Conceptual Structure of Continuum Physics

Roger Fosdick • David Owen • Paolo Podio-Guidugli
Editors

# Conceptual Structure of Continuum Physics

## In Recognition of the Fundamental Contributions of Walter Noll

Previously published in *Journal of Elasticity* Volume 135, Issues 1–2, 2019

Springer

*Editors*
Roger Fosdick
Aerospace Eng and Mech
University of Minnesota
Minneapolis, MN, USA

Paolo Podio-Guidugli
Accademia Nazionale dei Lincei
University of Rome TorVergata
Roma, Italy

David Owen
Mathematical Sciences
Carnegie-Mellon University
Pittsburgh, PA, USA

ISBN 978-94-024-1714-2



Printed on acid-free paper

This Springer imprint is published by the registered company Springer Nature B.V.
The registered company address is: Van Godewijckstraat 30, 3311 GX Dordrecht, The Netherlands

# Contents

J Elast (2019) 135:1–2
https://doi.org/10.1007/s10659-019-09729-9

# Foreword

**Roger Fosdick**[1]

Published online: 14 February 2019

**Walter Noll** (January 7, 1925–June 6, 2017)

This volume of the *Journal of Elasticity: The Physical and Mathematical Science of Solids* is dedicated to the memory of Professor Walter Noll. The collection consists of a biography, which includes a complete list of his publications, and 16 contributed papers.

✉ R. Fosdick
fosdick@aem.umn.edu

1 Dept. of Aerospace Engineering and Mechanics, University of Minnesota, Minneapolis, USA

Professor Walter Noll made major contributions to the foundations of continuum mechanics and continuum thermodynamics throughout the last half of the 20th century. He continued working and contributing with deep insight and characteristic scholarship until his death in 2017. Professor Noll affected directions of research with his conceptual understandings and descriptions of materials as a continuum governed by basic laws of physics, augmented with fundamental constitutive properties which characterize the behavioural differences among materials. His analysis, clarified by the application of fundamental principles, promoted both mathematical rigor and connection to the physical world. His depth of knowledge which made him conversant with philosophical issues concerning space, time, mass, force, and his understanding of relativity and developing field theories in physics, combined with his acumen to explain observations using mathematics, contributed to his abilities to isolate basic concepts upon which to build a foundation for continuum physics and its connection to kinetic and molecular theory. The physical and mathematical science community, especially mechanics, has greatly benefited from the depth and breadth of the contributions of Professor Walter Noll. His work has opened and made accessible a plethora of avenues worthy of exploration. Already, during his lifetime, many of his contributions have advanced our fundamental understanding of continuum physics as a way to characterize, understand and predict modern material behaviour. His publications have left a legacy of ideas and directions for future researchers and his presence as a man of principle and promoter of fundamental thinking as an essential element of scientific discovery will be evident for ages. His impact during a half century 'golden period' of advancement in continuum mechanics was great and will be lasting.

Roger Fosdick
Minneapolis, 2019

J Elast (2019) 135:3–72
https://doi.org/10.1007/s10659-019-09728-w

# Scientific Life and Works of Walter Noll

**Paolo Podio-Guidugli**[1,2] · **Epifanio G. Virga**[3]

Received: 1 February 2019 / Published online: 27 February 2019


**Abstract** Walter Noll (1925–2017) was an American mathematician of German birth who made lasting contributions to the foundations of continuum physics and the classical nonlinear field theory. This essay is an attempt to put in a broader perspective Noll's methods and achievements in the hope that young generations of researchers may find their inspiration in the talent and depth of the old. By no means should this be considered as a historical account on the development of continuum mechanics through the second half of the twentieth century. We are content to illuminate Noll's precious legacy.

**Keywords** Foundations of continuum Mechanics · Balance equations · Stress · Contact interactions · Constitutive equations · Fading memory · Thermodynamics · Rheology · Viscoelastic fluids · Molecular mechanics · Scale bridging · Theory of elasticity · Anisotropic media · Geometric elasticity · Media inhomogeneity

**Mathematics Subject Classification** 74A10 · 74A15 · 74A20 · 74A25 · 74A50 · 74B20 · 74D10 · 74E05 · 74E10 · 76A02 · 76A05 · 76A10 · 76A15 · 82B21 · 82C21

## Contents



✉ P. Podio-Guidugli
p.podioguidugli@gmail.com

E.G. Virga
eg.virga@unipv.it

1 Accademia Nazionale dei Lincei, Palazzo Corsini, Via della Lungara, 10-00165 Roma, Italy

2 Department of Mathematics, University of Rome TorVergata, Via della Ricerca Scientifica, 1-00133 Roma, Italy

3 Dipartimento di Matematica, Università di Pavia, Via Ferrata, 5-27100 Pavia, Italy

## 1 Introduction

Just as the rebirth from its own ashes for the Arab phoenix, a cleansed reiteration and accompanying renewal are essential for a scientific theory to achieve persistency and to enlarge its scope. In the early 1960 years, a new and authoritative narration of continuum mechanics was made available through two masterfully written treatises, namely, *The Classical Field Theories* [93], written by C. Truesdell and R.A. Toupin, with an appendix due to J.L. Ericksen, and the *The Non-Linear Field Theories of Mechanics* [B1], written by C. Truesdell and W. Noll.[1] The primary intention of this writing is to give an account of the scientific life and the works of the second author of the latter treatise, Walter Noll.

The life-long task that Noll assigned himself—his main legacy, in our opinion—has been to investigate the foundations of mechanics while furnishing and constantly renewing the mathematical structure and infrastructure of the various branches of that discipline [WA3]: continuum mechanics and thermodynamics *in primis*, but also classical mechanics [C1], [C2] and [C7], special relativity [WB1] and, through the work of his doctoral students, quantum mechanics.[2] Ever since he wrote his Ph.D. thesis, Noll made very clear what tools he intended to employ exclusively and systematically, as well as what subject he intended to address, with a method consisting in recreating and enlarging its scope by rethinking it afresh. He writes in [C15], the opening section of the third edition of [B1], that he was not "... very good at reading, digesting, and summarizing existing literature." But, adds he, "I cannot help rethinking and recreating whatever subject I wish to understand." To us, *wiping away the inessential* has been a distinctive and persistent trait of his method, his trademark as a researcher.

In view of our limited competencies, we will mostly focus here on Noll and continuum mechanics.[3] This self-limitation nothwithstanding, ours is a daunting task indeed, if not for other reasons because it is inevitably intermingled with a recounting of many of the most important developments of continuum mechanics from the early 1950s to date: although we both lived through a substantial and exciting part of those decades ourselves, we have no intention to sketch their history, now that those halcyon days are in the past and that most of those days' leading figures are either dead or retired. Nor do we intend to hint, if only indirectly, at an assessment of the present importance of continuum mechanics, let alone its future. Instead, our secondary intention is to let today researchers become aware of some older relevant contributions in that field.

[1] In his learned account on the history of continuum mechanics in the twentieth century, Maugin writes about these treatises [55, p. 61]: "Impressive encyclopedic contributions by Truesdell, Toupin and Noll were the lighthouses that 'illuminated' the world community of mechanics".

[2] We refer here to Lincoln E. Bragg's 1963 thesis "On Relativistic Worldlines and Motions, and on Non-Sentient Response"; to 1978 dissertation by Ray E. Artz, Jr., entitled "An Axiomatic Foundation for Finite-Dimensional Quantum Theory"; and to Vincent J. Matsko's thesis, entitled "Mathematical Structures of Special Relativity".

[3] Our choice also reflects our taste; in this, we cannot but agree with Truesdell [88, p. 105]:"It is tasteless to recommend one's own taste, but scarcely honest to recommend any other".

In preparation for our work, we have put together a comprehensive list of Noll's writings, to be found in Sect. 7, the last section of this article, distinguished in five categories:

- Books, such as [B1]
- Contributions to books, such as [C1]
- Articles in journals, such as [A18]
- Research reports, such as [R3]
- Manuscripts on the web, a category further subdivided in three sections:
  A. General, as for [WA6]
  B. Mathematical science, as for [WB1]
  C. Pure mathematics, as for [WC5]

We have also included a list of references to cited works, other than Noll's; they are numbered sequentially, such as [55].

The short biographical notes in Sect. 2 are meant to furnish a temporal and spatial framework for our account to follow of the scientific life and works of Walter Noll, as well as for the personal recollections to be found in Sect. 3. This section includes, besides each of our own, contributions by Marylin S. Marsh Noll, Brian Seguin, David R. Owen, Mario A. Bunge, and Mauro Ferrari, to whom all we are extremely grateful for sharing with us and our readers their memories of a multifaceted man and scientist.

Noll's scientific work is accounted for in Sects. 4 and 5. The former deals with those research themes, central to continuum mechanics and continuum thermodynamics, that recurrently attracted his attention throughout his professional life; the latter with a few other themes pertaining to the same disciplines, that he touched upon in an episodic manner. In order to give an idea of the contents of Sect. 4, we think it appropriate to list the titles of its subsections:

- 4.1 *On the Continuity of the Solid and Fluid States*
- 4.2 *Microscopic Counterparts of Macroscopic Balance Laws and Cauchy Stress*
- 4.3 *General Theory of Interactions*
  - 4.3.1 *Material Universes*
  - 4.3.2 *Interactions*
  - 4.3.3 *Fit Regions*
  - 4.3.4 *Stresses*
  - 4.3.5 *Balance Laws of Motion*
- 4.4 *Thermodynamics and Rheology*
  - 4.4.1 *Fading Memory*
  - 4.4.2 *Clausius-Duhem Inequality*
- 4.5 *Five Contributions to Natural Philosophy*
  - 4.5.1 *On the Illusion of Physical Space*
  - 4.5.2 *On Material Frame-Indifference*
  - 4.5.3 *Updating* The Non-Linear Field Theories of Mechanics
  - 4.5.4 *The Theory of Simple Semi-Liquids, a Conceptual Framework for Rheology*
  - 4.5.5 *Nematic Semi-Liquids*

The titles of Noll's papers dealt with in Sect. 5, and devoted to some episodic research interests of his, are:

- *"On the Rotation of an Incompressible Continuous Medium in Plane Motion"*
- *"On the Uniqueness and Non-existence of Stokes Flows"*
- *"On Exterior Boundary Value Problems in Linear Elasticity"*
- *"Materially Uniform Simple Bodies with Inhomogeneities"*

- *"Quasi-Invertibility in a Staircase Diagram"*
- *"On Certain Convex Sets of Measures and on Phases of Reacting Mixtures"*
- *"Representations of Certain Isotropic Tensor Functions"*
- *"Annihilators of linear differential operators"*
- *"The Geometry of Contact, Separation, and Reformation of Continuous Bodies"*

We delineate our spirits and provide some final thoughts on Walter Noll's legacy in the concluding Sect. 6.

## 2 Biographical Notes

Walter Noll was born in Berlin, Germany, on January 7, 1925; he died in Pittsburgh (PA), USA, on June 6, 2017 (see Fig. 1).

He studied at the the Technische Hochschule of Berlin, the Humboldt University, and the University of Paris. His first job, from 1951 to 1952, was as a Wissenschaftlicher Assistent at the Lehrsthul für Technische Mechanik, at the same Technische Hochschule where he studied. In the Fall of 1953, he arrived at the University of Indiana in Bloomington, where he was accepted as a graduate student and—in ten months!—he earned his Ph.D. in applied mathematics.[4] At the end of his one-year-long leave, he returned to his post at the Technische Hochschule, not before having put together a major paper [A2] where he cleansed and generalized the recent Irving & Kirkwood's derivation [46] from statistical mechanics of the main balance laws of continuum mechanics.

He immigrated to the U.S. in 1955, where he taught at the University of Southern California before joining in 1956 Carnegie Mellon University, at that time the Carnegie Institute of Technology, where he served as a professor of mathematics until his retirement from teaching, in 1993. He also served as a visiting professor at the Johns Hopkins University, the University of Karlsruhe, the Israel Institute of Technology, the École Politechnique in Nancy, the University of Pisa, the University of Pavia, and Oxford University.[5]

**Fig. 1** Walter Noll, in a photo taken at the opera ball in Pittsburgh on the 27th of October 1995

[4]A short personal account of his time in Bloomington is found in [C15] and in the autobiographical notes [WA1], which cover the years prior to 1956.

[5]These and other biographical matters are found in [79]. Another valuable source is [45]. The six first chapters of [45] are devoted to a meticulous biography of Noll, private and professional, based not only on documents but also on interviews of, and letters from, coworkers and colleagues; the remaining seven chapters attempt

## 3 Personal Recollections

We collect here a number of contributions from individuals who were either personally very close to Walter or, although remote from his acquaintance, were equally influenced by him in their own scientific practice or in their way of thinking science as a whole. Such an influence transcends the level of indirect impact and, in fact, attains the level of direct inspiration, which requires awareness. Thus, most of the words that follow in this section were written (or approved, when an editorial summary was imposed by space limitation) by witnesses who were so kind to contribute to our endeavor. These contributions also document how far-reaching Walter's thinking was, especially in disciplines foreign to some degree to his own.

Counting ourselves among those so influenced by Walter, we felt that we could not avoid the task of also recording here our own personal recollections and impressions, which is how we start this section, albeit very hesitantly. We hope that our readers will forgive our being the first in this list: if any primacy we wish thus to establish is that of taking responsibility for a bold gesture.

Whenever writing about a prominent figure with whom one had the good fortune to be acquainted, the greatest risk one faces is falling into the temptation of writing more about oneself and one's own involvement with the person in question than about the latter. We both promise that we shall resist this temptation, but from the start we ask for the readers' indulgence: it is in everybody's experience that being aware of a risk does not save one from overstepping it, as the fascination of prohibition teaches us all even too often.

### 3.1 Paolo Podio-Guidugli

The first time I met Walter was around the end of June 1968, when he came to Pisa and paid a long visit sponsored by the institutes (no departments during those times!) of mathematics and construction engineering of that university. He was accompanied by his family, his first wife Helga and his kids, Peter and Virginia. The last time I had the privilege to spend time with Walter was on the 23d of May 2017, a few days before he passed away, when he, his third wife Marilyn, Virginia, Mort Gurtin, Bill Williams, and myself, had a lovely dinner in a restaurant he liked on the outskirts of Pittsburgh, not far from where he always lived.

More and more along the five decades of our friendly relationship I got convinced of what I said in the opening of my invited talk at the XXXVII Meeting of the Society for Natural Philosophy on Mathematical Aspects of Continuum Mechanics, A Symposium in Honor of Walter Noll on the Occasion of his Retirement from Teaching (Pittsburgh, April 1993), and then I wrote in the dedication to him of my paper in this volume: that *Walter was the alchemist who distilled the quintessence of Continuum Mechanics.*

When Walter arrived in Pisa in 1968, he had been preceded by Clifford Truesdell and Bernard Coleman, and he was to be followed shortly after by other Carnegie-Mellon colleagues of his, Mort Gurtin, Bill Williams, David Owen, and by many other distinguished scientists from other research centers in USA, Europe, and Latin America. At the end of the 1960's and then for another fifteen years or so, due to the joint efforts of Gianfranco Capriz, Tristano Manacorda and Piero Villaggio (in alphabetical order), Pisa was the place

an account by themes of Noll's research work. In this book, appeared in print in 1996, Y.A. Ignatieff bravely took up an unusual endeavour for a historian of mathematics, namely, to write a scientific biography of a living mathematician. We are especially grateful to him for his bravery: his book has been one of our sources, to which we often turned.

to be in Europe, if you were interested in what Truesdell had termed Rational Continuum Mechanics. Where else would you have liked to be, if not where, in addition to world-class mechanicians, you could step every other day into such world-class mathematicians as Aldo Andreotti, Enrico Bombieri, Sergio Campanato, Ennio De Giorgi, Giovanni Prodi, Guido Stampacchia (again, in alphabetical order), and many others.

After his 1968 visit, Walter returned to Pisa, as well as to a number of other Italian universities (Bologna, Ferrara, Turin, Padua, Bari, Rome, Pavia, Milan, ...), many other times. His longest stay in Pisa took place in January-May 1985; he was then accompanied by his second wife Resie. Walter, who held a visiting professorship, was given an office at the Istituto di Scienza delle Costruzioni, where Villaggio, Epifanio Virga, and myself, had one as well—sheer luck for all of us, who enjoyed a unique daily opportunity to interact with him.

I had had many previous opportunities to do so, especially during two long visits of mine at Carnegie Mellon University in 1972 and 1979–1980, when I had attended courses and seminars given by him and frequently talked science with him, both in his office and, quite often, at the blackboard in the Math Department lounge, over what to me was a bucket of coffee (Walter never got to accept the fact that coffee in an Italian bar consists of few drops and is drunk in a few rapid sips). But, I had always been quite intimidated by his ability to nullify my attempts to add a bit to his Word or, worse, to deviate from it, if only slightly. During his extended 1985 stay in Pisa things changed, and Walter was kind enough to put together with me a note on the physical significance of discontinuous minimizers to equilibrium problems in finite elasticty [C13]. In the same span of time, the seeds were planted of his beautiful joint work with Virga, on space regions and function classes which are fit for a mathematically sound formulation of problems in continuum mechanics [A33], and on edge interactions [A34]. It is because of our treasured memories of that unique period of our lives that Epifanio and I have accepted to write this article, a job that others were certainly more qualified for than us.

Walter was the opposite of the solitary savant, disdainfully self-secluded in his own *turris eburnea*. His readiness for constructive engagement is well described in David Owen's recollections here below. To what David writes, I can add my memories of Walter as part of the audience in a number of seminars I held over the years at the Carnegie-Mellon math department, as well as my own first experience, in the Fall of 1972, as a speaker at the evening seminar series "Mathematical Foundations of Physics" that Noll helped organize. It was he who came to my rescue and eventually saved me with his constructive comments, after half an hour or so of crossfire from Bernie Coleman and others. Of the many lessons I learned that evening, Walter's was the most valuable one.

## 3.2 Epifanio G. Virga

My readers will not be surprised if, as an antidote to self-celebration, I start recounting my personal recollections with the admonitions that Walter used to repeat to me: "Do not jump to conclusions"; "Write clearly"; "Do state, not hide your assumptions". If I badly needed those warnings, he was a champion in the virtues I lacked.[6]

In his reasoning Walter never jumped to conclusions, even when they seemed to suggest themselves with compelling physical evidence. This is indeed the main and most ruinous

[6] Such a lack of virtues is perhaps the very reason why a good friend of mine recently said to me (with the generosity that only a true friend can afford) that "no Virga's scientific obituary would be justified" after my death. I can only assure the reader that I'll see that none will appear *before*.

pitfall for a natural philosopher: to be influenced by the urging reality of nature to the extent of making it prevail on the rigorous pace of deduction. So, as we are eager to show that either certain experimental quantitative measures or other qualitative features of a natural phenomenon support a cherished theoretical explanation or abstract principle of our own invention, we can easily indulge in considering established as a proved consequence what should at most be outlined as a mere possibility. Perhaps, we can involuntarily incur in such an error, not Walter, as his rigorous logic was unaffected by any sense of proximity to a thrilling physical discovery. He knew where to stop speculating, not to be fooled by premature, even if tempting conclusions, sure as he was of the value of his method.

Write clearly was another precept of Walter's scientific method. I remember vividly his characteristically pointed handwriting, rigorously in pencil. That were still pre-LaTex era and Walter, who in due time also became conversant with both TeX and LaTex, was used to draft all his notes by hand. *Pencil* was preferred because it invariably came with an *eraser*, the tool Walter treasured as eminently necessary for composing mathematics: "I cannot work without it," I often heard Walter saying, while frantically searching for one of the many on his desk. His prose was plain and effective, never pompous or affected: it remains for me an unparalleled example of clarity and conciseness, but I know that it did not come straight out of his pencil; it also resulted from his trained ability in the use of an eraser.

Assumptions are the life's blood of natural philosophy. Phrasing the most general assumptions within which a class of natural phenomena can be subsumed and modeled is at the basis of modern science as a deductive exercise. If this is perhaps all very clear, far less so is how axioms should be *posited*, what relates them to the phenomenological reality they are meant to describe, how to motivate them by building on an underlying finer (or coarser) description of matter. These are daunting tasks which can make whomever is not versed in modern epistemology grow more and more despondent. Walter's lesson was not to fear assumptions, but to state them clearly. Even if motivations or interpretations are possible for an assumption, the latter cannot be replaced by the former, no matter how sophisticated our apologetic skills may be. A single assumption is more valuable than the many ways there may be to motivate it: its truth only resides in the accountability of its logic consequences.

Maxims are in general reductive and should be avoided. However, in Walter's case, one which he used to repeat is adequately descriptive of his way of thinking: "Mathematics is the art of avoiding unnecessary calculations."[7] He hated calculations, especially lengthy ones, for fear of being baffled by the insipid complexity of algebra. This does not imply that he never computed anything or never used coordinates (as the myth has it). On the contrary, he performed only meaningful computations and employed coordinates adapted to the symmetry suggested by a preliminary group-theoretical analysis. On occasion, his papers contain diagrams and ingenious constructions, witnessing also the visual aspects of his geometric mind.

## 3.3 Marilyn S. Marsh Noll

Marilyn, Walter's third wife, assisted and supported Walter in his last years. She writes poetry and one of her poems, "Ordinary Tasks",[8] is dedicated to Walter. We also deemed it appropriate to record her words here, as a living testimony to Walter's last days. This is in tune with our endeavor in this obituary to illuminate Walter's commitment to mathematics as

[7] This is also in David Owen's recollection recorded in Sect. 3.5 below.

[8] This is also the title of the book [54] that collects some of her recent, minimalistic poems, seeking beauty in everyday life.

an absorbing style of living, also affecting the minute aspects of everyday life. What follows is taken from a letter to E.G.V. dated 4th November 2017.

> Thank you for your kind thoughts and information, and also relating to your last paragraph describing material you need for an obituary for Walter: "Especially, anything that could illuminate Walter's most recent thoughts on mathematics as a science and on mathematicians as professionals would be extremely useful: I would like our essay not to be confined to the undoubtedly high achievements of Walter's career, but also to encompass any recent contribution of his to the current scientific debate."
> As I think you know, Walter's hope was to put everything important of his on the Internet where everyone could reach it. I believe he accomplished a lot toward that end. But, during his last two years or so, he was not able to continue adding to that project.
> Today, I contacted Bill Hrusa, whom I thought might be able to help you because he actually visited with Walter several times over the past 1 1/2 years, bringing Juan Schäffer, Walter's longtime friend and colleague, with him. Because Walter's schedule was to sleep days and be up at least a few hours at night, these visits were confined to evening hours after 7 p.m.
> Walter would sit with them in our living room, and the three enjoyed good conversations, much relating to mathematics. Their last visit was on January 7, 2017, when I gave Walter a birthday dinner to which I invited Bill and Juan. This turned out to be a memorable event for all. Walter actually managed to sit at the dining room table to eat (which he hadn't done in a long while), and the conversation was unusually animated as they argued (Walter too) about their different views about mathematics, laughed and shared many of their funniest experiences. As you perhaps know, Juan passed away a month later (early February). Walter was badly affected by this loss, and went downhill soon afterwards. He died in early June.
> For a long time, although Walter hadn't been writing down material in his computer or anywhere, I know he was constantly thinking through his ideas and developing new ones. Late last year, when I inquired what he was thinking about those long hours in bed, he told me he wasn't always sleeping. He said he was constantly thinking about mathematics. "I'm never bored," he said, and seemed happy about that. Then, early this year, when I tried to wake him one morning (I seldom woke him in the morning, and can't remember why I did that day), he told me "I'm in the middle of teaching a relativity class." He obviously wanted to get back to that, so, of course, I let him sleep.

## 3.4 Brian Seguin

The text that follows was read by Brian, Noll's last doctoral student, on the occasion of his memorial.[9]

[9] A Service of Remembrance was held for Walter on the 21st of June 2017 at the Hunt Botanical Library of Carnegie Mellon University. The service was led by Pastor Chris Taylor of the Fox Chapel Presbyterian Church; the eulogy was read by Peter Noll, Walter's son; remembrances were given, in the order, by William O. Williams and David R. Owen, both Carnegie Mellon's Emeriti, and by William J. Hrusa, Professor at Carnegie Mellon University and Brian Seguin, Assistant Professor at the Loyola University of Chicago. The service was enriched by music: Marianne Cornetti, a world-renown mezzo-soprano (http://www.mariannecornetti.com/index.html), accompanied by pianist Mark Carver, sang *Pie Jesu* by Andrew Lloyd Webber, who was a composer very dear to Walter, and the *Ode to Joy* (from Beethoven's 9th Symphony) by Henry J. van Dyke. Marianne, who had been a student of Resie, Walter's late wife, has remained very close to Walter after Resie's death and has been most kind to Marilyn as well.

In some ways, it is odd for me to speak about Walter in this setting as I didn't meet him until after his 80th birthday. He had come to the University of Illinois to give a lecture in honor of Don Carlson's 65th birthday, and I was only a sophomore there at the time. However, even at this young age I was already partially aware of Walter's work on the foundations of continuum mechanics. A friend of mine had given me his biography and while the description of Walter's work in it was crude, it was enough to peak my interest. Don had encouraged me to attend Walter's lecture, not that I needed any encouraging, by saying something along the lines of "you don't know if you will get another chance to see Walter Noll lecture again." This is one of the few times that Don turned out to be wrong.

Several years later, I had been accepted to continuum mechanics as a graduate student. While chatting on the phone with Bill Hrusa to arrange my first visit, he asked if there was anyone in the department I would like to meet. I said that I had a copy of the Nonlinear Field Theories, of which Walter is a coauthor, that I would like to have signed by Walter, and Bill said he was sure that could be arranged. While I was visiting CMU, but before I met Walter, Bill mentioned that he would be willing to take me on as a PhD student if I was interested. This surprising me greatly as I knew Walter had retired several years ago. I was excited at the prospect. When I met Walter for dinner that evening, he gave me a packet of things to read. I recall I started to read them, trying to decrypt the Nollian terminology and notation, while I was on the plane going back to Chicago. Needless to say, I was hooked.

Walter was very generous with his time while he mentored me. Our schedule consisted of meeting once a week in the mid afternoon. (He wasn't much of a morning person, in fact he told me never to call him before noon.) These meeting usually began by him asking me if I had "solved all problems". Since I hadn't, we always had plenty to talk about. After several hours of working, we would go out to dinner, often joined by Marilyn. On several occasions, he invited me to his house for dinner. He said that Truesdell invited him out to dinner while he was a graduate student, so now he was returning the favor, and that I had to do the same when I had students. I promised I would. We would try to steer clear of math and physics topics for Marilyn's benefit, but this wasn't always possible. History was a common topic of discussion, especially his own history growing up in Germany and living through the War. I was particularly interested in this subject because while both of my grandfathers fought in the War, they died before I got a chance to talk to them about their experiences. Thus, in a way, Walter was a surrogate grandfather, passing down stories from a time that seemed long ago.

I know that I was very fortunate to have the opportunity to learn from Walter and hear him speak and think about mathematics and physics. On more than one occasion, people I've met at conferences have asked me what it was like to work with him. Because he was so influential in the field and was very opinionated, some people assumed that Walter would be difficult to talk to. Nothing could be further from the truth. When I went to conferences with Walter, he often engaged young students in conversation and enjoyed getting to know them.

Walter had a singular mind unlike any other mathematician I've met. He always stressed the importance of concepts. Only once these were understood, could one find the most appropriate mathematical structure to model them. This often lead to using abstract mathematics, but abstraction was never the objective. I would argue that his definition of a continuous body is one of the best examples of this. Walter once said that "intelligence is the ability to make distinctions." Well, according to this definition, he was very intelligent. I am happy that he shared not only this intelligence, but kindness and friendship with me and many others.

## 3.5 David R. Owen

We are grateful to David, who has agreed to reproduce here his thoughtful recollections of Walter as pronounced on the occasion of the Service of Remembrance (see footnote 9). We have kept his title as well as the splitting in short paragraphs that he suggested, as we find that they provide an effective guidance to the reader.

*Impressions and Recollections of Walter Noll over a Half-Century*

**Background** I met Walter in the Fall of 1967 when I arrived in the Department of Mathematics at Carnegie Mellon as a fresh-out-of-graduate school, newly appointed assistant professor. Walter already had been at Carnegie Mellon for about a decade and had established himself as a leader in research whose contributions set the stage for part of my doctoral research.

**Walter as a Clear-Thinking, Articulate Role-Model** From my earliest interactions with Walter it was apparent that he had the ability to state clearly and explicitly what I might only have sensed vaguely and incompletely. I learned from reading his papers and from discussions with him on mathematics that Walter was constantly searching for optimal mathematical concepts to describe physical systems and for optimal methods for analyzing the mathematical problems emerging from such descriptions.
As Walter often said in later years, "Mathematics is the art of avoiding unnecessary calculations." Walter pursued this art as an ongoing, ever-evolving process that led him to think and rethink the choices of concepts and methods used in his own research.
I recall an example of his ongoing revisions and rethinking: Walter suggested to a speaker at a scientific meeting that the speaker should use a different, better conceptual framework. The speaker responded by saying he was using the conceptual framework in one of Walter's papers. Walter responded that the framework in that paper was obsolete and that no one should use it.
Walter's ability to get to the heart of an issue emerged also in discussions on subjects other than mathematics. I recall over a period of time my having struggled to pin down the difference between scientific explanations of a phenomenon and religious explanations of the same phenomenon (for example, the emergence of life on earth) in a way that does not bias a priori one or the other type of explanation. I overheard a conversation in the mathematics lounge in which Walter gave the following definition: a scientific theory is a conceptual infrastructure for describing a natural system and for predicting its future behavior. This was the foothold that I needed, the laying down of a concise statement that revealed the main attributes of a scientific explanation through the notion of "scientific theory."

**His Readiness for Constructive Engagement** Walter always seemed ready and available to discuss issues, mathematical or not, either in person or by telephone. (I called him frequently enough over the years that I still know his home telephone number by heart.) During our conversations it was not unusual for him to say "What you said is wrong" or "What so-and-so said was false." Such criticisms from Walter were almost never an indictment of the person who committed the error. The absence in Walter's criticism of ad hominem statements was noteworthy at the time and is even more significant today when constructive dialogue is in short supply.

Walter's readiness for constructive engagement was evident in the classroom, particularly in the setting of the honors course, Mathematical Studies. The course was taught by a team of two instructors. Walter was paired frequently with Professor Juan Schäffer, his mathematical kindred spirit, and the students not only experienced the real-time, give-and-take of discussions between Walter and Juan, but were encouraged to participate in such discussions both within and outside the classroom.

Constructive engagement was also an important characteristic of Walter's written comments on students' homework and exams. Although grading papers was low on Walter's list of enjoyable activities (he said on more than one occasion that he'd rather do housework than grade papers), his grading system provided a uniform and objective guide that identified in an impersonal way the location and nature of errors and defects in the written work of students. Walter's system contained acronyms, for example WF, short for "wrong fruit", indicating that a student used attributes in the wrong context, or SS, short for "set the stage", indicating that the student had failed to describe fully the nature of the objects used in a particular argument.

Another example of Walter's readiness for constructive engagement was the evening seminar series "Mathematical Foundations of Physics" that he helped organize. The series featured speakers both from within and outside our faculty and had the twist that the speaker and any willing participants had dinner together before the seminar. The food and beverages set the stage for free-wheeling exchanges between speaker and audience members as well as among audience members. My first exposure to this format was as the speaker at one such seminar in the Fall of 1967, and I remember truly having had the floor only for the first five minutes of my talk. Thereafter there was a constant stream of questions from the audience, punctuated by disagreements between members of the audience. In the course of two and a half hours I did manage to communicate the main points that I had hoped to include, and the exchanges during the seminar led to significant improvements in the version that eventually appeared in print.

Walter's interactions with others were not all serious or all intellectual. In fact, his matter-of-fact style of discourse often led others to poke fun at him in a good-natured spirit to which Walter responded in kind. I remember a bus ride with Walter and others from a banquet at Fort Niagara back to our hotel in Buffalo. During the ride people for some reason were recalling at length their experiences riding horses. Walter listened silently and eventually stated simply and concisely "I once rode a horse," period. There was a long pause after which someone interjected "I am sure that the horse has never forgotten the experience," and I remember Walter laughing the loudest.

**His Sense of History** More than once Walter said to me that mathematics will be done in such and such a way 100 years from now. The fact that he said it more than once reminded me of the extent to which Walter thought of events in historical terms. This tendency is not surprising in view of the fact that he survived World War II and its aftermath in Germany and that he witnessed nearly a century of political, social and technological changes. Walter was strongly moved to record in short memoirs and essays his experiences during those times as well as his views on the costs and benefits of particular changes that he observed. I enjoyed his account of his experiences as a teenager in the German counterpart of the United States Civilian Conservation Corps, particularly since my father served as a physician in the CCC in California during the

1930's. I also enjoyed Walter's essay [WA4] that juxtaposed the historical role of a professor as one who guides students along the path of learning with the updated role of a professor as a principal investigator who teaches in his spare time.

**His Commitment to Professional and Personal Relationships** To me, Walter's day-to-day, rational bearing and demeanor initially masked his deep commitment to professional and personal relationships. However, over time I discovered how much effort he devoted in his support of family, friends and colleagues, as well as the unpretentious pride he felt for the United States of America, his adopted country. Perhaps Walter's commitment to these relationships will be the aspect of his life that I will remember the most going forward.

## 3.6 Mario A. Bunge

According to Ignatieff [45, p. 215], in a letter to G. Oravas dated 19th February 1969 Truesdell wrote "I agree that Bunge is a very good man. He is the only philosopher I know who has any contact with real science as it is practised". Such an appreciation is likely to have been corroborated by the acquaintance Truesdell had had with Bunge's work while editing the latter's book on the foundations of physics [9], which was published in 1967. For sure, Bunge's original views on the axiomatization of continuum mechanics were highly influenced by both Truesdell and Noll. It is instructive to hear from Bunge's own words his stance on the reduction of continuum mechanics to microphysics and his description of the state of foundation research in continuum mechanics prior to the advent on the scene of the "natural philosophy group". The following excerpts are taken from pp. 141–145 of [9].

> But the success of the reduction programme depends both on our mastery of microphysics and on a correct formulation of CM [Continuum Mechanics]: otherwise one would not know what should microphysics average to. ... In short CM must be studied by itself, on its own level, as well as in the light of microphycics if we want to understand it in statistical (but not purely corpuscular) terms. In other words, in addition to striving for a representational theory of matter we must secure a semiphenomenological theory of it. (Semiphenomenological rather than phenomenological, for the stress tensor is anything but an observable). And this both for the logical reason just mentioned and because the laws of bulk matter are different from those of its constituents and therefore entitled to a separate study, much as the laws of societies differ from those of their individual components. ...
> Unbelievable as it may seem, the foundations of CM were not seriously investigated until the 1950's. The whole interest in CM has died with the birth of relativity and the quantum. The subject was buried alive and has ever since been officially regarded as dead, to the point that no Pd.D.'s in physics are normally awarded in the field. But in the mid 1940's it was realized that not enough was known of CM even for engineering purposes: that not enough theorems were available, and that the existing theory did not account for newly discovered effects such as the swelling of a fluid emerging from a pipe and the climbing of a fluid up the walls of a rotating cylinder—not to speak of superfluids. A quick expansion followed and soon the subject began to be investigated in depth, until the foundations of CM were subjected to detailed revision by the "natural philosophy" group. The upshot was a gain in generality, mathematical rigor, and logical organization; in particular the bondage with respect to PM [Particle

> Mechanics] was shown to be fictitious. The gain was such that CM is now the best example of the benefits of FR [Foundations Research] for "hard" physics. (The fact that the subject is sometimes classified as a mathematical science and at other times as a branch of engineering only goes to show how thin our metascience can be.) In the following we shall borrow freely from that pioneer work ([82], [83], [C1], [93], [91]) Our treatment will be more schematic and mathematically less sophisticated, but on the other hand we shall add some assumptions necessary to make its physical content more apparent.

Bunge's and Noll's paths crossed not only ideally but also physically. In February 1965, Bunge chaired the Seminar on Foundations of Physics at the Department of Physics of the University of Delaware, where Noll delivered one of the main lectures, entitled "Space-Time Structures in Classical Mechanics" [C7]. Bunge's seminar was quite a cultural endeavour. We read from the preface to the book [8, p. vi] that collected the papers given at the seminar (written by Bunge himself):

> This volume collects the lectures on the foundations of physics given by eleven scientists at the University of Delaware. It is neither an anthology of disconnected items nor a smoothly running textbook but rather a progress report on a neglected yet vital area of basic physical research, namely foundations research. The investigation into the foundations of any branch of science is neither loose speculation nor popular science: it is an aspect of scientific research—in fact the deepest-searching part of basic research. Consequently it must be carried out by the scientist himself. Thus whether the time concept is a numerical variable or a function, whether particle mechanics is the primary mechanical theory or a particular case of continuum mechanics, whether transformation formulas are laws of nature or links between equivalent descriptions, whether thermodynamics has been fully reduced to statistical mechanics, whether the field concept is dispensable, whether the covariance principle is a law or a regulative principle, whether quantum mechanics is completely detached from classical physics or contains fragments of it, whether it has annihilated the physical object or given a more complex picture of it, and to which extent are the field variables measurable—all these are technical questions demanding a careful analysis of pieces of recent basic research. Yet all of these problems and indeed all questions in foundational research are philosophical as well as scientific.

This is at the same time an intellectual manifesto and a fairly complete summary of the diverse contributions heard at the meeting. An interesting story is told by Ignatieff [45, p. 47] about Bunge's harmonizing role in planning the Seminar. Apparently, Bunge had assigned a different topic to Noll; the way Noll asserts the reasons of his choice of topic and the decision to stick with it illuminates his vision for a (then) new axiomatic formulation of classical physics. From a letter of Noll to Bunge, dated 12th October 1964:

> I thought that my proposal to entitle my lecture "Space-Time Structures" should not be inconsistent with a content that deals mainly with the axiomatic foundations of classical mechanics. There is perhaps some ambiguity in the term "classical mechanics". As far as I am aware, the work by McKinsey, Suppes, and others deals mainly with the classical mechanics that is general enough to include continuum mechanics. It turns out that for this more inclusive "classical mechanics" (which has been developed extensively in the last 10 years or so) the classical space-time structure of Euclid and Newton is not as appropriate as another "neo-classical" space-time. It is this "neo-classical" space-time that I have worked on recently and that I should like to speak about.

Even if more than half a century has elapsed since these facts, we attempted to gain a personal insight into them directly from Mario Bunge, now aged 99. He was kind enough to answer our inquire in a letter to one of us, E.G.V., dated 2nd September 2018:

> My memories of Noll and Truesdell have become foggy after half a century. I only recall that both were very helpful an[d] supportive, and that I regarded both of them as my friends.

This is perhaps not very informative, but it witnesses the persistent impact that Noll's ideas have had in the philosophy of science.

### 3.7 Mauro Ferrari

From philosophy to medicine, you may say that it is a big jump and feel that influences of Noll's ideas might be hard to find there. Quite the opposite, as shown in great detail by the following witty and passionate words provided by Mauro Ferrari.[10]

> I was terrified as I walked to the podium, with the full notion that among those in the audience were some of the most accomplished mathematicians in mechanics of our age. Just a first-year graduate student at that time, the notion of speaking about my fledgling research project at the revered Department-of-it-all, at Carnegie Mellon, in front of Morton Gurtin, Clifford Truesdell, Bernard Coleman, and other distinguished thought leaders was downright scary.
> And then there was Walter Noll. I had studied his works in awe, and had had the privilege of hearing him speak just a few days prior, at a celebration event for Truesdell's 70th birthday. I was struck by the elegance and rigor, the inquisitiveness in his eloquence, the impression that with every door he was closing, by establishing theorems with his powerful mathematical reasoning, he was actually opening ten more—and loved every minute of it. And, I had appreciated his kindness toward students—which was a memory that emerged, very reassuring, as I was starting to speak, as in "Dr. Noll will save me, if I stumble today".
> It actually was probably "Gualtiero", rather than "Dr. Noll", since I had heard him ask the many Italian colleagues he interacted with to call him by the Italian translation of his name. There were many powerful collaborations between that group of scientists, and a cohort of Italian counterparts, including Epifanio Virga, Piero Villaggio, Paolo Podio-Guidugli, Gianpiero Del Piero, Cesare Davini, and others.
> As life would have it, in my career as a scientist I have pursued a different line and form of investigations from what was pursued by that community, but nevertheless I consider myself extremely fortunate, especially in my early formative stages as a scientist, to have had the opportunity to learn from their extraordinary rigor, methodologies, brilliance, commitment and unwavering dedication to scientific truth.
> *Gualtiero: I have only met you a handful of times, have only discussed my work with you perhaps in three, all-too-brief occasions, in which I have much benefited from your direct advice, but I am (undeservedly) writing this note to thank you effusively, for all you have done for science and for myself, with your monumental body of work.*
> Three specific examples emerge, as instances of the influence Dr. Noll had on my work—without of course affixing on him any blame or responsibility for the poverty

[10] M.F. is presently the president and CEO of the Houston Methodist Research Institute in Houston; among several other things, he also serves as executive vice president for Houston Methodist Hospital. https://en.wikipedia.org/wiki/Mauro_Ferrari.

of my personal scientific contributions! Chronologically, the first relates to what I personally consider one of the most important works in the mechanics literature: His celebrated 1963 paper with Bernard Coleman [A19]. In it, they established a rigorous, novel and exquisitely elegant procedure to exclude variables from constitutive equations, employing a statement of the Second Law of Thermodynamics (the Clausius-Duhem inequality), which is enforced for all processes described by the independent constitutive state variables. The results of the procedure were not surprising, but, rather, reassuring that prior phenomenological postulations were on solid theoretical ground. The foundational importance of the paper was on the methodology, which has since then impacted all constitutive law investigations in multi-physics mechanics. It provided a balancing perspective to the Truesdell's principle of equipresence.

The impact on my work was at least two-fold. First, it shaped my graduation thesis in mathematics from the University of Padova, which focused on thermoelasticity in media with microstructure, in particular those categorized by Ahmed Cemal Eringen as micropolar. For that thesis investigation, I employed a Coleman-Noll equivalent approach, reframing it in the form of an identically enforced equality, as opposed to an inequality, but with a positive-definite entropy generation term. The second impact was much more profound, pervasive and durable: The general, implicit admonition that less is better, when warranted, when developing constitutive frameworks. Both components of the admonition are essential: "Less is better" and the constraint "When warranted". Of course, this brings to mind Albert Einstein's quotable: "Make your theory as simple as possible—but not any simpler!".

Unavoidable doubt and confusion in identifying the boundaries of warranted simplicity are perhaps among the greatest challenges in science, and brute experimentation by itself is remarkably unsuitable to solve them. To elaborate on this perhaps counter-intuitive statement, I need to digress briefly to give an account of my 26 years of failures in the struggle to find a therapy that will cure cancer metastases to the lungs and liver—an objective that I have chosen to pursue in the early Nineties, largely for personal reasons. There is little doubt that it is an impactful goal, since it is estimated that about 80% of all cancer deaths results from metastases to lungs and liver, and the current therapeutic options (surgery, radiation, chemotherapy, biological macromolecular therapy) do not provide definitive cures in general, but rather only delay the inevitable—which is of course in itself a great success, but not the final objective. Our first attempt was with silicon microparticles carrying chemotherapeutic agents, and providing enhanced concentration of these in the target organs. Animal experimentation quickly showed that this approach would not work. Our work and that of a few other contemporaries pushed the boundaries of particle carrier manufacturing from the microscopic into the nanoscale domain—and thus, nanomedicine was born, with its primary applications to cancer. I had the privilege of leading in the formulation and the launch of the national funding plan of the USA in cancer nanotechnology, while I served at the National Cancer Institute (2003–2005). The Cancer Nanotechnology Plan is still in operation, currently funds hundreds of laboratories, and it has been estimated that between 5–10% of all cancer drugs currently used in clinical oncology are related to this effort in nanomedicine. But still, no cure for metastatic disease. Early in the first decade of the new millennium I became convinced that the "standard form" of nanomedicine (nanoscale spherical nanoparticle of one of many biomaterials; comprising one or more of many chemotherapeutic agents; possibly "targeted" by one or more of many different biological recognition biomolecules, such as monoclonal antibodies) would never succeed in curing visceral metastases.

Through abundant experimentation, both in-vitro and in in-vivo, we realized that the problem was—well, the mechanics of mass transport in the body, and its modifications (biobarriers) associated with the onset and growth of cancer lesions.
And here is the point, where brute force experimentation becomes futile, there just are not enough experimental mice in the world, and it is necessary to make recourse to mathematical insights, à la Noll. With several exceptional colleagues, we spent about ten years determining the mechanical framework of governing equations for the transport of nanoparticles (and other forms of biological and synthetic mass). We then used it as a "compass" to determine optimal physical design parameters for the nanocarriers, including non-spherical size, shape, surface charge, surface-expressed affinity and density, among others. No amount of experimentation would have sufficed to yield the conclusions that were reached by mathematical analysis. Less was better, when warranted, but the identification of the domains of warrant could only be attained by mathematics. That was the good news—but it came with bad news: the reconfirmation that the canonical model of nanomedicine would never be able to cure metastases.
Thus, the new field of Transport Oncophysics was born, but with the ominous recognition that the problem of curing metastases required a new paradigm, to see a successful conclusion. Toward it, we then developed and launched Multi-Stage Vectors, only to find that they were indeed useful in extending life of metastatic animals in multiple cancer types, but not curative of those pathologies. And, finally, the notion emerged of injectable Nano Particle Generators (iNPG), i.e., multistage carriers that deploy inside the biological microenvironment of lung and liver metastases the ingredients that are necessary to form exosome-like nanoparticles. If these are formulated in the right way (again, ten years of work in this sentence), the exosome-like particles that will be formed by the cancer itself will reach deep into the therapy-resistant cancer cells, across all biobarriers, and provide a definitive cure for a large portion of the animal models on which this has been tried. As we write, we have received extraordinary levels of financial support from government agencies (National Cancer Institute, Department of Defense Breast Cancer Program) and are expecting to initiate shortly the clinical trials for this new mechanics-based therapeutic approach. And how do we "convince" the cancer lesions we want to eliminate to commit hara-kiri, by forming lethal exosome-like nanoparticles that will pass their defenses and kill them? Well, in one word: Thermodynamics. The Second Law always wins. We have come full circle, thanks Gualtiero!
Truth be told, after these 26 years of failures I have become convinced that not only cancer manifests itself through the multiscale dis-regulation of mass transport, mediated largely by the evolution of biobarriers in carcinogenesis (as we postulated in Transport Oncophysics)—I now do believe that is what cancer actually is—a veritable disease of mechanical disruption. But, that is a topic for future presentation. Now it is time to address the third dimension of Noll's influence on our work. Back to micromechanics, which the readers who have not fallen asleep just yet might remember that I had entered as a domain of application of a modified form of the Coleman-Noll approach.
One key challenge of micromechanics is to determine the proper dimensional domains at which a homogeneous representation of matter is warranted. This is of course intrinsic to continuum mechanics itself, where a "dV" is imagined to correspond to a physical dimension, where that unavoidable, pesky discrete nature of matter can actually be ignored. Yet, it is not infrequent in engineering applications that structural and microstructural dimensions are not that far from each other, and a simple

continuum approach—let alone with assumption of materials homogeneity—can no longer be applied. Composite materials, granular media, multicrystals, and the whole field of nanotechnology come readily to mind, in this concern. Bottom-up approaches (from first principles up) give intractable formalisms for anything beyond a handful of atoms, while top-down approaches present an impossibly large number of parameters to be determined by an impossible experimental strategy. The need to meet in the middle between these two approaches requires suitable guidance tools—and here again, thermodynamics and a Coleman-Noll formalism are indispensable.

Facing these challenges, two strategies have been pursued. In the first (Doublet Mechanics) one discretizes matter with an overlaid array of representative points at finite distances, and develops a formalism that recovers continuum mechanics on one end, and lattice dynamics on the other—with many, not trivial scale theories in between. The constitutive equations for this approach were developed employing, naturally, a suitable form of the Coleman-Noll methodology.

The second approach is a homogenization theory, using a strain concentrator theory methodology, derived as an extension of Eshelby's exact results at extremely dilute conditions. Dubbed the "poly-inclusion theory", this approach has guided the development of hybrid, multiphase biomaterials, for the post-traumatic regeneration of bone and the surrounding soft tissues. The objective is the treatment of supercritical fractures, that is, lesions that are so large that amputation still remains the only option.

In conclusion, be it novel therapies for cancer metastases, or new approaches to save limbs that have suffered injuries beyond current repair, everything I have done in my career is based on the insights offered by fundamental principles of mechanics, and innovative methodologies for understanding their import on the problems of contemporary medicine. Throughout these efforts, Walter Noll and his exquisitely insightful approaches have provided me with invaluable guidance. *Grazie, Gualtiero!*

## 4 Recurrent Research Themes

The preceding recollections partly illuminate different aspects of Noll's scientific influence, as seen through the diverse lenses of personal perspectives; more about Noll's impact in science will be said wherever pertinent.

A purely chronological account of Noll's scientific production would be unnecessarily complicated, full of twists and turns.[11] Accordingly, Noll's system of notations—to which he attributed a special role, not just as a means to represent mathematical ideas, but as a part of their very creation—has been subject to a similar meandering evolution in time. This has made it impossible for us to use a single coherent notation throughout this section of our essay without betraying either the intention of our author or the historical truth. Our choice has been to sacrifice notational coherence to textual loyalty. Thus, we strive here to adhere to the original notation, explaining it wherever it occurs. The price for the reader is to tolerate that sometimes different symbols and fonts are used for the same physical quantity.

[11] For example, the expository paper [A29] is based on a series of lectures delivered at the Technion of Haifa, Israel, in the summer of 1972. It is typical of how Noll's thoughts progressed, through a sequence of successive refinements involving higher levels of abstraction. His earlier views about a comprehensive mathematical infrastructure of continuum mechanics and thermodynamics were first presented in 1965 in the *Bressanone Lectures* [C6]; seven years later, they had already become obsolete for him.

Thus, for example, the area measure could be either $\mathrm{da}r_x$ or $ds$, and the stress tensor could be either $T$ or $\boldsymbol{T}$, or even $\mathbf{S}$.

Noll's main research themes can be listed according to the following temporal partition of his professional life:

- **1953–1955** (firstly in Bloomington, then in Berlin)
  - microscopic counterparts of continuum-mechanics balance laws and Cauchy stress
  - outline of a general theory of simple material bodies
- **1955–1993** (mostly in Pittsburgh, as a professor at Carnegie Mellon University)
  - viscometric flows, rheology
  - foundational issues in classical and continuum mechanics
  - reformulations and refinements of a general theory of simple bodies
- **1994–2013** (in Pittsburgh, after his retirement from teaching)

  In this span of time, most of Noll's contributions are refinements and comments of his earlier writings; some also bear a controversial character and others are more of a philosophical nature.

A full list of Noll's publications is available in Sect. 7. We have renounced any attempt at a systematic account on each and all of them. Hereafter, we shall only refer to those actually required in developing our discourse about his recurrent research themes; others, which deal with episodic research themes, will be referred to in the next section.

On August 9, 1954, Noll defended is Ph.D. thesis, a some 80-page document soon to be published in article form in the *Journal of Rational Mechanics and Analysis* [A3]. We begin by [A3] because there one finds many of the research themes that he cultivated along his scientific life.

## 4.1 On the Continuity of the Solid and Fluid States

This subsection borrows the title, and Fig. 2 reproduces the table of contents, of [A3], the third of Noll's published works in the form of a journal article. This paper stands as an unprecedented, detailed and systematic exposition of a subject that still is the flagship theory of continuum mechanics, that is, the theory of simple material bodies.[12]

The Introduction is exemplar: to set the stage, the title is explained; then, a short but exhaustive summary is offered of the issues addressed, of the author's intentions, and of the main outcome of his efforts. Firstly, the state equation of a basic one-dimensional rheological model, Maxwell's, is laid down, and it is pointed out that the time-differentiated state equation of Hooke's model for a linearly elastic solid and the state equation of Newton's model for a linearly viscous fluid are obtained when the relaxation-time modulus $\tau$ tends to, respectively, infinity and zero (keeping the value of the viscosity modulus $\eta = \mu\tau$ fixed in the second instance):

$$\text{(Maxwell's model)}\quad \frac{ds}{dt} = \mu\frac{da}{dt} - \frac{s}{\tau} \Rightarrow \begin{cases} \frac{ds}{dt} = \mu\frac{da}{dt} & \text{(Hooke's model, } s = \mu a\text{)},\\ s = \eta\frac{da}{dt} & \text{(Newton's model)},\end{cases}$$

where $s =$ stress, $a =$ strain, and $\mu =$ elasticity modulus. The reader realizes to be in the presence of a smoothly $\tau$-parameterized family of materials, whose state equations embody

[12] A well-known exposition of this theory, including all developments up to 1965, is found in Chap. C of [B1], entitled "The general theory of material behavior". Some seven years later, Noll conceived and published [A28], that is, in his words from [WB8], "a general framework for the formulation of constitutive laws, using the concept of a state of a material element".

**TABLE OF CONTENTS**



**Fig. 2** Table of contents of [A3], the journal's article reproducing Noll's thesis

information both on their *aggregation* and on their *constitutive equations*,[13] from elastic solids to viscous fluids.[14] A three-dimensional nonlinear version of this one-dimensional linear exampling family of constitutive equations is proposed in Chap. II, where, among other things, in Sect. 6 Noll credits Maxwell [56] for "[t]he idea that solids and fluids are

[13] 'State' is a loaded term. The information about aggregation these state equations embody is singled out by performing, even only mentally, creep and relaxation experiments; the impact of Noll's and Coleman's work on rheology will be discussed in Sect. 4.4. The fortune of the term 'constitutive equation', which to our knowledge was used in this paper for the first time, has been such as to be eventually registered in the *Oxford English Dictionary* (http://www.oed.com/view/Entry/39861?rskey=iEMOLh&result=1&isAdvanced=true#eid8396069), making his coiner proud.

[14] "These two cases are also the extreme cases—pure fluidity and pure elasticity—in the landscape so beautifully described in his "continuity of states" by Walter Noll in 1955." [55, p. 2].

only limit cases of a general type of material" to be termed, he proposes, *hygrosteric*.[15] Accordingly, (pure) solidity and (pure) fluidity are viewed as extreme aggregation states.

Once a reasoned summary of contents of Chaps. III and IV is completed, the Introduction features a statement with which every mathematical modelist can only concur, namely, that constitutive equations may not "represent accurately the behavior of real material" but that "at least" they should lead to a mathematical theory that is "consistent, exact, and rigorous" (and as such capable of furnishing verifiable or falsifiable predictions, we add). Quite modestly, our author regards his as "a small step towards an axiomatic foundation of continuum mechanics", a statement that reminds us, *si parva licet componere magnis*, of other so-called 'small steps', like Mao Zedong's and Neil Armstrong's, and a statement that points, if only implicitly, to Hilbert's Sixth Problem (we shall return to the latter issue in Sect. 4.3). Finally, for the first time in a major paper on continuum mechanics, "coordinates, components and matrices" are banished in favor of "points, vectors and linear transformations", the corresponding objects in Euclidean geometry and modern linear algebra.

We see in Fig. 2 that paragraph A. Introduction is followed by two more preliminary paragraphs, B. Notations and C. Mathematical Preliminaries: it is in this third paragraph that we find a first exposition of those notions of algebra and analysis that Noll regarded as indispensable, at the time of this writing, to handle mechanical issues in a neat and conceptually economical manner and that he later implemented under form of a comprehensive mathematical syllabus of his taste [WC1]. He returned to these matters over and over again, in a generous attempt to put together a mathematical infrastructure for thermomechanics [WB8], [WB9] that not only mathematicians but "even ... physicists and engineering scientists" would eventually adopt. Unfortunately, this life-long effort was not successful. Worse, it brought him bitterness and disappointment, due to the charges of unreadability and 'Bourbakization' of his writings, charges he confuted in [WB8] (more on this is in our Sect. 6 below).

In Chap. I, where attention is explicitly confined to pure mechanics, general notions of the strain and stress involved in the formulation of constitutive equations for those materials bodies that will later be termed *simple* are introduced,[16] as well as the invariance principle guiding or rather restricting that formulation, that is, the principle of *isotropy of space*. Here is how Noll motivates and formulates this principle.

Just as in classical cosmology, [A3, p. 17],

> ... the physical space is [regarded as] homogeneous and isotropic ... there are no distinguished positions or directions ... the constitutive equations of a material can involve only the relative positions of the material points and not their absolute position in space. Therefore, if for a given body in one configuration the constitutive equation is satisfied, ... then the constitutive equation must be satisfied also for [a] second configuration [that] can be obtained from the first by a rigid displacement.

Being the constitutive equations regarded as "functional relations (*e.g.* differential or integral equations) between the stress $S(x,t)$, the density $\rho(x,t)$ and the motion $f(x,t)$" (*loc. cit.*),

[15] This adjective, an artefact combining the Greek adjectives ὑγρός for *fluid* and στερεός for *solid*, had some modest fortunes (e.g., references are made to 'Noll's theory of hygrosteric materials'), insufficient to make it to the *Oxford English Dictionary*. Neither they made it to the OED two terms, hereafter italicized, that became widely adopted though with acceptations slightly different from Noll's original ones, that is, *extra-stress* for the stress diminished by the pressure stress possibly to be found in a hygrosteric body *at ease* (more about this term later).

[16] We read in Sect. 14 of [B1] that a material is *simple* if "... the collection of kinematical facts needed to determine the stress reduces to the history of the deformation gradient alone."

isotropy of space is stated as the requirement that the constitutive equations be expressed as form-invariant relations about the transformations of those variables under whatever space isometry [A3, p. 19].

In the 1960 years, this principle was to be renamed the principle of *material frame-indifference*.[17] Writes Noll on p. VIII of [C15]:

> As a thesis topic, [Clifford Truesdell] gave me a paper written by S. Zaremba in 1903 and asked me to make sense of it. I found that, in order to do so, a general principle was needed, which I called 'The Principle of Isotropy of Space'. This principle also served to clarify many assertions in mechanics that had been obscure before (at least to me).

Decades later, on continuing his efforts to distill the essence of continuum mechanics, Noll devised a frame-free formulation of elasticity, to which such a principle was irrelevant, in fact alien (see [A28], [A37] and [WB2]; see also [A42] and [WB10]).

Noll has been repeatedly—and indeed rightly—faulted for not mentioning a relevant antecedent of his 1955 principle, which unfortunately had not been given a name by J.G. Oldroyd in his 1950 paper [58] devoted to the formulation of rheological equations of state in a form "independent of the frame of reference".[18] In fact, we see from the references in [A3] that Noll's sources, beside Zaremba's papers [106, 107], were Truesdell's 1952 and 1953 papers [82] and [83], where Oldroyd's contribution is overlooked.[19]

Today readers may find in Chaps. II and III, clearly defined and illustrated in detail, a panoply of concepts central to the mechanics of simple bodies, and appreciate the ample and appropriate vocabulary, much of which is still in use, adopted or developed by Noll to cope with those concepts.

---

[17] Other names were proposed, *in diminuendo*: principle of objectivity of material properties, principle of material objectivity, principle of objectivity (cf. paragrah $\zeta$ on p. 702 of [93] and Sect. 19A of [B1], where a history of the principle is told, from Hooke's 1675 *Philosophical-Scales* up to 1965, the year when [B1] appeared in print). We shall have more to say about material frame-indifference in Sect. 4.5.2.

[18] Curiously enough, in 1950 continuum mechanics was called *rheology* by Oldroyd and *hydrodynamics* by Irving and Kirkwood [46], as that discipline was probably referred to in their respective scientific communities.

[19] The only somehow relevant quotation is found in footnote 4 on p. 138 of [82]:

> Oldroyd bases his dynamical proposals upon an identification principle, in which the spatial and the material co-ordinates are made to coincide at a given instant and the numerical values of the components of a material tensor coincide with those of the corresponding components of an associated spatial tensor.

A more accurate reference to [58] is found both in footnote 3 on p. 703 of [93] and in Sect. 19A of [B1] (here [105], another relevant paper by Zaremba, is also cited; by the way, in [58] Zaremba is not mentioned at all), where the same quotation from [58] appears:

> The form of the completely general equations must be restricted by the requirement that the equations describe properties independent of the frame of reference ... *Moreover, only those tensor quantities need be considered which have a significance for the material element independent of its motion as a whole in space* [italics in the cited text].

Eventually, with characteristic intellectual honesty, Truesdell acknowledged in full the importance of Oldroyd's contribution. We read on p. 44 of [90], apropos of the criteria he had chosen to decide coverage and focus of his own papers [82, 83]:

> My failure to appreciate at its true value the fundamental memoir of Oldroyd [58] is a symptom of my failure to see then that what the field needed was *generality in order to reach simplicity and clarity* [italics in the cited text].

In [A3], $\rho$, $\mathbf{v}$, $D$, $W$, $S$, and $T = S + pI$, with $p$ the pressure and $I$ the identity tensor, denote mass density, motion velocity, symmetric and skew-symmetric parts of velocity gradient, stress, and extra-stress (recall footnote 15); a superposed dot denotes time differentiation. The general hygrosteric state equation (that is, Eq. (7.2)) reads

$$\dot{T} + TW - WT = \mathfrak{F}(D, T, \rho). \tag{1}$$

The following rather subtle distinction is made between three seemingly akin states of a material body before it is acted upon by external actions and/or confined in a space region: a material body is *at ease*, if it so happens that the fields $D$, $W$, and $\dot{S}$, have value zero; *at rest*, if the fields $\mathbf{v}$ and $\dot{S}$ have value zero; *in a natural state*, if the stress field has value zero and the density field has constant value. With this distinction in mind, we look at the diagram reproduced in Fig. 3 from Sect. 17 of [A3], which depicts the tree of mutually and progressively nested subclasses of the general class of hygrosteric materials. The first branching splits that class into *fluent* and *resilient* materials. A material body is fluent if it can be at ease only when the stress is a pressure; resilient if it can be at ease only when the stress takes values only in a specified range. Not surprisingly, all fluent material bodies are fluids by nature. Hypo-elastic materials, a constitutive class about to be proposed [85] at the time Noll was composing his thesis, are resilient, and such are elastic materials. In Sect. 15, some notions of particular interest are introduced: that of a stress distribution being *elastically relievable*; the opposite notion of *residual stress*; and the notion of a *dislocatable body*, that is to say, a body in a state of stress that can be piece-wise elastically relievable provided a finite number of appropriate cuts is performed.[20]

The final Chap. IV begins by associating the balance equations of mass and momentum to the constitutive equation (1) specifying hygrosteric response in general. The resulting system is shown to admit a number of special solutions: *pure stress relaxation*, in Sect. 19, consists in a three-dimensional version of the one-dimensional stress relaxation experiment of classical rheology; other solutions are valid when the stress field is spatially uniform (Sect. 20) and when the flow is linear and steady (Sect. 21), Poiseuille's (Sect. 22), and Couette's (Sect. 23). Noll's detailed mathematical study of all these classical flow problems is an obvious predecessor of the neat 1966 booklet [B2] on viscometric flows of non-Newtonian fluids he wrote with B.D. Coleman and H. Markowitz. We shall comment on Noll's and Coleman's joint work of direct relevance for rheology in Sect. 4.4 and we shall say more about [B2] in Sect. 4.5.4.

## 4.2 Microscopic Counterparts of Macroscopic Balance Laws and Cauchy Stress

We deal here with Noll's mathematical clarification and completion [A2], in 1955, of the 1950 pathbreaking paper [46] by J.H. Irving and J.G. Kirkwood. Interestingly, the "equations of hydrodynamics" of those authors, two physicists, became Noll's "Grundgleichungen der Thermomechanik der Kontinua". He returned to the subject in 2008 [WB7], in a presentation at the SIAM meeting in Philadelphia, and in 2010 [A43], on the occasion of the publication of an English translation of his 1955 paper [50]. Both this translation and Noll's paper appeared in volume 100 of the *Journal of Elasticity*, a thematic volume consisting of invited contributions and prompted by the renewal of interest that the success of Molecular Dynamics simulations had generated for a number of conceptual issues at the core of microscopic $\leftrightarrow$ macroscopic scale bridging.

[20] In this connection, Noll refers to Volterra's theory of dislocations, as exposed by Love in [53]. He tackled again this topic in [A23], see Sect. 5 below.

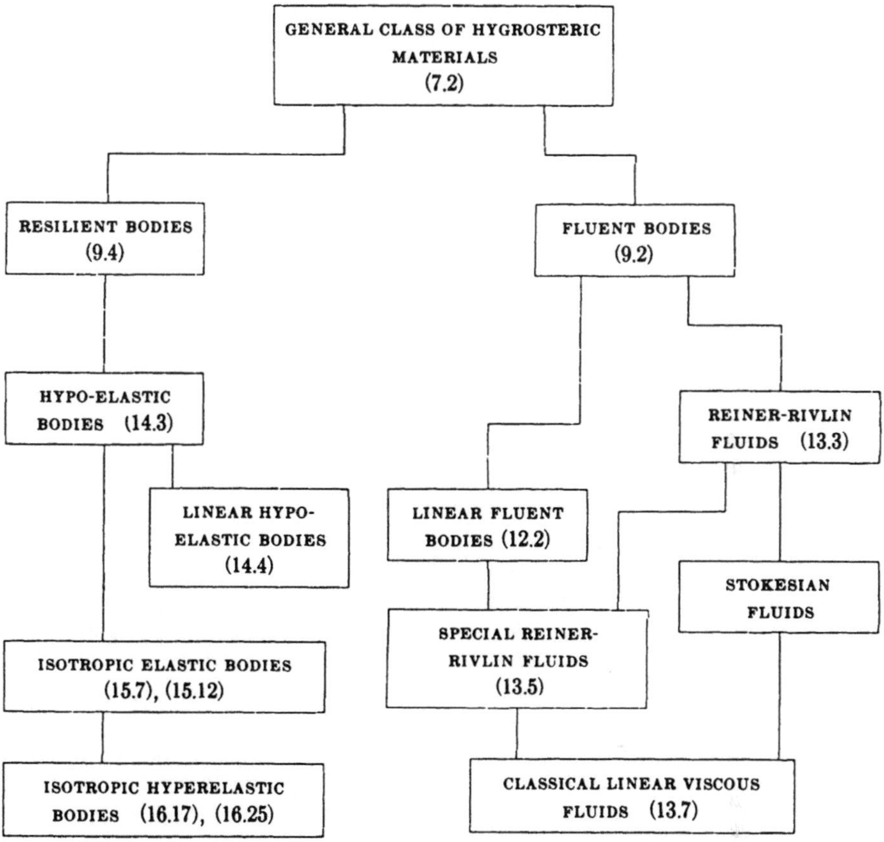


**Fig. 3** A diagram detailing the nested structure of the hygrosteric class of constitutive equations (numbers in boxes refer to equations in [A3])

One of those issues, perhaps the most important one for a convincing passage from discrete to continuum mechanics, was and is the construction of stress fields, perhaps more complex than Cauchy's, associated to molecular motions. In fact, while the only interactions among particles are at a distance, a continuous body is so modelled as to let its parts experience both distance and contact interactions; moreover, contact interactions and the associated stress fields carry the same information. Now, for a number of reasons, a direct microscopic interpretation for contact interactions is not viable, no matter how modest the complexity of particle interactions. But, if such interactions are the simplest elastic pair-wise interactions (or if they are reducible to pair-wise [1, 11, 12, 78]), then a microscopic interpretation *à la* Irving and Kirkwood for Cauchy stress can be achieved, whence, if one so wishes, an indirect microscopic interpretation of contact interactions follows.

Irving & Kirkwood's endeavor was in line with the XIX century views about the discrete-to-continuum passage. They worked in the framework of statistical mechanics to deduce from the Hamiltonian version of the Newtonian equations ruling the motion of a system of $N$ identical massy particles certain differential consequences that plausibly might serve as microscopic counterparts of the point-wise versions of mass, momentum, and energy, balance laws for simple material bodies. In particular, they individuated, to within a divergenceless addendum, a microscopic counterpart for the tensor whose divergence appears in momentum balance, and they proposed to interpret that tensor as the Cauchy stress.[21]

The key step of Irving & Kirkwood's deduction was the Liouvillean evolution of macroscopic observables obtained by coupling Dirac localization with state-space averaging of appropriately chosen microscopic observables: e.g., they took the microscopic counterpart of the current mass density field to be the expected value of the macroscopic observable

$$\sum_{J=1}^{N} m_J \delta(\boldsymbol{r}_J - \boldsymbol{r})$$

(here $m_J$ is the mass, and $\boldsymbol{r}_J$ the current position with respect to a fixed origin, of the typical particle; $\boldsymbol{r}$ is the position vector of the typical space point).

In [A2] Noll, apparently prompted to do so by Truesdell, cleaned up Irving & Kirkwood's developments by showing how to get rid of Dirac measures and their mathematically not palatable differentiation by series. Here we concentrate on the most subtle and creative of Noll's achievements in [A2] and [A43], the concept of a *reaching stress*: in his words, "...a concept of stress ...that describes interactions at a distance". For this very reason, that of reaching stress is a scale-bridging concept, akin to the familiar continuum mechanics notion of *contact stress* due to Cauchy but different from it. In fact, the latter notion need not to be given an interpretation in terms of microscopic interactions, whatever their nature. Instead, it describes the surface interaction related to contact forces exchanged by body parts at their imagined common boundary or exerted on a body's actual boundary by its environment; since both material surfaces and contact forces are intrinsically macroscopic notions, such is the notion of contact stress.

In the abstract of [A43] Noll states that "[p]erhaps contact stresses can be derived as limit cases of reaching stresses". This is, in our view, more a suggestion than a conjecture to be proved or disproved, a suggestion that to our knowledge has not been picked up so far.[22] However, reaching stresses can be written in a form that makes them strongly resemblant of contact stresses.

We premise a concise account of Noll's developments in [A43]. For $\mathcal{B}$ the region of the three-dimensional Euclidean space $\mathcal{E}$ occupied by a given material body, he denotes by $\boldsymbol{g}$ an integrable vector-valued mapping over $\mathcal{B} \times \mathcal{B}$ satisfying

$$\boldsymbol{g}(x, y) = -\boldsymbol{g}(y, x) \quad \text{for all } x, y \in \mathcal{B},$$

a skew-symmetry requirement crucial to the description of pair-wise particle interactions at a distance. Next, he extends $\boldsymbol{g}$ to null outside $\mathcal{B} \times \mathcal{B}$, and asks whether a tensor-valued field $\boldsymbol{R}$ over $\mathcal{B}$ can be found such that

$$\int_{\mathcal{P}} \left( \int_{\mathcal{E}} \boldsymbol{g}(x, y) dy \right) dx = \int_{\partial\mathcal{P}} \boldsymbol{R}\boldsymbol{n} \quad \text{for whatever part } \mathcal{P} \subset \mathcal{B}$$

(here $\partial\mathcal{P}$ and $\boldsymbol{n}$ denote the boundary surface of $\mathcal{P}$ and its outer normal), or rather, equivalently, such that

$$\int_{\mathcal{E}} \boldsymbol{g}(x, y) dy = \operatorname{div} \boldsymbol{R}(x) \quad \text{for all } x \in \mathcal{B}.$$

[21] For a discussion of this interpretation and an alternative to it, we refer the reader to [64].

[22] Similarly, it remains to be explored how reaching stresses can be related to both *contactors* and *semi-contactors*, which feature in the general theory of interactions, see Sect. 4.3.4.

The idea is to interpret $\boldsymbol{Rn}$ as a surface-interaction density over $\partial\mathcal{P}$ equivalent to the volume-interaction density $\boldsymbol{g}$ over $\mathcal{P}$; the task is to give a representation for $\boldsymbol{R}$ in terms of $\boldsymbol{g}$. Noll shows that such a representation is

$$\boldsymbol{R}(x) = \frac{1}{2}\int_{\mathcal{V}}\left(\int_0^1 \boldsymbol{g}\big(x-(1-\alpha)\boldsymbol{v}, x+\alpha\boldsymbol{v}\big)d\alpha\right)\otimes \boldsymbol{v}\,dv, \tag{2}$$

where $\mathcal{V}$ denotes the translation vector space of $\mathcal{E}$.[23]

As indicated in [64], on setting

$$\boldsymbol{r}(x,\boldsymbol{v}) := \frac{1}{2}\int_0^1 \boldsymbol{g}\big(x-(1-\alpha)\boldsymbol{v}, x+\alpha\boldsymbol{v}\big)d\alpha,$$

(2) becomes

$$\boldsymbol{R}(x) = \int_{\mathcal{V}} \boldsymbol{r}(x,\boldsymbol{v})\otimes \boldsymbol{v}\,dv; \tag{3}$$

to within appropriate specifications of densities, the vector $\boldsymbol{r}(x,\boldsymbol{v})$ may be interpreted as a reaching force per unit area on a surface through $x$ of normal $\boldsymbol{v}/|\boldsymbol{v}|$ and the tensorial integrand as the value at $x$ of an elementary reaching stress field. Representation (3) has a remarkable similarity with the well-known dyadic representation of contact stress due to Cauchy,

$$\boldsymbol{T}(x) = \sum_{i=1}^{3} \boldsymbol{c}(x,\boldsymbol{n}_i)\otimes \boldsymbol{n}_i,$$

where $\boldsymbol{c}(x,\boldsymbol{n}_i)$ denotes the contact force at point $x$ of a surface through $x$ of normal $\boldsymbol{n}_i$, provided the scalar triple product of vectors $\boldsymbol{n}_i$ is not null.

### 4.3 General Theory of Interactions

G. Hamel wrote to Truesdell in 1952 [90, p. 527], "in the concept of force lies the chief difficulty in the whole of mechanics", thus showing how the ambitious programme that Hilbert had conceived in 1900 was still in demand of achievement half a century later. The sixth of Hilbert's problems [43] reads as follows:[24]

> The investigations of the foundations of geometry suggest the problem: *To treat in the same manner, by means of axioms, those physical sciences in which mathematics already plays an important part; in the first rank are the theory of probabilities and mechanics. ...* [italics and ellipses in the cited text]

If mentioning geometry in connection with rigour sounds foreign to the modern ear, the following excerpt from [43] will clarify Hilbert's opinion about rigour and, by contrast, that of his contemporaries:[25]

---

[23] A step in the proof of this result consists in an application of Fubini's theorem, that is, of the Iterated Integration Theorem according to Walter, who used to advocate that theorems, laws of physics *et sim.* should be designated by alluding to their contents rather than be named after their discoverers.

[24] In Truesdell's translation [90, p. 507].

[25] Again in Truesdell's translation [90, p. 136].

> While I insist upon rigor in proof as a requirement for a perfect solution of a problem, I should like, on the other hand, to oppose the opinion that only the concepts of analysis, or even those of arithmetic alone, are susceptible of a fully rigorous treatment. This opinion ... I consider entirely mistaken. Such a one-sided interpretation of the requirement of rigor would soon lead us to ignore all concepts that derive from geometry, mechanics, and physics, to shut off the flow of new material from the outside world. ... What an important, vital nerve would be cut, were we to root out geometry and mathematical physics! On the contrary, I think that whenever mathematical ideas come up, where from the theory of knowledge or in geometry, or from the theories of natural science, the task is set for mathematics to investigate principles underlying these ideas and to establish them upon a simple and complete system of axioms in such a way that in exactness and in application to proof the new ideas shall be no whit inferior to the old arithmetical concepts.

Noll's theory of interactions, with all its ramifications, is the perfect example of how geometry and analysis can be brought together into a rigorous theory of mathematical physics. According to Truesdell, who credited Noll with the solution of the relevant part of Hilbert's sixth problem [90, p. 137]:[26]

> *The essence is conceptual analysis*, analysis not in the sense of the technical term but in the root meaning: logical criticism, dissection, and creative scrutiny. After comes the poetry of statement and the proof, the ornament of illuminating examples.

In Noll's mathematical approach, forces lose their dark character, which previous attempts to clarify this concept had contributed to accrue. In this connection, we find the following comment of Truesdell [89, p. 322] particularly illuminating:

> An *a priori* concept of force was, perhaps, Newton's main contribution, but in his writing it is rather illustrated than presented, explained, or defended explicitly, and at first only his own British disciples were converted to his way of looking at things. More than half a century after the publication of the *Principia*, D'Alembert, expressing the view that what we call force is only a manifestation of motion, spoke of Newtonian forces as "obscure and metaphysical beings, capable of nothing but spreading darkness over a science clear by itself." It was through the dozens of examples worked out by Daniel Bernoulli and Euler that the fruitfulness of taking forces as quantity given *a priori* came gradually to be accepted.

The right question is no longer: *What are forces?* They are undefined objects like points and lines in geometry. The right question is: *What are the properties of forces?* This is not a semantic, but a mathematical issue that can be settled through the adoption of (possibly different) sets of axioms, posited to represent the *mutual action* of material bodies, as every system of forces is ultimately reducible to a combination of elementary material *interactions*. The stark contrast between this approach and that still prevailing among Noll's contemporaries who knew about his work can easily be appreciated by reading the following excerpt from [76] published in 1974 (see also [90, p. 530]):

> The outstanding work of Walter Noll should also be mentioned, but his axioms on the concept of forces acting on bodies require a highly sophisticated and developed mathematical framework.

[26] Apart an earlier noteworthy attempt of Hamel in 1909 [40], as duly noted by Truesdell in his treatise [96, p. 6].

The notion of force, which Newton had introduced as a primitive concept (with no further explanation) and Euler systematically extended so as to include actions Newton had not envisaged, such as those arising from the contact between bodies, was not only fiercely rejected by D'Alembert.[27] Such a rejection of force was destined to grow in modern times, as noted by Wilszek in his witty column [100] (see also the sequels [101–103]), who quotes Tait and Russell, the latter going to the extreme of considering the "old notion of force" even "morally suspect". Despite the surge of much opposition, the "culture of force" remains undeniably strong. According to [100], this culture

> ...survives the competition, and continues to flourish, for one overwhelming good reason: It is much easier to work with.

In addition, we are reminded in [66] that also the quantum mechanics of molecules is imbued with forces, as tersely realized by Feynman [27]:

> Many of the problems of molecular structure are concerned essentially with forces. ...Usually these problems have been considered through the agency of energy, and its changes with changing configuration of the molecule. The reason for this indirect attack through energy, rather than the more qualitatively illuminating one, by considerations of force, is perhaps twofold. First it is probably thought that force is a quantity that is not easily described or calculated by wave mechanics, while energy is, and second, the first molecular problem to be solved is the analysis of band spectra, strictly a problem of energy as such. ... [F]orces are almost as easy to calculate as energies are ....

This passage of plain prose shows better than any grand philosophical discourse how the "cultures" of energy and force, which in modern physics are so often opposed to one another, are in the end not so different, and neither deserves to be banished.

Noll read both papers [66, 100] and wrote a short comment [WB5] about them. If Wilczek had opened his column saying that "Newton's second law of motion, $F = ma$, is the soul of classical mechanics". Noll closed his note saying that "[t]he soul of continuum mechanics is the analysis of contact forces". We shall return to this in Sect. 4.3.4 below.

We honestly cannot say whether Noll solved (and in whatever part) the sixth of Hilbert's problems.[28] Perhaps, the best we can do is to quote a passage from p. 48 of [C2]:

> Dans ce rapport je me propose de traiter la mécanique classique comme une structure mathématique. L'objectif de ce traitement est de donner des definitions mathématiques précises des concepts généraux de la mécanique classique en extrayant tout ce qui est commun à ses branches diverses: la mecanique des masses ponctuelles, la mécanique des corps rigides, la mécanique des fluides et des milieux élastiques, et la mécanique moderne des matériaux qui ne sont ni fluides ni élastiques dans le sens

[27] We also learn from [89, p. 113] that D'Alembert said that "...we shall be content to remark that the principle [of accelerating force], be it true or be it dubious, be it clear or be it obscure, is useless to mechanics and ought therefore to be banished from it."

[28] On a more general note, we cannot say either whether he was the only one who solved it, or, for that matter, whether the sport (so dear to popular science) of naming those who supposedly solved any of Hilbert's problems has any scientific significance. It is difficult to be in anyone else's mind, but perhaps Hilbert had not meant to start a mind contest; his only purpose might have been to set the scene for a modern, pervasive role of mathematics in science. A century later, we may at least say that this program has been successful and that that shift in perspective for the role of mathematics has been beneficial for all applied sciences, which have become more mathematical, even beyond their desire; this is the only lasting legacy of Hilbert's paper [43], and the merit is shared by the many mathematicians who have embraced the new century's mentality.

> classique (voir [A7] et les références données dans ce papier). Je crois que c'est une telle axiomatisation que Hilbert eut en vue quand il posa le sixième de ses célèbres problèmes en 1900 [43]. Je ne crois pas que la partie du sixième problème de Hilbert qui concerne la mécanique a trouvé une solution satisfaisante.

On the other hand, an excess of praise [WA2] may be counterproductive, as voiced with sarcasm in the following excerpt of Rivlin's review [67] of [B4]:

> For many years, Noll has been the beneficiary—or perhaps victim—of a propaganda campaign conducted on his behalf by Truesdell, his teacher and mentor. The present volume, a collection of 16 of Noll's papers with a panegyrical preface by Truesdell, is the latest thrust of this campaign. While it is undoubtedly true that Noll's influence upon research in the foundations of mechanics and thermodynamics has been considerable, much of it has been a product of Truesdell's campaign. Moreover, the extent to which this influence has been benign or malign may well be considered.

You may or may not agree with Rivlin, but the fact remains that an excess of praise, even if well-intended, invariably overshadows the real scientific merit, which is there and remains unaltered, no matter what we think. We strive in our account not to fall victim to the panegyric temptation and to offer to our readers a compass to navigate the sea of Noll's objective achievements through their continuous extensions and refinements.

### *4.3.1 Material Universes*

A fairly complete and systematic account of Noll's theory of interactions appeared in print in [A29] (we have already commented above on the origin of this expository paper). This theory builds upon the general notion of *material universe* $\Omega$. This is a set whose members $\mathcal{A}$, $\mathcal{B}$, $\mathcal{C}, \ldots$, are the *bodies* (or *material objects*, in Noll's original terminology). $\Omega$ is endowed with the structure of a Boolean algebra by a relation $\prec$, according to which $\mathcal{A} \prec \mathcal{B}$ means that "$\mathcal{A}$ is part of $\mathcal{B}$". $\Omega$ and $\prec$ satisfy 6 axioms, which are intuitively plausible and make rigorous and universal the different notions of *material system*, which different branches of mechanics describe vaguely to delimit their scope. Here are Noll's axioms, as they appear in [A29].[29]

(M1) $\mathcal{A} = \mathcal{B}$ if and only if $\mathcal{A} \prec \mathcal{B}$ and $\mathcal{B} \prec \mathcal{A}$.
(M2) $\mathcal{A} \prec \mathcal{B}$ and $\mathcal{B} \prec \mathcal{C}$ implies that $\mathcal{A} \prec \mathcal{C}$.

The partial ordering thus introduced in $\Omega$ warrants the uniqueness of the *greatest common part* $\mathcal{C}$ of two bodies $\mathcal{A}$ and $\mathcal{B}$. If such a $\mathcal{C}$ exists, it is called the *meet* of $\mathcal{A}$ and $\mathcal{B}$, and denoted $\mathcal{C} := \mathcal{A} \wedge \mathcal{B}$. Similarly, given $\mathcal{A}$ and $\mathcal{B}$ in $\Omega$, there can be at most one *least envelope* $\mathcal{D}$ of $\mathcal{A}$ and $\mathcal{B}$. If such a $\mathcal{D}$ exists, it is called the *join* of $\mathcal{A}$ and $\mathcal{B}$, and denoted $\mathcal{D} := \mathcal{A} \vee \mathcal{B}$.

(M3) There are two elements, $\varnothing$ and $\infty$ in $\Omega$, called the *material nothing* and the *material all*, respectively, such that $\varnothing \prec \mathcal{A} \prec \infty$ for all $\mathcal{A} \in \Omega$.

If $\mathcal{A} \wedge \mathcal{B} = \varnothing$, we say that $\mathcal{A}$ and $\mathcal{B}$ are *separate*.

(M4) For each $\mathcal{A} \in \Omega$ there is precisely an $\mathcal{A}^e \in \Omega$, called the *exterior* of $\mathcal{A}$, such that $\mathcal{A} \wedge \mathcal{A}^e = \varnothing$ and $\mathcal{A} \vee \mathcal{A}^e = \infty$.
(M5) If $\mathcal{A} \wedge \mathcal{B}^e = \varnothing$, then $\mathcal{A} \prec \mathcal{B}$.
(M6) For all $\mathcal{A}$ and $\mathcal{B}$ in $\Omega$, $\mathcal{A} \wedge \mathcal{B}$ exists.

[29] A more wordy presentation, with many illustrative examples, can be found in Sects. I.2 and I.3 of [96].

As proved in [C6], it follows from (M1)–(M6) that the operations $\wedge$ and $\vee$ define on $\Omega$ the structure of a Boolean algebra.

Of course, this adds little, if anything, to the theory of Boolean algebras,[30] though it is scholarly argued in [A39] that there is a non-trivial one-to-one correspondence between these and material universes. What is new here is the use of these algebraic structures to manipulate bodies and their parts, with the purpose of establishing on firm mathematical grounds a theory of material interactions between bodies. Any specific branch of mechanics would have its own material universe, which should be specified at the start in accord with the axioms (M1)–(M6) above. The theory of interactions that we shall outline below in Sect. 4.3.2 need not be adapted to the special material system (or model) under consideration. This theory not only applies to the bodies considered in Continuum Mechanics; it is far more general than that, as far more general is the very notion of body being evoked.[31] As is often the case, the gain in generality overweights the price paid to abstraction.

### *4.3.2 Interactions*

The only interactions between bodies in a material universe that Noll deemed admissible are defined for separate bodies.[32] Formally, a pair of separate bodies belongs to the set

$$(\Omega \times \Omega)_{\text{sep}} := \big\{(\mathcal{A}, \mathcal{B}) \in \Omega \times \Omega : \mathcal{A} \wedge \mathcal{B} = \varnothing\big\}. \tag{4}$$

It is often useful to consider the universe of all parts of a given body $\mathcal{B} \in \Omega$:

$$\Omega_{\mathcal{B}} := \{\mathcal{P} \in \Omega : \mathcal{P} \prec \mathcal{B}\}. \tag{5}$$

In [A29], Noll defines an *interaction* as being a mapping from the set $(\Omega \times \Omega)_{\text{sep}}$ of all pairs of separate bodies into a vector space $\mathcal{U}$ that satisfies the following axioms:

(I1) $I(\mathcal{C}_1 \vee \mathcal{C}_2, \mathcal{B}) = I(\mathcal{C}_1, \mathcal{B}) + I(\mathcal{C}_2, \mathcal{B})$.
(I2) $I(\mathcal{B}, \mathcal{C}_1 \vee \mathcal{C}_2) = I(\mathcal{B}, \mathcal{C}_1) + I(\mathcal{B}, \mathcal{C}_2)$.

A completely equivalent way to state (I1) and (I2) would be saying that both mappings

$$I(\cdot, \mathcal{B}^{\text{e}}) : \Omega_{\mathcal{B}} \to \mathcal{U} \quad \text{and} \quad I(\mathcal{B}^{\text{e}}, \cdot) : \Omega_{\mathcal{B}} \to \mathcal{U} \tag{6}$$

are *additive*.

We say that $I(\mathcal{A}, \mathcal{B})$ represents the "action that $\mathcal{B}$ exerts on $\mathcal{A}$", so that the "law of action and reaction" reduces to the requirement

$$I(\mathcal{A}, \mathcal{B}) + I(\mathcal{B}, \mathcal{A}) = 0, \tag{7}$$

which states that the interaction $I$ is *pairwise equilibrated*. A simple, but telling consequence of axioms (I1) and (I2) is that the interaction $I$ is pairwise equilibrated if and only if the mapping $\mathcal{A} \mapsto I(\mathcal{A}, \mathcal{A}^{\text{e}})$ is additive.

---

[30] For which a classical reference is the textbook [71].

[31] As an example, the reader may consult pp. 24–26 of [96] for an application of this theory to mass-point dynamics.

[32] As illustrated in [96, pp. 26–29], Rizzo later extended Noll's theory of interactions to interpenetrating bodies.

A special class of interactions are those called *balanced*, for which

$$I(\mathcal{A}, \mathcal{A}^e) = 0 \quad \forall \mathcal{A} \prec \Omega. \tag{8}$$

Clearly, (8) defines an additive mapping, and so we conclude that a balanced interaction is pairwise equilibrated (and obeys the law of action and reaction).

Typically, as shown below in Sects. 4.3.4 and 4.3.5, $I$ has the physical meaning of *force*, in which case (8) is the prototype of a balance law. However, Noll [A43] reminds us that his notion of interaction is far more general:

> The concept of an interaction is an abstraction. Its values may have the interpretation of forces, torques, or heat transfers. In most of the past literature these cases were treated separately even though much of the underlying mathematics is the same for all. Thus, this abstraction, like most others, is a labor saving device.

### *4.3.3 Fit Regions*

Continuum Mechanics requires more than a generic universe of bodies; it requires a notion of continuous body. In [C12], Noll conceives a *continuous body* as a set $\mathcal{B}$ endowed with structure by the prescription of a class of mappings, which he calls *placements*. The elements of $\mathcal{B}$ are *material points* and a placement puts them against a suitable background, or frame of reference, which is identified with a Euclidean space $\mathcal{E}$. Placements are accompanied by *displacements*,[33] which change a *shape* $\mathscr{B}$ of $\mathcal{B}$ (that is, the image of $\mathcal{B}$ in a placement) into another shape $\mathscr{B}'$ of $\mathcal{B}$. Noll, perhaps over-conservatively, took displacements to be regular, invertible mappings, obtained from $C^2$-diffeomorphisms between whole Euclidean spaces by restricting domain and codomain to suitable subsets.[34]

In continuum physics, we only encounter bodies in the shapes they take on in space, and so, as remarked by Truesdell [96, p. 90], specifications of the latter suffice "for efficient practice of the mathematical theory" of the former. Thus, the emphasis shifts from the universe of *bodies* to the universe of their *shapes*.

In the Introduction to [A33], Noll acknowledges that since his first attempts to use precise mathematical structures to analyze basic concepts of continuum physics [C1], it was clear to him that a concept of *fit region*[35] in a Euclidean space was utterly necessary. Only such regions could be occupied by continuous bodies. One may at first deem this to be a simple issue to settle. For example, one might define a fit region to be a set with piecewise smooth boundary. However, explicit, simple examples in [C12] and [A33] show that sets with piece-

[33] As was typical in Noll's evolving terminology, displacements, were later to be renamed *transplacements* (in [A33], for example). They are commonly called *deformations* in most contemporary textbooks, such as [35], for example. Noll felt always strongly about notation and terminology, to the point of apologizing, in a rather peculiar way, for his (frequent) changes of mind, as in the following footnote, taken from [A42]:

> In 1958, Noll introduced the unfortunate terms "configuration" and "deformation" for what are called "placement" and "transplacement" here. He apologizes. Since these old terms were used in [91], they were widely accepted, and we are now in the ironic position to fight against a terminology that Noll introduced.

[34] Noll's interest for *category theory* [WC2] also permeated his views on the foundations of continuum mechanics [C12]. In the language of that theory, the class of all displacements is the class of *morphisms* of a category whose objects are pairs $(\mathscr{B}, \mathcal{E})$, where $\mathscr{B}$ is a subset of $\mathcal{E}$. Likewise, a morphism from $(\mathscr{B}, \mathcal{E})$ to $(\mathscr{B}', \mathcal{E}')$ is an inverbible mapping from $\mathscr{B}$ onto $\mathscr{B}'$ that can be extended to a $C^2$-diffeomorphism from $\mathcal{E}$ onto $\mathcal{E}'$.

[35] What came to be designated as a fit region had already been called a *nice* region [C12].

wise smooth boundaries may have intersections which fail to be so, thus jeopardizing the stability of the whole class.

As pointed out in [C12], the class of fit regions should possess a number of properties to ensure that what was common practice in continuum physics can be justified in precise mathematical terms. This, which was so typically Noll's attitude, should not be mistaken for the frigid attitude of the medical examiner, who knows everything but invariably comes too late. This was—as already said—the only way Noll could understand Mechanics (or any other branch of Mathematics, for that matter), by redoing it afresh in his own mind and with his own terminology. In our essay, we strive to show by example that this attitude, which has only too often been bitterly criticized, is far from having proved sterile.

Fit regions should first obey the axioms (M1)–(M6) that define a material universe.[36] Second, they should be closed under deformations.Third, each fit region should have a *surface-like* boundary for which a form of the Integral-Gradient Theorem[37] should be valid. Moreover, and this is perhaps a distinctive feature of [A33]:

> It is also desirable that the class of fit regions include all that can possibly be imagined by an engineer but exclude those that can be dreamt up only by an ingenious mathematician.

We already know that the first requirement is not met by the class of all regions with piecewise smooth boundary. As shown in [A33], the second requirement makes it illusory to achieve a satisfactory class of fit regions by considering subclasses of regions with piecewise smooth boundary, if simple regions, such as rectangles, are to be fit, as they should.[38] Quite on the contrary, it had become clear that the right strategy was enlarging that class, rather than restricting it, by allowing it to encompass sets of *finite perimeter*.[39] The fit regions proposed in [A33] are subsets of a Euclidean space $\mathcal{E}$ that are (i) *bounded*, (ii) *regularly open*[40] of (iii) *finite perimeter*, and with (iv) *negligible boundary*.[41]

---

[36]More precisely, the fit regions that are parts of a given fit region $\mathscr{B}$ should be materially ordered by inclusion. They thus constitute the universe $\Omega_{\mathscr{B}}$, for which $\varnothing$ is the empty set and $\infty$ is $\mathscr{B}$ itself.

[37]As already remarked above, Noll advocated for the use of self-descriptive names to be given to theorems and principles. This theorem, in particular, is best known in the Western literature as the Gauss-Green Theorem.

[38]A simple counterexample is constructed in [A33] as follows. Think of a rectangle deformed in such a way that one side becomes the graph of a smoothly oscillating function, such as $f(x) = \exp(-1/x^2)\sin(1/x)$, say for $0 < x < 1$. Then take the intersection with an adjacent, undeformed rectangle, so as to chop off all protruding humps. The result is a countable union of chops, which cannot be regarded as a set with piecewise smooth boundary.

[39]These are also known as *Caccioppoli sets*; see, for example, Chap. 1 of [33]. Perhaps, Banfi and Fabrizio [4] were the first to realize the role played by these sets in continuum physics (see also [5]). Other applications can be found in the papers [108], [72], and [39], all published not long before [A33]. However, the class of regions proposed in [4], besides being unnecessarily large, did not obey the requirement of constituting a material universe. All three requirements put forward in [A33] were instead satisfied by the class proposed by Gurtin, Williams, and Ziemer [39], but their development, being based on subtle measure-theoretic concepts, was unnecessarily sophisticated.

[40]A *regularly* open set is an open set that coincides with the interior of its closure. The reason for this condition is to prevent the boundary of a fit region from being other than its *outer* delimiter. Similarly, a *regularly closed* set is a closed set that coincides with the closure of its interior. It makes no substantial difference whether one uses the class of regularly open or regularly closed sets as an ambient for fit regions, because the process of taking the closure is a natural one-to-one correspondence from the former class to the latter, its inverse being the process of taking the interior.

[41]A subset of $\mathcal{E}$ is *negligible* if for every $\varepsilon > 0$ it can be covered by a finite collection of balls, the sum of whose volumes does not exceed $\varepsilon$. Thus a negligible set has zero volume-measure. Conversely, a compact

Axioms (M1)–(M6) are satisfied for any two fit regions $\mathscr{P}$ and $\mathscr{Q}$ being subsets of a given fit region $\mathscr{B}$ and such that

$$\mathscr{P} \wedge \mathscr{Q} = \mathscr{P} \cap \mathscr{Q}, \qquad \mathscr{P} \vee \mathscr{Q} = \overline{\overset{\circ}{\mathscr{P} \cup \mathscr{Q}}}, \qquad \mathscr{P}^{\mathrm{e}} = \overline{\overset{\circ}{\mathscr{B} \setminus \mathscr{P}}}.$$

The most stringent requirement for a fit region is perhaps that it possess a surface-like boundary. That this property is actually enjoyed by the regions introduced in [A33] follows from deeper aspects of the theory of sets of finite perimeter, mostly due to the work of De Giorgi [32] and admirably[42] recounted in Chaps. 4 and 5 of the book [97]. It is interesting that the surface-like boundary of a fit region $\mathscr{B}$ according to [A33] is *not* its *topological* boundary $\partial\mathscr{B}$, but the *reduced* boundary $\partial^*\mathscr{B}$, which enjoys three important (and non-trivial) properties: (i) the area-measure of $\partial^*\mathscr{B}$ is the perimeter of $\mathscr{B}$, (ii) $\partial\mathscr{B}$ is the topological closure of $\partial^*\mathscr{B}$, and (iii) $\partial^*\mathscr{B}$ differs from the union of a countable collection of compact subsets of $C^1$-surfaces only by a set of area-measure zero.

As shall be recalled in Sect. 4.3.4 below, it is indeed the reduced boundary of fit regions that will play a crucial role in the later developments of the theory of *contact* interactions, whose general principles had already been refreshed in [C12].

### 4.3.4 Stresses

In 1957, which saw the birth of Noll's first attempt at a rigorous theory on the foundations of classical mechanics, it was generally believed that any such attempt was not worth the time and effort. Noll was perfectly aware of such a state of affairs, and was equally unaffected by that. He wrote [C1]:

> It is a widespread belief even today that classical mechanics is a dead subject, that its foundations were made clear long ago, and that all that remains to be done is to solve special problems. This is not so. It is true that the mechanics of systems of a finite number of mass points has been on a sufficiently rigorous basis since Newton. Many textbooks on theoretical mechanics dismiss continuous bodies with the remark that they can be regarded as the limiting case of a particle system with an increasing number of particles. They cannot. The erroneous belief that they can had the unfortunate effect that no serious attempt was made for a long period to put classical continuum mechanics on a rigorous axiomatic basis. ... A clarification of these foundations is of importance also for the following reason. It is known that continuous matter is really made up of elementary particles. The basic laws governing the elementary particles are those of quantum mechanics. The science that provides the link between these basic laws and the laws describing the behavior of gross matter is statistical mechanics. At the present time this link is quite weak, partly because the mathematical difficulties are formidable, and partly because the basic laws themselves are not yet completely clear. A rigorous theory of continuum mechanics would give at least some precise information on what kind of gross behavior the basic laws ought to predict.[43]

set with zero volume-measure is negligible. Since the boundary of a bounded set is compact, condition (iv) could be replaced by "a fit region has boundary of zero volume-measure".

[42] Truesdell [96, p. 88] praises this book saying "Be it noted that this clear, excellent, and compact book is written by and for engineers."

[43] The reader should recall Bunge's words in Sect. 3.6.

We mention this here because perhaps the most striking new result of [C1] was the proof that the *Cauchy Stress Principle*, which constituted the classical basic assumption to deduce the existence of the stress tensor, could rather be derived from more basic assumptions.[44]

Establishing the weakest assumptions that warrant the existence of the stress tensor has been another recurrent theme in Noll's research. He returned over and over again on this issue, seeking the most general mathematical infrastructure that would guarantee the tersest simplicity. In [WB4], which was posted on the web in 2005,[45] he found the last setting for his theory in the class of fit regions. Here, we shall briefly outline the contents of [WB4], avoiding all unnecessary technical details, but pausing instead on some of Noll's own critiques, which show to what extent he was still dissatisfied with his own infrastructure and conceived his journey towards perfection as being *ad infinitum*.

The paper [WB4] is concerned with *surface interactions*, which are a subclass of *contact interactions*. These latter were properly defined in [A34]; in words, they are interactions that vanish whenever the *contact* between the interacting bodies is empty. Now, the contact between bodies, which is whatever part their closures have in common, is to be distinguished from the *reduced* contact, which is the part of the contact where a normal can be defined. A *surface interaction* is defined as an interaction that vanishes whenever the *area* of the reduced contact vanishes. *Edge interactions*, which were studied in [A34] and will be briefly recalled shortly below, are examples of contact interactions that are not surface interactions.

In the material universe $\Omega_{\mathscr{B}}$ of all fit regions that are parts of a fit region $\mathscr{B}$, Noll defines in [WB4] a *directed contact* $\mathrm{no}_{(\mathscr{P},\mathscr{Q})}$ as the restriction of the reduced contact between parts $\mathscr{P}$ and $\mathscr{Q}$ of the outer normal field defined on the reduced boundary $\partial^*\mathscr{P}$ of $\mathscr{P}$.

A special role in the theory is played by the set $\mathrm{Dc}_{\Omega_{\mathscr{B}}}$ of all directed contacts for $\Omega_{\mathscr{B}}$. Noll proves that for every surface interaction $I$ of $(\Omega_{\mathscr{B}} \times \Omega_{\mathscr{B}})_{\mathrm{sep}}$ onto the linear space $\mathscr{U}$ there is exactly one mapping $\mathbf{C} : \mathrm{Dc}_{\Omega_{\mathscr{B}}} \to \mathscr{U}$ such that

$$\mathbf{C}(\mathrm{no}_{(\mathscr{P},\mathscr{Q})}) = I(\mathscr{P},\mathscr{Q}) \quad \forall(\mathscr{P},\mathscr{Q}) \in (\Omega_{\mathscr{B}} \times \Omega_{\mathscr{B}})_{\mathrm{sep}}. \tag{9}$$

The mapping $\mathbf{C}$ is called the *contact flux*. Given an appropriate subset $\Gamma$ of $\mathrm{Dc}_{\Omega_{\mathscr{B}}}$, two special instances of contact fluxes may arise. In the first one, $I$ admits a *proto-contactor*[46] $\mathbf{s} : \mathscr{B} \times \mathrm{Usph}\,\mathcal{V} \to \mathscr{U}$ relative to $\Gamma$, which represents $\mathbf{C}$ through the formula

$$\mathbf{C}(\mathbf{n}) = \int_{\mathscr{S}} \mathbf{s}\big(x, \mathbf{n}(x)\big)\mathrm{dar}_x \quad \forall \mathbf{n} \in \Gamma, \tag{10}$$

where $\mathscr{S}$ is the domain in $\mathscr{B}$ where $\mathbf{n}$ is defined. When (10) applies, $\mathbf{s}$ represents the density per unit area of the bodies' reduced contact; it is the *traction field* of most traditional treatments, albeit phrased in a rather abstract and rarefied setting, which is fully appreciated only by recalling that the domain of a mapping is itself part of its definition.[47]

[44] This quote also illuminates Noll's unwearying attitude towards science since his youth.

[45] This paper is based on a lecture that Walter gave in April 1993 at the meeting of the Society for Natural Philosophy that took place at Carnegie Mellon University on the occasion of his retirement from teaching.

[46] Here, we adopt Noll's notation $\mathrm{Usph}\,\mathcal{V}$ for the unit sphere in the translation space $\mathcal{V}$. It is admittedly rather unusual.

[47] This specifically applies here to the outward unit normal field.

In the second special instance, a contact flux not only admits a proto-contactor as density,[48] but this also depends linearly on the outer unit normal $\mathbf{n}$, that is,

$$\mathbf{s}\big(x, \mathbf{n}(x)\big) = \mathbf{S}(x)\mathbf{n}(x), \tag{11}$$

where the linear transformation $\mathbf{S}$ is the *contactor*, which in the case where $I$ represents surface contact forces is Cauchy's *stress tensor*.

In this language, Noll [WB4] easily summarizes both Cauchy's seminal result and his own first extension:

> In 1823, Cauchy proved, in essence, a theorem of the following form: If $I$ admits a proto-contactor relative to a suitable class $\Gamma$ and if suitable assumptions are made then $I$ admits a contactor relative to $\Gamma$. Cauchy simply assumed the existence of a proto-contactor. This assumption is often called "the Cauchy Stress Principle". The description of the class $\Gamma$ and the assumptions needed for the proof of his theorem were quite vague by modern standards.
> In 1957 (see Theorem IV on p. 275 in [C1]), I proved a theorem of the following form: if suitable assumptions are made then $I$ admits a proto-contactor relative to an appropriate class $\Gamma$. Again the description of the class $\Gamma$ and the assumptions needed for the proof of my theorem were not very precise. One reason was that, at that time, the requisite conceptual infrastructure was missing.

In the same language, Noll then phrases in a way that he deems appropriate to modern standards the list of conditions that would guarantee the existence of both proto-contactors and contactors (that is, generalized tractions and stresses, respectively), thus turning the special instances of contact fluxes into reality.

(a) The resultant $I(\mathscr{P}, \mathscr{P}^{\mathrm{e}})$ is *locally volume-bounded.* This means that the total action exerted on $\mathscr{P}$ by its exterior is bounded by the volume of $\mathscr{P}$ and thus vanishes when the latter does.
(b) $I$ is *locally area-bounded.* This means that $I(\mathscr{P}, \mathscr{Q})$ is bounded by the area of the reduced boundary that $\mathscr{P}$ and $\mathscr{Q}$ have in common. Thus, when the area of the reduced contact between $\mathscr{P}$ and $\mathscr{Q}$ vanishes, so also does $I(\mathscr{P}, \mathscr{Q})$.
(c) For every $\mathbf{n} \in \mathrm{Dc}_{\Omega_{\mathscr{B}}}$, the contact flux $\mathbf{C}$ has a continuous density $\mathbf{s}_{\mathbf{n}}$. This does *not* mean that there is a proto-contactor in the class $\mathrm{Dc}_{\Omega_{\mathscr{B}}}$ of all possible directed contacts between parts of $\Omega_{\mathscr{B}}$, as the mapping $\mathbf{s}_{\mathbf{n}}$, which is different for different $\mathbf{n}$'s, is far more general than the mapping $\mathbf{s}$ in (10).

Clearly, these conditions are not completely independent. In particular, (c) implies (b). It remains an open problem to see under what additional hypotheses (b) implies (c) [WB4]:

> Adding some additional hypotheses to (b) may insure that a modified (b) implies (c), but it is not clear to me how such additional hypotheses should be formulated.

Combining these conditions, Noll [WB4] proves two fundamental theorems that subsume his life-long search for the neatest setting allowing for the definition of stress.

The first theorem establishes that if conditions (a) and (c) are assumed to be valid then the interaction $I$ admits a proto-contactor relative to the smooth class $\mathrm{SDc}_{\Omega_{\mathscr{B}}}$ of directed contacts having support on surfaces of class $C^1$.

[48] It is perhaps not completely useless to stress that a proto-contactor is indeed a special form of surface density for a contact flux.

The second theorem establishes that if the interaction $I$ satisfies condition (a) and admits a *continuous* proto-contactor $\mathbf{s}$ relative to the smooth class $\mathrm{SDc}_{\Omega_{\mathscr{B}}}$, then $I$ admits a continuous contactor $\mathbf{S}$ relative to $\mathrm{SDc}_{\Omega_{\mathscr{B}}}$.

As Noll remarked, a number of issues remain unresolved. These are at least three:

(i) It is not clear whether conditions (a) and (c) alone are sufficient to insure the existence of a *continuous* proto-contactor.

(ii) If the interaction $I$ does admit a continuous proto-contactor and hence a continuous contactor, then $I$ is clearly pairwise equilibrated on all smooth directed contacts (supported on $C^1$-surfaces). However, if conditions (a) and (c) are satisfied but the proto-contactor, which exists, fails to be continuous, it is not clear whether $I$ is still pairwise equilibrated.

(iii) It is also unclear what happens if the class of smooth directed contacts $\mathrm{SDc}_{\Omega_{\mathscr{B}}}$ is replaced by a larger subclass.

This is the state of affairs in Noll's view in 2005.

Similar issues had already been addressed in the past. For example, attempts at dispensing with the assumption of continuity were made since the early days of Noll's original theory. We only cite the works [36–38], as they are likely to have been directly influenced by Noll's physical proximity to the authors at CMU.

We renounce following up in the literature the further partial answers possibly given to Noll's queries in the past few years; we should not be too far from truth in saying that these must be quite scant. These themes are no longer fashionable. But fashion goes in cycles, so in due time the readers will find here a digest of Noll's latest theory on stress, which is part of his legacy to future generations.

We close this section by recalling from [A34] Noll & Virga's treatment of edge interactions, which we have already mentioned above as possible, though less usual forms of contact interactions.

The idea that forces can be distributed along the edges of a body had already occurred in Toupin's theory of second-grade elasticity [81]. As remarked by Truesdell [96, p. 196], that treatment was affected by the same limitations from which suffered the theory of interaction prior to 1960's, that is, it was only concerned with resultants "without bringing into the open the system of forces giving rise to these resultants". The paper [A34] is an attempt to reconcile Toupin's theory with the general theory of interactions sketched here. A notion of edge interactions as special contact interactions that supplement surface interactions and contribute with those to the total balance is developed in [A34] with details that cannot be reproduced here. However, the universe that serves as ambient for edge interactions is not as general as the one employed above for surface interactions; the regions adopted as shapes of bodies are *not* fit regions, but *bi-regular* regions, which constitute a subclass of Kellogg's *regular* regions [48, p. 112].

An edge interaction between two internal parts of a body is *not* to be regarded as the analog of external actions concentrated along curves; the theory of edge interactions has nothing in common with the theories of stress concentration. An edge interaction between two internal parts of a body manifests itself only when a side of one part is in "everywhere-cusped" contact with a side of the other part along a common edge. Cusps play a very special role in this theory, which also proved capable of incorporating surface tension into its abstract scheme.[49]

[49] In this context, surface tension arises from the balance equation obtained for a special constitutive choice of the tensor that plays the role of an *edge stress*. Traditionally, that very same balance equation is obtained

In two respects the theory presented in [A34] does not equal the level of generality attained in the theory of surface interactions. First, a number of assumptions were made, which postulated special representations for edge interactions. Second, the shapes of the interacting bodies were subject to a number of requirements. In other words, the class of admissible shapes was defined through the properties it was requested to have, as fit regions had been before they were actually identified. How the class of fit regions should be modified to accommodate edge interactions is a difficult problem, which to the best of our knowledge is still open.

### 4.3.5 Balance Laws of Motion

When the interaction between bodies is realized by a system of forces, this is delivered by a vector $\mathbf{f}$ in the translation space $\mathcal{V}$ of a frame of reference $\mathcal{E}$ (see Sect. 4.3.3). Noll's specific assumption for forces is as follows: *The systems of forces are frame-indifferent.* That is, if $\mathbf{f}$ and $\mathbf{f}^*$ denote the same system of forces in the reference frames $\mathcal{E}$ and $\mathcal{E}^*$, then

$$\mathbf{f}^* = \mathbf{Q}\mathbf{f}, \tag{12}$$

where $\mathbf{Q}$ is the (possibly time dependent) orthogonal tensor that changes $\mathcal{E}$ into $\mathcal{E}^*$.

In [C2], Noll posited what Truesell [90, p. 546] called Noll's *Axiom of Dynamics*:[50]

> *For every assignment of forces to pairs of bodies, the* working *of the system of forces acting on each body is frame-indifferent, no matter what be the motion.*

Here *working* is the rate at which the system of forces expends power in a motion.

The preceding axiom is quite a general statement. Once the transformation rules for all quantities involved are declared, its consequences can become explicit. Noll proved in [C2] that necessary and sufficient conditions for his axiom to hold are the following:

(i) the resultant force on every body vanishes;
(ii) the resultant torque on every body vanishes.

Thus, Noll's Axiom of Dynamics is *equivalent* to Euler's balance laws of motion, with a yet unspecified inertia (see [96, p. 151]).[51]

Noll's approach to inertia is also innovative. As regard his Axiom of Dynamics he writes in [C2]:

> Cet axiome est compatible avec les réalités physiques si l'on regarde les forces d'inertie comme des forces véritables qui sont les interactions entre les corps dans notre système solaire et la totalité des objets dans le reste de l'univers. Les repères inertiaux n'entrent plus dans la partie générale de la nouvelle axiomatisation. La loi d'inertie est regardée comme un postulat constitutif. Les repères privilégiés sont liés explicitement à la position du reste de l'univers.

---

by assuming that the boundary of the body behaves like a material skin. Such an assumption was *not* made in [A34], where the surface tension emerges, as it were, from *within* the body.

[50] We also learn from [90, p. 546] that this postulate dates from 1959 but was only published in 1963.

[51] Revisitations and extensions of Noll's Axiom of Dynamics and its links with the evolution equations of various mechanical systems (classical and generalized alike) are too numerous to be mentioned here. We just refer the reader to [16], which we take as a representative for many others.

As explained in great detail in [96, p. 64], the inertia of a body is interpreted as the action exerted on the body by the bodies outside the *great system*.[52] The *first axiom of inertia* then becomes the statement about the existence of a reference frame, called *inertial*, where the linear momentum of a body vanishes if and only if the body has no inertia. The *second axiom of inertia* further identifies the inertia of a body in an inertial frame of reference with the opposite to the rate of change of the linear momentum of the body. This assertion, combined with the frame-indifference of forces (12) allows one to determine the inertia of bodies in all frames of reference (see [96, p. 68]).[53]

### 4.4 Thermodynamics and Rheology

Most of Noll's work on thermodynamics and rheology was made in collaboration with Bernard D. Coleman, a senior fellow of Mellon Institute in Pittsburgh, PA, from 1957 to 1988 and a professor at Carnegie-Mellon University from 1967 to 1988.[54]

From 1958 to 1966, Coleman and Noll published a number of research papers that, in the words of Tanner and Walters, "...had an immense influence, particularly amongst workers in theoretical continuum mechanics", whose "impact on experimental rheologists was also significant" [79, pp. 93–94]. We find it convenient to distinguish those papers in two categories: those having to do with making thermodynamics a chapter of continuum physics in the 'rational' spirit that C.A. Truesdell advocated; and those concerning rheological issues, a streak somehow ended by *Viscometric Flows of Non-Newtonian Fluids*, the slim book [B2] we already referred to in Sect. 4.1 that Coleman and Noll coauthored with another researcher affiliated to the Mellon Institute, Herschel Markowitz.[55] The papers comprising this second category are [A10], [A11], [A14], [A15], [A16], [A17], and [A18]; they were published between 1959 and 1962, in various primary journals whose readership was physics-oriented. [A9], the first paper in the first category, appeared in 1959; it has to do with equilibrium at negative temperatures, an issue in classical statistical mechanics. The other papers are, in chronological order, [A12], [A13], [A19], and [A21]; they were all published in the *Archive for Rational Mechanics and Analysis*, the journal that, at that time and for long afterwards, was the forum where new matters of presumably persistent value and of, in a broad sense, common interest for mechanicians and mathematical analysts were presented and debated.[56]

It would be way beyond the space limits and the purpose of our present writing to give an account, no matter how much abridged, of all the papers quoted above. We shall content

---

[52]Einstein [18] called *Mach's principle* the (generic) assertion that makes the distribution of matter throughout the universe responsible for the inertia of local matter (see, for example, [2, 3]).

[53]In a different, related perspective, the connection between inertia and invariance properties has been explored in [62].

[54]B.D. Coleman, "an extraordinary figure of the 20th century continuum mechanics and thermodynamics[, who] made a number of deep and permanent contributions that changed the way these sciences are now understood, presented, and applied" [73], passed away on August 2nd, 2018.

[55]This book features a limpid exposition of explicit solutions to motion problems that, in addition to being interesting per se, are of direct rheological interest, in that they point at a feasible experimental program to achieve the constitutive characterisation in terms of three *viscometric functions* of the shear rate of a class of incompressible Rivlin-Ericksen fluids whose extra-stress is sensitive to the past history of the deformation gradient. A preliminary presentation of the method of viscometric functions on which this book is based was given at the *First International Symposium on Pulsatile Blood Flow*, held in 1963 at the Presbyterian Hospital in Philadelphia [C4], see also [45, p. 128].

[56]In particular, [A12] appeared in the fourth volume of this journal. In [89, p. 328], Truesdell tells an interesting story about the origin of this paper and its connection with a critical retrospective view on the history of mechanics:

ourselves to comment on the two, [A13] and [A14], where the until then vague notion of *fading memory* was carefully formalized and applied; and on [A19], where it was shown how to make use of the *Clausius-Duhem inequality* to select thermodynamically admissible constitutive equations from which the superfluous information had been previously trimmed off.

### 4.4.1 Fading Memory

The title of Coleman & Noll's 1960 paper [A13]—*An Approximation Theorem for Functionals with Applications in Continuum Mechanics*—is centred on the use of an analytical tool the authors develop to capture memory effects, whenever "the present value of a physical quantity ... is determined by the values ... of a second quantity at all times $s$ in the past ... [via] a functional relationship". Let

$$T = \underset{s=0}{\overset{\infty}{\mathfrak{F}}}\big(F(s)\big), \tag{13}$$

be such a relationship, positing that the present value of the stress $T$ in a simple material body is determined by all values $F(s)$ of the deformation gradient "$s$ time units ago" via the functional $\mathfrak{F}$. The analytical tool is a norm

$$\|F\| = \left(\int_0^\infty \big(|F(s)|h(s)\big)^p ds\right)^{1/p}, \quad 1 \le p < \infty, \tag{14}$$

over a function space consisting of *histories*

$$F^t(s) := F(t-s), \quad s \ge 0, \tag{15}$$

a norm that "assign[s] a greater influence to the values of $[F]$ for small $s$ (recent past) than for very large $s$ (distant past)", because of the *influence function* $h$,

> a [nonnegative] real-valued function which approaches zero rapidly as $s \to \infty$.... The physical idea that the memory of the system is fading corresponds to the mathematical assumption that the functional [in (13)] is *continuous* with respect to convergence in the norm [(14)] of the function space of histories [(15)]. The influence function $h$ characterizes the rapidity with which the memory is fading.

The properties a functional $\mathfrak{F}$ as in (13) is required to have in order to be called a *memory functional* are carefully listed in Sect. 4 of [A13]; the most important one is continuity and $n$-times Fréchet differentiability at the null history $F^t(0) \equiv I$. The theory is shown: (i) to cover the case of *simple fluids*, a concept introduced in Sect. 6; (ii) to allow for $n$-order approximations of the response, e.g., *incompressible second-order fluids* generalizing Newtonian fluids (Sect. 7); and (iii) to handle simple materials in general (Sect. 8).

We regard the 1961 paper [A14] entitled *Foundations of Linear Viscoelasticity* (a citation hit of Coleman and Noll's) as the main application of the theoretical framework developed

---

This paper is directed explicitly toward a problem that originated in historical study. Noticing the thermodynamic method of Duhem for deriving the inequalities satisfied by the linear elasticities and linear viscosities in a lecture at Berlin in 1955 I had pointed out that corresponding considerations for the theory of finite elastic strain were not known, and I proposed the problem of finding the restrictions necessarily satisfied by the strain-energy function of an elastic material [86].

in [A13]. In line with the institutional policy of the prestigious journal where it appeared, *Reviews of Modern Physics*, which addresses a large audience, this paper takes a long route to show how the theory developed in [A13] accomodates not only the standard linear theory of small-strain viscoelastic deformations but also a more general theory which is given a somewhat oxymoronic name, *finite linear viscoelasticity*.

Section 2 is a primer of continuum kinematics, where among other things $W$ and $E$, the skew and symmetric parts of $H := F - I$, and the deformation measure $C := F^T F$ are introduced. Section 3 furnishes a complete user-friendly presentation of the mathematisation of the fading-memory notion from [A13]. Slight notational differences apart, Eq. (3.4) reads

$$T(t) = \mathfrak{h}\big(C(t)\big) + \mathop{\mathfrak{K}}_{s=0}^{\infty}\big(C(t-s) - I; C(t)\big), \quad \mathop{\mathfrak{K}}_{s=0}^{\infty}\big(0; C(t)\big) = 0, \tag{16}$$

a specification of (13) where the stress response of a simple material to a deformation history is split into the instantaneous response to the current value $C(t)$, ruled by a continuously differentiable function $\mathfrak{h}$, plus the response to the past history $0 < s \mapsto C(t-s)$, ruled by a memory functional $\mathfrak{K}$; the latter "vanishes when the material has always been at rest" as $(16)_2$ prescribes and is defined in a neighborhood of, and Fréchet-differentiable at, the null history; both $\mathfrak{h}$ and $\mathfrak{K}$ have symmetric second-order tensors as values. The introduction of an influence function allows for a well-defined linearisation of functional $\mathfrak{K}$ at the null history; moreover, thanks to a standard integral representation theorem for continuous linear functional over Hilbert spaces, the constitutive equation (16) is approximated by

$$T(t) = \mathfrak{h}\big(C(t)\big) + \int_0^{\infty} \boldsymbol{\Gamma}(s, C)\big(C(t-s) - I\big) ds \tag{17}$$

(cf. Eq. (3.14) of [A14]). Section 4 is devoted to the deduction of the constitutive equation of *infinitesimal viscoelasticity*, namely,

$$T(t) - T_r = W(t)T_r - T_r W(t) + \big[\boldsymbol{\Omega} + \boldsymbol{\Phi}(0)\big]E(t) + \int_0^{\infty} \dot{\boldsymbol{\Phi}}(s)E(t-s)ds. \tag{18}$$

Here $T_r := \mathfrak{h}(I)$ is the *residual stress*, i.e., "the stress the material would sustain if it had been held in the reference configuration at all times in the past"; $\boldsymbol{\Omega} := \partial_C \mathfrak{h}|_{C=I}$ is such that

$$\mathfrak{h}\big(C(t)\big) = T_r + \boldsymbol{\Omega} E(t); \tag{19}$$

and the integral term in (17) is replaced by

$$\boldsymbol{\Phi}(0)E(t) + \int_0^{\infty} \dot{\boldsymbol{\Phi}}(s)E(t-s)ds, \tag{20}$$

where

$$\boldsymbol{\Phi}(s) := -2\int_s^{\infty} \boldsymbol{\Gamma}(\sigma)d\sigma, \quad \lim_{s\to\infty} \boldsymbol{\Phi}(s) = 0;$$

both (19) and (20) hold true to within smaller order terms; both fourth-order tensors $\boldsymbol{\Omega}$ and $\boldsymbol{\Phi}(0)$ are tacitly endowed with the algebraic symmetries typical of an elasticity tensor. In Sect. 5, (17) is identified with the constitutive equation upon which the theory of finite linear viscoelasticity is based (cf. Eq. (5.1) of [A14]), a theory that, provided all past deformation histories have small norm and depending on the choice of the reference configuration, includes as special cases standard linear viscoelasticity, the theory of infinitesimal elastic deformations superposed on large deformations, and classical linear elasticity.

### 4.4.2 Clausius-Duhem Inequality

Coleman & Noll's 1963 paper [A19], perhaps the most quoted one of all papers dealing with continuum thermodynamics, is rather short, some ten printed pages in all. The three sections of its first half integrate a quick exposition of that subject, intended as a physical theory susceptible of a precise mathematical setting within which to model the known phenomenology carefully and, in addition, to make rigorous quantitative and qualitative predictions about the yet-to-be-observed behavior of large classes of material bodies.

In the first two lines of Sect. 1, the "basic physical concepts of classical continuum mechanics" are listed: they are the concepts of a *body*, "regarded to be a smooth manifold whose elements are the material points", of a "*configuration* of a body, and [of a] *force system* acting on a body, ... resolved into body forces with a mass density and contact forces with a surface density".[57] Later, "five new basic concepts" are added, those of *temperature*, specific *internal energy* and *entropy*, *heat flux* and *heat supply*. Moreover, to the "general principles" of mechanics, "such as the principle of objectivity[58] or the law of balance of linear momentum", "two new principles" are added: "the *first law* of thermodynamics, *i.e.* the law of balance of energy, and the *second law*, which for continua takes the form of the Clausius-Duhem inequality". In Sect. 2, the notion of *thermodynamic process* is stated: a set of eight time-dependent fields over the body manifold such as to satisfy, for every part $\mathcal{B}$ of it, both the balance law of linear momentum

$$\int_{\mathcal{B}} \ddot{\boldsymbol{x}}\, dm = \int_{\mathcal{B}} \boldsymbol{b}\, dm + \int_{\partial\mathcal{B}} \boldsymbol{T}\boldsymbol{n}\, ds \tag{21}$$

and the balance law of energy

$$\frac{1}{2}\frac{d}{dt}\int_{\mathcal{B}} \dot{\boldsymbol{x}}\cdot\dot{\boldsymbol{x}}\, dm + \int_{\mathcal{B}} \dot{\varepsilon}\, dm = \int_{\mathcal{B}} (\dot{\boldsymbol{x}}\cdot\boldsymbol{b} + r) dm + \int_{\partial\mathcal{B}} (\dot{\boldsymbol{x}}\cdot\boldsymbol{T}\boldsymbol{n} - \boldsymbol{q}\cdot\boldsymbol{n}) ds, \tag{22}$$

where the current *position vector* of a typical body point is $\boldsymbol{x}$, symmetric *stress* $\boldsymbol{T}$, external non-inertial *body force* $\boldsymbol{b}$, *internal energy* $\varepsilon$, *heat flux* $\boldsymbol{q}$, external radiative *heat supply* $r$, *entropy* $\eta$, and *temperature* $\vartheta$, "assumed to be always positive"; and where $dm = \rho\, dv$, with $\rho$ the current mass density, while $dv$ and $ds$ denote the volume and area measures. The local versions of these two balance laws are, respectively,

$$\operatorname{div}\boldsymbol{T} - \rho\ddot{\boldsymbol{x}} = -\rho\boldsymbol{b} \tag{23}$$

and

$$\operatorname{tr}\{\boldsymbol{T}\boldsymbol{L}\} - \operatorname{div}\boldsymbol{q} - \rho\dot{\varepsilon} = -\rho r, \tag{24}$$

where $\boldsymbol{L}$ is the velocity gradient. It is appropriate to specify here that the position at time $t$ of a body point whose referential position vector is $\boldsymbol{X}$ is delivered by a so-called "deformation function" $\boldsymbol{x} = \boldsymbol{\chi}(\boldsymbol{X}, t)$, a vector-valued mapping which is supposed to be as smooth

[57] In the parlance of 'rational' continuum mechanicists, a physical concept is termed 'basic' when it is regarded as 'primitive' and, as such, it is defined by the "mathematical entities" chosen to represent and manipulate it. As is not infrequent and indeed healthy when a theory is in a phase of development and consolidation, both the list of basic concepts and the properties of their mathematical counterparts were to change, not always slightly, in later expositions of continuum theories, beginning with the publication of volume III/3 of the Handbuch der Physik [B1], coauthored by Noll with Tuesdell [C15].

[58] Recall footnote 17.

as desirable in both its arguments and such that $\chi(\cdot,t)$ is invertible at each time $t$; that $F=\nabla x := \partial_X \chi$ denotes the deformation gradient and $v=\dot{x} := \partial_t \chi$ the velocity; and that, when $v$ is seen as a time-dependent field over the current configuration, then for the velocity gradient we have that $L = \operatorname{grad} v := \partial_x v$.

The constitutive assumptions in Sect. 3 delineate the class of "elastic materials with heat conduction and viscosity", a special and important case considered in [A13] and in [A14]. A *thermodynamic process* obtains when a smooth time dependence is chosen for both "the deformation function $\chi$ and the entropy distribution $\eta$", so that deformation, velocity and their gradients $F$ and $L$ are computable, as is the time rate of entropy. Next, the fields of internal energy, temperature, and stress, are determined with the use of the constitutive equations (3.1)–(3.3), namely,

$$\varepsilon = \hat{\varepsilon}(F,\eta), \qquad \vartheta = \hat{\vartheta}(F,\eta), \qquad T = \hat{T}(F,\eta) + \hat{\mathfrak{l}}(F,\eta)[L] \tag{25}$$

(here the "viscous stress" $\hat{\mathfrak{l}}(F,\eta)[L]$ is the result of the linear action of the fourth-order tensor $\mathfrak{l}(F,\eta)$ over the velocity gradient); moreover, as soon as the temperature field is available via $(25)_2$, the constitutive equation (3.4) in [A19],

$$q = \hat{q}(F, \vartheta, \operatorname{grad}\vartheta), \tag{26}$$

determines the heat flux. "Finally $b$ and $r$ can be chosen so that [(23)] and [(24)] hold."

The key point of the paper—indeed, its main element of novelty—comes at the beginning of Sect. 4 with the definition of the *entropy production* in a thermodynamic process,

$$\Gamma = \int_{\mathcal{B}} \gamma\, dm = \frac{d}{dt}\int_{\mathcal{B}} \eta\, dm - \int_{\mathcal{B}} \frac{1}{\vartheta} r\, dm + \int_{\partial\mathcal{B}} \frac{1}{\vartheta} q\cdot n\, ds, \tag{27}$$

and with the following

**Postulate** *For every process admissible in a body consisting of a given material and for every part* $\mathcal{B}$ *of this body the inequality*

$$\Gamma \geq 0 \tag{28}$$

*is valid.*

Write Coleman and Noll:

> The inequality [(28)] is called the *Clausius-Duhem inequality, and the postulate above is a statement of the second law of thermodynamics within the framework presented here. The main purpose of the present paper is to derive logical consequences from the postulate* [italics in the cited text].

As remarked elsewhere [63], this postulate is an implicit confession of ignorance about the internal causes of entropy changes: it is formulated as an inequality—at variance with what we did for energy changes with (22), a balance law—because neither we can exclude the existence of internal causes (i.e., internal entropy sources) nor we know how to model them explicitly.

Note that the integrands of the third and fourth integrals in (27), that is, the entropy supply and the entropy flux, have been presumed proportional, respectively, to the radiative heat supply and the heat flux, in a manner scaled by temperature. This proportionality presumption encodes in a precise mathematical form various experimental evidences of the fact that a heat inflow has less and less pronounced entropic effects as temperature gets bigger and bigger. It also conforms to Helmholtz's identification of *coldness*, the inverse of temperature, as the integrating factor needed to prove his celebrated Heat Theorem, with which he strove to establish a representation for the mechanical analogue of entropy (see, e.g., [29]).[59] Note also that, as stated in [A19], an entropy imbalance postulate can legitimately be meant to concern material classes different from that specified by the constitutive prescriptions (25)–(26). Such a generality in scope has been one of the two reasons why the reduction procedure exemplified by Coleman and Noll has been adopted so often ever since their paper was published. The other reason has been the facile, at bottom algebraic, argument expedient to reach physically relevant conclusions when adopting the Coleman-Noll procedure: the use and abuse of that procedure has been the main justification for a good many of authors to claim 'consistency with thermodynamics' for their theories.

Here is how Coleman and Noll impose consistency with the entropy postulate (28) of the constitutive equations (25). We reproduce the original procedure step by step, because in our mind it has kept its flavour, no matter how many repetitions with variants of it have been made since 1963.

Firstly, with the use of the local version (24) of internal energy balance and granted positivity of temperature, a local equivalent of Clausius-Duhem inequality is stated,

$$\rho\vartheta\gamma = \rho(\vartheta\dot{\eta} - \dot{\varepsilon}) + \operatorname{tr}\{\boldsymbol{T}\boldsymbol{L}\} - \frac{1}{\vartheta}\boldsymbol{q}\cdot\operatorname{grad}\vartheta \geq 0. \tag{29}$$

Next, since $(25)_1$ implies that

$$\dot{\varepsilon} = \operatorname{tr}\big\{\partial_{\boldsymbol{F}}\hat{\varepsilon}(\boldsymbol{F},\eta)\dot{\boldsymbol{F}}\big\} + \partial_\eta\hat{\varepsilon}(\boldsymbol{F},\eta)\dot{\eta},$$

and since $\dot{\boldsymbol{F}} = \boldsymbol{L}\boldsymbol{F}$, inequality (29) is given the form

$$\rho\big(\vartheta - \partial_\eta\hat{\varepsilon}(\boldsymbol{F},\eta)\big)\dot{\eta} + \operatorname{tr}\big\{\big(\boldsymbol{T} - \rho\boldsymbol{F}\partial_{\boldsymbol{F}}\hat{\varepsilon}(\boldsymbol{F},\eta)\big)\boldsymbol{L}\big\} - \frac{1}{\vartheta}\boldsymbol{q}\cdot\operatorname{grad}\vartheta \geq 0. \tag{30}$$

Consideration of *homogeneous* deformation-and-entropy processes, that is, processes in which "$\boldsymbol{F}$ and $\eta$ are independent on the material point and depend only on time"—and hence $\vartheta$ also does, in view of $(25)_2$—permits to write (30), with the use of $(25)_3$, as

$$\begin{aligned}&\rho\big(\vartheta - \partial_\eta\hat{\varepsilon}(\boldsymbol{F},\eta)\big)\dot{\eta} + \operatorname{tr}\big\{\big(\hat{\boldsymbol{T}}(\boldsymbol{F},\eta) - \rho\boldsymbol{F}\partial_{\boldsymbol{F}}\hat{\varepsilon}(\boldsymbol{F},\eta)\big)\boldsymbol{L}\big\}\\ &\quad + \operatorname{tr}\big\{\big(\hat{\mathbb{I}}(\boldsymbol{F},\eta)[\boldsymbol{L}]\big)\boldsymbol{L}\big\} \geq 0.\end{aligned} \tag{31}$$

For (31) to hold for all continuations of any given homogeneous process such that $\dot{\boldsymbol{F}}$—and hence $\boldsymbol{L}$—is null and $\dot{\eta}$ has an arbitrary value, the *temperature relation*

$$\vartheta = \hat{\vartheta}(\boldsymbol{F},\eta) = \partial_\eta\hat{\varepsilon}(\boldsymbol{F},\eta) \tag{32}$$

must hold. Moreover, in view of this first conclusion, (31) also implies that

$$\alpha\operatorname{tr}\big\{\big(\hat{\boldsymbol{T}}(\boldsymbol{F},\eta) - \rho\boldsymbol{F}\partial_{\boldsymbol{F}}\hat{\varepsilon}(\boldsymbol{F},\eta)\big)\boldsymbol{L}\big\} + \alpha^2\operatorname{tr}\big\{\big(\hat{\mathbb{I}}(\boldsymbol{F},\eta)[\boldsymbol{L}]\big)\boldsymbol{L}\big\} \geq 0$$

[59] As shown in Sect. 1.8.1 of [63], life gets much harder (but not impossible) if the proportionality presumption is renounced. Various later papers of Noll (e.g., [A41], [A42], and [WB8]) testify to his faithfulness to that presumption.

whatever the number $\alpha$, that is, equivalently, that both the *stress relation*

$$\hat{T}(F,\eta) = \rho F \partial_F \hat{\varepsilon}(F,\eta) \tag{33}$$

and the *dissipation inequality*

$$\operatorname{tr}\{(\hat{\mathfrak{l}}(F,\eta)[L])L\} \geq 0 \tag{34}$$

hold true, the latter for all tensors $L$. In conclusion, for the constitutive class preliminarily defined by (25)–(26), consistency with the entropy-production inequality (27) implies that the temperature and stress relations are both specified by the constitutive equation delivering the internal energy; moreover, inequality (30) reduces to

$$\rho\vartheta\gamma = \operatorname{tr}\{(\mathfrak{l}(F,\eta)[L])L\} - \frac{1}{\vartheta} q \cdot \operatorname{grad}\vartheta \geq 0. \tag{35}$$

But this is not all: under the assumption that the temperature relation is invertible for each $F$ fixed so as to give

$$\eta = \hat{\eta}(F,\vartheta), \tag{36}$$

consideration of processes such that the deformation gradient is a spacetime constant and the temperature distribution is time independent, so that entropy is uniquely determined by (36) and constant in time, induces the following consequence of inequality (30)

$$-\hat{q}(F,\eta,\operatorname{grad}\vartheta) \cdot \operatorname{grad}\vartheta \geq 0, \tag{37}$$

whatever value the vector grad $\vartheta$ may take. Coleman and Noll offer an argument to show that it follows from (37) that

$$\hat{q}(F,\eta,\mathbf{0}) = \mathbf{0}, \tag{38}$$

a conclusion "expressing the non-existence of a piezo-caloric effect", according to the terminology introduced by Pipkin and Rivlin [61].[60] In summary, given the reasonable proposition of characterising the constitution of a material class by laying down the equation set (25)–(26), the Coleman-Noll procedure for testing the compatibility of that proposition with the Clausius-Duhem inequality has led to signal an information redundancy in it, because a choice of mapping $\hat{\varepsilon}$ determines both $\hat{\vartheta}$ and $\hat{T}$, and to specify the two dissipation mechanisms, mechanical and thermal in nature, encoded by mappings $\hat{\mathfrak{l}}$ and $\hat{q}$.

Other equivalent ways to reach the same conclusions can be conceived. A systematic customary use of the Clausius-Duhem inequality is made in [35], where the advantages of introducing another function of state, Helmhholtz free energy $\psi := \varepsilon - \vartheta\eta$, beside internal energy and entropy are demonstrated. No matter how formulated, as remarked by Gurtin [34, p. 53], a procedure of Coleman-Noll type

[60] In fact, more can be deduced from (37), namely, that a thermodynamically compatible generalisation of Fourier law of heat conduction is

$$\hat{q}(F,\eta,\operatorname{grad}\vartheta) = \hat{K}(F,\eta,\operatorname{grad}\vartheta)\operatorname{grad}\vartheta,$$

with the *conductivity tensor* $\hat{K}$ such that

$$\operatorname{grad}\vartheta \cdot \hat{K}(F,\eta,\operatorname{grad}\vartheta)\operatorname{grad}\vartheta \geq 0 \quad \text{for all } (F,\eta)$$

(see [63], Sect. 1.6).

> ... is based on the premise that the second law be satisfied in all conceivable processes, irrespective of the difficulties involved in producing such processes in the laboratory. The rational application of this procedure requires external forces and supplies [such as $\boldsymbol{b}$ in (23) and $r$ in (24)] that may be assigned arbitrarily to ensure satisfaction of the underlying balances in all processes. This may seem artificial, but it is no more artificial than theories based on virtual power, a paradigm that requires arbitrary variations, not guaranteed to be consistent with the resulting evolution equations, granted a constitutive description. The Coleman-Noll procedure makes explicit the external fields needed to support the 'virtual processes' used, and in so doing ensures that these external fields, whether virtual or not, enter the theory in a thermodynamically consistent manner.

## 4.5 Five Contributions to Natural Philosophy

In 2004, Noll published a booklet entitled *Five Contributions to Natural Philosophy*[61] [WB2], where he collected his latest reflections on the recurrent themes of his lifelong scientific activity and shaped a coherent programme for further refinements and extensions. It is not clear whether he had planned to pursue this programme himself; certainly, he felt the (felicitous for us) urge to re-systematize his thoughts. These *Contributions* serve well our purposes here, as they constitute a critical retrospective from which we also learn what he would have improved in his own foundational theories, what he would have redone afresh, and in what mathematical language. We list below all five *Contributions*, commenting briefly on the features that most interested us.

### *4.5.1 On the Illusion of Physical Space*

The notion of a *frame of reference* (not to be confused with a system of coordinates!) is a necessary prerequisite for the description of any physical event.[62] In the mathematical language of his 1987 treatise *Finite-Dimensional Spaces* [B5] and [WC1], Noll prefers to construct a *frame-space*; this is a Euclidean point space hosting all *locations* where events unravel. Only in one such space, geometric concepts such as straight line, direction, angle, etc., acquire a definite sense and *vectors* become members of the translation-space associated with the frame-space.

Although a frame-space is necessary to do physics, there is no need for a physical space to exist:

> It should be clear ... that the existence of a physical space, apart from any frame of reference, is an illusion. The mystery is that so many intelligent people, Newton included, fell victim to this illusion. ... I was also a victim of the illusion of physical space until the early nineteen sixties. In my doctoral thesis, in 1954, I proposed what I called "the principle of isotropy of space". Later I realized that there is no such thing as physical space and that the "principle of isotropy of space" should be replaced by what is now called the *principle of frame-indifference*.

[61] This collection of works is dedicated to the memory of Clifford Truesdell, his mentor, who had revived the use of *Natural Philosophy* to designate the endeavour to understand nature by using conceptual mathematical tools. See also [A38], for a further discussion on the modern use of this old term.

[62] See also [A45], which reproduces a lecture given at the *Seminario Matematico e Fisico di Milano* on October 4, 2012.

> I submit that a solution of the mystery of the illusion of physical space does not come from physics but from psychology and neural science, and in particular from the way our brain processes visual information. A brilliant analysis of this way is given in the Chapter entitled *The Mind's Eye* in the book [60] by Steven Pinker.

In classical physics, space is accompanied by time, but their mingling is not problematic. The *pre-classical spacetime*[63] defined in [WB1][64] is a mathematical structure with the following ingredients: (i) a set called the *eventworld*, (ii) a relation on the eventworld called *precedence*, (iii) a *timelapse* function, which assigns to each pairs of events, with the first preceding the second, the time lapse between them, and (iv) a *distance* function, which assigns to each pair of simultaneous events the distance between them.

Simultaneity of events and time lapse measures have an absolute meaning in classical physics, but they become problematic whenever the finite, non-vanishing time of propagation for any type of information is taken into account. Then space and time need to mingle in perhaps a counterintuitive manner to accommodate the fact that the transmission of information cannot be regarded as instantaneous. A full account on relativistic space-time is given in [WB1].

The appropriate mathematical structure to describe the space-time of special relativity is Minkowskian chronometry, which Noll had already phrased in his own axiomatic terms in 1964 [A20];[65] the same conceptual framework of this latter paper is still used in [WB1]. We find it useful to comment briefly[66] on the genesis and impact of [A20]. This paper apparently originated by a question posed by Suppes [75]: "Is it possible to find natural axioms which fix a direction of time?" Suppes adds that "Robb's meticulous axiomatization [68] in terms of the notion of *after* provides no answer."

Szekeres [77] criticized Noll's work. In particular, he considered it "inappropriate to build up kinematics from axioms which make use of assumed properties of 3-space, either explicitly or under the disguise of an inner-product space."[67] In essence, this critique targets the assumption about the existence of an inner-product; this implicitly amounts to assuming the existence of spaces of simultaneity, which would be desirable to obtain as a derived concept.[68] In 1973, Schutz [69] proposed yet another axiomatization of special relativity; here, Noll's work is only cited in passing in the Introduction alongside with Suppes'. We found no later influence of [A20],[69] unlike Robb's work [68] which is still widely known,

---

[63] This is nothing but the *neo-classical space-time* introduced in [C7], the paper that Noll read at the *Delaware Seminar in the Foundations of Physics* organized by Bunge in 1965, about which we have said already in Sect. 3.6 above.

[64] This booklet, written with Vincent J. Matsko who was Noll's student, grew out of a course that Noll taught every two or three years since 1961. The manuscript was based on handwritten notes by V.J.M. who had taken the course as a graduate student in 1987. V.J.M. contributed many original ideas, especially for Chap. 3, provided details of proofs, and produced virtually all the exercises.

[65] Ideas contained in this paper were further developed in two papers published in 1977 and 1978 with J.J. Schäffer [A31], [A32].

[66] A fuller account, to which we refer the interested reader, is given in Chap. 16 of [45].

[67] Thus, Maugin [55, p. 75] can say "... Toupin [and] Noll also tried their hands at the relativistic framework, but without much response and enthusiasm from relativists."

[68] We are indebted to M. Dafermos for this comment.

[69] The fate of oblivion did not strike only Noll's work. Ignatieff [45, p. 232] has a list of early studies on the axiomatics of special relativity, mainly from the German literature (and one also by Carathéodory [10]), which have been almost completely forgotten.

as it is credited for having introduced for the first time notions of *causality* that play a fundamental role in today's relativity theories.[70]

### 4.5.2 On Material Frame-Indifference

Here, the principle of material frame-indifference, the way we now know it, is traced back to its first appearance:

> The term "principle of material frame-indifference" was introduced in 1965 by C. Truesdell and me in our contribution [B1] to the Encyclopedia of Physics. Earlier, I had used the term "principle of objectivity", and some people use this term to this day. I meant "objectivity" to express independence of the "observer", but Truesdell disliked the term as being too easily misinterpreted. In fact, "observer" is also easily misinterpreted; a much better term is "frame of reference", or "frame" for short. ... Of course, the principle has been applied implicitly for a long time without the use of an explicit name or formulation.

This illustrates well both the need for a good choice of names in science and the need for clear statements for what may be considered to be implicit, but it is instead only vague.

The principle of *material frame-indifference* is formulated thus:

> The constitutive laws governing the internal interactions between the parts of the system should not depend on whatever external frame of reference is used to describe them.

It applies *only* to *internal* interactions, not to actions of the environment on the system, since these latter are usually tied to the frame of reference being adopted. In particular, *inertia* should always be considered as an external action, and thus exonerated from the obedience to the principle (see also Sect. 4.3.5 above). Similarly, the principle applies only to *external* frames of reference, that is, to the frames related to the environment, not to those constructed from the material system itself.

A quite original application of these ideas is made to the continuum theory of liquid crystals. These are anisotropic fluids endowed with an extra unit vector field $\mathbf{n}$, called the *director*, which describes the average orientation of the elongated molecules that constitute the material. Oseen [59] and Frank [28] had proposed to describe the elastic free-energy density $\sigma$ as a function $\sigma(\mathbf{n}, \nabla\mathbf{n})$, which measures the energetic cost associated with a distortion from the ground state characterized by *any* uniform director field, $\nabla\mathbf{n} \equiv \mathbf{0}$. The following invariance requirement is typically asked for $\sigma$ [24],

$$\sigma\big(\mathbf{Q}\mathbf{n}, \mathbf{Q}\nabla\mathbf{n}\mathbf{Q}^{\mathsf{T}}\big) = \sigma(\mathbf{n}, \nabla\mathbf{n}), \tag{39}$$

where $\mathbf{Q}$ is any orthogonal tensor (for *nematic* liquid crystals) or any *proper* orthogonal tensor (for *cholesteric* liquid crystals). Noll does not regard (39) as a consequence of his principle of material frame-indifference, but rather as

> a certain kind of material symmetry. Roughly, it states that the director-field interacts isotropically (or hemitropically in the cholesteric case) with the underlying body. In other words, the body has no implicit preferred directions in addition to the explicit one given by the director-field itself.

[70] We have also learned this from M. Dafermos. Maybe Noll was not familiar with Robb's work, but he must have known about it, because in Suppes' paper [75], which was one of Noll's motivations, that work is cited.

In tune with this interpretation, he deems as *internal* the frames constructed from liquid crystals not subject to flows, and thus exempt from the principle of material frame-indifference.

To give an elementary application of the principle, Noll considers the case of two mass-points interacting through a force $\mathbf{f}(x, y)$ that depends only on the positions $x$ and $y$ of the two particles. He shows, with an argument which is essentially the same first presented in 1957 [R1], that the principle of material frame-indifference demands that $\mathbf{f}(x, y)$ acts along the line joining $x$ and $y$, and it is odd with respect to the exchange of $x$ and $y$. This latter property is usually called the *law of action and reaction*; it is instructive to see how it derives from a general principle of invariance.

In 1972, Noll [A28] had proposed a theory for *simple* materials, which among other things made the principle of material frame-indifference vacuously satisfied by using an intrinsic mathematical formulation that does not use any frame-space. In retrospect, Noll casts doubts on the viability of this approach:

> However, the mathematics that is needed to do this, although not necessarily complicated, is not familiar to many people and hence resisted by some as being "too abstract". Also, it seems that the action of the environment on a system cannot be described without using a frame of reference, and hence one must introduce such a frame in the end when dealing with specific problems.

We seem to hear similar doubts voiced even in the most orthodox circles of the continuum mechanics community.

This contribution also features the construction of a *frame-space* from a suitable rigid material system. Mathematically, this amounts to construct a Euclidean space from a given metric set.

In accord with the preceding *Contribution*, Noll also comments on the possible extension of the principle to the theory of relativity:

> There is no simple notion of distance, and hence one cannot define unambiguously what is meant by a rigid system. Therefore, there is no relativistic counterpart of a frame of reference. ...In view of these remarks, it is not clear what a relativistic counterpart to the principle of material frame-indifference would be. The only proposal for such a counterpart that I know of is the principle of non-sentient response of Bragg [7].

### *4.5.3 Updating the* The Non-Linear Field Theories of Mechanics

The treatise [B1], which was first published in 1965 after a long labor well described in [A36],[71] is with no doubt a landmark for modern continuum mechanics. It served well two distinct purposes: first, to bring order in the conceptual and notational chaos that reined in the discipline; second, to account for all relevant contributions that had been published in the literature prior to its appearance. After nearly 40 years from the first publication, Noll felt the need for a radical revision, which does not overshadow his pride for the role he had played in the first place:

> Clifford Truesdell was my thesis advisor when I was a graduate student at Indiana University in 1954. Later he invited me to become a co-author in a treatise which he initiated: *The Non-Linear Field Theories of Mechanics* [B1]. ...Over the years [B1] became a basic reference treatise for continuum mechanics. It was reprinted

[71] See also [C15] and [WA5].

> in 1992, and a translation into Chinese appeared in 2002. A third edition, with a preface by Stuart Antman, appeared in 2004. Starting with my thesis, I introduced a mathematical infrastructure, mainly based on coordinate-free linear algebra, into continuum mechanics. I believed that this structure would provide better clarity and conceptual insight. Unfortunately, this infrastructure was familiar to almost nobody working in the field of continuum mechanics and was scornfully dismissed by some as lacking any value. Clifford Truesdell, on the other hand, had an open mind and soon began to see the value of my approach. Thus, [B1], which appeared in 1965, was largely written using this new mathematical infrastructure. ... [B1] is in many respects obsolete and perhaps should be updated after almost 40 years of its original publication. I believe that such an update should be very different from the original, and I would like to present here some guidelines.

Already in 1992, when a second edition of [B1] was published as a standalone book, Truesdell and Noll had identified the major faults of their original work. With admirable intellectual honesty, they wrote:

> The "General theory of material behavior" presented in Chap. C, although still useful, can no longer be regarded as the final word. The "Principle of Determinism for the Stress" stated on p. 56 has only limited scope. It should be replaced by a more inclusive principle, using the concept of "state" rather than a history of infinite duration, as a basic ingredient. In fact, forcing the theory of "materials of the rate type" into the general framework of the treatise as is done on p. 95 must now be regarded as artificial at best, and unworkable in general.

The mathematical infrastructure that Noll outlines in this third *Contribution* surpasses the new theory of simple materials he had proposed in 1972 [A28], which was the first attempt to replace the concept of history of infinite duration with that of state in defining simple material elements.[72] As remarked in [A40], the new theory of simple materials in [A28] had two basic ingredients:

(a) With each material element one associates a state space. A state, i.e., an element of this space, determines the present configuration of the element and the present stress in it.
(b) Given the initial state of a material element and subjecting it to an arbitrary deformation process, the final state is determined.

In Noll's revisitation of [A28], two major pillars should form the foundation for an update of [B1]: these are a *continuous body system* and a *state-space system*. The continuous body system[73] is based on two ingredients: (i) a class of subsets in three-dimensional Euclidean spaces eligible to be occupied by a body system when placed in a frame of reference, and (ii) a class of mappings which are candidates for changing the placement of a body system from one frame to another.

For (i), Noll still deems it appropriate to use the *fit regions* of [A33]. The state-space system (ii) is a more complicated structure based on seven ingredients: (a) a set of *conditions*, (b) a set of *processes*, (c) a set of *states*, called the *state-space*, (d) a mapping from states into conditions, (e) an *evolution functional*, delivering a final state after applying a process to

[72] Noll himself acknowledges that a similar notion of state, although based on different axioms, had been introduced by Willems [104].

[73] This is a new term for what Noll used to call simply a body, a concept including a no longer required connectedness.

an initial state, (f) a set of *responses*, and (g) a mapping delivering a response when applied to a state.

Processes and evolution functionals are then subject to a number of *natural* assumptions, which would be too complicated to explain here in details.[74] Suffice it to say that the resulting axiomatic structure forms the basis of the new theory of simple materials first presented in [A28]. In particular, for a special class of materials called *semi-elastic*, this theory is just the reformulation of the older theory of simple materials with memory first formulated in [A7] and reprised in [B1].

This *Contribution* goes beyond the new theory of simple materials. Indeed, not all materials fall in this abstract category. Having in mind liquid crystals as special examples, Noll introduces *second-grade* body elements and formulates an axiomatic, frame-free theory to describe them, which builds heavily on his lecture notes on *Geometry of Differential Manifolds* [WC6]. This theory assigns a Riemannian structure to a second-grade body element, which not only consists of a *metric tensor* but also brings in a *symmetric affine connection*.[75]

### *4.5.4 The Theory of Simple Semi-Liquids, a Conceptual Framework for Rheology*

In this *Contribution*, Noll applies the new theory of simple materials to the study of *monotonous states*, which include the state of *simple shearing* as a special case. Monotonous states are related to *monotonous processes*, which have played a crucial role in the rheology of simple fluids.

The first idea of what was to become a monotonous process was introduced in 1962 by Coleman [13], but he used for it the term "substantially stagnant motion", which was shortly after replaced by Noll [A17] with "motion with constant stretch history". It had also been introduced in [13] the term "viscometric flow", which has enjoyed a much longer life. Monotonous flows were first described in terms of tensorial exponentials in [A17], and a rigorous proof of a fundamental representation theorem for monotonous processes was given in [A30]. It would be a bit too long to dwell here on a formal definition of a monotonous process and the details of the relevant representation theorem. We find instead illuminating reproducing the analogy with curves in a three-dimensional Euclidean space established by Noll. If a process has its analog in a curve in 3-space, then a monotonous process would be represented by a *monotonous curve*, that is, a curve which has constant curvature and torsion. The only such curves are straight lines, circles, or helices, to which there correspond three distinct classes of monotonous processes (and three *viscometric functions* [B2], see also Sects. 4.1 and 4.4 above).

A *semi-liquid* is a material characterized by being incompressible and by having its symmetries described by the *unimodular group*.[76] This *Contribution* shows that the entire theory

---

[74]These assumptions are meant to replace the ones that are already common in continuum mechanics textbooks, namely, the principles of *determinism*, *local action*, and *material frame-indifference* (see, for example, Sect. IV.2 of [96], for a detailed account).

[75]In this regard, the following quote from [A45] is appropriate:

> In about 1970, at a conference in Germany, I was asked by a Physics professor whether General Relativity could be made coordinate-free. I said yes, but it took me decades to develop the necessary background. The reason was that a fully coordinate-free reformulation of the theory of differentiable manifolds had to be created first.

[76]The unimodular group comprises all (invertible) tensors whose determinant has modulus 1. In [A22] Noll gave an original, alternative proof of the fact that the unimodular group is the *maximal* subgroup of the orthogonal group.

of viscometric flows treated in [B2] can be extended to semi-liquids; thus, this new theory should also be the basis for a radical update of [B2] (not only of Chap. E of [B1]).

The examples of Poiseuille and Couette flows are analyzed explicitly, whereas other examples of viscometric flows, such as helical, cone and plate, and torsional flows, are left for the reader to elaborate on. With the support of these applications, Noll may rightly say:

> I believe that the theory of semi-liquids is a good mathematical model for many materials considered by rheologists, toothpaste being a familiar example.

However, the new theory of simple semi-liquids presented here allows the existence of plugs (regions of constant velocity) in viscometric flows, while the old theory did not. A thorough study of plugs is carried out in [WB6].

### *4.5.5 Nematic Semi-Liquids*

This *Contribution* is rather technical in character; it contains a special application of the theory outlined in Sect. 4.5.4 to a class of semi-liquids that have some connections with nematic liquid crystals.

The attempts to frame liquid crystals (in, at least, some of their motions) within the general theory of simple fluids have an interesting, long history. This starts perhaps with Wang [98], who defines a simple *subfluid* as a simple material whose isotropy group contains a *dilatation* group.[77] Coleman [14] called *simple liquid crystal* a simple material whose symmetry group is neither the full unimodular group nor any subgroup of the orthogonal group. He identified two special cases of simple liquid crystals according to this definition. First, when the material symmetry group is the group of all unimodular tensors that perserve or reverse all vectors in a one-dimensional or two-dimensional subspace. Second, when the material symmetry group is the group of all unimodular tensors mapping a one-dimensional or two-dimensional subspace into itself. While the second case is included in Wang's theory, the first is not.

All these theories, envisaging liquid crystals as special simple fluids, allowed neither for couple stresses nor for a non-symmetric stress tensor, as instead did the more general, early theories developed by Ericksen in a series of papers [21, 22, 24]. For a special constitutive choice, lucidly discussed in [21], the stress tensor is symmetric, although it depends on the nematic director $\mathbf{n}$ (here not necessarily of fixed length) besides the stretching tensor $\mathbf{D}$. The balance equation for linear momentum is supplemented in this theory by an evolution equation for $\mathbf{n}$, appropriately made invariant under change of frame.[78]These equations are reproduced in Sects. 127 and 129 of [B1]; in this final *Contribution*, Noll provides for them a full frame-free derivation within his new abstract theory of simple materials. Noll also retraces in this formal setting Ericksen's results for the corresponding viscometric functions [23], reproduced in Eqs. (129.6) of [B1], where he now also spots a small error: "$\lambda_1$ in the first line should be replaced by $\lambda_2$".[79]

[77] A dilatation group is the group of all unimodular tensors having three fixed linearly independent vectors as their eigenvectors.

[78] Ericksen [21] explicitly renounces seeking inspiration for his equations in the molecular theories for suspensions of ellipsoidal [47] or dumbell particles [65], which had been derived by Jeffery and Prager, respectively. This task was undertaken by Hand [41, 42], who proposed a theory for anisotropic fluids with a symmetric stress tensor and an evolution equation for a supplementary symmetric tensor representing the alignment of suspended particles. Hand showed that his theory includes as special cases Jeffery's, Prager's, and Ericksen's.

[79] The history of the dynamics of liquid crystals has seen more than the attempts to represent them at all costs as simple fluids. Ericksen [25] reformulated in a continuum mechanics context the static theories of

## 5 Episodic Research Themes

Besides the recurrent research themes described in the preceding section, Noll touched, somehow episodically, upon a number of other subjects, which were either completely independent of the former, or only tangentially related to them. These works are generally not very well-known, some perhaps also testimony to the publish-or-perish fever that afflicted the young Noll too [45, p. 201]; precisely these features make some of them not less interesting than those amply discussed in Sect. 4. We give below a short account of a few of them, extracted (in chronological order) from the list in Sect. 7 of the articles published in English in regular journals. Each paragraph below has the same title as the corresponding paper. Occasionally, we borrow freely from the introductions and abstracts to describe the contents of the selected papers.

**"On the Rotation of an Incompressible Continuous Medium in Plane Motion" [A4]** Noll considers a plane motion of an incompressible medium with homogeneous density. The plane region occupied by the medium is assumed to be bounded by a finite number of smooth curves. If the region is finite, all but one of the closed boundary curves move rigidly. If it extends to infinity, all finite closed boundary curves move rigidly. This motion is labeled by I. Another motion, labeled by II, is obtained from I by superposing a rotation with constant angular velocity about a fixed axis perpendicular to the plane of the motion. The paper shows that motion II is dynamically possible whenever I is, and that the stresses in II differ from those in I only by a pressure field.

G.I. Taylor [80] had already proved this theorem for the special case of the irrotational motion of an inviscid fluid caused by rigid cylindrical bodies moving in an infinite container. Noll's paper points out that Taylor's theorem remains valid for any material whatsoever (it may even be an anisotropic solid). The essential assumptions are that the medium be incompressible and that the motion be plane.

**"On the Uniqueness and Non-existence of Stokes Flows" [A5] (with R. Finn)** It has been known since the time of Stokes that under suitable assumptions on the behavior of the velocity field at infinity, there exists no steady two-dimensional flow of a viscous fluid past an obstacle, in which the velocity components are infinitesimals of the first order. On the other hand, flows of this type in three dimensions are explicitly known.

An explanation sometimes advanced for the discrepancy in results is that in the known flow past a sphere, the assumption on the ratio of inertial to viscous forces which is used to derive the equations of motion becomes violated in a neighborhood of infinity. In the view of the authors, this phenomenon brings into serious question both the physical and the mathematical validity of the known results, and makes it imperative to investigate the sense

Oseen [59] and Frank [28]. Unfortunately, however, these Authors' theory could not be obtained as the static limit of Ericksen's theory of anisotropic fluids recalled above. Leslie [52] succeeded in extending an earlier constitutive theory of his [51] consistent with Ericksen's theory of anisotropic fluids, thus arriving at what has become known as the Ericksen-Leslie theory for the dynamics of liquid crystals, perfectly compatible with Ericksen's static theory, and blessed with an impressive number of experimental confirmations [30, Chap. 5]. In this theory, liquid crystals are no longer regarded as simple fluids. This confirms the suspicion that Coleman and Noll [A16] had already raised:

> It is possible that the theory of simple fluids cannot be successfully applied to the physics of liquid crystals. My colleague and I, however, are not sufficiently familiar with the mechanical behavior of such substances to hazard a categorical statement on this.

in which the boundary problem is correctly set. More precisely, it is natural to ask whether there exist two-dimensional flows in which the velocity tends to its limit more slowly than has in the past been assumed, and whether a three-dimensional flow is unique in a (physically reasonable) class of flows for which the usual uniqueness proofs break down.

This paper proves that the answer to the first question is no, whereas to the second is yes. In fact, it is shown that there are no two-dimensional flows for which the velocity is bounded, and the uniqueness of a three-dimensional flow is proven under the single assumption that the velocity tends to a limit at infinity.

**"On Exterior Boundary Value Problems in Linear Elasticity" [A8]** The questions considered in this note were suggested by the paper [A5] on the uniqueness and non-existence of Stokes flow. The equations of linear isotropic elasticity are only slightly more complicated than the equations governing Stokes flow, and it proves possible to extend the treatment of [A5] to analogous elastic problems.

**"Materially Uniform Simple Bodies with Inhomogeneities" [A23]** Here Noll presents a detailed study of the structure of bodies that consist of a uniform simple material, yet are not necessarily homogeneous (see also [C8]). The concept of simple material employed here is more inclusive than the one described in [A7] (and reprised in Chap. C III of [B1]) because it can be appropriate not only to mechanical material properties, but also to thermal, optical, electrical, magnetic, or any other type of material properties. A body may be simple with respect to any particular such material property or to any combination of them. The physical theory relevant to these properties need not be made explicit.

A precise definition of a *materially uniform simple body* is given here. The nature of the coherence of a uniform body with respect to the local material properties under consideration is described mathematically either in terms of a *material uniformity* or in terms of a *uniform reference*. In general, neither of these is uniquely determined by the simple body structure, but the degree of non-uniqueness can be delimited precisely. There may or may not exist uniform references that are gradients of global configurations. If they do exist, the body is homogeneous, and the theory becomes trivial.

A material uniformity is equivalent to an affine connection. The Cartan torsion of this connection describes locally the deviation from homogeneity and is therefore called the *inhomogeneity* of the given material uniformity. Associated with each smooth uniform reference is also a Riemannian structure on the body, whose relation to the material connection is studied in detail. The difference between the Riemannian connection and the material connection determines the *contortion* of the given uniform reference. Contortion and inhomogeneity determine one another.

A special type of non-homogeneity, called *contorted aeolotropy* generalizes the more familiar *curvilinear aeolotropy*. In contorted aeolotropy, the deviation from homogeneity is given by a distribution of rotations on a suitable global configuration, and the contortion describes the local behavior of this distribution.

For applications to materially uniform but inhomogeneous bodies, a new version of Cauchy's balance equation is much more suitable than the usual one. This new version gives rise, for example, to a definite differential equation for the theory of inhomogeneous but materially uniform elastic bodies.

This paper embodies Noll's contribution to the theory of continuous distributions of dislocations; it adds to a vast literature, of which he is also aware.[80] Here are his comments on the features that in his opinion distinguish his work from the preceding ones:

> Motivated by heuristic considerations, mostly concerning lattice defects in crystals, these authors lay down a priori certain geometric structures to describe distributions of dislocations. These geometric structures are formally of the same type as some of those occurring in the present paper. The conceptual status of the theory presented here, however, is very different. I show that once a constitutive assumption defining a materially uniform simple body is laid down, the geometric structures of the body are determined. The geometry is thus the natural outcome, not the first assumption, of the theory. Since the underlying constitutive assumption is very general, the real materials to which the theory can be expected to apply need be neither crystalline, nor elastic, nor solid.

Similar remarks can be found in Sect. 34 of [B1], which contains a previous version of the theory presented here (a version which Noll considers superseded by [A23], but which highly influenced Wang's work [99] on this subject). In [B1] (see, in particular, p. 89) is another remark, which we also find worth recalling:

> In one respect, however, Noll's theory is less general than some of those above referenced, since couple-stresses are excluded from the start. That couple-stresses are really needed for an adequate description of the behavior of real solids has not yet been shown, in our opinion.

In 1990, Epstein and Maugin [19, 20] provided a meaningful connection between Noll's theory of material inhomegeneity and Eshelby's material stress [26], which had been introduced in 1951 (see also [55, p. 254]).

**"Quasi-Invertibility in a Staircase Diagram" [A24]** This is a technical note that deals with objects and morphisms in an Abelian category.

It starts with the definition of a *quasi-invertible* morphism. In particular, for module-homomorphisms, quasi-invertibility means that both Kernel and Image are direct summands. For monomorphisms, quasi-inverses coincide with left inverses, i.e., monomorphisms are quasi-invertible if and only if they are left-invertible. For epimorphisms, quasi-inverses coincide with right inverses, i.e., epimorphisms are quasi-invertible if and only if they are right-invertible.

Properties of semi-invertible morphisms are proved which were needed in the paper [A27] on annihilators of differential operators, but may as well have other applications.

**"On Certain Convex Sets of Measures and on Phases of Reacting Mixtures" [A25]** The paper starts with a quote of Gibbs' memoir on the equilibrium of mixtures [31, p. 153]:

> It does not seem probable that $r$ can ever exceed $n+2$ [$n=$ number of components, $r=$ number of coexistent phases]. An example of $n=1$ and $r=3$ is seen in the coexistent solid, liquid, and gaseous forms of any substance of invariable composition.

On which Noll comments:

[80] Noll knew about the works of Kondo, Nye, Bilby, Bullough, Smith, Seeger, Kröner, and Günther, possibly through the expository accounts of Kröner [49], Bilby [6], and Seeger [70], see also [B1, p. 88].

> It is interesting that Gibbs was not certain about the matter. It seems to me that this is so because the mathematical tools necessary to render the basic concepts of his theory completely unambiguous were not available at the time. These tools are available now, and it is perhaps worthwhile to develop a modern version of Gibbs' theory. This paper is intended to be a step toward such a version.

The subject concerns chemically reacting mixtures of fluids. External influences such as gravity are assumed to be absent and surface tension effects are neglected. The problem is to characterize the stable equilibrium states of such mixtures. The set of possible states of a mixture are described mathematically as a certain convex set of Radon measures. The stable states are those that maximize a certain functional (the entropy) on this set of measures.

The general theory is rather abstract and formulated in Banach spaces, but it applies seamlessly to the (finite-dimensional) case of reacting mixtures.[81] It turns out that Gibbs' assertion is correct for what are here called strictly stable states ("stable equilibrium" in Gibbs' terminology, see [31, p. 110]). It is not necessarily correct for states that are stable but not strictly stable ("neutral equilibrium" in Gibbs' terminology).

**"Representations of Certain Isotropic Tensor Functions" [A26]** Two types of isotropic functions are considered in this paper. First, scalar-valued functions depending on a symmetric second-order tensor and a vector. Second, vector-valued functions of the same arguments. For both types of functions, Noll gives general and elegant representation formulae.

**"Annihilators of Linear Differential Operators" [A27] (with H.D. Dombrowski)** Elementary kinematic problems in potential theory and linear elasticity can be subsumed under the following general problem. Given a differential operator $f$, the "integrability conditions" of the differential equation $f(x) = y$ are described by differential operators $g$ with the property $g \circ f = 0$. Such differential operators are called *annihilators* of $f$.

This paper deals with the following problem: Given any linear differential operator, find all of its annihilators. This problem has a constructive solution if one can find an annihilator $g_*$ that is generating, in the sense that every annihilator $g$ of $f$ is of the form $g = l \circ g_*$, where $l$ is some differential operator. Thus, the following particular question is addressed: Under which conditions does a differential operator admit a generating annihilator, and how can it be determined if it exists?

To answer this question, two characteristics of $f$ turn out to be of crucial importance: the *characteristic polymorphism* of $f$ and the *sequence of curvatures* of $f$. These notions are closely related to others known in the literature: these are the "characteristic polynomial" of a linear differential operator and the "curvature" of a covariant differential, respectively.

The main result of the paper enables one to find a generating annihilator of a differential operator $f$ provided one knows a generating annihilator of the characteristic polymorphism of $f$ and provided almost all curvatures of $f$ vanish.

[81] The Gibbsian stability of phases was also to be later considered in [17], among other issues. The authors of this paper had, however, missed [A25], presumably on the account of the mathematical language it was clad in. In a note added in proof they say:

> We believe that this interesting paper has not gained wide recognition because of the non-standard although elegant mathematical formalism it employs to examine the morphology of material phases. In fact, it was only after our work was completed that we came to appreciate Noll's contribution.

In retrospect, it is perhaps fair to consider Noll's paper [A25] as the precursor of the growth of interest in the variational approach to phase transformations that was to follow in the mathematical literature.

**"The Geometry of Contact, Separation, and Reformation of Continuous Bodies" [A35]** Here is Noll's own motivation for this study:

> Separation and contact of real objects are everyday phenomena. Real objects may develop cracks, may split, or may shatter to pieces. Fluids may cavitate or mingle. Real objects may stick together, merge, or be pieced together to form composites. The laws that govern these phenomena have been of great interest to engineers for a long time and there has been renewed interest in the discovery and study of these laws in recent years.

The paper intends to create a mathematical infrastructure which would facilitate the study of these phenomena and create a satisfactory mathematical description of changes of *coherence* for such bodies. The programme that Noll conceives here is quite ambitious (and, as far as we know it has not yet been carried out); this paper just moves the first step:

> It includes only what I call the "geometry" of changes of coherence. A next step would be a description of "kinematics" of such changes, i.e., of changes of the geometry of coherence with time. Next would come a study of the forces involved in such changes, and finally one would like to develop a general theory of the constitutive laws governing the phenomena governing such changes.

One of the problems with the description of separation and contact of continuous bodies is that the material points on the surfaces of separation or contact disappear or are created. One may ask, therefore, what retains its identity during such processes. The answer given here is that not individual material points of a body are preserved and retain their identity, but the whole materially ordered set of the subbodies of the body and the associated Boolean algebra. The material points are determined by filters (in the sense of Boolean algebra) of such subbodies. What may change as a result of separation or contact is the set of filters that determine material points.

Changes of coherence of a continuous body can come about not only by separation and contact, but also by processes that can be understood as a separation followed by a simultaneous contact or vice versa. Sliding would be an example. If such separation and simultaneous contact preserve material points, the process is called a "reformation". Such reformations may serve to obtain insight into phenomena such as phase transitions, acceleration waves, and shock waves.

## 6 Conclusion

One may think that the most natural and appropriate way to close our journey through the scientific life of Walter Noll would be to state what we expect his lifetime effort will pass on to the next generations of researchers and what we believe will be his legacy. On reinforcing Mark Twain's warning [74, p. 408], we prefer to abstain from prophecies altogether. Our reflections are thus bound to the past, where Walter's action has already been more than effective.

Maybe the best and most objective praises came from those who would not consider themselves as his followers or have been critical towards him. For example, Maugin says [55, p. 69]:

> It is clear that he was more familiar than Truesdell with modern mathematics (in particular topology and geometry). This he demonstrated with maestria in his mathematical theory of simple continua [A7], [A28] and in his approach to material uniformity

> and inhomogneities [A23]. He is very much responsible for the extensive use of an intrinsic notation avoiding tensorial indices in contemporary continuum physics.

The latter statement is possibly stressed by Antman, who in the preface to the third edition [92] of [B1] wrote:

> It is interesting to note that Truesdell and Noll's system of notation, symbols, and terminology has been widely adopted, even by the scientific descendents of critics of their enterprise.

As a matter of fact, while working at providing a new and more encompassing paradigm for continuum physics, two by no means secondary tools were for Noll appropriate notations and terminology.

This is a time when many seem convinced that quality is a quantity, hence it can be measured. So metrics have replaced judgment in many fields, including science. It is not for us to explore the sociological motivations and implications of such a state of affairs; we deem it to be a cold matter of fact, a common experience, which can hardly be disputed. However, if anything, our essay has shown that Walter Noll's scientific life is the perfect testimony to the delusion of the quantifiability of quality. How many of us use concepts, names, and notation that Walter introduced with not a single citation to his work?

One may think that, if not citations, such a plethora of indirect, often unaware admirers would have at least given him recognition. Unfortunately, this was not always the case, especially in the last part of his life. In [WB2, p.3] Walter writes:

> A letter from the editor of the *Reviews*,[82] written in 1988, informed me that a 1961 paper by Bernard Coleman and me had become a *citation classic* and that they would welcome receiving other papers from me.[83] However, my paper[84] was rejected even though the editor conceded that "the article is clearly written". Here are some quotes from the reviewer:
>
>> I enjoyed reading this paper and very much would like to see it published. I am afraid, however, that the Reviews of Modern Physics is not the appropriate place. I believe that the overwhelming majority of the readers of the journal will consider the paper unreadable. Not because the material presented is intrinsically difficult, but rather because the author's individual form of the 'Bourbakian' style is far removed from anything that physicists are willing to digest. ... Professor Noll is highly respected in the mathematical community and has more than once proved himself to be ahead of his time. ...

There is something to this final comment, however. Noll's thesis in 1955 was so ahead of his time that his advisor had requested that all its formulas were transcribed in the then more popular indicial form, but it was published, this notwithstanding. His thoughts of nearly half a century later could not find a home in most traditional publishing venues, whatever their prestige and circulation. You may say: what a striking contrast!

Quite the contrary: the contrast is only illusory. With his thesis, Noll began a life-long effort to codify continuum mechanics. At the beginning of his career, this effort provided the concepts and notation that were to be absorbed so deeply and to raise persistently the level

[82] That is, the *Reviews of Modern Physics*.

[83] The paper Noll refers to here is [A14], see Sect. 4.4.1.

[84] This paper, which had been submitted in 1995, was an earlier version of the Contribution *On Material Frame Indifference*, which eventually appeared in [WB2], see Sect. 4.5.2.

of exposition and thought in the field. But, this effort was accompanied by a compulsion for repeated reexamination and reformulation of the formal structure and the mathematical infrastructures he himself had proposed. In the myopic view of many, there was no need for that. While Walter's purpose was to best capture the essence and to widen the scope of continuum mechanics, his compulsion originated a quite widespread rejection of the formats he was proposing as improvements to the accepted Nollian syllabus; eventually, "[this compulsion] drove him far out on a limb with respect to the accessibility of his ideas and their potential influence on the field."[85]

There is a striking similarity between the first and the last years of Noll's scientific life: Walter was as lonely at the start of his journey into discovery as he was at the end. He trusted to the same degree his *free* thinking and he was always equally unwilling to give it up. The only difference was that in his last years the web made it easier for his voice to be heard [WA9].[86] Indeed, he knew only too well that expressing one's own thoughts is part of thinking, as free expression and free thinking are but the two sides of one and the same coin.

This is the ultimate lesson that Walter has taught us: to persevere in free thinking through our whole scientific lives. A formidable task indeed: we might not be up to it, but we must at least try.

## 7 Publications of Walter Noll

Here we list all works by Walter Noll that we were able to retrace both in the scientific literature and on the web. They are organized in a number of self-explanatory categories.

### Books

[B1] The Non-Linear Field Theories of Mechanics, vol. III/3 of Encyclopedia of Physics. Springer, Berlin, 1965. 2nd edn. 1992, Chinese transl. 2000, 3rd edn. 2004 edited by S.S. Antman (with C.A. Truesdell).

[B2] Viscometric Flows of Non-Newtonian Fluids, Theory and Experiment, vol. 5 of Springer Tracts in Natural Philosophy. Springer, Berlin, 1966. (with B.D. Coleman & H. Markovitz).

[B3] Continuum Theory of Inhomogeneities in Simple Bodies: A Reprint of Six Memoirs. Springer, New York, 1968. (with R. Toupin & C.C. Wang).

[B4] The Foundations of Mechanics and Thermodynamics: Selected Papers. Springer, Berlin, 1974.

[B5] Finite-Dimensional Spaces: Algebra, Geometry, and Analysis, vol. I. Nijhoff, Dordrecht, 1987.

### Contributions to Books

[C1] The foundations of classical mechanics in the light of recent advances in continuum mechanics. In: L. Henkin, P. Suppes, A. Tarski (eds.), The Axiomatic Method, with Special reference to Geometry and Physics, vol. 27 of Studies in Logic and the Foundations of Mathematics, pp. 266–281. North-Holland, Amsterdam, 1959.

[85] We here quote verbatim a passage from a message by D.R. Owen to us, whose contents we incorporated in the above considerations.

[86] His late writings posted on the web are a treasure trove of reflections and worth perusing conjectures.

Proceedings of an International Symposium on the Axiomatic Method, held at the University of California, Berkeley, December 27, 1957–January 4, 1958. Reprinted in [B4], pp. 32–47.

[C2] La mécanique classique, basée sur un axiome d'objectivité. In: A. Châtelet, J. Destouches, F. Aeschlimann (eds.), La Méthode Axiomatique dans les Mécaniques Classiques et Nouvelles, pp. 47–56. Gauthier-Villar, Paris, 1963. Actes du 4e Colloque International de Logique et Philosophie des Sciences tenu sous la présidence d'Albert Châtelet, Institut Henri-Poincaré, Paris, Septembre 1959. Reprinted in [B4], pp. 135–144.

[C3] Simple fluids with fading memory. In: M. Reiner, D. Abir (eds.), Second-Order Effects in Elasticity, Plasticity and Fluid Dynamics, pp. 530–552. Macmillan/Pergamon, Oxford, 1964. Proceedings of an International Symposium, Haifa, Israel, April 23–27, 1962 (with B.D. Coleman).

[C4] Modern theories of flow in tubes. In: E.O. Attinger (ed.), Pulsatile Blood Flows, pp. 77–83. McGraw-Hill, New York, 1964. Proceedings of the First International Symposium on Pulsatile Blood Flow, held at the Presbyterian Hospital in Philadelphia, April 11–13, 1963.

[C5] The equations of finite elasticity. In: R. Finn (ed.), Applications of Nonlinear Partial Differential Equations in Mathematical Physics, vol. 17 of Proceedings of Symposia in Applied Mathematics, pp. 93–101. American Mathematical Society, Providence, RI, 1965. Lecture notes prepared for the American Mathematical Society short course on Applications of Nonlinear Partial Differential Equations in Mathematical Physics held in New York, April 20–23, 1964.

[C6] The foundations of mechanics. In: G. Grioli, C. Truesdell (eds.), Non-linear Continuum Theories, vol. 36 of CIME Summer Schools, pp. 159–200. Spinger, Heidelberg, 2011. Lectures given at a Summer School of the Centro Internazionale Matematico Estivo (C.I.M.E.) held in Bressanone (Bolzano), Italy, May 31–June 9, 1965. Reprint of the 1st edn. C.I.M.E., Ed. Cremonese, Roma, 1966.

[C7] Space-time structures in classical mechanics. In: M. Bunge (ed.), Delaware Seminar in the Foundations of Physics, pp. 28–34. Springer, Berlin, 1967. Reprinted in [B4], pp. 204–210.

[C8] Inhomogeneities in materially uniform simple bodies. In: E. Kröner (ed.), Mechanics of Generalized Continua, pp. 239–246. Springer, Berlin, 1968. Proceedings of the IUTAM-Symposium on The Generalized Cosserat Continuum and the Continuum Theory of Dislocations with Applications, Freudenstadt and Stuttgart (Germany), 1967. Reprinted in [B3], pp. 1–8.

[C9] On the concept of the symmetry group of a physical system. In: P.G. Glockner, M.C. Singh (eds.), Symmetry, Similarity, and Group Theoretic Methods in Mechanics, pp. 83–99. University of Calgary Press, Calgary, 1974. Proceedings of a Symposium held in Calgary, August 19–21, 1974.

[C10] The role of the second law of thermodynamics in classical continuum physics. In: B. Gal-Or (ed.), Modern Developments in Thermodynamics: An Interdisciplinary Collective Treatise, pp. 117–119. Wiley, New York, 1974.

[C11] A general framework for problems in the statics of finite elasticity. In: G.M. De La Penha, L.A.J. Medeiros (eds.), Contemporary Developments in Continuum Mechanics and Partial Differential Equations, vol. 30 of North-Holland Mathematics Studies, pp. 363–387. North-Holland, Amsterdam, 1978.

[C12] Continuum mechanics and geometric integration theory. In: F.W. Lawvere, S.H. Schanuel (eds.), *Categories in Continuum Physics: Lectures given at a Workshop held at SUNY, Buffalo 1982*, vol. 1174 of Lecture Notes in Mathematics, pp. 17–29. Springer-Verlag, Berlin, 1986.

[C13] Discontinuous displacements in elasticity. In: Proceedings of the Meeting on Finite Thermoelasticity (Roma, 1985), vol. 76 of Contributi del Centro Linceo Interdisciplinare di Scienze Matematiche e Loro Applicazioni, pp. 151–163. Accademia Nazionale dei Lincei, Roma, 1986 (with P. Podio-Guidugli).

[C14] On cracks in elastic bodies. In: Beitrage zur Mechanik: Festschrift zum 65. Geburtstag von Prof. Dr. Rudolf Trostel, pp. 249–253. Technische Universitat Berlin, Berlin, 1993.

[C15] The Genesis of the Non-Linear Field Theories of Mechanics. In: S.S. Antman (ed.), The Non-Linear Field Theories of Mechanics, vol. III/3 of Encyclopedia of Physics, 3rd edn., pp. VII–XII, 2004, see [92]. See also [A36].

## Articles in Journals

[A1] Eine Bemerkung zur Schwarzschen Ungleichheit. Mathematische Nachrichten **7**(1952): 55–59 (with E. Mohr).

[A2] Die Herleitung der Grundgleichungen der Thermomechanik der Kontinua aus der statistischen Mechanik. Journal of Rational Mechanics and Analysis **4**(1955): 627–646.

[A3] On the continuity of the solid and fluid states. Journal of Rational Mechanics and Analysis **4**(1955): 3–81. Reprinted in [94].

[A4] On the rotation of an incompressible continuous medium in plane motion. Quarterly of Applied Mathematics **15**(1957): 317–319.

[A5] On the uniqueness and non-existence of Stokes flows. Archive for Rational Mechanics and Analysis **1**(1957): 97–106 (with R. Finn).

[A6] Verschiebungsfunktionen für elastische Schwingungsprobleme. ZAMM—Zeitschrift für Angewandte Mathematik und Mechanik [Journal of Applied Mathematics and Mechanics] **37**(1957): 81–87.

[A7] A mathematical theory of the mechanical behavior of continuous media. Archive for Rational Mechanics and Analysis **2**(1958): 197–226. Reprinted in [B4], pp. 1–30, in [B3], pp. 25–54, and in [94].

[A8] On exterior boundary value problems in linear elasticity. Archive for Rational Mechanics and Analysis **2**(1958): 191–196 (with R.J. Duffin).

[A9] Conditions for equilibrium at negative absolute temperatures. Physical Review **115**(1959): 262–265 (with B.D. Coleman).

[A10] Helical flow of general fluids. Journal of Applied Physics **30**(1959): 1508–1512. Reprinted in [94] (with B.D. Coleman).

[A11] On certain steady flows of general fluids. Archive for Rational Mechanics and Analysis **3**(1959): 289–303. Reprinted in [B4], pp. 49–63, and in [94] (with B.D. Coleman).

[A12] On the thermostatics of continuous media. Archive for Rational Mechanics and Analysis **4**(1959): 97–128. Reprinted in [B4], pp. 65–96, and in [95] (with B.D. Coleman).

[A13] An approximation theorem for functionals, with applications in continuum mechanics. Archive for Rational Mechanics and Analysis **6**(1960): 355–370. Reprinted in [B4], pp. 97–112, and in [94] (with B.D. Coleman).

[A14] Foundations of linear viscoelasticity. Reviews of Modern Physics **33**(1961): 239–249. Reprinted in [B4], pp. 113–123, and in [95] (with B.D. Coleman).

[A15] Normal stresses in second-order viscoelasticity. Transactions of the Society of Rheology **5**(1961): 41–46 (with B.D. Coleman).

[A16] Recent results in the continuum theory of viscoelastic fluids. Annals of the New York Academy of Sciences **89**(1961): 672–714 (with B.D. Coleman).
[A17] Motions with constant stretch history. Archive for Rational Mechanics and Analysis **11**(1962): 97–105. Reprinted in [B4], pp. 125–133.
[A18] Steady extension of incompressible simple fluids. The Physics of Fluids **5**(1962): 840–843 (with B.D. Coleman).
[A19] The thermodynamics of elastic materials with heat conduction and viscosity. Archive for Rational Mechanics and Analysis **13**(1963): 167–178. Reprinted in [B4], pp. 145–156 (with B.D. Coleman).
[A20] Euclidean geometry and Minkowskian chronometry. The American Mathematical Monthly **71**(1964): 129–144. Reprinted in [B4], pp. 183–198.
[A21] Material symmetry and thermostatic inequalities in finite elastic deformations. Archive for Rational Mechanics and Analysis **15**(1964): 87–111. Reprinted in [B4], pp. 157–181 (with B.D. Coleman).
[A22] Proof of the maximality of the orthogonal group in the unimodular group. Archive for Rational Mechanics and Analysis **18**(1965): 100–102. Reprinted in [B4], pp. 200–202.
[A23] Materially uniform simple bodies with inhomogeneities. Archive for Rational Mechanics and Analysis **27**(1967): 1–32. Reprinted in [B4], pp. 211–242, and in [B3], pp. 55–86. See also errata in Archive for Rational Mechanics and Analysis **31**(1968): 401; **38**(1970): 405–406.
[A24] Quasi-invertibility in a staircase diagram. Proceedings of the American Mathematical Society **23**(1969): 1–4.
[A25] On certain convex sets of measures and on phases of reacting mixtures. Archive for Rational Mechanics and Analysis **38**(1970): 1–12.
[A26] Representations of certain isotropic tensor functions. Archiv der Mathematik **21**(1970): 87–90.
[A27] Annihilators of linear differential operators. Journal d'Analyse Mathématique **24**(1971): 207–284 (with H.D. Dombrowski).
[A28] A new mathematical theory of simple materials. Archive for Rational Mechanics and Analysis **48**(1972): 1–50. Reprinted in [B4], pp. 243–292.
[A29] Lectures on the foundations of continuum mechanics and thermodynamics. Archive for Rational Mechanics and Analysis **52**(1973): 62–92. Reprinted in [B4], pp. 294–324.
[A30] The representation of monotonous processes by exponentials. Indiana University Mathematics Journal **25**(1976): 209–214.
[A31] Orders, gauge, and distance in faceless linear cones; with examples relevant to continuum mechanics and relativity. Archive for Rational Mechanics and Analysis **66**(1977): 345–377. See also erratum in Archive for Rational Mechanics and Analysis **68**(1978): 383–384 (with J.J. Schäffer).
[A32] Order-isomorphisms in affine spaces. Annali di Matematica Pura ed Applicata **117**(1978): 243–262 (with J.J. Schäffer).
[A33] Fit regions and functions of bounded variation. Archive for Rational Mechanics and Analysis **102**(1988): 1–21 (with E.G. Virga).
[A34] On edge interactions and surface tension. Archive for Rational Mechanics and Analysis **111**(1990): 1–31 (with E.G. Virga).
[A35] The geometry of contact, separation, and reformation of continuous bodies. Archive for Rational Mechanics and Analysis **122**(1993): 197–212.
[A36] The genesis of Truesdell's Nonlinear Field Theories of Mechanics. Journal of Elasticity **70**(2003): 23–30. See also [C15].

[A37] A frame-free formulation of elasticity. Journal of Elasticity **83**(2006): 291–307. See also [WB3].
[A38] On the past and future of natural philosophy. Journal of Elasticity **84**(2006): 1–11. See also [WA7].
[A39] Monoids, Boolean algebras, materially ordered sets. International Journal of Pure and Applied Mathematics **37**(2007): 187–202. See also [WC8] (with B. Seguin).
[A40] Plugs in viscometric flows of simple semi-liquids. Nihon Reoroji Gakkaishi (Journal of the Society of Rheology, Japan) **37**(2009): 1–10. See also [WB6] (with B. Seguin).
[A41] Thoughts on thermomechanics. Journal of Thermal Stresses **33**(2009): 15–28.
[A42] Basic concepts of thermomechanics. Journal of Elasticity **101**(2010): 121–151 (with B. Seguin).
[A43] Thoughts on the concept of stress. Journal of Elasticity **100**(2010): 25–32. See also [WB7].
[A44] A conceptual framework for problems in elastostatics. Journal of Elasticity **105**(2011): 107–116. See also [WB11].
[A45] Physics and mathematics without coordinates. Milan Journal of Mathematics **81**(2013): 259–264.

## Research reports

[R1] On the foundations of the mechanics of continuous media. Technical Report 17, Department of Mathematics, Carnegie Institute of Technology, 1957. 68 pages.
[R2] Isocategories and tensor functors. Research Report 92-142, Department of Mathematical Sciences, Mellon College of Science, 1992. 19 pages. http://repository.cmu.edu/cgi/viewcontent.cgi?article=1481&context=math.
[R3] Linearly induced mappings between cones of quadratic forms. Research Report 93-156, Department of Mathematical Sciences, Mellon College of Science, 1993. 23 pages. http://repository.cmu.edu/cgi/viewcontent.cgi?article=1024&context=math (with R.E. Artz).

## Manuscripts on the Web

The material published on the web is organized following closely the structure given by Noll to his webpage http://www.math.cmu.edu/~wn0g/noll/. Each entry is supplemented by a short abstract or description, mostly written in Noll's own words, possibly slightly edited. Hence the use of the pronoun "I" occurring at some places.

*A. General*

[WA1] Autobiographical Notes. 1988. 17 pages. This is a description of my life from my birth in 1925 and my arrival in Pittsburgh in 1956. The emphasis is on my education in Germany, France, and the United States and on the development of my interest in mathematics. http://www.math.cmu.edu/~wn0g/noll/AB.pdf.
[WA2] The Naturalization and Apotheosis of Walter Noll. 1993. 18 pages. This is the text the lecture given by Clifford Truesdell in April 1993 at the meeting of the Society for Natural Philisophy on the occasion of my retirement from teaching. The emphasis is on the situation in the Graduate Institute of Applied Mathematics at Indiana University both before and after my arrival there in 1953. http://www.math.cmu.edu/~wn0g/noll/TL.pdf (by C. Truesdell).

[WA3] The Conceptual Infrastructure of Mathematics. 1995. 5 pages. By conceptual infrastructure of mathematics I mean the concepts, the terminology, the symbols, and the notations that mathematicians and people who apply mathematics use in their daily professional activities. This infrastructure was and still is being developed in four stages. Here I make an attempt to describe these stages. http://www.math.cmu.edu/~wn0g/noll/CIM.pdf.

[WA4] The Role of the Professor. 1997. 3 pages. This essay is intended not only to help professors better understand their own role, but also to help the public at large better appreciate this role. Although the essay is written from the point of view of a professor of mathematics, its essence should apply to professors in any field. http://www.math.cmu.edu/~wn0g/RP.pdf.

[WA5] The genesis of the "Non-Linear Field Theories of Mechanics". 2002. 9 pages. I was co-author, with Clifford Truesdell, of The Non-Linear Field Theories of Mechanics. It was first published in 1965 and has become the standard reference work in the field. It was reprinted in 1992, translated into Chinese in 2000, and again reprinted in 2004. This is an account of the efforts that went into the creation of this work. This paper was published in [A36]. A part of it is included in the beginning of the 2004 edition of [B1]. http://www.math.cmu.edu/~wn0g/noll/GEN.pdf.

[WA6] Mathematics should not be boring. 2003. 18 pages. This is the text of a lecture before the mathematics teachers of the Catholic Schools in the Diocese of Pittsburgh. First, I make the distinction between arithmetic and true mathematics, starting with geometry and algebra. For the latter, rote memorization is deadly while conceptual understanding and problem solving ability are essential. I illustrate this insight by a section entitled The art of avoiding unnecessary calculation and a section on The Theorem of Pythagoras. http://www.math.cmu.edu/~wn0g/noll/mnb.pdf.

[WA7] On the Past and Future of Natural Philosophy. 2005. 13 pages. The term "Natural Philosophy" was used from the 17th to the middle of the 19th century for what is now called "Natural Science". About 50 years ago, the term was revived by Clifford Truesdell. I describe why he revived this term. Here is his description of the meaning of the term: "In modern natural philosophy, the physical concepts themselves are made mathematical at the outset, and mathematics is used to formulate theories". Truesdell provided the leadership that led to the foundation of the "Society for Natural Philosophy" in 1963. Recently, I wrote a booklet entitled "Five Contributions to Natural Philosophy" [WB2], dedicated to the memory of Clifford Truesdell. I describe its purpose and my hopes for the future of natural philosophy. http://www.math.cmu.edu/~wn0g/noll/PFNP.pdf. See also [A38].

[WA8] What is Mathematics all about. 2006. 9 pages. I list 8 aspects of mathematics and give my take on their relative importance. Two of these are the creation of new concepts by abstraction and the clarification of old ones. The former is illustrated by the creation of the concept of a monoid and the latter by the clarification of the concept of volume. http://www.math.cmu.edu/~wn0g/noll/WMA.pdf.

[WA9] The Future of Scientific Publication. 2009. 6 pages. This is a proposal for a system, based on the internet and websites, that is faster, cheaper, and more efficient than the traditional one. http://www.math.cmu.edu/~wn0g/noll/future%20of%20sc.pub.pdf.

*B. Mathematical Science*

[WB1] Mathematical Structures of Special Relativity. 1993. 239 pages. This textbook is based on a course given several times by Walter Noll from about 1970 to 1993. The approach presented is intended to convey a deeper understanding of the subject and an intuition for relativistic phenomena. Such an understanding can be gained only by the modern mathematics of sets, mappings, relations, and mathematical structures. Here are the chapters of the book:

Introduction
1. Eventworlds http://www.math.cmu.edu/~wn0g/noll/ch1.pdf
2. Timed Eventworlds http://www.math.cmu.edu/~wn0g/noll/ch2.pdf
3. Flat Eventworlds http://www.math.cmu.edu/~wn0g/noll/ch3.pdf
4. Classical Spacetimes http://www.math.cmu.edu/~wn0g/noll/ch4.pdf
5. Minkowskian Spacetimes http://www.math.cmu.edu/~wn0g/noll/ch5.pdf
6. World Momentum http://www.math.cmu.edu/~wn0g/noll/ch6.pdf
7. Electromagnetism http://www.math.cmu.edu/~wn0g/noll/ch7.pdf

Appendices http://www.math.cmu.edu/~wn0g/noll/appsidx.pdf
(with V.J. Matsko).

[WB2] Five Contributions to Natural Philosophy. 2004. 73 pages. This booklet provides a blueprint for updating "The Non-Linear Field Theories of Mechanics" [B1] and hence a guideline for future developments in the field. It should be available next to "The Non-Linear Field Theories of Mechanics" in every scientific library. *Contents*: [0] Introduction; [1] On the Illusion of Physical Space; [2] On Material Frame Indifference; [3] Updating The Non-Linear Field Theories of Mechanics; [4] The Theory of Simple Semi-Liquids, a Conceptual Framework for Rheology; [5] Nematic Semi-Liquids. http://www.math.cmu.edu/~wn0g/noll/FC.pdf.

[WB3] A Frame-Free Formulation of Elasticity. 2005. 16 pages. As I pointed out at the end of Sect. 4 in part [2] of the booklet [WB2], it should be possible to make the principle of material frame-indifference vacuously satisfied by using an intrinsic mathematical frame-work that does not use an external frame-space at all when describing the internal interactions of a physical system. In Sects. 2 and 3 below I do just that for the classical theory of elasticity by specializing the treatment given in. In Sect. 4 I do the same for hyperelasticity, i.e., elasticity based on a strain-energy function. In Sect. 5 I comment on possible restrictions on the intrinsic response functions defined in Sects. 2 and 4. http://www.math.cmu.edu/~wn0g/noll/FFFE.pdf. See also [A37].

[WB4] On the Theory of Surface Interactions. 2005. 14 pages. I use the term *interaction* as an abstraction that applies, for example, to force-interactions, torque-interactions, and heat-transfers. If an interaction is a *surface-interaction* it can be characterized, under suitable conditions, by a *contactor*. In the case of force-interactions, torque-interactions, and heat-transfers, the contactor is the field of *stress-tensors*, *couplestress-tensors*, and *heatflux-vectors*, respectively. The first to prove, making suitable assumptions, the existence of such a contactor was Cauchy in 1823, and there has been a large literature on this subject since then. Here I describe what I consider to be the present situation. http://www.math.cmu.edu/~wn0g/noll/Surface%20Interactions.pdf.

[WB5] On the Concept of Force. 2007. 4 pages. This essay is a reaction to an article by Frank Wilczek [100] in the October 2004 issue of Physics Today followed by the comments [66, 102] in the August 2005 issue. http://www.math.cmu.edu/~wn0g/noll/Force.pdf.

[WB6] Plugs in Viscometric Flows of Simple Semi-Liquids. 2007. 14 pages. Here we apply the theory of simple semi-liquids described in part [4] of the booklet [WB2] above in order to analyze all the viscometric flows treated in the book

[B2]. Only simple liquids are treated in this book. When dealing with simple semi-liquids, plugs can occur. This paper gives a detailed analysis of such plugs. http://www.math.cmu.edu/~wn0g/noll/Plugs_.pdf (with B. Seguin). See also [A40].

[WB7] Thoughts on Stress. 2008. 8 pages. This is the content of a lecture given at the SIAM meeting in Philadelphia in May 2008. http://www.math.cmu.edu/~wn0g/noll/TOS.pdf. See also [A43].

[WB8] Thoughts on Thermomechanics. 2008. 8 pages. First, I give a short description of the history of the interpretation of the second law of thermo-dynamics as a restriction on the kind of material properties that physical systems occurring in nature can have. Second, I describe the desirability of developing a framework, using the concept of a state of a material element, for developing a general thermomechanical theory of simple materials. Third, I give a taste of the mathematical infrastructure needed for such a theory. Finally, I give an outline of the paper [A41] by Brian Seguin and myself which may be a model for the first few chapters of future textbooks on continuum thermomechanics. http://www.math.cmu.edu/~wn0g/noll/Thoughts%20on%20Thermomechanics.pdf.

[WB9] Basic Concepts of Thermomechanics. 2009. 21 pages. This paper is intended to serve as a model for the first few chapters of future textbooks on continuum mechanics and continuum thermomechanics. http://www.math.cmu.edu/~wn0g/noll/BCTM+.pdf (with B. Seguin).

[WB10] Frame-Free Thermomechanics. 2010. 100 pages. This is Brian Seguin's doctoral thesis, submitted in May 2010. http://www.math.cmu.edu/~wn0g/noll/Seguin%20Thesis.pdf (by B. Seguin).

[WB11] A Conceptual Framework for Problems in Elastostatics. 2011. 14 pages. In finite elasticity, the traditional local boundary-value problems are rarely appropriate. More often one has to consider non-local environmental loads. This paper gives a general description of schemes for formulating such non-local problems. Interesting examples are cavity problems, special cases of which are what I call basketball problems and tire problems. The actual solution to these problems I leave to future work. http://www.math.cmu.edu/~wn0g/noll/Elastostatics.c.pdf. See also [A44].

*C. Pure Mathematics*

[WC1] Finite-Dimensional Spaces, Algebra, Geometry, and Analysis, vol. I. 1987. 393 pages. This treatise is not a run-of-the-mill mathematics book. It offers a fresh point of view on some of the basic infrastructure of mathematics. It unifies all aspects of finite-dimensional spaces by stressing the interplay between algebra, geometry, topology, and analysis. Most of the proofs are original and many concepts and results cannot be found elsewhere. The approach is uncompromisingly coordinate-free and $\mathbb{R}^n$-free when dealing with concepts. However, there is a chapter on how to handle general coordinate systems efficiently when dealing with special situations. This treatise cannot be understood by "sampling" miscellaneous sections or chapters. The reader must first become familiar with the basic terminology and notation presented in the Introduction and the preliminary Chap. 0. This treatise should be a suitable textbook for core courses at the advanced undergraduate and beginning graduate level. Also, it should be of use to theoretically inclined scientists and engineers who wish to use contemporary mathematics as a conceptual tool. The first volume of this treatise was first published in 1987, see

[B5]. The present version, posted on my website in 2006, is a corrected reprinting. The second volume was never completed, but preliminary versions of various chapters are posted and frequently updated on this website.

Introduction

0. Basic Mathematics http://www.math.cmu.edu/~wn0g/noll/FDSI.Ch.0..pdf

1. Linear Spaces http://www.math.cmu.edu/~wn0g/noll/2ch1a.pdf

2. Duality, Bilinearity http://www.math.cmu.edu/~wn0g/noll/2ch2a.pdf

3. Flat Spaces http://www.math.cmu.edu/~wn0g/noll/2ch3a.pdf

4. Inner-Product Spaces, Euclidean Spaces http://www.math.cmu.edu/~wn0g/noll/2ch4a.pdf

5. Topology http://www.math.cmu.edu/~wn0g/noll/FDSI.Ch%205.pdf

6. Differential Calculus http://www.math.cmu.edu/~wn0g/noll/2ch6a.pdf

7. Coordinate Systems http://www.math.cmu.edu/~wn0g/noll/2ch7a.pdf

8. Spectral Theory http://www.math.cmu.edu/~wn0g/noll/2ch8a.pdf

9. The Strucure of General Lineons http://www.math.cmu.edu/~wn0g/noll/2ch9a.pdf

Index http://www.math.cmu.edu/~wn0g/noll/FDS1%20ind.pdf.

[WC2] Isocategories and Tensor Functors. 1992. 19 pages. In this paper, I show how the concepts of an *isocategory* (category all of whose morphisms are isomorphisms) and the corresponding concept of an *isofunctor* can be used to improve the conceptual infrastructure of many branches of mathematics. The crucial new idea is that of a natural assignment, a modification of the idea of a natural tansformation introduced by Eilenberg and Mac Lane. Isofunctors that involve the isocategory LIS of all linear isomorphism of finite-dimensional linear spaces are called *tensor functors*, because they can be used to clarify most uses of the term "tensor" in the literature of mathematics and physics. Of particular importance are the "*analytic tensor functors*", which can serve to be the basis for a completely coordinate-free presentation of the theory of differentiable manifolds in [WC6]. http://www.math.cmu.edu/~wn0g/noll/Iso.%20and%20Tensor.pdf, also published in [R2].

[WC3] Monoids. 1992. 5 pages. These are sketchy lecture notes intended to show that the concepts of divisibility, prime elements and prime-decompositions are most efficiently treated in the framework of monoids. The results then apply both to the multiplicative monoid of the natural numbers and that of the monic polynomials over a field. http://www.math.cmu.edu/~wn0g/noll/monoids.pdf.

[WC4] Linearly Induced Mappings between Cones of Quadratic Forms. 1993. 24 pages. This paper deals with mappings between cones of positive quadratic forms which are induced by linear mappings between the underling vector spaces, *i.e.*, the spaces which are the domains of the forms. Three fundamental results are proved, two of which were previously announced by the second author [Walter Noll]. The first result states that there is an inclusion-reversing one-to-one correspondence between the lattice of subspaces of a given space and the lattice of faces of the cone of quadratic forms on that space. The second result states that all cone-isomorphisms between cones of quadratic forms are induced by linear isomorphisms between the underlying spaces. The third result states that a given cone-linear mapping $F$ between cones of quadratic forms is induced by some linear mapping between the underlying spaces if and only if both $F$ and its transpose preserve faces. http://www.math.cmu.edu/~wn0g/noll/LIM_2007+.pdf (with R.E. Artz).

[WC5] Finite-Dimensional Spaces, Algebra, Geometry, and Analysis, vol. II. 1994. The plan was to publish this volume in its entirety. Here are preliminary manuscripts of the Chaps. 0, 1, 2, 3, and 4.

0. Preliminaries (11 pages) http://www.math.cmu.edu/~wn0g/noll/FDSII,Ch.0.pdf

1. Determinants, Invariants, Covariants (18 pages) http://www.math.cmu.edu/~wn0g/noll/FDSII,Ch.1.pdf

2. Differential Calculus II (17 pages) http://www.math.cmu.edu/~wn0g/noll/FDSII,Ch.2.pdf

3. Manifolds in a Euclidean Space (33 pages) http://www.math.cmu.edu/~wn0g/noll/FDSII,Ch.3.pdf

4. Volume Integrals (47 pages) http://www.math.cmu.edu/~wn0g/noll/FDSII,%20Ch.4.pdf.

[WC6] Geometry of Differentiable Manifolds. 1994. 95 pages, preliminary version. These lecture notes are unusual because the topic is treated in a totally coordinate-free manner. Included are the theory of connectors and connections, Riemannian and symplectic manifolds, and Lie groups. Also, a coordinate-free version of Einstein's equation of general relativity is presented.

Contents http://www.math.cmu.edu/~wn0g/noll/contents.pdf

1. Preliminaries http://www.math.cmu.edu/~wn0g/noll/book1.pdf

2. Manifolds and Bundles http://www.math.cmu.edu/~wn0g/noll/book2.pdf

3. Connections http://www.math.cmu.edu/~wn0g/noll/book3.pdf

4. Gradients http://www.math.cmu.edu/~wn0g/noll/book4.pdf

5. Geometric Structures http://www.math.cmu.edu/~wn0g/noll/book5.pdf

Bibliography http://www.math.cmu.edu/~wn0g/noll/bib.pdf

(with S.-M. Chiou).

[WC7] The Chain Rule for Higher Derivatives. 1995. 8 pages. A Chain Rule of Order $n$ should state, roughly, that the composite of two functions that are $n$ times differentiable is again $n$ times differentiable, and it should give a formula for the $n$th derivative of this composite. Many textbooks contain a Chain Rule of order 2 and perhaps 3, but I do not know of a single one that contains a Chain Rule of arbitrary order $n$ with an explicit and useful formula. The main purpose of this paper is to derive such a formula. http://www.math.cmu.edu/~wn0g/noll/CHR.pdf.

[WC8] Monoids, Boolean Algebras, Materially Ordered Sets. 2007. 11 pages. In this paper, the interplay between certain mathematical structures is elucidated. First, it is shown that there is a one-to-one correspondence between bounded half-lattices and commutative idempotent monoids (c.i.-monoids). Adding certain additional structural ingredients and axioms, such c.i.-momoids become Boolean algebras. There is a non-trivial one-to-one correspondence between these and what we call *materially ordered sets*, which are half-lattices that satisfy certain additional axioms. Such materially ordered sets can serve as mathematical models for certain physical systems. The correspondence between materially ordered sets and Boolean algebras can be used to show, for example, that the law of action and reaction (Newton's Third Law) is not an independent axiom but a consequence of fundamental balance laws. http://www.math.cmu.edu/~wn0g/noll/MBM.pdf. See also [A39] (with B. Seguin).

[WC9] On the Conjugacy of Orthogonal Groups. 2008. 10 pages. This paper presesnts some insights that should be of interest to anybody who ever heard of orthogonal groups. http://www.math.cmu.edu/~wn0g/noll/orthogonal%20groups.pdf.

[WC10] Scalar Algebras and Quaternions: An approach Based on the Algebra and Topology of Finite-Dimensional Real Linear Spaces, Part 1. 2009. 36 pages. This expository paper addresses the category of scalar algebras, *i.e.*, real finite-dimensional division algebras. An original proof shows that these algebras possess natural inner products. The inner products are then used as an essential ingredient in a proof of the Frobenius classification according to which every scalar algebra is isomorphic to one of the algebras $\mathbb{R}$, $\mathbb{C}$, and $\mathbb{H}$. http://www.math.cmu.edu/~wn0g/noll/qu1.pdf. (by R.E. Artz).



## References

1. Admal, N.C., Tadmor, E.B.: A unified interpretation of stress in molecule systems. J. Elast. **100**(1), 63–143 (2010)
2. Assis, A.K.T.: On Mach's principle. Found. Phys. Lett. **2**, 301–318 (1989)
3. Assis, A.K.T., Graneau, P.: Nonlocal forces of inertia in cosmology. Found. Phys. **26**, 271–283 (1996)
4. Banfi, C., Fabrizio, M.: Sul concetto di sottocorpo nella meccanica dei continui. Atti Accad. Naz. Lincei Cl. Sci. Fis. Mat. Natur. Rend. **66**(2), 136–142 (1979). Available at http://www.bdim.eu/item?id=RLINA_1979_8_66_2_136_0
5. Banfi, C., Fabrizio, M.: Global theory for thermodynamic behaviour of a continuous medium. Ann. Univ. Ferrara **27**, 181–199 (1981)
6. Bilby, B.A.: Continuous Distributions of Dislocations pp. 329–398. North-Holland, Amsterdam (1960)
7. Bragg, L.E.: On relativistic worldlines and motions, and on non-sentient response. Arch. Ration. Mech. Anal. **18**, 127–166 (1965)
8. Bunge, M. (ed.): Delaware Seminar in the Foundations of Physics. Studies in the Foundations Methodology and Philosophy of Science, vol. 1. Springer, Berlin (1967)
9. Bunge, M.: Foundations of Physics. Springer Tracts in Natural Philosophy, vol. 10. Springer, New York (1967)
10. Carathéodory, C.: Zur Axiomatik der Speziellen Relativitätstheorie. Sitz. Preuss. Akad. Wiss., Phys. Math. Kl. **5**, 12–27 (1924)
11. Di Carlo, A.: A major serendipitous contribution to continuum mechanics. Mech. Res. Commun. **93**, 41–46 (2018)
12. Di Carlo, D.: Continuum mechanics as a computable coarse-grained picture of molecular dynamics. J. Elast. (2019, in press)
13. Coleman, B.D.: Kinematical concepts with applications in the mechanics and thermodynamics of incompressible viscoelastic fluids. Arch. Ration. Mech. Anal. **9**, 273–300 (1962)
14. Coleman, B.D.: Simple liquid crystals. Arch. Ration. Mech. Anal. **20**, 41–58 (1965)
15. Coleman, B.D., Feinberg, M., Serrin, J. (eds.): Analysis and Thermomechanics: A Collection of Papers Dedicated to W. Noll on His Sixtieth Birthday. Springer, Berlin (1987)
16. Del Piero, G.: An axiomatic framework for the mechanics of generalized continua. Rend. Lincei Mat. Appl. **29**, 31–61 (2018)
17. Dunn, J.E., Fosdick, R.L.: The morphology and stability of material phases. Arch. Ration. Mech. Anal. **74**, 1–99 (1980)
18. Einstein, A.: Prinzipielles zur allgemeinen Relativitätstheorie. Ann. Phys. **55**, 241–244 (1918)
19. Epstein, M., Maugin, G.A.: The energy-momentum tensor and material uniformity in finite elasticity. Acta Mech. **83**, 127–133 (1990)
20. Epstein, M., Maugin, G.A.: Sur le tenseur de moment matériel d'Eshelby en élasticité non linéaire. C. R. Acad. Sci. Paris **310**, 675–678 (1990)
21. Ericksen, J.L.: Anisotropic fluids. Arch. Ration. Mech. Anal. **4**, 231–237 (1959)
22. Ericksen, J.L.: Theory of anisotropic fluids. Trans. Soc. Rheol. **4**, 29–39 (1960)
23. Ericksen, J.L.: Transversely isotropic fluids. Kolloid Z., 117–122 (1960)
24. Ericksen, J.L.: Conservation laws for liquid crystals. Trans. Soc. Rheol. **5**, 23–34 (1961)
25. Ericksen, J.L.: Hydrostatic theory of liquid crystals. Arch. Ration. Mech. Anal. **9**, 371–378 (1962)
26. Eshelby, J.D.: The force on an elastic singularity. Philos. Trans. R. Soc. Lond. A **244**, 87–112 (1951)
27. Feynman, R.P.: Forces in molecules. Phys. Rev. **56**, 340–343 (1939)

28. Frank, F.C.: On the theory of liquid crystals. Discuss. Faraday Soc. **25**, 19–28 (1958)
29. Gallavotti, G.: Statistical Mechanics: a Short Treatise. Texts and Monographs in Physics. Springer, Berlin (1999)
30. de Gennes, P.G., Prost, J.: The Physics of Liquid Crystals, 2nd edn. Clarendon, Oxford (1993)
31. Gibbs, J.W.: On the equilibrium of heterogeneous substances. In: The Collected Works, vol. I. Trans. Conn. Acad., vol. 3, pp. 108–248. Longmans, New York (1875–1878). 343–524, 55–353, 1928, Also available at https://gallica.bnf.fr/ark:/12148/bpt6k95192s/f82.image
32. De Giorgi, E., Colombini, F., Piccinini, L.C.: Frontiere Orientate di Misura Minima e Questioni Collegate. Quaderni Scuola Normale Superiore, vol. 1. Editrice Tecnico Scientifica, Pisa (1972)
33. Giusti, E.: Minimal Surfaces and Functions of Bounded Variation. Monographs in Mathematics, vol. 80. Springer, New York (1984)
34. Gurtin, M.: Configurational Forces as Basic Concepts of Continuum Physics. Applied Mathematical Sciences, vol. 137. Springer, Berlin (2000)
35. Gurtin, M.E., Fried, E., Anand, L.: The Mechanics and Thermodynamics of Contiuna. Cambridge University Press, Cambridge (2010)
36. Gurtin, M.E., Martins, L.C.: Cauchy's theorem in classical physics. Arch. Ration. Mech. Anal. **60**, 305–324 (1976)
37. Gurtin, M.E., Mizel, V.J., Williams, W.O.: A note on Cauchy's stress theorem. J. Math. Anal. Appl. **22**, 398–401 (1968)
38. Gurtin, M.E., Williams, W.O.: An axiomatic foundation for continuum thermodynamics. Arch. Ration. Mech. Anal. **26**, 83–117 (1967)
39. Gurtin, M.E., Williams, W.O., Ziemer, W.P.: Geometric measure theory and the axioms of continuum thermodynamics. Arch. Ration. Mech. Anal. **92**, 1–22 (1986), reprinted in [15]
40. Hamel, G.: Über die Grundlagen der Mechanik. Math. Ann. **66**, 350–397 (1908)
41. Hand, G.L.: A theory of dilute suspensions. Arch. Ration. Mech. Anal. **7**, 81–86 (1961)
42. Hand, G.L.: A theory of anisotropic fluids. J. Fluid Mech. **13**, 33–46 (1962)
43. Hilbert, D.: Mathematische probleme. Arch. Math. Phys. **1**, 213–217 (1901), reprinted in [44]
44. Hilbert, D.: Gesammelte Abhandlungen, vol. 3. Springer, Berlin (1970)
45. Ignatieff, Y.A.: The Mathematical World of Walter Noll. A Scientific Biography. Springer, Berlin (1996)
46. Irving, J.H., Kirkwood, J.G.: The statistical mechanical theory of transport processes. IV. The equations of hydrodynamics. J. Chem. Phys. **18**(6), 817–829 (1950)
47. Jeffery, G.B.: The motion of ellipsoidal particles immersed in a viscous fluid. Proc. R. Soc. Lond. A **102**, 161–179 (1922)
48. Kellogg, O.D.: Foundations of Potential Theory. Springer, Berlin (1929)
49. Kröner, E.: Kontinuumstheorie der Versetzungen und Eigenspannungen. Springer, Berlin (1958)
50. Lehoucq, R.B., Von Lilienfeld-Toal, A.: Translation of Walter Noll's "Derivation of the fundamental equations of continuum thermodynamics from statistical mechanics". J. Elast. **100**, 5–24 (2010)
51. Leslie, F.M.: Some constitutive equations for anisotropic fluids. Q. J. Mech. Appl. Math. **19**, 357–370 (1966)
52. Leslie, F.M.: Some constitutive equations for liquid crystals. Arch. Ration. Mech. Anal. **28**, 265–283 (1968)
53. Love, A.E.H.: A Treatise on the Mathematical Theory of Elasticity, 4th edn. Cambridge University Press, Cambridge (1927)
54. Marsh Noll, M.: Ordinary Tasks. Madbooks, Pittsburgh (2016)
55. Maugin, G.A.: Continuum Mechanics through the Twentieth Century. A Concise Historical Perspective. Solid Mechanics and Its Applications, vol. 196. Springer, Dordrecht (2013)
56. Maxwell, J.C.: On the dynamical theory of gases, IV. Philos. Trans. R. Soc. Lond. **157**, 49–88 (1867). Reprinted in [57], pp. 26–78
57. Niven, W.D. (ed.): The Scientific Papers of James Clerk Maxwell, vol. 2. Dover, New York (1965)
58. Oldroyd, J.G.: On the formulation of rheological equations of state. Proc. R. Soc. Lond. A **200**(1063), 523–541 (1950)
59. Oseen, C.W.: The theory of liquid crystals. Trans. Faraday Soc. **29**(4), 883–899 (1933)
60. Pinker, S.: How the Mind Works. Norton, New York (1997)
61. Pipkin, A.C., Rivlin, R.S.: The formulation of constitutive equations in continuum physics. Tech. Rep. DA 4531/4, Department of the Army, Ordnance Corps (1958)
62. Podio-Guidugli, P.: Inertia and invariance. Ann. Mat. Pura Appl. **172**, 103–124 (1997)
63. Podio-Guidugli, P.: Continuum Thermodynamics. SISSA Lecture Notes, vol. 1. Springer, Berlin (2019)
64. Podio-Guidugli, P.: On the mechanical modeling of matter, molecular and continuum. J. Elast. (2019). https://doi.org/10.1007/%2Fs10659-018-9709-y
65. Prager, S.: Stress-strain relations in a suspension of dumbbells. Trans. Soc. Rheol. **1**, 53–62 (1957)

66. Ravi, R.: Comments on the culture of the force. Phys. Today **58**(8), 15–16 (2005)
67. Rivlin, R.S.: Review of "The foundations of mechanics and thermodynamics: selected papers. W. Noll". Am. Sci. **64**, 100–101 (1976), the book reviewed is [B4]
68. Robb, A.A.: Geometry of Time and Space. A Theory of Time and Space. Cambridge University Press, Cambridge (1936). This is the second edition of a book published in 1914 with the title
69. Schutz, J.W.: Foundations of Special Relativity: Kinematic Axioms for Minkowski Space-Time. Lecture Notes in Mathematics, vol. 361. Springer, Berlin (1973)
70. Seeger, A.: Recent advances in the theory of defects in crystals. Phys. Status Solidi **1**, 669–698 (1961)
71. Sikorsky, R.: Boolean Albegras, 3rd edn. Springer, Berlin (1969)
72. Šilhavý, M.: The existence of the flux vector and the divergence theorem for general Cauchy fluxes. Arch. Ration. Mech. Anal. **90**, 195–212 (1985). Reprinted in [15]
73. Šilhavý, M.: The scientific work of B.D. Coleman. J. Math. Mech. Solids (2019, in press)
74. Smith, H.E. (ed.): Autobiography of Mark Twain, vol. 1. University of California Press, Berkeley (2010)
75. Suppes, P.: Axioms for relativistic kinematics with or without parity. In: Henkin, L., Suppes, P., Tarski, A. (eds.) The Axiomatic Method, Studies in Logic and the Foundations of Mathematics, vol. 27, pp. 291–307. Elsevier, Amsterdam (1959)
76. Suppes, P.: The axiomatic method in the empirical sciences. In: Henkin, L. (ed.) Proceedings of the Turski Symposium. Proceedings of Symposia in Pure Mathematics, vol. 25, pp. 465–479. American Mathematical Society, Providence (1960)
77. Szekeres, G.: Kinematic geometry; an axiomatic system for Minkowski space-time: M.L. Urquhart in memoriam. J. Aust. Math. Soc. **8**, 134–160 (1968)
78. Tadmor, E.B., Miller, R.E.: Modeling Materials: Continuum, Atomistic and Multiscale Techniques. Cambridge University Press, Cambridge (2011)
79. Tanner, R.I., Walters, K.: Rheology: An Historical Perspective. Rheology Series, vol. 7. Elsevier, Amsterdam (1998)
80. Taylor, G.I.: Motion of solids in fluids when the flow is not irrotational. Proc. R. Soc. Lond. A **93**, 99–113 (1917)
81. Toupin, R.A.: Elastic materials with couple-stresses. Arch. Ration. Mech. Anal. **11**, 385–414 (1962)
82. Truesdell, C.: The mechanical foundations of elasticity and fluid dynamics. J. Ration. Mech. Anal. **1**, 125–300 (1952). Extensive additions and corrections appeared in [83] and further minor misprints were corrected in [84]. A corrected reprint of this essay incorporating all previous amendments and some other slips was published as a book in 1966 [87]
83. Truesdell, C.: Corrections and additions to "The mechanical foundations of elasticity and fluid dynamics". J. Ration. Mech. Anal. **2**, 593–616 (1953)
84. Truesdell, C.: Corrigenda. J. Ration. Mech. Anal. **3**, 801–802 (1954)
85. Truesdell, C.: Hypo-elasticity J. Ration. Mech. Anal. **4**, 83–133 (1955). 1019–1020
86. Truesdell, C.: Das ungelöste Hauptproblem der endlichen Elastizitätstheorie. Z. Angew. Math. Mech. **36**, 97–103 (1956)
87. Truesdell, C.: The Mechanical Foundations of Elasticity and Fluid Dynamics. International Science Review Series, vol. VIII, Part 1. Gordon & Breach, New York (1966)
88. Truesdell, C.: Six Lectures on Modern Natutal Philosophy. Springer, Berlin (1966)
89. Truesdell, C.: Essays in the History of Mechanics. Springer, Berlin (1968)
90. Truesdell, C.: An Idiot's Fugitive Essays on Science. Springer, New York (1984)
91. Truesdell, C., Noll, W.: The Non-Linear Field Theories of Mechanics, 2nd edn. Springer, Berlin (1992), see [92] for an annotated edition
92. Truesdell, C., Noll, W.: The Non-Linear Field Theories of Mechanics, 3rd edn. Springer, Berlin (2004), edited by S.S. Antman
93. Truesdell, C., Toupin, R.: The Classical Field Theories. In: Flügge, S. (ed.) Encyclopedia of Physics, vol. III/1. Springer, Berlin (1960)
94. Truesdell, C.A. (ed.): Continuum Mechanics II: The Rational Mechanics of Materials. International Science Review Series, vol. VIII, Part 2. Gordon & Breach, New York (1965)
95. Truesdell, C.A. (ed.): Continuum Mechanics III, Foundations of Elasticity Theory. International Science Review Series, vol. VIII, Part 3. Gordon & Breach, New York (1965)
96. Truesdell, C.A.: A First Course in Rational Continuum Mechanics, vol. 1, 2nd edn. Academic Press, San Diego (1991)
97. Vol'pert, A.I., Hudjaev, S.I.: Analysis in Classes of Discontinuous Functions and Equations of Mathematical Physics. Mechanics: Analysis, vol. 8. Nijhoff, Dordrecht (1985)
98. Wang, C.C.: A general theory of subfluids. Arch. Ration. Mech. Anal. **20**, 1–40 (1965)
99. Wang, C.C.: On the geometric structures of simple bodies, a mathematical foundation for the theory of continuous distributions of dislocations. Arch. Ration. Mech. Anal. **27**, 33–94 (1967)

100. Wilczek, F.: Whence the force of $F = ma$? I: Culture shock. Phys. Today **57**(10), 11–12 (2004)
101. Wilczek, F.: Whence the force of $F = ma$? II: Rationalizations. Phys. Today **57**(12), 10–11 (2004)
102. Wilczek, F.: Comments on the culture of the force. Phys. Today **58**(8), 17 (2005)
103. Wilczek, F.: Whence the force of $F = ma$? III: Cultural diversity. Phys. Today **58**(7), 10–11 (2005)
104. Willems, J.C.: Dissipative dynamical systems, part I: general theory. Arch. Ration. Mech. Anal. **45**, 321–351 (1972)
105. Zaremba, S.: Le principe des mouvements relatifs et les équations de la mécanique physique. Bull. Int. Acad. Sci. Cracovie, 614–621 (1903). Available at https://www.biodiversitylibrary.org/item/47339#page/665/mode/1up
106. Zaremba, S.: Sur une forme perfectionée de la théorie de la relaxation. Bull. Int. Acad. Sci. Cracovie, 594–614 (1903). Available at https://www.biodiversitylibrary.org/item/47339#page/665/mode/1up
107. Zaremba, S.: Sur une conception nouvelle des forces intérieures dans un fluide en mouvement. Mém. Sci. Math. Gauthier-Villars, No. 82 (1937)
108. Ziemer, W.P.: Cauchy flux and sets of finite perimeter. Arch. Ration. Mech. Anal. **84**, 189–201 (1983)

J Elast (2019) 135:73–89
https://doi.org/10.1007/s10659-019-09735-x

# Rotationally Symmetric Motions and Their Blowup for Incompressible Nonlinearly Elastic and Viscoelastic Annuli

**Stuart S. Antman**[1]

*In memoriam Walter Noll*

Received: 25 September 2018 / Published online: 27 March 2019


**Abstract** This paper treats rotationally symmetric motions of incompressible, transversely isotropic, nonlinearly elastic and viscoelastic annuli subjected to the live loads of (time-dependent) centrifugal force and hydrostatic pressure. The incompressibility is responsible for some simplification of the analysis of such motions, which is nevertheless more complicated than that for radially symmetric motions. The theory is illustrated with blowup theorems, which have proofs different from those used for compressible media.

**Keywords** Rotationally symmetric motion of nonlinearly elastic and viscoelastic annuli · Blowup theorems

**Mathematics Subject Classification** 34A34 · 35B44 · 35Q74 · 74A05 · 74B20 · 74D10 · 74H35

## 1 Introduction

In 1960 Knowles [18] proved that *radial* motions of *incompressible* transversely isotropic nonlinearly elastic cylindrical shells are governed by ordinary differential equations rather than by the quasilinear partial differential equations governing such motions of such compressible shells (see [23]). This work was soon extended to radial motions of spherical shells [15, 19]. (The motions of these bodies and of those discussed below are described by the full equations of motion of 3-dimensional continuum mechanics and their specializations to 2-dimensional problems, and not by the equations of any shell theory.) Thus specific nonlinear initial-boundary-value problems for such motions could be given detailed qualitative analyses, which furnish the first substantive treatments of dynamical problems for nonlinearly elastic materials (see [24, 25]). Calderer used phase-plane methods to study blowup phenomena for incompressible elastic shells in [8], and analyzed such problems for inelastic materials with nonlinear constitutive equations in [9, 10]. Fosdick, Ketema, & J.-H. Yu

✉ S.S. Antman
ssa@math.umd.edu

1 Department of Mathematics, Institute for Physical Science and Technology, Institute for Systems Research, University of Maryland, College Park, MD 20742, USA

[13] analyzed free and forced radial motions of elastic and viscoelastic spherical shells with nonconvex stored energies. The papers [20, 21] treated combined twisting and breathing motions for special simple constitutive equations for incompressible elastic cylindrical shells.

This paper shows that the analysis of *rotationally* symmetric motions of incompressible transversely isotropic solids can be simplified (but not rendered trivial) by virtue of the incompressibility. The theory is illustrated with blowup theorems, which have proofs different from those used for compressible media [4, 6–10, 16, 17, 23].

Our annulus is a model for the (plane-strain) problem of an infinitely long, thick circular cylindrical tube and for the (plane-stress) problem of an annular plate.

**Notation.** Lower-case Greek subscripts assume values 1 and 2. Repeated Greek indices are summed over 1,2. Vectors in 2- and 3-dimensional Euclidean space are denoted by lower-case bold-face italic letters, $\boldsymbol{u}, \boldsymbol{v}, \dots$, and second-order tensors (linear transformations) from 2-dimensional Euclidean space $\mathbb{E}^2$ to itself are denoted by upper-case bold-face italic letters, $\boldsymbol{A}, \boldsymbol{B}, \dots$. The dot product of vectors $\boldsymbol{u}$ and $\boldsymbol{v}$ is denoted as usual by $\boldsymbol{u} \cdot \boldsymbol{v}$.

Since we must carefully balance three different coordinate systems, we employ Gibbs dyadic notation throughout: The value of a tensor $\boldsymbol{A}$ at a vector $\boldsymbol{u}$ is denoted $\boldsymbol{A} \cdot \boldsymbol{u}$ (rather than $\boldsymbol{A}\boldsymbol{u}$). The dyad $\boldsymbol{b}\boldsymbol{a}$ is the tensor taking a vector $\boldsymbol{c}$ to the vector $(\boldsymbol{b}\boldsymbol{a}) \cdot \boldsymbol{c} \equiv (\boldsymbol{a} \cdot \boldsymbol{c})\boldsymbol{b}$. If $\{\boldsymbol{a}_1, \boldsymbol{a}_2\}$ and $\{\boldsymbol{b}_1, \boldsymbol{b}_2\}$ are orthonormal bases for $\mathbb{E}^2$, then the sum $\boldsymbol{b}_\alpha \boldsymbol{a}_\alpha$ of dyads is the tensor taking the $\boldsymbol{a}_\beta$ to the $\boldsymbol{b}_\beta$, and bases for tensors on $\mathbb{E}^2$ are $\{\boldsymbol{a}_\alpha \boldsymbol{a}_\beta\}$, $\{\boldsymbol{a}_\alpha \boldsymbol{b}_\beta\}$, $\{\boldsymbol{b}_\alpha \boldsymbol{a}_\beta\}$, $\{\boldsymbol{b}_\alpha \boldsymbol{b}_\beta\}$. If the $\boldsymbol{a}_\alpha, \boldsymbol{b}_\alpha, \boldsymbol{c}_\alpha, \boldsymbol{d}_\alpha$ are each bases, and if $\boldsymbol{A} = A_{\alpha\beta} \boldsymbol{a}_\alpha \boldsymbol{b}_\beta$, $\boldsymbol{B} = B_{\gamma\delta} \boldsymbol{c}_\gamma \boldsymbol{d}_\delta$, then the inner product $:$ on the space of tensors is defined by $\boldsymbol{A} : \boldsymbol{B} := A_{\alpha\beta} B_{\gamma\delta} (\boldsymbol{a}_\alpha \cdot \boldsymbol{c}_\gamma)(\boldsymbol{b}_\beta \cdot \boldsymbol{d}_\delta)$, which reduces to $A_{\alpha\beta} B_{\alpha\beta}$ when $\boldsymbol{a}_\alpha = \boldsymbol{c}_\alpha$, $\boldsymbol{b}_\beta = \boldsymbol{d}_\beta$. The transpose of $\boldsymbol{A} = A_{\alpha\beta} \boldsymbol{a}_\alpha \boldsymbol{b}_\beta$ is $\boldsymbol{A}^* = A_{\alpha\beta} \boldsymbol{b}_\beta \boldsymbol{a}_\alpha$. The inverse of $\boldsymbol{A}$ is denoted $\boldsymbol{A}^{-1}$ and the transpose of the inverse (which is the inverse of the transpose) is denoted $\boldsymbol{A}^{-*}$.

Partial and ordinary derivative are often denoted by subscripts or by $\partial$ bearing an appropriate subscript. E.g., $\frac{\partial \psi}{\partial t}(s,t) \equiv \psi_t(s,t) \equiv \partial_t \psi_t(s,t)$, $\frac{du}{dt}(t) \equiv u_t(t)$.

In our analysis of the governing equations we simplify some formulas by letting $c$ and $C$ denote computable positive constants whose meaning can change in each appearance, even in the same equation. We use without comment the Cauchy-Bunyakovskiĭ-Schwarz inequality and the elementary inequality that $2xy \le \delta^2 x^2 + \delta^{-2} y^2$ for real $x, y, \delta$ with $\delta \neq 0$.

## 2 Kinematics

Let $\{\boldsymbol{i}, \boldsymbol{j}, \boldsymbol{k}\}$ be the standard orthonormal basis for Euclidean 3-space. Introduce the basis $\{\boldsymbol{\jmath}_1(\phi), \boldsymbol{\jmath}_2(\phi)\}$ for the $\{\boldsymbol{i}, \boldsymbol{j}\}$-plane by

$$\boldsymbol{\jmath}_1(\phi) := \cos\phi\, \boldsymbol{i} + \sin\phi\, \boldsymbol{j}, \qquad \boldsymbol{\jmath}_2(\phi) := -\sin\phi\, \boldsymbol{i} + \cos\phi\, \boldsymbol{j}, \tag{2.1}$$

so that

$$\boldsymbol{\jmath}_{1\phi} = \boldsymbol{\jmath}_2, \qquad \boldsymbol{\jmath}_{2\phi} = -\boldsymbol{\jmath}_1. \tag{2.2}$$

The natural reference configuration of an annulus in the $\{\boldsymbol{i}, \boldsymbol{j}\}$-plane consists of all the material points with scaled positions

$$s\boldsymbol{\jmath}_1(\phi), \quad a \le s \le 1, \quad 0 \le \phi < 2\pi \tag{2.3}$$

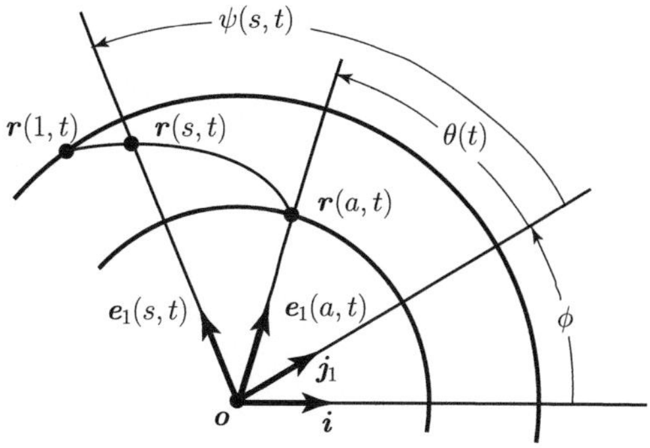


**Fig. 1** The geometry of the rotationally symmetric deformation. Here $\boldsymbol{r}(s,t)=v(s,t)\mathbf{e}_1(s,t)$, $\boldsymbol{r}(a,t)=u(t)\mathbf{e}_1(a,t)$, $\theta(t)=\psi(a,t)$. Unit vectors with the subscript 2, which are not shown, are rotated $\frac{\pi}{2}$ radians counter-clockwise from the corresponding unit vectors with subscript 1

with $a>0$. These material points are identified by the polar coordinates $(s,\phi)$. To describe rotationally symmetric motions introduce a rotating basis

$$\begin{aligned}\mathbf{e}_1(s,t;\phi)&:=\cos\psi(s,t)\boldsymbol{j}_1(\phi)+\sin\psi(s,t)\boldsymbol{j}_2(\phi),\\ \mathbf{e}_2(s,t;\phi)&:=-\sin\psi(s,t)\boldsymbol{j}_1(\phi)+\cos\psi(s,t)\boldsymbol{j}_2(\phi),\\ \text{with}\quad \psi(a,t)&=:\theta(t).\end{aligned}\tag{2.4}$$

Thus

$$\begin{aligned}\mathbf{e}_{1s}(s,t;\phi)&=\psi_s(s,t)\mathbf{e}_2(s,t;\phi), &\mathbf{e}_{2s}(s,t;\phi)&=-\psi_s(s,t)\mathbf{e}_1(t;\phi),\\ \mathbf{e}_{1\phi}(s,t;\phi)&=\mathbf{e}_2(s,t;\phi), &\mathbf{e}_{2\phi}(s,t;\phi)&=-\mathbf{e}_1(s,t;\phi),\\ \mathbf{e}_{1t}(s,t;\phi)&=\psi_t(s,t)\mathbf{e}_2(s,t;\phi), &\mathbf{e}_{2t}(s,t;\phi)&=-\psi_t(s,t)\mathbf{e}_1(s,t;\phi).\end{aligned}\tag{2.5}$$

In a general rotationally symmetric motion at time $t$ the material point $(s,\phi)$ has position

$$\boldsymbol{r}(s,t;\phi)=v(s,t)\mathbf{e}_1(s,t;\phi),\quad v(a,t)=:u(t).\tag{2.6}$$

See Fig. 1. (In anticipation of our particular initial-boundary-value problems, we have privileged the inner radius by the definitions of $\theta$ and $u$. A variant of these formulas would privilege the outer radius.) Henceforth we suppress the appearance of $\phi$ wherever it plays no essential role.

The deformation tensor is

$$\begin{aligned}\boldsymbol{F}&=\boldsymbol{r}_s\boldsymbol{j}_1+s^{-1}\boldsymbol{r}_\phi\boldsymbol{j}_2\\ &=\big[v_s(s,t)\mathbf{e}_1(s,t)+v(s,t)\psi_s(s,t)\mathbf{e}_2(s,t)\big]\boldsymbol{j}_1+s^{-1}v(s,t)\mathbf{e}_2(s,t)\boldsymbol{j}_2.\end{aligned}\tag{2.7}$$

An incompressible (locally area-preserving) deformation is defined by

$$\begin{aligned}1=\det\boldsymbol{F}&\equiv\big[(\boldsymbol{F}\cdot\boldsymbol{j}_1)\times(\boldsymbol{F}\cdot\boldsymbol{j}_2)\big]\cdot\boldsymbol{k}\\ &=s^{-1}\big\{[v_s\mathbf{e}_1+v\psi_s\mathbf{e}_2]\times[v\mathbf{e}_2]\big\}\cdot\boldsymbol{k}=s^{-1}vv_s,\end{aligned}\tag{2.8}$$

whence

$$v(s,t)=\sqrt{u(t)^2+s^2-a^2},$$

$$\begin{gathered}v_s(s,t)=sv(s,t)^{-1},\quad v(s,t)v_t(s,t)=u(t)u_t(t),\\ v^3v_{ss}=v^2-s^2\equiv u(t)^2-a^2,\end{gathered}\tag{2.9}$$

$$v_t(s,t)^2+v(s,t)v_{tt}(s,t)=u_t(t)^2+u(t)u_{tt}(t),$$

$$\begin{aligned}v_{tt}(s,t)&=u_t(t)^2v(s,t)^{-1}+u(t)u_{tt}(t)v(s,t)^{-1}-u(t)^2u_t(t)^2v(s,t)^{-3}\\&\equiv s^{-1}\partial_s\big[u_t{}^2v+uu_{tt}v+u^2u_t{}^2v^{-1}\big]=:s^{-1}\partial_s\zeta(s,t).\end{aligned}\tag{2.10}$$

(The function $\zeta$ is an acceleration potential.)

These identities, direct consequences of incompressibility, play a fundamental role in simplifying the analysis of rotationally symmetric motions. We often use these identities without comment. Thus the rotationally symmetric motion of a incompressible annulus is determined by the scalar function $u$ of the single variable $t$ and the scalar function $\psi$ of $s$ and $t$. (Note that incompressibility ensures that $u(t)\equiv v(a,t)$ and likewise $v(1,t)$ each determine $v(s,t)$.) Taylor's theorem implies the obvious geometrical fact that

$$v(1,t)-v(a,t)\sim\frac{1-a^2}{2v(a,t)}\sim\frac{1-a^2}{2v(1,t)}\to 0\quad\text{as } v(a,t)\to\infty.\tag{2.11}$$

In summary,

$$\boldsymbol{r}_t=v_t\boldsymbol{e}_1+v\psi_t\boldsymbol{e}_2,\tag{2.12}$$

$$\boldsymbol{r}_{tt}=\big[v_{tt}-v\psi_t^2\big]\boldsymbol{e}_1+[2v_t\psi_t+v\psi_{tt}]\boldsymbol{e}_2,\tag{2.13}$$

$$\boldsymbol{F}=(v_s\boldsymbol{e}_1+v\psi_s\boldsymbol{e}_2)\boldsymbol{j}_1+s^{-1}v\boldsymbol{e}_2\boldsymbol{j}_2=\big(sv^{-1}\boldsymbol{e}_1+v\psi_s\boldsymbol{e}_2\big)\boldsymbol{j}_1+s^{-1}v\boldsymbol{e}_2\boldsymbol{j}_2=:F_{\alpha\beta}\boldsymbol{e}_\alpha\boldsymbol{j}_\beta,\tag{2.14}$$

$$\boldsymbol{F}^{-1}=s^{-1}v\boldsymbol{j}_1\boldsymbol{e}_1+\boldsymbol{j}_2(-v\psi_s\boldsymbol{e}_1+v_s\boldsymbol{e}_2)=s^{-1}v\boldsymbol{j}_1\boldsymbol{e}_1+\boldsymbol{j}_2\big(-v\psi_s\boldsymbol{e}_1+sv^{-1}\boldsymbol{e}_2\big),\tag{2.15}$$

$$\boldsymbol{F}^{-*}=\boldsymbol{e}_1\big(s^{-1}v\boldsymbol{j}_1-v\psi_s\boldsymbol{j}_2\big)+v_s\boldsymbol{e}_2\boldsymbol{j}_2=\boldsymbol{e}_1\big(s^{-1}v\boldsymbol{j}_1-v\psi_s\boldsymbol{j}_2\big)+sv^{-1}\boldsymbol{e}_2\boldsymbol{j}_2,\tag{2.16}$$

$$\begin{aligned}\boldsymbol{C}:=\boldsymbol{F}^*\cdot\boldsymbol{F}&=\big[\boldsymbol{j}_1(v_s\boldsymbol{e}_1+v\psi_s\boldsymbol{e}_2)+s^{-1}v\boldsymbol{j}_2\boldsymbol{e}_2\big]\cdot\big[(v_s\boldsymbol{e}_1+v\psi_s\boldsymbol{e}_2)\boldsymbol{j}_1+s^{-1}v\boldsymbol{e}_2\boldsymbol{j}_2\big]\\&=\big(s^2v^{-2}+v^2\psi_s{}^2\big)\boldsymbol{j}_1\boldsymbol{j}_1+s^{-1}v^2\psi_s(\boldsymbol{j}_1\boldsymbol{j}_2+\boldsymbol{j}_2\boldsymbol{j}_1)+s^{-2}v^2\boldsymbol{j}_2\boldsymbol{j}_2\\&=:C_{\alpha\beta}\boldsymbol{j}_\alpha\boldsymbol{j}_\beta.\end{aligned}\tag{2.17}$$

$\boldsymbol{C}$ is the Green deformation tensor.

## 3 Stress and Constitutive Equations

Set

$$\dot{\boldsymbol{C}}:=\dot{\boldsymbol{F}}^*\cdot\boldsymbol{F}+\boldsymbol{F}^*\cdot\dot{\boldsymbol{F}}\tag{3.1}$$

Let

$$\boldsymbol{T} \equiv T_{\alpha\beta}\,\boldsymbol{e}_\alpha \boldsymbol{j}_\beta, \qquad \boldsymbol{S} \equiv \boldsymbol{F}^{-1} \cdot \boldsymbol{T} \equiv S_{\alpha\beta}\boldsymbol{j}_\alpha \boldsymbol{j}_\beta \tag{3.2}$$

be the first and second Piola-Kirchhoff stresses at $(s, \phi, t)$. As a consequence of the equation of rotational motion, the tensor $\boldsymbol{S}$ is required to be symmetric. For the treatment of problems for incompressible media, decompose $\boldsymbol{T}$ as

$$\boldsymbol{T}(s,\phi,t) = -p(s,\phi,t)\boldsymbol{F}(s,\phi,t)^{-*} + \boldsymbol{T}^{\mathrm{A}}(s,\phi,t) \tag{3.3}$$

where $p$ is a pressure (Lagrange multiplier), an unknown function, maintaining the constraint of incompressibility, and $\boldsymbol{T}^{\mathrm{A}}(s,\phi,t)$ is the active (or extra) stress. Thus

$$T_{11} = -ps^{-1}v + T_{11}^{\mathrm{A}}, \quad T_{12} = pv\psi_s + T_{12}^{\mathrm{A}}, \quad T_{21} = T_{21}^{\mathrm{A}}, \quad T_{22} = -psv^{-1} + T_{22}^{\mathrm{A}}. \tag{3.4}$$

We study problems in which material properties vary only with the radius $s$. Then the general constitutive equation for an incompressible nonlinearly viscoelastic material of strain-rate type has the form

$$\boldsymbol{T}(s,\phi,t) = -p(s,\phi,t)\boldsymbol{F}(s,\phi,t)^{-*} + \hat{\boldsymbol{T}}^{\mathrm{A}}\big(\boldsymbol{F}(s,\phi,t), \boldsymbol{F}_t(s,\phi,t), s\big) \tag{3.5}$$

where

$$\hat{\boldsymbol{T}}^{\mathrm{A}}(\boldsymbol{F}, \dot{\boldsymbol{F}}, s) = \hat{\boldsymbol{T}}^{\mathrm{E}}(\boldsymbol{F}, s) + \hat{\boldsymbol{T}}^{\mathrm{D}}(\boldsymbol{F}, \dot{\boldsymbol{F}}, s) = \boldsymbol{F} \cdot \hat{\boldsymbol{S}}^{\mathrm{A}}(\boldsymbol{C}, \dot{\boldsymbol{C}}, s) \tag{3.6}$$

with

$$\hat{\boldsymbol{T}}^{\mathrm{E}}(\boldsymbol{F}, s) := \hat{\boldsymbol{T}}^{\mathrm{A}}(\boldsymbol{F}, \boldsymbol{O}, s) = \boldsymbol{F} \cdot \hat{\boldsymbol{S}}^{\mathrm{A}}(\boldsymbol{C}, \boldsymbol{O}, s) = \frac{\partial \hat{W}}{\partial \boldsymbol{F}}(\boldsymbol{F}, s) \equiv \frac{\partial \hat{W}^+}{\partial \boldsymbol{F}}\big(\boldsymbol{F}^* \cdot \boldsymbol{F}, s\big), \tag{3.7}$$

where $\hat{\boldsymbol{S}}^{\mathrm{A}}$ is symmetric, and where $\hat{W}$ and $\hat{W}^+$ are stored-energy functions. (Here we adopt the convention that the superposed dot has no operational significance, but merely identifies a constitutive argument that is to be occupied by a time derivative.) The superposed carets identify constitutive functions, distinguishing them from functions with arguments $(s, \phi, t)$. $\hat{\boldsymbol{T}}^{\mathrm{E}}$ is the *equilibrium response function* and $\hat{\boldsymbol{T}}^{\mathrm{D}}$ is the dissipative *viscoelastic response function*. Condition (3.6) ensures that these constitutive equations are frame-indifferent.

The symmetry of $\hat{\boldsymbol{S}}^{\mathrm{A}}$ says that

$$\begin{aligned}
&\boldsymbol{F}^{-1} \cdot \hat{\boldsymbol{T}}^{\mathrm{A}} \\
&\quad \equiv \big[s^{-1}v\boldsymbol{j}_1\boldsymbol{e}_1 + \boldsymbol{j}_2(-v\psi_s\boldsymbol{e}_1 + v_s\boldsymbol{e}_2)\big] \cdot [\hat{T}_{11}^{\mathrm{A}}\boldsymbol{e}_1\boldsymbol{j}_1 + \hat{T}_{12}^{\mathrm{A}}\boldsymbol{e}_1\boldsymbol{j}_2 + \hat{T}_{21}^{\mathrm{A}}\boldsymbol{e}_2\boldsymbol{j}_1 + \hat{T}_{22}^{\mathrm{A}}\boldsymbol{e}_2\boldsymbol{j}_2] \\
&\quad = (\hat{\boldsymbol{T}}^{\mathrm{A}})^* \cdot \boldsymbol{F}^{-*} \\
&\quad \equiv [\hat{T}_{11}^{\mathrm{A}}\boldsymbol{j}_1\boldsymbol{e}_1 + \hat{T}_{12}^{\mathrm{A}}\boldsymbol{j}_2\boldsymbol{e}_1 + \hat{T}_{21}^{\mathrm{A}}\boldsymbol{j}_1\boldsymbol{e}_2 + \hat{T}_{22}^{\mathrm{A}}\boldsymbol{j}_2\boldsymbol{e}_2] \cdot \big[\boldsymbol{e}_1\big(s^{-1}v\boldsymbol{j}_1 - v\psi_s\boldsymbol{j}_2\big) + v_s\boldsymbol{e}_2\boldsymbol{j}_2\big],
\end{aligned} \tag{3.8}$$

whence the constitutive equations for the first Piola-Kirchhoff stress must satisfy the *symmetry condition*:

$$sv^{-1}\hat{T}_{21}^{\mathrm{A}} - s^{-1}v\hat{T}_{12}^{\mathrm{A}} = v\psi_s\hat{T}_{11}^{\mathrm{A}}, \quad \big(v_s = sv^{-1}\big). \tag{3.9}$$

For our study of rotationally symmetric motions we naturally require a corresponding rotational symmetry in constitutive response:

**1.** The material of (3.6), (3.7) is transversely isotropic in the sense that the annulus responds the same way to shearing deformations in the clockwise and counter-clockwise senses. Thus $\jmath_1 \cdot \hat{\boldsymbol{S}}^{\mathrm{A}} \cdot \jmath_2$ changes sign when $\jmath_1 \cdot \boldsymbol{C} \cdot \jmath_2$ and $\jmath_1 \cdot \dot{\boldsymbol{C}} \cdot \jmath_2$ change sign. This condition implies that $\hat{W}^+$ depends on $\jmath_1 \cdot \boldsymbol{C} \cdot \jmath_2$ through its square, so that $\hat{W}^+$ depends on $\psi_s$ only through its square. The constitutive functions are independent of the material coordinate $\phi$.

We impose natural growth conditions on the constitutive functions:

**2.** The equilibrium response function $\hat{\boldsymbol{T}}^{\mathrm{E}}$ satisfies an analog of the Strong Ellipticity Condition adapted to incompressible nonlinearly elastic materials [2, Sect. 13.3], which provides a monotone response of the first Piola-Kirchhoff stress with respect to the deformation tensor in directions tangent to the constraint manifold $\det \boldsymbol{F} = 1$:

$$\begin{aligned} &\boldsymbol{ab} : \frac{\partial \hat{\boldsymbol{T}}^{\mathrm{E}}}{\partial \boldsymbol{F}} : \boldsymbol{ab} > 0 \quad \forall\, \boldsymbol{F} \text{ with } \det \boldsymbol{F} = 1 \\ &\forall\, \boldsymbol{a} \equiv \cos\alpha\, e_1 + \sin\alpha\, e_2, \qquad \boldsymbol{b} \equiv \cos\beta\, \jmath_1 + \sin\beta\, \jmath_2 \end{aligned} \tag{3.10}$$

such that

$$\begin{aligned} \boldsymbol{b} \cdot \boldsymbol{F}^{-1} \cdot \boldsymbol{a} &= \boldsymbol{b} \cdot \left[ s^{-1} v \jmath_1 e_1 + \jmath_2 \left( -v\psi_s e_1 + s v^{-1} e_2 \right) \right] \cdot \boldsymbol{a} \\ &\equiv s^{-1} v \cos\alpha \cos\beta - \psi_s \cos\alpha \sin\beta + s v^{-1} \sin\alpha \sin\beta = 0 \\ \iff \quad \tan\beta &= \frac{s^{-1} v}{v\psi_s - s v^{-1} \tan\alpha} \end{aligned} \tag{3.11}$$

by virtue of (2.15). (The first line of (3.11) is a corrected consequence of the last equality of formula [2, Eq. 13.3.42], which is missing the power $-1$ on $\boldsymbol{F}$.) In particular, for $\boldsymbol{a} = e_2$, $\boldsymbol{b} = \jmath_1$, so that $\alpha = \frac{\pi}{2}$, $\beta = 0$, (3.10) becomes

$$\frac{\partial \hat{T}^{\mathrm{E}}_{21}}{\partial F_{21}} \equiv \frac{\partial (e_2 \cdot \boldsymbol{T}^{\mathrm{A}} \cdot \jmath_1)}{\partial (e_2 \cdot \boldsymbol{F} \cdot \jmath_1)} > 0. \tag{3.12}$$

(Taking the complementary $\beta = \frac{\pi}{2}$, $\alpha = 0$ reduces (3.11) to the overly special $\psi_s = 0$, so that (3.10) gives $\partial T^{\mathrm{E}}_{12} / \partial F_{12} > 0$ if $\psi_s = 0$. Recall that $F_{12} = 0$ for the deformations that we study. Inequality (3.12) seems to be the only simple and nontrivial consequence of (3.10) and (3.11).)

**3.** The components $\hat{T}^{\mathrm{E}}_{11}, \hat{T}^{\mathrm{E}}_{21}, \hat{T}^{\mathrm{E}}_{22}$ of the equilibrium response function are respectively strictly increasing functions of the corresponding components $F_{11}, F_{21}, F_{22}$ of $\boldsymbol{F}$. (This condition is not unlike Ball's Polyconvexity Condition [5], the 3-dimensional version of which says that the stored-energy function has the form $W^{\ddagger}(\boldsymbol{F}, (\det \boldsymbol{F})\boldsymbol{F}^{-1}, \det \boldsymbol{F}, s)$ where $W^{\ddagger}$ is a convex function of its first two tensor arguments and its third scalar argument. For our 2-dimensional problem for incompressible materials, these arguments reduce to $\boldsymbol{F}, \boldsymbol{F}^{-1}, 1$ with $\boldsymbol{F}$ and $\boldsymbol{F}^{-1}$ having exactly the same set of components.)

**4.** The constitutive equations should penalize a total radial compression, characterized by $v_s \to 0$ and a total azimuthal compression characterized by $v \to 0$. Since $v_s = s v^{-1}$, such penalizations simultaneously penalize total radial and azimuthal extensions, characterized by $v_s \to \infty$ and $v \to \infty$.

**5.** In analogy with condition (3.10) the dissipative part $\hat{\boldsymbol{T}}^{\mathrm{D}}$ of $\hat{\boldsymbol{T}}^{\mathrm{A}}$ and therefore $\hat{\boldsymbol{T}}^{\mathrm{A}}$ itself is a strictly monotone function of $\dot{\boldsymbol{F}}$ for $\dot{\boldsymbol{F}}$ tangent to the constraint manifold for incompressible nonlinearly viscoelastic materials of strain-rate type:

$$\dot{\boldsymbol{H}}:\frac{\partial\hat{\boldsymbol{T}}^{\mathrm{D}}}{\partial\dot{\boldsymbol{F}}}(\boldsymbol{F},\dot{\boldsymbol{F}},s):\dot{\boldsymbol{H}}>0\quad\forall\,\boldsymbol{F},\dot{\boldsymbol{F}},\dot{\boldsymbol{H}}$$
$$\text{with }\det\boldsymbol{F}=1,\quad \boldsymbol{F}^{-*}\cdot\dot{\boldsymbol{H}}=\boldsymbol{O}. \tag{3.13}$$

(The derivation of this condition follows the derivation of (3.10), (3.11).) Let $\dot{\boldsymbol{H}}=\dot{H}_{\alpha\beta}\,\mathbf{e}_\alpha\boldsymbol{j}_\beta$. Then (2.16) converts the last equation of (3.13) to

$$\begin{aligned}\boldsymbol{F}^{-*}\cdot\dot{\boldsymbol{H}}&=\left[\mathbf{e}_1\left(s^{-1}v\boldsymbol{j}_1-v\psi_s\boldsymbol{j}_2\right)+sv^{-1}\mathbf{e}_2\boldsymbol{j}_2\right]:\dot{H}_{\alpha\beta}\,\mathbf{e}_\alpha\boldsymbol{j}_\beta\\&=s^{-1}v\dot{H}_{11}-v\psi_s\dot{H}_{12}+sv^{-1}\dot{H}_{22}=0.\end{aligned}\tag{3.14}$$

The choice of $\dot{\boldsymbol{H}}=\dot{H}_{21}\,\mathbf{e}_2(a,t)\boldsymbol{j}_1$ trivially satisfies this restriction, so that its use in (3.13) yields a complement of (3.12):

$$\frac{\partial\hat{T}^{\mathrm{A}}_{21}}{\partial\dot{F}_{21}}\equiv\frac{\partial\hat{T}^{\mathrm{D}}_{21}}{\partial\dot{F}_{21}}>0. \tag{3.15}$$

**6.** In analogy with paragraph 3 is the condition that the components $\hat{T}^{\mathrm{D}}_{11},\hat{T}^{\mathrm{D}}_{21},\hat{T}^{\mathrm{D}}_{22}$ of the dissipative part of the stress are respectively strictly increasing functions of the corresponding nonzero components $\dot{F}_{11},\dot{F}_{21},\dot{F}_{22}$ of $\boldsymbol{F}$.

**The Mooney-Rivlin material.** A popular constitutive equation for rubber is the Mooney-Rivlin equation for isotropic incompressible elastic materials [2, (13.5.3)], [24, (95.1)], the use of which can simplify the analysis of specific problems. Its material formulation is

$$\hat{\boldsymbol{T}}^{\mathrm{E}}=\hat{\boldsymbol{T}}^{\mathrm{MR}}=-p\boldsymbol{F}^{-*}+\tfrac12\mu(s)\left[1+\gamma(s)\right]\boldsymbol{F}-\tfrac12\mu(s)\left[1-\gamma(s)\right]\boldsymbol{F}^{-*}\cdot\boldsymbol{F}^{-1}\cdot\boldsymbol{F}^{-*}. \tag{3.16}$$

Here $\mu$ is a positive-valued function and $\gamma$ is a function with values in [0, 1]. (The special case of (3.16) for $\gamma=1$ defines the neo-Hookean material, which for our work is degenerate.) The componential form of (3.16) is

$$\begin{aligned}\hat{T}^{\mathrm{MR}}_{11}&=\tfrac12\mu(1+\gamma)v_s-\tfrac12\mu(1-\gamma)s^{-1}v^3\left(s^{-2}+\psi_s{}^2\right),\\\hat{T}^{\mathrm{MR}}_{12}&=\tfrac12\mu(1-\gamma)v\psi_s\left[s^{-2}v^2+\left(v_s{}^2+v^2\psi_s{}^2\right)\right],\\\hat{T}^{\mathrm{MR}}_{21}&=\tfrac12\mu(1+\gamma)v\psi_s+\tfrac12\mu(1-\gamma)v\psi_s\equiv\mu v\psi_s,\\\hat{T}^{\mathrm{MR}}_{22}&=\tfrac12\mu(1+\gamma)s^{-1}v-\tfrac12\mu(1-\gamma)v_s\left(v_s{}^2+v^2\psi_s{}^2\right).\end{aligned}\tag{3.17}$$

(These formulas of course satisfy the symmetry condition (3.9).) Since $v_s=sv^{-1}$, when $\gamma\in[0,1)$ (so that the material is not neo-Hookean), the formulas for all these stresses except $T^{\mathrm{A}}_{21}$ account for an expected extreme behavior when $v$ and consequently $u$ approach either of the extreme values 0 and $\infty$:

$$\begin{aligned}&\hat{T}^{\mathrm{MR}}_{11}\to\infty &&\text{as}\quad v_s\to\infty, &&\text{i.e., as}\quad v\searrow 0,\\&\hat{T}^{\mathrm{MR}}_{11}\to-\infty &&\text{as}\quad v\to\infty, &&\text{i.e., as}\quad v_s\searrow 0,\\&\hat{T}^{\mathrm{MR}}_{22}\to\infty &&\text{as}\quad v\to\infty, &&\text{i.e., as}\quad v_s\searrow 0,\\&\hat{T}^{\mathrm{MR}}_{22}\to-\infty &&\text{as}\quad v_s\to\infty, &&\text{i.e., as}\quad v\searrow 0.\end{aligned}\tag{3.18}$$

We adopt (3.18) as a concrete realization of constitutive restriction 4 for the equilibrium response $\hat{\boldsymbol{T}}^{\mathrm{E}}$ for all materials that we treat.

The stored-energy function for the Mooney-Rivlin material has the form

$$\hat{W}^{\mathrm{MR}} = \tfrac{1}{4}(1+\gamma)\mu\big[s^2 v^{-2} + s^{-2}v^2 + v^2 {\psi_s}^2\big] + \text{const}, \tag{3.19}$$

which is comparable to

$$T_{\alpha\beta}^{\mathrm{MR}} F_{\alpha\beta} = \mu\gamma\big[s^2 v^{-2} + s^{-2}v^2 + v^2 {\psi_s}^2\big]. \tag{3.20}$$

These constitutive equations satisfy properties 1–4. (But this $\hat{W}^{\mathrm{MR}}$ is a convex function of $F_{11}, F_{22}, F_{21}$ only for $\gamma \in [\frac{1}{2}, 1]$.) At the end of Sect. 3 we shall see that system (3.17) just barely meets a condition ensuring good behavior of solutions. Thus it, like many special constitutive equations, can be misleading about the general behavior of other materials. It, however, admits blowup phenomena, so its use in this setting could signal dangerous instabilities.

## 4 Equations of Motion and Boundary Conditions

Let $\varrho(s)$ be the density per unit reference area at any material point with radial reference coordinate $s$. Assume that $\varrho$ is bounded and has a positive lower bound. Under the assumption that there are no true body forces—we shall encounter effective body forces coming from rotating coordinate systems—the equations of motion are

$$\nabla \cdot \boldsymbol{T}^* \equiv \nabla \cdot \big(\boldsymbol{T}^{\mathrm{A}*} - p\boldsymbol{F}^{-1}\big) \equiv \big(\boldsymbol{j}_1\partial_s + s^{-1}\boldsymbol{j}_2\partial_\phi\big)\cdot\big(\boldsymbol{T}^{\mathrm{A}*} - p\boldsymbol{F}^{-1}\big) = \varrho(s)\boldsymbol{r}_{tt}, \tag{4.1}$$

with

$$\begin{aligned}\nabla \cdot \boldsymbol{T}^{\mathrm{A}*} &= \big[\boldsymbol{j}_1\partial_s + s^{-1}\boldsymbol{j}_2\partial_\phi\big]\cdot T^{\mathrm{A}}_{\alpha\beta}\boldsymbol{j}_\beta\,\boldsymbol{e}_\alpha = \partial_s(T^{\mathrm{A}}_{\alpha 1}\boldsymbol{e}_\alpha) + s^{-1}T^{\mathrm{A}}_{\alpha\beta}\boldsymbol{j}_2\partial_\phi\cdot(\boldsymbol{j}_\beta\,\boldsymbol{e}_\alpha)\\ &= T^{\mathrm{A}}_{\alpha 1 s}\boldsymbol{e}_\alpha + T^{\mathrm{A}}_{11}\psi_s\boldsymbol{e}_2 - T^{\mathrm{A}}_{21}\psi_s\boldsymbol{e}_1 + s^{-1}\big[(T^{\mathrm{A}}_{11} - T^{\mathrm{A}}_{22})\boldsymbol{e}_1 + (T^{\mathrm{A}}_{12} + T^{\mathrm{A}}_{21})\boldsymbol{e}_2\big],\end{aligned} \tag{4.2}$$

$$\begin{aligned}\nabla\cdot\big(-p\boldsymbol{F}^{-1}\big) &= \big[\boldsymbol{j}_1\partial_s + s^{-1}\boldsymbol{j}_2\partial_\phi\big]\cdot\big\{-p\big[s^{-1}v\boldsymbol{j}_1\boldsymbol{e}_1 + \boldsymbol{j}_2\big(-v\psi_s\boldsymbol{e}_1 + sv^{-1}\boldsymbol{e}_2\big)\big]\big\}\\ &= -\big(ps^{-1}v\big)_s\boldsymbol{e}_1 - ps^{-1}v\psi_s\boldsymbol{e}_2 - s^{-2}vp\boldsymbol{e}_1\\ &\quad + s^{-1}p\big(v\psi_s\boldsymbol{e}_2 + sv^{-1}\boldsymbol{e}_1\big) - s^{-1}p_\phi\big(-v\psi_s\boldsymbol{e}_1 + sv^{-1}\boldsymbol{e}_2\big).\end{aligned} \tag{4.3}$$

Then (4.2) and (4.3) convert the equation of motion (4.1) to

$$\begin{aligned}&-\big(ps^{-1}v\big)_s\boldsymbol{e}_1 - ps^{-1}v\psi_s\boldsymbol{e}_2 - s^{-2}vp\boldsymbol{e}_1\\ &\quad + s^{-1}p\big(v\psi_s\boldsymbol{e}_2 + sv^{-1}\boldsymbol{e}_1\big) - s^{-1}p_\phi\big(-v\psi_s\boldsymbol{e}_1 + sv^{-1}\boldsymbol{e}_2\big)\\ &\quad + T^{\mathrm{A}}_{\alpha 1 s}\boldsymbol{e}_\alpha + T^{\mathrm{A}}_{11}\psi_s\boldsymbol{e}_2 - T^{\mathrm{A}}_{21}\psi_s\boldsymbol{e}_1 + s^{-1}\big[(T^{\mathrm{A}}_{11} - T^{\mathrm{A}}_{22})\boldsymbol{e}_1 + (T^{\mathrm{A}}_{12} + T^{\mathrm{A}}_{21})\boldsymbol{e}_2\big]\\ &= \varrho\big\{\big[v_{tt} - v\psi_t^2\big]\boldsymbol{e}_1 + [v\psi_{tt} + 2v_t\psi_t]\boldsymbol{e}_2\big\},\end{aligned} \tag{4.4}$$

the componential versions of which are

$$-s^{-1}vp_s + s^{-1}p_\phi v\psi_s + T^{\mathrm{A}}_{11s} - T^{\mathrm{A}}_{21}\psi_s + s^{-1}(T^{\mathrm{A}}_{11} - T^{\mathrm{A}}_{22}) = \varrho\big[v_{tt} - v{\psi_t}^2\big], \tag{4.5}$$

$$- p_\phi v^{-1} + T^{\mathrm{A}}_{21s} + T^{\mathrm{A}}_{11}\psi_s + s^{-1}(T^{\mathrm{A}}_{12} + T^{\mathrm{A}}_{21})$$
$$= \varrho[v\psi_{tt} + 2v_t\psi_t] \equiv \varrho v^{-1}\left(v^2\psi_t\right)_t \equiv \varrho(s)v^{-1}\left[\left(u^2 + s^2 - a^2\right)\psi_t\right]_t. \tag{4.6}$$

**Boundary conditions.** Since the annuli under study are intact (and not subject to dislocations), we naturally require that

$$\boldsymbol{r}(s, 2\pi) = \boldsymbol{r}(s, 0), \quad \psi(s, 2\pi) = \psi(s, 0), \quad p(s, 2\pi) = p(s, 0), \quad \boldsymbol{T}^{\mathrm{A}}(s, 2\pi) = \boldsymbol{T}^{\mathrm{A}}(s, 0) \tag{4.7}$$

(and consequently we are free to extend the functions $\boldsymbol{r}(s, \cdot)$, $\psi(s, \cdot)$, $p(s, \cdot)$, $\boldsymbol{T}^{\mathrm{A}}(s, \cdot)$ to $\mathbb{R}$ as functions of $\phi$ with period $2\pi$).

We shall illustrate the analysis of rotationally invariant motions for a specific problem in which internal pressure and centrifugal force play central roles: Let the outer boundary be traction-free:

$$\begin{aligned} o = \boldsymbol{T}(1,t) \cdot \boldsymbol{j}_1 &\equiv \left[-p(1,t)\boldsymbol{F}(1,t)^{-*} + \boldsymbol{T}^{\mathrm{A}}(1,t)\right] \cdot \boldsymbol{j}_1 \\ &\equiv -p(1,t)v(1,t)\mathrm{e}_1(1,t) + \boldsymbol{T}^{\mathrm{A}}(1,t) \cdot \boldsymbol{j}_1, \end{aligned} \tag{4.8}$$

so that

$$T^{\mathrm{A}}_{21}(1,t) = 0, \qquad T^{\mathrm{A}}_{11}(1,t) = p(1,t)v(1,t). \tag{4.9}$$

Let the normal component of traction on the inner boundary be a hydrostatic pressure of intensity $q(t)$ per *actual* length:

$$q(t)\frac{u(t)}{a} = \mathrm{e}_1(a,t)_1 \cdot \boldsymbol{T}(a,t) \cdot (-\boldsymbol{j}_1) = -T_{11}(a,t) = p(a,t)\frac{u(t)}{a} - T^{\mathrm{A}}_{11}(a,t). \tag{4.10}$$

($q$ is positive for the natural case of a true inflational hydrostatic pressure.) To formulate a tractable problem in which centrifugal effects intervene when $u$ is not fixed, let this boundary be kept circular and be spun about the origin with prescribed angular speed $\theta_t$:

$$\psi_t(a,t) \equiv \theta_t(t). \tag{4.11}$$

Below we discuss how to effect such a condition. We assume that $q$ and $\theta_{tt}$ are continuous and bounded.

Since constitutive property 1 of Sect. 3 prevents the constitutive functions from depending on $\phi$, the only term in (4.6) that could depend on $\phi$ is $p_\phi$. Thus $p_\phi$ must be independent of $\phi$, so $p$ must be linear in $\phi$. The continuity (periodicity) requirement $(4.7)_3$ implies that

$$p_\phi = 0. \tag{4.12}$$

Equation $(2.9)_1$ and the symmetry condition (3.9) then reduce (4.6) to

$$(svT^{\mathrm{A}}_{21})_s = s\varrho\left(v^2\psi_t\right)_t. \tag{4.13}$$

Since the form of $v$ is known by virtue of (2.9), the substitution of the constitutive functions into (4.13) converts it to a partial differential equation for $\psi$ parametrized by (the $t$-dependent) $u$.

Integrate (4.5) from $a$ to $s$ subject to boundary condition (4.10) and simplify the integrand to represent $p$:

$$\begin{aligned} p(s,t) &= q(t) + T_{11}^{A}(s,t)sv(s,t)^{-1} \\ &\quad + \int_a^s v(\xi,t)^{-1}\big\{T_{11}^{A}(\xi,t)\xi^2 v(\xi,t)^{-2} - T_{22}^{A}(\xi,t) - \xi T_{21}^{A}(\xi,t)\psi_s(\xi,t) \\ &\quad - \varrho(\xi)\xi\big[v_{tt}(\xi,t) - v(\xi,t)\psi_t(\xi,t)^2\big]\big\}\,d\xi. \end{aligned} \tag{4.14}$$

Evaluating this equation at $s = 1$ and using the boundary condition (4.9) yield the important equation

$$q(t) = \int_a^1 \left[-\frac{s^2}{v^3}T_{11}^{A} + \frac{1}{v}T_{22}^{A} + \frac{s}{v}T_{21}^{A}\psi_s + s\varrho\left(\frac{v_{tt}}{v} - \psi_t^2\right)\right] ds. \tag{4.15}$$

The form of (4.13) shows that motions with straight sections, characterized by $\psi_s = 0$, must be of a very restricted class: In this case transverse isotropy, which causes $T_{21}^{A} = 0$ when $\psi_s = 0$, reduces (4.13) to $(v^2\psi_t)_t = 0$. Thus there is a function $g$ such that $v(s,t)^2\psi_t(t) = g(s)$. In view of the form (2.9) of $v$, two $s$-derivatives of this equation yield $2\psi_t(t) = g_{ss}(s)$, so that $\psi_t = \omega$ (const). Then there are constants $A, B$ such that $v(s,t)^2\omega = g = A + Bs + \frac{1}{2}\omega s^2$, so that $v$ and therefore $u$ must be independent of $t$.

For the simpler analogous planar problems described by the special Cosserat theory of rods, in which the configuration of a rod consists of a plane curve equipped with cross sections, there is a rich theory of breathing and shearing motions in which the cross sections, perforce constrained to remain straight, have oscillating rotations, while concentric material circles, perforce constrained to remain circular, oscillate radially [1]. This paper does not model incompressibility, for which there is a satisfactory theory of rods [3].

The form of (4.6) also gives conditions under which the shape of a deformed ray does not vary with time: $\psi_t = 0$. In this case the right-hand side of (4.6) vanishes, so that the left-hand side would be $t$-dependent, unless $u(t)$ is constant. Under these conditions, the vanishing of the left-hand side of (4.6) becomes a quasilinear ordinary differential equation for $\psi$ as a function of $s$. Boundary-value problems for it subject to the monotonicity condition and suitable growth conditions can be readily analyzed by standard variational methods [2, Chap. 7].

Since $\theta_t$ is prescribed, the Principle of Virtual Power says that the shear component $T_{21}(a,t)$ cannot be prescribed. This traction is responsible for the net torque $2\pi u(t)T_{21}(a,t)$ on the annulus, which must equal the time-derivative of its angular momentum:

$$u(t)T_{21}(a,t) = \frac{d}{dt}\int_a^1 s\varrho(s)\boldsymbol{k}\cdot\big[\boldsymbol{r}(s,t)\times\boldsymbol{r}_t(s,t)\big]ds \equiv \frac{d}{dt}\int_a^1 s\varrho(s)v(s,t)^2\psi_t(s,t)\,ds. \tag{4.16}$$

This condition, of course, also follows from an integration of the equation of motion (4.6) over $[a, 1]$, the use of the boundary condition $(4.9)_1$, and the use of the symmetry condition (3.9).

Since $u$ is not prescribed, the mechanical prescription of $\theta_t$ in a way that permits nontrivial rotationally symmetric motions requires some artifice: Suppose that the annulus has a thin flange near $s = a$. Such a motion could be approximately effected by attaching the annulus to a central rotating rigid disk with lubricated radially disposed rigid spokes passing only through the flange. Cf. [21]. (Since we are limiting our attention to rotationally symmetric motions we need not impose an additional constraint to maintain the circularity of the inner boundary.)

If the inner radius were fixed, then incompressibility would cause the outer radius to be bounded. On the other hand, when (4.8) and (4.10) hold, there is a possibility, to be investigated below, for the annulus to expand radially, possibly without bound, under the effects of centrifugal force and internal pressure.

## 5 The Energy Equation

The energy equation plays a fundamental role in the analysis of blowup problems for unconstrained solids [4, 6–10, 23]. Here we use it, however, merely to ensure that solutions have a natural level of regularity.

Condition (2.12) implies that the kinetic energy density is $\frac{1}{2}\varrho({v_t}^2 + v^2{\psi_t}^2)$. The constitutive equations (3.3)–(3.7) say that stress power density is

$$\hat{\boldsymbol{T}}:\boldsymbol{F}_t = -p\boldsymbol{F}^{-*}:\boldsymbol{F}_t + \hat{W}_{\boldsymbol{F}}:\boldsymbol{F}_t + \hat{\boldsymbol{T}}^{\mathrm{D}}:\boldsymbol{F}_t. \tag{5.1}$$

Since (2.14) and (2.16) imply that $\boldsymbol{F}^{-*}:\boldsymbol{F}_t = 0$, the integration by parts of the dot product of the equation of motion (4.1) with the velocity $\boldsymbol{r}_t$ over the area of the annulus and the time interval $[0, t]$ yields the energy equation

$$K(t) + \Phi(t) + D(t) + B_1(t) - B_a(t) = E \tag{5.2}$$

where

$$\begin{aligned}
K(t) &:= \frac{1}{2}\int_a^1 s\varrho\big[v_t(s,t)^2 + v(s,t)^2\psi_t(s,t)^2\big]\,ds,\\
\Phi(t) &:= \int_a^1 s\hat{W}\big(\boldsymbol{F}(s,t),s\big)\,ds,\\
D(t) &:= \int_0^t\int_a^1 \hat{\boldsymbol{T}}^{\mathrm{D}}\big(\boldsymbol{F}(s,\tau),\boldsymbol{F}_t(s,\tau),s\big):\boldsymbol{F}_t(s,\tau)\,s\,ds\,d\tau,\\
B_1(t) &:= \int_0^t \big[\boldsymbol{T}(1,\tau)\cdot\boldsymbol{j}_1\big]\cdot\boldsymbol{r}_t(1,\tau)\,d\tau,\\
B_a(t) &:= \int_0^t a\big[\boldsymbol{T}(a,\tau)\cdot\boldsymbol{j}_1\big]\cdot\boldsymbol{r}_t(a,\tau)\,d\tau
\end{aligned} \tag{5.3}$$

are the kinetic energy function, the stored-energy function, the work of the dissipative stresses, the work of the traction on the outer boundary, and the negative of the work of the traction on the inner boundary. $E$ is the constant initial energy.

Condition (4.8) causes $B_1 = 0$. Conditions (2.12), (2.14), $(2.16)_3$, (3.3), (4.10) cause the integrand of $B_a$ to be

$$\begin{aligned}
&a\boldsymbol{r}_t(a,\tau)\cdot\big[\boldsymbol{T}(a,\tau)\cdot\boldsymbol{j}_1\big]\\
&= a\big[u_t(\tau)\boldsymbol{e}_1(a,\tau) + u(\tau)\theta_t(\tau)\boldsymbol{e}_2(a,\tau)\big]\cdot\big[-p(a,\tau)\boldsymbol{F}(a,\tau)^{-*}\cdot\boldsymbol{j}_1 + T^{\mathrm{A}}_{\alpha 1}(a,\tau)\boldsymbol{e}_\alpha(a,\tau)\big]\\
&= a\big[u_t(\tau)\boldsymbol{e}_1 a,(\tau) + u(\tau)\theta_t(\tau)\boldsymbol{e}_2(a,\tau)\big]\cdot\big\{-p(a,\tau)a^{-1}u(\tau)\boldsymbol{e}_1(a,\tau) + T^{\mathrm{A}}_{\alpha 1}(a,\tau)\boldsymbol{e}_\alpha(a,\tau)\big\}\\
&= -q(\tau)u(\tau)u_t(\tau) + aT^{\mathrm{A}}_{21}(a,\tau)u(\tau)\theta_t(\tau).
\end{aligned} \tag{5.4}$$

For an elastic material, for which $\boldsymbol{T}^{\mathrm{D}}=\boldsymbol{O}$, these boundary conditions reduce the energy equation (5.2) to

$$K(t)+\Phi(t)-\int_0^t\big[q(\tau)u(\tau)u_t(\tau)+aT_{21}^{\mathrm{A}}(a,\tau)u(\tau)\theta_t(\tau)\big]\,d\tau=E. \tag{5.5}$$

Replacing $T_{21}^{\mathrm{A}}(a,\tau)u(\tau)$, which is not prescribed, with its formula from (4.16) yields

$$\begin{aligned}
&K(t)+\Phi(t)-E-\int_0^t q(\tau)u(\tau)u_t(\tau)\,d\tau\\
&=a\int_0^t\theta_t(\tau)\frac{d}{d\tau}\bigg[\int_a^1 s\varrho(s)v(s,\tau)^2\psi_t(s,\tau)\,ds\bigg]d\tau\\
&\equiv a\theta_t(t)\int_a^1 s\varrho(s)v(s,t)^2\psi_t(s,t)\,ds-a\theta_t(0)\int_a^1 s\varrho(s)v(s,0)^2\psi_t(s,0)\,ds\\
&\quad-a\int_0^t\theta_{tt}(\tau)\int_a^1 s\varrho(s)v(s,\tau)^2\psi_t(s,\tau)\,ds\,d\tau.
\end{aligned} \tag{5.6}$$

There is a significant simplification if $q$ and $\theta_t$ are constant:

$$\begin{aligned}
&K(t)+\Phi(t)-E-\tfrac{1}{2}q\big[u(t)^2-u(0)^2\big]\\
&\quad=a\theta_t\int_a^1 s\varrho(s)v(s,t)^2\psi_t(s,t)\,ds-a\theta_t\int_a^1 s\varrho(s)v(s,0)^2\psi_t(s,0)\,ds.
\end{aligned} \tag{5.7}$$

For brevity, we limit our attention to this case, which, of course, excludes an unbounded $\theta_t$.

The Cauchy-Bunyakovskiĭ-Schwarz inequality implies that $\int_a^1 s\varrho v^2\psi_t\,ds$, appearing in (5.6) and (5.7), satisfies

$$\int_a^1 s\varrho v^2|\psi_t|\,ds\le\sqrt{\int_a^1 s\varrho v^2\,ds}\sqrt{\int_a^1 s\varrho v^2{\psi_t}^2\,ds}\le\frac{1}{2\beta}\int_a^1 s\varrho v^2\,ds+\frac{\beta}{2}\int_a^1 s\varrho v^2{\psi_t}^2\,ds \tag{5.8}$$

for any $\beta>0$. The last integral appears in the kinetic energy. Substitute (5.8) into (5.7) to obtain

$$\frac{1}{2}\int_a^1 s\varrho {v_t}^2\,ds+\frac{1-a|\theta_t|\beta}{2}\int_a^1 s\varrho v^2{\psi_t}^2\,ds+\int_a^1 s\hat{W}\,ds\le E_2+\tfrac{1}{2}qu^2+\frac{a|\theta_t|}{2\beta}\int_a^1 s\varrho v^2\,ds \tag{5.9}$$

where $E_2$ is another constant. Now choose $\beta$ so small that $1-a|\theta_t|\beta\ge 0$, so that there is a constant $C>0$ such that

$$\int_a^1 s\hat{W}\,ds\le E_2+\tfrac{1}{2}qu^2+C\int_a^1 s\varrho v^2\,ds. \tag{5.10}$$

We get the same inequality when the dissipative stress power, the integrand of $D$ in (5.3), satisfies the natural condition that it be positive.

**Energetic preclusion of total compression.** For our ensuing study of blowup, we need a proof of the expected property that $u(t)$ is bounded below by a positive number for all $t$ when the boundary conditions are reasonable: Suppose that the elastic response satisfies

$$W(\boldsymbol{F},s)\ge Av^{-\kappa}-B\quad\text{if }|\boldsymbol{F}|<\infty \tag{5.11}$$

for appropriate positive constants $\kappa$ and $A$ and arbitrary constant $B$. Then (5.10) and (5.11) imply that

$$A\int_a^1 s\big[u^2+s^2-a^2\big]^{-\kappa/2}\,ds-(1-a)B\le\int_a^1 sW\,ds\le E_2+\tfrac{1}{2}qu^2+C\int_a^1 s\varrho v^2\,ds. \quad (5.12)$$

For $\kappa\ge 2$ (a condition ensured by the requirement that

$$\hat{W}^+\ge AC_{11}-B, \quad (5.13)$$

cf. (2.9), (3.7)) the left-hand side of inequality (5.12) approaches $\infty$ if $u\to 0$, violating (5.12). The stored-energy function (3.19) for the Mooney-Rivlin material just satisfies (5.11) with $\kappa=2$. In summary, if $q$ and $\theta_t$ are constants and if (5.11) holds with $\kappa\ge 2$, then $u$ has a positive lower bound:

$$u(s,t)\ge c. \quad (5.14)$$

**A bound on shear.** If (5.13) holds, then (2.17) and the energy estimate (5.10) imply that

$$\int_a^1 sv^2{\psi_s}^2\,ds\le C, \quad (5.15)$$

so that (5.14) then implies that

$$\int_a^1 s{\psi_s}^2\,ds\le C. \quad (5.16)$$

## 6 Blowup

We assume that $q$ is bounded and that inequalities (5.14)–(5.16) hold (whether or not the special conditions in Sect. 5 ensuring these inequalities hold). We limit our attention to classical solutions, for which $v$ and $\psi$ are twice continuously differentiable on $(a,1)\times(0,\infty)$ and continuously differentiable on $[a,1]\times[0,T]$ for all finite $T$. A key role in the study of blowup is played by the acceleration term $\int_a^1 s\varrho v^{-1}v_{tt}\,ds$ in (4.15). If $\varrho$ is constant, then this integral can be evaluated in closed form by the use of $(2.10)_2$.

$$\begin{aligned}\int_a^1 sv^{-1}v_{tt}\,ds&=\int_a^1\left\{\frac{s({u_t}^2+uu_{tt})}{u^2+s^2-a^2}-\frac{su^2{u_t}^2}{(u^2+s^2-a^2)^2}\right\}ds\\&\equiv\tfrac{1}{2}\big(u_t^2+uu_{tt}\big)\ln\big(1+\big(1-a^2\big)u^{-2}\big)+\tfrac{1}{2}\big(1-a^2\big)u^2{u_t}^2. \end{aligned} \quad (6.1)$$

It is more convenient, however, for the ensuing analysis, even for constant $\varrho$, to evaluate this integral by the Mean-Value Theorem: There is a $b(t)\in[a,1]$ such that

$$\int_a^1 s\varrho v^{-1}v_{tt}\,ds=\beta(t)^{-1}\tilde{v}(t)^{-1}v_{tt}\big(b(t),t\big)\equiv\beta(t)^{-1}\left\{\frac{[u(t)^2]_{tt}}{2\tilde{v}(t)^2}-\frac{u(t)^2u_t(t)^2}{\tilde{v}(t)^4}\right\}$$

$$\text{where }\beta(t):=\big[(1-a)b(t)\varrho\big(b(t)\big)\big]^{-1},\quad \tilde{v}(t):=v\big(b(t),t\big)\equiv\sqrt{u(t)^2+b(t)^2-a^2}, \quad (6.2)$$

and where (2.10) was used. Then (4.15) becomes

$$(u^2)_{tt} = 2\frac{u^2 u_t(t)^2}{\tilde{v}(t)^2} + 2\beta\tilde{v}^2\left[q + \int_a^1 s\varrho\psi_t{}^2\,ds + \int_a^1\left(\frac{s^2}{v^3}T_{11}^{\mathrm{A}} - \frac{1}{v}T_{22}^{\mathrm{A}} - \frac{s}{v}T_{21}^{\mathrm{A}}\psi_s\right)ds\right]. \quad (6.3)$$

Let us first study this equation for the Mooney-Rivlin material of (3.17), for which

$$\begin{aligned}
&\int_a^1\left(\frac{s^2}{v^3}T_{11}^{\mathrm{A}} - \frac{1}{v}T_{22}^{\mathrm{A}} - \frac{s}{v}T_{21}^{\mathrm{A}}\psi_s\right)ds \\
&\quad\equiv \int_a^1 \{[\tfrac{1}{2}\mu(1+\gamma)s^3v^{-4} - \tfrac{1}{2}\mu(1-\gamma)s(s^{-2}+\psi_s{}^2)] \\
&\qquad - [\tfrac{1}{2}\mu(1+\gamma)s^{-1} - \tfrac{1}{2}\mu(1-\gamma)sv^{-2}(s^2v^{-2}+v^2\psi_s{}^2)] - s\mu\psi_s{}^2\}\,ds \\
&\quad\equiv \int_a^1 \mu(s^3v^{-4} - s^{-1} - s\psi_s{}^2)\,ds. \qquad (6.4)
\end{aligned}$$

The function $s \mapsto s^3 v(s,t)^{-4}$ in (6.4) (which can be explicitly integrated [14, 2.175.8]) readily admits simple two-sided estimates:

$$\begin{aligned}
&\int_a^1 \mu(s)s^3v^{-4}\,ds \equiv \int_a^1 \mu(s)s^3(s^2+u^2-a^2)^{-2}ds \\
&\left\{\begin{matrix}\le & \max\mu \\ \ge & a^2\min\mu\end{matrix}\right\}\int_a^1 s(s^2+u^2-a^2)^{-2}ds = \left\{\begin{matrix}\max\mu \\ a^2\min\mu\end{matrix}\right\}\frac{1}{2}\left(\frac{1}{u^2} - \frac{1}{u^2+1-a^2}\right),
\end{aligned} \qquad (6.5)$$

each of which approaches $\infty$ as $u \searrow 0$. But assumption (5.14) precludes this possibility. Condition (5.16), giving the integrability of $\psi_s{}^2$, then ensures that the right-hand side of (6.4) is bounded.

Our basic theorem, which illustrates the typicality of the Mooney-Rivlin material and which will be interpreted after its statement, is

**Blowup Theorem.** *If the term in brackets in* (6.3) (*for any material*) *has a positive lower bound $c$ on any time interval of the form* $[\tau, \infty)$, *then there are initial conditions such that the radius $u$ blows up to infinity exponentially fast.*

**Proof.** Inequality (6.3) yields

$$(u^2)_{tt} \ge 2\bar{\beta}cu^2 =: \lambda^2 u^2 \quad \text{where } \bar{\beta} := [(1-a)\max\varrho]^{-1}(\le\beta) \qquad (6.6)$$

Thus there is a positive-valued function $f$ such that (6.6) is equivalent to

$$(u^2)_{tt} = \lambda^2 u^2 + f(t), \qquad (6.7)$$

the general solution of which is

$$u(t)^2 = A\cosh\lambda t + B\sinh\lambda t + \frac{1}{\lambda}\int_0^t f(\tau)\sinh\lambda\tau\,d\tau. \qquad (6.8)$$

Thus there are initial conditions for which $u$ grows exponentially fast. □

Since (6.4) is bounded below for the Mooney-Rivlin material, there is exponentially fast blowup when the internal pressure $q$, a control, is large enough so that the term in brackets in (6.3) is positive. We make the angular velocity $\theta_t$ an analogous control under the regularity assumption that $\psi_{st}$ is integrable, for in this case

$$|\psi_t(s,t) - \theta_t(t)| \le \int_a^1 |\psi_{st}(\xi,t)|\,d\xi \le \text{const}. \tag{6.9}$$

Thus if the prescribed $|\theta_t|$ is big enough, then $|\psi_t|$ has a positive lower bound $c$, and the first integral within the braces of (6.3) can be made arbitrarily large by taking $|\theta_t|$ sufficiently large. In particular, a sufficiently large angular speed $|\theta_t|$ and its accompanying centrifugal force can dominate the effect of a negative $q$, i.e., can dominate the effect of a tensile traction on the inner boundary.

All that is needed to extend this analysis to a large family of constitutive equations for elastic annuli or for annuli of any material is for these equations to ensure that the second integral in (6.3) is bounded below. Such a bound, however, may not be available for various materials that are not elastic. Using an exceedingly simple model of viscoelasticity we indicate what can be done:

In (6.3) now take

$$T_{11}^{\mathrm{A}} = T_{11}^{\mathrm{MR}}, \quad T_{22}^{\mathrm{A}} = T_{22}^{\mathrm{MR}} + \kappa s^{-1} v_t, \quad T_{21}^{\mathrm{A}} = T_{21}^{\mathrm{MR}} \tag{6.10}$$

where $\kappa$ is a positive number. Here the only constitutive equation for the stress modified from the Mooney-Rivlin version (3.17) is that for $T_{22}^{\mathrm{A}}$, which now accounts for viscous dissipation in the circumferential direction, the direction in which the largest stretch can be expected to occur.

In this case, the right-hand sides of (6.3) and (6.4) are modified by the addition of

$$\begin{aligned} &-2\beta\kappa\tilde{v}^2 \int_a^1 s^{-1} v^{-1} v_t\,ds \equiv -2\beta\kappa\tilde{v}^2 \int_a^1 s^{-1} v^{-2} u u_t\,ds \\ &\equiv -2\beta\kappa \int_a^1 s^{-1} u u_t \frac{u^2 + b(t)^2 - a^2}{u^2 + s^2 - a^2}\,ds \equiv -\left(u^2\right)_t h(t), \\ &c \le h(t) := \beta\kappa \int_a^1 s^{-1} \frac{u^2 + b(t)^2 - a^2}{u^2 + s^2 - a^2}\,ds \le C. \end{aligned} \tag{6.11}$$

For any positive number $\delta$,

$$-\left(u^2\right)_t h \equiv -2uu_t h \ge -\left(\delta^2 u^2 + \delta^{-2} {u_t}^2\right) C. \tag{6.12}$$

Now take $\delta$ so large that

$$\delta^{-2} {u_t}^2 C \le 2u^2 \tilde{v}^{-2} \le 2 \tag{6.13}$$

(see (6.3)). Now take $q$ and $\theta_t$ so large that

$$\delta^2 u^2 C \le q + \int_a^1 s\varrho {\psi_t}^2\,ds. \tag{6.14}$$

Under these conditions we reproduce the proof of the Blowup Theorem to show that *there are initial conditions for which solutions of the equations of motion for this simple viscoelastic material blow up exponentially fast.* These techniques can clearly handle a much richer collection of viscoelastic materials.

## 7 Comments

There are several interesting open problems for rotationally symmetric motions:
**Existence of solutions to initial-boundary-value problems for incompressible annuli consisting of nonlinearly viscoelastic materials of strain-rate type.** The analysis of such problems for unconstrained media relies greatly on the presence of a strong dissipative mechanism to kill shocks and to prevent a total compression [23]. Whether (3.13) and (3.15) suffice for the treatment of incompressible media is not clear. That the governing equations have but a single independent spatial variable should greatly simplify the analysis. An obvious approach is to use the Faedo method [12], which employs the weak formulation of the governing equations (just like the Principle of Virtual Power) to convert the analysis to that for a sequence of systems of time-dependent ordinary differential equations. This weak formulation must account for the constraint of incompressibility, a task for which it is eminently well suited.

**Analysis of initial-boundary-value problems of nonlinear elasticity.** These problems present some novelties from the viewpoint of the available theory of quasilinear hyperbolic partial differential equations [11] (which typically admit shocks). The chief of these is the coupling of the function $\psi$ which satisfies a second-order hyperbolic partial differential equation with the function $u$, the presence of which makes the governing equations non-autonomous in $t$, and thereby moves them away from the standard theory.

**Stability.** There is a need to analyze the stability of rotationally invariant motions, which arise uniquely from rotationally invariant initial and boundary conditions. (Cf. [22].)

**Twisting motions.** It would be interesting to extend the studies of twisting motions of infinite incompressible elastic cylinders by [20, 21] to a much richer class of materials. (My preliminary studies indicate that accounting for the coupling with longitudinal shearing destroys the simplicity of the motions.)

**Acknowledgement** I am indebted to a referee who discovered a serious gaffe in the first version of this work.



## References

1. Antman, S.S.: Breathing oscillations of rotating nonlinearly elastic and viscoelastic rings. In: Durban, D., Givoli, D., Simmonds, J.G. (eds.) Advances in the Mechanics of Plates and Shells, pp. 1–16. Kluwer Academic, Dordrecht (2001)
2. Antman, S.S.: Nonlinear Problems of Elasticity, 2nd edn. Springer, New York (2005)
3. Antman, S.S., Schuricht, F.: Incompressibility in rod and shell theories. Math. Model. Numer. Anal. **33**, 289–304 (1999)
4. Antman, S.S., Ulusoy, S.: Blowup of solutions for the planar motions of rotating nonlinearly elastic rods. Int. J. Non-Linear Mech. **94**, 28–35 (2017)
5. Ball, J.M.: Convexity conditions and existence theorems in nonlinear elasticity. Arch. Ration. Mech. Anal. **63**, 337–403 (1977)
6. Ball, J.M.: Remarks on blow-up and nonexistence theorems for nonlinear evolution equations. Q. J. Math. **28**, 473–486 (1977)
7. Ball, J.M.: Finite time blow-up in nonlinear problems. In: Crandall, M.G. (ed.) Nonlinear Evolution Equations, pp. 189–205. Academic Press, San Diego (1978)
8. Calderer, M.C.: The dynamic behavior of nonlinearly elastic spherical shells. J. Elast. **13**, 17–47 (1983)

9. Calderer, M.C.: Finite time blow-up and stability properties of materials with fading memory. J. Differ. Equ. **63**, 289–305 (1986)
10. Calderer, M.C.: The dynamic behavior of viscoelastic spherical shells. Math. Methods Appl. Sci. **9**, 13–34 (1987)
11. Dafermos, C.M.: Hyperbolic Conservation Laws in Continuum Physics, 3rd edn. Springer, Berlin (2010)
12. Evans, L.C.: Partial Differential Equations, 2nd edn. AMS, Providence (2010)
13. Fosdick, R., Ketema, Y., Yu, J-H.: Dynamics of a viscoelastic spherical shell with a nonconvex strain energy function. Q. Appl. Math. **56**, 221–244 (1998)
14. Gradstein, L.S., Ryzhik, M.: Tables of Integrals, Series, and Products. Academic Press, San Diego (1980)
15. Guo, Z.-h., Solecki, R.: Free and forced finite-amplitude oscillations of an elastic thick-walled hollow sphere made of incompressible material. Arch. Mech. Stosow. **15**, 427–433 (1963)
16. Knops, R.: Logarithmic convexity and other techniques applied to problems in continuum mechanics. In: Knops, R. (ed.) Symposium on Non-Well-Posed Problems and Logarithmic Convexity. Lect. Notes Math., vol. 316, pp. 31–54. Springer, Berlin (1973)
17. Knops, R., Levine, H.A., Payne, L.E.: Non-existence, instability, and growth theorems for solutions of a class of abstract nonlinear equations with applications to nonlinear elastodynamics. Arch. Ration. Mech. Anal. **55**, 52–72 (1974)
18. Knowles, J.K.: Large amplitude oscillations of a tube of incompressible elastic material. Q. Appl. Math. **18**, 71–77 (1960)
19. Knowles, J.K., Jakub, M.T.: Finite dynamic deformations of an incompressible elastic medium containing a spherical cavity. Arch. Ration. Mech. Anal. **18**, 376–387 (1965)
20. Novozhilova, L., Pence, T.J., Urazhdin, S.: Exact solutions for axially varying three-dimensional twist motion in a neo-Hookean solid. Q. J. Mech. Appl. Math. **56**, 123–138 (2003)
21. Nowinski, J.L., Wang, S.D.: Finite radial oscillations of a spinning thick-walled cylinder. J. Acoust. Soc. Am. **40**, 1548–1553 (1966)
22. Rabier, P., Oden, J.T.: Bifurcation in Rotating Bodies. Masson, Paris (1989)
23. Stepanov, A.B., Antman, S.S.: Radially symmetric motions of nonlinearly viscoelastic bodies under live loads. Arch. Ration. Mech. Anal. **226**, 1209–1247 (2017)
24. Truesdell, C., Noll, W.: Non-linear Field Theories of Mechanics, 3rd edn. Springer, Berlin (2004)
25. Wang, C.-C., Truesdell, C.: Introduction to Rational Elasticity. Noordhoff, Groningen (1973)

J Elast (2019) 135:91–115
https://doi.org/10.1007/s10659-019-09727-x

# Interactions in a Multi-scale Representation of Sparse Media: From Mechanics to Thermodynamics

**Gianfranco Capriz**[1] · **Paolo Maria Mariano**[2]

Received: 7 July 2018 / Published online: 18 February 2019


**Abstract** We develop further a discrete-to-continuum approach to sparse media for which we do not grant the common axiom of permanent identification of the material elements. After refining our previous work on this topic, we focus our attention on the identification of stresses and self-actions at continuum scale in terms of the molecular interactions. Moreover, we introduce and discuss a notion of tensorial temperature by considering a grand canonical statistical ensemble constituted by molecules freely flowing in and out a control volume.

**Keywords** Continuum thermodynamics · Multi-scale models · Invariance · Kinetic theory · Granular matter · Discrete-to-continuum schemes

**Mathematics Subject Classification** 05C38 · 15A15 · 05A15 · 15A18

## 1 Introduction

We carry on in this paper with a program aiming at the construction of a continuum model of sparse media. They are clusters of *molecules* (a word used here in a generic sense, at times to be considered as a synonymous of *grains*) free to flow in space. They can interact at short distance with one another and, possibly, the environment, at least through collisions, which preserve the identity of molecules. For such media we construct a multi-scale continuum representation of the pertinent mechanics by relating values of continuum fields to statistical properties of molecular clusters in a control volume, called a *loculus*.

We consider non-negligible fluctuations in the molecular motion, which render arduous, if not impossible, a permanent choice of specific molecular clusters to be considered as

✉ P.M. Mariano
paolo.mariano@unifi.it

G. Capriz
gianfranco.capriz@mac.com

1 Dipartimento di Matematica, Università di Pisa, Accademia dei Lincei, Rome, Italy

2 DICeA, Università di Firenze, via Santa Marta 3, 50136 Firenze, Italy

material elements in a continuum representation. The circumstance forces us to disregard the axiom of permanent identification of material elements, an axiom which underpins the traditional continuum descriptions of bodies. This is a key aspect of our theoretical proposal.

We presume to know a distribution of places and velocities of molecules within the loculus at the instant $\tau$. It allows us to construct useful averages, through which we decompose the molecular motion within the loculus into elements pertaining to its mass center, an affine motion around it, and fluctuations. To these last two elements of the molecular motion, we associate a tensor moment of momentum and a Reynolds tensor, respectively. They account for the possible anisotropy in the moving cluster within the loculus (such an anisotropy plays a crucial role in various flows, such as turbulent states in which the homogeneous mixing assumption of fluid particle trajectories does not apply, as in heavy rainfall under appreciable wind). Also, they satisfy tensor balances of interactions, which we derive by imposing the invariance, under isometric changes in observers, of the external power, at continuum scale, on a generic part of the macroscopic region that we presume as actual place of the sparse medium in its continuum Eulerian representation, the one we propose.

Spurious fluctuations remain, left out by the averaging procedure used to construct mechanics. We attribute to them thermal effects. In our view, in fact, the boundary between mechanical and thermal phenomena depends on the level of detail that we reach in describing mechanics (a similar point of view on this boundary has been adopted by Grmela [22]). Roughly speaking, we could say that we decide a representation of what we consider essential features of the motion pertaining to some classes of bodies (or related electromagnetic effects), leaving to heat what remains out of that description.

In our work [12], we have already proposed a definition of temperature based on fluctuations, a scalar based on the distribution density of the fluctuating kinetic energy, and we refine such a definition here. However, this is not the sole possible definition. We show here, in fact, that even a tensor definition of temperature, at least for a special class of distributions, makes sense and can be defined in terms of statistical mechanics, besides the circumstance that the Reynolds tensor introduced here can be considered itself a tensor representation of temperature, a choice that we do not adopt here, leaving the Reynolds tensor associated with fluctuations in the representation of mechanics. Such a variety of possibilities suggests that the boundary (borderland, if you want) between mechanics and thermodynamics is (let us say) 'flexible', at times with fragmented or fluctuating appearance.

Our work contributes to foundational discussions on the justification of discrete-to-continuum representations for the mechanical behavior of condensed matter. An analysis of this topic, by now canonical, has been carried out by Irving and Kirkwood [26], prosecuted and deepened by Noll [36] (see also [28]), and further discussed by Murdoch [33], Pitteri [42, 43], and Peters [41] (see also the book [34] for further references and analyses). Differences with our approach rest on the whole structure emerging from our attention to a multi-scale representation, which leads to a multi-field continuum scheme, in contrast to Irving-Kirkwood's single-scale view, related to a standard Cauchy-type continuum. In addition, in Inving-Kirkwood's approach and related works, a Langevin-type equation for the distribution density plays a role. At variance, we do not exploit from the beginning that equation. Such a choice does not allow us to derive the expression of the tensors representing in average (and at continuum level) the interactions within the loculus and between neighboring loculi, as in the single scale Irving-Kirkwood's and Noll's approaches. The question remained open in our work [12]. Here we tackle it and identify the tensors representing various types of actions in the balance equations that we propose.

There is something more, however. Besides the foundational need of analyzing the link between continuum representations of matter and his discrete appearance at microscopic scales, there is also a technological interest for that relation. Fully atomistic simulations could predict with details the behavior of a material before its manufacture, just by resorting to first-principles-based-simulations and avoiding the introduction of phenomenological information (which require to have already at disposal the material in laboratory). However, fully atomistic computations on macroscopic specimens have prohibitive costs. The circumstance has pushed the development of schemes coupling molecular dynamics simulations at microscale with continuum mechanics approaches, linked with finite-element computations. Delicate questions pertain the matching between the two view; they have been tackled variously (see, e.g., [17, 32, 46, 51, 52]). In short we can roughly distinguish between *partitioned-domain* and *hierarchical* methods (see the book [48] for a comprehensive review). The former class incorporates atomistic (or molecular dynamics) analyses just in critical regions, those in which the presence of defects requires a closer look at the details of local phenomena and related interactions (an example is the analysis of partially fragmented unstable regions around crack tips in presence of cohesive effects). Hierarchical methods adopt the bottom-up approach over the whole body, exploiting microscale discrete analyses to supply constitutive information for the continuum scheme. Cauchy-Born's rule assures the link between the two scales. It refers to a cell (the reservoir of molecular dynamics calculations) which has been considered variously from a statistical viewpoint: an ensemble with isostress-isothermal or isostress-isenthalpic, canonical or microcanonical character (see [1, 25, 38, 40, 45, 50], for the consequences of specific choices on the nature of the cell).

We take also a cell in space and adopt on it a statistical view, but, differently on the quoted approaches, we consider that cell as a grand canonical ensemble for we allow it to exchange mass with the surrounding environment; and we adopted the word *loculus*, instead of the more common *cell*, to remind the circumstance. In fact, standard molecular dynamics methods emerge commonly with the aim of investigating the behavior of solids, while our approach comes from thinking of gases and granular matter. However, our view makes sense also for solids, particularly those suffering diffusion of particles or defects. Also, our approach can be reduced to a more traditional fashion by inserting appropriate constraints, such as the one avoiding mass exchange with the environment, which is not the sole point of difference with proposals already quoted above, because inside the loculus we adopt a refined-scale-type view, by investigating local statistical properties on small neighborhoods of points within the loculus. This is like we have a macroscopic scale, a mesoscopic one (the loculus), and a microscopic one (neighborhoods of places *within* the loculus).

Another difference with common hierarchical methods of molecular dynamics is that we consider a version of the Cauchy-Born rule, which we reinterpret in two-scale sense and write in terms of velocities. Our recourse to a multi-field approach motivates the choice of a two-scale version of the Cauchy-Born rule (we record a single-scale expression of that rule in terms of velocities in [2], while we find a displacement-based two-scale version of it already in [31], although with reference to a different setting, but always multi-field; extended analyses of the rule in its traditional single-scale expression in terms of displacements are in [18–20, 47]).

In the next section we summarize and refine the peculiar features of our approach (further details are in references [4–12]). In Sect. 3, we identify actions at continuum scale in terms of intermolecular or molecule-environment interactions. We discuss in Sect. 4 boundary conditions pertinent to our continuum modeling: this is a particularly delicate issue because it involves the need of considering the way in which we model the environment. Eventually, we dedicate Sect. 5 to the discussion of possible scalar and tensor definition of temperature on statistical basis—our approach does not rest on the assumption of a definite temperature as, e.g., in the isostress-isothermal approach to molecular dynamics. Additional remarks appear in the last section.

## 2 Basic Definitions and Derived Properties

We consider the point space $\mathcal{E}$—which we take to be three-dimensional after defining its translation space and assigning it Euclidean structure—as an ambient for the cluster of flowing molecules constituting the body under analysis here. In $\mathcal{E}$, the coarse image of the body that we consider at an instant $\tau$ is a bounded, arcwise connected region $\mathcal{B}_\tau$, endowed with surface-like smooth boundary, oriented by the normal $n$. We attribute to every point $x \in \mathcal{B}_\tau$ statistical entities constructed by analyzing the population of molecules within a loculus. We write $\mathsf{e}(x)$ to indicate the loculus pertaining to $x$ among all (with the same shape but potentially different content of molecules) possible loculi, at the instant $\tau$. $\mathsf{e}(x)$ is a cube with side length $\delta$, a value to be prescribed as a multiple of the mean free path in the flowing cluster considered, to allow the possible presence of several colliding molecules within $\mathsf{e}(x)$. The cluster in $\mathsf{e}(x)$ should be so populous to assure the possibility of statistical estimates beyond averages.

We carry on such estimates by using a distribution $\Theta(\tau, x; y, w)$, which furnishes at instant $\tau$ the number of molecules per unit volume in $\mathsf{e}(x)$ with velocities within the infinitesimal ball $(w + \mathrm{d}w)$, which are around a point, indicated by the vector $y$ issued by $x$, in a neighborhood of it, say a ball with radius $\mathrm{d}y$. Notation $\mathrm{d}w$ indicates the vector measure in *the real vector space* $\mathcal{V}$ of velocities. By assumption $\Theta(\tau, x; y, w)$ is sufficiently regular to justify the pertinent analyses.

In what follow, for the sake of simplicity, we refer our analyses to flat metrics in the ambient space.

Our multi-scale view has a structure resting on a trivial bundle with basis $\mathcal{E}$ and typical fiber $\mathcal{E}_x$, where the subscript $x$ indicates that $\mathcal{E}_x$ contains $\mathsf{e}(x)$. In other words, we attach at every point $x$ of $\mathcal{E}$ a copy $\mathcal{E}_x$ of $\mathcal{E}$ itself. The choice allows us to point out that we consider distinguished and non-overlapping loculi. It formalizes the naive idea of considering two spatial scales, one larger than $\delta$, the other smaller. However, although we find it convenient to exploit such a structure, events occur just in the physical space. To account for this aspect, *when we impose on some entity independence of the observer, rotation tensors entering the requirement are the same at both scales.*

### 2.1 Gross Fields

The total number $\omega(\tau, x)$ of molecules in $\mathsf{e}(x)$ at time $\tau$ is given by

$$\omega(\tau, x) := \int_{\mathsf{e}(x)} \int_{\mathcal{V}} \Theta(\tau, x; y, w)\, \mathrm{d}w\, \mathrm{d}y.$$

The averages

$$w_*(\tau, x; y) := \theta^{-1} \int_{\mathcal{V}} w \Theta(\tau, x; y, w)\, \mathrm{d}w$$

and

$$v(\tau, x) := \delta^{-3} \int_{\mathsf{e}(x)} \theta w_*(\tau, x; y)\, \mathrm{d}y,$$

where

$$\theta(\tau, x; y) := \int_{\mathcal{V}} \Theta(\tau, x; y, w)\, dw,$$

play essential roles in the representation of motion that we propose here. The vector $w_*$ is the average velocity around $y$ in $\mathsf{e}(x)$, while $v$ is the average velocity of the whole set of grains within $\mathsf{e}(x)$. We decompose $w_*$ into the sum

$$w_*(\tau, x; y) = v(\tau, x) + B(\tau, x)y + c(\tau, x; y),$$

where $B(\tau, x)$ is *second-rank tensor* and $c(\tau, x; y)$ a *vector*. The product $By$ defines an affine motion within $\mathsf{e}(x)$, while $c$ is the fluctuation of $w_*$ at $y$, with respect to $v + By$.

We define $B$ to be such that

$$K = YB^T \tag{1}$$

(the superscript $T$ indicates transposition), with

$$Y(\tau, x) := \omega^{-1} \int_{\mathsf{e}(x)} \theta y \otimes y\, dy \quad \text{and} \quad K(\tau, x) := \omega^{-1} \int_{\mathsf{e}(x)} \theta y \otimes (w_* - v)\, dy.$$

$Y$ is the instantaneous inertia binor of the cluster of molecules in $\mathsf{e}(x)$. $K$ is the instantaneous tensor moment of momentum per unit mass, again for all molecules in $\mathsf{e}(x)$. The definition of $K$ imposes

$$\int_{\mathsf{e}(x)} \theta y \otimes c\, dy = 0 \quad \text{and} \quad \int_{\mathsf{e}(x)} \theta c\, dy = 0.$$

We use the symbol $L$ to indicate the spatial gradient of the averaged velocity $v$. The difference $\sigma := \operatorname{tr} L - \operatorname{tr} B$ measures the instantaneous relative rate of change in volume across the scale determined by $\delta$.

We also define a linear mapping $G$ to be solution to the differential equation

$$\dot{G} + \frac{\sigma}{2} G = BG. \tag{2}$$

We interpret $G$ as the gradient of a deformation inside $\mathsf{e}(x)$, which links the current binor $Y$ of $\mathsf{e}(x)$ to a binor $Y_\star$ of a reference (although fictitious) arrangement of molecules (see [11, 12]). Fixed $\sigma$ and $B$, once we know $G$, the binor $Y_\star$ is such that $Y = GY_* G^T$. The choice implies

$$\dot{Y} = \left(BG - \frac{\sigma}{2} G\right) Y_* G^T + GY_* \left(BG - \frac{\sigma}{2} G\right)^T = BY + YB^T - \sigma Y. \tag{3}$$

Finally, we define a Reynolds tensor $H$ associated with fluctuations by

$$H(\tau, x) := \omega^{-1} \int_{\mathsf{e}(x)} \theta c \otimes c\, dy.$$

## 2.2 Objective Time Derivative

The rates $v$ and $B$ refer to phenomena at different spatial scales, distinguished by $\delta$ but interacting. To account for the interaction, we choose to define a time derivative following the affine motion around $x$ with rate $B$. We denote it by a superposed circle. For $A$ a second-rank contravariant tensor with both components in the current ambient space, as $Y$, $K$, and $H$ are, according to reference [30], we define $\mathring{A}$ by

$$\mathring{A} := \dot{A} - BA - AB^T + \sigma A, \tag{4}$$

where $\dot{A} := \frac{\partial A}{\partial \tau} + (\nabla A)v$. When $\sigma = 0$, $\mathring{A}$ is the Oldroyd derivative of $A$.

We compute

$$\mathring{Y} := \dot{Y} - BY - YB^T + \sigma Y = 0,$$

so that the *inertia tensor* $\mathcal{I} = I \operatorname{tr} Y - Y$ of $\mathsf{e}$, with $I$ the second-rank identity, vanishes in the flat metric space considered here.

For a linear operator like $G$, which has only one component in the current ambient space, the other belonging to the reference space where we place the loculus with inertia binor $Y_\star$, we write $\mathring{G}$ as

$$\mathring{G} = \dot{G} - BG + \frac{\sigma}{2} G,$$

while, for $v$, we have

$$\mathring{v} = \dot{v} + \frac{\sigma}{2} v.$$

These derivatives emerge on the basis of a local integration backwards in time of both $v$ and $By$, as shown in references [11] and [12].

## 2.3 Local Form of the Balance of Mass

We assume that all the molecules have mass $\mu$ each. At continuum scale, we define the density of mass, $\rho$, at $x$ and time $\tau$, to be $\rho(\tau, x) := \mu\delta^{-3}\omega(\tau, x)$; it is a global quantity for $\mathsf{e}(x)$; its time variation depends on the change of mass within the loculus, i.e., $\dot{\rho} = -\rho \operatorname{tr} B$ because $\operatorname{tr} B$ indicates the rate of change in volume of the molecular cluster in $\mathsf{e}(x)$, due to the affine velocity $By$. The algebraic sign minus refers to the inflow of mass. The previous equation can be written also as

$$\mathring{\rho} + \rho \operatorname{div} v = 0, \tag{5}$$

where $\mathring{\rho} = \dot{\rho} - \sigma\rho$ (see [11] and [12] for further remarks and related proofs).

## 2.4 Observers and Rules of Their Changes

Let $\mathcal{O}$ and $\mathcal{O}'$ be two observers (each a reference frame) in physical space, differing one another by a rigid-body motion. $\mathcal{O}$ records the velocity $v$, while $\mathcal{O}'$ measures $v'$. When we rotate-back the velocity $v'$ into the frame $\mathcal{O}$, the resulting vector $v^\diamond$ is given by

$$v^\diamond = v + \hat{\mathsf{c}} + \hat{\mathsf{W}}(x - x_0), \tag{6}$$

with $x_0$ an arbitrary point in space, $\hat{\mathsf{c}}$ a (relative) translational velocity, and $\hat{\mathsf{W}}$ a second-rank skew-symmetric tensor of the relative rotational speed.

By indicating by $B^\diamond$ the counterpart of $B$, with meaning analogous to $v^\diamond$, we consider a rule of changes in observers given by

$$B^\diamond = B + \hat{\mathsf{W}}. \tag{7}$$

$c$ is observer independent, being a relative velocity, inside the loculus.

We also assume objectivity for $\Theta$.

## 2.5 External Power, Invariance, and Balance

We call a part $\mathrm{b}_\tau$ of $\mathcal{B}_\tau$ any inner its subset with the same regularity and non-null volume measure. With reference to a generic part, we call a power of the external actions over $\mathrm{b}_\tau$ the functional $\mathcal{P}^{ext}_{\mathrm{b}_\tau}(v, B)$ defined by

$$\mathcal{P}^{ext}_{\mathrm{b}_\tau}(v, B) := \int_{\mathrm{b}_\tau} \left(\rho b^\ddagger \cdot v + \rho O^\ddagger \cdot B\right) \mathrm{d}x + \int_{\partial \mathrm{b}_\tau} (\mathsf{t} \cdot v + \hat{\mathsf{M}} \cdot B)\, \mathrm{d}\mathcal{H}^2(x), \tag{8}$$

where $\mathrm{d}\mathcal{H}^2(x)$ is the surface measure over $\partial\mathrm{b}_\tau$. Thinking of the events at a spatial scale below $\delta$, in writing the second integral above, we are imagining that loculi pertaining to points on $\partial\mathrm{b}_\tau$ include just molecules belonging to $\mathrm{b}_\tau$ at the instant $\tau$. The definition of $\mathcal{P}^{ext}_{\mathrm{b}_\tau}(v, B)$ involves bulk and surface actions conjugated with the kinematic mechanisms associated with $v$ and $B$. By assumption, the volume actions $b^\ddagger$ and $O^\ddagger$, a vector and a second-rank tensor, respectively, both depending on space and time, admit additive decomposition into inertial (indicated by a superscript *in*) and non-inertial components: $b^\ddagger = b + b^{in}$ and $O^\ddagger = O + O^{in}$. Also, besides space and time, we assume that the contact actions $\mathsf{t}$ and $\hat{\mathsf{M}}$, vector and second-rank tensor, respectively, depend on the normal $n$ to the boundary $\partial\mathrm{b}_\tau$, i.e., $\mathsf{t} = \tilde{\mathsf{t}}(\tau, x, n)$ and $\hat{\mathsf{M}} = \tilde{\mathsf{M}}(\tau, x, n)$.

**Theorem 1** *If we assume, as a foundational axiom, that $\mathcal{P}^{ext}_{\mathrm{b}_\tau}(v, B)$ is invariant under the rigid-body-type class of changes in observers described by* (6) *and* (7), *i.e.,*

$$\mathcal{P}^{ext}_{\mathrm{b}_\tau}\left(v^\diamond, B^\diamond\right) = \mathcal{P}^{ext}_{\mathrm{b}_\tau}(v, B)$$

*for any choice of $\hat{\mathsf{c}}$ and $\hat{\mathsf{W}}$, and the part $\mathrm{b}_\tau$, the following propositions hold true:*

- *If $\sup_{x\in\mathcal{B}_\tau} |\rho b^\ddagger| < +\infty$ and $\tilde{\mathsf{t}}(\tau, \cdot, n)$ is continuous, the contact action $\mathsf{t}$ satisfies the action-reaction principle*

$$\tilde{\mathsf{t}}(\tau, x, n) = -\tilde{\mathsf{t}}(\tau, x, -n) \tag{9}$$

  *and is linear in the normal $n$, i.e.,*

$$\tilde{\mathsf{t}}(\tau, x, n) = T(\tau, x)n \tag{10}$$

- *If the bulk action $\rho b^\ddagger$ is continuous in space and the second-rank tensor stress field $T$ is $C^1(\mathcal{B}_\tau) \cap C(\bar{\mathcal{B}}_\tau)$, the following pointwise balance holds:*

$$\rho b^\ddagger + \operatorname{div} T = 0. \tag{11}$$

– *Under the boundedness of the bulk actions assumed above, if* $\sup_{x \in \mathcal{B}_\tau} |\rho O^{\ddagger}| < +\infty$ *and* $\hat{\mathsf{M}}(\tau, \cdot, n)$ *is continuous in space, the contact action* $\hat{\mathsf{M}}$ *satisfies the action-reaction principle*

$$\mathrm{skw}\big(\tilde{\mathsf{M}}(\tau, x, n) - \tilde{\mathsf{M}}(\tau, x, -n)\big) = 0, \tag{12}$$

*and is linear in the normal* $n$*, i.e.,*

$$\tilde{\mathsf{M}}(\tau, x, n) = \mathbf{m}(\tau, x)n. \tag{13}$$

– *In addition, if* $\rho O^{\ddagger}$ *is continuous in space and the third-rank tensor stress field* $\mathbf{m}$ *is* $C^1(\mathcal{B}_\tau) \cap C(\bar{\mathcal{B}}_\tau)$*, there exists a second-rank tensor field* $A$*, depending on space and time, such that*

$$\mathrm{skw} T = \mathrm{skw} A, \tag{14}$$

*and*

$$\rho O^{\ddagger} - A + \mathrm{div}\, \mathbf{m} = 0. \tag{15}$$

The proof of this theorem is in reference [12]. We could also reduce the assumptions. In fact, the boundedness of bulk actions would allow us to prove that the contact ones depend on the oriented boundary just through the normal, by using a Hamel-Noll-like argument, rather than assuming such a dependence, as we do here for the sake of conciseness. As a matter of fact, the proof of this theorem rests on the assumed invariance of the *external power alone* under changes in observers that are isometric in the ambient space where we place the gross configuration of the body. This view is an extension to multi-field theories of the analogous proposal by Noll for simple bodies [37]. The extension for continua with microstructure (in the multi-field setting) has been proposed first by Capriz and Mazzini in 1998 [13], restricted to the case in which the special orthogonal group $SO(3)$ may act directly on the manifold of microstructural shapes. In 2002 [29], independently, Mariano proposed an extension overcoming that restriction (see also [30]).

To identify the inertial components of the bulk actions, we refer to a reduced tensor kinetic energy per unit mass, $\mathcal{W}_*$, defined by

$$\mathcal{W}_* = \frac{1}{2\omega} \int_{\mathsf{e}(x)} \theta w_\star \otimes w_\star \, \mathrm{d}y.$$

Instead of involving the true molecular velocity $w$, the tensor $\mathcal{W}_*$ refers only to the average $w_*$ of $w$ over $\mathcal{V}$. By using the decompositions $w_* = v + By + c$, we rewrite $\mathcal{W}_*$ as

$$\mathcal{W}_* = \frac{1}{2}\big(v \otimes v + BYB^T + H\big).$$

Then, we *postulate* the following tensor inertial balance [12]:

$$\int_{\mathfrak{b}_\tau} \rho \mathring{\mathcal{W}}_* \, \mathrm{d}x + \int_{\mathfrak{b}_\tau} \rho \mathrm{sym}\big(v \otimes b^{in} + BO^{in}\big) \, \mathrm{d}x$$
$$- \frac{1}{2} \int_{\mathfrak{b}_\tau} (\rho J - Z) \, \mathrm{d}x - \frac{1}{2} \int_{\partial \mathfrak{b}_\tau} \mathfrak{J} \, \mathrm{d}\mathcal{H}^2(x) = 0, \tag{16}$$

where the symmetry of $\mathcal{W}_*$ suggests to choose the symmetric component of the tensor power density $v \otimes b^{in} + BO^{in}$; we add bulk sources per unit volume, $\rho J$ and $Z$, due to the environment and events inside the loculus, respectively, and fluxes $\hat{\mathtt{J}}$, all compensating the time variation of $H$, i.e., tensor kinetic energy fluctuations with respect to the motion described by $v + By$.

We compute

$$\mathring{\mathcal{W}}_* = \mathrm{sym}(v \otimes \mathring{v}) + \mathrm{sym}(B\mathring{K}) + \frac{1}{2}\dot{H} - \mathrm{sym}(BH) + \frac{1}{2}\sigma H,$$

where the last three addenda constitute $\mathring{H}$ (recall the symmetry of $H$) [12].

We assume that the inertial balance (16) holds for any choice of $\mathrm{b}_\tau$ and the rate fields $v$ and $B$. The identifications

$$\rho b^{in} = -\rho\dot{v} \quad \text{and} \quad \rho O^{in} = -\rho\mathring{K} + \rho H$$

are compatible with it; the integral balance

$$\int_{\mathrm{b}_\tau} \rho(\dot{H} + \sigma H)\,\mathrm{d}x - \int_{\mathrm{b}_\tau} (\rho J - Z)\,\mathrm{d}x - \int_{\partial\mathrm{b}_\tau} \hat{\mathtt{J}}\,\mathrm{d}\mathcal{H}^2(x) = 0,$$

remains. We characterize its elements by further assumptions:

(1) The flux $\hat{\mathtt{J}}$ depends on the normal $n$ to $\partial\mathrm{b}_\tau$, besides space and time: $\hat{\mathtt{J}} = \tilde{\mathtt{J}}(\tau, x, n)$.
(2) Bulk actions determined by the fluctuations inside the loculus are bounded: $\sup_{x\in\mathcal{B}_\tau} |\rho(\dot{H} + \sigma H) - (\rho J - Z)| < +\infty$.
(3) $\tilde{\mathtt{J}}(\tau, \cdot, n)$ is continuous; so are the other integrands.

Assumption (1) could be derived from the previous integral balance and the presumed boundedness of bulk actions, through a Hamel-Noll-like argument. However, we accept it here as an assumption for the sake of conciseness. Also, bulk action boundedness and flux continuity imply for $\hat{\mathtt{J}}$ an action-reaction property: $\tilde{\mathtt{J}}(\tau, x, n) = -\tilde{\mathtt{J}}(\tau, x, -n)$. Such a result and the standard tetrahedron argument imply the linearity of $\hat{\mathtt{J}}$ with respect to $n$: $\hat{\mathtt{J}} := \tilde{\mathtt{J}}(\tau, x, n) = \tilde{\mathbf{j}}(\tau, x)n$, with $\mathbf{j} = \tilde{\mathbf{j}}(\tau, x)$ the value of a third-rank tensor field. Finally, when $\rho\dot{H} - (\rho J - Z)$ is continuous in space and $\tilde{\mathbf{j}}(\tau, \cdot)$ is $C^1(\mathcal{B}_\tau) \cap C(\bar{\mathcal{B}}_\tau)$, we get the pointwise balance

$$\rho(\dot{H} + \sigma H) = \rho J - Z + \mathrm{div}\,\mathbf{j}. \tag{17}$$

The weak version of the resulting integral balances reads

$$\begin{aligned}\int_{\mathrm{b}_\tau} \rho\mathring{\mathcal{W}}_*\,\mathrm{d}x = {} & \int_{\mathrm{b}_\tau} \rho\left(\mathrm{sym}(v \otimes b + BM) + \frac{1}{2}J\right)\mathrm{d}x \\ & + \int_{\partial\mathrm{b}_\tau} \left(\mathrm{sym}\big(v \otimes Tn + \mathbf{m}(Bn)\big) + \frac{1}{2}\mathbf{j}n\right)\mathrm{d}\mathcal{H}^2(x) \\ & - \int_{\mathrm{b}_\tau} \left(\mathrm{sym}\big(LT^T + BA + \mathbf{b}\mathbf{m}^T\big) + \frac{1}{2}Z\right)\mathrm{d}x,\end{aligned}$$

where $\mathbf{b} := \nabla B$ and the superscript $T$ indicates the major transposition for the third-rank tensor $\mathbf{m}$ (an obvious misprint on the transposition is in reference [12]). The term

$$\mathrm{sym}\big(LT^T + BA + \mathbf{b}\mathbf{m}^T\big) + \frac{1}{2}Z \tag{18}$$

indicates a tensor power density of internal actions.

### 2.6 A Prominent Constrained Motion

When the dynamics (in particular fluctuations with respect to the average velocity) do not impose discrepancy between the backward integration of $v$ and $By$, so that we get $F = G$ or $B = L$, for which $\sigma$ vanishes, we may treat the identity between the two gradients as an internal constraint. Consequently, by following the traditional approach to constraints, we suppose that all measures of stresses be sums of an active term, specified by a constitutive rule and indicated below by a subscript $a$, and a reactive term. All such latter terms are collectively such as to lead to a vanishing power for all straining permitted by the constraint. Balance equations result reduced to

$$\frac{\partial \rho}{\partial \tau} + \operatorname{div}(\rho v) = 0,$$

$$\rho\left(\frac{\partial v}{\partial \tau} + Lv\right) + \operatorname{div}\left(\rho\left(H - \left(\frac{\partial L}{\partial \tau} + (\nabla L)v\right)Y\right)\right)$$

$$= \rho b - \operatorname{div}(\rho O^T) + \operatorname{div}\left(\operatorname{sym}(T_a + A_a) - (\operatorname{div} \mathbf{m}_a)^T\right),$$

$$\rho\left(\frac{\partial H}{\partial \tau} + (\nabla H)v\right) = \rho J$$

(see reference [10] for the pertinent proofs). Actually, the active components $T_a$ and $A_a$ act together as a sum, and $\mathbf{m}_a$ plays the role of an hyperstress of the type appearing in strain-gradient continuous approaches.

## 3 Actions at Continuum Scale in Terms of Molecular Interactions

### 3.1 Stresses and Self-actions

The stress tensors introduced above and the companion self-actions (namely $A$ and $Z$) represent interactions at continuum scale.

In particular, the standard stress tensor, $T(x, \tau)$, represents at macroscopic scale (the one beyond $\delta$) interactions between neighboring loculi, due to the spatial variation of the average velocity $v$.

The tensor self-action, $A(x, \tau)$, describes interactions among particles within the loculus at $x$, not accounting for the activity in the neighboring loculi.

The (third-rank) microstress tensor, $\mathbf{m}(x, \tau)$, represents the interactions between neighboring loculi, due to variations of the pertinent affine motion.

To identify them from the presumed knowledge of molecular interactions, first we construct the discrete internal interaction tensor power $P_{dis}^{int}(\tau, x)$ within the loculus and across neighboring loculi in terms of molecule-to-molecule interactions. Then, we identify such an expression with its continuum counterpart, requiring that *the identification should hold for any choice of the velocity fields involved.*

- We consider only pairwise interactions.
- A sufficiently smooth vector function $\tilde{\mathfrak{f}}$, with values

$$\mathfrak{f} := \tilde{\mathfrak{f}}(\tau, x; y, w, y', w'),$$

represents the *interaction*, at the instant $\tau$, within the loculus $\mathsf{e}(x)(x)$, between a molecule at $y$ and one at $y'$, characterized by velocities $w$ and $w'$, respectively. We assume

$$\tilde{\mathsf{f}}(\tau, x; y, w, y, w) = 0 \tag{19}$$

to exclude *molecular self-interactions*. Objectivity requirements may restrict further the dependence of $\tilde{\mathsf{f}}$ on its entries. Here, we do not explore such an aspect because our aim is just to render explicit in terms of molecular actions the way we consider $T$, $A$, $\mathbf{m}$, and $Z$ at continuum scale.

– Although in our molecular approach we consider the molecules as mass points, *in reality they are extended in space* and if we say that a molecule is at $y$ in $\mathsf{e}$, we refer to the placement of mass center. To account indirectly for the molecular radius, otherwise not included in our modeling, and the possibility of having interactions at *short* distance (as already mentioned), we include into the averages presented below an influence function, $f$, which we write both as $f(y, y')$ and $f(x, x')$, considering *the same structure* for it at both scales involved. Such a function is at least integrable; in any case, we consider it as smooth as necessary in the following developments. Its choice is a constitutive ingredient and has to be determined time to time in every specific application. In any case, however, we assume that it is compactly supported and, fixed $x$, its support must be such that, all points in it satisfy the relations

$$v(\tau, x') - v(\tau, x) = L(x' - x)$$

and

$$B(\tau, x')y - B(\tau, x)y = (\mathbf{b}y)(x' - x).$$

Of course, when the body undergoes homogeneous macroscopic straining, all $x'$ different from $x$ satisfy at least the first condition above. However, this does not mean that the domain of $f$ might cover the entire body. It is limited to a *small* compact neighborhood of $x$, the adjective taking meaning—we repeat—in each specific circumstance (here, we essentially speak about a meta-model, in fact, so that we do not go, if not just marginally, into constitutive questions). In any case, we consider the two conditions above as a version of the Cauchy-Born rule re-interpreted in two-scale sense and written in terms of velocities.

By averaging velocities, we get

$$\bar{\mathsf{f}}(\tau, x; y, y') := \theta^{-2} \int_{\mathcal{V}} \Theta(\tau, x; y, w) \int_{\mathcal{V}} \Theta(\tau, x; y', w') \mathsf{f}(\tau, x; y, w, y', w') \, \mathrm{d}w' \, \mathrm{d}w.$$

Then, a space average over $\mathsf{e}(x)$ allows us to define

$$\mathsf{f}_*(\tau, x; y) := \delta^{-3} \int_{\mathsf{e}(x)} \theta f(y, y') \bar{\mathsf{f}}(\tau, x; y, y') \, \mathrm{d}y',$$

which is the force on $y$, within $\mathsf{e}(x)$ at the instant $\tau$, exerted by the molecules in a small neighborhood defined by $f(y, y')$. Its tensor power performed on $w_*$ is given by

$$\begin{aligned} \delta^{-3} \int_{\mathsf{e}(x)} \theta w_* \otimes \mathsf{f}_* \, \mathrm{d}y &= v \otimes \delta^{-3} \int_{\mathsf{e}(x)} \theta \mathsf{f}_* \, \mathrm{d}y + \delta^{-3} B \int_{\mathsf{e}(x)} \theta y \otimes \mathsf{f}_* \, \mathrm{d}y + \delta^{-3} \int_{\mathsf{e}(x)} \theta c \otimes \mathsf{f}_* \, \mathrm{d}y \\ &= v \otimes \bar{\mathsf{t}}(\tau, x) + \delta^{-3} B \int_{\mathsf{e}(x)} \theta y \otimes \mathsf{f}_* \, \mathrm{d}y + \delta^{-3} \int_{\mathsf{e}(x)} \theta c \otimes \mathsf{f}_* \, \mathrm{d}y, \end{aligned}$$

where

$$\bar{\mathsf{t}}(\tau, x) := \delta^{-3} \int_{\mathsf{e}(x)} \theta \mathsf{f}_*(\tau, x; y)\, \mathrm{d}y.$$

Such a tensor power refers to $\mathsf{e}(x)$. It does not include interactions among neighboring loculi. To account for them with a view towards the aim of constructing the discrete analog of the continuum version (18) of the inner power, we consider two further vector fields with values

$$\mathsf{t} := \tilde{\mathsf{t}}(\tau, x, x')$$

and

$$\mathsf{k} := \tilde{\mathsf{k}}(\tau, x, x'; y).$$

$\tilde{\mathsf{t}}$ may depend on $v$ and $v'$, i.e.,

$$\tilde{\mathsf{t}}(\tau, x, x') = \hat{\mathsf{t}}(\tau, x, x', v(\tau, x), v'(\tau, x')),$$

so it can be also for $\tilde{\mathsf{k}}$, although we leave implicit such a dependence for the sake of conciseness. $\tilde{\mathsf{t}}$ indicates interaction at macroscopic scale between two neighboring loculi, due to possible variations of the averaged velocity $v$ from $x$ to $x'$, and is such that

$$\tilde{\mathsf{t}}(\tau, x, x) = \bar{\mathsf{t}}(\tau, x).$$

$\mathsf{k}$ represents inter-loculus interactions due to the variation of $B$ from loculus to loculus. We assume

$$\tilde{\mathsf{k}}(\tau, x, x; y) = 0.$$

The sum

$$\tilde{\mathsf{t}}(\tau, x, x') + \int_{\mathsf{e}(x)} \theta(\tau, x, x'; y) \tilde{\mathsf{k}}(\tau, x, x'; y)\, \mathrm{d}y$$

is the overall interaction between $\mathsf{e}(x)$ and $\mathsf{e}(x')$.

We write the discrete internal interaction tensor power $P_{dis}^{int}$ as

$$P_{dis}^{int}(\tau, x) := P_{dis[x]}^{int}(\tau, x) + P_{dis[B]}^{int}(\tau, x) + \delta^{-3} B \int_{\mathsf{e}(x)} \theta y \otimes \mathsf{f}_* \,\mathrm{d}y + \delta^{-3} \int_{\mathsf{e}(x)} \theta c \otimes \mathsf{f}_* \,\mathrm{d}y$$

where

$$\begin{aligned} P_{dis[x]}^{int}(\tau, x) &:= \left|\mathrm{supp}_x f(x, x')\right|^{-1} \int_{\mathcal{B}} f(x, x')(v' - v) \otimes \tilde{\mathsf{t}}(\tau, x, x')\, \mathrm{d}x' \\ &= \left|\mathrm{supp}_x f(x, x')\right|^{-1} L \int_{\mathcal{B}} f(x, x')(x' - x) \otimes \tilde{\mathsf{t}}(\tau, x, x')\, \mathrm{d}x', \end{aligned}$$

with $\mathrm{supp}_x f(x, x')$ the support of $f(x, \cdot)$ around $x$, and

$$P_{dis[B]}^{int}(\tau, x) := \left|\mathrm{supp}_x f(x, x')\right|^{-1} \nabla B \int_{\mathcal{B}} f(x, x') \int_{\mathsf{e}(x)} (x' - x) \otimes y \otimes \tilde{\mathsf{k}}(\tau, x, x'; y)\, \mathrm{d}y \mathrm{d}x'.$$

Then, we *impose* as an axiom the identity

$$P_{dis}^{int} = LT^T + BA + \mathbf{b}\mathbf{m}^T + \frac{1}{2}Z,$$

presuming that *it holds for any choice* of $L$, $B$, and $\nabla B$. As a consequence, we find

$$T(\tau, x) = \left|\mathrm{supp}_x f(x, x')\right|^{-1} \int_{\mathcal{B}} f(x, x')\tilde{\mathsf{t}}(\tau, x, x') \otimes (x' - x)\,\mathrm{d}x',$$

$$\mathbf{m}(\tau, x) = \left|\mathrm{supp}_x f(x, x')\right|^{-1} \int_{\mathcal{B}} f(x, x') \int_{\mathsf{e}(x)} \tilde{\mathsf{k}}(\tau, x, x'; y) \otimes y \otimes (x' - x)\,\mathrm{d}y\mathrm{d}x',$$

$$A(\tau, x) = \delta^{-3} \int_{\mathsf{e}(x)} \theta y \otimes \mathsf{f}_*(\tau, x; y)\,\mathrm{d}y,$$

$$Z(\tau, x) = 2\delta^{-3} \int_{\mathsf{e}(x)} \theta c \otimes \mathsf{f}_*(\tau, x; y)\,\mathrm{d}y.$$

The results in Theorem 1 impose also that $\mathsf{t} = \tilde{\mathsf{t}}(\tau, x, x')$ and $\mathsf{f}_*(\tau, x, x')$ be such that

$$\mathrm{skw}\left(\int_{\mathcal{B}} f(x, x')\tilde{\mathsf{t}}(\tau, x, x') \otimes (x' - x)\,\mathrm{d}x'\right) = \mathrm{skw}\left(\int_{\mathsf{e}(x)} \theta y \otimes \mathsf{f}_*(\tau, x; y)\,\mathrm{d}y\right),$$

which is obviously satisfied at least when $(x' - x) = y$, $\mathsf{t} = \mathsf{f}_*$, $\mathrm{supp}\, f(x, x') \cap \mathcal{B} = \mathsf{e}(x)$, and $f(x, x')$ is the indicator function of $\mathsf{e}(x)$, i.e., when we reduce ourselves to a single-scale representation.

Moreover, if we would consider a jammed cluster of grains at rest, focusing our attention on a grain with center of mass at $x$, a grain in contact with those around it, and indicate by $\mathsf{t}$ the contact force, in single-scale representation, the expression of $T$ obtained above, reduces to the one commonly used in the analysis of granular packing (see, e.g., references [3, 23, 24, 44]). In this case, the influence function is just the indicator of a region referring to the grains in contact with the one with center of mass at $x$; the difference $x' - x$ refers to the centers of mass pertaining to the grains in contact; the integral reduces to a sum over all contacts in the volume considered. In particular, when the setting is two-dimensional, $f(x, x')$ becomes the indicator function of Edwards' quadrons (see, e.g., [3]).

## 3.2 External Actions

We write

$$\ell := \tilde{\ell}(\tau, x; y, w)$$

for a possible external action on a molecule crossing $y$ in $\mathsf{e}(x)$ with velocity $w$. By assumption, as a function of $w$, such a covector-valued map (identified with a vector because we refer to orthonormal frames of reference) has sufficient regularity for the related developments listed below. Its average over the space of velocities is then

$$\ell_* := \tilde{\ell}_*(\tau, x; y) := \theta^{-1} \int_{\mathsf{e}(x)} \Theta(\tau, x; y, w)\tilde{\ell}(\tau, x; y, w)\,\mathrm{d}w.$$

By recalling the tensor inertial balance (16), we postulate that the power density $w_* \otimes \ell_*(\tau, x; y)$, integrated over the loculus, equals the power of non-inertial bulk actions $b$

and $O$, supplemented by the external source and fluxes appearing in the balance (17) *for any choice of the velocity fields involved*, namely

$$\delta^{-3}\mu\omega(\tau,x)\int_{\mathsf{e}(x)}\theta w_* \otimes \ell_* \,\mathrm{d}y = v\otimes\rho b + \rho B O + \frac{1}{2}(\rho J + \operatorname{div}\mathbf{j}) \tag{20}$$

We compute

$$w_* \otimes \ell_* = v\otimes \ell_* + By\otimes \ell_* + c\otimes \ell_*.$$

Then, the arbitrariness of $v$ and $B$ implies from the assumed balance (20) the following identities

$$\rho b = \delta^{-3}\mu\omega(\tau,x)\int_{\mathsf{e}(x)}\theta\tilde{\ell}_*(\tau,x;y)\,\mathrm{d}y,$$

$$\rho O = \delta^{-3}\mu\omega(\tau,x)\int_{\mathsf{e}(x)}\theta y\otimes\tilde{\ell}_*(\tau,x;y)\,\mathrm{d}y,$$

$$\rho J + \operatorname{div}\mathbf{j} = 2\delta^{-3}\mu\omega(\tau,x)\int_{\mathsf{e}(x)}\theta c\otimes\tilde{\ell}_*(\tau,x;y)\,\mathrm{d}y,$$

which clarify further the meaning of bulk actions appearing in the balance equations derived at continuum level from a requirement of invariance for the external power under changes in observers and the validity of a tensor inertial integral balance.

Finally, due to the identifications above, the integral balance (17) imposes

$$\operatorname{skw}\left(\mu\omega(\tau,x)\int_{\mathsf{e}(x)}\theta c\otimes\tilde{\ell}_*(\tau,x;y)\,\mathrm{d}y - 2\int_{\mathsf{e}(x)}\theta c\otimes \mathsf{f}_*(\tau,x;y)\,\mathrm{d}y\right) = 0.$$

## 4 Boundary Conditions

Boundary conditions depend on the way we model the body *and* the environment, i.e., whether the environment is modeled as the body. In traditional issues of continuum mechanics, we implicitly assume that the environment has the same description of the body under scrutiny, so that Dirichlet and Neumann conditions may be naturally prescribed, and the latter may also be nonzero because we may imagine appropriate loading devices. When the environment has a different description from the body, actions at the boundary may be nonzero when they have a counterpart in the description of the environment.

In what we describe here, a question is whether we discuss boundary conditions by looking directly at continuum scale, i.e., taking the balance equations in Sect. 3, or we aim at reconstructing boundary conditions for the macroscopic fields from an analysis of microscopic scales, by considering (let us say) near-wall particle-type interactions. In the latter case, we should think of a container, posing the question of determining near-wall distributions. For a rapid flow of colliding grains, assuming that the distribution be Maxwellian would lead us to closed form classical solutions dating back to Maxwell himself and later re-discussed (see, e.g., [27]). However, such an assumption can be considered just a low order approximation because the presence of a wall containing, e.g., a gas changes the distribution at least in a boundary layer. The analysis of this circumstance can be tackled rather systematically on the basis of a linearized Boltzmann equation, supplemented by a correction of super-Burnett order (a critical review of this method is in reference [21]). However, such an approach is of single-scale-type and actually we are unable to propose a version of it adapted to our multi-scale view. By directly looking at the continuum scheme, we obviously recognize that the boundary $\partial\mathcal{B}_\tau$—which we may assume to be smooth for the sake of simplicity—is an interface between the cluster of molecules considered and the environment. At continuum scale, our approach pictures the cluster in $\mathcal{B}_\tau$ as a complex body described by at least two fields (if we leave $H$ a part). Consequently, if we describe the environment as a simple body, to write Neumann-type boundary conditions (the only conditions that we try to discuss here), we have the problem of writing down the balance of actions across the interface between these two different types of bodies (or better, different descriptions of them). By indicating by a superscript "−" the inner trace of a given field to the boundary $\partial\mathcal{B}_\tau$, oriented by $n$, we could write

$$T^{-}n = \bar{h}$$

for the macroscopic stress, with $\bar{h}$ the value of the trace of the analogous (Cauchy) stress in the environment, applied to $n$. The resulting relation is standard and is commonly simplified from a notational viewpoint to $Tn = \bar{h}$. However, if we want to apply the same reasoning to the third-rank tensor $\mathbf{m}$, in the description of the environment (say a wall), assumed to be a simple body, we do not find any counterpart, which could be, at least ideally, generated by a loading device. Consequently, the only natural thing to be written should be

$$\mathbf{m}^{-}n = 0.$$

A similar reasoning applies also to $\mathbf{j}$, so that we could write

$$\mathbf{j}^{-}n = 0,$$

unless we have reasons to describe the environment with the same scheme adopted for the molecular cluster in $\mathcal{B}_\tau$. However, $\mathbf{j}$ describes a flux of (tensor) kinetic energy associated with fluctuations. In principle, we could have some external actions (e.g., wind on a sand pile) allowing such a flux although we describe the environment as a simple body. In that case, we could imagine a boundary condition of the type

$$\mathbf{j}^{-}n = \hat{\mathbf{h}},$$

with the right-hand side term a prescribed flux. Also, if the molecules would undergo polarization—in that case the present model should be extended to account for the phenomenon—charged walls of a container could also induce a flux of moment of momentum, which could imply a nonzero value of $\mathbf{m}^{-}n$.

In any case, although these considerations could appear reasonable, they are just approximations, depending on the type of interactions among particles and the nature of the wall they meet.

For a gas in a container, e.g., we know that, for $1 - \bar{e}^2 \ll 1$, with $\bar{e}$ the degree of restitution in the collisions, the initial distribution converges rapidly (i.e., after a few collisions) to equilibrium or equilibrium-like distributions. Even in these cases, and also presumably with higher probability when we are beyond them, the common Knudsen layer—or more intricate versions of it in complicated situations—could be considered at a macroscopic continuum scale as something justifying the description of the body boundary as a structured surface, i.e., an interface with its own peculiar energy. In this case the previous (tentative) Neumann-type boundary conditions should include the divergence of a surface stress or microstress (see [16] for the analysis of structured interfaces in complex materials). Here, we limit ourselves just to the indication of the possibility.

Various options are possible, in fact, including those in which we may adopt a description of the environment analogous to the one of the cluster considered, a possibility already mentioned above, if acceptable for some reasons. In essence, something desirable in continuum mechanics would be a general boundary layer theory for interactions, which should include what we know of boundary layers in simple fluids, going beyond that special (although prominent) case (specific problems to be tackled in the possible enterprise are evidenced in some chapters of reference [14]).

The variety of circumstances notwithstanding, once we make a choice, a question is the 'weight' of what we neglect, a weight that we are not always able to evaluate, above all before performing specific computations and comparing them with the results of experiments, which require, to be designed, a preliminary theoretical view on the class of phenomena under analysis.

## 5 From Mechanics to Thermodynamics: Scalar and Tensor Temperatures

### 5.1 Scalar Temperature

We have already discussed in reference [11] a possible notion of temperature, which emerges from the analysis of the molecular motion in the present scheme, by attributing the responsibility of what we macroscopically perceive as thermal phenomena to a fluctuation velocity $\hat{c}$ defined by

$$\hat{c} := w - v(\tau, x) - B(\tau, x)y.$$

However, the (averaged) fluctuation velocity $c := w_* - v(\tau, x) - B(\tau, x)y$ used to define $H$, carries with it at least partial information on $\hat{c}$, for $w_*$ is the average of $w$ on the space $\mathcal{V}$ of velocities. For this reason, we find it appropriate to indicate a possible alternative to our previous proposal in reference [11] (see also [7]), by relating the notion of temperature to the residual fluctuation velocity $\bar{c}$ defined by

$$\bar{c} := \hat{c} - c.$$

We report below the related definition of scalar temperature, which does not diverge from the one in reference [11] for its formal appearance but it involves a different ingredient: $\bar{c}$.

However, before going on, to justify subsequent choices, we find it useful recalling that, for $\Omega$ an open subset of $\mathcal{V}$ and $\mathcal{D}(\Omega)$ the space of $C^\infty$ maps into a linear space, compactly supported over $\Omega$, we define a distribution $\mathcal{T}$ to be a linear form on $\mathcal{D}(\Omega)$ such that $\lim_i \mathcal{T}(\varphi_i) = \varphi$, for any sequence $\{\varphi_i\}$ converging towards $\varphi$ in $\mathcal{D}(\Omega)$. As $\mathcal{T}$ belongs to the dual space of $\mathcal{D}(\Omega)$, namely $\mathcal{D}^*(\Omega)$, we use the common notation $\langle \mathcal{T}, \varphi \rangle$ for the value $\mathcal{T}(\varphi)$. Moreover, if $\mathcal{T}$ is an integrable map on every compact subset of $\Omega$, in short $\mathcal{T} \in L^1_{loc}(\Omega)$, $\mathcal{T}$ defines a distribution through the formula

$$\langle \mathcal{T}, \varphi \rangle = \int_\Omega \mathcal{T}(u) \cdot \varphi(u)\, \mathrm{d}u,$$

with $\mathcal{T}(u)$ belonging to the dual space of $\varphi(u)$.

Such remarks (see [15]) suggest to us to adopt an operational view on $\Theta$, which can be considered depending on $\tau$, $x$, $y$, $\bar{c}$, once we leave the determination of $\upsilon$ and $B$ to the balance equations.

By distinguishing symbols to underline the present view on $\Theta$, leaving unspecified $\tau$, $x$, and $y$ for the sake of conciseness, we write

$$\Theta(u) = \hat{\mathrm{t}} \cdot u,$$

with $\hat{\mathrm{t}}$ an appropriate *covector* (identified with a *vector* because we refer to orthonormal frames of reference) depending on $\tau$, $x$, $y$, and $\bar{c}$.

We assume *isotropy* of $\Theta$ with respect to $\bar{c}$. Such an (arbitrary) assumption implies

$$\hat{\mathrm{t}} := \bar{\alpha}(\tau, x; y, \xi')\bar{c},$$

with $\bar{\alpha}$ a scalar-valued function, assumed to have a sufficiently smooth dependence on its entries, among which we include by assumption the residual fluctuation molecular kinetic energy

$$\xi' := \frac{1}{2}|\bar{c}|^2$$

per unit mass of the molecules, instead of $\bar{c}$.

With these notations, the residual fluctuation kinetic energy $\kappa$ per unit mass of all molecules in $\mathsf{e}$, is given by

$$\kappa = \omega^{-1} \int_0^{+\infty} \left( \int_{\mathsf{e}} \bar{\alpha}\, \mathrm{d}y \right) \xi' \mathrm{d}\xi'.$$

By defining

$$\xi = \kappa^{-1}\xi',$$

and

$$\alpha = \kappa\omega^{-1} \int_{\mathsf{e}} \bar{\alpha}\, \mathrm{d}y,$$

we find

$$\Theta(u) = \omega\kappa^{-1}\alpha\bar{c} \cdot u, \quad \forall u \in \mathcal{V},$$

with the constraints

$$\int_0^{+\infty} \alpha(\tau, x; \xi)\mathrm{d}\xi = 1, \qquad \int_0^{+\infty} \xi\alpha(\tau, x; \xi)\mathrm{d}\xi = 1. \tag{21}$$

The function $\alpha(\cdot)$ can be interpreted as the probability (frequency of occurrence) of finding in a ball of radius $\mathrm{d}y$ centered at $y \in \mathsf{e}(x)$ molecules with spurious fluctuation kinetic energy $\xi$ per unit molecular mass. Essentially, $\alpha$ is an additional descriptor of the microscopic fluctuations within the loculus. For this reason we consider it to be governed by an independent rule. More precisely, we consider a version of $\alpha(\tau, x, \xi)$ defined by

$$\hat{\alpha}(\tau, x; \xi) := \kappa\alpha(\tau, x; \xi)$$

and presume that its total time variation $D_\tau \hat{\alpha}$ be given by

$$D_\tau \hat{\alpha} = \mathcal{F}(\hat{\alpha}),$$

where the right-hand side term a functional to be specified. In particular, by imagining to refer to Maxwell distribution, we specialize the previous equation by writing

$$\frac{\partial \hat{\alpha}}{\partial \tau} + \frac{\partial \hat{\alpha}}{\partial x} \cdot v = \frac{1}{\tau_*}\left(\beta + \frac{\partial \psi}{\partial \xi} + \operatorname{div} p\right), \tag{22}$$

where $\tau_*$ is a constant with the dimension of time, the scalar $\beta$ a source, $p$ a vector associated with a flux, $\frac{\partial \psi}{\partial \xi}$ the derivative of a potential with respect to $\xi$. The source $\beta$ has to be specified by a constitutive rule, which should determine the energy transfer inducing thermal effects. As for the previous equation, with nonzero right-hand side term, a constant is not a solution, we propose below to take the average of $\alpha$ over its support. A special analysis of Eq. (22) is in our previous work [12]. Here, we just mention that, once we know $\hat{\alpha}$, in the portion of its range where $\ln \hat{\alpha}$ is defined, we can consider as an absolute temperature the scalar $\vartheta$ defined by

$$\vartheta = -\frac{\rho \delta^3}{\omega k_B}\left(\frac{\partial \ln \hat{\alpha}}{\partial \xi}\right)^{-1},$$

with $k_B$ the Boltzmann constant.

We could also select as a putative temperature the scalar $\vartheta_*$ defined by

$$\vartheta_* = \frac{1}{\omega_*}\left(\frac{\ln \alpha(\xi_0) - \ln \alpha(\xi_1)}{\xi_1 - \xi_0}\right)^{-1},$$

where

$$\omega_* = \frac{\omega k_B}{\rho \delta^3},$$

and we have imagined $\hat{\alpha}$ to be supported and monotonous over the interval $[\xi_0, \xi_1]$, where we have averaged the reciprocal of $\vartheta$.

Such a freedom in the possible definitions of (scalar) temperature underlines further the variegate nature of the body of concepts that we call thermodynamics. Some specific examples may give further light on the nature of $\vartheta_*$.

*Example 5.1* Consider the case in which

$$\alpha(\xi) = (1 - \alpha_0)\xi_0^{-1} \exp\bigl(\alpha_0(\xi_0 - \xi_1)\bigr),$$

where $\alpha_0$ satisfies the relation

$$\alpha_0 - 1 = \alpha_0 \xi \left(1 - \exp(\alpha_0 \xi_0)\right)^{-1}.$$

In these conditions, we write

$$\omega \vartheta_* = \xi_0 \left(\ln\left(\frac{\alpha(0)}{\alpha(\xi_0)}\right)\right)^{-1},$$

where, with the specific choices above,

$$\alpha(0) = (1-\alpha_0)\xi_0^{-1}\exp\alpha_0\xi_0$$

and

$$\alpha(\xi_0) = (1-\alpha_0)\xi_0^{-1},$$

so that

$$\omega_*\vartheta_* = \alpha_0^{-1} > 1.$$

*Example 5.2* Let us take

$$\alpha(\xi) = \begin{cases} 0, & \xi < \xi_0, \\ (1-\xi)^{-1}\exp(\frac{\xi_0-\xi}{1-\xi_0}), & \text{otherwise.} \end{cases}$$

We compute

$$\ln\alpha(\xi_0) = -\ln(1-\xi_0)$$

and

$$\ln\alpha(\xi_0) = -\ln(1-\xi_0) + \frac{\xi_0-\xi_1}{1-\xi_0},$$

so that

$$\omega_*\vartheta_* = 1-\xi_0.$$

When we leave a part exponential type distributions, we may find unexpected behaviors, which could limit, in a sense, the proposal of $\theta_*$.

*Example 5.3* Consider Pareto's distribution over the nonnegative real line ($\xi_0 = 0$ and $\xi_1 = +\infty$), i.e., for $m > 2$

$$\alpha_m(\xi) = \frac{(m-1)(m-2)^{m-1}}{(m-2+\xi)^m}.$$

For $m = 3$, we have

$$\alpha_3(\xi) = \frac{2}{(1+\xi)^m},$$

so that

$$\alpha_3(\xi_0) = 2, \qquad \alpha_3(\xi_1) = 0.$$

For $m > 3$, we have

$$\alpha_m(\xi_0) = \frac{m-1}{m-2}, \qquad \lim_{\xi_1\to+\infty}\alpha_m = 0.$$

Consequently, we find

$$\theta_* \to +\infty.$$

## 5.2 Tensorial Temperature

Other options for a definition of temperature are possible, even with tensorial character.

For example, we find in literature pertinent notions of scalar and tensor temperatures for force and torque balanced granular clusters, jammed in a container with volume $V$, after shaking it at a given frequency and amplitude for a given interval of time (see, e.g., [3, 23, 24, 39]). We summarize their essential aspects *to mark the difference with what we propose here*, beyond the obvious fact that we refer to full dynamics and *do not include* just jammed configurations of grain clusters, although such bodies fall within our proposal. Consequently, we restrict attention for a while on the above mentioned container, with volume $V$, in which we imagine $N_V$ grains at a jammed state. Consider the $i$-th grain and imagine it extended in space. It has $z_i$ contacts, each assumed to be exerted at a single point. Let us indicate one of such points on the boundary of the $i$-th grain by a vector $r_{ij}$ issued by the grain center of mass; $j$ refers to the $j$-th grain in contact with the $i$-th one. At the contact point, the two grains exchange a force $\mathtt{f}_{ij}$. Consider a single-scale representation of this body (a common choice in the analyses we are summarizing here). The total stress $\hat{T}$ of the cluster is given by $\hat{T}^T := V^{-1} \sum_i^{N_V} \sum_j^{z_i} r_{ij} \otimes \mathtt{f}_{ij}$. It is exerted on the container walls and is balanced. Assigned volume $V$, number of grains $N_V$, and total stress $\hat{T}$, we may have several configurations compatible with that values. Each configuration allows us to define a state $\varsigma$, characterized by different contact points and grain-to-grain interactions. If we assume that all states are equally likely (Edwards' microcanonical assumption; see [39]), the probability $P_\varsigma$ that a subsystem of volume $\bar{V} < V$ be in a given state $\varsigma$ is given by $P_\varsigma = \exp(-\frac{\bar{V}}{\chi(V)})$. The scalar factor $\chi(V)$, dubbed the compactivity, is the inverse of a temperature. Also, the state space can be factorized into subsets $\mathcal{M}_{\hat{T}}$, each containing all possible (balanced) jammed states with total stress $\hat{T}$. We can define then the number $\varrho_{N_V}(\hat{T})$ of states pertaining to $\mathcal{M}_{\hat{T}}$. By imposing local dynamics (say, transient repeated shaking) we can connect states in the same sector leaving to the introduction of further external actions the transition to different sectors. However, dynamics are in general not ergodic, so to define the density of states in the state space, indicated here by $X(\bar{T}, V, N_V)$, we need to account for the frequency $\beta_\varsigma^{dyn}(\bar{T}, V, N_V)$ with which each state is accessed by the dynamics chosen to create it. We take $X(\bar{T}, V, N_V) := \sum_{\varsigma \in \mathcal{M}_{\hat{T}}} \beta_\varsigma^{dyn}(\bar{T}, V, N_V)$ (details are in reference [23]). In this way, by looking at portions of material in the container as subsystems exchanging actions with the surrounding ones, we can think in terms of canonical ensemble with states parameterized by the stress instead of the energy. Consequently, we can accept, in accord to the Boltzmann ensemble, an entropy for this system given by $\ln X(\bar{T}, V, N_V)$. Then, a *temperature tensor* $\Xi$ emerges as defined by $\Xi := \frac{\partial \ln X(\bar{T}, V, N_V)}{\partial \bar{T}}$ (see [24]). When the cluster is isotropic, states are parameterized only by the trace of $\bar{T}$ so that we recover a scalar temperature given by $\frac{\partial \ln X(\operatorname{tr} \bar{T}, V, N_V)}{\partial \operatorname{tr} \bar{T}}$ (as defined in reference [24]). The inverse of $\Xi$ is strongly related to Edwards' angoricity tensor, defined as the derivative of $\bar{T}$ with respect to entropy [3]. In all these statistical analyses, beyond analogous formal treatments, a substantial difference with thermally activated systems is that jammed granular matter does not change state spontaneously by fluctuations at grain scale; it needs an action exerted by the environment.

By coming back to the view proposed here, since we focus attention on the dynamics of a molecular system not (or not necessarily) in a jammed state, we cannot resort to statistical view based on the canonical ensemble constructed in some sense (unless we are in a jammed state).

In the common single scale representation of fast granular flows, a temperature tensor appear as given by the dyad $(w - v) \otimes (w - v)$ (see, e.g., [21], and this point of view is also marginally taken in [49]), i.e., it is determined by the fluctuation velocity with respect to the only average $v$ considered. In our multi-scale setting, we have different fluctuations, so that we can hardly consider just the dyad $(w-v)\otimes(w-v)$ as a basic ingredient for a possible temperature tensor. We have inserted both $v$ and $By$ in the representation of mechanics so that we could consider $H$, defined by using the fluctuation velocity $c$, as a putative temperature tensor. The choice, however, could be questionable because, as shown in Sect. 3, $H$ enters the inertial terms of the balance of tensor moment of momentum, declaring in this way a mechanical nature.

Consequently, by looking at the ground of our choices for a notion of scalar temperature, as discussed above, we could start considering the dyad $\bar{c}\otimes\bar{c}$, which we could call upon to play the role of temperature tensor (see, once again, how the temperature tensor is considered in [21]) or even to use it as an ingredient to construct other notions of temperature tensor.

By replacing $\xi$ with $\bar{c}\otimes\bar{c}$, in fact, we could obtain a temperature tensor along the same path followed in Sect. 5.1. Besides formal aspects—which are rather straightforward—such an approach would underline the *instability of states with negative temperature*, those associated with an oblate tensor; in fact, if it would be axially symmetric, a small variation of any its eigenvalues would render prolate it.

Another issue to be stressed is that the choice of function space for the distribution has constitutive character (itself matter of modeling) and is decisive in the above recalled analyses. For example, if the distribution would be sum of two functions obtained as limits of half a Gaussian distribution, we would not recover the above mentioned instability.

Also, by looking at a special class of distributions, we may obtain another notion of temperature tensor (based on analyses in [35]), which *underlines the character of grand canonical ensemble that we attribute to the loculus*. To this aim, we take at the instant $\tau$ a loculus $\mathsf{e}(x)$ imagining that it contains a number $\omega(\tau,x)$ of molecules. For it, we define a *residual fluctuation tensor kinetic energy* as

$$\frac{1}{2}\bar{c}\otimes\bar{c},$$

and consider a special family of distribution densities of the type

$$\Theta(\tau,x;y,\bar{c}\otimes\bar{c})=\lambda(\tau,x;y)\exp\bigl(-\Lambda\cdot(\bar{c}\otimes\bar{c})\bigr),$$

where $\Lambda$ is a second-rank symmetric tensor and $\lambda$ an appropriate real-valued function. Then, we define the *averaged residual fluctuation kinetic energy* $\Gamma_\omega$ by

$$\Gamma_\omega(\tau,x;\Lambda)=\frac{1}{2\delta^3}\mu\int_{\mathsf{e}(x)}\lambda(\tau,x;y)\exp\bigl(-\Lambda\cdot(\bar{c}\otimes\bar{c})\bigr)(\bar{c}\otimes\bar{c})\,dy.$$

A standard requirement in the statistical mechanics of grand canonical ensembles (see, e.g., [35]) prescribes that to assure thermodynamic stability, there should exist a differentiable entropy function $\eta_\omega(\tau,x;\Gamma_\omega)$ such that

$$\Lambda=\frac{\partial\eta_\omega}{\partial\Gamma_\omega}.$$

When $\det\Lambda\neq 0$, we may accept

$$\Upsilon:=\Lambda^{-1}$$

as a (second-rank) temperature tensor.

However, the content of molecules within e varies because the loculus is open to mass exchange. The number of molecules should also be connected with temperature, so with entropy. A question is how to consider the number $\omega$ of molecules as referred to the loculus. If we maintain the view that the molecules are mass points, physics would suggest us to take $\omega$ as natural-number-valued. However, when we want to consider, even indirectly, that, beyond our modeling representation, molecules are in reality extended in space, we may allow $\omega(\tau, x)$ to take arbitrarily values on the positive real line, so that, for example, $\omega = 3.5$ would mean that in $\mathsf{e}(x)$ there are 3 molecules and half of a fourth molecule has entered at the instant $\tau$ the cube e. Moreover, since the loculus has finite size, if we want to account indirectly the spatial extension of the molecules, we should imagine $\omega$ on an interval $[0, k]$, rather than on the entire nonnegative real line. This aspect plays a role later. For the moment, we consider the entire nonnegative real line and define on it a probability with density $\pi_\omega$, satisfying the normalization condition

$$\int_0^{+\infty} \pi_\omega \, \mathrm{d}\omega = 1. \tag{23}$$

By using $\pi_\omega$, and leaving unexpressed $\tau$ and $x$ for the sake of notational conciseness, we define the second-rank tensor

$$\Gamma(\Lambda, \pi) := \int_0^{+\infty} \pi_\omega \Gamma_\omega(\Lambda) \, \mathrm{d}\omega,$$

the (scalar) entropy

$$\eta(\Lambda, \pi) := \int_0^{+\infty} \pi_\omega \eta_\omega\big(\Gamma_\omega(\Lambda)\big) \, \mathrm{d}\omega - \int_0^{+\infty} \pi_\omega \ln \pi_\omega \, \mathrm{d}\omega,$$

and the average number of molecules $\bar{\omega}$ within e, namely

$$\bar{\omega} := \int_0^{+\infty} \pi_\omega \omega \, \mathrm{d}\omega.$$

The integral

$$\int_0^{+\infty} \pi_\omega \ln \pi_\omega \, \mathrm{d}\omega$$

is Shannon's entropy, which accounts for the loss of information pertaining to $\pi_\omega$. To determine $\pi_\omega$, we define first the Lagrangian

$$\int_0^{+\infty} \mathcal{L}(\pi_\omega) \, \mathrm{d}\omega := \eta(\Lambda, \pi) - \Lambda \cdot \Gamma(\Lambda, \pi) + \Lambda \cdot M \bar{\omega} - (\gamma - 1) \int_0^{+\infty} \pi_\omega \, \mathrm{d}\omega,$$

where $\gamma - 1$ is the Lagrangian multiplier through which we impose the constraint (23) and $M$ is a second-rank tensor playing here the role of chemical potential. Then, we impose the stationarity condition

$$\frac{\mathrm{d}\mathcal{L}}{\mathrm{d}\pi_\omega} = 0.$$

By computing the derivative, we get

$$\eta_\omega\big(\Gamma_\omega(\Lambda)\big) - \ln \pi_\omega - \Lambda \cdot \Gamma_\omega(\Lambda) + \Lambda \cdot M\omega - \gamma = 0.$$

The normalization condition (23) implies

$$\gamma = \ln \int_0^{+\infty} \exp\big(\eta_\omega\big(\Gamma_\omega(\Lambda)\big) - \Lambda \cdot \Gamma_\omega(\Lambda)\big) \exp(\Lambda \cdot M\omega)\, \mathrm{d}\omega$$

and

$$\pi_\omega = \frac{1}{\mathcal{Z}} \exp\big(\eta_\omega\big(\Gamma_\omega(\Lambda)\big) - \Lambda \cdot \Gamma_\omega(\Lambda)\big) \exp(\Lambda \cdot M\omega),$$

where

$$\mathcal{Z} := \exp(\gamma).$$

Let us write for the sake of conciseness

$$\gamma = \ln \int_0^{+\infty} \exp\big(\tilde{a}(s)\big)\, \mathrm{d}s = \lim_{r\to\infty} \ln \int_0^r \exp\big(\tilde{a}(s)\big)\, \mathrm{d}s.$$

The boundedness of $\mathtt{e}(x)$ and the finite extension in space of the molecules allow us to consider the functions of $\omega$ defined above, as compactly supported over the nonnegative real line, precisely, defined over $[0, k]$, where $k$ is the number of molecules pertaining to the higher packing of a jammed state in $\mathtt{e}(x)$. Also, we should also presume that $\tilde{a}(s)$ be bounded so that the integral converges. Consequently, we write

$$\ln \int_0^r \exp\big(\tilde{a}(s)\big)\, \mathrm{d}s \le \ln\Big(r \sup_{[0,r]} \exp\big(\tilde{a}(s)\big)\Big)$$

and *for large r* we get

$$\frac{\ln \int_0^r \exp(\tilde{a}(s))\, \mathrm{d}s}{\ln r} \le 1 + \frac{\ln(\sup_{[0,r]} \exp(\tilde{a}(s)))}{\ln r}.$$

Consequently, under the boundedness assumptions above, $\gamma$ *behaves asymptotically as* $\ln \omega$.

In accord to standard thermodynamic notions, reinterpreted here in tensorial sense, we may define a tensor grand canonical potential as

$$\Psi(\Lambda, M) = -\Upsilon \ln \mathcal{Z},$$

determining further tensor versions of common thermodynamic relations.

A question left here open is the kind of evolution equation the temperature tensor $\Upsilon$ should satisfy.

## 6 Additional Remarks

We have already discussed in our previous work [12] limits and advantages of our approach. Here we want just to add ancillary remarks. In fact, the identification of the stresses and self-actions at continuum scale in terms of molecular interactions underlines explicitly the nature of the 'measures' of actions appearing in general multi-field descriptions of materials (or bodies, if you want), called complex to remind the intricate influence of microstructural events on the gross behavior, an influence not completely representable in terms of the standard format of continuum mechanics. Also, our analyses underline—we repeat—how 'fluctuating' (let us say) the border between mechanics and thermodynamics may be.

**Acknowledgement** This work has been developed within the programs of the research group in 'Theoretical Mechanics' of the "Centro di Ricerca Matematica Ennio De Giorgi" of the Scuola Normale Superiore in Pisa. The support of GNFM-INDAM is acknowledged.



## References

1. Andersen, H.C.: Molecular dynamics simulations at constant pressure and/or temperature. J. Chem. Phys. **72**, 2384–2393 (1980)
2. Bacci, M., Mariano, P.M.: Protein dynamics: an approach based on the Cauchy-Born rule. Physica E **61**, 69–82 (2014)
3. Blumenfeld, R., Edwards, S.F.: On granular stress statistics: compactivity, angoricity, and some open issues. J. Phys. Chem. **113**, 3981–3987 (2009)
4. Brocato, M., Capriz, G.: Clockwork, ephemeral and hybrid continua. Phys. Mesomech. **14**, 124–144 (2011)
5. Capriz, G.: Elementary preamble to a theory of granular gases. Rend. Mat. Univ. Padova **110**, 179–198 (2003)
6. Capriz, G.: Pseudofluids. In: Capriz, G., Mariano, P.M. (eds.) Material Substructures in Complex Bodies: From Atomic Level to Continuum, pp. 238–261. Elsevier, Amsterdam (2006)
7. Capriz, G.: A quest for an 'extended' continuum mechanics. Note Mat. **27**, 27–41 (2007)
8. Capriz, G.: On ephemeral continua. Phys. Mesomech. **11**, 285–298 (2008)
9. Capriz, G., Giovine, P.: Hypocontinua. In: Albers, B., Kuczma, M. (eds.) Continuum Media with Microstructure 2, pp. 23–43. Springer, Berlin (2016)
10. Capriz, G., Giovine, P.: Classes of ephemeral continua. Math. Methods Appl. Sci. **43**, 1175–1196 (2017)
11. Capriz, G., Mariano, P.M.: Objective fluxes in a multiscale continuum description of sparse medium dynamics. Physica A **415**, 354–365 (2014)
12. Capriz, G., Mariano, P.M.: Multi-scale kinetic description of granular clusters: invariance, balance, and temperature. Contin. Mech. Thermodyn. **30**, 1323–1342 (2018)
13. Capriz, G., Mazzini, G.: Invariance and balance in continuum mechanics. In: Nonlinear Analysis and Continuum Mechanics, pp. 27–35. Springer, New York (1998)
14. Capriz, G., et al. (eds.): Whence Boundary Conditions in Continuum Physics? Accademia Nazionale dei Lincei, Roma (2004)
15. Chipot, M.M.: Elements of Nonlinear Analysis. Birkhäuser, Basel (2000)
16. de Fabritiis, C., Mariano, P.M.: Geometry of interactions in complex bodies. J. Geom. Phys. **54**, 301–323 (2005)
17. E, W., Huang, Z.: Matching conditions in atomistic-continuum modeling of materials. Phys. Rev. Lett. **87**(1–4), 135501 (2001)
18. E, W., Ming, P.: Cauchy-Born rule and the stability of crystalline solids: static problems. Arch. Ration. Mech. Anal. **183**, 241–297 (2007)
19. E, W., Ming, P.: Cauchy-Born rule and the stability of crystalline solids: dynamic problems. Acta Math. Appl. Sin. Engl. Ser. **529**, 529–550 (2007)
20. Ericksen, J.L.: On nonlinear elasticity theory of crystal defects. Int. J. Plast. **14**, 9–24 (1998)
21. Goldhirsch, I.: Theory of rapid granular flows. In: Mehta, A. (ed.) Granular Physics, pp. 176–208. Cambridge University Press, Cambridge (2007)
22. Grmela, M.: Externally driven macroscopic systems: dynamics versus thermodynamics. J. Stat. Phys. **166**, 282–316 (2017)
23. Henkes, S., Chakraborty, B.: Statistical mechanics framework for static granular matter. Phys. Rev. E **79**, 061301 (2009)
24. Henkes, S., O'Hern, C.S., Chakraborty, B.: Entropy and temperature of a static granular assembly: an ab initio approach. Phys. Rev. Lett. **99**, 038002 (2007)
25. Hoover, W.G.: Canonical dynamics: equilibrium phase-space distributions. Phys. Rev. A **31**, 1695–1697 (1985)
26. Irving, J.H., Kirkwood, J.G.: The statistical mechanical theory of transport processes. IV. The equations of hydrodynamics. J. Chem. Phys. **18**, 817–829 (1950)
27. Jenkins, J.T.: Boundary conditions for rapid granular flow: flat, frictional walls. J. Appl. Mech. **114**, 120–127 (1992)

28. Lehoucq, R.B., von Lilienfeld-Tonal, A.: Translation of Walter Noll's "Derivation of the fundamental equations of continuum thermodynamics from statistical mechanics". J. Elast. **100**, 5–24 (2010)
29. Mariano, P.M.: Multifield theories in mechanics of solids. Adv. Appl. Mech. **38**, 1–93 (2002)
30. Mariano, P.M.: Mechanics of material mutations. Adv. Appl. Mech. **XX**, 1–92 (2014)
31. Mariano, P.M., Stazi, F.L.: Strain localization in elastic microcracked bodies. Comput. Methods Appl. Mech. Eng. **190**, 5657–5677 (2001)
32. Memarnahavandi, A., Larsson, F., Runesson, K.: A goal-oriented adaptive procedure for the quasi-continuum method with cluster approximation. Comput. Mech. **55**, 617–642 (2015)
33. Murdoch, A.I.: A corpuscular approach to continuum mechanics: basic considerations. Arch. Ration. Mech. Anal. **88**, 291–321 (1985)
34. Murdoch, A.I.: Physical Foundations of Continuum Mechanics. Cambridge University Press, Cambridge (2012)
35. Naudts, J., van der Straeten, E.: The grand canonical ensemble in generalized thermostatistics. J. Stat. Mech. Theory Exp. **2004**, P12002 (2004)
36. Noll, W.: Die Herleitung der Grundgleichungen der Thermomechanik der Kontinua aus der statistichen Mechanik. J. Ration. Mech. Anal. **4**, 627–646 (1955)
37. Noll, W.: La Mécanique classique, basée sur une axiome d'objectivité. In: La Méthode Axiomatique dans les Mécaniques Classiques et Nouvelles, Colloque International, Paris, 1959, pp. 47–56. Gauthier-Villars, Paris (1963)
38. Nosé, S.: A unified formulation of the constant temperature molecular-dynamics methods. J. Chem. Phys. **81**, 511–519 (1984)
39. Oakeshott, R.B.S., Edwards, S.F.: Theory of powders. Physica A **157**, 1080–1090 (1989)
40. Parrinello, M., Rahman, A.: Polymorphic transitions in single crystals: a new molecular dynamics method. J. Appl. Phys. **52**, 7182–7190 (1981)
41. Peters, M.H.: Completing Irving and Kirkwood's molecular theory of transport processes: nonequilibrium entropy conservation. Ind. Eng. Chem. Res. **48**, 166–171 (2009)
42. Pitteri, M.: Continuum equations of balance in classical statistical mechanics. Arch. Ration. Mech. Anal. **94**, 291–305 (1986). Corrigendum published as "Correction of an error in my paper: "Continuum equations of balance in classical statistical mechanics" [Arch. Rational Mech. Anal. 94 (1986), no. 4, 291–305]". Arch. Rational Mech. Anal. **100**, 315–316 (1988)
43. Pitteri, M.: On a statistical-kinetic model for generalized continua. Arch. Ration. Mech. Anal. **111**, 99–120 (1990)
44. Puckett, J.G., Daniels, K.E.: Equilibrating temperaturelike variables in jammed granular subsystems. Phys. Rev. Lett. **110**, 058001 (2013)
45. Ray, J.R., Rahman, A.: Statistical ensemble and molecular dynamics studies of anisotropic solids. II. J. Chem. Phys. **82**, 4243–4247 (1985)
46. Shenoy, V.B., Miller, R., Tadmor, E.B., Rodney, D., Phillips, R., Ortiz, M.: An adaptive finite element approach to atomic-scale mechanics—the quasicontinuum method. J. Mech. Phys. Solids **47**, 611–647 (1999)
47. Stackgold, I.: The Cauchy relations in a molecular theory of elasticity. Q. Appl. Math. **8**, 169–186 (1950)
48. Tadmor, E.B., Miller, R.E.: Modeling Materials: Continuum, Atomistic and Multiscale Techniques. Cambridge University Press, Cambridge (2011)
49. Truesdell, C.A., Muncaster, R.G.: Fundamentals of Maxwell's Kinetic Theory of a Simple Monatomic Gas. Treated as a Branch of Rational Mechanics. Academic Press, New York (1980)
50. Ulz, M.H.: A multiscale molecular dynamics method for isothermal dynamic problems using the seamless heterogeneous multiscale method. Comput. Methods Appl. Mech. Eng. **295**, 510–524 (2015)
51. Wurm, P., Ulz, M.H.: A stochastic approximation approach to improve the convergence behavior of hierarchical atomistic-to-continuum multiscale models. J. Mech. Phys. Solids **95**, 480–500 (2016)
52. Xu, M., Gracie, R., Belytschko, T.: A continuum-to-atomistic bridging domain method for composite lattices. Int. J. Numer. Methods Eng. **81**, 1635–1658 (2010)

J Elast (2019) 135:117–148
https://doi.org/10.1007/s10659-018-9697-y

# "Reality" and Representation in Mechanics: The Legacy of Walter Noll

**Gianpietro Del Piero**[1,2]

Received: 12 July 2018 / Published online: 20 September 2018


**Abstract** The foundations of the mechanics of generalized continua are revisited in the light of the theoretical progress made in the last decades. The paper includes a summary of the scientific activity of W. Noll, to whom a large part of this progress is due.

**Keywords** Foundations of mechanics · Generalized continua · Two-scale deformations · Reference frames

**Mathematics Subject Classification** 74AXX · 74A05 · 74A10 · 74A20 · 74A60 · 74C15

## 1 Introduction

A large part of the progress of mechanics in the last decades is due to W. Noll. The present paper is a personal revisitation, dedicated to his memory as an act of homage, admiration, and gratitude. My purpose is to drive the reader's attention to the fundamentals of mechanics, nowadays scarcely investigated, a common opinion being that they are firmly established since long time. On the contrary, I believe that the restless advance of knowledge tends to make them obsolete, and that thorough re-discussion and updating are necessary.

My first concern is to make a clear distinction between physical reality and its representation. In Sect. 2, I distinguish between the *observation* of a physical body and its *placement* on a *support* endowed with the properties of the Newtonian *absolute space*. In this spirit, in Sect. 3, when defining the basic objects necessary to describe the geometry of a classical continuum, I distinguish between a *configuration* and its *placements* on the support, and between a *reference system* and its representatives on the support, the *reference frames*. I also insist on the question of the changes of reference frame, which has always been a source of great confusion.

Dedicated to the memory of Walter Noll

✉ G. Del Piero
dlpgpt@unife.it

1 Dipartimento di Ingegneria, Università di Ferrara, 44100 Ferrara, Italy

2 International Research Center M&MoCS, Cisterna di Latina, Italy

Much of the current research on material behavior deals with *gradient theories*, for which the geometry of the classical continuum is insufficient, and at least two length scales, micro- and macroscopic, are required. A geometry of continua with two length scales, or *two-scale continua*, is developed in Sect. 4, where precise definitions of two-scale configurations, placements, and deformations are given.

The rest of the paper is devoted to more strictly mechanical aspects. In the short Sect. 5, some traditional views on the nature of forces and on the basic postulates of mechanics are substantially modified. There has always been some reluctance about taking energy as a primitive object of mechanics. In the past, this was due to the idea that this can be properly done only for conservative systems. Today, there is plenty of beautiful theories including both conservative and dissipative systems, and the time is ready for a change. In Sect. 5.3 the traditional axioms, the Euler balance laws of linear and angular momentum, are replaced by two energetic postulates, the *conservation principle* and the *dissipation principle*, which are mechanical versions of the two laws of thermodynamics. In the next Sect. 6.1 the traditional axioms are shown to be consequences of the indifference of the energy to changes of placement within the same configuration, a result proved by Noll in 1959 [35].

Section 6 deals with *classical continua*, that is, continua for which the external energy is a function of the macroscopic deformation alone. For such continua, the existence proof of the stress tensor does not involve the internal energy. By consequence, the traditional *equation of virtual power* is not an equality between external and internal powers, but only an identity between equivalent forms of the external power. The equality between the time derivatives of the internal and external energies becomes the true equation of virtual power, and the constitutive equations become relations between the partial derivatives in space of the energy densities. In the last part of this section it is shown that in the presence of dissipation the two basic postulates produce an *activation condition* for the dissipative strain rate and a *flow rule* determining its direction.

Section 7 deals with *generalized continua*. For such continua, the indifference restrictions do not provide an adequate number of field equations. The missing equations come from the *pseudobalance equations*, which are a consequence of the conservation principle. These equations allow to deduce constitutive equations, activation conditions, and flow rules, using the Coleman-Noll procedure. The comparison with the gradient theory of plasticity made in Sect. 7.4 shows the greater theoretical consistency and conceptual simplicity of the proposed framework.

Throughout the paper, Noll's ideas are inextricably mingled with my own views. To somehow separate the two, I devoted Sect. 8.1 to a summary of Noll's original achievements. In Sect. 8.2 his formulation of the constitutive relations in terms of state variables is discussed. This formulation replaced his earlier formulation in terms of infinite deformation histories, which is recalled in Sect. 8.3. In the following Sect. 8.4 the pseudobalance equations are compared with the *microbalance equations* proposed as supplementary field equations for generalized continua. Finally, Sect. 8.5 deals with the problem of inertia, a typical case of confusion between physical reality and representation, which gave origin to veritable intellectual monsters, like "changes of observers" and "apparent forces".

We assume that the reader is familiar with the definitions of Euclidean space and of the associated vector space $\mathscr{V}$, as given e.g. in [37]. By a *tensor* we mean a linear map on $\mathscr{V}$, and for a homogeneous map $\mathfrak{h} : \mathscr{V} \to \mathbb{R}$ of order one we write $\mathfrak{h} \triangleright v$ in place of $\mathfrak{h}(v)$. As pointed out in [13], the operator $(\triangleright)$ shares with the inner product the property

$$\mathfrak{h} \triangleright (\lambda v) = \lambda \mathfrak{h} \triangleright v \quad \forall \lambda \geq 0,$$

but is clearly not symmetric, and in general it is not additive with respect to $v$.

## 2 "Reality" and Representation

The *real world* is a terribly complicated object, which, nevertheless, seems to obey very precise laws. The discovery of these laws is the purpose of physics. In the past, it was a common belief that a complete understanding of the laws of Nature was only a question of time.[1] Today, more modestly, most scientists tend to be satisfied with approximate descriptions of specific aspects of reality.

Mechanics is the part of physics which studies the laws of the equilibrium and motion of physical objects. To get an idea of how such objects are represented, very instructive is the analogy with the procedure followed to represent a portion of the terrestrial surface:

- to draw a geographic map, an operator has to go on the spot and measure distances between points. The distances are communicated to another operator, which places the points on a plane surface, for example, paper or blackboard, with the measured distances modified by a suitable *scale factor*. The two operators are the *observer* and the *placer*, and the plane surface is the *support* of the representation.

This operation is never exact. Indeed,

- it is hardly believable that an observer can measure distances between more than a finite number of points. To represent, for example, a coastal line, all he can do is to select a finite number of points on the coast, and measure their mutual distances,
- the placer places images of these points on the support. Joining the points with a line, he gets a representation of the coast. But this representation is not unique, since (i) there are infinite ways of placing on paper a set of points at given distances, and (ii) once the points are placed, there are infinite ways of joining them with a line,[2]
- since the terrestrial surface is curved, its representation on a plane is only approximate. However, the error is negligible if a sufficiently small portion of the earth's surface is represented.

In mechanics the objects of the representation, the *bodies*, are located in the *physical space*, which we perceive as three-dimensional. But, at least at my knowledge, it is not known whether or not this space is curved, and, if so, there is no idea of the order of magnitude of its curvature. Like in the case of the earth's surface, the error made by neglecting the curvature of space is negligible when conveniently small portions of the universe are considered. This is why in classical mechanics the question of the curvature of space is usually ignored.

To represent a body, the procedure is the same adopted for representing the earth's surface: an observer measures the mutual distances of a finite set of *body particles*. The distances are communicated to a placer, who places a corresponding set of points on a support. Like in the case of the geographic map, this support is *material*. But now it is three-dimensional. It is a sort of "three-dimensional blackboard" made of material points, whose positions are considered immutable.

Immutability is a property of an *absolute space*. In Newtonian mechanics this property was attributed to the physical space, which was imagined to be a sort of big three-dimensional blackboard made of transparent, imponderable *ether particles*.[3] Later, this idea was gradually abandoned. What is left is the possibility of *representing* the physical space as an absolute space.

---

[1] See, e.g., Laplace's introduction to his "Essai philosophique sur les probabilités".

[2] For instance, the line can be piecewise linear, or piecewise polynomial of different orders.

[3] In Newton's words, "Absolute space, in its own nature and without anything external, always remains similar and immovable". Cited in Mach [29], p. 226.

## 3 The Representation of a Body as a Continuum

In classical mechanics, to represent a body the observer measures the distances between a finite number of body particles, and the placer places their images on the support. An image of the body as a continuum is constructed by "filling" the spaces between the placed points. This passage from discrete to continuum is a crucial point which deserves a careful description.

### 3.1 Configurations and Placements

Let $\Xi^N$ be a set of $N$ particles $X^\alpha, X^\beta \ldots$, and let $d(X^\alpha, X^\beta)$ be their distances measured by the observer. The set $\Xi^N$ endowed with the distance function $d$ is a finite-dimensional metric space called a *configuration* of $\Xi^N$. A *placement* on $\mathscr{E}$ of a configuration $(\Xi^N, d)$ is a set $\mathfrak{X}^N$ of $N$ points $x^\alpha, x^\beta$ of $\mathscr{E}$, whose Euclidean distances coincide with the distances $d(X^\alpha, X^\beta)$[4]

$$\left|x^\alpha - x^\beta\right| = d\left(X^\alpha, X^\beta\right) \quad \forall x^\alpha,\ x^\beta \in \mathfrak{X}^N. \tag{1}$$

The choice of a special distance function $d_R$ defines a *reference configuration* $(\Xi^N, d_R)$. Among all possible placements of the reference configuration, the placer selects a *reference placement* $\mathfrak{X}^N_R$. Then, he constructs a three-dimensional region $\Omega_R$ of $\mathscr{E}$ by "filling" the finite set $\mathfrak{X}^N_R$.[5] This region is a representation of the body *as a continuum*. It is said to be a *reference placement* of the body. Note that, since a finite set can well be extracted from bodies with a discrete, for example, atomistic, structure, it does not really matter whether or not the body is itself a continuum. In other words, *the distinction between continuous and discrete bodies is a matter of representation and does not depend on physical observation.*[6]

The region $\Omega_R$ is not an exact representative of the body, because only the points $x^\alpha_R$ of $\mathfrak{X}^N_R$ are exact images of body particles. For the same reason, the boundary of $\Omega_R$ is not an exact representative of the physical boundary of the body. Therefore, *the regularity of the boundary also comes from the representation and not from physical observation.*[7]

### 3.2 Deformations

Let $\mathfrak{X}^N_R$ and $\mathfrak{X}^N$ be placements of $(\Xi^N, d_R)$ and $(\Xi^N, d)$ on $\mathscr{E}$, and let $f^N$ be the discrete map from the points $x^\alpha_R$ of $\mathfrak{X}^N_R$ to the corresponding points $x^\alpha$ of $\mathfrak{X}^N$

$$f^N\left(x^\alpha_R\right) = x^\alpha. \tag{2}$$

[4]The coincidence of the distances is possible only if the physical space has itself the structure of an Euclidean space. Otherwise the coincidence is only approximate, just as for the distances between points on the earth's surface and between their images on the support.

[5]The "filling" operation may consist, for example, in joining the neighboring points of $\mathfrak{X}^N_R$ with a finite-element mesh, and taking as $\Omega_R$ the region covered by the mesh, possibly with some regularization at the boundary.

[6]After receiving a finite number of distances between particles, it is up to the placer to decide between a discrete and a continuous representation. The origins of continuum mechanics go back to the deliberate choice of the continuous representation made by Cauchy [4].

[7]In many cases, it is not even clear what the physical boundary of a body is. The regularity to be attributed to the boundary was a major concern of Noll [39, 43]. The properties of what he called *fit regions* are in fact restrictions on the choices to be made by the placer.

Let $\Omega_R$ be the reference placement obtained by "filling" the set $\mathfrak{X}_R^N$, and let $f$ be an extension of $f^N$ from $\mathfrak{X}_R^N$ to $\Omega_R$. We say that the extended map $f$ is a *deformation from* $\Omega_R$ if it obeys some conditions of physical plausibility. Minimal plausibility conditions are that $f$ be

(i) injective,
(ii) orientation-preserving.

Condition (i) excludes interpenetration of matter, and (ii) excludes placements which are the mirror image of each other. Additional regularity can be imposed, more for convenience than for physical reasons. For example, differentiability is assumed because it makes possible the use of the instruments of differential calculus.[8] The maximum order of differentiability is also decided by the placer. By consequence, *the theories involving higher-order deformation gradients have no direct relation with physical reality*.[9]

Due to the injectivity assumption, every deformation $f$ admits an *inverse*, which is the deformation from $f^{-1} : f(\Omega_R) \to \Omega_R$ such that

$$f^{-1}\big(f(x_R)\big) = x_R \quad \forall x_R \in \Omega_R. \tag{3}$$

Moreover, if $f_\emptyset$ and $f$ are deformations from $\Omega_R$, the composition

$$f_\emptyset \circ f^{-1} \tag{4}$$

is the deformation which with each $x \in f(\Omega_R)$ associates the point $f_\emptyset(f^{-1}(x))$ of $f_\emptyset(\Omega_R)$. Thus, a deformation between regions of $\mathscr{E}$ can be defined in terms of deformations $f$, $f_\emptyset$ from the reference placement.

The properties (i) and (ii) being preserved in the composition, the set of all deformations between regions of $\mathscr{E}$ has the algebraic structure of a *group*.[10] Additional regularity assumptions are considered admissible only if they preserve this algebraic structure.[11] Therefore, the *set of all admissible deformations* of a body is a set of maps between regions of $\mathscr{E}$ with the properties (i), (ii), plus some "admissible" regularity properties.

Once a reference placement $\Omega_R$ has been fixed, a *configuration of the body* is identified with the metric space $(\Omega_R, d)$, and a *placement* of $(\Omega_R, d)$ is a region $f(\Omega_R)$, with $f$ an admissible deformation from $\Omega_R$ to $\mathscr{E}$ such that

$$\big|f(x_R) - f(x_{\emptyset R})\big| = d(x_R, x_{\emptyset R}) \quad \forall x_R,\ x_{\emptyset R} \in \Omega_R. \tag{5}$$

Though every configuration admits an infinity of placements, a one-to-one correspondence between configurations and placements can be obtained by introducing a *reference system* in the physical space and a corresponding *reference frame* in $\mathscr{E}$.

[8] For example, for differentiable functions $f$ the preservation of the orientation and the local injectivity at $x_R$ are ensured by the analytical condition $\det \nabla f(x_R) > 0$.

[9] For example, a *second-order continuum* need not represent a body with a twice differentiable structure. The body's structure may even be discrete. But, by a decision of the placer, only twice differentiable deformations are considered admissible. An example is Ambrosio and Tortorelli's model for fracture [2], in which cracked configurations are represented by differentiable functions with very large values of the deformation gradient inside the cracked zone.

[10] Noll [33].

[11] For example, as shown by (4), differentiability is preserved under composition if all deformations from the reference placement are differentiable and have a differentiable inverse.

### 3.3 Reference Systems and Reference Frames

A *reference system* is a discrete set of points of the physical space, whose distances from any point of the body determine uniquely the position of that point on the support. The advantage of using a reference system is that the observer, instead of measuring the mutual distances between all points of $\Xi^N$, can measure only the distance of each point of $\Xi^N$ from the points of the reference system.

In the example of the earth's surface, the construction of a reference system starts with the selection of an *observation point* $X^0$ and of a *direction* $E^1$. The point can be any point on the earth's surface, and the direction is conventionally the *north*, shown by the pole star or by the hand of a compass. The observer is supposed to recognize the points of the surface which are *aligned with* $X^0$ along the direction $E^1$. Among them he selects a point $X^1$, and measures the distance $d(X^1, X^0)$. The placer places $X^0$ and $X^1$ at two points $x^0, x^1$ of the support, at the distance $|x^1 - x^0| = d(X^1, X^0)$.[12]

For any other point $X$ of the surface, the observer measures the distances $d(X^0, X)$ and $d(X^1, X)$. When placing $X$ on the support, the placer realizes that, because of the two possible *orientations* of the triple $\{X^0, X^1, X\}$, there are two points whose distances from $x^0$ and $x^1$ are $d(X^0, X)$ and $d(X^1, X)$. To remove this ambiguity the observer selects a second direction $E^2$, for example, the *east*, takes a point $X^2$ aligned with $X^0$ along $E^2$, and measures the distances $d(X^2, X^0)$ and $d(X^2, X^1)$. The position of $X$ on the support is now uniquely determined by its distances from the points $X^0, X^1, X^2$. They form a *reference system* for the observed portion of the earth's surface.

For bodies in the physical space all goes much in the same way, except that, the dimension of the support being three instead of two, three independent directions $E^i$ are required. Therefore, a reference system for the physical space is a metric space $(\Xi^*, d)$, where $\Xi^*$ is a quadruple $\{X^0, X^i\}$ and $d$ is a distance function. A *reference frame* is a placement of the reference system on $\mathscr{E}$, that is, a quadruple $\mathfrak{X}^* = \{x^0, x^i\}$ of points of $\mathscr{E}$ such that

$$\left|x^i - x^0\right| = d\left(X^i, X^0\right), \qquad \left|x^i - x^j\right| = d\left(X^i, X^j\right). \tag{6}$$

To prove that the choice of a reference frame determines uniquely the position in $\mathscr{E}$ of any body particle $X$, we start from the distance conditions

$$|x - x^0| = d\left(X, X^0\right), \qquad |x - x^i| = d\left(X, X^i\right), \tag{7}$$

to be satisfied by the image $x$ of $X$, for the measured distances $d(X, X^0)$, $d(X, X^i)$. Then we set

$$x^i = x^0 + d\left(X^i, X^0\right)e^i, \tag{8}$$

where the unit vectors $e^i$ are the images of the directions $E^i$ in the selected reference frame. Since the $e^i$ are three linearly independent vectors and $\mathscr{E}$ is three-dimensional, the vector $(x - x_0)$ is a linear combination of these vectors

$$x - x^0 = \sum_j \xi^j e^j. \tag{9}$$

[12] Modified by the scale factor, here taken equal to one for simplicity.

The coefficients $\xi^j$ are determined by the identity

$$d(X, X^i) = |x - x^i| = |(x - x^0) + (x^0 - x^i)| = \left|\sum_j \xi^j e^j - d(X^i, X^0)e^i\right|, \tag{10}$$

which, squared, leads to the linear system[13]

$$d^2(X, X^i) = d^2(X, X^0) - 2d(X^i, X^0)\sum_j (e^j \cdot e^i)\xi^j + d^2(X^i, X^0). \tag{11}$$

Thus, for every fixed reference frame $\mathfrak{X}^*$, the position $x$ of any body particle $X$ is determined by its distances from the points $X^0$, $X^i$ of the reference system.

### 3.4 Evolutions

Usually, when drawing a geographic map one is not interested in how the terrestrial surface changes in time. This is not the same in mechanics, whose objects are *deformable bodies*. For them, the changes of distances with time, called *evolutions*, are of primary importance. We specify that by *time* we mean any monotonic scalar function of the physical time, and that the term *motion* will be reserved to evolutions with respect to the physical time.

Let $d_k$, $k = 0, 1 \ldots M$, be distance functions on a finite set $\Xi^N$ of body particles, obtained by observations made at $M$ instants $t_k$. By "filling" the discrete set $\{t_k\}$, a family $t \mapsto d_t$ of distance functions such that

$$d_t = d_k \quad \text{at} \quad t = t_k, \tag{12}$$

is constructed over the interval $[t_0, t_M]$ of the real line.[14] With each $t$ the placer associates a discrete deformation $f_t^N$ from the reference placement $\mathfrak{X}_R^N$ of $\Xi^N$, such that the distance conditions (10), now rewritten in the form

$$d_t(X^\alpha, X^\beta) = |f_t^N(x_R^\alpha) - f_t^N(x_R^\beta)| \quad \forall X^\alpha,\ X^\beta \in \Xi^N, \tag{13}$$

are satisfied. This deformation is uniquely determined by the choice of a family $t \mapsto \mathfrak{X}_t^N$ of reference frames. The reference system being the same for all $t$, all reference frames have the same distances $|x_t^0 - x_t^i|$, $|x_t^i - x_t^j|$ and the same orientation.

Finally, the discrete deformations $f_t^N$ are extended to "admissible" deformations $f_t$ from $\Omega_R$. A family $t \mapsto f_t$ constructed in this way is an *evolution* of the body, or rather, of the continuous representative $\Omega_R$ of the body, relative to the family $t \mapsto \mathfrak{X}_t^N$ of reference frames.

### 3.5 Changes of Reference Frame

Let $x_{\emptyset R}$ be a fixed point of $\mathscr{E}$, let $x_R$ be any point of $\Omega_R$, and let $x^*, x_\emptyset^*$ and $x, x_\emptyset$ be their positions according to two reference frames $\mathfrak{X}^*$, $\mathfrak{X}$. Then the distances $|x^* - x_\emptyset^*|$ and $|x - x_\emptyset|$ are the same, that is, there is an orthogonal tensor $Q$ such that

$$x^* - x_\emptyset^* = Q[x - x_\emptyset] \quad \forall x_R \in \Omega_R. \tag{14}$$

[13] The solution of this system is unique because the vectors $e^i$ are linearly independent. I omit the proof that the solution is independent of $i$.

[14] Just as the regularity in space, the regularity in time can be decided by the placer, possibly without any physical motivation.

This is the transformation rule for the points of $\Omega_R$ under a change $\mathfrak{X} \to \mathfrak{X}^*$ of the reference frame. Moreover, since a difference of points is an element of the vector space $\mathscr{V}$ associated with $\mathscr{E}$, this is also the transformation rule for the vectors of $\mathscr{V}$.

Now consider an evolution of $\Omega_R$, and assume that the reference system is the same for all $t$, while the reference frame may vary with $t$.[15] Then if $t \mapsto f_t^*$ and $t \mapsto f_t$ are the descriptions of the same evolution with respect to two families $t \mapsto \mathfrak{X}_t^*$ and $t \mapsto \mathfrak{X}_t$ of reference frames, the positions of the points $x_R, x_{\emptyset R}$ of $\Omega_R$ at the instant $t$ are

$$x_t^* = f_t^*(x_R), \quad x_{\emptyset t}^* = f_t^*(x_{\emptyset R}), \qquad x_t = f_t(x_R), \quad x_{\emptyset t} = f_t(x_{\emptyset R}). \tag{15}$$

Since the distances $|x_t^* - x_{\emptyset t}^*|$ and $|x_t - x_{\emptyset t}|$ are the same, for every $t$ there is an orthogonal tensor $Q_t$ such that $(x_t^* - x_{\emptyset t}^*) = Q_t[x_t - x_{\emptyset t}]$, that is,

$$f_t^*(x_R) = f_t^*(x_{\emptyset R}) + Q_t\big[f_t(x_R) - f_t(x_{\emptyset R})\big]. \tag{16}$$

This is the transformation rule for the deformations. By differentiation with respect to $x_R$ we get the transformation rule for the deformation gradients

$$\nabla f_t^*(x_R) = Q_t \nabla f_t(x_R), \tag{17}$$

and by differentiation with respect to $t$ we get the transformation rule for the velocities

$$v_t^*(x_R) = v_t^*(x_{\emptyset R}) + Q_t\big[v_t(x_R) - v_t(x_{\emptyset R})\big] + Q_t Y_t\big[f_t(x_R) - f_t(x_{\emptyset R})\big], \tag{18}$$

with $Y_t$ the skew-symmetric tensor resulting from the expansion

$$Q_{t+\varepsilon} = Q_t\big(I + \varepsilon Y_t + o(\varepsilon)\big), \tag{19}$$

and with $f_t^*$, $f_t$ and $v_t^*$, $v_t$ the positions and velocities relative to the two reference frames.

The definition of the velocity $v_t$ as the time derivative of $x_t$ has no physical meaning, because the positions $x_t$ are chosen by the placer. The only velocities with a physical meaning are those of *the body particles relative to the reference system*, which are deduced from distance measurements made by the observer at different times. By consequence, the only physically meaningful velocities in a representation on $\mathscr{E}$ *are those relative to the image of the reference system in* $\mathscr{E}$.

A vector field is said to be *indifferent* if under a change of reference frame it transforms like the vectors of $\mathscr{V}$, that is, with the law (14). According to this definition, the vector field $(f_t^*(x_R) - f_t^*(x_{\emptyset R}))$ of the relative positions is indifferent, the velocity $v_t^*(x_R)$ is not indifferent, and the relative velocity $(v_t^*(x_R) - v_t^*(x_{\emptyset R}))$ is indifferent only if $Y_t = 0$, that is, only if the reference frame does not depend on $t$.

It is a common belief that *the velocity is not indifferent because it depends on the observer*. The underlying idea is that the observer observes different velocities when changing his position in the physical space. But there are no positions if the space is not absolute. Then, unless the physical space is an absolute space, the above statement is meaningless.

[15] According to ([51], Sect. 17), a reference system is "a set of objects whose mutual distances change comparably little in time, like the walls of a laboratory, the fixed stars, or the wooden horses on a merry-go-round". To assume that the reference system does not change with $t$ does not mean that these objects "occupy a fixed position in the physical space", since only an absolute spaces has "positions", and we are not supposing that the physical space is an absolute space. We are only supposing that *the mutual distances* of the points of the reference system do not change during the evolution.

The observer measures distances between material particles and points of the reference system. Whether or not he is "moving", whatever this means, is irrelevant. The non-indifference of velocity is not because it depends on the observer, but *because it depends on the reference frame*. That is, non-indifference is not a physical fact, but a problem of representation.[16]

# 4 Two-Scale Representations

Going back to the example of the geographic map assume that, after representing a portion of the earth's surface, the observer becomes interested in a more detailed representation. For this, he selects additional body particles $X^{\alpha\beta}$ in the vicinity of each $X^\alpha$ of $\Xi^N$, and measures their distances from a local reference system. In this way he gets a refined representation, which may look very different from the previous one.

Further refinements can be obtained by taking additional body particles in the vicinity of the $X^{\alpha\beta}$ and reiterating the process. The resulting representation of the body is based on measurements made on smaller and smaller portions of the body, that is, at different *scales*. Note that a *multi-scale representation* does not reflect any physical status of the body. It only reflects the number of scales at which the measurements have been made. In what follows we restrict our analysis to *two-scale continua*, that is, to continua constructed on distance measurements made on two scales, the *macroscale* and the *microscale*.

## 4.1 Two-Scale Placements

At the macroscale, just as in a single-scale representation, the observer selects a set $\Xi^N$ of $N$ body particles $X^\alpha$, and measures their distances $d^{\alpha 0}, d^{\alpha i}$ from the points $X^0$, $X^i$ of the given reference system $\Xi^*$. The placer selects a reference frame $\mathfrak{X}^*$, and places $\Xi^N$ on the *macroscopic placement* $\mathfrak{X}^N$ whose points $x^\alpha$ are determined by the distances

$$\left|x^\alpha - x^0\right| = d^{\alpha 0}, \qquad \left|x^\alpha - x^i\right| = d^{\alpha i}, \tag{20}$$

from the points $x^0, x^i$ of $\mathfrak{X}^*$.

At the microscale, for each $X^\alpha \in \Xi^N$ the observer takes a *local reference system* $\Xi^{\alpha *} = (X^{\alpha 0}, X^{\alpha i})$, with local observation point $X^{\alpha 0} = X^\alpha$ and with the $X^{\alpha i}$ aligned with $X^\alpha$ along the same directions $E^i$ of the macroscopic reference system. Then in the vicinity of $X^\alpha$ he selects a set $\Xi^{N_\alpha}$ of $N^\alpha$ body particles $X^{\alpha\beta}$ and measures their distances $d^{\alpha\beta 0}, d^{\alpha\beta i}$ from the points of the local reference system $\Xi^{\alpha *}$.

The placer places $\Xi^{\alpha *}$ into the local reference frame $\mathfrak{X}^{\alpha *} = (x^{\alpha 0}, x^{\alpha i})$, with

$$x^{\alpha 0} = x^\alpha, \qquad x^{\alpha i} = x^{\alpha 0} + d\left(X^{\alpha i}, X^{\alpha 0}\right) e^i, \tag{21}$$

where $x^\alpha$ is the image of $X^\alpha$ in the macroscopic placement and the unit vectors $e^i$ are the directions of the vectors $(x^i - x^0)$ in the macroscopic reference frame $\mathfrak{X}^*$. Finally, he places $\Xi^{N_\alpha}$ on the *microscopic placement* $\mathfrak{X}^{N_\alpha}$ whose points $x^{\alpha\beta}$ are determined by the distances

$$\left|x^{\alpha\beta} - x^{\alpha 0}\right| = d^{\alpha\beta 0}, \qquad \left|x^{\alpha\beta} - x^{\alpha i}\right| = d^{\alpha\beta i}. \tag{22}$$

[16]The identification of observers with reference frames, see, e.g., Truesdell and Noll [51], Sect. 17, is an example of confusion between "reality" and representation, which may lead to incorrect conclusions. See the discussion on the law of gravitation in Sect. 8.5 below.

The macroscopic placement $\mathfrak{X}^N$ and the $N$ microscopic placements $\mathfrak{X}^{N_\alpha}$ form the *two-scale placement* $(\mathfrak{X}^N, \{\mathfrak{X}^{N_\alpha}\})$ of the two-scale set $(\Xi^N, \{\Xi^{N_\alpha}\})$, relative to the two-scale reference frame $(\mathfrak{X}^*, \{\mathfrak{X}^{\alpha *}\})$.

A *two-scale reference placement* of the finite set $(\Xi^N, \{\Xi^{N_\alpha}\})$ is a pair $(\mathfrak{X}_R^N, \{\mathfrak{X}_R^{N_\alpha}\})$, where the $N$ points $x_R^\alpha$ which form $\mathfrak{X}_R^N$ are images of the body particles of $\Xi^N$ determined by the distances $d_R^{\alpha 0}, d_R^{\alpha i}$, and the $N_\alpha$ points $x_R^{\alpha\beta}$ which form each $\mathfrak{X}_R^{N_\alpha}$ are images of the body particles of $\Xi^{N_\alpha}$ determined by the distances $d_R^{\alpha\beta 0}, d_R^{\alpha\beta i}$.

Just as in the one-scale representation, the finite set $\mathfrak{X}_R^N$ can be extended to a three-dimensional region $\Omega_R$ of $\mathscr{E}$. Moreover, each set $\mathfrak{X}_R^{N_\alpha}$ can be extended to a three-dimensional neighborhood $\mathscr{N}_R^\alpha$ of the corresponding $x_R^\alpha$. The region $\Omega_R$ is again a representative on $\mathscr{E}$ of the body imagined as a continuum. With the subsets $\mathscr{N}_R^\alpha$ determined by distance measurements at the microscopic scale, this region becomes a *reference placement* of the body in a two-scale representation.

## 4.2 Two-Scale Deformations

Given a reference placement $\Omega_R$, a two-scale placement $(\mathfrak{X}^N, \{\mathfrak{X}^{N_\alpha}\})$ of $(\Xi^N, \{\Xi^{N_\alpha}\})$ can be described by a pair $(f^N, \{f^{N_\alpha}\})$, where $f^N$ is a deformation from $\mathfrak{X}_R^N$ and the $f^{N_\alpha}$ are deformations from the $\mathfrak{X}_R^{N_\alpha}$. By "filling" procedures, $f^N$ can be extended to a differentiable function $f$ over $\Omega_R$, and the $f^{N_\alpha}$ can be extended to differentiable functions $f^\alpha$ over neighborhoods $\mathscr{N}_R^\alpha$ of the points $x_R^\alpha$. If these neighborhoods are sufficiently small, the deformations $f^\alpha$ are conveniently described by their gradients $\nabla f^\alpha$ at $x_R^\alpha$. The $N$ tensors

$$F^\alpha\left(x_R^\alpha\right) = \nabla f^\alpha\left(x_R^\alpha\right) \tag{23}$$

can be extended to a continuous tensor field $F$ over $\Omega_R$. This field is a description of the microscopic deformation of the body, and the pair $(f, F)$ is a *two-scale deformation* from $\Omega_R$.[17]

In a two-scale deformation from $\Omega_R$, a point $x_R$ of $\mathscr{N}_R^\alpha$ is mapped into the point $f(x_R)$ by the macroscopic deformation and into the point $f^\alpha(x_R)$ by the microscopic deformation. Then the vector $e_R = x_R - x_R^\alpha$ is mapped into $\nabla f(x_R^\alpha)[e_R]$ and into $F(x_R^\alpha)[e_R]$, respectively, and the two vectors are different in general. Since the directions $e^i$ associated with the physical directions $E^i$ are the same in the macroscopic and in the local reference frames, this difference reveals a misalignment between the same physical direction represented at the two scales. This difference will be called the *distortion* of the two-scale deformation at $x_R$. In the special case $F = \nabla f$ the distortion is zero, and the two-scale deformation $(f, \nabla f)$ becomes indistinguishable from the one-scale deformation $f$. Therefore, a one-scale deformation $f$ can be viewed as a two-scale deformation of the particular form $(f, \nabla f)$.

The pair $(f(\Omega_R), F(\Omega_R))$ is the *two-scale placement* of $\Omega_R$ generated by the two-scale deformation $(f, F)$. A two-scale placement generated by a two-scale deformation of the form $(f, \nabla f)$ will be called *undistorted*. In particular, the reference placement $\Omega_R$ can be identified with the undistorted placement generated by the identical two-scale deformation $(\iota, I)$, where $\iota$ is the identical deformation $\iota(x) = x$ and $I$ is the identity tensor.

[17] Since $f^\alpha$ is differentiable, the conditions (i), (ii) imply that the tensor $\nabla f^\alpha(x^\alpha)$ is invertible and with positive determinant. Then the field $F$ is made of invertible second-order tensors with positive determinant. However, being the gradient of a different function $f^\alpha$ at each $x_R^\alpha$, $F$ is not expected to be itself a gradient.

## 4.3 Inverses and Compositions of Two-Scale Deformations

Let $(f, F)$ be a two-scale deformation from an undistorted placement $\Omega_R$, and let $f^{-1}$ and $F^{-1}$ be the inverses of $f$ and $F$. Then $(f^{-1}, F^{-1})$ is the two-scale deformation which brings the two-scale placement $(f(\Omega_R), F(\Omega_R))$ back to $\Omega_R$. This is the *inverse map* of $(f, F)$.

Now let $(f_1, F_1)$ and $(f_2, F_2)$ be two-scale deformations from $\Omega_R$. Since $(f_1, F_1)^{-1}$ maps $(f_1(\Omega_R), F_1(\Omega_R))$ into $\Omega_R$ and $(f_2, F_2)$ maps $\Omega_R$ into $(f_2(\Omega_R), F_2(\Omega_R))$, the composition

$$(f, F) = (f_2, F_2) \circ (f_1, F_1)^{-1} = \left(f_2 \circ f_1^{-1}, F_2 F_1^{-1}\right) \tag{24}$$

maps $(f_1(\Omega_R), F_1(\Omega_R))$ into $(f_2(\Omega_R), F_2(\Omega_R))$. Thus, a two-scale deformation between any two placements can be defined by a composition of deformations and inverse deformations from an *intermediate undistorted placement*.

The pair $(\iota, F)$ is a two-scale deformation in which all changes in the positions of the points of $\Omega_R$ occur at the microscopic level. A deformation of this type is called a *purely microscopic deformation*. On the contrary, $(f, \nabla f)$ is a *purely macroscopic deformations*, that is, a two-scale deformation in which the changes in the positions at the microscopic and at the macroscopic level are the same. As said above, a purely macroscopic deformation can be identified with a single-scale deformation. Special cases of (24) are decomposition

$$(f, F) = (f, \nabla f) \circ \left(\iota, F^{-1} \nabla f\right)^{-1} = \left(f \circ \iota, \nabla f \left(\nabla f^{-1} F\right)\right) \tag{25}$$

of a two-scale deformation $(f, F)$ into a purely microscopic deformation followed by purely macroscopic deformation, and the decomposition

$$(f, F) = \left(\iota, F \nabla f^{-1}\right) \circ (f, \nabla f) = (\iota \circ f) \circ \left(\left(F \nabla f^{-1}\right) \nabla f\right) \tag{26}$$

into a purely macroscopic deformation followed by a purely microscopic deformation. Both involve *multiplicative decompositions*

$$\nabla f = F \tilde{F}^d = \hat{F}^d F, \qquad \tilde{F}^d = F^{-1} \nabla f, \qquad \hat{F}^d = \nabla f F^{-1}, \tag{27}$$

of the macroscopic deformation gradient $\nabla f$ into the microscopic deformation tensor $F$ preceded, or followed, by tensors $\tilde{F}^d$, $\hat{F}^d$ which measure the *disarrangements* occurring at the microscopic scale.[18] The *additive decomposition* of the total deformation into a part without disarrangements plus a part due to the disarrangements

$$\nabla f = F + F^d \tag{28}$$

is usually considered appropriate only to infinitesimal deformations. Its physical relevance for finite deformations emerged only recently.[19] Note that the identities

$$\nabla f - F^d = \hat{F}^{d-1} \nabla f = \nabla f \tilde{F}^{d-1}, \tag{29}$$

establish a one-to-one correspondence between $F^d$, $\hat{F}^d$, $\tilde{F}^d$ for every fixed $\nabla f$.

---

[18] For an account of the literature on the subject see Deseri and Owen [15] and Owen [45]. In plasticity, the decompositions in (27) are called the *Kröner-Lee* and the *Clifton decomposition*, respectively. In the paper [11] they are compared with the additive decomposition (28).

[19] Its relevance is due to the form of the energy of a structured deformation, see Sect. 5.2 below.

### 4.4 Two-Scale Evolutions

Let $k \mapsto (f_k, F_k)$ be a family of two-scale deformations from an undistorted placement $\Omega_R$, based on distance measurements made at a finite number of instants $t_k$. By "filling" the discrete set $\{t_k\}$, a family $t \mapsto (f_t, F_t)$ of two-scale deformations can be constructed. This is a *two-scale evolution* of the body.

The construction involves the choice of a fixed reference system $(X^0, X^i)$, of its representative $(x_R^0, x_R^i)$ in $\mathscr{E}$, and of a family $t \mapsto (x_t^0, x_t^i)$ of reference frames, all with the same distances and orientation of the reference system. Let us see the effects of a change of family of reference frames.

Let $t \mapsto \mathfrak{X}_t^*$ and $t \mapsto \mathfrak{X}_t$ be two families of reference frames, and let $t \mapsto (f_t^*, F_t^*)$ and $t \mapsto (f_t, F_t)$ be the corresponding representations of the evolution. The transformation rules for the positions $f_t^*(x_R)$ and for the velocities $v_t^*(x_R)$ are the same as in (16), (18). It remains to determine the transformation rules for the microscopic deformations $F_t^*(x_R)$ and for their velocities.

For the microscopic deformations, the preservation of the lengths of the vectors $|F_t^*[e_R]|$ implies that there is an orthogonal tensor $\mathcal{Q}_t$ such that $F_t^* = \mathcal{Q}_t F_t$. Moreover, the preservation of the distortions implies

$$F_t^* e \cdot \nabla f_t^* e = F_t e \cdot \nabla f_t e \tag{30}$$

for all directions $e$ and, therefore, $\nabla f_t^{*T} F_t^* = \nabla f_t^T F_t$. Then, from (17),

$$\nabla f_t^T F_t = \nabla f_t^{*T} F_t^* = \nabla f_t^T Q_t^T \mathcal{Q}_t F_t. \tag{31}$$

This implies $Q_t^T \mathcal{Q}_t = I$, that is, $\mathcal{Q}_t = Q_t$. Then the transformation rule for the microscopic deformation is

$$F_t^* = Q_t F_t, \tag{32}$$

with the same $Q_t$ appearing in (16) and (18). Denoting by $V_t^*$, $V_t$ the time derivatives of $F_t^*$, $F_t$ and recalling that by (19) the time derivative of $Q_t$ is $Q_t Y_t$, by time differentiation we get

$$V_t^* = Q_t (V_t + Y_t F_t), \tag{33}$$

which is the transformation rule for the microscopic strain rates $V$.

## 5 The Causes of Deformation

In the previous sections the geometric aspects of the representation of physical bodies have been analyzed. Now it is time to explore the *causes of deformation*, and to investigate the relations between causes and effects.

### 5.1 External Energy and External Power

A deformation can be produced by several physical agents, such as contact with other bodies, gravitation, heat, electric and magnetic fields, and other factors acting either on the macroscale or on microscales of different orders. These agents are usually assumed to act through vectorial quantities called *forces* or *actions*. In the past, forces were considered as physical entities. Today, the prevailing view is that a body deforms because it receives some *energy* from the exterior, and that forces are derived quantities related to energy.[20]

For a body represented by the region $\Omega_R$, the *external energy* is assumed to be a function of the deformations from $\Omega_R$, which for a continuum with a two-scale geometry are the pairs $(f, F)$.[21]

$$\mathscr{E}_{ext}(\Omega_R) = \mathscr{E}_{ext}(\Omega_R, f, F). \tag{34}$$

It is currently assumed that there are two ways for supplying energy, *at distance* and *by contact*. In the first case energy is supplied to the interior of the body, and in the second case it is supplied through the boundary. In an evolution $t \mapsto (f_t, F_t)$, an external energy of the form

$$\mathscr{E}_{ext}(\Omega_R, f_t, F_t) = \int_{\Omega_R} \beta\big(f_t(x_R), F_t(x_R)\big)\,\mathsf{dv} + \int_{\partial\Omega_R} \sigma\big(f_t(x_R), F_t(x_R)\big)\,\mathsf{da}, \tag{35}$$

is assumed, with the volume density $\beta$ and the surface density $\sigma$ differentiable functions of the deformation $(f_t, F_t)$.[22]

By differentiation with respect to $t$, the *external power*

$$\mathscr{P}_{ext}(\Omega_R, v_t, V_t) = \int_{\Omega_R} (b_{Rt} \cdot v_t + B_{Rt} \cdot V_t)\,\mathsf{dv} + \int_{\partial\Omega_R} (s_{Rt} \cdot v_t + S_{Rt} \cdot V_t)\,\mathsf{da} \tag{36}$$

is obtained, where the *body forces* $b_{Rt}$, $B_{Rt}$ and the *surface tractions* $s_{Rt}$, $S_{Rt}$ are the partial derivatives of $\beta$ and $\sigma$, respectively. Therefore, the external power is the product of the *external actions* $b_{Rt}, B_{Rt}, s_{Rt}, S_{Rt}$ by the velocities $v_t$, $V_t$. Energy is a primitive concept, and the external actions are related quantities intervening in the description of the energy changes in an evolution. Since (36) holds for all evolutions, the velocities can be those of any possible evolution, that is, they can be any pair $(v_t, V_t)$ of *virtual velocities* from $(f_t(\Omega_R), F_t(\Omega_R))$.

## 5.2 Internal Energy and Internal Power

The *internal energy* of the body is supposed to depend on the two parts $F$, $F^d$ of the additive decomposition (28) of $\nabla f$, and on their gradients

$$\mathscr{E}_{int}(\Omega_R) = \mathscr{E}_{int}\big(\Omega_R, F, F^d, \nabla F, \nabla F^d\big). \tag{37}$$

The choice of $F$ and $F^d$ as independent variables is motivated by the theory of *structured deformations*, with which the two-scale deformations $(f, F)$ can be identified. It has been proved that each structured deformation is the limit of uniformly convergent sequences $n \mapsto f_n$ of piecewise continuous one-scale deformations,[23] and that the relaxed limit of the

[20] It has been said that "*force* has the same status in science as the notion of *epicycles* in astronomy; although the use of these terms may lead to correct results, it should not be supposed that they are part of nature itself" (attributed to G. Berkeley by Jammer [27], p. 204), and that forces should be rather considered as "*methodological intermediates* (between physical reality and representation) ... comparable to the so-called middle term in the traditional syllogism" (*ibid.*, preface to the Dover Edition, 1999).

[21] More in general, a dependence on additional *state variables* can be assumed, see Sect. 8 below.

[22] In fact there are energies, such as the frictional contact energy, which are only Gâteaux differentiable, see footnote 27. Here we assume differentiability just for simplicity. Note that $b_{Rt}(x_R)$ and $s_{Rt}(x_R)$ are vectors, while $B_{Rt}(x_R)$ and $S_{Rt}(x_R)$ are second-order tensors.

[23] Del Piero and Owen [14].

energies of such sequences is the sum of two parts depending on $F$ and $F^d$, respectively.[24] The subsequent extension to second-order structured deformations, to be identified with two-scale deformations, led to the additional dependence on $\nabla F$.[25] Here, the dependence on $\nabla F^d$ has been added.[26]

We make a distinction between the *free part* and the *dissipative part* of the energy. With the idea that dissipation is due to the disarrangements, we assume that the energy density associated with $F$ is a differentiable function $\varphi$, while the energy density associated with $F^d$ is the sum of a differentiable part $\varphi^d$ plus a *dissipation potential* $\phi^d$.[27] Therefore, the assumed form of the internal energy is

$$\mathscr{E}_{int}(\Omega_R, F, F^d, \nabla F, \nabla F^d) = \int_{\Omega_R} \big(\varphi(F, \nabla F) + \varphi^d(F^d, \nabla F^d) + \phi^d(F^d, \nabla F^d)\big)\,\mathsf{dv}, \tag{38}$$

where $(\varphi + \varphi^d)$ is the free part and $\phi^d$ is the dissipative part of the energy density. By differentiation with respect to $t$, the *internal power*

$$\begin{aligned}\mathscr{P}_{int}(\Omega_R, V, V^d, \nabla V, \nabla V^d) = \int_{\Omega_R} &(\nabla_1\varphi(F, \nabla F)\cdot V + \nabla_2\varphi(F, \nabla F)\cdot \nabla V\\ &+ \nabla_1\varphi^d(F^d, \nabla F^d)\cdot V^d + \nabla_2\varphi^d(F^d, \nabla F^d)\cdot \nabla V^d\\ &+ \check{\nabla}_1\phi^d(F^d, \nabla F^d) \triangleright V^d + \check{\nabla}_2\phi^d(F^d, \nabla F^d) \triangleright \nabla V^d)\,\mathsf{dv},\end{aligned} \tag{39}$$

is obtained, with $\nabla_1, \nabla_2$ partial derivatives with respect to $F$, $F^d$ and $\nabla F$, $\nabla F^d$, with $V$, $V^d$ virtual velocities for $F$ and $F^d$, and with

$$\check{\nabla}_1\phi^d(F^d, \nabla F^d) \triangleright V^d + \check{\nabla}_2\phi^d(F^d, \nabla F^d) \triangleright \nabla V^d > 0 \tag{40}$$

for all $V^d \neq 0$.

## 5.3 The Two Basic Postulates

The *law of conservation of energy* states that in every evolution of a body and at every $t$ the difference between the internal and the external energies is constant. Moreover, a part of the internal energy is assumed to be *free*, that is, fully recoverable by the application of suitable external actions, while the remaining part is irreversibly dissipated. This *conservation principle* and this *dissipation principle* are the two basic axioms of mechanics. In terms of powers, they can be stated as follows: in every evolution of a body and at every $t$,

[24] Choksi and Fonseca [5], Owen [44], Del Piero [8].

[25] Barroso et al. [3], Theorem 3.2.

[26] This additional dependence is generally assumed, see, e.g., the theory of Gurtin and Anand discussed below. Perhaps, in the theorem proved in [3] this dependence is excluded by some restrictive hypothesis.

[27] We recall that a dissipation potential is a positive Gâteaux differentiable function, and that a function $\phi$ is *Gâteaux differentiable* if the directional derivative

$$\check{\nabla}\phi(F^d) \triangleright V = \lim_{\varepsilon\to 0^+} \frac{\phi(F^d + \varepsilon V) - \phi(F^d)}{\varepsilon}$$

exists at all $F^d$ and for all directions $V$. For homogeneous functions $\mathfrak{h}$ of degree one, the notation $\mathfrak{h}(V) = \mathfrak{h} \triangleright V$ was introduced in [13], Sect. 2.2, to which we refer for further details.

– *the internal power is equal to the external power,*
– *the internal power is the sum of a free part and a dissipative part.*

The two principles can be viewed as *purely mechanical versions* of the first two laws of thermodynamics.[28] From them, a fairly general axiomatic framework for mechanics can be deduced under an auxiliary assumption, the *cut principle* of Euler and Cauchy. None of the traditional axioms, such as the balance laws of linear and angular momentum or the indifference of the constitutive equations, is necessary. Indeed, as it will be shown below, both are consequences of the indifference of energy.

## 6 Classical Continua

A *classical continuum* is a continuum with $B = S = 0$. That is, a continuum in which the external power depends only on the macroscopic deformation

$$\mathscr{P}_{ext}(\Omega_R, v_t) = \int_{\Omega_R} b_{Rt} \cdot v_t \,\mathsf{dv} + \int_{\partial\Omega_R} s_{Rt} \cdot v_t \,\mathsf{da}. \tag{41}$$

The absence of the virtual velocity $V_t$ notably simplifies the scheme introduced above.

### 6.1 Indifference and Balance Laws

We recall that $v_t$ is the velocity of points placed on $\mathscr{E}$ according to a reference frame chosen by the placer. Though the placer may choose a reference frame variable with $t$, it is just matter of common sense that energy and power cannot depend on such choices. Therefore, once a reference system has been chosen, if $t \mapsto f_t$ and $t \mapsto f_t^*$ are the representations of the same evolution with respect to two families $t \mapsto \mathfrak{X}_t$, $t \mapsto \mathfrak{X}_t^*$ of reference frames, not only the virtual velocities $v_t$ and $v_t^*$ are different, but also $b_{Rt}$ and $s_{Rt}$ must transform in such a way that

$$\int_{\Omega_R} b_{Rt}^* \cdot v_t^* \,\mathsf{dv} + \int_{\partial\Omega_R} s_{Rt}^* \cdot v_t^* \,\mathsf{da} = \int_{\Omega_R} b_{Rt} \cdot v_t \,\mathsf{dv} + \int_{\partial\Omega_R} s_{Rt} \cdot v_t \,\mathsf{da}. \tag{42}$$

By (18), the transformation law for $v_t$ is

$$v_t^*(x_R) = Q_t\big(v_t(x_R) + Y_t[x_t - x_{\emptyset t}] + a_t\big), \qquad x_t = f_t(x_R),\ x_{\emptyset t} = f_t(x_{\emptyset R}), \tag{43}$$

with $Q_t$ an orthogonal tensor, $Y_t$ a skew-symmetric tensor, and $a_t$ a vector. Substituting into (42) and omitting for simplicity the reference to $t$, we get

$$\begin{aligned} &\int_{\Omega_R} \big(b_R^* \cdot Q\big(v + Y[x - x_\emptyset] + a\big) - b_R \cdot v\big) \,\mathsf{dv} \\ &\quad + \int_{\partial\Omega_R} \big(s_R^* \cdot Q\big(v + Y[x - x_\emptyset] + a\big) - s_R \cdot v\big) \,\mathsf{da} = 0, \end{aligned} \tag{44}$$

for all $Q, Y, a$ and for all virtual velocities $v$. In particular, for $Y = 0$ and $a = 0$, from the

[28] See, e.g., Gurtin et al. [25], Sect. 29.

arbitrariness of $v$ we deduce the transformation laws of the external actions under a change of reference frame[29]

$$b_R^* = Qb_R, \qquad s_R^* = Qs_R. \tag{45}$$

Then all terms involving $v$ cancel from (44), which by this way reduces to

$$\int_{\Omega_R} b_R \cdot \big(Y[x - x_\emptyset] + a\big)\,\mathsf{dv} + \int_{\partial\Omega_R} s_R \cdot \big(Y[x - x_\emptyset] + a\big)\,\mathsf{da} = 0. \tag{46}$$

Since $a$ and $Y$ are mutually independent, this results in the separate conditions

$$\mathscr{P}_{ext}(\Omega_R, a) = 0, \qquad \mathscr{P}_{ext}\big(\Omega_R, Y[x - x_\emptyset]\big) = 0. \tag{47}$$

Using the representation $Yu = y \times u$ for all vectors $u$, from the arbitrariness of $a$ and $y$ the *Euler balance laws of linear and angular momentum*

$$\begin{aligned} \int_{\Omega_R} b_R\,\mathsf{dv} + \int_{\partial\Omega_R} s_R\,\mathsf{da} = 0, \\ \int_{\Omega_R} (x - x_\emptyset) \times b_R\,\mathsf{dv} + \int_{\partial\Omega_R} (x - x_\emptyset) \times s_R\,\mathsf{da} = 0, \end{aligned} \tag{48}$$

follow.[30] Therefore, *the balance laws are consequences of the indifference of the external power.*[31]

A current interpretation of conditions (47) is that *the power of all rigid velocities is zero*, and this is called the *principle of independence of the power on superimposed rigid motions*. This *principle* met large success, because it agrees with "engineering" intuition. In reality, the meaning of conditions (47) is deeper. It says that the energy depends on the *configurations* of the body, but not on the placements. Placements belonging to the same configuration have the same energy, and the power is zero in all evolutions involving placements belonging to the same configuration. By consequence, to get conditions (47) there is no need of *superimposing* any *rigid motion*, whatever this is intended to mean.[32]

### 6.2 The Cut Principle and the Stress Tensor

The *cut principle of Euler and Cauchy* is the assumption that an external power of the form (41) can be defined for any part $\Pi_R$ of $\Omega_R$. In particular, surface tractions $s_{\partial\Pi_R}$ can be defined at every interior surface $\partial\Pi_R$ of the body in the same way in which they were defined for $\partial\Omega_R$.[33] Then, omitting for simplicity the reference to $x_R$ and $t$, for every part $\Pi_R$ of $\Omega_R$ and

[29] Therefore, the external actions are *indifferent*, in the sense that they obey the transformation law (14) for vector fields.

[30] Note that the "moment of momentum" is evaluated at the points $x_t$ of the region $f_t(\Omega_R)$ and not at the points $x_{Rt}$ of $\Omega_R$.

[31] This deduction of the balance laws from indifference is due to Noll [35]. It answers the questions raised by Truesdell about the nature of this law [49].

[32] Since the property (47) is a direct consequence of the conservation principle, to call it *principle* is inappropriate. In Noll's words, "the principle of material frame-indifference is not a law of physics… it is merely a prescription for avoiding nonsense" [42].

[33] Truesdell [50], p. 154. This assumption excludes, for example, the presence of different surface effects at the physical boundary and at the interior surfaces of the body.

for every $t$ we have

$$\mathscr{P}_{ext}(\Pi_R, v) = \int_{\Pi_R} b_R \cdot v \,\mathsf{dv} + \int_{\partial \Pi_R} s_{\partial \Pi_R} \cdot v \,\mathsf{da}, \tag{49}$$

with the same body force $b_R$ as in (41), with $s_{\partial \Pi_R}$ the contact action on $\partial \Pi_R$, and with $v$ any virtual velocity.

Consider a family $\varepsilon \mapsto \Pi_{R\varepsilon} = \varepsilon \Pi_R$ of homothetic regions, and for each $\Pi_{R\varepsilon}$ write the balance law $(48)_1$. For $\varepsilon \to 0$, the volume integral tends to zero faster than the area of $\Pi_{R\varepsilon}$. Then, in the limit, the area integral also tends to zero faster than $\mathsf{a}(\Pi_{R\varepsilon})$

$$\lim_{\varepsilon \to 0+} \frac{1}{\mathsf{a}(\Pi_{R\varepsilon})} \int_{\partial \Pi_{R\varepsilon}} s_{\partial \Pi_{R\varepsilon}} \,\mathsf{da} = 0. \tag{50}$$

From this property, taking regions $\Pi_R$ of particular shapes it has been proved that the contact action $s_{\partial \Pi_R}$ at a boundary point of $\Pi_R$ depends only on the exterior unit normal to $\partial \Pi_R$ at that point,[34] and that this dependence is linear.[35] That is, that there is a tensor field $T_R$ over $\Omega_R$ such that[36]

$$s_{\partial \Pi_R}(x_R) = T_R(x_R) n_R \tag{51}$$

for all regions $\Pi_R$ for which $x_R$ is a boundary point. The tensor $T_R(x_R)$ is the *Piola stress* at $x_R$.

Substituting into (49) and using the divergence theorem, the external power takes the form of a volume integral

$$\mathscr{P}_{ext}(\Pi_R, v) = \int_{\Pi_R} \big((\operatorname{div} T_R + b_R) \cdot v \,\mathsf{dv} + T_R \cdot \nabla v\big) \,\mathsf{dv}, \tag{52}$$

where, by the indifference condition $(47)_1$,

$$\operatorname{div} T_R + b_R = 0. \tag{53}$$

Moreover, recalling that $x - x_\emptyset = f(x_R) - f(x_{\emptyset R})$, we have

$$T_R \cdot \nabla\big(Y[x - x_\emptyset]\big) = T_R \cdot Y \nabla f = T_R \nabla f^T \cdot Y, \tag{54}$$

and from condition $(47)_2$ we deduce the symmetry of the tensor $T_R \nabla f^T$.[37] In conclusion, the external power (52) takes the *reduced form*

$$\mathscr{P}_{ext}(\Pi_R, v) = \int_{\Pi_R} T_R \cdot \nabla v \,\mathsf{dv}, \qquad \big(T_R \nabla f^T\big)^W = 0, \tag{55}$$

where $A^W$ denotes the skew-symmetric part of a tensor $A$.

---

[34] This was a conjecture of Cauchy, proved by Noll in [34].

[35] This is the famous *tetrahedron theorem* of Cauchy.

[36] In particular, for opposite directions $n_R$ and $-n_R$ this implies

$$s_R(x_R, n_R) = -s_R(x_R, -n_R),$$

which is the local form of Newton's *law of action and reaction*.

[37] We recall that $(\det F)^{-1} T_R \nabla f^T$ is the Cauchy stress tensor. Therefore, this condition states the symmetry of the Cauchy tensor.

In classical mechanics, the balance laws are usually taken as basic postulates. From them, the representation (51) of the contact actions is deduced using Noll's theorem (formerly, *Cauchy's postulate*) and the tetrahedron theorem of Cauchy. Then, the local forms (53), $(55)_2$ of the balance laws follow from the divergence theorem. Since this approach leads to the same results of the approach based on energy conservation, a choice could be considered as matter of taste. However, as we shall see, there are some difficulties in extending the approach based on balance laws to generalized continua.

## 6.3 The "Equation of Virtual Power"

Equating the expressions (49) and $(55)_1$ of $\mathscr{P}_{ext}(\Pi_R, v)$, the equation

$$\int_{\Pi_R} T_R \cdot \nabla v \, \mathrm{dv} = \int_{\Pi_R} b_R \cdot v \, \mathrm{dv} + \int_{\partial \Pi_R} s_{\partial \Pi_R} \cdot v \, \mathrm{da} \tag{56}$$

is obtained. This equation is currently interpreted as an equality between the external power (49) and the internal power, which is supposed to have the form $(55)_1$. But $(55)_1$ is in fact an alternative expression of the external power, following from indifference and from the cut principle. Therefore, (56) is *an identity between two different forms of the external power*. It has nothing to do with the internal power. By consequence, $T_R$ *is not an internal action*, as it is commonly believed. It is a *tensorial representative of the external action*.

There have been attempts of taking (56) as the basic postulate of mechanics.[38] This has some advantages. Indeed, the existence of the stress tensor need not be proved anymore, since it is assumed from the beginning. Moreover, the balance laws and their local expressions easily follow from (56). A further reason in favor of this approach is that, as we shall see in Sect. 7, equation (56) can be easily extended to generalized continua.

## 6.4 The Constitutive Equation and the Activation Condition

For a classical continuum, the energy densities $\varphi$, $\varphi^d$ are assumed to be independent of $\nabla F$ and $\nabla F^d$, respectively.[39] Then the equality between the internal power (39) and the external power $(55)_1$ required by the conservation principle reduces to

$$\int_{\Omega_R} \big( \nabla\varphi(F) \cdot \big( \nabla v - V^d \big) + \nabla\varphi^d\big(F^d\big) \cdot V^d + \check{\nabla}\phi^d\big(F^d\big) \triangleright V^d \big) \, \mathrm{dv} = \int_{\Omega_R} T_R \cdot \nabla v \, \mathrm{dv}, \tag{57}$$

where $\nabla\varphi$, $\nabla\varphi^d$ stay in place of $\nabla_1\varphi$, $\nabla_1\varphi^d$ and $\nabla v - V^d$ stays in place of $V$. This is the real *equation of virtual power*, to be satisfied for all virtual velocities $v$, $V^d$. It expresses mathematically both the conservation and the dissipation principle.[40]

In particular, for $V^d = 0$ we get the *constitutive equation*

$$T_R = \nabla\varphi(F), \tag{58}$$

[38] Germain [20, 21].

[39] See Choksi and Fonseca [5].

[40] It is remarkable that a single equation collects the information traditionally given by two separate conditions, the energy balance and the entropy imbalance, see, e.g., [25], Sects. 26, 27. Note that in the present analysis it is assumed that the dissipation potential is positively homogeneous of order one. In reality, as explained in the *post-post scriptum* at the end of the paper, alternative options are possible.

which relates the stress tensor $T_R$ with the microscopic deformation $F$. There is a remarkable conceptual change in the way in which this equation has been deduced. Conventionally, a constitutive equation is conceived as a relation between internal actions and deformations. Here, (58) is a relation between the tensorial measure $T_R$ of the external actions and the corresponding internal action $\nabla\varphi$, which, in turn, is a function of the deformation $F$. This has the consequence that the stress measures, being determined by the external actions which appear in the external power, cannot be anymore chosen arbitrarily.[41]

Substituting (58) into (57) the terms in $\nabla v$ cancel, and (57) reduces to

$$\int_{\Omega_R} \left(T_R - \nabla\varphi^d(F^d)\right) \cdot V^d \,\mathsf{dv} = \int_{\Omega_R} \check{\nabla}\phi^d(F^d) \rhd V^d \,\mathsf{dv}. \tag{59}$$

In view of the dissipative character of $\phi^d$, by localization we get

$$\left(T_R - \nabla\varphi^d(F^d)\right) \cdot V^d = \check{\nabla}\phi^d(F^d) \rhd V^d \geq 0. \tag{60}$$

The deduction of (58) and (60) from the two basic postulates is a purely mechanical counterpart of the *Coleman-Noll procedure* for the deduction of constitutive relations in thermodynamics.[42] In particular, the relation (60) plays a double role: it restricts the deformations $(f, F)$ at which a non-null strain rate $\delta F^d$ can take place, and determines its direction. Indeed, when restricted to the unit tensors $N$

$$\left(T_R - \nabla\varphi^d(F^d)\right) \cdot N = \check{\nabla}\phi^d(F^d) \rhd N, \qquad |N| = 1, \tag{61}$$

this relation tells us that a non-null $\delta F^d$ can take place in a direction $N$ only if the projection of $(T_R - \nabla\varphi^d(F^d))$ along $N$ is equal to the directional derivative of $\varphi^d$ along the same direction. In an evolution with $T_R = \nabla\varphi^d(F^d) = 0$ at $t = 0$, for sufficiently small $t > 0$ we have

$$\left(T_{Rt} - \nabla\varphi^d(F^d_t)\right) \cdot N < \check{\nabla}\phi^d(F^d_t) \rhd N \qquad \forall N, \tag{62}$$

and therefore the relation (60) is satisfied only by $V^d = 0$. If at some $t > 0$ the equality is reached at $N = N^d$, at this $t$ a non-null strain rate $\delta F^d$ can take place in the direction $N^d$.[43] Thus, the relation (61) is both an *activation condition* for $\delta F^d$ and a *flow rule* which specifies its direction.[44]

---

41 As done, for example, in the gradient plasticity theory of Sect. 7.4.

42 Coleman and Noll [7]. In particular, (60) is a mechanical counterpart of their relation (5.1), and the inequality on the right is a mechanical counterpart of the Clausius-Duhem inequality ([51], Sect. 79). However, while in [7] the stress tensor $T_R$ is regarded as an internal action, and for it a constitutive equation is postulated, here $T_R$ is an internal action, and the constitutive equation (58) is *deduced* from the conservation principle.

43 Here and in the following for the "time" derivative I use the symbol $\delta$, reserving the more familiar superimposed dot to the case in which "time" is the physical time.

44 In the paper [13] it has been shown that $\check{\nabla}\phi^d$ is the *bounding map* which determines the *elastic region* in which the tensor $(T_R - \nabla\varphi^d(F^d))$ must lie, and that $\nabla\varphi^d(F^d)$ is the *backstress tensor* which appears, for example, in the kinematical hardening model. If the boundary of the elastic region has corner points, the determination of $\delta F^d$ at such points is not unique.

### 6.5 Determination of $\delta F^d$

The direction $N^d$ of $\delta F^d$ is determined, not uniquely in general, by condition (61). To determine its magnitude, consider the "time" derivative of (58)

$$\delta T_R = \nabla^2\varphi(F)[\delta F], \tag{63}$$

which, after setting

$$\nabla^2\varphi(F) = \mathbb{C}, \quad \delta F = \delta(\nabla f - F^d), \quad \delta\nabla f = \nabla u, \quad \delta F^d = \lambda^d N^d, \tag{64}$$

takes the form

$$\delta T_R = \mathbb{C}[\nabla u] - \lambda^d \mathbb{C}[N^d]. \tag{65}$$

Then consider the derivative of (61) at $N = N^d$

$$\big(\delta T_R - \delta\nabla\varphi^d(F^d)\big)\cdot N^d = \delta\breve{\nabla}\phi^d(F^d) \triangleright N^d, \tag{66}$$

which, using (65) and setting[45]

$$\begin{aligned} &\delta\nabla\varphi^d(F^d) = \nabla^2\varphi^d(F^d)[\delta F^d], && \delta\breve{\nabla}\phi^d(F^d) = \breve{\nabla}^2\phi^d(F^d)\{\delta F^d\}, \\ &\mathbb{C}^d = \nabla^2\varphi^d(F^d), && \breve{\mathbb{K}} = \breve{\nabla}^2\phi^d(F^d), \end{aligned} \tag{67}$$

takes the form

$$\mathbb{C}[\nabla u]\cdot N^d = \lambda^d\big((\mathbb{C}+\mathbb{C}^d)[N^d]\cdot N^d + \breve{\mathbb{K}}\{N^d\} \triangleright N^d\big). \tag{68}$$

If the sum on the right is positive, the system made of this equation plus the constraint $\lambda^d \geq 0$ has the unique solution

$$\lambda^d = \frac{(\mathbb{C}[\nabla u]\cdot N^d)^+}{(\mathbb{C}+\mathbb{C}^d)[N^d]\cdot N^d + \breve{\mathbb{K}}\{N^d\} \triangleright N^d}, \tag{69}$$

where $a^+ = \max\{a, |a|\}$ is the positive part of a real number $a$. This solution is a positively homogeneous function of degree one of $\mathbb{C}[\nabla u]\cdot N^d$.

## 7 Generalized Continua

A *generalized continuum* is a continuum whose external energy does not depend only on the macroscopic deformation $f$. Here, the dependence (34) on the two-scale deformation $(f, F)$ is considered. The dependence on additional state variables is briefly discussed in Sect. 8.

---

[45] Equation $(67)_2$ implies the existence of the second directional derivative

$$\breve{\nabla}^2\phi^d(A)\{H\} \triangleright K = \lim_{\varepsilon\to 0^+} \frac{\breve{\nabla}\phi^d(A+\varepsilon H) \triangleright K - \breve{\nabla}\phi^d(A) \triangleright K}{\varepsilon}.$$

### 7.1 The External Power

Let us consider generalized continua in which the external power has the form (36) and the internal power has the form (39). Recalling the transformation law (33) for $V$, the indifference condition (44) is now replaced by

$$\int_{\Omega_R} \big(b_R^* \cdot Q\big(v + Y[x - x_\emptyset] + a\big) - b_R \cdot v + B_R^* \cdot Q(V + YF) - B_R \cdot V\big)\,\mathsf{dv}$$

$$+ \int_{\partial\Omega_R} \big(s_R^* \cdot Q\big(v + Y[x - x_\emptyset] + a\big) - s_R \cdot v + S_R^* \cdot Q(V + YF) - S_R \cdot V\big)\,\mathsf{da} = 0. \quad (70)$$

For $Y = 0$ and $a = 0$, we find the transformation laws

$$b_R^* = Qb_R, \qquad B_R^* = QB_R, \qquad s_R^* = Qs_R, \qquad S_R^* = QS_R, \quad (71)$$

and substitution into (70) yields

$$\mathscr{P}_{ext}(\Omega_R, a, 0) = 0, \qquad \mathscr{P}_{ext}\big(\Omega_R, Y[x - x_\emptyset], YF\big) = 0. \quad (72)$$

The first condition again provides the balance law $(48)_1$ of linear momentum. Using the cut principle, from this condition we deduce the representation (51) for the contact actions $s_{\partial\Pi_R}$ and the local balance law (53). The external power (36) then reduces to

$$\mathscr{P}_{ext}(\Omega_R, v, V) = \int_{\Omega_R} (T_R \cdot \nabla v + B_R \cdot V)\,\mathsf{dv} + \int_{\partial\Omega_R} S_R \cdot V\,\mathsf{da}. \quad (73)$$

Thus, for generalized continua the indifference of power does not reduce the external power to a volume integral as in the case of classical continua. However, this reduction is obtained directly, as a consequence of the conservation of energy.

### 7.2 The Pseudobalance Equation

A consequence of energy conservation is the equation of virtual power. By extending the cut principle to the external actions $B_R$, $S_R$, this equation can be written for any part $\Pi_R$ of $\Omega_R$. In particular, the external power over $\Pi_R$ has the expression (73), with the same $T_R$ and $B_R$, and with $S_R$ replaced by the contact actions $S_{\partial\Pi_R}$ at the boundary of $\Pi_R$. Equating to the corresponding internal power (29) and taking $V^d = 0$ and $\nabla v = V$ an arbitrary constant tensor field, we get

$$\int_{\partial\Pi_R} S_{\partial\Pi_R}\,\mathsf{da} = \int_{\Pi_R} (\nabla_1\varphi - T_R - B_R)\,\mathsf{dv}. \quad (74)$$

As shown by the balance law $(48)_1$ applied to $\Pi_R$, the contact actions $s_{\partial\Pi_R}$ of the classical continuum have the volume density $b_R^c = -b$.[46] Now, (74) shows that *the contact actions*

[46] That is, the volume density for the contact actions is the opposite of the volume density for the distance actions. The assumption (74) was made in 1973 by Noll, who called it *area-volume continuity* ([37], p. 78). The contact actions with this property were called *weakly balanced Cauchy fluxes* by Gurtin and Martins [23] and by Šilhavý [47], who used assumption (74) to weaken the regularity hypotheses made in the standard proof of the tetrahedron theorem.

$S_{\partial\Pi_R}$ *also have a volume density*, and that this density is $B_R^c = \nabla_1\varphi - T_R - B_R$. The difference is that, while for the contact actions $s_{\partial\Pi_R}$ the existence of the volume density follows from the indifference conditions (47), here it follows from energy conservation.

Due to its resemblance to the balance equation of linear momentum, equation (74) is called a *pseudobalance equation*.[47] This equation *looks like, but is not, a balance equation.* Indeed, while a balance equation involves an internal and an external action, (74) involves two representatives, surface and volume, of the same external action.

Taking a family $\varepsilon \mapsto \Pi_{R\varepsilon} = \varepsilon\Pi_R$ of homothetic regions, we see that the volume integral in (74) tends to zero faster than the area of $\Pi_{R\varepsilon}$. Then the contact actions $S_{\partial\Pi_R}$ have the property

$$\lim_{\varepsilon\to 0^+} \frac{1}{\mathsf{a}(\Pi_{R\varepsilon})} \int_{\partial\Pi_{R\varepsilon}} S_{\partial\Pi_{R\varepsilon}}\,\mathsf{da} = 0, \tag{75}$$

that is, the same property (50) of the contact actions $s_{\partial\Pi_R}$. Therefore, using the theorems of Noll and Cauchy, it can be proved that $S_{\partial\Pi_R}$ depends only on the normal $n_R$ to $\partial\Pi_R$, and that this dependence is linear. That is, there exists a third-order tensor $\mathbb{T}_R$ such that

$$S_{\partial\Pi_R} = \mathbb{T}_R n_R, \qquad (S_{ij} = \mathbb{T}_{ijk} n_k). \tag{76}$$

By substitution into (73) and use of the divergence theorem, the external power takes the form of a volume integral

$$\mathscr{P}_{ext}(\Pi_R, v, V) = \int_{\Pi_R} \big(T_R \cdot \nabla v + (\operatorname{div}\mathbb{T}_R + B_R)\cdot V + \mathbb{T}_R \cdot \nabla V\big)\,\mathsf{dv}, \tag{77}$$

for which the indifference condition $(72)_2$ becomes[48]

$$Y \cdot \int_{\Pi_R} \big(T_R \nabla f^T + (\operatorname{div}\mathbb{T}_R + B_R) F^T + \mathbb{T}_R \nabla F^T\big)\,\mathsf{dv} = 0, \tag{78}$$

for all skew-symmetric tensors $Y$. By the arbitrariness of $\Pi_R$ and $Y$ we conclude that the integrand function must be a symmetric tensor, that is, that

$$\big(T_R \nabla f^T + \big(\operatorname{div}\mathbb{T}_R + B_R\big) F^T + \mathbb{T}_R \nabla F^T\big)^W = 0. \tag{79}$$

### 7.3 The Coleman-Noll Procedure

By the conservation principle, the external power (77) must be equal to the internal power (39). Then setting $\nabla v = V + V^d$ and omitting the arguments $(F, \nabla F)$, $(F^d, \nabla F^d)$, we get

$$\begin{aligned}&\int_{\Pi_R} \big(T_R \cdot \big(V + V^d\big) + (\operatorname{div}\mathbb{T}_R + B_R)\cdot V + \mathbb{T}_R \cdot \nabla V\big)\,\mathsf{dv}\\ &= \int_{\Pi_R} \big(\nabla_1\varphi \cdot V + \nabla_2\varphi \cdot \nabla V + \nabla_1\varphi^d \cdot V^d + \nabla_2\varphi^d \cdot \nabla V^d + \check{\nabla}_1\phi^d \triangleright V^d + \check{\nabla}_2\phi^d \triangleright \nabla V^d\big)\,\mathsf{dv},\end{aligned} \tag{80}$$

[47] Del Piero [9]. Equation (74) suggests that the contact actions can be represented as distance actions and vice versa. The founders of continuum mechanics were well aware of this fact, see, e.g., Maxwell [30] and (Love [28], Note B of the Appendix).

[48] In components, $(\operatorname{div}\mathbb{T}_R F^{dT})_{ih} = \mathbb{T}_{ijk,k} F^d_{hj}$ and $(\mathbb{T}_R \nabla F^{dT})_{ih} = \mathbb{T}_{ijk} F^d_{hj,k}$.

and since $V$ and $\nabla V$ can be varied independently, we conclude that

$$\begin{gathered} T_R + \operatorname{div} \mathbb{T}_R + B_R = \nabla_1 \varphi, \quad \mathbb{T}_R = \nabla_2 \varphi, \\ \left(T_R - \nabla_1 \varphi^d\right) \cdot V^d - \nabla_2 \varphi^d \cdot \nabla V^d = \breve{\nabla}_1 \phi^d \triangleright V^d + \breve{\nabla}_2 \phi^d \triangleright \nabla V^d \geq 0. \end{gathered} \tag{81}$$

These relations, plus the local balance law (53), are the result of the Coleman-Noll procedure. For the particular generalized continuum considered here, this procedure does not determine uniquely the constitutive equations, the field equations, and the activation condition, as it occurred for the classical continuum. Indeed, in the present case in the expression (36) of the external power there are two tensorial measures, $T_R$ and $\mathbb{T}_R$, of external actions. While for $\mathbb{T}_R$ the field equation is $(81)_1$ and the constitutive equation is $(81)_2$, any constitutive equation and field equation compatible with condition $(81)_3$ is admissible for $T_R$.

If for $T_R$ we choose the constitutive equation (58) of the one-scale continuum, we have the constitutive equations

$$T_R = \nabla_1 \varphi, \qquad \mathbb{T}_R = \nabla_2 \varphi, \tag{82}$$

and from (53) and $(81)_1$ we have the field equations

$$\operatorname{div} T_R + b_R = 0, \qquad \operatorname{div} \mathbb{T}_R + B_R = 0. \tag{83}$$

Moreover, taking $\varphi^d$ and $\phi^d$ independent of $\nabla F^d$ in accordance with [3], the activation condition for $\delta F^d$ reduces to

$$\left(T_R - \nabla_1 \varphi^d\right) \cdot V^d = \breve{\nabla}_1 \phi^d \triangleright V^d \geq 0. \tag{84}$$

Finally, the indifference condition (79) reduces to

$$\left(T_R \nabla F^T + \mathbb{T}_R \nabla F^T\right)^W = 0. \tag{85}$$

As said above, this is not the only possible choice. A well known alternative choice is made in the gradient theory of plasticity, which we discuss in the next section.

## 7.4 Comparison with the Gradient Theory of Plasticity

Several theories of gradient plasticity have been formulated since the initial idea of Aifantis.[49] Here we consider the gradient theory of Gurtin and Anand [24] in the version given in [25], and compare it with our previous results. Following our notation for the virtual velocities and superimposing a hat on Gurtin and Anand's external actions, the expressions of the internal and external power assumed in equations (90.8) and (90.12) of [25] become

$$\begin{aligned} \mathscr{P}^d_{int}\left(\Pi_R, V, V^d\right) &= \int_{\Pi_R} \left(\hat{T}_R \cdot V + \hat{T}_R^d \cdot V^d + \hat{\mathbb{T}}_R \cdot \nabla V^d\right) \mathsf{dv}, \\ \mathscr{P}^d_{ext}\left(\Pi_R, v, V^d\right) &= \int_{\Pi_R} \hat{b}_R \cdot v \,\mathsf{dv} + \int_{\partial \Pi_R} \left(\hat{s}_{\partial \Pi_R} \cdot v + \hat{S}_{\partial \Pi_R} \cdot V^d\right) \mathsf{da}, \end{aligned} \tag{86}$$

with $\hat{T}_R^d$ and $\hat{\mathbb{T}}_R$ *microscopic stress* and *microscopic hyperstress* measures assumed in addition to the macroscopic stress $\hat{T}_R$. Equating the two powers, using the relations

$$\hat{s}_{\partial \Pi_R} = \hat{T}_R n_R, \qquad \hat{S}_{\partial \Pi_R} = \hat{\mathbb{T}}_R n_R, \tag{87}$$

[49] Aifantis [1], Fleck and Hutchinson [18], Gudmundson [22], Fleck and Willis [19], etc.

the divergence theorem, and the arbitrariness of $v$ and $V^d$, they get their field equations (90.18) and (90.23)

$$\operatorname{div} \hat{T}_R + \hat{b}_R = 0, \qquad \operatorname{div} \hat{\mathbb{T}}_R + \hat{T}_R - \hat{T}_R^d = 0. \tag{88}$$

A dissipative term $\hat{\delta}$ is introduced in the *local free energy imbalance equation* (90.27)

$$\hat{T}_R \cdot V + \hat{T}_R \cdot V^d + \hat{\mathbb{T}}_R \cdot \nabla V^d - \delta\psi = \hat{\delta} \geq 0. \tag{89}$$

For the free energy the constitutive choice

$$\psi = \psi\big(F, \nabla F^d\big) \tag{90}$$

is made, from which equation (90.36)

$$\delta\psi = \hat{T}_R \cdot V + \hat{\mathbb{T}}_R^{en} \cdot \nabla V^d \tag{91}$$

follows, with $\hat{\mathbb{T}}_R^{en}$ the "energetic part" of $\hat{\mathbb{T}}_R$. By substitution into (89) they get

$$\hat{T}_R \cdot V^d + \hat{\mathbb{T}}_R^{diss} \cdot \nabla V^d = \delta \geq 0, \tag{92}$$

where $\hat{\mathbb{T}}_R^{diss} = \hat{\mathbb{T}}_R - \hat{\mathbb{T}}_R^{en}$ is the "dissipative part" of $\hat{\mathbb{T}}_R$.

Comparing with the results of the previous section, we first observe that the external power $(86)_1$ involves the virtual velocity $V^d$ instead of $V$ as in (36). Thus, a preliminary point to verify is whether, at least in some cases, the two expressions of the external power are equivalent. For this purpose, rewrite the power (36) in the form

$$\begin{aligned}
&\int_{\Pi_R} \big(b_R \cdot v + B_R \cdot \big(\nabla v - V^d\big)\big)\,\mathsf{dv} + \int_{\partial\Pi_R} \big(T_R n_R \cdot v + \mathbb{T}_R n_R \cdot \big(\nabla v - V^d\big)\big)\,\mathsf{da} \\
&\quad = \int_{\Pi_R} \big((b_R - \operatorname{div} B_R) \cdot v - B_R \cdot V^d + \operatorname{div} \mathbb{T}_R \cdot \nabla v + \mathbb{T}_R \cdot \nabla^2 v\big)\,\mathsf{dv} \\
&\qquad + \int_{\partial\Pi_R} \big((B_R + T_R) n_R \cdot v - \mathbb{T}_R n_R \cdot V^d\big)\,\mathsf{da}.
\end{aligned} \tag{93}$$

If

$$\mathbb{T}_R \cdot \nabla^2 v = 0, \tag{94}$$

we get the power $(86)_2$, with

$$\hat{b}_R = b_R - \operatorname{div} B_R, \qquad \hat{B}_R = -B_R, \qquad \hat{s}_{\partial\Pi_R} = (B_R + T_R) n_R, \qquad \hat{S}_{\partial\Pi_R} = -\mathbb{T}_R n_R. \tag{95}$$

In (86) $\hat{B}_R$ is assumed to be zero. Then $B_R = 0$, and from (87) we get

$$\hat{b}_R = b_R, \qquad \hat{B}_R = 0, \qquad \hat{s}_{\partial\Pi_R} = -s_{\partial\Pi_R}, \qquad \hat{S}_{\partial\Pi_R} = -S_{\partial\Pi_R}. \tag{96}$$

Thus, the two expressions of the external power are equivalent if condition (94) is satisfied for all $\nabla^2 v$, that is, if $\mathbb{T}_R$ is skew-symmetric with respect to the last two indices. In Gurtin and Anand's theory, this assumption is indeed verified, as follows from their equations (90.34), (90.46).

We now consider the internal power $(86)_1$, for which the comparison with (36) leads to the identifications

$$\begin{aligned} &\nabla_1\varphi = \hat{T}_R, \qquad \nabla_2\varphi = 0, \qquad \nabla_1\varphi^d = \hat{T}_R^d, \qquad \nabla_2\varphi^d = \hat{\mathbb{T}}_R^{en}, \\ &\check{\nabla}_1\phi^d = 0, \qquad \check{\nabla}_2\phi^d = \hat{\mathbb{T}}_R^{diss}. \end{aligned} \tag{97}$$

Thus, the model of Gurtin and Anand is a particular generalized continuum of the type analyzed in the previous section, with $\mathbb{T}_R$ skew-symmetric with respect to the last two indices and with $\varphi$ and $\phi^d$ independent of $\nabla v$ and $V^d$, respectively.

As to the procedure followed to construct the model, we point out that the equation of the free energy imbalance (89) is redundant, since it is sufficient to specify from the beginning that $\phi^d$ is the dissipative part of the energy (37). Then assumption (91) can be replaced by the declaration that, of the three terms which compose the power $(86)_1$, the first is free, the second is dissipative, and the third is the sum of a free and of a dissipative part.[50]

# 8 Walter Noll, and Beyond

The mechanics of two-scale continua developed in the previous sections suggests broader ideas in the direction of the axiomatization of continuum mechanics. A large part of the recent progress in this direction is due a single scientist, Walter Noll. I devote this final section to a quick review of his contributions and of their extensions to generalized continua.

## 8.1 A Summary of Noll's Achievements

"Walter Noll's influence upon research into the foundations of mechanics and thermodynamics in the past twenty years is plain, everywhere acknowledged". This *incipit* of the preface to the volume of Noll's selected papers [38] was written by C. Truesdell as early as the mid 1970's. Almost half a century later, many signs, including the massive dependence of the present paper on Noll's work, show that his influence is still far from fading.

Just to mention a few items, to him is due the sharp distinction between general laws of physics and constitutive equations describing the behavior of particular classes of materials,[51] as well as the redemption from the slavery of coordinates, now replaced by *intrinsic*, that is, coordinate-free, formulations.[52] To him and to B.D. Coleman is also due the *Coleman-Noll procedure*, which has a large impact on the growing theory of material behavior.

To him is also due the clarification of the concept of *configuration*, a term previously used rather informally, since only a few were aware of the precise meaning of *assemblage of relative positions* attributed to it by Maxwell [31]. Noll's formal definitions, in which *extrinsic* and *intrinsic configurations* were identified with placements and distance functions, led to frame-free definitions of deformation and evolution. With the consequent definitions of *intrinsic deformation* and *intrinsic stress*, they made possible a frame-free description of the body's evolutions [36].

[50] Note that it is incorrect to write $\hat{\mathbb{T}}_R^{diss} \triangleright \nabla V^p$ as an inner product as done in (90.48) of [25], since $\hat{\mathbb{T}}_R^{diss}$ is the map $\check{\nabla}_2\phi^d$, which is not linear but only homogeneous of order one.

[51] Truesdell and Noll [51], Sect. 1.

[52] See Noll's comments in [32], p. 8, [51], Sect. 15, [36], p. 2. For an intrinsic formulation of elasticity see [41], and for an extension to thermomechanics see Seguin's thesis [46].

About extrinsic configurations or *placements*, Noll's views evolved during the years. Initially, a *reference frame* was for him a set of objects in the physical space, determining the position and velocity of a body particle relative to that set. Later, he identified reference frames either with Euclidean spaces or with the associated translation spaces. This led to the ideas of *placement* and *motion* of the body in a *frame space*, by which the replacement of the "real" body with its representative on an Euclidean "absolute space" was eventually achieved ([36], p. 11).

Noll's activity covered other specific aspects of material behavior, such as the theory of *materials with memory*, later replaced by the *new theory* involving *state variables*. Both theories are discussed in the next two sections. Finally, among his contributions to the foundations, of primary importance are his deduction of the balance equations from the indifference of power [35], and his re-interpretation of the law of inertia [35, 40]. They are discussed in the final Sects. 8.4, 8.5.

## 8.2 State Variables

For some time after the formulation of the theory of constitutive equations, the research influenced by Noll's ideas was almost exclusively concentrated on simple materials with fading memory. Other models, like plasticity or damage, were considered rough engineering approximations, without a precise theoretical collocation. A turning point came with the "New Theory", with the introduction of the concept of *state variable*, already familiar in other fields of physics.[53] Roughly, a state variable is a physical agent which produces a deformation of the body. In the New Theory, the stress is a function of the current values of the deformation and of the state variables, and the latter are determined by the initial state and by the evolution $t \mapsto f_t$ of the deformation.

Noll's theory is restricted to classical continua with a one-scale geometry. In a continuum with a two-scale geometry it can be assumed that each state variable $\rho^k$ determines a microscopic deformation $F^k$, whose dependence on $\rho^k$ is specified by geometry if $\rho^k$ is a geometric variable, and by a *physical law* of experimental origin otherwise.[54] Just as for a two-scale continuum without state variables, the macroscopic deformation gradient $\nabla f$ can be decomposed into the sum

$$\nabla f = F + \sum_k F^k \quad k = 1 \ldots n, \tag{98}$$

of a microscopic deformation $F$ not due to the state variables plus $n$ deformations $F^k$ due to the state variables. Comparing with the decomposition (28), we see that the sum of the $F^k$ can be identified with the deformation $F^d$ of a two-scale continuum

$$F^d = \sum_k F^k. \tag{99}$$

[53] The paper [36] in which the New Theory was formulated did not receive an adequate attention. This had the unfortunate consequence that most authors refer to Noll's initial views rather than to their more mature developments. This is the case of the theory of fading memory which, as said in the next section, was criticized when it had already been set aside by the New Theory.

[54] For example, this is the case of microscopic deformations due to variations of temperature or of the electric or magnetic field.

In the previous sections, a reversible internal energy $\varphi^d$ and a dissipation potential $\phi^d$ were associated with $F^d$. Alternatively, a pair $(\varphi^k, \phi^k)$ can be associated with each $F^k$.[55] For all dissipation potentials which are homogeneous of order one, there is an activation law of the type (61).

According to the definition given in Sect. 6, in a classical continuum the external energy is independent of the state variables, while in a generalized continuum there is some dependence of the external energy on the state variables. We call *structural* the state variables which contribute to the external energy, and *constitutive* the state variables which appear in the internal energy and not in the external energy. For example, for a continuum characterized by the equation of virtual power (57) there are no structural state variables, the virtual velocity $v$ can be regarded as the only geometric variable, and $V^d$ can be regarded as a constitutive state variable.

## 8.3 Continua with Memory

A *simple material* is a material in which the current stress is determined by the past history of the deformation gradient.[56] Denoting by $\nabla f^t$ an evolution of $\nabla f$ over the infinite time interval $(-\infty, t)$, the constitutive equation of a simple material is

$$T_{Rt}(x_R) = \mathfrak{G}\big(\nabla f_t(x_R), \nabla f^t(x_R)\big). \tag{100}$$

Since it is difficult to believe that an observer can extend his measurements to the infinite past, the influence on the constitutive functional $\mathfrak{G}$ of deformations occurred in the far past was supposed to be negligible.[57] This is the basic assumption of the theory of *materials with fading memory*.[58] This theory was criticized in a series of papers by Fichera,[59] in which he showed that the behavior of $\mathfrak{G}$ at infinity is crucial in determining the well-posedness of boundary-value problems, in contrast with the fact that the deformations occurred in an infinitely far past are inaccessible to experiments. But long before their formulation, Fichera's objections were set aside by the New Theory, in which the introduction of state variables made possible to replace infinite with finite past histories. That is, to replace (100) by a constitutive equation

$$T_{Rt}(x_R) = \mathfrak{g}\big(\nabla f_t(x_R), F_t^k(x_R)\big), \tag{101}$$

in which the current values $F_t^k$ of the deformations $F^k$ associated with the state variables are determined by *evolution equations* of the form

$$F_t^k(x_R) = \mathscr{G}\big(F_{t_0}^k(x_R), \nabla f_{(t_0,t)}(x_R)\big). \tag{102}$$

These equations involve the finite histories $\nabla f_{(t_0,t)}$, which are restrictions of $\nabla f^t$ to the finite intervals $(t_0, t)$.

[55] This is the case in the New Theory and in the theory of *generalized standard materials* of Halphen and Nguyen [26].

[56] Noll [33], Truesdell and Noll [51], Sect. 28.

[57] Noll [33], Truesdell and Noll [51], Sect. 38.

[58] Coleman and Noll [6].

[59] See [16, 17], and other papers cited therein.

For a classical continuum for which $F^d$ is the unique state variable, the constitutive equation (101) takes the form (58)

$$T_{Rt} = \nabla\varphi\big(\nabla f_t - F_t^d\big), \tag{103}$$

and the evolution equation is obtained by time integration of the incremental relation $(64)_4$

$$\delta F_t^d = \lambda_t^d N_t^d, \tag{104}$$

with $N_t^d$ a solution of (61) and $\lambda_t^d$ the positively homogeneous function of order one of $\mathbb{C}[\nabla u_t] \cdot N_t^d$ given by (69). The extension to generalized continua will not be considered here for brevity.

### 8.4 Balance Equations and Pseudobalance

In the traditional formulations of continuum mechanics, the balance equations (48) of linear and angular momentum are taken as basic postulates, and from them the existence of the stress tensor and the local forms (53) and $(55)_2$ of the balance equations are deduced. To extend this approach to generalized continua, supplementary balance equations are required for the microscopic deformation and for the structural state variables. *Microbalance* equations were indeed proposed by several authors, but their motivations, if any, were not convincing.[60]

Neither the discovery of Noll that the balance equations are a consequence of the indifference of the external power may help in this circumstance. Indeed, the indifference conditions do not produce the desired supplementary equations, but only transform the symmetry restriction (54) into (78). The missing equations are provided by the pseudobalance equation (74) for the microscopic deformation and by a similar equation for each of the state variables [9].

### 8.5 Inertia, changes of observers, and apparent forces

The criticisms of the Newtonian concepts of motion and inertia have a long history, which starts with Leibnitz in the 17-th century, and continues with many authors, among which Berkeley in the 18-th, Mach in the 19-th, and Noll in the 20-th.[61] Their criticisms produced large, perhaps not yet completely perceived, impacts on the traditional axiomatic layout of mechanics.

*Inertia* is a phenomenon involving a particular external energy, the *kinetic energy*, with a volume density of the form

$$\varphi^{(in)}(x_R) = -\tfrac{1}{2}\rho(x_R)\big|\bar{v}_t(x_R)\big|^2, \tag{105}$$

where $\rho$ is the *mass density* in the reference placement, and

(i) $t$ is the physical time,
(ii) $\bar{v}_t$ is the physical velocity at the time $t$,

[60] See my comments in Sects. 1 and 7 of [10] and in Sect. 4 of [12].

[61] Berkeley's objections are summarized in Jammer [27], Chap. 11. For Mach see [29], Chapt. II.VI.8 and Appendix XX. For Noll see [35, 40].

the *physical velocity* being the velocity of the body particles with respect to the reference system.[62]

According to Noll, the kinetic energy is the gravitational energy received by the body from the "rest of the universe".[63] In extreme simplification, a "universe" can be viewed as a set made of two elements, the body and the "rest of the universe", and the kinetic energy can be supposed proportional to the square of the relative velocity of the two elements.[64]

The kinetic energy is a part of the volume density $\beta(f(x_R))$ in the expression (35) of the external energy.[65] The corresponding external power is[66]

$$-\rho(x_R)\dot{\bar{v}}_t(x_R)\cdot\bar{v}_t(x_R), \tag{106}$$

and the corresponding external action is

$$b_{Rt}^{(in)}(x_R) = -\rho(x_R)\dot{\bar{v}}_t(x_R). \tag{107}$$

By recording the positions of the body particles with respect to the reference system and the physical time, the observer observes a *physical motion* of the body. In the Euclidean space $\mathscr{E}$, this motion is represented by an evolution $t \mapsto f_t$ in which the deformations are relative to a family $t \mapsto \mathfrak{X}_t$ of reference frames. The velocity $v_t = \dot{f}_t$ in the representation corresponds to the physical velocity $\bar{v}_t$, in the sense that $|v_t| = |\bar{v}_t|$, only if the reference frame is the same for all $t$. That is, to represent a *motion* and not simply an evolution, the placer is still free of choosing any reference frame, but this frame must be the same for all $t$.

By consequence, if $\mathfrak{X}$ and $\mathfrak{X}^*$ are two reference frames, the change of frame is represented by a constant orthogonal tensor $Q$. For $Q_t = Q$ we have $Y_t = 0$, and the transformation law (18) reduces to

$$v_t^*(x_R) - v_t^*(x_{\emptyset R}) = Q\big[v_t(x_R) - v_t(x_{\emptyset R})\big]. \tag{108}$$

Keeping $x_{\emptyset R}$ fixed and varying $x_R$ we find that $v_t^*(x_R) = Qv_t(x_R)$, and therefore

$$\big|v_t^*(x_R)\big| = \big|v_t(x_R)\big|. \tag{109}$$

Then the kinetic energy (105) evaluated with respect to the two constant reference frames is indifferent.

Looking at the transformation law (18) for the velocity, it is generally asserted that *the kinetic energy is not indifferent, because it depends on the observer*, and that, by consequence, the expression (107) of the inertia action holds only for a privileged class of observers, called *inertial* or *Galilean*. A non-inertial observer, besides the inertial force (107), should observe

62 In a two-scale continuum, the kinetic energy can be augmented with a term depending on the microscopic velocity $V$ and proportional to a microscopic equivalent of the mass. In a multi-scale continuum this leads to "generalized" conceptions of kinetic energy and mass, see e.g. [48].

63 ... "on regarde les forces d'inertie comme des forces véritables qui sont les interactons entre les corps dans notre syst'eme solaire et la totalité des objets dans le reste de l'univers" [35].

64 In general, the "fixed stars" are taken as the reference system representing the "rest of the universe". For experiments in which the gravitational effect of the earth is considered predominant, the "rest of the universe" can be identified with the earth, and the walls of the laboratory can be taken as the reference system. "I have remained to the present day the only one who insists upon referring the law of inertia to the earth, and in the case of motions of great spatial and temporal extent, to the fixed stars", Mach [29], p. 568.

65 Usually the kinetic energy is placed, with the plus sign, in the integral representation (37) of the *internal* energy, though the inertia force (107) is generally regarded as an external action.

66 The superimposed dot denotes the derivative with respect to the physical time.

some supplementary *apparent forces*. I think that believing in apparent forces requires a *sacrificium intellectus* even greater than believing in the *luminiferous ether* of 18-th century physics. In reality, everything is much simpler.

According to the definition given in Sect. 3, an observer measures only distances between body particles and points of the reference system. Then there is no room for "privileged observers". Moreover, since two observers provide the same distances at the same instants, one of them is redundant, and we can well assume that there is just one observer. So, there are no "changes of observer". Neither there are "apparent forces" in the physical world. They show up only in the representation, when the placer takes a reference frame variable with $t$. But this is a bad representation, because a variable reference frame does not represent the physical velocity relative to the reference system. Therefore, apparent forces have nothing to do with "reality", they are only the effect of a bad representation.[67]

According to Noll, the inertia force $b_R^{in}$ is an external action exerted on the body by the rest of the universe, and the gravitation law $b_R^{in} = -m_R \dot{v}$ is the corresponding constitutive equation. One of the "large, perhaps not yet completely perceived, impacts on the traditional layout of mechanics", of which I was speaking at the beginning of this section, is that in this way the law of inertia is downgraded from a keystone principle of mechanics to the more modest status of constitutive law for a particular class of external actions.[68]

This change suggests a thorough revision of the textbooks, which almost unanimously consider the study of motion as the main object of mechanics. Inertia forces should not anymore be given a special status in the face of other distance actions, motions should not anymore be regarded as different from, and more general than, the so-called *quasi-static* evolutions, and physical time and physical velocity should not anymore be given a special status with respect to other measures of time and velocity.[69]

**Post Scriptum** In the time elapsed between the submission and acceptance of the present paper, it occurred to me to read over again Noll's note "On the illusion of physical space".[70] I was surprised about how many of my views, which I believed to be original, are already there. However, there is a point in which my views differ from his, and this is the concept of *physical space*, about which he writes: "the existence of physical space, apart from the frame of reference, is an illusion".

I believe that it makes no sense to say that the existence of the physical space is an illusion,[71] or, even worse, to put the existence of a reference frame as a condition for the existence of the physical space. Probably, Noll meant that the existence of physical space *as an absolute space* cannot be conceived without a reference frame. This is a clear example of confusion between "reality" and representation. The physical space is "reality". Its representation on a support having the properties of an absolute space requires the choice of a reference system in the "reality" and of a reference frame in the support.

[67] This criticism of apparent forces is mine. There is no trace of it in Noll's papers. Perhaps the reason is that, according to his updated definition of placement, each placement of the body belongs to a different Euclidean space [36, 37]. Therefore, the idea of a "variable reference frame" makes no sense. On the contrary, this idea is meaningful when all placements belong to the same $\mathscr{E}$.

[68] For the *revolutionary character* of this change see a more detailed comment in Sect. 3.7 of my paper [12].

[69] This agrees with Noll's belief that "the basic concepts of mechanics should *not* include items such as momentum, kinetic energy, and angular momentum, because they are relevant only when inertia is important", [40], paper N2.

[70] [40], paper N1.

[71] Noll himself quotes a dictionary's definition saying that the physical space is "the unlimited and indefinitely three-dimensional expanse in which all material objects are located and all events occur". He seems to accept this definition, "vague and ambiguous as it may be".

**Post-post Scriptum** Just before delivering the manuscript to the printer, I realized that my presentation in Sect. 6.4 is too particular, since not all continua have evolutions governed by an activation condition and a flow rule. My presentation was influenced by the case of plasticity, in which the dissipation potential is a positively homogeneous function of order one, and therefore is only Gâteaux differentiable at the origin. But this is not the only possible choice. For example, a positively homogeneous function of order two provides a dissipation potential which is differentiable everywhere and, as it will be shown in a forthcoming paper, requires no activation condition and no flow rule.



## References


1. Aifantis, E.C.: On the microstructural origin of certain inelastic models. ASME J. Eng. Mater. Technol. **106**, 326–330 (1984)
2. Ambrosio, L., Tortorelli, V.M.: Approximation of functionals depending on jumps by elliptic functionals via $\Gamma$-convergence. Commun. Pure Appl. Math. **43**, 999–1036 (1990)
3. Barroso, A.C., Matias, J., Morandotti, M., Owen, D.R.: Second-order structured deformations: relaxation, integral representation and applications. Arch. Ration. Mech. Anal. **225**, 1025–1072 (2017)
4. Cauchy, A.-L.: Sur l'équilibre et le mouvement intérieur des corps considérés comme des masses continues. Œuvres complètes Sér. **9**(2), 342–369 (1829)
5. Choksi, R., Fonseca, I.: Bulk and interfacial energy densities for structured deformations of continua. Arch. Ration. Mech. Anal. **138**, 37–103 (1997)
6. Coleman, B.D., Noll, W.: An approximation theorem for functionals with applications in continuum mechanics. Arch. Ration. Mech. Anal. **6**, 97–112 (1960)
7. Coleman, B.D., Noll, W.: The thermodynamics of elastic materials with heat conduction and viscosity. Arch. Ration. Mech. Anal. **13**, 167–178 (1963)
8. Del Piero, G.: The energy of a one-dimensional structured deformation. Math. Mech. Solids **6**, 387–408 (2001)
9. Del Piero, G.: On the method of virtual power in continuum mechanics. J. Mech. Mater. Struct. **4**, 281–292 (2009)
10. Del Piero, G.: Nonclassical continua, pseudobalance, and the law of action and reaction. Math. Mech. Complex Syst. **2**, 71–107 (2014)
11. Del Piero, G.: On the decomposition of the deformation gradient in plasticity. J. Elast. **131**, 111–124 (2018)
12. Del Piero, G.: An axiomatic framework for the mechanics of generalized continua. Rend. Lincei Mat. Appl. **29**, 31–61 (2018)
13. Del Piero, G.: The variational structure of classical plasticity. Math. Mech. Complex Syst. **6**, 137–180 (2018)
14. Del Piero, G., Owen, D.R.: Structured deformations of continua. Arch. Ration. Mech. Anal. **124**, 99–155 (1993)
15. Deseri, L., Owen, D.R.: Toward a field theory for elastic bodies undergoing disarrangements. J. Elast. **70**, 197–236 (2003)
16. Fichera, G.: I dificili rapporti tra l'analisi funzionale e la fisica matematica. Rend. Semin. Mat. Univ. Padova **68**, 245–259 (1982)
17. Fichera, G.: Sul principio della memoria evanescente. Rend. Mat. Accad. Lincei, Suppl. **9**(1), 161–170 (1990)
18. Fleck, N.A., Hutchinson, J.W.: A phenomenological theory for strain gradient effects in plasticity. J. Mech. Phys. Solids **41**, 1825–1857 (1993)
19. Fleck, N.A., Willis, J.R.: A mathematical basis for strain-gradient plasticity theory—Part II: tensorial plastic multiplier. J. Mech. Phys. Solids **57**, 1045–1057 (2009)
20. Germain, P.: La méthode des puissances virtuelles en mécanique des milieux continus. Première partie: théorie du second gradient. J. Méc. **12**, 235–274 (1973)
21. Germain, P.: The method of virtual power in continuum mechanics. Part 2: microstructure. SIAM J. Appl. Math. **25**, 556–575 (1973)

22. Gudmundson, P.: A unified treatment of strain gradient plasticity. J. Mech. Phys. Solids **52**, 1379–1404 (2004)
23. Gurtin, M.E., Martins, L.C.: Cauchy's theorem in classical physics. Arch. Ration. Mech. Anal. **60**, 305–324 (1976)
24. Gurtin, M.E., Anand, L.: A theory of strain-gradient plasticity for isotropic, plastically irrotational materials. Part II: finite deformations. Int. J. Plast. **21**, 2297–2318 (2005)
25. Gurtin, M.E., Fried, E., Anand, L.: The Mechanics and Thermodynamics of Continua. Cambridge University Press, Cambridge (2010)
26. Halphen, B.: Nguyen Quoc Son, sur les matériaux standards généralisés. J. Méc. **14**, 39–63 (1975)
27. Jammer, M.: Concepts of Force. Dover, Mineola (1999)
28. Love, A.E.H.: A Treatise on the Mathematical Theory of Elasticity, 4th edn. Cambridge University Press, New York (1927). Reprinted by Dover Publisher, 1944
29. Mach, E.: Die Mechanik in ihrer Entwickelung historisch-kritisch dargestellt. Brockhaus, Leipzig (1883). 2nd English edn. "The Science of Mechanics. A Critical and Historical Account of Its Development". The Open Court publisher, Chicago and London (1919)
30. Maxwell, J.C.: On action at a distance. Proc. R. Inst. G. B. VII **44**(54), 311–323 (1875). Also in: Scientific Papers **2**, 311–323
31. Maxwell, J.C.: Matter and Motion. Society for Promoting Christian Knowledge, London (1878). Dover edition, New York (1952)
32. Noll, W.: On the continuity of the solid and fluid states. J. Ration. Mech. Anal. **4**, 3–81 (1955)
33. Noll, W.: A mathematical theory of the mechanical behavior of continuous media. Arch. Ration. Mech. Anal. **2**, 197–226 (1958). Reprinted in [38]
34. Noll, W.: The foundations of classical mechanics in the light of recent advances in continuum mechanics. In: The Axiomatic Method, with Special Reference to Geometry and Physics, Symposium at Berkeley, 1957, pp. 266–281. North-Holland, Amsterdam (1959). Reprinted in [38]
35. Noll, W.: La mécanique classique, basée sur un axiome d'objectivité. In: La Méthode Axiomatique dans les Mécaniques Classiques et Nouvelles, Symposium in Paris, 1959, pp. 47–56. Gauthier-Villars, Paris (1963). Reprinted in [38]
36. Noll, W.: A new mathematical theory of simple materials. Arch. Ration. Mech. Anal. **48**, 1–50 (1972). Reprinted in [38]
37. Noll, W.: Lectures on the foundations of continuum mechanics and thermodynamics. Arch. Ration. Mech. Anal. **52**, 62–92 (1973). Reprinted in [38]
38. Noll, W.: The Foundations of Continuum Mechanics and Thermodynamics. Selected Papers of W. Noll. Springer, Berlin (1974)
39. Noll, W.: Continuum mechanics and geometric integration theory. In: Lawvere, F.W., Schnauel, S.H. (eds.) Categories in Continuum Physics. Springer Lecture Notes in Mathematics, vol. 1174, pp. 17–29. Springer, Berlin (1986)
40. Noll, W.: Five contributions to natural philosophy (2004). http://repository.cmu.edu/math
41. Noll, W.: A frame-free formulation of elasticity. J. Elast. **83**, 291–307 (2006)
42. Noll, W.: On the past and future of natural philosophy. J. Elast. **84**, 1–11 (2006)
43. Noll, W., Virga, E.G.: Fit regions and functions of bounded variation. Arch. Ration. Mech. Anal. **102**, 1–21 (1988)
44. Owen, D.R.: Structured deformations—part two. In: Del Piero, G., Owen, D.R. (eds.) Lecture Notes to the Ravello Summer School of the Gruppo Nazionale di Fisica Matematica. Quaderni dell'Istituto Nazionale di Alta Matematica, vol. 58 (2000)
45. Owen, D.R.: Elasticity with gradient-disarrangements: a multiscale geometrical perspective for strain-gradient theories of elasticity and plasticity. J. Elast. **127**, 115–150 (2017)
46. Seguin, B.: Frame-free continuum thermomechanics. Ph.D. thesis, Carnegie Mellon University (2010) http://repository.cmu.edu/math/dissertations
47. Šilhavý, M.: The existence of the flux vector and the divergence theorem for general Cauchy fluxes. Arch. Ration. Mech. Anal. **90**, 195–212 (1985)
48. Šilhavý, M.: Mass, internal energy, and Cauchy's equations in frame-indifferent thermodynamics. Arch. Ration. Mech. Anal. **107**, 1–22 (1989)
49. Truesdell, C.: Whence the law of moment of momentum? In: Essays in the History of Mechanics. Springer, New York (1968)
50. Truesdell, C.: A First Course in Rational Continuum Mechanics, vol. 1, 2nd edn. Academic Press, Boston (1991)
51. Truesdell, C., Noll, W.: The non-linear field theories of mechanics. In: Flügge, S. (ed.) Handbuch der Physik, Vol. III/3. Springer, Berlin (1965)

J Elast (2019) 135:149–182
https://doi.org/10.1007/s10659-018-9707-0

# Elasticity with Hierarchical Disarrangements: A Field Theory That Admits Slips and Separations at Multiple Submacroscopic Levels

**Luca Deseri**[1,2,3,4,5] · **David R. Owen**[6]

*To Walter Noll, whose writings set the foundations of continuum mechanics and whose commitment to colleagues, friends, and family endures in our memory*

Received: 8 June 2018 / Published online: 12 December 2018


**Abstract** The complexity and variety of geometrical changes in physical systems at submacroscopic levels has led to various approaches to the broadening of the classical theory of finite elasticity. One approach, the field theory "elasticity with disarrangements", employed the multiscale geometry of structured deformations in order to incorporate the effects of disarrangements such as slips and separations at a single submacroscopic level on the macroscopic response of a continuous body. This article extends that field theory by enriching the underlying geometry so as to include the effects of disarrangements at more than one submacroscopic level. The resulting field theory broadens the scope of this approach, sharpens the description of the physical nature of dissipative mechanisms that can arise, and increases the variety of systems of contact forces that can serve as boundary loadings for a body that evolves via multiscale geometrical processes.

**Keywords** Structured deformations · Hierarchies · Multiscale geometry · Elasticity · Disarrangements · Field theory

**Mathematics Subject Classification** 74A · 74B20 · 74C15 · 74M25

✉ D.R. Owen
do04@andrew.cmu.edu

[1] DICAM-Mechanical, Civil and Environmental Engineering, University of Trento, 38123 Trento, Italy

[2] MEMS-Mechanical Engineering and Materials Sciences, Swanson School of Engineering, University of Pittsburgh, Pittsburgh PA 15206, USA

[3] Department of Civil and Environmental Engineering, Carnegie Mellon University, Pittsburgh PA 15213, USA

[4] Department Mechanical Engineering, Carnegie Mellon University, Pittsburgh PA 15213, USA

[5] THMR-Department of Nanomedicine-Houston Methodist Hospital, Houston TX 77030, USA

[6] Department of Mathematical Sciences, Carnegie Mellon University, Pittsburgh PA 15213, USA

## 1 Introduction

In the article [18] we provided a first step in a program to employ structured deformations of continua [12] in order to obtain a field theory capable of describing, in the context of dynamics and large isothermal deformations, the evolution of bodies that undergo smooth deformations at the macroscopic level, that can experience piecewise smooth deformations at submacroscopic levels, and that can not only store energy but can also dissipate energy during such multilevel geometrical changes. The field theory [18] that we formulated incorporated the effects at the macrolevel of smooth deformations and of non-smooth deformations (disarrangements) at one submacroscopic level by providing field relations that govern the (time-dependent) macroscopic deformation $g$ of the body as well as the tensor field $G$ of deformations without disarrangements at the submacroscopic level. Among the field relations in [18] that govern the structured motions $(g, G)$ of the body is the accommodation inequality that includes the relation $0 < \det G \leq \det \nabla g$, i.e., the volume changes caused by smooth deformations at the submacroscopic level cannot exceed the volume changes at the macroscopic level. The accommodation inequality is the basis for the Approximation Theorem: at each time there exist piecewise smooth, injective mappings $f_n$ on the body that tend to $g$ and whose gradients $\nabla f_n$ tend to $G$ as $n$ tends to $\infty$. The passage to $\infty$ can be interpreted as the process of "zooming out" from the generally non-smooth geometrical changes at the submacroscopic level to the generally smoother geometrical fields $g$ and $G$. In other words, $g$ and $G$ provide at the macrolevel some (but generally not all) of the geometrical information available at the submacroscopic level. The additional information carried in the field $G$, the deformation without disarrangements, is complemented by the information carried in the field $M = \nabla g - G$, which can be shown to capture at the macrolevel the geometrical effects of slips and separations in the approximating deformations $f_n$ in the limit as $n$ tends to $\infty$.

The main obstacle in formulating a field theory in terms of $g$ and $G$ is to provide relations between these fields in addition to the algebraically obvious additive decomposition

$$\nabla g = G + M.$$

In [18] we provided additional relations among these fields by obtaining for the stress $S$ in the macroscopic reference configuration (the Piola-Kirchhoff stress) both an additive decomposition $\det((\nabla g)^{-1}G)S = S_\backslash + S_d$ and a multiplicative decomposition $\det((\nabla g)^{-1}G)S = S_\backslash G^T(\nabla g)^{-T}$. The theory of structured deformations permits us to identify the field $S_\backslash$ as the stress without disarrangements and the field $S_d$ as the stress due to disarrangements, and the two decompositions of $S$ show that, given the structured deformation $(g, G)$, the two fields $S_\backslash$ and $S_d$ cannot be prescribed independently. Thus, there is a "consistency relation"

$$S_\backslash G^T(\nabla g)^{-T} = S_\backslash + S_d \tag{1}$$

that the two refined stress fields must satisfy. This consistency relation is universal, i.e., it is valid for every body undergoing structured deformations in the presence of a given system of contact forces that determine $S$, without regard to the material comprising that body [23]. The constitutive assumptions in [18] then allow us through the response function $(G, M) \longmapsto \Psi(G, M)$ for the free energy density $\psi$ to interpret the refined stress $S_\backslash$ already defined in terms of $S$ and $(g, G)$, as a "driving stress" corresponding to the geometrical field $G$ and to interpret the refined stress $S_d$ as a "driving stress" corresponding to the geometrical field $M$. These constitutive relations express the dependence of both $S_\backslash$ and $S_d$ upon $G$

**Fig. 1** An example of three-level hierarchies: large stack (macroscale) of bundled papers (intermediate level). The finest scale is represented by each paper sheet (reproduced with permission of the photographer Mr. M. Schmorgan)

and $M$, and the consistency relation then provides the desired further restriction on the three fields $\nabla g$, $G$, and $M$. Moreover, the additive decompositions both of $S$ and of $\nabla g$ as well as our constitutive assumptions allow us to calculate the internal dissipation $\Gamma$ through the formula

$$\Gamma = S \cdot \nabla \dot{g} - \dot{\psi} = (S_{\backslash} \cdot \dot{M} + S_d \cdot \dot{G}) / \det\big((\nabla g)^{-1} G\big) = D_G \Psi \cdot \dot{M} + D_M \Psi \cdot \dot{G}.$$

These formulas give us a precise physical as well as mathematical insight into the sources of dissipation in this multilevel setting and led to a necessary departure from the standard approach [9] for assuring compatibility of constitutive relations with the Second Law.

Our field theory [18] and its refined geometry have been applied (i) to distinguish among the geometrical phenomena of contact, of intermingling, and of mixing of elastic constituents and to identify the additional forces and moments exerted by a mixture of elastic bodies on each constituent [34], (ii) to define and apply a notion of submacroscopically stable equilibrium of elastic bodies that selects among the many phases available to the body those that dissipate energy faster than they store energy in quasistatic, purely submacroscopic processes [19], and (iii) to define and identify a related notion of stable disarrangement phases of elastic bodies that, in the context of aggregates of small elastic bodies, provides a setting for the emergence of no-tension materials with nonlinear response in compression [11, 21].

These applications indicate the variety of possibilities afforded by structured deformations for enriching through effects at a single submacroscopic level the purely macroscopic field theory of non-linear elasticity. Nevertheless, many (i) *natural* and (ii) *man-made* physical systems have a rich enough geometrical structure to permit the identification of hierarchies consisting of more than one physically meaningful submacroscopic level. In each of the above categories both (a) *soft* and (b) *hard* materials can be singled out (see Figs. 1–3).

*Natural* soft tissues exhibiting hierarchies include muscles, cartilage, cornea, tendons (see, e.g., [28, 29, 32, 42]), while natural hard tissues with hierarchical structures include bone, nacre, enamel, etc. (see, e.g., [4, 6, 28, 38, 43]). In [32] and [38] one finds evidence that submacroscopic disarrangements in hierarchical materials can occur at various levels of the hierarchy and can be influenced by the organization of the hierarchy. This evidence is provided with particular regard to muscles and enamel.

Furthermore, in [6, 29] it is suggested that toughening mechanisms, characterized by distributions of submacroscopic separations at the various levels of the hierarchies, do closely follow the internal arrangements of such materials. For such mechanisms a statistical model

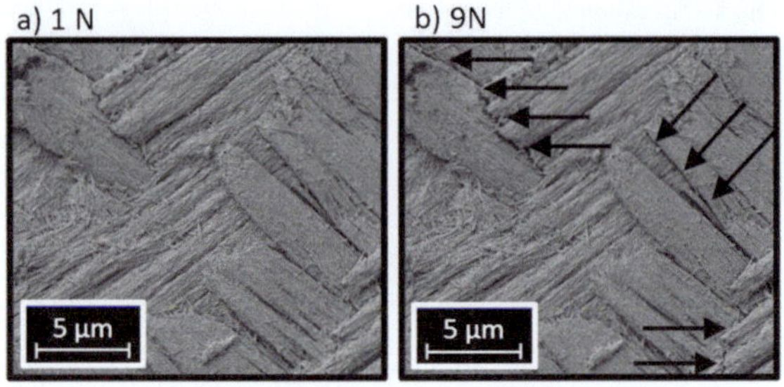


**Fig. 2** Disarrangements at different levels of hierarchy: SEM-image of an *enamel specimen* (**a**) after pre-loading, showing very little submacroscopic cracks opening, (**b**) at a load level of 9–N, displaying more opening (reproduced from [38] with permission of Prof. Dr. rer. nat. G.A. Schneider)

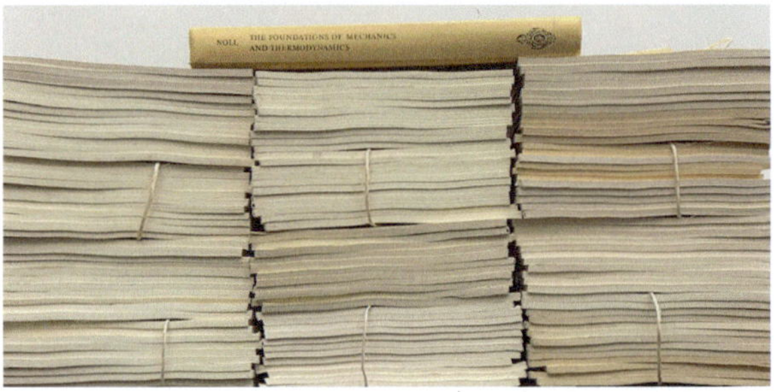

**Fig. 3** Zoom-in across the scales: the presence of disarrangement at two different levels is evident because of the separation between (**i**) bundled papers and (**ii**) sheets

based on bundles of fibers has been extensively used in this context [10, 31, 39]. Among natural materials exhibiting hierarchical structures and analogous modes of disarrangements, one finds wood, bamboo [6, 24, 27], and plants in general, forming a span of hard-to-soft materials.

*Man-made* materials with similar features are primarily bioinspired composites [6], nowadays available thanks to the evolving advancements in additive manufacturing [25, 44].

Our goal here is to take a further step in our program by broadening the field theory [18] to include the effects of geometrical changes at more than one submacroscopic level. To simplify the explanations and to provide focus, we restrict our attention to the case of two submacroscopic levels: submacroscopic level 2, the finer submacroscopic level that represents in this case the most magnified view of geometrical changes, and submacroscopic level 1, a level intermediate to submacroscopic level 2 and to the macroscopic level (level 0). Thus, we provide not only the macroscopic deformation field $g$ at each time, but also two tensor fields $G_1$ and $G_2$ on the body that ultimately provide the effects at the macrolevel of geometrical changes without disarrangements at each of the submacrocscopic levels.

In Sect. 2.1 we describe the refined geometry of three-level structured deformations $(g, G_1, G_2)$ based partly on discussions with Gianpietro Del Piero in the late 1990′s. The three-level analogues, (3) and Theorem 1 in the Appendix, of the accommodation inequality and the Approximation Theorem identify $G_2$ as a gradient followed by two limit operations that correspond to zooming out through both submacroscopic levels. As such, $G_2$ incorporates at the macrolevel nothing of the slips and separations at either submacroscopic level. By contrast, $G_1$ is identified as a limit (zooming out from submacroscopic level 2 to level 1) followed by a gradient followed by a second limit (zooming out from submacroscopic level 1 to the macrolevel), and so incorporates the effects both of smooth and of non-smooth changes at level 2, but only the effects of smooth deformations at level 1. The three-level hierarchy is completed with the field $G_0 := \nabla g$, consisting first of two limits that span both the submacroscopic levels, followed by the gradient operation. Consequently, $G_0$ incorporates both types of geometrical changes at both submacroscopic levels.

The analysis in Sect. 2.2 includes a discussion of the algebraically trivial additive decomposition $G_0 = G_2 + (G_1 - G_2) + (G_0 - G_1)$, in which each of the differences $G_{i-1} - G_i$ on the right-hand side represents the effect of interchanging the order of a "limit operator" with the gradient operator and, consequently, incorporates only the effects of disarrangements at level $i$. Our definition $M_i = G_{i-1} - G_i$ of the disarrangement tensor at level $i$ recasts the last additive decomposition in the form

$$\nabla g = G_0 = G_2 + M_2 + M_1 \tag{2}$$

and so resolves the difference $G_0 - G_2$, spanning both submacroscopic levels, separately into the effects $M_2$ of disarrangements at level two only, added to the effects $M_1$ of disarrangements at level one only. This is the source of the term "hierarchical disarrangements" and motivates the preferred choice of refined geometrical variables $G_2$, $M_2$, $M_1$ in most of our treatment. We show also in Sect. 2.2 how factoring the invertible tensor field $G_2$ from either the left of the sum $G_2 + M_2 + M_1$ or from the right yields two multiplicative decompositions of $\nabla g = G_0$ in both of which each factor has a definite interpretation supported by limit formulas derived from the Approximation Theorem. Moreover, a notion of composition of three-level structured deformations permits us to factor via composition an arbitrary three-level structured deformation into a succession of three three-level structured deformations, each of which describes geometrical changes occurring at exactly one of the three levels under consideration.

For each three-level structured deformation and for each tensor field $N$ ("tensor flux") on the macroscopically deformed body, we introduce in Sect. 2.3, for $i = 0, 1, 2, \ldots$ various refined versions $N_i$, $N_{i-1.\backslash}$ and $N_{i-1.d}$ of the field $N$ that depend upon the submacroscopic or macroscopic level $i$ and depend upon whether disarrangements at that level are to be incorporated. Our development focuses on the example $N = T$, the Cauchy stress (stress in the deformed configuration), in which case $T_0$ is the (first) Piola-Kirchhoff stress (stress in the reference configuration). However, the treatment here could also apply to other tensor fluxes such as the momentum flux. The interpretations of the refined versions $N_i$, $N_{i-1.\backslash}$ and $N_{i-1.d}$ are justified on the basis of the Approximation Theorem, and we provide additive decompositions that parallel the decomposition of $\nabla g$ in (2) as well as multiplicative decompositions. The availability of universal additive and of multiplicative decompositions of tensor fluxes at each level permits us in Sect. 2.4 to derive a consistency relation for the tensors $N_{i-1.\backslash}$ and $N_{i-1.d}$ at each submacroscopic level analogous to (1) for the case of the stress and for only one submacroscopic level.

In Sect. 3 the notion of a structured motion as a time-parameterized family of three level structured deformations is introduced. In this context, one can use the three-term additive decomposition (2) of $G_0$ and the corresponding decomposition of tensor fluxes $N$ to compute a nine-term additive decomposition of the inner product $N_0 \cdot \dot{G}_0$, where $\dot{G}_0$ is the time derivative of $G_0$. When $N$ is taken to be the Cauchy stress field $T$ defined on the deformed configuration of the body at each time, then $T_0 \cdot \dot{G}_0$ is the stress power per unit volume in the reference configuration of the body. Three of the nine terms in the expression for the stress power contain a "matching" pair of factors, indicating that the location of the corresponding tractions are in proximity to the resulting geometrical changes both in the sense of spatial location and in the sense of hierarchical level. The remaining six terms of $T_0 \cdot \dot{G}_0$ each contains a mismatched pair of factors: *either* the two hierarchical levels of the refined stress and of the rate of deformation measure do not match *or* the disarrangement status "without disarrangements" versus "due to disarrangements" of the two factors do not match. This remote character of the stress and deformation rates in the mismatch terms lead us in subsequent sections to assign a special thermodynamical status to these six terms.

Section 4 contains the conceptual background for the specific constitutive assumptions that we make in Sect. 5. This background is needed, since it provides a procedure for assuring that constitutive assumptions are in accord with the Second Law of Thermodynamics that is less restrictive than the now standard procedure of Coleman and Noll [9]. The key observation already employed in [18] that we use here in Sect. 4 is the following: constitutive assumptions identify a constitutive class, i.e., a subset $\mathcal{C}$ of the collection of all dynamical processes available to a body, and the requirement that the Second Law of Thermodynamics be satisfied also identifies a set $\mathcal{T}$ of dynamical processes; compatibility of the constitutive assumptions with the Second Law is simply the requirement that the first set be included in the second: $\mathcal{C} \subset \mathcal{T}$. For a given free energy response function, our approach provides in Sect. 5 constitutive relations defined in terms of that response function and its partial derivatives, including an inequality involving the fields $G_2$, $M_2$, $M_1$ and their time derivatives. These relations define a constitutive class that we denote by $\mathcal{E}_{hd}$, and we verify the inclusion $\mathcal{E}_{hd} \subset \mathcal{T}$. An advantage of our particular choice $\mathcal{E}_{hd}$ is the particular physical interpretation that it affords for the internal dissipation $\Gamma = T_0 \cdot \dot{G}_0 - \dot{\psi}$ in dynamical processes: $\Gamma$ turns out to be the sum of the six "mismatch" terms in the expression for the stress power $T_0 \cdot \dot{G}_0$, and internal dissipation is thereby identified with lack of proximity of applied tractions to the sites of occurrence of the geometrical changes that they work against. Here, "proximity" means that the tractions and the rates of deformation must correspond to the same (submacroscopic or macroscopic) level and to the same disarrangement status ("without disarrangements" or "due to disarrangements").

In Sect. 5.1 we detail the choice of $\mathcal{E}_{hd}$, the constitutive class mentioned in the previous paragraph. One of our constitutive equations requires that the free energy density at any point and time be determined through a given response function $\Psi$ by the values at that point and time of $G_2$, $M_2$, and $M_1$. Three additional constitutive equations relate the values of the refined stress measures $T_{1.\backslash}$, $T_{1.d}$, and $T_{0.d}$ at each point and time to the values of $G_2$, $M_2$, and $M_1$ by means of the partial derivatives $D_{G_2}\Psi$, $D_{M_2}\Psi$, and $D_{M_1}\Psi$ of $\Psi$, respectively. Doing so permits us to interpret each of these stress measures as a "driving force (per unit area)" corresponding to a refined geometrical variable. The final constitutive assumption is the requirement that the power expended by the "mismatch" terms in the stress power $T_0 \cdot \dot{G}_0$ be non-negative. This inequality is a restriction on dynamical processes that, together with the other constitutive assumptions, assures that the constitutive class $\mathcal{E}_{hd}$ is compatible with the Second Law. In Sect. 5.2 we rewrite with $N = T$ the two consistency relations by substituting into them the constitutively assigned expressions for the refined stress measures in terms of the derivatives of the free energy response $\Psi$. We so obtain two tensorial consistency equations that directly restrict the three fields $G_2$, $M_2$, and $M_1$ and that eventually provide closure to the field theory under consideration. We also record the "stress relation" that gives the Piola-Kirchhoff stress $S = T_0$ in terms of $G_2$, $M_2$, and $M_1$ as the sum of the three partial derivatives of $\Psi$. The stress relation then permits one to write the balance of linear momentum equation as a restriction on the fields $g$, $G_1 = G_2 + M_2$, and $G_2$. Additional stress relations for the referential stresses $T_1$ and $T_2$ at submacroscopic levels 1 and 2 also are obtained in terms of the partial derivatives of $\Psi$.

We are aware that the accommodation inequality $0 < \det G \leq \det F$ could be treated as an internal constraint and, hence, appropriate reaction stresses would arise. This would lead to an enrichment of the refined measures for the stress and, ultimately, of the stress measures themselves (see [35] for details in the context of ordinary disarrangements and of "gradient disarrangements").

The refined geometry of three-level structured deformations permits the determination of transformation rules under change of observer for the new tensor fields that enter into our constitutive relations, and we note in Sect. 5.3 that $G_1$, $G_2$, $M_2$, $M_1$, $T_i$ for $i = 0, 1, 2$, as well as $T_{i-1.\backslash}$ and $T_{i-1.d}$ for $i = 1, 2$ all transform in the same manner under change of observer as does $G_0 = \nabla g$. Frame-indifference in the present context is the statement that the constitutive class $\mathcal{E}_{hd}$ chosen in Sect. 5.1 be closed under change of observer, i.e., a dynamical process is in $\mathcal{E}_{hd}$ if and only if all the dynamical processes obtained from it by a change of observer also are in $\mathcal{E}_{hd}$. It is interesting that we are able to show in Sect. 5.3 that frame-indifference here is equivalent to the usual invariance property of the free energy response $\Psi$ together with the symmetry of the Cauchy stress $T$. Consequently, balance of angular momentum is guaranteed on dynamical processes in $\mathcal{E}_{hd}$, once it is required that $\mathcal{E}_{hd}$ be closed under change of observer.

In Sect. 5.4 we extend to elasticity with hierarchical disarrangements the notion of "stable disarrangement phase" already defined and studied ([19, 21, 22]) in the context of elasticity with disarrangements at only one submacroscopic level [18]. In the present context, for a given tensor $F = G_0$, the decomposition $F = G_2 + M_2 + M_1$, the consistency relations for stress at submacroscopic levels 1 and 2 and a symmetry condition arising from frame indifference, along with the accommodation inequalities, provide four tensorial equations and three inequalities that restrict the tensor variables $G_2$, $M_2$, and $M_1$. A triple $G_2$, $M_2$, $M_1$ that satisfies these equations and inequalities is called a disarrangement phase corresponding to $F$. We pointed out in the context of elasticity with disarrangements ([19, 21, 22]) that multiple disarrangement phases corresponding to a given $F$ are to be expected, and the same assertion applies in the present, richer context. Accordingly, we follow [21] and define a stable disarrangement phase $G_2$, $M_2$, $M_1$ corresponding to $F$ to be a disarrangement phase corresponding to $F$ with the lowest free energy among all disarrangement phases corresponding to $F$. Because of the constraints imposed by (i) the consistency relations and frame indifference, (ii) the decomposition for $F$, and (iii) the accommodation inequalities, a stable disarrangement phase need not be a stationary point of the free energy and, therefore, the stress need not vanish at a stable disarrangement phase and (as examples in the two-level case show) need not be hydrostatic.

We collect together in Sect. 5.5 all of the geometrical and constitutive information from previous sections and record the system of field relations for elasticity with hierarchical disarrangements. The field relations amount to thirty-three scalar equations and four inequalities to be satisfied by the thirty components of $g$, $G_2$, $M_2$, and $M_1$. (Three of the scalar equations express the symmetry of the Cauchy stress and can be omitted in cases where a particular free energy response function automatically provides the required symmetry.) We show how under particular circumstances these field relations in the context of two submacroscopic levels reduce to those for elasticity with disarrangements, in which only one submacroscopic level is admitted, or, under other particular circumstances reduce to the field relations of finite elasticity, in which no submacroscopic levels are admitted.

In Sect. 6 the case where the response function $\Psi$ does not depend upon the disarrangement tensors $M_1$ and $M_2$ is considered. This special circumstance bars the storage of energy in the body through non-smooth geometrical changes, and, following terminology introduced in the two-level case [20], we call this the case of purely dissipative disarrangements. The field relations simplify considerably in the case of purely dissipative disarrangements, and we point out that among the field relations, all but the accommodation inequalities are invariant under interchange of the fields $M_1$ and $M_2$. Examples are provided in Sect. 6 that illustrate how the accommodation inequalities serve in the case of purely dissipative disarrangements to determine the particular level at which given disarrangements may arise.

The refined geometry of hierarchical structured deformations provides, in addition to the Piola-Kirchhoff stress $T_0 = S$, the submacroscopic reference stresses $T_1$ and $T_2$ introduced in Sect. 2.3. Each incorporates to a different extent the disarrangements occurring at submacroscopic levels one and two. In Sect. 7 we use the reference stresses $T_0$, $T_1$, and $T_2$ to refine and broaden the types of boundary conditions available for the fields $g$, $G_1$, $G_2$ that determine structured deformations satisfying the field relations in Sect. 5.5.

The paper concludes with an appendix in which a complete statement and proof of the Approximation Theorem for three-level structured deformations are provided.

# 2 Hierarchical Structured Deformations

## 2.1 Definition and Approximation Theorem

The geometrical basis for our present considerations is the multilevel geometry of hierarchical structured deformations that we describe here for the case of three levels: the macroscopic level along with two submacroscopic levels. The case of a multilevel geometry with only one submacroscopic level was developed in a series of papers on structured deformations ([1, 2, 7, 12, 13, 36, 40]) from which followed a number of applications (e.g., [3, 5, 8, 14–16, 18, 21, 22, 30, 34, 37, 41]) that include a field theory of elasticity with (submacroscopic) disarrangements [18] and a broadened field theory that includes the effects of gradient disarrangements [35]. The case with more than two submacroscopic levels can be inferred easily from the present treatment. For present purposes, we may define a *three-level structured deformation* to be a triple $(g, G_1, G_2)$ in which the injective, piecewise continuously differentiable point-valued mapping $g : \mathcal{B} \longrightarrow \mathcal{E}$ from the body $\mathcal{B}$ into Euclidean space $\mathcal{E}$ is called the *macroscopic deformation*, and the piecewise continuous tensor-valued mappings $G_1$ and $G_2$ from $\mathcal{B}$ into $Lin\mathcal{V}$ are called the *deformation without disarrangements at submacroscopic level one* and *the deformation without disarrangements at submacroscopic level two*, respectively. The triple $(g, G_1, G_2)$ is required to satisfy the *accommodation inequalities* throughout $\mathcal{B}$ (excepting the points of discontinuity of $\nabla g$, $G_1$ and $G_2$):

$$c < \det G_2(X) \leq \det G_1(X) \leq \det \nabla g(X) \tag{3}$$

with $c$ a positive constant that may depend upon the given triple but not upon the points $X$ in $\mathcal{B}$. (The notions of piecewise continuous differentiability and piecewise continuity are made precise in [12] through the notion of "simple deformations" and "piecewise-fit regions"; the set $\kappa$ that appears in the definitions of simple and structured deformations in [12] is here taken to be the empty set.) Here, $\nabla$ denotes the classical derivative of a smooth mapping or of the restriction of a piecewise smooth mapping to points of differentiability. When the body has the regularity of a piecewise-fit region [12], the Approximation Theorem for three-level structured deformations provides a double sequence $(n_1, n_2) \longmapsto f_{n_1 n_2}$ of injective, piecewise continuously differentiable mappings from $\mathcal{B}$ into $\mathcal{E}$ satisfying

$$\lim_{n_1\to\infty} \lim_{n_2\to\infty} f_{n_1 n_2} = g, \tag{4}$$

so that the deformation gradient $\nabla g$ is given by

$$G_0 := F := \nabla g = \nabla \lim_{n_1\to\infty} \lim_{n_2\to\infty} f_{n_1 n_2}, \tag{5}$$

as well as satisfying

$$G_1 = \lim_{n_1\to\infty} \nabla \lim_{n_2\to\infty} f_{n_1 n_2}, \tag{6}$$

$$G_2 = \lim_{n_1\to\infty} \lim_{n_2\to\infty} \nabla f_{n_1 n_2}. \tag{7}$$

Here, each of the limits $\lim_{n_1\to\infty}$, $\lim_{n_2\to\infty}$ separately may be taken in the sense of $L^\infty$-convergence on the body, i.e., in the sense of essential uniform convergence, although other settings may require different choices of regularity of the fields $g$, $G_1$, and $G_2$ and possibly weaker notions of convergence (see [1, 7, 8, 40] for the case of two-level structured deformations). A complete statement of the Approximation Theorem for three-level structured deformations and a proof are provided in the Appendix.

It is clear from (5)–(7) that $G_0$, $G_1$, and $G_2$ capture the effects of applying the gradient operator $\nabla$ following, between, or prior to applying the limit operations $\lim_{n_1\to\infty}$ and $\lim_{n_2\to\infty}$, and we explore now in more depth the significance of these relations. We interpret each injective mapping $f_{n_1 n_2}$ in (4)–(7) to be a description of the geometrical changes occurring at the finest level (highest magnification), i.e., at the submacroscopic level 2. The non-smooth changes (disarrangements) may consist of slips and separations of small pieces of the body that, individually, deform smoothly. The classical gradient $\nabla f_{n_1 n_2}$ then reflects only the smooth changes occurring at level 2 in each of the pieces. The limits $\lim_{n_2\to\infty} f_{n_1 n_2}$ and $\lim_{n_2\to\infty} \nabla f_{n_1 n_2}$ reflect the process of "zooming out" from submacroscopic level 2 to submacroscopic level 1: $\lim_{n_2\to\infty} f_{n_1 n_2}$ reflects the effects at level 1 of both disarrangements occurring at level 2 and of smooth changes occurring at level 2; on the contrary, $\lim_{n_2\to\infty} \nabla f_{n_1 n_2}$ reflects the effects at level 1 only of deformations without disarrangements occurring at level 2. The process $\lim_{n_1\to\infty}$ of zooming out from level 1 to the macrolevel, level 0, along with the relation (7), then justifies the terminology already introduced: "deformation without disarrangements at level 2" for the tensor field $G_2$. In the same spirit, for each $n_1$ the field $\nabla \lim_{n_2\to\infty} f_{n_1 n_2}$ only reflects the smooth part of the deformations $\lim_{n_2\to\infty} f_{n_1 n_2}$ at level 1, so that the iterated limit $\lim_{n_1\to\infty} \nabla \lim_{n_2\to\infty} f_{n_1 n_2}$ reflects the effects at the macrolevel only of the smooth part of the deformations $\lim_{n_2\to\infty} f_{n_1 n_2}$ at level 1. The relation (6) then justifies the terminology already introduced: "deformation without disarrangements at level 1" for the tensor field $G_1$.

As an example we consider the *three-level shear,* an adaptation to the present context of the two-level shear for two-level structured deformations [12, 14]. Let $u, v$ be orthogonal unit vectors, let $\mu_0, \mu_1, \mu_2$ be scalars, and define for each $X$ in $\mathcal{B}$

$$g(X) = X + \mu_0 \big(v \cdot (X - o)\big)\, u, \tag{8}$$

$$G_1(X) = I + \mu_1 u \otimes v, \quad G_2(X) = I + \mu_2 u \otimes v, \tag{9}$$

with $o$ a preassigned point in $\mathcal{B}$, the unit cube with one vertex at $o$ and two edges parallel to $u$ and $v$. The macroscopic deformation $g$ is a simple shear of amount $\mu_0$ in the direction $u$ whose shearing plane has normal $v$, while the angle of shearing is $\tan^{-1}\mu_0$. Because $G_0(X) = \nabla g(X) = I + \mu_0\, u \otimes v$, we have $\det G_i = 1$ for $i = 0, 1, 2$, so that the accommodation inequalities (3) are satisfied with $c = 1/2$ and with equality in the last two.

An approximating sequence $(n_1, n_2) \longmapsto f_{n_1 n_2}$ for the three-level shear that satisfies (4)–(7) in the Approximation Theorem is described geometrically as follows. The mapping $f_{n_1 n_2}$ first slices $\mathcal{B}$ into $n_1$ pieces ("cards") by means of $n_1 - 1$ parallel planes with normal the unit vector $v$ that are separated by amount $1/n_1$. It leaves the bottom card fixed and translates each higher card in the deck, relative to the one below it, by the amount $(\mu_0 - \mu_1)/n_1$ in the direction $u$, and successively shears the $i^{th}$ translated card via a simple shear of amount $\mu_1$, again in the direction $u$ with normal orientation $v$. The resulting deck of $n_1$ translated and sheared cards approximates the range $g(\mathcal{B})$ of the original mapping $g$, and this sequence of injective, piecewise smooth approximations indexed on $n_1$ would converge to $g$ uniformly, while its gradients would converge uniformly to $G_1$, if we at this point would let $n_1$ tend to infinity. In fact, the description up to this point is precisely that of a two-level shear and its approximates at submacroscopic level 1 [12]. The further action of $f_{n_1 n_2}$ involves slicing each of the $n_1$ translated and sheared level 1 cards into $n_2$ level 2 cards by means of $n_2 - 1$ planes all with normal $v$ as before. Within the $i$th level 1 card, the $j$th level 2 card is translated relative to the card below it by amount $(\mu_1 - \mu_2)/n_2$ in the direction $u$ and sheared via a simple shear of amount $\mu_2$ with direction $u$ and normal orientation $v$. The resulting sequence of multilevel shears $f_{n_1 n_2}$ is easily shown to satisfy (4)–(7) for the specific three-level structured deformation defined in (8) and (9).

It is worth pointing out in the example of three-level shears that the differences

$$G_0 - G_1 = (\mu_0 - \mu_1)u \otimes v \quad \text{and} \quad G_1 - G_2 = (\mu_1 - \mu_2)u \otimes v$$

involve the differences $\mu_0 - \mu_1$ and $\mu_1 - \mu_2$ that appear in the numerators of the expressions for the tangential discontinuities $(\mu_0 - \mu_1)/n_1$ and of $(\mu_1 - \mu_2)/n_2$ caused by the approximating sequence of piecewise smooth, injective deformations $f_{n_1 n_2}$. This suggests that $G_0 - G_1$ captures the deformation due to disarrangements at submacroscopic level 1, while $G_1 - G_2$ captures the deformation due to disarrangements at submacroscopic level 2. An analysis analogous to that given in [13] for general two-level structured deformations shows that for arbitrary three-level structured deformations this interpretation of the differences $G_{i-1} - G_i$ for $i = 1, 2$ is justified. In fact, for each three-level structured deformation $(g, G_1, G_2)$ and for each sequence $f_{n_1 n_2}$ satisfying (4)–(7) in the Approximation Theorem, we have for every point $X$ and $r > 0$:

$$\int_{B_r(X)} (G_1 - G_2)(y) dV_y = \lim_{n_1 \to \infty} \lim_{n_2 \to \infty} \int_{J(f_{n_1 n_2}) \cap B_r(X)} [f_{n_1 n_2}](y) \otimes \nu_y \, dA_y \tag{10}$$

and

$$\int_{B_r(X)} (G_0 - G_1)(y) dV_y = \lim_{n_1 \to \infty} \int_{J(\lim_{n_2 \to \infty} f_{n_1 n_2}) \cap B_r(X)} \left[ \lim_{n_2 \to \infty} f_{n_1 n_2} \right](y) \otimes \nu_y \, dA_y. \tag{11}$$

In these relations, $B_r(X)$ denotes the ball of radius $r$ centered at the point $X$ in $\mathcal{B}$, while $J(h)$ denotes the jump set of a function $h$, $\nu_y$ the normal to $J(h)$ at the point $y$, and $[h](y)$ its jump at $y$. While (6) and (7) show that $G_1$ and $G_2$ capture the effects at the macrolevel of smooth submacroscopic deformations at levels 1 and 2, respectively, the relations (10) and (11) show that the tensor

$$M_1 := G_0 - G_1 \tag{12}$$

captures the effects at the macrolevel of disarrangements (discontinuous deformations) at submacroscopic level 1 and that

$$M_2 := G_1 - G_2 \tag{13}$$

captures the effects at the macrolevel of disarrangements at submacroscopic level 2. We call $M_1$ and $M_2$ the *disarrangement tensors* for the three-level structured deformation $(g, G_1, G_2)$. Their role from (10) and (11) in capturing separately the non-smooth deformations that occur at different levels motivates the term "hierarchical disarrangements" in the title of this paper. It is of mathematical interest to note from (5)–(7), (12) and (13) that each of the disarrangement tensors $M_1$ and $M_2$ captures quantitatively the effect of interchanging once the operation "$\nabla$", the classical gradient operator, and the operation "lim" that here describes zooming out one level.

## 2.2 Additive Decompositions, Multiplicative Decompositions, and Factorizations

We record using (5), (10), and (11) the algebraically simple but geometrically significant relation

$$\nabla g = G_0 = G_2 + M_2 + M_1 \tag{14}$$

that provides an additive decomposition of the macroscopic deformation gradient as a sum of the disarrangement tensors $M_1$ and $M_2$ at levels one and two, respectively, and of the tensor without disarrangements at level two $G_2$. Thus, all of the local geometrical changes captured by $G_0 = \nabla g$ are described in a simple way in terms of the deformation without disarrangements $G_2$ at level two and the deformations due to disarrangements $M_1$ and $M_2$ at levels one and two. Similarly, we may write for $i = 1, 2$:

$$G_{i-1} = G_i + M_i \tag{15}$$

to obtain additive decompositions for $G_0$ and for $G_1$ into a part without disarrangements and a part due to disarrangements. While each of these decompositions is algebraically trivial, the identification relations (6), (7), (10), and (11) fully justify the interpretation and the terminology that we provide for each term in the sums on the right-hand sides of (14) and (15).

We emphasize that the additive decompositions (14) and (15) are valid whatever the size of the underlying deformations. The additive decomposition (14) for $\nabla g$ provides immediately two multiplicative decompositions:

$$\nabla g = G_2\big(G_2^{-1}M_1 + G_2^{-1}M_2 + I\big) =: G_2 M_r, \tag{16}$$

$$\nabla g = \big(M_1 G_2^{-1} + M_2 G_2^{-1} + I\big)G_2 =: M_l G_2. \tag{17}$$

In many theories of inelastic behavior of materials, particularly in theories of plasticity, analogous decompositions are introduced either as constitutive assumptions or as geometrical assumptions that rest on the existence of intermediate configurations that are not attainable in general by a three-dimensional body. Here, the multiplicative decompositions are derived from additive decompositions, and the factors in each multiplicative decomposition have definite interpretations obtained from the identification relations (6), (7), (10), and (11). Moreover, each of the multiplicative factors $M_r$ and $M_l$ itself has an additive decomposition whose terms reflect the presence of disarrangements at two submacroscopic levels:

$$M_r = G_2^{-1}M_1 + G_2^{-1}M_2 + I, \tag{18}$$

$$M_l = M_1 G_2^{-1} + M_2 G_2^{-1} + I. \tag{19}$$

(See [12, 17] for detailed discussions of the corresponding additive and multiplicative decompositions in the case of two-level structured deformations.)

In spite of the usefulness of multiplicative decompositions of the form (16) and (17), these decompositions do not reflect factorizations of the deformations of the body as a whole. We record here a definition of *composition of three-level structured deformations* (see [12] for the two-level counterpart) that does reflect geometrical changes of the body as a whole:

$$(g, G_1, G_2) \circ (\tilde{g}, \tilde{G}_1, \tilde{G}_2) = \big(g \circ \tilde{g}, (G_1 \circ \tilde{g})\tilde{G}_1, (G_2 \circ \tilde{g})\tilde{G}_2\big). \tag{20}$$

In this definition we use the same symbol "∘" on the left to denote "composition of structured deformations" as we do on the right for "composition of tensor fields and point mappings", but no confusion should arise from this practice. This definition leads immediately to the following factorization of an arbitrary three-level structured deformation:

$$(g, G_1, G_2) = (g, G_0, G_0) \circ \left(\iota, G_0^{-1}G_1, G_0^{-1}G_1\right) \circ \left(\iota, I, G_1^{-1}G_2\right) \tag{21}$$

where $\iota$ denotes the identity mapping $X \longmapsto \iota(X) = X$, and where $I$ is the constant tensor field whose only value is the identity tensor $u \longmapsto Iu = u$. The factorization is easily verified by evaluating the (associative) operation ∘ twice on the right-hand side of this relation, and we show below (22) that each factor on the right-hand side is, itself, a three-level structured deformation.

The following table records, for each of the three factors on the right-hand side of (21), the deformation without disarrangements at level two, "$G_2$", as well as the disarrangement tensors "$M_1$" and "$M_2$" at levels one and two:

| Factor in (21) | "$G_2$" | "$M_1$" | "$M_2$" |
|---|---|---|---|
| $(g, G_0, G_0)$ | $G_0$ | 0 | 0 |
| $(\iota, G_0^{-1}G_1, G_0^{-1}G_1)$ | $G_0^{-1}G_1$ | $I - G_0^{-1}G_1$ | 0 |
| $(\iota, I, G_1^{-1}G_2)$ | $G_1^{-1}G_2$ | 0 | $I - G_1^{-1}G_2$ |

The table shows that the first factor $(g, G_0, G_0)$ causes no disarrangements at either submacroscopic level, and we call $(g, G_0, G_0)$ a *classical deformation* or a *purely macroscopic deformation*, since in classical treatments of geometrical changes no disarrangements are included and only deformations at the macrolevel are considered. The second factor $(\iota, G_0^{-1}G_1, G_0^{-1}G_1)$ produces no geometrical changes at the macrolevel, since $i(X) = X$ for all $X \in \mathcal{B}$, and causes no disarrangements at submacroscopic level 2. The third factor $(\iota, I, G_1^{-1}G_2)$ produces no geometrical changes at either the macrolevel or at the submacroscopic level 1, since $\iota(X) = X$ for all $X \in \mathcal{B}$ and since $Iu = u$ for all $u \in \mathcal{V}$, and $(i, I, G_1^{-1}G_2)$ causes no disarrangements at submacroscopic level 1.

It is convenient to use the notation $K_i = G_{i-1}^{-1}G_i$ for $i = 1, 2$ and to write the factorization (21) in the form

$$(g, G_1, G_2) = (g, G_0, G_0) \circ (\iota, K_1, K_1) \circ (\iota, I, K_2). \tag{22}$$

The accommodation inequalities (3) for the three-level structured deformation $(g, G_1, G_2)$ tells us that at each point $X$ both $\det K_1(X)$ and $\det K_2(X)$ lie in the interval $(0, 1]$, and we refer to $\det K_i$ as the volume fraction associated with the purely level-$i$ factor, for $i = 1, 2$. Moreover, it is clear that each of the factors on the right-hand side of (22) is, itself, a three-level structured deformation, because each has the required regularity and satisfies the accommodation inequalities (3) in one of the following forms:

$$0 < c_0 < \det G_0(X) = \det G_0(X) = \det \nabla g(X) \quad \text{for } (g, G_0, G_0),$$

$$0 < c_1 < \det K_1(X) = \det K_1(X) \leq 1 = \det \nabla \iota(X) \quad \text{for } (\iota, K_1, K_1),$$

$$0 < c_2 < \det K_2(X) \leq 1 = \det I = \det \nabla \iota(X) \quad \text{for } (\iota, I, K_2).$$

In these relations, $c_j$ for $j = 0, 1, 2$, can be chosen using the fact that $\det G_j$ has a strictly positive lower bound and a finite upper bound on the closure of the body.

## 2.3 Decomposition of Tensor Fluxes

For a given three-level structured deformation $(g, G_1, G_2)$ and for a smooth tensor field $N$: $g(\mathcal{B}) \longrightarrow Lin\mathcal{V}$, we call $N$ a *tensor flux on the deformed configuration*, and we define for $i = 0, 1, 2$ the *reference flux at level* $i$ to be the tensor field $N_i$: $\mathcal{B} \longrightarrow Lin\mathcal{V}$ given by

$$N_i = (N \circ g)G_i^*, \tag{23}$$

where, for each invertible tensor $A$, $A^* = (\det A)A^{-T}$. We also refer to $N_0$ as the macroscopic reference flux. For example, when $N = T$ is the Cauchy stress field on $g(\mathcal{B})$, then $T_0 = (T \circ g)G_0^* = (T \circ g)(\nabla g)^*$ is the Piola-Kirchhoff stress, usually denoted by $S$. The macroscopic reference stress $T_0$ as well as the submacroscopic reference stresses $T_i = (T \circ g)G_i^*$ for $i = 1, 2$ can be rewritten using (4), (6), and (7) as

$$\begin{aligned} S = T_0 &= (T \circ g)\Big(\nabla \lim_{n_1\to\infty}\lim_{n_2\to\infty} f_{n_1 n_2}\Big)^*, \\ T_1 &= (T \circ g)\lim_{n_1\to\infty}\Big(\nabla \lim_{n_2\to\infty} f_{n_1 n_2}\Big)^*, \\ T_2 &= (T \circ g)\lim_{n_1\to\infty}\lim_{n_2\to\infty}(\nabla f_{n_1 n_2})^*. \end{aligned} \tag{24}$$

This is the primary example of tensor flux that we shall use here, and the reader may without loss read all of the following with $N$ replaced by $T$. Later in this section we justify calling the submacroscopic reference stress $T_1$ the part of the macroscopic reference stress $T_0 = S$ without level one disarrangements, and we justify calling the submacroscopic reference stress $T_2$ the part of the submacroscopic reference stress $T_1$ without level two disarrangements (see relations (29)).

For a general flux $N$, the reference fluxes $N_1$ and $N_2$ at the submacroscopic levels one and two are intended to capture the idea that the geometry of three-level structured deformations allows us to identify refined measures of fluxes defined on the body $\mathcal{B}$ arising from the given flux $N$ on the deformed body $g(\mathcal{B})$. In order to develop further this idea, we use the tensor fields $K_i$ defined above (22) to write for $i = 1, 2$ the following product rule

$$\det K_i \operatorname{div} N_{i-1} = \operatorname{div}\big((\det K_i)N_{i-1}\big) - N_{i-1}\nabla \det K_i$$

for the divergence of the product $(\det K_i)N_{i-1}$. By adding and subtracting on the right-hand side the field $\operatorname{div}(N_{i-1}K_i^*)$, where $K_i^* = (\det K_i)K_i^{-T}$, we obtain for $i = 1, 2$

$$\det K_i \operatorname{div} N_{i-1} = \operatorname{div}\big(N_{i-1}K_i^*\big) + \operatorname{div}\big((\det K_i)N_{i-1} - N_{i-1}K_i^*\big) - N_{i-1}\nabla \det K_i. \tag{25}$$

Arguments given in [18] in the context of two-level structured deformations and summarized in [21] show, by applying the Approximation Theorem to the purely level-$i$ factor in the factorization (22), $(i, K_1, K_1)$ for $i = 1$ or $(i, I, K_2)$ for $i = 2$, that the left-hand side $\det K_i \operatorname{div} N_{i-1}$ can be interpreted as a volume density of the flux of $N_{i-1}$ taking into account both the sites of disarrangements and the preassigned surfaces across which the flux of $N_{i-1}$ is measured. Moreover, the term $\operatorname{div}(N_{i-1}K_i^*)$ on the right-hand side can be interpreted as a volume density of the flux of $N_{i-1}$, taking into account only preassigned surfaces. Consequently, the remaining terms $\operatorname{div}((\det K_i)N_{i-1} - N_{i-1}K_i^*) - N_{i-1}\nabla \det K_i$ can be interpreted as the volume density of the flux of $N_{i-1}$, taking into account only the sites of disarrangements at level $i$, and we may call (25) the *decomposition at level $i$ of the volume* *density of the flux of* $N_{i-1}$. It is shown further in [18] that the tensor field $N_{i-1}K_i^*$ accounts for all of the volume density of the flux without level $i$ disarrangements, while the tensor field $(\det K_i)N_{i-1} - N_{i-1}K_i^*$ accounts for only the flux due to level $i$ disarrangements.

The trivial algebraic identity for $i = 1, 2$,

$$(\det K_i)N_{i-1} = N_{i-1}K_i^* + \big((\det K_i)N_{i-1} - N_{i-1}K_i^*\big), \tag{26}$$

and the identification just above of each term on the right-hand side of (25) motivates the following terminology: $N_{i-1}K_i^*$ is *the part of* $N_{i-1}$ *without level i disarrangements*, and $(\det K_i)N_{i-1} - N_{i-1}K_i^*$ is *the part of* $N_{i-1}$ *due to level i disarrangements*. For conciseness, we use the following notation for the two parts of $N_{i-1}$:

$$N_{i-1.\backslash} := N_{i-1}K_i^* \quad \text{and} \quad N_{i-1.d} := (\det K_i)N_{i-1} - N_{i-1}K_i^*, \tag{27}$$

In the expression $N_{i-1}K_i^*$ the subscript $i-1$ on $N$ indicates the level at which the reference flux is measured, while the subscript $i$ on $K_i$ and $K_i^*$ indicates the level at which disarrangements are being considered.

We rewrite (26) in the shorter form for $i = 1, 2$:

$$(\det K_i)N_{i-1} = N_{i-1.\backslash} + N_{i-1.d}. \tag{28}$$

The appearance of $\det K_i$ multiplying $N_{i-1}$ on the left-hand sides of (26) and (28) is natural, in light of the fact that the adjugate $K_i^*$ of $K_i$ also contains that factor. (One could divide the decomposition (26) through by $\det K_i$ and so obtain a decomposition of $N_{i-1}$, itself, but maintaining the presence of $\det K_i$ has advantages that are apparent in what follows.) One may view the role of $\det K_i$ in front of $N_{i-1}$ in (28) as adjusting for the possible reduction in volume due to the "purely level-$i$ factor" in the factorization (22). One also may regard the decomposition (28) as the counterpart for fluxes of the geometrical decomposition $G_{i-1} = G_i + M_i$ that results from the definition of $M_i$ as the difference $G_{i-1} - G_i$.

We have argued for $i = 1, 2$ that $N_{i-1.\backslash} = N_{i-1}K_i^*$ represents the part of the reference flux $N_{i-1}$ without disarrangements at level $i$, and it is interesting that this part of the reference flux $N_{i-1}$ at level $i-1$ actually recovers all of the reference flux $N_i$ at the next level, level $i$, i.e., we have at each of the two submacroscopic levels:

$$N_i = N_{i-1.\backslash} \quad \text{for } i = 1, 2. \tag{29}$$

For example, the reference stress $T_1$ represents the part of the Piola-Kirchhoff stress $T_0 = S$ without disarrangements at level 1, while $T_2$ represents the part of $T_1$ without disarrangements at level 2. To establish (29) we note that $G_i = G_{i-1}K_i$, so that $G_i^* = G_{i-1}^*K_i^*$ and therefore we have

$$N_i = (N \circ g)G_i^* = (N \circ g)G_{i-1}^*K_i^* = N_{i-1}K_i^* = N_{i-1.\backslash}.$$

Inasmuch as the additive decomposition (28) is the counterpart for fluxes of the geometrical decomposition $G_{i-1} = G_i + M_i$, it is natural to seek a counterpart for fluxes of the full geometrical decomposition (14): $G_0 = G_2 + M_2 + M_1$. In fact, it is not difficult to arrive at the corresponding full decomposition of fluxes in the following form:

$$(\det K_1)(\det K_2)\, N_0 = N_{1.\backslash} + N_{1.d} + (\det K_2)N_{0.d}. \tag{30}$$

Indeed, the decomposition (28) of fluxes at level i with $i = 2$, the definition $(27)_2$ with $i = 1$, and the identification (29) of $N_i$ with $i = 1$ permit us to write

$$N_{1.\backslash} + N_{1.d} + (\det K_2)N_{0.d} = (\det K_2)N_1 + (\det K_2)\big((\det K_1)N_0 - N_{0.\backslash}\big)$$
$$= (\det K_2)N_1 + (\det K_2)\big((\det K_1)N_0 - N_1\big)$$

and this verifies (30). We refer to the decompositions (14) of $G_0$ and (30) of $N_0$ as *parallel decompositions*, because corresponding terms on the right-hand sides refer to the same disarrangement status and to the same level at which disarrangements occur.

## 2.4 Universal Consistency Relations for Tensor Fluxes at Submacroscopic Levels

For $i = 1, 2$ the definition $(27)_1$ of $N_{i-1.\backslash}$ is a multiplicative relation, while the decomposition (28) is an additive relation. Together these two relations yield:

$$N_{i-1.\backslash} + N_{i-1.d} = (\det K_i)N_{i-1} = N_{i-1.\backslash}K_i^T.$$

Multiplying on the right the left-most and the right-most members by $K_i^{-T} = (G_i^{-1}G_{i-1})^{-T} = G_i^T G_{i-1}^{-T}$ we find that

$$N_{i-1.\backslash}\big(G_i^T G_{i-1}^{-T} - I\big) + N_{i-1.d}G_i^T G_{i-1}^{-T} = 0.$$

Multiplying both sides on the right by $G_{i-1}^T$ and using the definition $M_i = G_i - G_{i-1}$ we obtain the *consistency relation at submacroscopic level i*:

$$N_{i-1.\backslash}M_i^T + N_{i-1.d}G_{i-1}^T = 0 \quad \text{for } i = 1, 2. \tag{31}$$

The consistency relations (31) show that, for a given three-level structured deformation and a given tensor field $N$ on $g(\mathcal{B})$, the parts of $N_{i-1}$ with and without disarrangements at each submacroscopic level $i$ cannot be assigned independently. In the context of two-level structured deformations and when $N$ is taken to be the Cauchy stress tensor $T$, the consistency relation for $i = 1$ can be rewritten in terms of $S = T_0$ and reads

$$S_\backslash M_1^T + S_d(\nabla g)^T = 0.$$

It was derived in [18] and provides an essential relation among the field relations that govern elastic bodies undergoing disarrangements at one submacroscopic level.

It is important to keep in mind that the consistency relations (31) are universal, in the sense that apply to every body, whatever its material composition, undergoing a structured deformation $(g, G_1, G_2)$ in the presence of a given tensor flux $N$. The universality property as well as a uniqueness property of the associated decompositions (27) and (28) were studied in the context of two-level structured deformations in [23].

It is helpful to rewrite the two consistency relations (31) in terms of the three tensor fluxes $N_{1.\backslash}$, $N_{1.d}$, and $N_{0.d}$ that appear in the full decomposition (30) of the reference flux $N_0$ at the macroscopic level. Putting $i = 1$ in (31) and multiplying both sides by $\det K_2$ we have

$$0 = (\det K_2)N_{0.\backslash}M_1^T + (\det K_2)N_{0.d}G_0^T$$
$$= (\det K_2)N_1 M_1^T + (\det K_2)N_{0.d}G_0^T$$
$$= (N_{1.\backslash} + N_{1.d})M_1^T + (\det K_2)N_{0.d}G_0^T,$$

where we have used (29) and (28) in the second and third relations. This relation together with (31) for $i = 2$ and the relation $G_1 = G_2 + M_2$ yield the desired form of the consistency relations:

$$N_{1.\backslash} M_1^T + N_{1.d} M_1^T + (\det K_2) N_{0.d} G_0^T = 0, \tag{32}$$

$$N_{1.\backslash} M_2^T + N_{1.d} M_2^T + N_{1.d} G_2^T = 0. \tag{33}$$

## 3 Motions and a Decomposition of the Stress-Power

We follow some earlier treatments of motions in the context of two-level structured deformations ([17, 18, 33]) by requiring that, at each time $t$ in a given interval, there is specified a three-level structured deformation $(g(\cdot,t), G_1(\cdot,t), G_2(\cdot,t))$ of the body. We require for every $X$ in $\mathcal{B}$ that the field $t \longmapsto g(X,t)$ is twice continuously differentiable, while $t \longmapsto G_i(X,t)$ for $i = 0, 1, 2$ is continuously differentiable. Thus, structured motions here may be regarded as smoothly time-parameterized families of three-level structured deformations. (See [33] for an alternative approach in the context of two-level structured deformations in which time-like disarrangements are admitted.) It is convenient in what follows to denote derivatives with respect to $t$ via Newtonian notation (superposed dots) and, without risk of ambiguity, to omit frequently the explicit writing of the variable $t$.

We may now use the parallel decompositions (14) of $G_0 = \nabla g$ and the full decomposition of fluxes (30) with $N$ equal to $T$, the Cauchy stress, to compute $(\det K_1)(\det K_2) T_0 \cdot \dot{G}_0$, the *stress-power* $\mathcal{P}$ in a given motion (measured per unit volume at submacroscopic level two):

$$\begin{aligned} \mathcal{P} &:= (\det K_1)(\det K_2) T_0 \cdot \dot{G}_0 \\ &= \left\{ T_{1.\backslash} + T_{1.d} + (\det K_2) T_{0.d} \right\} \cdot \{ \dot{G}_2 + \dot{M}_2 + \dot{M}_1 \} \\ &= T_{1.\backslash} \cdot \dot{G}_2 + T_{1.d} \cdot \dot{M}_2 + (\det K_2) T_{0.d} \cdot \dot{M}_1 + \mathcal{P}_{mix}. \end{aligned} \tag{34}$$

The stress-power $\mathcal{P}$ is an inner product of two factors, each with three terms, and so can be written as a sum of nine terms. Here, each of the first three terms displayed explicitly in the last expression in (34) contains a stress measure and a measure of rate of deformation that match both in terms of submicroscopic levels and in terms of whether or not the effects of disarrangements are captured. For example, in the inner product $T_{1.\backslash} \cdot \dot{G}_2$, the factor $T_{1.\backslash}$ is the part of $T_1$ without level 2 disarrangements and $\dot{G}_2$ is the rate of change of deformation without level 2 disarrangements. The sum of the remaining six terms is designated in (34) by $\mathcal{P}_{mix}$. These six terms include, for example, the terms $T_{1.\backslash} \cdot \dot{M}_2$ and $T_{1.d} \cdot \dot{M}_1$, and each such term has the property that there is at least one "mismatch" between its factors with regard to submacroscopic level or with regard to the inclusion or not of effects of disarrangement. Thus, in the term $T_{1.\backslash} \cdot \dot{M}_2$ in the sum $\mathcal{P}_{mix}$, $T_{1.\backslash}$ is the part of $T_1$ *without* level two disarrangements, while $\dot{M}_2$ is the rate of deformation *due to* level two disarrangements. Consequently, we follow the terminology introduced in [18] in the context of two-level structured motions and call $\mathcal{P}_{mix}$ the *mixed power* in the given motion (again measured per unit volume at submacroscopic level two). The refined geometry of hierarchical structured deformations here permits us to distinguish the mixed power $\mathcal{P}_{mix}$ from the total stress power $\mathcal{P}$, and this distinction will allow us to identify explicitly, in both physical and mathematical terms, the sources of dissipation that arise through the constitutive prescriptions that we make in Sect. 5.1.

## 4 Dynamical Processes, Constitutive Classes, and the Dissipation Inequality

We adapt the discussion in [18] to specify here a dynamical process by giving the fields $g$, $G_1$, and $G_2$ associated with a structured motion as well as the Cauchy stress field $T \circ g$, the volume density $\psi$ of the Helmholtz free energy in the macroscopic reference configuration, and the mass density $\rho_0$ in that configuration. The fields $g$, $G_1$, $G_2$, $T \circ g$, and $\psi$ are defined on pairs $(X, t)$ with $X \in \mathcal{B}$ and $t$ in a time interval that can depend upon $g$, while $\rho_0$ is defined on points $X \in \mathcal{B}$. Henceforth, we consider the density $\rho_0$ as fixed and omit it when listing the quantities associated with dynamical processes. The formulas from above (22) for $K_i = G_{i-1}^{-1} G_i$ with $i = 1, 2$, accompanied by the relations (23) and (27) with $N = T$, show that the stresses $T_{i-1.\backslash}$ and $T_{i-1.d}$ with and without level $i$ disarrangements are also specified once a dynamical process is specified. The same is true for the stress power $\mathcal{P}$ and the mixed power $\mathcal{P}_{mix}$ (see the formula (34)), as well as the time derivative $\dot{\psi}$ of the free energy density.

A constitutive class for the body [26] is simply a collection $\mathcal{C}$ of dynamical processes of the body, and a particular choice of constitutive class then limits the dynamical processes that are to be considered. In practice, a constitutive class is specified by giving a list of response functions and requiring that the fields $g$, $G_1$, $G_2$, $T \circ g$, and $\psi$ in each dynamical process be related through equations and/or inequalities involving the response functions in the given list.

Another limitation on the collection of dynamical processes in the present isothermal setting is thermodynamical in nature and here takes the form of a dissipation inequality:

$$\dot{\psi}(X, t) \leq T_0(X, t) \cdot \dot{G}_0(X, t), \tag{35}$$

asserting that the rate of change of free energy density cannot exceed the stress power in the macroscopic reference configuration. We denote by $\mathcal{T}$ the collection of dynamical processes that satisfy the dissipation inequality, and we impose the requirement $\mathcal{C} \subset \mathcal{T}$ on possible choices of constitutive classes $\mathcal{C}$. This requirement suffices to assure that our constitutive relations are compatible with the Second Law. The standard, Coleman-Noll procedure for assuring compatibility of constitutive relations with the Second Law here proves to be too restrictive in that it rules out non-zero internal dissipation in smooth processes. (For a discussion of this shortcoming of the Coleman-Noll procedure in the context of elastic bodies undergoing two-level structured deformations, see Sects. 6 and 7 of the article [18].)

## 5 Elasticity with Hierarchical Disarrangements

### 5.1 Constitutive Assumptions

The constitutive assumptions that we make here in the context of three-level structured deformations are adaptations of those made for the field theory elasticity with disarrangements [18] in the context of two-level structured deformations. It is important to keep in mind that, other than the free energy density $\psi$ itself, all of the fields that enter into our constitutive relations have been identified through the multilevel geometry of structured deformations and through the decompositions of deformations and of fluxes that this geometry provides. Consequently, we can avoid the pitfall of confusing, say, a constitutive equation containing a particular stress field with a relation that defines that field.

We first assume that the density $\psi$ of free energy per unit volume in the macroscopic reference configuration is determined at each time $t$ and point $X$ in $\mathcal{B}$ by the values of the fields $G_2$, $M_2$, and $M_1$:

$$\psi(X,t) = \Psi\big(G_2(X,t), M_2(X,t), M_1(X,t)\big), \tag{36}$$

where $\Psi$ is a smooth function of its three tensor arguments. Next, we consider the decomposition (30) again with $N = T$, and we make constitutive assumptions on the stresses $T_{1.\backslash}$, $T_{1.d}$, and $T_{d.1}$ appearing in that decomposition:

$$T_{1.\backslash}(X,t) = (\det K_1)(\det K_2)\ |_{(X,t)}\ D_{G_2}\Psi\big(G_2(X,t), M_2(X,t), M_1(X,t)\big), \tag{37}$$

$$T_{1.d}(X,t) = (\det K_1)(\det K_2)\ |_{(X,t)}\ D_{M_2}\Psi\big(G_2(X,t), M_2(X,t), M_1(X,t)\big), \tag{38}$$

$$T_{0.d}(X,t) = (\det K_1)\ |_{(X,t)}\ D_{M_1}\Psi\big(G_2(X,t), M_2(X,t), M_1(X,t)\big). \tag{39}$$

These three constitutive assumptions reflect the refined geometry of three-level structured deformations by linking the submacroscopic level and the disarrangement status of a given stress field on the left-hand side with the variable with respect to which the free energy is differentiated on the right-hand side. This feature of the constitutive relations (37)–(39) tell us in language found in the physics literature that the previously defined stress fields $T_{1.\backslash}$, $T_{1.d}$, and $T_{0.d}$ now play the role of "driving forces" corresponding to the geometrical fields $G_2$, $M_2$, and $M_1$. (As we remarked at the beginning of this section, the stresses $T_{1.\backslash}$, $T_{1.d}$, and $T_{0.d}$ were not *defined* to be driving forces.)

The factors $(\det K_1)(\det K_2)$ and $(\det K_1)$ involving determinants of $K_1$ and $K_2$ in (37)–(39) adjust for the fact that $\Psi$ measures free energy per unit volume in the macroscopic reference configuration, while the stresses $T_{1.\backslash}$, $T_{1.d}$ refer to level two disarrangements and $T_{0.d}$ refers to level one disarrangements. Moreover, the factors $\det K_1$ and $\det K_2$ in the three constitutive relations (37)–(39) facilitate the following computation:

$$\begin{aligned}(\det K_1)(\det K_2)\dot{\psi} &= (\det K_1)(\det K_2)\, D_{G_2}\Psi(G_2, M_2, M_1)\cdot \dot{G}_2 \\ &\quad + (\det K_1)(\det K_2)\, D_{M_2}\Psi(G_2, M_2, M_1)\cdot \dot{M}_2 \\ &\quad + (\det K_1)(\det K_2)\, D_{M_1}\Psi(G_2, M_2, M_1)\cdot \dot{M}_1 \\ &= T_{1.\backslash}\cdot \dot{G}_2 + T_{1.d}\cdot \dot{M}_2 + \det K_2 T_{0.d}\cdot \dot{M}_1 \\ &= \mathcal{P} - \mathcal{P}_{mix},\end{aligned} \tag{40}$$

where $\mathcal{P}$ and $\mathcal{P}_{mix}$ are the stress-power and the mixed stress-power defined in the previous section. Because $\mathcal{P} = (\det K_1)(\det K_2)T_0 \cdot \dot{G}_0$, we conclude that the relation (40) is equivalent to

$$T_0 \cdot \dot{G}_0 - \dot{\psi} = \frac{\mathcal{P}_{mix}}{(\det K_1)(\det K_2)}. \tag{41}$$

Thus, our constitutive assumptions (37)–(39) imply that the *rate of internal dissipation* per unit volume in the macroscopic reference configuration,

$$\Gamma := T_0 \cdot \dot{G}_0 - \dot{\psi} = \frac{\mathcal{P}_{mix}}{(\det K_1)(\det K_2)}, \tag{42}$$

is given, to within the positive factor $(\det K_1)(\det K_2)$, by the six "mismatch" terms that comprise the mixed power $\mathcal{P}_{mix}$:

$$\begin{aligned}(\det K_1)(\det K_2)\Gamma &= T_{1.\backslash}\cdot \dot{M}_2 + T_{1.\backslash}\cdot \dot{M}_1 + T_{1.d}\cdot \dot{G}_2 + T_{1.d}\cdot \dot{M}_1 \\ &\quad + (\det K_2)\, T_{0.d}\cdot \dot{G}_2 + (\det K_2)\, T_{0.d}\cdot \dot{M}_2. \end{aligned} \tag{43}$$

The first two mismatch terms $T_{1.\backslash}\cdot \dot{M}_2 + T_{1.\backslash}\cdot \dot{M}_1$ represent the power expended by stresses $T_{1.\backslash}$ without level two disarrangement against rates of change of deformation $\dot{M}_2$ and $\dot{M}_1$ due to level 2 and to level 1 disarrangements, respectively. The third and fourth terms $T_{1.d}\cdot \dot{G}_2 + T_{1.d}\cdot \dot{M}_1$ represent the power expended by stresses $T_{1.d}$ due to level two disarrangements against a rate of change of deformation without (level one or level two) disarrangements $\dot{G}_2$ or against a rate of deformation $\dot{M}_1$ due to level one disarrangements. Finally, the fifth and sixth mismatch terms $(\det K_2)\, T_{0.d}\cdot \dot{G}_2 + (\det K_2)\, T_{0.d}\cdot \dot{M}_2$ represent stresses $T_{0.d}$ due to level one disarrangements expending power against rates of deformation $\dot{G}_2$ and $\dot{M}_2$ at level 2. Thus, *our constitutive assumptions tell us that internal dissipation arises only when the contact forces and the geometrical changes against which they do work are mismatched in terms of submacroscopic level or in terms of whether or not disarrangements are tracked.*

We note from the formula (43) and from the three constitutive relations (37)–(39) that we may write the rate of internal dissipation $\Gamma$ in the following form:

$$\begin{aligned}\Gamma &= D_{G_2}\Psi\cdot \dot{M}_2 + D_{G_2}\Psi\cdot \dot{M}_1 + D_{M_2}\Psi\cdot \dot{G}_2 + D_{M_2}\Psi\cdot \dot{M}_1 \\ &\quad + D_{M_1}\Psi\cdot \dot{G}_2 + D_{M_1}\Psi\cdot \dot{M}_2, \end{aligned} \tag{44}$$

and this formula renders transparent the mismatches described above. In view of (44) our final *constitutive assumption* takes the form of a restriction on the fields $G_2$, $M_2$, $M_1$, and on their time derivatives $\dot{G}_2$, $\dot{M}_2$, $\dot{M}_1$:

$$0 \le \Gamma(X,t) \tag{45}$$

for every $X \in \mathcal{B}$ and for every time $t$. Of course, the dissipation inequality (45) just imposed is a sufficient condition on the dynamical processes available to the body for satisfaction of the second law of thermodynamics in the present, isothermal context.

Our constitutive assumptions (36), (37)–(39), and (45) thus specify a collection $\mathcal{E}_{hd}$ of dynamical processes (namely those that satisfy (36), (37)–(39), and (45)) and, thereby, specify an elastic body undergoing disarrangements at two submacroscopic levels. The notation $\mathcal{E}_{hd}$ and the term *elasticity with hierarchical disarrangements* are used to distinguish the present choice of the constitutive class from our choice $\mathcal{E}_d$ made when only one submacroscopic level was considered [18]. As we pointed out in the previous paragraph, our constitutive assumptions guarantee that all of the (isothermal) dynamical processes available to such a body also are compatible with the second law of thermodynamics, so that our constitutive class is included in the collection of dynamical processes satisfying the second law, i.e., $\mathcal{E}_{hd} \subset \mathcal{T}$, as required in the previous section.

The decomposition (30) with $N = T$ along with the constitutive equations (37)–(39) imply

$$\begin{aligned}(\det K_1)(\det K_2)T_0 &= T_{1.\backslash} + T_{1.d} + (\det K_2)T_{0.d} \\ &= (\det K_1)(\det K_2)D_{G_2}\Psi + (\det K_1)(\det K_2)D_{M_2}\Psi \\ &\quad + (\det K_2)(\det K_1)D_{M_1}\Psi \end{aligned}$$

and thereby yield the *stress relation* for the Piola-Kirchhoff stress $S = T_0$:

$$T_0 = D_{G_2}\Psi + D_{M_2}\Psi + D_{M_1}\Psi. \tag{46}$$

An analogous stress relation

$$T_1 = (\det K_1)(D_{G_2}\Psi + D_{M_2}\Psi) \tag{47}$$

for the reference stress $T_1$ at level 1 follows from the relations :

$$(\det K_2)T_1 = T_{1.\backslash} + T_{1.d} = (\det K_1)(\det K_2)D_{G_2}\Psi + (\det K_1)(\det K_2)D_{M_2}\Psi.$$

Finally, the stress relation

$$T_2 = (\det K_1)(\det K_2)D_{G_2}\Psi \tag{48}$$

follows from the constitutive assumption (37) and from (29) with $i = 2$ and with $N = T$.

We note also from the relations (23), (29) with $i = 2$, and from (37) that the Cauchy stress field is given by

$$\begin{aligned} T \circ g &= (\det G_2)^{-1} T_2 G_2^T = (\det G_2)^{-1} T_{1.\backslash} G_2^T \\ &= (\det K_1 \det K_2)(\det G_2)^{-1} D_{G_2}\Psi\, G_2^T \\ &= (\det G_0)^{-1} D_{G_2}\Psi\, G_2^T. \end{aligned} \tag{49}$$

Similarly, we have

$$\begin{aligned} T \circ g &= (\det G_1)^{-1} T_1 G_1^T = (\det G_1)^{-1}(\det K_1)(D_{G_2}\Psi + D_{M_2}\Psi)(G_2 + M_2)^T \\ &= (\det G_0)^{-1}(D_{G_2}\Psi + D_{M_2}\Psi)(G_2 + M_2)^T \end{aligned} \tag{50}$$

and

$$T \circ g = (\det G_0)^{-1} T_0 G_0^T = (\det G_0)^{-1}(D_{G_2}\Psi + D_{M_2}\Psi + D_{M_1}\Psi)(G_2 + M_2 + M_1)^T. \tag{51}$$

## 5.2 Universal Consistency Relations for Stresses Revisited

As a consequence of the constitutive equations (37)–(39), the consistency relations (32) and (33) with $N = T$ now take the form

$$(D_{G_2}\Psi + D_{M_2}\Psi)\mid_{(G_2,M_2,M_1)} M_1^T + D_{M_1}\Psi \mid_{(G_2,M_2,M_1)} G_0^T = 0, \tag{52}$$

$$D_{G_2}\Psi \mid_{(G_2,M_2,M_1)} M_2^T + D_{M_2}\Psi \mid_{(G_2,M_2,M_1)} G_1^T = 0. \tag{53}$$

These relations are two tensorial equations that, in view of the decompositions $G_1 = G_2 + M_2$ and $G_0 = G_1 + M_1 = G_2 + M_2 + M_1$, restrict the tensors of deformation $G_2$, $M_2$, and $M_1$ at each point $X$ in the body and at each time $t$. Alternatively, the tensors of deformation $G_2$, $M_2$, and $M_1$ can be expressed in terms of the original tensors $G_2$, $G_1$, and $G_0 = \nabla g$ associated with the three-level structured deformation $(g, G_1, G_2)$, so that the consistency relations (52) and (53) provide two tensorial relations that relate the fields $G_1$ and $G_2$, associated with the two submacroscopic levels, to the gradient $\nabla g$ of the macroscopic deformation field $g$.

We note that an alternative derivation of the consistency relations for stresses (52), (53) is provided by using the three different equations (49)–(51) for $T \circ g$ in terms of the partial derivatives of $\Psi$. In fact, subtracting (49) from (50) yields the consistency relation (53), while subtracting (50) from (51) recovers (52). This alternative derivation should provide the simplest derivation of the consistency relations applicable when more than two submacroscopic levels are to be considered.

### 5.3 Material Frame Indifference

At this point, we have shown that the response of an elastic body undergoing hierarchical disarrangements is completely determined by the smooth response function $(G_2, M_2, M_1) \longmapsto \Psi(G_2, M_2, M_1)$ whose values provide the free energy per unit volume in the macroscopic reference configuration at each pair $(X, t)$:

$$\psi(X,t) = \Psi\big(G_2(X,t), M_2(X,t), M_1(X,t)\big).$$

In order to guarantee that the response of such a body is frame-indifferent, we first must consider how three-level structured motions may transform under a change of observer $g \longrightarrow g^{obs}$, where the motions $g$ and $g^{obs}$ are related by

$$g^{obs}(X,t) = x(t) + Q(t)\big(g(X,t) - X_0\big) \tag{54}$$

with $t \longmapsto x(t)$ a smooth point-valued mapping and $t \longmapsto Q(t)$ a smooth, orthogonal tensor-valued mapping. The definition of change of observer (54) and the chain rule tell us that the macroscopic deformation gradient $G_0 = \nabla g$ satisfies the transformation rule $G_0 \to QG_0$ under change of observer. The identification relations (6) and (7) for $G_1$ and $G_2$ tell us that these quantities have the same transformation property as $G_0$, and the definitions of $M_1$ and $M_2$ as differences formed from $G_0$, $G_1$, and $G_2$ yield the transformation rules:

$$G_i \to QG_i \quad \text{for } i = 0, 1, 2, \tag{55}$$

$$M_i \to QM_i \quad \text{for } i = 1, 2. \tag{56}$$

Thus, *the deformations with and without disarrangements at all levels transform in the same manner under change of observer*. Consequently, we say that the three-level structured motion $(g^{obs}, G_1^{obs}, G_2^{co})$ is obtained from $(g, G_1, G_2)$ by means of a change of observer if $g^{obs}$ is given by (54) and

$$G_i^{obs}(X,t) = Q(t)G_i(X,t) \quad \text{for all pairs } (X,t) \text{ and for } i = 1, 2. \tag{57}$$

We use the notation $(g, G_1, G_2) \to (g^{obs}, G_1^{obs}, G_2^{obs})$ to indicate that the three-level structured motion $(g^{obs}, G_1^{obs}, G_2^{obs})$ is obtained from $(g, G_1, G_2)$ by means of a change of observer.

It is standard in continuum physics to require that the free energy density $\psi(X,t)$ be unchanged and that the Cauchy stress $T(g(X,t),t)$ be replaced by $Q(t)T(g^{obs}(X,t),t)Q(t)^T$ under a change of observer. We adopt these transformation rules here and so arrive at the transformation rule for dynamical processes under change of observer:

$$(g, G_1, G_2, \psi, T \circ g) \to \big(g^{obs}, G_1^{obs}, G_2^{obs}, \psi^{obs}, T^{obs} \circ g^{obs}\big), \tag{58}$$

where for all $X$ and $t$: $\psi^{obs}(X,t) = \psi(X,t)$, $(T^{obs} \circ g^{obs})(X,t) = Q(t)T(g(X,t),t)Q(t)^T$, and where $(g, G_1, G_2) \to (g^{obs}, G_1^{obs}, G_2^{obs})$. We recall the definition $\Gamma := T_0 \cdot \dot{G}_0 - \dot{\psi}$ of the rate of internal dissipation in a dynamical process, and we may now determine the transformation rule for $\Gamma$ under a change of observer:

$$\begin{aligned}
\Gamma^{co} &= T_0^{obs} \cdot \left(G_0^{obs}\right)^{\cdot} - \left(\psi^{obs}\right)^{\cdot} \\
&= \left(T^{obs} \circ g^{obs}\right)\left(G_0^{obs}\right)^* \cdot \left(G_0^{obs}\right)^{\cdot} - \dot{\psi} \\
&= Q(T \circ g)Q^T (QG_0)^* \cdot (QG_0)^{\cdot} - \dot{\psi} \\
&= Q(T \circ g)Q^T \left(QG_0^*\right) \cdot (\dot{Q}\, G_0 + Q\, \dot{G}_0) - \dot{\psi} \\
&= T_0 \cdot \dot{G}_0 - \dot{\psi} + T_0 G_0^T \cdot Q^T \dot{Q} = \Gamma + T_0 G_0^T \cdot Q^T \dot{Q}.
\end{aligned}$$

Thus we have: $\Gamma \to \Gamma + T_0 G_0^T \cdot Q^T \dot{Q}$ under a change of observer, and, because $Q^T \dot{Q}$ is skew-valued, we may also write

$$\Gamma \to \Gamma + sk\left(T_0 G_0^T\right) \cdot Q^T \dot{Q}, \tag{59}$$

where $sk(A) := (1/2)(A - A^T)$ is the skew part of a tensor $A$. When $N$ is taken to be the Cauchy stress $T$, the following additional transformation rules of the type (55) and (56) follow from the definitions and relations established in Sect. 2.3 on tensor fluxes:

$$\begin{aligned}
(T_i)^{obs} &= QT_i && \text{for } i = 0, 1, 2, \\
(T_{i-1.\backslash})^{obs} &= QT_{i-1.\backslash} && \text{for } i = 1, 2, \\
(T_{i-1.d})^{obs} &= QT_{i-1.d} && \text{for } i = 1, 2.
\end{aligned} \tag{60}$$

We say that the response of an elastic body undergoing hierarchical disarrangements is *frame-indifferent* if the constitutive class $\mathcal{E}_{hd}$ is "closed under a change of observer", i.e., if for every change of observer there holds

$$(g, G_1, G_2, \psi, T \circ g) \in \mathcal{E}_{hd} \implies \left(g^{obs}, G_1^{obs}, G_2^{obs}, \psi^{obs}, T^{obs} \circ g^{obs}\right) \in \mathcal{E}_{hd}. \tag{61}$$

In other words, modification of a dynamical process in $\mathcal{E}_{hd}$ by a change of observer cannot produce a dynamical process lying outside of $\mathcal{E}_{hd}$. To obtain the consequences of frame-indifference (61) we first note that the constitutive relation (45) tells us that for every dynamical process in $\mathcal{E}_{hd}$, both $\Gamma$ and $\Gamma^{obs}$ must be non-negative. Because the orthogonal-valued mapping $t \longmapsto Q(t)$ is arbitrary and $Q^T(t)\dot{Q}(t)$ is skew, the formula (59) is equivalent to the symmetry of the tensor field $T_0 G_0^T = (\det G_0)(T \circ g)$, i.e., the Cauchy stress $T(X, t)$ is symmetric for every $X$ and $t$ in every dynamical process in $\mathcal{E}_{hd}$. According to the relation (49), the preservation of the dissipation inequality (45) under change of observer is equivalent to the symmetry condition

$$\left(D_{G_2}\Psi(G_2, M_2, M_1)G_2^T\right)^T = D_{G_2}\Psi(G_2, M_2, M_1)G_2^T \tag{62}$$

for all three-level structured motions. Similarly, the constitutive relation (36) and the transformation rules (55), (56) tell us that the invariance of the free energy under change of observer is equivalent to the condition:

$$\Psi(QG_2, QM_2, QM_1) = \Psi(G_2, M_2, M_1) \tag{63}$$

for all three-level structured motions and for all orthogonal-valued mappings $t \longmapsto Q(t)$. If we restrict this condition to the case of fields that are constant in space and time, then $Q$

becomes an arbitrary orthogonal tensor, and $G_2$, $M_2$, and $M_1$ are tensors satisfying

$$0 < \det G_2 \leq \det(G_2 + M_2) \leq \det(G_2 + M_2 + M_1). \quad (64)$$

If we restrict attention to tensors $G_2$, $M_2$, and $M_1$ for which all the inequalities in (64) are strict and assume that $\Psi$ is smooth, then we may differentiate both sides of (63) in turn with respect to $G_2$, $M_2$, and $M_1$ to find that

$$D_A\Psi(QG_2, QM_2, QM_1) = QD_A\Psi(G_2, M_2, M_1) \quad (65)$$

for $A = G_2, M_2, M_1$. Continuity of the derivatives of $\Psi$ then permits us to extend (65) to tensors satisfying (64), including the equalities. Therefore, we have obtained transformation rules for the partial derivatives of $\Psi$ under change of observer, and they agree with the rules (60) for the stresses $T_{1.\backslash}$, $T_{1.d}$, and $T_{0.d}$. Because the tensors $K_i = G_{i-1}^{-1}G_i$ are unchanged under a change of observer, it follows that each of the three remaining constitutive relations (37)–(39) is satisfied on a dynamical process if and only if it is satisfied on every dynamical process obtained from it by means of a change of observer. These considerations establish the following characterization of frame-indifference in the present context: *the constitutive class $\mathcal{E}_{hd}$ is closed under changes of observer if and only if the response function $\Psi$ satisfies (62) and (63) for all tensors $G_2$, $M_2$, and $M_1$ satisfying (64) and for all orthogonal tensors $Q$*. In view of the formula (49), the condition (62) is equivalent to the symmetry of the Cauchy stress and, hence, implies the balance of angular momentum in local form.

## 5.4 Stable Disarrangement Phases

Among the geometrical and constitutive requirements that we have imposed on a given elastic body with free energy response function $\Psi$, we focus in this subsection on the accommodation inequalities (3), on the consistency relations (52) and (53), and on the frame-indifference condition (62). The relation (3) is a system of inequalities restricting the determinants of the geometrical fields $G_0$, $G_1$, and $G_2$, while (52), (53), and (62) are tensor equations in terms of the response function $\Psi$ that restrict $G_0$, $G_1$, and $G_2$. For a point $X$ in the body and a time $t$, we denote by $F$ the macroscopic deformation gradient $G_0(X,t) = \nabla g(X,t)$ which we consider for the moment as fixed. By virtue of (3), (14), (52), (53), and (62) we then seek tensors $G_2$, $M_2$ and $M_1$ satisfying the relations

$$G_2 + M_2 + M_1 = F, \quad (66)$$

$$0 < \det G_2 \leq \det(G_2 + M_2) \leq \det F, \quad (67)$$

$$D_{G_2}\Psi(G_2, M_2, M_1)M_1^T + D_{M_2}\Psi(G_2, M_2, M_1)M_1^T + D_{M_1}\Psi(G_2, M_2, M_1)\,F^T = 0, \quad (68)$$

$$D_{G_2}\Psi(G_2, M_2, M_1)\,M_2^T + D_{M_2}\Psi(G_2, M_2, M_1)M_2^T + D_{M_2}\Psi(G_2, M_2, M_1)G_2^T = 0, \quad (69)$$

$$\left(D_{G_2}\Psi(G_2, M_2, M_1)G_2^T\right)^T = D_{G_2}\Psi(G_2, M_2, M_1)G_2^T. \quad (70)$$

For a given macroscopic deformation gradient tensor $F$, these four tensor equations and three inequalities are the requirements that our notion of elastic body imposes on the deformation without disarrangements $G_2$ at level 2 and on the deformations due to disarrangements $M_2$ and $M_1$ at levels 2 and 1. (Of course, the equation (66) can be used to eliminate say $G_2$ from the remaining relations leaving only the disarrangement tensors $M_1$ and $M_2$ to be determined in terms of $F$, but we prefer to retain the full system (66)–(70) for present considerations.)

Given a macroscopic deformation gradient $F$, a triple $(G_2, M_2, M_1)$ that satisfies the system (66)–(70) is called a *(three-level) disarrangement phase corresponding to* $F$. In the context of three-level structured deformations, only the three-level geometry and the free energy response function $\Psi$ are employed in determining the system (66)–(70). It is clear from the consistency relations (68) and (69) and from the symmetry condition (70) that the disarrangement phases for $F$ include all triples $(G_2, M_2, M_1)$ that satisfy the geometrical conditions (66), (67) and also that are stationary points of $\Psi$, i.e., all three partial derivatives of $\Psi$ vanish at $(G_2, M_2, M_1)$.

We call a disarrangement phase $(G_2, M_2, M_1)$ corresponding to $F$ *stable* if

$$\Psi(G_2, M_2, M_1) \le \Psi\big(G_2', M_2', M_1'\big)$$

for all disarrangement phases $(G_2', M_2', M_1')$ corresponding to $F$. The problem of determining stable disarrangement phases then becomes the problem of minimizing the free energy subject to the constraints (66)–(70) imposed by the additive decomposition of $F$, the accommodation inequalities, and the consistency relations. It is important to note that, because of the constraints included in (66)–(70), a stable disarrangement phase need not be a stationary point of the free energy and, therefore, the stress need not vanish at a stable disarrangement phase and (as examples in the two-level case show) need not be hydrostatic.

We note that "disarrangement phase" and "stable disarrangement phase" are constitutively based concepts that do not refer to balance laws. Such notions of material stability were introduced and studied for two-level structured deformations in the context of elasticity with disarrangements ([19, 21, 22]) with a view toward applications to granular media and, more generally, to "elastic aggregates", i.e., continuous bodies that submacroscopically are composed of a large number of elastic bodies. In the present context of three-level structured deformations, the notions of disarrangement phase and stable disarrangement phase permit the description of the loss of material stability during the course of a dynamical process as well as the identification of the submacroscopic level at which destabilizing disarrangements arise.

### 5.5 Field Relations

We collect together and record now the full system of field relations that govern motions $(X, t) \longmapsto (g(X, t), G_1(X, t), G_2(X, t))$ of an elastic body undergoing disarrangements at two submacroscopic levels with free energy response $(G_2, M_2, M_1) \longmapsto \Psi(G_2, M_2, M_1)$ that is invariant under change of observer:

$$\rho_0 \ddot{g} = \operatorname{div}(D_{G_2}\Psi + D_{M_2}\Psi + D_{M_1}\Psi) + b, \tag{71}$$

$$D_{G_2}\Psi\, M_1^T + D_{M_2}\Psi\, M_1^T + D_{M_1}\Psi\, (\nabla g)^T = 0, \tag{72}$$

$$D_{G_2}\Psi\, M_2^T + D_{M_2}\Psi\, M_2^T + D_{M_2}\Psi\, G_2^T = 0, \tag{73}$$

$$\big(D_{G_2}\Psi\, G_2^T\big)^T = D_{G_2}\Psi\, G_2^T, \tag{74}$$

$$0 < \det G_2 \le \det(G_2 + M_2) \le \det(G_2 + M_2 + M_1), \tag{75}$$

$$\nabla g = G_2 + M_2 + M_1, \tag{76}$$

$$\begin{aligned} 0 \le{}& D_{G_2}\Psi \cdot \dot{M}_2 + D_{G_2}\Psi \cdot \dot{M}_1 + D_{M_2}\Psi \cdot \dot{G}_2 + D_{M_2}\Psi \cdot \dot{M}_1 \\ &+ D_{M_1}\Psi \cdot \dot{G}_2 + D_{M_1}\Psi \cdot \dot{M}_2. \end{aligned} \tag{77}$$

Here, $\rho_0$ denotes the density of the body in the macroscopic reference configuration (level 0), (71) is the balance of linear momentum in that configuration, (72) and (73) are the consistency relations for stresses at submacroscopic levels 1 and 2, (74) expresses the invariance of the internal dissipation under change of observer or, equivalently, the symmetry of the Cauchy stress $T$ which, in turn, guarantees that the balance of angular momentum is satisfied, (75) contains the accommodation inequalities, (76) is the additive decomposition of $\nabla g$, and (77) is the dissipation inequality. Of course, the field $G_1$ needed to specify the three level structured deformation $(g, G_1, G_2)$ can be recovered from the fields $G_2$ and $M_2$ from the relation $G_1 = G_2 + M_2$.

The system of field relations (71)–(77) amounts to thirty-three scalar equations and two sets of inequalities that are to be satisfied by the thirty scalar components of $g$, $G_2$, $M_2$, and $M_1$. When $\Psi$ does not depend upon $M_1$ and $M_2$ and when one sets $M_1 = M_2 = 0$, then $G_1 = G_2 = \nabla g$, so that only classical deformations are admitted, and the field relations (71)–(77) then reduce to those of finite elasticity:

$$\rho_0 \ddot{g} = \operatorname{div}\bigl(D\Psi(\nabla g)\bigr) + b, \tag{78}$$

$$\bigl(D\Psi(\nabla g)(\nabla g)^T\bigr)^T = D\Psi(\nabla g)(\nabla g)^T, \tag{79}$$

where the remaining relations are satisfied identically. In particular, the inequalities are satisfied with equality throughout, while the assumed invariance of $\Psi$ under change of observer in the special setting of finite elasticity implies that (79) is satisfied. If we require $D_{M_2}\Psi$ is identically zero and consider motions in which $M_2 = 0$, but in which $M_1$ need not vanish, then the field relations (71)–(77) for the three-level case reduce to those of elasticity with disarrangements in the context of two-level structured deformations [18]. Consequently, it would be appropriate in studies of the field relations (71)–(77) to consider counterparts of the notions of submacroscopically stable equilibria [21] and of moving phase boundaries [20] already investigated for the two-level case.

## 6 The case of Purely Dissipative Disarrangements

In general, disarrangements such as slips and separations at submacroscopic levels 1 and 2 can contribute to the free energy through dependence of $\Psi$ upon $M_1$ and $M_2$. For example, the disarrangements at both submacroscopic levels in the three-level shear introduced in Sect. 2 can be captured via approximating sequences in terms of the slips between parallel faces of small pieces of the body ("cards"). In single crystals, such slips can arise across crystallographic planes and may produce a dependence of energy stored on amount of slip that is periodic [8]. A significantly different dependence of energy upon disarrangements arises in the case of elastic aggregates, i.e., continua that are comprised of many small elastic bodies, where we can consider interactions between different pieces of the aggregate as they move relative to one another, while maintaining the same level of deformation in the individual pieces. Such interactions correspond to short-range forces across boundaries of the pieces and may or may not result in additional changes in free energy of the aggregate. In the absence of a crystalline structure of the pieces or other sources of cohesive forces between pieces of the aggregate, one expects that disarrangements will not alter the free energy of the aggregate, and it is appropriate now to adapt a notion of "purely dissipative disarrangements" (introduced earlier for two-level structured deformations of elastic bodies [20, 21]) to the present three-level setting.

A physical context in which purely dissipative, hierarchical disarrangements might arise would be an aggregate consisting of tiny bundles, each of which is an aggregate of thin, rectangular elastic sheets held loosely together by an inextensible string (see Fig. 1). Assuming then that the only mechanism for storage of energy is the deformation of the elastic sheets, the collection of tiny bundles would conform to the notion of a body undergoing purely dissipative disarrangements at two submacroscopic levels, because the relative motion of the elastic sheets within a bundle and the relative motion of different bundles would, themselves, not store energy. Thus we are led to consider the case in which the constitutive relation (36) takes the special form

$$\psi(X,t) = \Psi\big(G_2(X,t)\big), \tag{80}$$

while we continue to allow the disarrangement fields $M_1$ and $M_2$ to be non-zero. In this case, we have $D_{M_2}\Psi = D_{M_1}\Psi = 0$ and $D_{G_2}\Psi = D\Psi$, so that the field relations (71)–(74) simplify as follows:

$$\rho_0\ddot{g} = \operatorname{div}\big(D\Psi(G_2)\big) + b, \tag{81}$$

$$D\Psi(G_2)\, M_1^T = 0, \tag{82}$$

$$D\Psi\ (G_2) M_2^T = 0, \tag{83}$$

$$0 < \det G_2 \leq \det(G_2 + M_2) \leq \det(G_2 + M_2 + M_1), \tag{84}$$

$$\nabla g = G_2 + M_2 + M_1, \tag{85}$$

$$0 \leq D\Psi\ (G_2) \cdot (\dot{M}_1 + \dot{M}_2). \tag{86}$$

We note that the last relation (74) in (71)–(74) automatically is satisfied in the present case in view of (80) and in view of the assumed invariance (63) of free energy under change of observer. In the present case of purely dissipative disarrangements, the field relations amount to thirty scalar equations and two sets of inequalities for thirty unknown components of the fields $g$, $G_2$, $M_2$, $M_1$. In this context the Piola-Kirchhoff stress $S = T_0$ in the stress relation (46) is given by

$$S = D\Psi(G_2) = D\Psi(\nabla g - M_1 - M_2). \tag{87}$$

An interesting aspect of the special field relations (81)–(86) is the following: the equations (81), (82), (83), (85), as well as the stress relation (87) and the dissipation inequality (86) are unchanged under interchange of $M_1$ and $M_2$, while only the system of inequalities (84) changes under that interchange. Consequently, if for given $g$ and $G_2$, the unchanged relations are satisfied for the pair of disarrangement fields $(M_1, M_2)$, then they are satisfied by the pair $(M_2, M_1)$, and the level at which the disarrangements associated with $M_1$ and $M_2$ can occur is controlled soley by which, if any, of the following two systems of inequalities may be satisfied:

$$\det G_2 \leq \det(G_2 + M_2) \leq \det(\nabla g) = \det(G_2 + M_2 + M_1), \tag{88}$$

$$\det G_2 \leq \det(G_2 + M_1) \leq \det(\nabla g) = \det(G_2 + M_1 + M_2). \tag{89}$$

If only the first system of inequalities (88) is satisfied, then the disarrangements corresponding to the field $M_2$ can occur at submacroscopic level two but not at level one, and the field relations all are satisfied by the structured motion $(g, G_2 + M_2, G_2)$. Similarly, if only the second system (89) is satisfied, then the disarrangements associated with the field $M_1$ can occur at level two but not at level one, and the field relations are satisfied by the structured motion $(g, G_2 + M_1, G_2)$. If both systems of inequalities are satisfied then either set of disarrangements may occur at either level, and the field relations are satisfied by both of the motions $(g, G_2 + M_2, G_2)$ and $(g, G_2 + M_1, G_2)$.

For an example of the last type in which either disarrangement field may correspond to either hierarchical level, we put $G_2 = I$ and assume also that the derivative $D\Psi(I)$ of the free energy response at the identity has non-trivial nullspace. We choose a unit vector $v$ in the nullspace of $D\Psi(I)$ and a unit vector $u$ perpendicular to $v$, we put $M_1 = -u \otimes v$, $M_2 = u \otimes v$, and we choose $g$ to be a homogeneous deformation with $\nabla g = I + M_1 + M_2 = I$ to obtain, as is easily verified, a time-independent solution of the field relations (81)–(86). The two systems of inequalities (88) and (89) become

$$1 \leq \det(I + u \otimes v) \leq \det(I - u \otimes v + u \otimes v) = 1 \tag{90}$$

and

$$1 \leq \det(I - u \otimes v) \leq \det(I - u \otimes v + u \otimes v) = 1. \tag{91}$$

Thus, both systems are satisfied (with equality), and we may assign either the submacroscopic level *one* to the disarrangements associated with $u \otimes v$ or the submacroscopic level *two* to the disarrangements associated with $u \otimes v$. If say we assign level one to $u \otimes v$, then the resulting solution of the field equations corresponds to the three-level shear $(g, G_1, G_2)$ defined in (8), (9) with $\mu_0 = \mu_2 = 0$ and $\mu_1 = 1$, so that $G_1 = I + u \otimes v$, $G_2 = I$. The alternative assignment of level two to the disarrangements associated with $u \otimes v$ results in the three-level shear $(g, G_1, G_2)$ in (8), (9) with $\mu_0 = \mu_2 = 0$ and $\mu_1 = -1$, where now we simply replace $u$ by $-u$ in all of the formulas for the previous choice.

In order to provide an example in which only one of the two systems of inequalities (88) and (89) is satisfied, we assume now that $D\Psi(I)$ has nullspace of dimension at least two, and we choose perpendicular unit vectors $u$ and $v$ in the nullspace. We let $\varepsilon$ with $0 < \varepsilon < 1/2$ be given, we put $G_2 = I$, and define

$$\begin{aligned} M_1 &:= -\varepsilon u \otimes u - (1 - \varepsilon) v \otimes u, \\ M_2 &:= u \otimes v. \end{aligned} \tag{92}$$

As above, we choose $g$ to be a time-independent homogeneous deformation with gradient $G_2 + M_2 + M_1$. Again, the equations (81), (82), (83), (85) and the dissipation inequality (86)—all invariant under interchange of $M_1$ and $M_2$—are satisfied for the time-independent three-level structured motion $(g, G_1, G_2)$. The two systems of inequalities (88) and (89) now read

$$\det I \leq \det(I + u \otimes v) \leq \det\big(I + -\varepsilon u \otimes u - (1 - \varepsilon) v \otimes u - \varepsilon u \otimes u - (1 - \varepsilon) v \otimes u\big)$$

and

$$\det I \leq \det\big(I - \varepsilon u \otimes u - (1 - \varepsilon) v \otimes u\big) \leq \det\big(I + u \otimes v - \varepsilon u \otimes u - (1 - \varepsilon) v \otimes u\big).$$

The first system reduces to $1 \leq 1 \leq 2(1 - \varepsilon)$, which validates the choice $u \otimes v$ for the level 2 disarrangement tensor because $0 < \varepsilon < 1/2$. The second reads $1 \leq 1 - \varepsilon \leq 2(1 - \varepsilon)$ which eliminates the choice $-\varepsilon u \otimes u - (1 - \varepsilon) v \otimes u$ for the level 2 disarrangement tensor, since the first inequality is false. The elimination of the second choice shows that the choice $-\varepsilon u \otimes u - (1-\varepsilon)v \otimes u$ for the level 2 disarrangement tensor to construct the triple $(g, G_1, G_2)$ eliminates the possibility of approximating that triple by *injective*, piecewise smooth mappings in the sense of the Approximation Theorem. In more geometrical terms, the possibility of zooming out from level two to level one by means of the combination of shrinkage and shearing associated with the disarrangement tensor $-\varepsilon u \otimes u - (1-\varepsilon)v \otimes u$ is eliminated by the accommodation inequalities, whereas the same combination may occur in zooming out from level 1 to level 0 if the tensor $u \otimes v$ accounts for the disarrangements in zooming out from level 2 to level 1.

## 7 Boundary Conditions

We have obtained a detailed system of field relations (71)–(77) for the fields $g$, $G_1$, and $G_2$ that determine a structured motion in an elastic body undergoing hierarchical disarrangements at two submacroscopic levels. These relations are formulated in terms of the fields $g$, $G_2$, $M_2$, and $M_1$, because these fields provide the refined decomposition $\nabla g = G_2 + M_2 + M_1$ of the macroscopic deformation gradient as well as the parallel decomposition

$$S = T_0 = D_{G_2}\Psi + D_{M_2}\Psi + D_{M_1}\Psi = (\det K_1 \det K_2)^{-1}\big(T_{1.\backslash} + T_{1.d} + (\det K_2)T_{0.d}\big)$$

of the Piola-Kirchhoff stress (see Sect. 5.1).

The formulation of boundary conditions of place for (71)–(77) appears to be straightforward, because, even in the refined geometry of three-level structured deformations, the only field available for the computation of position is the macroscopic deformation field $g$. The refined decomposition $\nabla g = G_2 + M_2 + M_1$ does not provide a decomposition of $g$, itself, and therefore does not provide refined versions of boundary conditions of place. Thus, we single out a portion $\partial\mathcal{B}_p$ of the boundary of the body and require for all times $t$ and $X \in \partial\mathcal{B}_p$:

$$g(X,t) = g_p(X,t), \tag{93}$$

where $g_p$ is a given field on $\partial\mathcal{B}_p$.

Formulation of boundary conditions of traction is more interesting, and the starting point for our discussion is the well-known fact that the contact force $\int_{g(\mathcal{S},t)} T(x,t)n(x)dA_x$ on an oriented surface $g(\mathcal{S},t)$ in the deformed configuration is given by

$$\int_{\mathcal{S}} T_0(X,t)N(X)dA_X = \int_{\mathcal{S}} T\big(g(X,t),t\big)G_0^*(X,t)N(X)dA_X \tag{94}$$

where $N(X)dA_X$ is the vector element of area at the point $X$ in the surface $\mathcal{S}$ in $\mathcal{B}$ (and $n(x)dA_x$ is the corresponding vector element of area for $g(\mathcal{S},t)$ at the point $g(X,t)$ in $g(\mathcal{S},t)$). In other words, the reference stress $T_0$ carries the same information about contact forces supported by the smooth surface $\mathcal{S}$ in $\mathcal{B}$ as does the Cauchy stress $T$ for the smooth surface $g(\mathcal{S},t)$ in the deformed configuration $g(\mathcal{B},t)$. This observation simply restates the equivalence of the two stress measures $T_0$ and $T$ of contact forces in continuum theories.

In the present context of three-level structured motions $(g, G_1, G_2)$, there are two additional measures of stress obtained when $i = 1, 2$ from the definition (23) with $N = T$: $T_1 = (T \circ g)G_1^*$ and $T_2 = (T \circ g)G_2^*$. Like $T_0$, both are defined at each time $t$ on $\mathcal{B}$ and both allow us to compute for each smooth surface $\mathcal{S}$ in $\mathcal{B}$ the surface integrals:

$$\int_{\mathcal{S}} T_i(X,t)N(X)dA_X = \int_{\mathcal{S}} T\big(g(X,t),t\big)G_i^*(X,t)N(X)dA_X \quad \text{for } i = 1, 2. \tag{95}$$

The integrals on the right-hand sides of (94) and (95) differ only by the oriented element of area $G_i^*(X,t)N(X)dA_X$ in the integrand. The continuity of the adjugate operation $^*$ and the formulas (6) and (7) here yield for every double sequence $(n_1, n_2) \longmapsto f_{n_1 n_2}$ as in the Approximation Theorem:

$$G_1^*(X,t)N(X) = \Big(\Big(\lim_{n_1\to\infty} \nabla \lim_{n_2\to\infty} f_{n_1 n_2}\Big)(X,t)\Big)^* N(X), \tag{96}$$

$$G_2^*(X,t)N(X) = \Big(\Big(\lim_{n_1\to\infty} \lim_{n_2\to\infty} \nabla f_{n_1 n_2}\Big)(X,t)\Big)^* N(X), \tag{97}$$

while (5) yields

$$G_0^*(X,t)N(X) = \Big(\nabla \lim_{n_1\to\infty} \lim_{n_2\to\infty} f_{n_1 n_2}\Big)^* N(X). \tag{98}$$

Based on our discussion of (6)–(5) in Sect. 2, we may assert that the choice of index $i$ then influences the measure of contact force $\int_{\mathcal{S}} T_i(X,t)N(X)dA_X = \int_{\mathcal{S}} T(g(X,t),t)G_i^*(X,t) \times N(X)dA_X$ by adjusting the element of area on $\mathcal{S}$ to account for no disarrangements whatsoever at sites on the given surface ($i = 2$), disarrangements only at submacroscopic level 2 ($i = 1$), or disarrangements at both submacroscopic levels 1 and 2 ($i = 0$). We conclude that the reference stresses $T_1$ and $T_2$ at the submacroscopic levels each carries different information about contact forces supported by the smooth surface $\mathcal{S}$ in $\mathcal{B}$ than does $T_0$, and the difference is accounted for by whether or not the effect of disarrangements at the submacroscopic levels is included in the element of area on $\mathcal{S}$.

We observe that prescription of the traction vector $\tau(X,t)$ at each point on a portion $\partial\mathcal{B}_\tau$ leads, for example, to the following three traction boundary conditions: choose $i \in \{0, 1, 2\}$ and require

$$T_i(X,t)N(X) = \tau(X,t) \quad \text{for every } X \in \partial\mathcal{B}_\tau \text{ and for every } t. \tag{99}$$

Here, $N(X)$ denotes the unit outward normal to $\partial\mathcal{B}_\tau$ at $X$, and we call (99) the *level i traction boundary condition*. The equivalence of $T_0$ and $T$ as measures of contact forces assures us that the level 0 traction boundary condition is equivalent to specifying a traction boundary condition on a portion of $g(\mathcal{B},t)$ using the Cauchy stress field. Moreover, the stress relation (46) permits us to rewrite the level 0 traction boundary condition in terms of the response function $\Psi$: for every $X \in \partial\mathcal{B}_\tau$ and for every $t$

$$(D_{G_2}\Psi + D_{M_2}\Psi + D_{M_1}\Psi)\,|_{(X,t)}\, N(X) = \tau(X,t). \tag{100}$$

Similarly, (47) permits us to rewrite the level 1 traction condition as: for every $X \in \partial\mathcal{B}_\tau$ and for every $t$

$$(\det K_1)(D_{G_2}\Psi + D_{M_2}\Psi)\,|_{(X,t)}\, N(X) = \tau(X,t), \tag{101}$$

while (48) yields for the level 2 traction condition: for every $X \in \partial\mathcal{B}_\tau$ and for every $t$

$$(\det K_1)(\det K_2)D_{G_2}\Psi\,|_{(X,t)}\, N(X) = \tau(X,t). \tag{102}$$

We recall from Sect. 2, $K_i := G_{i-1}^{-1}G_i$ for $i = 1, 2$. Relation (102) shows that satisfaction of the level 2 traction boundary condition amounts to the statement that only the driving traction associated with $G_2$ is activated in balancing the applied traction $\tau$, while (100) tells us that the driving tractions corresponding to $G_2$, $M_2$ and $M_1$ add together in balancing $\tau$. For the intermediate, level 1 boundary condition (101), the two driving tractions corresponding to $G_2$ and $M_2$ combine to balance $\tau$.

Up to this point, we have not provided any guidance as to which of the traction conditions (100)–(102) should be employed to impose traction boundary conditions. The formulation of our field theory in terms of a free energy response that depends upon $G_2$, $M_2$, and $M_1$ or, equivalently, in terms of $G_i$ for $i = 0, 1, 2$ presupposes knowledge for the given body about the number of submacroscopic levels available to the body and the manner in which the body stores energy in response to geometrical changes with and without disarrangements at these levels. It is consistent with this presupposition to suppose available some information about the device that provides the tractions $\tau$ to the body on the surface $\partial\mathcal{B}_\tau$.

For example, if the body is an aggregate composed of small elastic grains, then it is natural to identify the submacroscopic level 2 with a collection of such grains large enough to justify calculating the gradients $\nabla f_{n_1,n_2}$ that approximate $G_2$ by averaging deformations of individual grains in such a collection. We presume then that it is known whether or not the tractions $\tau$ provided by a given loading device can vary appreciably over a surface of area comparable to the cross section of such a collection. If so, it is natural to impose the level 2 traction condition (102). On the other hand, if $\tau$ varies appreciably only over surfaces of area comparable to the cross section of larger submacroscopic features such as shear bands or such as load-supporting, geometrically irregular chains of grains, then it is natural to impose the level 1 traction condition (101). Finally, if the imposed traction $\tau$ varies only over surfaces of area large compared to the cross sections of the larger submacroscopic features, then it is natural to impose the level 0 traction condition (100). We conclude that the choices of traction boundary condition available in this multilevel approach allow us to match the choice of level of the traction condition to characteristics of the loading device, and this matching capability provides flexibility beyond that afforded by a single level approach.

While there is no qualitative theory yet available for the field equations of our field theory, we indicate here initial conditions that appear natural in view of the form of the field equations. Consequently, for a given initial time $t_I$ we suggest that specification of the fields $g(\cdot, t_I)$, $\dot{g}(\cdot, t_I)$, $G_1(\cdot, t_I)$, and $G_2(\cdot, t_I)$ on the body $\mathcal{B}$ provides appropriate initial conditions to supplement the boundary conditions considered above in order to determine the evolution of the body in the form of a three-level structured motion.

**Acknowledgement** The partial support from the grant ERC-2013-ADG-340561-INSTABILITIES is gratefully acknowledged.

**Publisher's Note** Springer Nature remains neutral with regard to jurisdictional claims in published maps and institutional affiliations.

## Appendix: Proof of the Approximation Theorem for Three-Level Structured Deformations

In this appendix we use the terms "simple deformation" and "piecewise-fit region" in the sense of [12]. Roughly speaking, a simple deformation is a piecewise smooth, injective mapping, and a piecewise-fit region is a finite union of regions without unopened cracks and with finite surface area.

**Theorem 1** *For each three-level structured deformation* $(g, G_1, G_2)$ *from a piecewise-fit region* $\mathcal{B}$ *there exists a double sequence* $(n_1, n_2) \longmapsto f_{n_1,n_2}$ *of simple deformations from* $\mathcal{B}$

*for which*

$$\lim_{n_1\to\infty}\lim_{n_2\to\infty} f_{n_1,n_2} = g,$$

$$\lim_{n_1\to\infty}\nabla \lim_{n_2\to\infty} f_{n_1,n_2} = G_1,$$

$$\lim_{n_1\to\infty}\lim_{n_2\to\infty} \nabla f_{n_1,n_2} = G_2,$$

*where each of the iterated limits* $\lim_{n_1\to\infty}$ *and* $\lim_{n_2\to\infty}$ *is taken in the sense of* $L^\infty$-*convergence.*

*Proof* Let a three-level structured deformation $(g, G_1, G_2)$ and a positive integer $n_1$ be given. The three-level accommodation inequality (3) implies that the pair $(g, G_1)$ is a (two-level) structured deformation. By the Approximation Theorem for two-level structured deformations and properties of the determinant mapping, we may choose a constant $\bar{C} > 0$ (that depends only upon the field $G_1$) and a simple deformation $f_{n_1}$ such that

$$\|g - f_{n_1}\|_\infty < \frac{1}{n_1}, \qquad \|G_1 - \nabla f_{n_1}\|_\infty < \frac{\bar{C}}{n_1}, \quad \text{and} \quad \|\det G_1 - \det \nabla f_{n_1}\|_\infty < \frac{1}{n_1}, \tag{103}$$

where $\|\cdot\|_\infty$ denotes the $L^\infty$ norm. By the three-level accommodation inequality (3) and (103) we conclude that

$$\det \nabla f_{n_1} > \det G_1 - \frac{1}{n_1} \geq \det G_2 - \frac{1}{n_1} > c - \frac{1}{n_1}, \tag{104}$$

and we take from this point on $n_1 > c^{-1}$. (The inequality (104) can be written in terms of fields because the accommodation inequality holds at almost every point in $\mathcal{B}$ for one and the same positive number $c$.) We define

$$\varphi(n_1) := 2\left(1 - \left(1 - \frac{1}{n_1 c}\right)^{\frac{1}{3}}\right) > 0 \tag{105}$$

and note that

$$\lim_{n_1\to\infty} \varphi(n_1) = 0, \tag{106}$$

as well as $\varphi(n_1) > 1 - (1 - \frac{1}{n_1 c})^{\frac{1}{3}}$, so that

$$1 - \frac{1}{n_1 c} > \left(1 - \varphi(n_1)\right)^3. \tag{107}$$

We infer from (104), (105), and (107) that

$$\begin{aligned}\det \nabla f_{n_1} &> \left(1 - \frac{1}{n_1 \det G_2}\right) \det G_2 \\ &> \left(1 - \frac{1}{n_1 c}\right) \det G_2 > \left(1 - \varphi(n_1)\right)^3 \det G_2 \\ &= \det\big(\left(1 - \varphi(n_1)\right) G_2\big).\end{aligned} \tag{108}$$

By (106) and the inequality $\det G_2 > c$, we may choose $N_1 > c^{-1}$ such that

$$\left(1-\varphi(n_1)\right)^3 > \frac{1}{2} \quad \text{for all } n_1 > N_1.$$

The estimate (108) then implies

$$\det \nabla f_{n_1} > \det\left(\left(1-\varphi(n_1)\right)G_2\right) > \frac{c}{2} \quad \text{for all } n_1 > N_1,$$

and we conclude that the pair $(f_{n_1}, (1-\varphi(n_1))G_2)$ satisfies the accommodation inequality for (two-level) structured deformation for all $n_1 > N_1$. Thus, for each $n_1 > N_1$ and positive integer $n_2$, the Approximation Theorem for (two-level) structured deformations permits us to choose a simple deformation $f_{n_1,n_2}$ such that

$$\| f_{n_1,n_2} - f_{n_1} \|_\infty < \frac{1}{n_2} \quad \text{and} \quad \left\| \nabla f_{n_1,n_2} - \left(1-\varphi(n_1)\right)G_2 \right\|_\infty < \frac{1}{n_2}.$$

Therefore, for each $n_1 > N_1$ we may let $n_2$ tend to $\infty$ in the last two inequalities to obtain

$$\lim_{n_2\to\infty} f_{n_1,n_2} = f_{n_1} \quad \text{and} \quad \lim_{n_2\to\infty} \nabla f_{n_1,n_2} = \left(1-\varphi(n_1)\right)G_2,$$

and, in view of (106), we may then let $n_1$ tend to $\infty$ in each of the last two relations to conclude from (103)

$$\begin{aligned}
\lim_{n_1\to\infty} \lim_{n_2\to\infty} f_{n_1,n_2} &= \lim_{n_1\to\infty} f_{n_1} = g, \\
\lim_{n_1\to\infty} \nabla \lim_{n_2\to\infty} f_{n_1,n_2} &= \lim_{n_1\to\infty} \nabla f_{n_1} = G_1, \\
\lim_{n_1\to\infty} \lim_{n_2\to\infty} \nabla f_{n_1,n_2} &= \lim_{n_1\to\infty} \left(\left(1-\varphi(n_1)\right)G_2\right) = G_2.
\end{aligned}$$

In all of these relations, the limits are taken in the sense of $L^\infty$. □

## References


1. Baia, M., Matias, J., Santos, P.M.: A relaxation result in the framework of structured deformation in a bounded variation setting. Proc. R. Soc. Edinb. A **142**(2), 239–271 (2012)
2. Barroso, A.C., Matias, J., Morandotti, M., Owen, D.R.: Second-order structured deformations: relaxation, integral representation and applications. Arch. Ration. Mech. Anal. **225**, 1025–1072 (2017)
3. Barroso, A.C., Matias, J., Morandotti, M., Owen, D.R.: Explicit formulas for relaxed disarrangement densities arising from structured deformations. Math. Mech. Complex Syst. **5**, 163–189 (2017)
4. Bertoldi, K., Bigoni, D., Drugan, W.J.: Nacre: an orthotropic and bimodular elastic material. Compos. Sci. Technol. **68**, 1363–1375 (2008)
5. Carita, G., Matias, J., Morandotti, M., Owen, D.R.: Dimension reduction in the context of structured deformations. J. Elast. (2018). https://doi.org/10.1007/s10659-018-9670-9
6. Chen, Q., Pugno, N.M.: Bio-mimetic mechanisms of natural hierarchical materials: a review. J. Mech. Behav. Biomed. Mater. **19**, 3–33 (2013)
7. Choksi, R., Fonseca, I.: Bulk and interfacial energy densities for structured deformations of continua. Arch. Ration. Mech. Anal. **138**, 37–103 (1997)
8. Choksi, R., Del Piero, G., Fonseca, I., Owen, D.R.: Structured deformations as energy minimizers in some models of fracture and hysteresis. Math. Mech. Solids **4**, 321–356 (1999)
9. Coleman, B.D., Noll, W.: The thermodynamics of elastic materials with heat conduction and viscosity. Arch. Ration. Mech. Anal. **13**, 167–178 (1963)

10. Daniels, H.E.: The statistical theory of the strength of bundles of threads I. Proc. R. Soc. A **183**(995) (1945). https://doi.org/10.1098/rspa.1945.0011
11. Del Piero, G.: Limit analysis and no-tension materials. Int. J. Plast. **14**, 259–271 (1998)
12. Del Piero, G., Owen, D.R.: Structured deformations of continua. Arch. Ration. Mech. Anal. **124**, 99–155 (1993)
13. Del Piero, G., Owen, D.R.: Integral-gradient formulae for structured deformations. Arch. Ration. Mech. Anal. **131**, 121–138 (1995)
14. Del Piero, G., Owen, D.R.: Structured Deformations, XXII Scuola Estiva di Fisica Matematica, Ravello, Settembre 1997. Quaderni dell' Istituto Nazionale di Alta Matematica. Gruppo Nazionale di Fisica Matematica, Firenze (2000)
15. Deseri, L., Owen, D.R.: Active slip-band separation and the energetics of slip in single crystals. Int. J. Plast. **16**, 1411–1418 (2000)
16. Deseri, L., Owen, D.R.: Energetics of two-level shears and hardening of single crystals. Math. Mech. Solids **7**, 113–147 (2002)
17. Deseri, L., Owen, D.R.: Invertible structured deformations and the geometry of multiple slip in single crystals. Int. J. Plast. **18**, 833–849 (2002)
18. Deseri, L., Owen, D.R.: Toward a field theory for elastic bodies undergoing disarrangements. J. Elast. **70**, 197–236 (2003)
19. Deseri, L., Owen, D.R.: Submacroscopically stable equilibria of elastic bodies undergoing dissipation and disarrangements. Math. Mech. Solids **15**, 611–638 (2010)
20. Deseri, L., Owen, D.R.: Moving interfaces that separate loose and compact phases of elastic aggregates: a mechanism for drastic reduction or increase in macroscopic deformation. Contin. Mech. Thermodyn. **25**, 311–341 (2013)
21. Deseri, L., Owen, D.R.: Stable disarrangement phases of elastic aggregates: a setting for the emergence of no-tension materials with non-linear response in compression. Meccanica **49**, 2907–2932 (2014)
22. Deseri, L., Owen, D.R.: Stable disarrangement phases arising from expansion/contraction or from simple shearing of a model granular medium. Int. J. Eng. Sci. **96**, 111–130 (2015)
23. Deseri, L., Owen, D.R.: Submacroscopic disarrangements induce a unique, additive and universal decomposition of continuum fluxes. J. Elast. **122**, 223–230 (2016)
24. Gibson, L.: The hierarchical structure and mechanics of plant materials. J. R. Soc. Interface **9**(76), 2749–2766 (2012)
25. Gu, G.X., Takaffoli, M., Buehler, M.J.: Hierarchically enhanced impact resistance of bioinspired composites. Adv. Mater. **29**(28), 1700060 (2017)
26. Gurtin, M.E.: An Introduction to Continuum Mechanics. Academic Press, New York (1981)
27. Habibi, M.K., Lu, Y.: Crack propagation in bamboo's hierarchical cellular structure. Sci. Rep. **4**, 5598 (2014). https://doi.org/10.1038/srep05598
28. Lakes, R.: Materials with structural hierarchy. Nature **361**, 511–515 (1993)
29. Launey, M.E., Buehler, M.J., Ritchie, R.O.: On the mechanistic origins of toughness in bone. Annu. Rev. Mater. Res. **40**, 25–53 (2010)
30. Matias, J., Morandotti, M., Zappale, E.: Optimal design of fractured media with prescribed macroscopic strain. J. Math. Anal. Appl. **449**, 1094–1132 (2017)
31. Newman, W.I., Gabrielov, A.M.: Failure of hierarchical distributions of fibre bundles I. Int. J. Fract. **50**(1), 1–14 (1991)
32. Oliver, K., Seddon, A., Trask, R.S.: Morphing in nature and beyond: a review of natural and synthetic shape-changing materials and mechanisms. J. Mater. Sci. **51**, 10663–10689 (2016)
33. Owen, D.R.: Elasticity with disarrangements. In: Multiscale Modeling in Continuum Mechanics and Structured Deformations. CISM Courses and Lectures, vol. 447. Springer, Berlin (2004)
34. Owen, D.R.: Field equations for elastic constituents undergoing disarrangements and mixing. In: Šilhavý, M. (ed.) Mathematical Modelling of Bodies with Complicated Bulk and Boundary Behavior. Quaderni di Matematica, vol. 20. Aracne, Rome (2007)
35. Owen, D.R.: Elasticity with gradient disarrangments: a multiscale perspective for strain-gradient theories of elasticity and of plasticity. J. Elast. **127**, 115–150 (2017)
36. Owen, D.R., Paroni, R.: Second-order structured deformations. Arch. Ration. Mech. Anal. **155**, 215–235 (2000)
37. Owen, D.R., Paroni, R.: Optimal flux densities for linear mappings and the multiscale geometry of structured deformations. Arch. Ration. Mech. Anal. **218**, 1633–1652 (2015)
38. Ozcoban, H., Yilmaz, E.D., Schneider, G.A.: Hierarchical microcrack model for materials exemplified at enamel. Dent. Mater. **34**(1), 69–77 (2018)
39. Pugno, N., Bosia, F., Abdalrahman, T.: Hierarchical fiber bundle model to investigate the complex architectures of biological materials. Phys. Rev. E **85**, 011903 (2012)

40. Šilhavý, M.: On the approximation theorem for structured deformations from BV(Ω). Math. Mech. Complex Syst. **3**, 83–100 (2015)
41. Šilhavý, M.: The general form of the relaxation of a purely interfacial energy for structured deformations. Math. Mech. Complex Syst. **5**, 191–215 (2017)
42. Simonini, I., Pandolfi, A.: Customized finite element modelling of the human cornea. PLoS ONE **10**(6), e0130426 (2015)
43. Wang, R., Gupta, H.S.: Deformation and fracture mechanisms of bone and nacre. Annu. Rev. Mater. Res. **41**, 41–73 (2011)
44. Wegst, U.G.K., Bai, H., Saiz, E., Tomsia, A.P., Ritchie, R.O.: Bioinspired structural materials. Nat. Mater. **14**, 23–36 (2015)

J Elast (2019) 135:183–235
https://doi.org/10.1007/s10659-019-09734-y

# Continuum Mechanics as a Computable Coarse-Grained Picture of Molecular Dynamics

**Antonio DiCarlo**[1]

Received: 4 August 2018 / Published online: 5 April 2019


**Abstract** In this paper, I determine the minimum amount of information that continuum mechanics needs to obtain from Newtonian molecular dynamics, in order to avail itself of stress-strain responses uniformly valid for a vast range of macroscopic regimes, being quantitatively determined by microscopic physical properties. Described from the opposite, bottom-up point of view, the procedure I put forward uses the basic kinematic and dynamical machinery of continuum mechanics to upscale molecular dynamics to the macroscopic level in a practicable and efficient way.

**Keywords** Molecular dynamics · Continuum mechanics · Space-time averaging · Streaming velocity · Thermal velocity · Stress · Hierarchical multiscale modelling

**Mathematics Subject Classification** 82C22 · 82C21 · 74A25 · 74A10 · 82C80

> Let me say one other thing which I keep saying at conferences when I get the chance. The significance is not the long-time tail. The significance is that hydrodynamics applies at the microscopic scale. This is the real philosophical break through. I mean when you listened to Uhlenbeck in those days, he would say that the kinetic time of a collision which for hard spheres is the time between collisions, and is $10^{-14}$ seconds, and there is the hydrodynamic time scale, which is seconds. I don't know what people imagined between $10^{-14}$ and seconds, but the big push philosophically, is that hydrodynamics applies on almost the kinetic time scale, that is it merges with the kinetic time scale. And from what I can remember historically, that dates back to an argument between Boltzmann and Mach. Mach, of course, did not believe in kinetic and molecular theory, and Boltzmann did. Mach was, of course, a great hydrodynamicist, but did not believe there was any connection. That break between kinetic theory and continuum hydrodynamics was deeply inbred in every statistical

✉ A. DiCarlo
adicarlo@mac.com

[1] CECAM–IT–SIMUL Node, Rome, Italy

mechanician. That, to me, is the real importance. Once you realize that hydrodynamics applies to a molecule moving in a fluid, it's obvious that you get this long-time tail. Nobody dared believe that until we did the computer experiments.

B.J. Alder interviewed by D. Mac Kernan et al. (2002) [3]

L'objet de cette Leçon paraîtra sans doute fort peu important aux ingénieurs qui s'intéressent spécialement à l'équilibre d'élasticité. Mais, outre qu'il est souvent nécessaire d'étudier l'éffet des vibrations sur certaines constructions, le temps n'est-il pas venu de se demander si l'état moléculaire des corps dont le repos nous paraît le mieux établi, est bien réellement un état statique ; s'il n'est pas, au contraire, le résultat de vibrations très rapides, et qui ne s'arrêtent jamais ? Tout porte à penser, en effet, que le repos relatif des molécules d'un corps n'est qu'un cas très-exceptionnel, une pure abstraction, une chimère peut-être.

G. Lamé (1852) [15, p. 136]

## 1 Introduction

Nearly a decade ago, this same journal devoted another special issue (volume 100, issue 1–2, June 2010) to "Statistical Mechanics, Molecular Modeling, and the Notion of Stress." Its main feature was a late English translation [16] of the 1955 paper [24] in which Noll had straightened out and simplified the formal calculations of the seminal hydrodynamical paper by Irving and Kirkwood, published five years earlier [14]. With other contributions I will refer to later, that small collection contained the first published paper, authored by Podio-Guidugli [28], where the molecular dynamics method by Parrinello and Rahman [25, 26] was discussed and analysed from a continuum mechanical perspective.

Much more recently, in [11] I brought to the fore the latent theoretical potential of that method, maintaining that its foundational idea—if properly reinterpreted and extended—yields a completely satisfactory solution to the long-standing problem of determining the microscopic, particle-wise correlative of the continuum notion of stress, thus providing a quantitative atomistic foundation to macroscopic stress analysis. Such a solution, outlined in [11], is presented and discussed thoroughly in this paper. My main tool is *averaging on mesoscopic space-time cells*, each of which should be thought of as a mesoscopically sized space window kept open for a mesoscopic time interval to monitor a mesoscopic sample of the underlying macroscopic particle system.

As I argued in [11], I think that the ensemble-averaging approach by Irving & Kirkwood and Noll is less effective, for reasons I will just summarise here.[1] First and foremost, their procedure does not identify the stress directly, but through its divergence. This is an unavoidable consequence of relating continuum and particle quantities according to the role they play in the force balance equation. Secondly, their approach is intrinsically limited to binary interactions and requires the interaction potential energy to be artificially localised in space. Finally, I disagree with the primacy accorded by Irving and Kirkwood to ensemble averages over space-time averages, even though the purpose and function of spatial and temporal averaging are explicitly recognised in [14], where ensemble averages are correlated with experimental repetitions, and space-time averages with the mesoscopic spatial resolution and relaxation time of measuring devices. Agreeing with Irving and Kirkwood's

[1] The reader interested in this discussion is referred to the first two sections of [11].

**Table 1** Typical length ($L$) and time ($T$) scales

| | Length scale | Time scale |
|---|---|---|
| Macroscopic | $L_\mathrm{M} \gtrapprox 1\ \mu\mathrm{m}$ | $T_\mathrm{M} \gtrapprox 1\ \mu\mathrm{s}$ |
| Mesoscopic | $L_\mathrm{m} \approx 10\ \mathrm{nm}$ | $T_\mathrm{m} \approx 1\ \mathrm{ns}$ |
| Microscopic | $L_\mu \approx 1\ \mathrm{nm}$ | $T_\mu \approx 1\ \mathrm{fs}$ |

physical interpretation of averages of different kinds, I disagree with their attitude towards them: in my view, space-time averages are primary, being inescapable in each and every macroscopic observation, independently of whether it is repeated many times or not at all.

The peculiar feature that puts my space-time averaging procedure apart from other similar techniques is that it extracts not just the average velocity of a bunch of particles, but the mesoscopically best spatially *affine* approximation to their velocities—which I call *local streaming field*. Correspondingly, microscopic velocity fluctuations are not gauged by the difference between the velocity of individual particles and their average translational velocity, but by the difference between the velocity of individual particles and the value of the local streaming field at their respective instantaneous positions—a difference I call *thermal* velocity (cf. Sect. 2.2).

The very possibility of coarse-graining the underlying particle dynamics into a continuum field theory depends upon an adequate separation of space and time scales. In comparison with ensemble averaging, space-time averaging has the merit of: (i) making quantitatively explicit the different scales involved; (ii) providing specific conditions under which a continuum picture is feasible; (iii) emphasising the essential role played by well-tuned mesoscopic tolerances in constructing and using the macroscopic theory. In regard to the first item, three pairs of well-separated length and time scales—*micro-*, *meso-* and *macroscopic*—are introduced here and exploited in Sect. 2. For definiteness, their typical orders of magnitude are shown in Table 1.[2] The microscopic length scale $L_\mu$ is set by the effective range of short-ranged particle interactions $R$ (cf. Sect. 2.1) and the microscopic time scale $T_\mu$ by the precision required to resolve the fine details of particle motion. The mesoscopic length scale $L_\mathrm{m}$ and space scale $T_\mathrm{m}$ correspond, respectively, to the linear size of the computational cell employed and the time span covered by a typical molecular-dynamics simulation aimed at extracting robust statistical properties of a molecular sample. As will be made precise in Sect. 2.1, the size of a mesoscopic space cell should be $\approx L_\mathrm{m}$, and the duration of a mesoscopic time interval should be $\approx T_\mathrm{m}$. The macroscopic space scale $L_\mathrm{M}$ and time scale $T_\mathrm{M}$ characterise the coarseness of a typical sampling grid in space-time that is able to provide a satisfactory reconstruction of the macroscopic fields of interest. In most applications of continuum mechanics, the relevant values of $L_\mathrm{M}$ and $T_\mathrm{M}$ are much larger than those given in Table 1, possibly pertinent to micromechanical analyses.

The feasibility conditions mentioned in item (ii) emerge naturally in Sect. 2.3, where the macroscopic fields of velocity and mass density are constructed by assigning the values of cell-wise average densities (momentum per unit mass, mass per unit volume) to the cell centres. The twofold requirement is, first, that cells should be densely populated by particles and, second, that the fields so derived should vary slowly both in space and time with respect to the mesoscopic space and time scales $L_\mathrm{m}$ and $T_\mathrm{m}$.[3] The required smoothness of the macro-

[2]Here and in what follows, $a \approx b$ means that $a$ and $b$ are of the same order of magnitude; $a \gtrapprox b$ that $a$ is at least of the order of magnitude of $b$.

[3]These two conditions are often violated: the first fails in a rarefied gas approaching the Knudsen-flow regime; the second in a mesoscopic neighbourhood of shocks and sharp material interfaces.

scopic fields plays a pivotal role in proving that the local streaming fields have a first-order contact with the macroscopic velocity field at the respective cell centres (cf. Sect. 2.4). The smoothness of the velocity field also ensures that, at a conventional initial time, a unique time-parameterised pathline passes through each place where the macroscopic velocity is defined. This pathline is interpreted as the trajectory of a *macroscopic point*—an essential contrivance of continuum mechanics—whose current velocity approximates the average velocity of the throng of particles swarming through any mesoscopic space-time cell centred at its current position. As an unavoidable consequence, at different times the velocity of the very same macroscopic point encapsulates the average of (slightly or largely) different sets of particles—a conundrum intrinsic to any continuum caricature of an underlying atomistic picture.[4] The collection of such macroscopic points forms a deformable medium, whose motion is defined by their time evolution. *Co-moving space cells*, which I introduce next (cf. Sect. 2.5), should be thought of as mesoscopically small pieces of this macroscopic medium. Thermal degrees of freedom can then be attributed to all particles that are currently in a given co-moving space cell, by pulling back their current positions to the corresponding reference cell—i.e., to the snapshot of that co-moving cell taken at the conventional initial time. This allows me to give a fully referential representation of the motion of that bunch of particles, in terms of (a) slow collective degrees of freedom, associated with the continuum evolution, and (b) fast peculiar degrees of freedom, parameterising the microscopic thermal motion within the cell (cf. Sect. 2.6). The way in which the local streaming field is selected ensures that both momentum *and* moment of momentum of thermal velocities average to zero on each mesoscopic space-time cell—a far-reaching property of microscopic velocity fluctuations. In particular, this property is used in Sect. 2.7 to give a microscopic justification to the macroscopic mass conservation equation.

As regards item (iii), from Sect. 2.3 onwards the incongruity of macroscopic descriptors and microscopic precision becomes apparent. Far from being an imperfection, the introduction of suitable mesoscopic tolerances is intrinsic to any coarse-grained model of a point particle system. It should be borne in mind that the very gist of any upscaling procedure is precisely to extract macroscopic observables by blurring out the unavoidable microscopic noise—hopefully, without loosing mesoscopically significant information. One should accordingly be aware that demanding microscopic precision of a continuum theory is nonsensical. In fact, it is just by giving up on this level of precision—illusory at a macroscopic scale—that we are allowed to replace huge collections of line Dirac measures concentrated on point particle paths with *mesoscopically equivalent* Lebesgue-dominated measures.[5]

The above sketched construction of continuum fields, based on complete information about the microscopic details of the underlying particle system, just amounts to post-processing of molecular dynamics data. Such an a posteriori construction, while useful for summarising the collective kinematic features of the intricate microscopic evolution, is far from satisfactory. First of all, it is disproportionately expensive, if we are mainly interested in large-scale kinematic features, and altogether unfeasible on truly macroscopic space and time scales. Secondly, it gives us no information on macroscopic dynamical quantities, such as stress. However, the results presented in Sect. 2 form the basis of a viable multiscale method (reminiscent of the Parrinello-Rahman dynamics [25, 26]), elaborated

[4] In my own view, *all* continua are essentially "ephemeral," to use the expressive terminology coined by Capriz [7, 8]. My mesoscopic space cells are akin to his "loculi."

[5] My mesoscopic averages are integrals on a space-time cell—the Cartesian product of a space cell and a time interval—of product measures $\mu_1 \times \mu_2$, with $\mu_1$ a finite sum of (possibly vector- or tensor-valued) Dirac measures on the ambient three-dimensional space and $\mu_2$ the Lebesgue measure on the time line (cf. Sect. 2.2).

in Sect. 3. This method hinges on the idea of splitting a priori the dynamics of the system into a macroscopic component, parameterised by the slow degrees of freedom of macroscopic points, and a microscopic component, parameterised by the fast thermal degrees of freedom of individual particles. The two components, while evolving in parallel, are coupled cell-wise on the intermediate, mesoscopic space-time scale, where the continuum field theory—stripped of all macroscopic stress-strain recipe—informs particles of the prevailing hydrodynamical conditions and, in return, particle dynamics provides the continuum theory with molecularly-based stress data, via mesoscopic space-time averages.

The macroscopic configuration of the system can be adequately sampled at a finite number of points, which are spaced at distances of order $L_{\mathrm{M}}$ and whose degrees of freedom evolve significantly on the $T_{\mathrm{M}}$ time scale. A co-moving space cell of size $L_{\mathrm{m}}$, populated by a cloud of representative particles, is associated with each of these representative macroscopic points. While the collective degrees of freedom of each cell evolve on the $T_{\mathrm{M}}$ time scale, being slaved to the macroscopic degrees of freedom of the system, the thermal degrees of freedom of the particles evolve on the $T_{\mu}$ time scale. However, time histories of duration $T_{\mathrm{m}}$ of these microscopic degrees of freedom suffice to obtain the stress value in each cell—and hence to construct the macroscopic stress field and its divergence, which drives the evolution of the macroscopic configuration. Therefore, the fast microscopic dynamics needs to be computed only in bouts of mesoscopic duration, spaced by pauses of macroscopic duration, corresponding to the time-stepping intervals at which the slow macroscopic dynamics is evolved. Moreover, this computation is naturally parallel, since different clouds interact with each other only through their collective (streaming) degrees of freedom, so that the thermal degrees of freedom of a cloud are segregated from those of all the other clouds.

The mathematical machinery operating within each co-moving cell is developed in Sects. 3.1–3.6. Lagrangian forces dual to the collective and thermal degrees of freedom are defined in Sect. 3.1. In each of the subsequent four sections, the contributions to these generalised forces from the microscopic forces of inertia, external forces, short-range interaction forces, and reactive forces are separately presented. Based on the these results, Sect. 3.6 gives the Lagrangian equations of motion governing the dynamics of each co-moving cell. The concerted macroscopic-microscopic dynamics is discussed in Sect. 3.7.

The final Section 4 is a composite commentary addressing a variety of issues. The equations ruling the coupled dynamics of a co-moving space cell and of the cloud of point particles associated with it are compared and contrasted with the ones given by Parrinello and Rahman [25] and by Ribarsky and Landman [32]. Further arguments for and against different averaging techniques are put forward. The correct and efficient exploitation of scale separation in space and time is elaborated on. Doubts are raised on the possibility of giving a straightforward atomistic justification of higher-gradient continuum theories. Lastly, the main limitations of this work are pointed to.

The paper is supplemented by three appendices. Appendix A lists and explains the symbols and notations used. A basic account of Newtonian particle mechanics is given in Appendix B: despite its elementary nature, I found it necessary, both to set notations and to present coherently the concepts and tools needed in the core sections of the paper. Appendix C collects the most laborious proofs of the main results given in Sect. 3.

## 2 Coarse-Graining Via Mesoscopic Space-Time Averages

In this section, I play Laplace's God and imagine to know in microscopic, particle-wise detail the motion of a material system of macroscopic extent $\mathscr{M}$ (whose particles I label with the first $N$ natural numbers) over a macroscopic time interval. For notations the reader is referred to Appendices A and B. Following and extending Murdoch's ideas [19–22], I propound the use of space-time averaging to organise and summarise the overwhelming quantity of sheer molecular information.

## 2.1 Mesoscopic Space-Time Cells

I assume that all interactions are *short-ranged*, meaning by this that the interaction force $\mathbf{f}_i^{\mathsf{int}}(\tau)$ applied to the general $i$-th particle by the other particles in $\mathscr{M}$ (cf. Appendix B.7) may be safely evaluated by simply ignoring all particles whose current distance from $\mathbf{q}_i(\tau)$ is larger that a given *cutoff radius* $R$, which sets the microscopic length scale $L_\mu$:[6]

$$L_\mu := R. \tag{1}$$

To stress this assumption—which will prove essential in what follows—interaction forces will be relabelled explicitly as short-ranged:

$$\mathbf{f}_i^{\mathsf{int}} \equiv \mathbf{f}_i^{\mathsf{shr}}. \tag{2}$$

A space-time cell is, by definition, the Cartesian product $\mathscr{C} \times \mathscr{I}$ of a space cell $\mathscr{C}$ and a time interval $\mathscr{I}$. I define a *convex open* set $\mathscr{C} \subset \mathscr{E}$ to be a *mesoscopic space cell* if (and only if) its inner and outer diameters

$$D_{\mathsf{in}} := 2\sup\{r \mid \mathscr{B}_r(\mathbf{x}) \subset \mathscr{C}; \mathbf{x} \in \mathscr{E}\}, \qquad D_{\mathsf{out}} := 2\inf\{r \mid \mathscr{B}_r(\mathbf{x}) \supset \mathscr{C}; \mathbf{x} \in \mathscr{E}\} \tag{3}$$

are of the same order of magnitude $L_{\mathsf{m}}$, which sets the mesoscopic length scale, larger than the microscopic length scale $L_\mu$ defined in (1):

$$D_{\mathsf{in}} \approx D_{\mathsf{out}} \approx L_{\mathsf{m}} \gg L_\mu \tag{4}$$

(as is standard, in (3) $\mathscr{B}_r(\mathbf{x})$ denotes the space ball of radius $r$ centred at $\mathbf{x}$). A *mesoscopic time cell* is simply an open time interval

$$\mathscr{I} := ]t - T_{\mathsf{m}}/2, t + T_{\mathsf{m}}/2[ \tag{5}$$

whose duration $T_{\mathsf{m}}$ has order of magnitude larger than the microscopic time scale $T_\mu$ (cf. Table 1):

$$T_{\mathsf{m}} \gg T_\mu. \tag{6}$$

Mesoscopic space-time cells are the Cartesian products of mesoscopic space cells and time cells. Of course, we have at our disposal an uncountably infinite number of mesoscopic space-time cells, and we shall take advantage of all of them—albeit only in principle.

[6]This rough characterisation, sufficient for my present purposes, covers also *weak* long-range interactions, as defined in [6]. Coarse-graining *strong* long-range interactions is an altogether different story, since they generically produce non-additive collective quantities, such as energy.

## 2.2 Local Affine Approximation to Particle Velocities

For each space-time cell $\mathscr{C} \times \mathscr{I}$, I wish to determine the affine vector field

$$\begin{aligned} \alpha : \mathscr{E} &\to \mathsf{V}\mathscr{E} \\ \mathbf{x} &\mapsto \mathbf{v} + \mathbf{L}(\mathbf{x} - \mathbf{o}) \qquad (\mathbf{o} \in \mathscr{E}) \end{aligned} \tag{7}$$

(having physical dimension velocity) that minimises the quadratic functional[7]

$$\alpha \mapsto \int_{\mathscr{C}\times\mathscr{I}} \sum_i m_i \left\| \dot{\mathbf{q}}_i(\tau) - \alpha\big(\mathbf{q}_i(\tau)\big) \right\|^2 \delta_{\mathbf{q}_i(\tau)} \times d\tau, \tag{8}$$

where

$$\delta_{\mathbf{q}_i(\tau)}(\mathscr{C}) = \begin{cases} 1 & \text{if } \mathbf{q}_i(\tau) \in \mathscr{C}, \\ 0 & \text{otherwise}. \end{cases} \tag{9}$$

The rationale for (8) is that the minimiser provides the best spatially affine, constant-in-time approximation to the microscopic velocities, in the sense of the *mesoscopic kinetic energy seminorm*

$$\|\mathfrak{v}\|_{\mathscr{C}\times\mathscr{I}} := \frac{1}{T_{\mathsf{m}}} \int_{\mathscr{C}\times\mathscr{I}} \sum_i m_i \|\mathbf{v}_i(\tau)\|^2 \delta_{\mathbf{q}_i(\tau)} \times d\tau. \tag{10}$$

In principle, the sum over $i$ in (8) and in all that follows runs over all particles from 1 to $N$. However, only an extremely small minority of them visit the mesoscopic space cell $\mathscr{C}$ during the mesoscopic time interval $\mathscr{I}$. Let us denote by $\mathscr{S}_{\mathscr{C}\times\mathscr{I}}$ the small material subsystem comprising these particles, and by $\mathscr{Q}_{\mathscr{C}\times\mathscr{I}} \subset \mathscr{Q}$ its configuration manifold. When restricted to the trivial subbundle $\mathsf{T}\mathscr{Q}_{\mathscr{C}\times\mathscr{I}}$, (10) is a bona fide Euclidean norm. The minimiser $\alpha_{\mathscr{C}\times\mathscr{I}}$ is pinpointed by the stationarity conditions with respect to the vector $\mathbf{v}$ and the tensor $\mathbf{L}$ parameterising the field $\alpha$ in (8):

$$M_{\mathscr{C}\times\mathscr{I}}\alpha_{\mathscr{C}\times\mathscr{I}}(\mathbf{o}) + \mathbf{L}_{\mathscr{C}\times\mathscr{I}}\left(\frac{1}{T_{\mathsf{m}}} \int_{\mathscr{C}\times\mathscr{I}} \sum_i m_i \big(\mathbf{q}_i(\tau) - \mathbf{o}\big)\delta_{\mathbf{q}_i(\tau)} \times d\tau\right) = \mathbf{m}_{\mathscr{C}\times\mathscr{I}}, \tag{11a}$$

$$\left(\frac{1}{T_{\mathsf{m}}} \int_{\mathscr{C}\times\mathscr{I}} \sum_i m_i \big(\mathbf{q}_i(\tau) - \mathbf{o}\big)\delta_{\mathbf{q}_i(\tau)} \times d\tau\right) \otimes \alpha_{\mathscr{C}\times\mathscr{I}}(\mathbf{o}) + \mathbf{L}_{\mathscr{C}\times\mathscr{I}}\mathbf{E}_{\mathbf{o}} = \mathbf{M}_{\mathbf{o}}^{\mathsf{mom}}, \tag{11b}$$

where the scalar quantity

$$M_{\mathscr{C}\times\mathscr{I}} := \frac{1}{T_{\mathsf{m}}} \int_{\mathscr{C}\times\mathscr{I}} \sum_i m_i \delta_{\mathbf{q}_i(\tau)} \times d\tau \tag{12}$$

is the mesoscopically time-averaged cell mass (strictly positive if at least one particle visits $\mathscr{C}$ during $\mathscr{I}$), the vector quantity

$$\mathbf{m}_{\mathscr{C}\times\mathscr{I}} := \frac{1}{T_{\mathsf{m}}} \int_{\mathscr{C}\times\mathscr{I}} \sum_i m_i \dot{\mathbf{q}}_i(\tau) \delta_{\mathbf{q}_i(\tau)} \times d\tau \tag{13}$$

[7]Given a Radon measure $\mu$ on space-time, I denote its integral on $\mathscr{C} \times \mathscr{I}$ by $\int_{\mathscr{C}\times\mathscr{I}} \mu$.

is the mesoscopically time-averaged cell momentum, and the tensor quantities

$$\mathbf{E_o} := \frac{1}{T_{\mathrm{m}}} \int_{\mathscr{C}\times\mathscr{I}} \sum_i m_i \big(\mathbf{q}_i(\tau) - \mathbf{o}\big) \otimes \big(\mathbf{q}_i(\tau) - \mathbf{o}\big) \delta_{\mathbf{q}_i(\tau)} \times d\tau, \tag{14}$$

$$\mathbf{M}_{\mathbf{o}}^{\mathrm{mom}} := \frac{1}{T_{\mathrm{m}}} \int_{\mathscr{C}\times\mathscr{I}} \sum_i m_i \big(\mathbf{q}_i(\tau) - \mathbf{o}\big) \otimes \dot{\mathbf{q}}_i(\tau) \delta_{\mathbf{q}_i(\tau)} \times d\tau \tag{15}$$

are, respectively, the mesoscopically time-averaged cell Euler tensor with respect to the origin $\mathbf{o}$ and cell moment of momentum with respect to the same origin.[8] The mesoscopically time-averaged first moment of mass, appearing in both (11a) and (11b), vanishes if the origin $\mathbf{o}$ coincides with the mesoscopically time-averaged centre of mass

$$\mathbf{c}_{\mathscr{C}\times\mathscr{I}} = \arg\min\left( \mathscr{E} \ni \mathbf{x} \mapsto \frac{1}{T_{\mathrm{m}}} \int_{\mathscr{C}\times\mathscr{I}} \sum_i m_i \|\mathbf{q}_i(\tau) - \mathbf{x}\|^2 \delta_{\mathbf{q}_i(\tau)} \times d\tau \right), \tag{16}$$

which is uniquely defined if $M_{\mathscr{C}\times\mathscr{I}} > 0$: cf. (12). Henceforth, I will assume $\mathbf{o} = \mathbf{c}_{\mathscr{C}\times\mathscr{I}}$, and define the current *central radius vector* $\mathbf{r}_i(\tau)$ as

$$\mathbf{r}_i(\tau) := \mathbf{q}_i(\tau) - \mathbf{c}_{\mathscr{C}\times\mathscr{I}}, \tag{17}$$

implying that

$$\int_{\mathscr{C}\times\mathscr{I}} \sum_i m_i \mathbf{r}_i(\tau) \delta_{\mathbf{q}_i(\tau)} \times d\tau = \mathbf{0}. \tag{18}$$

Equations (11a)–(11b) take then the simpler form

$$M_{\mathscr{C}\times\mathscr{I}} \mathbf{v}_{\mathscr{C}\times\mathscr{I}} = \mathbf{m}_{\mathscr{C}\times\mathscr{I}}, \qquad \mathbf{L}_{\mathscr{C}\times\mathscr{I}} \mathbf{E}_{\mathscr{C}\times\mathscr{I}} = \mathbf{M}_{\mathscr{C}\times\mathscr{I}}^{\mathrm{mom}}, \tag{19}$$

where

$$\mathbf{v}_{\mathscr{C}\times\mathscr{I}} := \alpha_{\mathscr{C}\times\mathscr{I}}(\mathbf{c}_{\mathscr{C}\times\mathscr{I}}) \tag{20}$$

is the *mesoscopic central velocity*, and

$$\mathbf{E}_{\mathscr{C}\times\mathscr{I}} := \frac{1}{T_{\mathrm{m}}} \int_{\mathscr{C}\times\mathscr{I}} \sum_i m_i \mathbf{r}_i(\tau) \otimes \mathbf{r}_i(\tau) \delta_{\mathbf{q}_i(\tau)} \times d\tau, \tag{21a}$$

$$\mathbf{M}_{\mathscr{C}\times\mathscr{I}}^{\mathrm{mom}} := \frac{1}{T_{\mathrm{m}}} \int_{\mathscr{C}\times\mathscr{I}} \sum_i m_i \mathbf{r}_i(\tau) \otimes \dot{\mathbf{q}}_i(\tau) \delta_{\mathbf{q}_i(\tau)} \times d\tau \tag{21b}$$

are the mesoscopically time-averaged cell central Euler tensor and cell central moment of momentum, respectively. If $M_{\mathscr{C}\times\mathscr{I}} > 0$ and the symmetric tensor $\mathbf{E}_{\mathscr{C}\times\mathscr{I}}$ is positive definite, hence invertible, then the minimiser

$$\alpha_{\mathscr{C}\times\mathscr{I}} : \mathbf{x} \mapsto \mathbf{v}_{\mathscr{C}\times\mathscr{I}} + \mathbf{L}_{\mathscr{C}\times\mathscr{I}}(\mathbf{x} - \mathbf{c}_{\mathscr{C}\times\mathscr{I}}) \tag{22}$$

[8] My definition of choice of the tensor product is such that $(\mathbf{a} \otimes \mathbf{b})\mathbf{u} = (\mathbf{a} \cdot \mathbf{u})\mathbf{b}$ for all $\mathbf{a}$, $\mathbf{b}$, $\mathbf{u} \in \mathsf{V}\mathscr{E}$. This definition implies that $(\mathbf{La}) \cdot \mathbf{b} = \mathbf{L} \cdot (\mathbf{a} \otimes \mathbf{b})$ and $(\mathbf{La}) \otimes \mathbf{b} = (\mathbf{a} \otimes \mathbf{b})\mathbf{L}^\top$ for all $\mathbf{L} \in \mathsf{V}\mathscr{E} \otimes \mathsf{V}\mathscr{E}$. Note that in Cartesian index notation $(\mathbf{a} \otimes \mathbf{b})_{ij} = a_j b_i$. The moment of momentum $\mathbf{M}_{\mathbf{o}}^{\mathrm{mom}}$ as defined in (15) is *not* the rotational momentum, which equals $2\,\mathrm{skw}\,\mathbf{M}_{\mathbf{o}}^{\mathrm{mom}}$ (cf. (197) in Appendix B.3). More in general, it is instructive to compare the time-averaged cell quantities (12), (13), (16), (21a), and (21b) with the analogous mass-related quantities defined instantaneously for a fixed subsystem of particles in Appendix B.3.

is uniquely defined by

$$\mathbf{v}_{\mathscr{C}\times\mathscr{I}} = (1/M_{\mathscr{C}\times\mathscr{I}})\mathbf{m}_{\mathscr{C}\times\mathscr{I}}, \qquad \mathbf{L}_{\mathscr{C}\times\mathscr{I}} = \mathbf{M}^{\mathrm{mom}}_{\mathscr{C}\times\mathscr{I}}\mathbf{E}^{-1}_{\mathscr{C}\times\mathscr{I}}. \tag{23}$$

Note, in particular, that the mesoscopic central velocity $\mathbf{v}_{\mathscr{C}\times\mathscr{I}}$ equals the mass-averaged particle velocity over $\mathscr{C}\times\mathscr{I}$, and that the actual velocity $\dot{\mathbf{q}}_i(\tau)$ of the general $i$-th particle in $\mathscr{S}_{\mathscr{C}\times\mathscr{I}}$ decomposes into the sum of the *streaming* velocity

$$\alpha_{\mathscr{C}\times\mathscr{I}}\big(\mathbf{q}_i(\tau)\big) = \mathbf{v}_{\mathscr{C}\times\mathscr{I}} + \mathbf{L}_{\mathscr{C}\times\mathscr{I}}\big(\mathbf{q}_i(\tau) - \mathbf{c}_{\mathscr{C}\times\mathscr{I}}\big) \tag{24a}$$

and the *thermal* velocity

$$\mathbf{w}_i(\tau) := \dot{\mathbf{q}}_i(\tau) - \alpha_{\mathscr{C}\times\mathscr{I}}\big(\mathbf{q}_i(\tau)\big). \tag{24b}$$

As long as a particle is in $\mathscr{C}$ during $\mathscr{I}$, its streaming velocity only depends on its current position, once the spatially affine, timewise constant streaming velocity field $\alpha_{\mathscr{C}\times\mathscr{I}}$ is given. In turn, the field $\alpha_{\mathscr{C}\times\mathscr{I}}$ depends on the microscopic particle vagaries only through the mesoscopic space-time averages $\mathbf{c}_{\mathscr{C}\times\mathscr{I}}$, $\mathbf{v}_{\mathscr{C}\times\mathscr{I}}$, and $\mathbf{L}_{\mathscr{C}\times\mathscr{I}}$.

Restricting the scope of the minimisation procedure (8) by putting $\mathbf{L} = \mathbf{0}$ in (7), i.e., by reducing (8) to

$$\mathbf{v} \mapsto \frac{1}{T_{\mathrm{m}}}\int_{\mathscr{C}\times\mathscr{I}} \sum_i m_i \|\dot{\mathbf{q}}_i(\tau) - \mathbf{v}\|^2 \delta_{\mathbf{q}_i(\tau)} \times d\tau,$$

would extract far less information from the same pool of microscopic data, yielding just the mass-averaged particle velocity $(23)_1$ (or, in terms of fields, a spatially constant streaming velocity). Put this way, it seems unwise not to take advantage of the extra information provided by (8), at the marginal extra cost of computing $(23)_2$. However, this is exactly what is done in most, if not all,[9] averaging procedures, performed either on space-time cells or on a statistical ensemble. Correspondingly, the thermal velocity is customarily identified with the (position-independent) *peculiar* velocity

$$\mathbf{w}_i(\tau) := \dot{\mathbf{q}}_i(\tau) - \mathbf{v}_{\mathscr{C}\times\mathscr{I}}.$$

I find definition (24b) more proper and more fecund than this one. Both definitions imply that thermal velocities have zero mesoscopic momentum:

$$\int_{\mathscr{C}\times\mathscr{I}} \sum_i m_i \mathbf{w}_i(\tau)\delta_{\mathbf{q}_i(\tau)} \times d\tau = \mathbf{0}. \tag{25}$$

Under definition (24b), also its mesoscopic moment of momentum is null:

$$\int_{\mathscr{C}\times\mathscr{I}} \sum_i m_i \mathbf{r}_i(\tau) \otimes \mathbf{w}_i(\tau)\delta_{\mathbf{q}_i(\tau)} \times d\tau = \mathbf{0}. \tag{26}$$

Equalities (25) *and* (26) hold if and only if

$$\int_{\mathscr{C}\times\mathscr{I}} \sum_i m_i \alpha\big(\mathbf{q}_i(\tau)\big) \cdot \mathbf{w}_i(\tau)\delta_{\mathbf{q}_i(\tau)} \times d\tau = 0 \tag{27}$$

for *all* affine velocity field (7), i.e., for *all* local streaming velocities.

[9] I consider the 2010 paper by Rossi and Testa [33] a notable, though partial, exception: cf. the last paragraph of Sect. 4.3 below.

## 2.3 Macroscopic Density and Velocity Fields

In principle, the construction leading to (12) and (23) may be repeated for all conceivable mesoscopic space-time cells. For all place $\mathbf{x}$ and time $t$ a mesoscopic neighbourhood of which is thick with particles, one can construct a mesoscopic space cell $\mathscr{C}$—in fact, infinitely many of them—such that, for a mesoscopic time interval $\mathscr{I}$ centred at $t$,

$$\mathbf{c}_{\mathscr{C}\times\mathscr{I}} = \mathbf{x} \quad \text{to within a preset } \textit{mesoscopically negligible} \text{ deviation,}$$

a statement which I condense in the approximate equality

$$\|\mathbf{c}_{\mathscr{C}\times\mathscr{I}} - \mathbf{x}\| \underset{\text{m}}{\cong} 0. \tag{28}$$

Here and in what follows, $a \underset{\text{m}}{\cong} b$ means that the quantities $a$ and $b$ are *mesoscopically indistinguishable*. Analogously, $a \underset{\text{m}}{>} b$ means that $a - b > c$ for all $c \underset{\text{m}}{\cong} 0$ or, in words, that $a$ is *mesoscopically larger* than $b$. In particular, the positivity requirements put on the mesoscopic mass (12) and on the mesoscopic central Euler tensor (21a) are to be strengthened into[10]

$$M_{\mathscr{C}\times\mathscr{I}} \underset{\text{m}}{>} 0, \qquad \mathbf{E}_{\mathscr{C}\times\mathscr{I}} \underset{\text{m}}{>} 0, \tag{29}$$

in order to ensure that the time-averaged centre of mass $\mathbf{c}_{\mathscr{C}\times\mathscr{I}}$, the mesoscopic *mass density*

$$\varrho_{\mathscr{C}\times\mathscr{I}} := \frac{M_{\mathscr{C}\times\mathscr{I}}}{V_{\mathscr{C}}} \tag{30}$$

(with $V_{\mathscr{C}} := H^3(\mathscr{C}) \underset{\text{m}}{>} 0$ the space cell volume[11]), the mesoscopic central velocity $\mathbf{v}_{\mathscr{C}\times\mathscr{I}}$, and the linear part $\mathbf{L}_{\mathscr{C}\times\mathscr{I}}$ of the local streaming velocity are robust against microscopic fluctuations. Then, I plan to construct macroscopic fields out of intensive quantities extracted from space-time cell averages by positing

$$\varrho(\mathbf{x}, t) := \varrho_{\mathscr{C}\times\mathscr{I}}, \qquad \mathbf{v}(\mathbf{x}, t) := \mathbf{v}_{\mathscr{C}\times\mathscr{I}}, \qquad \mathbf{L}(\mathbf{x}, t) := \mathbf{L}_{\mathscr{C}\times\mathscr{I}}. \tag{31}$$

These definitions are legitimate, provided that *any* other mesoscopic space-time cell $\mathscr{C}' \times \mathscr{I}'$ centred at the (mesoscopically determined) time $t$ and sharing the same centre of mass $\mathbf{x}$ (to within a mesoscopically negligible deviation) assigns the same values (31) to the pair $(\mathbf{x}, t)$, to within preset mesoscopic tolerances:

$$\varrho_{\mathscr{C}'\times\mathscr{I}'} \underset{\text{m}}{\cong} \varrho_{\mathscr{C}\times\mathscr{I}}, \qquad \mathbf{v}_{\mathscr{C}'\times\mathscr{I}'} \underset{\text{m}}{\cong} \mathbf{v}_{\mathscr{C}\times\mathscr{I}}, \qquad \mathbf{L}_{\mathscr{C}'\times\mathscr{I}'} \underset{\text{m}}{\cong} \mathbf{L}_{\mathscr{C}\times\mathscr{I}}. \tag{32}$$

In terms of the notations introduced in (31), the fundamental velocity decomposition (24a)–(24b) may be rewritten as

$$\dot{\mathbf{q}}_i(\tau) = \mathbf{v}(\mathbf{x}, t) + \mathbf{L}(\mathbf{x}, t)\mathbf{r}_i(\tau) + \mathbf{w}_i(\tau). \tag{33}$$

The necessary robustness of definitions (31) against mesoscopically legitimate variations in the sampling procedure cannot be attained unless the quantities so obtained vary slowly both

[10] Inequality $(29)_2$ is to be construed in the sense of quadratic forms: $(\mathbf{E}_{\mathscr{C}\times\mathscr{I}}\mathbf{n}) \cdot \mathbf{n} \underset{\text{m}}{>} 0$ for all unit vector $\mathbf{n}$.

[11] $H^3$ is the volume measure on $\mathscr{E}$.

in space and time with respect to the mesoscopic space and time scales $L_{\mathrm{m}}$ and $T_{\mathrm{m}}$. In other words, filtering out wavelengths shorter than a macroscopic length

$$L_{\mathrm{M}} \gg L_{\mathrm{m}} \tag{34a}$$

and periods shorter than a macroscopic time span

$$T_{\mathrm{M}} \gg T_{\mathrm{m}} \tag{34b}$$

should leave the fields

$$(\mathbf{x}, t) \mapsto \big(\varrho(\mathbf{x}, t), \mathbf{v}(\mathbf{x}, t), \mathbf{L}(\mathbf{x}, t)\big) \tag{35}$$

mesoscopically equivalent. Under this assumption, mesoscopic space-time cells can be left in the background, after having been instrumental in constructing the macroscopic space-time fields (35). Wherever and whenever either the *denseness requirement* (29) or the *smoothness requirement* (34a)–(34b) fails, continuum fields cannot be properly defined.[12] Hence, in general, the domain of definition $\mathscr{D}_{\mathrm{M}}$ of fields (35) is a *proper* open subset of space-time, typically non connected. At any place-time pair $(\boldsymbol{\xi}, \tau) \in \mathscr{D}_{\mathrm{M}}$, on the contrary, the values taken by these fields can be obtained—at least in principle—from averages computed on the product of a mesoscopic space cell centred at $\boldsymbol{\xi}$ times a mesoscopic time cell $\mathscr{I}_\tau$ centred at $\tau$:

$$\mathscr{I}_\tau := ]\tau - T_{\mathrm{m}}/2, \tau + T_{\mathrm{m}}/2[. \tag{36}$$

With this notation, $\mathscr{I} \equiv \mathscr{I}_t$ (cf. (5)).

## 2.4 Macroscopic Kinematic Compatibility

Owing to the smoothness requirement (34a)–(34b) on the fields $\mathbf{v}, \mathbf{L}$ parameterising the mesoscopic streaming velocity fields (24a), $\mathbf{L}$ equals—to within a mesoscopically negligible error—the *spatial gradient* $\mathbf{G}$ of the macroscopic velocity field $\mathbf{v}$:

$$\mathbf{L}(\mathbf{x}, t) \underset{\mathrm{m}}{\cong} \nabla \mathbf{v}|_{(\mathbf{x},t)} =: \mathbf{G}(\mathbf{x}, t). \tag{37}$$

To prove this approximate equality, for any given mesoscopic space cell $\mathscr{C}_-$, let us pick another mesoscopic space cell $\mathscr{C}_+$ such that the intersection $\mathscr{C}_- \cap \mathscr{C}_+ =: \mathscr{C}$ is also a mesoscopic space cell (Fig. 1). Since the set of open convex subsets of $\mathscr{E}$ is closed under intersection, taking for $\mathscr{C}_+$ any sufficiently small perturbation of $\mathscr{C}_-$ that keeps it open and convex will work. Let $\mathbf{c}_-$, $\mathbf{c}_+$, and $\mathbf{c}$ be, respectively, the $T_{\mathrm{m}}$-averaged centres of mass of $\mathscr{C}_-$, $\mathscr{C}_+$, and $\mathscr{C}$ at time $t$, and define (cf. Fig. 1)

$$\mathbf{u}_- := \mathbf{c}_- - \mathbf{c}, \qquad \mathbf{u}_+ := \mathbf{c}_+ - \mathbf{c}, \qquad \mathbf{u} := \mathbf{c}_+ - \mathbf{c}_- = \mathbf{u}_+ - \mathbf{u}_-. \tag{38}$$

In principle, we can take the perturbation as small as we please. However, if we take it too small, place differences (38) will be polluted, and ultimately dominated, by microscopic noise. Hypothesis (34a)–(34b) implies that the restrictions of the streaming velocity fields $\alpha_{\mathscr{C}_- \times \mathscr{I}}$ and $\alpha_{\mathscr{C}_+ \times \mathscr{I}}$ to $\mathscr{C}$, namely

$$\alpha_- := \alpha_{\mathscr{C}_- \times \mathscr{I}}|_{\mathscr{C}}, \qquad \alpha_+ := \alpha_{\mathscr{C}_+ \times \mathscr{I}}|_{\mathscr{C}}, \tag{39}$$

[12] As pointed out in footnote 3, requirement (29) fails in a rarefied gas approaching the Knudsen-flow regime, and requirement (34a)–(34b) in a mesoscopic neighbourhood of shocks and sharp material interfaces.

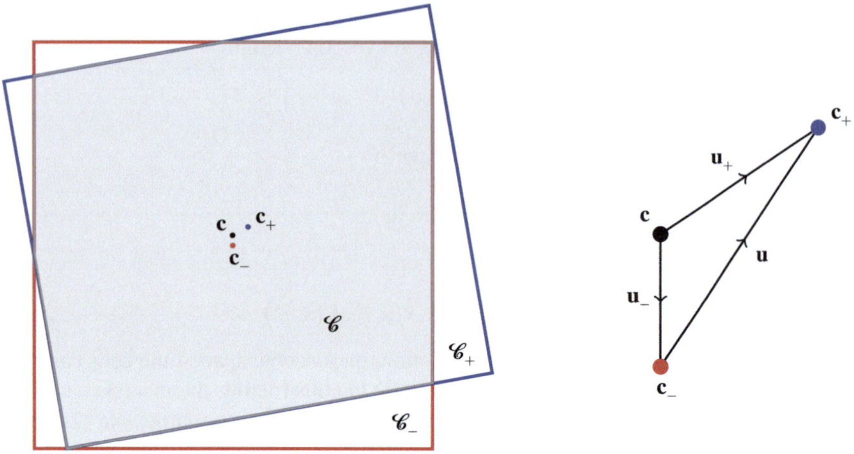


**Fig. 1** Left: Two-dimensional cartoon of two mesoscopic space cells $\mathscr{C}_-$ (contoured in red), $\mathscr{C}_+$ (contoured in blue), such that also their intersection $\mathscr{C} = \mathscr{C}_- \cap \mathscr{C}_+$ (coloured in gray) qualifies as a legitimate mesoscopic space cell. Their respective $T_{\mathrm{m}}$-averaged centres of mass are labelled $\mathbf{c}_-$ (red dot), $\mathbf{c}_+$ (blue dot), and $\mathbf{c}$ (black dot). Right: Central area of $\mathscr{C}$ magnified to illustrate vectors $\mathbf{u}_-$, $\mathbf{u}_+$, and $\mathbf{u}$ as defined in (38)

are mesoscopically indistinguishable, in the sense that the mesoscopic kinetic energy norm (10) of their difference $\alpha_+ - \alpha_-$ is mesoscopically negligible:

$$\frac{1}{T_{\mathrm{m}}} \int_{\mathscr{C}\times\mathscr{I}} m_i \left\| \alpha_+\big(\mathbf{q}_i(\tau)\big) - \alpha_-\big(\mathbf{q}_i(\tau)\big) \right\|^2 \delta_{\mathbf{q}_i(\tau)} \times d\tau \underset{\mathrm{m}}{\cong} 0. \tag{40}$$

Definitions (24a) and (38) imply

$$\alpha_+\big(\mathbf{q}_i(\tau)\big) - \alpha_-\big(\mathbf{q}_i(\tau)\big) = \mathbf{v}_+ - \mathbf{v}_- - \mathbf{L}_+\mathbf{u}_+ + \mathbf{L}_-\mathbf{u}_- + (\mathbf{L}_+ - \mathbf{L}_-)\big(\mathbf{q}_i(\tau) - \mathbf{c}\big), \tag{41}$$

where

$$\mathbf{v}_\mp := \mathbf{v}(\mathbf{c}_\mp, t), \qquad \mathbf{L}_\mp := \mathbf{L}(\mathbf{c}_\mp, t). \tag{42}$$

On account of (12) and (21a), substituting (41) into (40) yields

$$M_{\mathscr{C}\times\mathscr{I}} \|\mathbf{v}_+ - \mathbf{v}_- - \mathbf{L}_+\mathbf{u}_+ + \mathbf{L}_-\mathbf{u}_-\|^2 + \big((\mathbf{L}_+ - \mathbf{L}_-)\mathbf{E}_{\mathscr{C}\times\mathscr{I}}\big) \cdot (\mathbf{L}_+ - \mathbf{L}_-) \underset{\mathrm{m}}{\cong} 0. \tag{43}$$

Since both terms on the left side of (43) are nonnegative, their sum is mesoscopically negligible if and only if each term is. Since $\mathbf{E}_{\mathscr{C}\times\mathscr{I}} \underset{\mathrm{m}}{>} 0$ (cf. footnote 10), then

$$\mathbf{L}(\mathbf{c}_+, t) \underset{\mathrm{m}}{\cong} \mathbf{L}(\mathbf{c}_-, t), \tag{44}$$

and, on account of (38), (43) reduces to

$$M_{\mathscr{C}\times\mathscr{I}} \|\mathbf{v}_+ - \mathbf{v}_- - \mathbf{L}_-\mathbf{u}\|^2 \underset{\mathrm{m}}{\cong} 0. \tag{45}$$

Since $M_{\mathscr{C}\times\mathscr{I}} \underset{\mathrm{m}}{>} 0$, then

$$\mathbf{v}(\mathbf{x} + \mathbf{u}, t) \underset{\mathrm{m}}{\cong} \mathbf{v}(\mathbf{x}, t) + \mathbf{L}(\mathbf{x}, t)\,\mathbf{u}, \tag{46}$$

where, to increase readability and conform with the notations in (37), $\mathbf{c}_-$ has been renamed $\mathbf{x}$. Since (46) holds for any sufficiently small—but *not* necessarily mesoscopically negligible—vector **u**, statement (37) is proved. In other terms, the local streaming velocity field (22) is mesoscopically equivalent to the restriction of the macroscopic velocity field **v** to the space-time cell where it is defined:

$$\alpha_{\mathscr{C}\times\mathscr{I}} \underset{\mathrm{m}}{\cong} \mathbf{v}|_{\mathscr{C}\times\mathscr{I}}. \tag{47}$$

This approximate equality, combined with (24a)–(24b), yields an approximate decomposition of particle velocity into the sum of the macroscopic velocity prevailing at its current position and its thermal velocity:

$$\dot{\mathbf{q}}_i(\tau) \underset{\mathrm{m}}{\cong} \mathbf{v}\big(\mathbf{q}_i(\tau)\big) + \mathbf{w}_i(\tau). \tag{48}$$

## 2.5 Macroscopic Motion; Co-Moving Mesoscopic Space Cells

Due to the assumed piecewise smoothness of the macroscopic velocity field

$$\mathbf{v} : \mathscr{D}_{\mathsf{M}} \to \mathsf{V}\mathscr{E}, \tag{49}$$

for each pair $(\mathbf{X}, t_0) \in \mathscr{D}_{\mathsf{M}}$ the ordinary differential equation

$$\dot{\chi}(\mathbf{X}, t) = \mathbf{v}\big(\chi(\mathbf{X}, t), t\big) \tag{50a}$$

has a unique, well-behaved local solution[13] under the initial condition

$$\chi(\mathbf{X}, t_0) = \mathbf{X}. \tag{50b}$$

By construction, the smooth *placement* map $\chi(\cdot, t)$ is bijective on its image for each time $t$ for which it is defined. Consequently, the *deformation gradient*

$$\mathbf{F} := \nabla\chi \tag{51}$$

takes values in the special linear group $\mathsf{SL}(\mathsf{V}\mathscr{E})$ (i.e., tensors with positive determinant). For each pair $(\mathbf{X}, t_0) \in \mathscr{D}_{\mathsf{M}}$, the map $\mathbf{F}(\mathbf{X}, \cdot)$ is the unique solution to the tensor-valued ordinary differential equation

$$\dot{\mathbf{F}}(\mathbf{X}, t) = \mathbf{G}\big(\chi(\mathbf{X}, t), t\big)\mathbf{F}(\mathbf{X}, t), \tag{52a}$$

where **G** is the *spatial velocity gradient* (37), under the initial condition

$$\mathbf{F}(\mathbf{X}, t_0) = \mathbf{I}, \tag{52b}$$

with **I** the identity on $\mathsf{V}\mathscr{E}$. The place $\chi(\mathbf{X}, t)$ is interpreted as the position at the current time $t$ of the *macroscopic point* supposedly placed in **X** at the initial time $t_0$, and the vector $\dot{\chi}(\mathbf{X}, t)$ is interpreted as the current macroscopic velocity of such a point, labelled with **X**. Let **x** be the position where the macroscopic motion $\chi$ brings the **X**-labelled macroscopic point at time $t$, i.e.,

$$\mathbf{x} = \chi(\mathbf{X}, t). \tag{53}$$

[13] Existence, uniqueness and behaviour of global solutions crucially depend on jump conditions between different connected components of $\mathscr{D}_{\mathsf{M}}$. Such conditions make substantial extensions to continuum mechanics per se, and would demand case-specific microscopic foundations.

Since the placement $\chi(\cdot, t)$ is bijective on its image whenever it is defined, then the macroscopic point placed in $\mathbf{x}$ at time $t$ is uniquely defined:[14]

$$\mathbf{X} = \chi^{-1}(\mathbf{x}, t). \tag{54}$$

Moreover, since the placement varies slowly with respect to the mesoscopic space scale $L_{\mathrm{m}}$, the inverse image under $\chi(\cdot, t)$ of a mesoscopic space cell $\mathscr{C}$ centred at $\mathbf{x}$ is mesoscopically indistinguishable from its inverse image under the *affine* approximation to $\chi(\cdot, t)$ at $\mathbf{X}$:

$$\mathscr{C}_0 := \left\{\mathbf{X} + \mathbf{F}(\mathbf{X}, t)^{-1}\mathbf{u} \mid \mathbf{x} + \mathbf{u} \in \mathscr{C}\right\} \underset{\mathrm{m}}{\cong} \chi^{-1}(\mathscr{C}, t). \tag{55}$$

Equivalently, the space cell $\mathscr{C}$ may be regarded as (mesoscopically equivalent to) the value taken at time $t$ by the *co-moving space cell* $\mathscr{C}_\chi$, defined as the $\chi$-induced cell-valued map

$$t \mapsto \mathscr{C}_\chi(t) := \left\{\chi(\mathbf{X}, t) + \mathbf{F}(\mathbf{X}, t)\mathbf{u} \mid \mathbf{X} + \mathbf{u} \in \mathscr{C}_0\right\} \underset{\mathrm{m}}{\cong} \chi(\mathscr{C}_0, t). \tag{56}$$

With this notation,

$$\mathscr{C}_0 \equiv \mathscr{C}_\chi(t_0), \qquad \mathscr{C} \equiv \mathscr{C}_\chi(t). \tag{57}$$

Since invertible affine transformations of $\mathscr{E}$ map open convex sets into open convex sets, $\mathscr{C}_\chi(\tau)$ $(t_0 \le \tau \le t)$ is a legitimate mesoscopic space cell, provided that the (symmetric positive definite) *change-of-metric* tensor associated with the deformation gradient $\mathbf{F}(\mathbf{X}, \tau)$, namely

$$\mathbf{C}(\mathbf{X}, \tau) := \mathbf{F}(\mathbf{X}, \tau)^\top \mathbf{F}(\mathbf{X}, \tau), \tag{58}$$

has eigenvalues of the order of magnitude of 1.[15]

### 2.6 (Approximate) Complete Motion: Macroscopic Plus Thermal

The peculiar motion $\tau \mapsto \mathbf{r}_i(\tau)$ of the $i$-th particle jiggling within $\mathscr{C}$ during $\mathscr{I}$ is conveniently pulled back to $\mathscr{C}_0$ by introducing the corresponding *reference* radius vector[16]

$$\mathbf{s}_i(\tau) := \mathbf{F}(\mathbf{X}, t)^{-1}\mathbf{r}_i(\tau). \tag{59}$$

Since $\mathscr{C}$ is centred at $\mathbf{x}$, (17) may hence be rewritten as

$$\mathbf{q}_i(\tau) = \chi(\mathbf{X}, t) + \mathbf{r}_i(\tau) \underset{\mathrm{m}}{\cong} \chi(\mathbf{X}, \tau) + \mathbf{F}(\mathbf{X}, \tau)\mathbf{s}_i(\tau), \tag{60}$$

[14] In what follows, the one-to-one *time-dependent* relationship between $\mathbf{x}$ and $\mathbf{X}$ will often be understood without explicit notation. Care should be taken when taking time derivatives.

[15] Hence the practical necessity of subdividing the time interval $[t_0, t]$ into subintervals $[t_{j-1}, t_j]$, $j = 1, \dots, k$, with $t_k = t$, such that $\mathbf{F}(\mathbf{X}, t_{j-1+})^{-\top}\mathbf{C}(\mathbf{X}, t_{j-})\mathbf{F}(\mathbf{X}, t_{j-1+})^{-1}$ has eigenvalues of the order of magnitude of 1, for all $j = 1, \dots, k$. At each time cut $t_j$ $(j < k)$, the molecular pattern within the (parallelepipedal) space cell, extended by periodic replications, is sampled by a new space cell, mesoscopically equivalent to the old one, constructed in such a way that $\mathbf{C}(\mathbf{X}, t_{j+}) \approx \mathbf{I}$. The transition across such time cuts should be seamless. The original "fictitious Lagrangian" (quoted verbatim from [40]) introduced by Parrinello and Rahman [25, 26] was rightly criticised [10, 32, 40] on the ground that it does not satisfy this condition.

[16] This vector corresponds to the "scaled coordinate" vector introduced by Andersen [4] and generalised by Parrinello & Rahman [25, 26].

the second (approximate) equality stemming from the fact that $\chi(\mathbf{X},\cdot)$ and $\mathbf{F}(\mathbf{X},\cdot)$ vary slowly on the $T_{\mathrm{m}}$ time scale:

$$\chi(\mathbf{X},\tau) \underset{\mathrm{m}}{\cong} \chi(\mathbf{X},t), \qquad \mathbf{F}(\mathbf{X},\tau) \underset{\mathrm{m}}{\cong} \mathbf{F}(\mathbf{X},t) \quad \text{over } \mathscr{I}. \tag{61}$$

Equation (60) provides a (mesoscopically accurate) fully referential representation of particle positions, in terms of the Lagrangian degrees of freedom $\chi(\mathbf{X},\cdot)$, $\mathbf{F}(\mathbf{X},\cdot)$ and $\mathbf{s}_i(\cdot)$. In turn, the velocity decomposition (33) may be rewritten as

$$\begin{aligned}\dot{\mathbf{q}}_i(\tau) &\underset{\mathrm{m}}{\cong} \dot{\chi}(\mathbf{X},t) + \dot{\mathbf{F}}(\mathbf{X},t)\mathbf{s}_i(\tau) + \mathbf{w}_i(\tau)\\ &\underset{\mathrm{m}}{\cong} \dot{\chi}(\mathbf{X},\tau) + \dot{\mathbf{F}}(\mathbf{X},\tau)\mathbf{s}_i(\tau) + \mathbf{w}_i(\tau),\end{aligned} \tag{62}$$

where the first approximate equality uses (37), (50a), (52a) and (59), and the second relies on the fact that

$$\dot{\chi}(\mathbf{X},\tau) \underset{\mathrm{m}}{\cong} \dot{\chi}(\mathbf{X},t), \qquad \dot{\mathbf{F}}(\mathbf{X},\tau) \underset{\mathrm{m}}{\cong} \dot{\mathbf{F}}(\mathbf{X},t) \quad \text{over } \mathscr{I}. \tag{63}$$

Taking the time derivative of (60) and comparing it with (62), one finally obtains

$$\mathbf{w}_i(\tau) \underset{\mathrm{m}}{\cong} \mathbf{F}(\mathbf{X},\tau)\dot{\mathbf{s}}_i(\tau) \quad \text{at all time } \tau. \tag{64}$$

This approximate equality, substituted back into (62), provides a fully referential representation of particle velocities:

$$\dot{\mathbf{q}}_i(\tau) \underset{\mathrm{m}}{\cong} \dot{\chi}(\mathbf{X},\tau) + \dot{\mathbf{F}}(\mathbf{X},\tau)\mathbf{s}_i(\tau) + \mathbf{F}(\mathbf{X},\tau)\dot{\mathbf{s}}_i(\tau) \quad \text{at all time } \tau, \tag{65a}$$

where the pair $(\dot{\chi}(\mathbf{X},\cdot), \dot{\mathbf{F}}(\mathbf{X},\cdot))$ parameterises the streaming (macroscopic) velocity, and $\dot{\mathbf{s}}_i$ parameterises the thermal (microscopic) velocity. On account of (50a), (52a), and (53), the spatial counterpart of (65a)–(65b) reads

$$\dot{\mathbf{q}}_i(\tau) \underset{\mathrm{m}}{\cong} \mathbf{v}(\boldsymbol{\xi},\tau) + \mathbf{G}(\boldsymbol{\xi},\tau)\mathbf{r}_i(\tau) + \mathbf{w}_i(\tau), \tag{65b}$$

with

$$\boldsymbol{\xi} = \chi(\mathbf{X},\tau). \tag{66}$$

Both representations (65a)–(65b) will be used in Sect. 3. The initial condition (52b) obviously implies $\mathbf{s}_i(t_0) = \mathbf{r}_i(t_0)$. In general, however,

$$\mathbf{s}_i(t) \underset{\mathrm{m}}{\ncong} \mathbf{r}_i(t_0), \tag{67}$$

unless the *Cauchy-Born rule* is enforced, thus making the thermal velocity mesoscopically negligible:

$$\mathbf{r}_i(t) = \mathbf{F}(\mathbf{X},t)\mathbf{r}_i(t_0) \quad \Longrightarrow \quad \mathbf{w}_i(t) \underset{\mathrm{m}}{\cong} \mathbf{0}. \tag{68}$$

### 2.7 Mesoscopic Properties of Mass Distribution; Mass Conservation

The smoothness requirement (34a)–(34b) on the macroscopic field $\varrho$ of mesoscopic mass density entails that, for all mesoscopic space-time cell $\mathscr{C} \times \mathscr{I} \subset \mathscr{D}_{\mathrm{M}}$, the mesoscopically averaged centre of mass (16) is mesoscopically indistinguishable from the *centroid* $\overline{\mathbf{c}}_{\mathscr{C}}$ of $\mathscr{C}$:

$$\mathbf{c}_{\mathscr{C}\times\mathscr{I}} \underset{\mathrm{m}}{\cong} \overline{\mathbf{c}}_{\mathscr{C}} := \arg\min\left(\mathscr{E} \ni \boldsymbol{\xi} \mapsto \int_{\mathscr{E}} \mathbf{1}_{\mathscr{C}}(\mathbf{z})\|\mathbf{z}-\boldsymbol{\xi}\|^2 H^3\right), \tag{69}$$

where $\mathbf{1}_{\mathscr{C}}$ is the indicator function of $\mathscr{C}$ and $\mathbf{z} \in \mathscr{E}$. Consequently, (26) implies

$$\int_{\mathscr{C}\times\mathscr{I}} \sum_i m_i \bar{\mathbf{r}}_i(\tau) \otimes \mathbf{w}_i(\tau)\delta_{\mathbf{q}_i(\tau)} \times d\tau \underset{\mathrm{m}}{\cong} \mathbf{0}, \tag{70}$$

with $\bar{\mathbf{r}}_i(\tau)$ the current centroidal radius vector of the $i$-th particle:[17]

$$\mathbf{q}_i(\tau) - \bar{\mathbf{c}}_{\mathscr{C}} =: \bar{\mathbf{r}}_i(\tau) \underset{\mathrm{m}}{\cong} \mathbf{r}_i(\tau). \tag{71}$$

Taking the trace of (70), and defining the radial component of the thermal velocity

$$\overline{w}_i(\tau) := \mathbf{w}_i(\tau) \cdot \frac{\bar{\mathbf{r}}_i(\tau)}{\|\bar{\mathbf{r}}_i(\tau)\|}, \tag{72}$$

we find that

$$\int_{\mathscr{C}\times\mathscr{I}} \sum_i m_i \|\bar{\mathbf{r}}_i(\tau)\| \, \overline{w}_i(\tau)\delta_{\mathbf{q}_i(\tau)} \times d\tau \underset{\mathrm{m}}{\cong} 0. \tag{73}$$

This holds, in particular, for any spherical space cell, i.e., for any ball $\mathscr{B}_r$ such that $r \approx L_{\mathrm{m}}$. Hence, considering two concentric balls $\mathscr{B}_r$ and $\mathscr{B}_{r+\varepsilon} \supset \mathscr{B}_r$, we find that the same result holds for the spherical shell $\mathscr{B}_{r+\varepsilon} \setminus \mathscr{B}_r$:

$$\int_{(\mathscr{B}_{r+\varepsilon}\setminus\mathscr{B}_r)\times\mathscr{I}} \sum_i m_i \|\bar{\mathbf{r}}_i(\tau)\| \, \overline{w}_i(\tau)\delta_{\mathbf{q}_i(\tau)} \times d\tau \underset{\mathrm{m}}{\cong} 0. \tag{74}$$

Since (74) holds for any small (but not necessarily mesoscopically negligible) $\varepsilon$, this implies that the *net* mesoscopic mass flux due to the thermal velocity across the boundary of any spherical space cell $\mathscr{B}_r$ is mesoscopically negligible:

$$\int_{\partial\mathscr{B}_r\times\mathscr{I}} \sum_i m_i \overline{w}_i(\tau)\delta_{\mathbf{q}_i(\tau)} \times d\tau \underset{\mathrm{m}}{\cong} 0, \tag{75}$$

Since (75) holds for any small (but not necessarily mesoscopically negligible) $r$, this proves that the thermal momentum is a *mesoscopically divergence-free vector measure*. Hence, the *net* mesoscopic mass flux due to the thermal velocity across the boundary of *any* mesoscopic space cell $\mathscr{C}$ is mesoscopically negligible:

$$\int_{\partial\mathscr{C}\times\mathscr{I}} \sum_i m_i \mathbf{w}_i(\tau) \cdot \mathbf{n}_{\partial\mathscr{C}}\big(\mathbf{q}_i(\tau)\big)\delta_{\mathbf{q}_i(\tau)} \times d\tau \underset{\mathrm{m}}{\cong} 0, \tag{76}$$

with $\mathbf{n}_{\partial\mathscr{C}}$ the unit outer normal field to $\partial\mathscr{C}$.

In turn, the approximate equality (48) implies that only the thermal velocity affects the mesoscopic mass flux through the boundary of a *co-moving* space cell (56). In fact, (48) establishes that the velocity *relative* to $\mathscr{C}_\chi$ of the general $i$-th particle lying on $\partial\mathscr{C}_\chi(\tau)$ at time $\tau$ is mesoscopically equivalent to its current thermal velocity:

$$\dot{\mathbf{q}}_i(\tau) - \mathbf{v}\big(\mathbf{q}_i(\tau)\big) \underset{\mathrm{m}}{\cong} \mathbf{w}_i(\tau). \tag{77}$$

[17] The relation $a \underset{\mathrm{m}}{\cong} b$ between *microscopic* quantities, as in (71), should be construed as meaning that taking $a = b$ has mesoscopically negligible effects on all mesoscopic integral quantities computed from either $a$ or $b$.

Now, let

$$M_{\mathscr{C}_\chi}(t) := M_{\mathscr{C}\times\mathscr{I}} \tag{78}$$

be the *t-centred mesoscopic mass content* of the co-moving cell $\mathscr{C}_\chi$ (cf. definition (12) and identity (57)). Since an invariable mass is carried by each particle, and particles are not created nor destroyed, (76) and (77) imply that $M_{\mathscr{C}_\chi}$ is *mesoscopically stationary*:

$$\int_{\mathscr{I}} \dot{M}_{\mathscr{C}_\chi}(\tau)d\tau \underset{\mathrm{m}}{\cong} 0 \quad \Rightarrow \quad M_{\mathscr{C}_\chi}(t) \underset{\mathrm{m}}{\cong} M_{\mathscr{C}_\chi}(t_0) \quad \text{at all time } t. \tag{79}$$

Then, combining the definition (30) of mesoscopic mass density and the relation between current and reference volumes of $\mathscr{C}_\chi$ (cf. identity (57))

$$V_{\mathscr{C}} \underset{\mathrm{m}}{\cong} \det\big(\mathbf{F}(\mathbf{X},t)\big) V_{\mathscr{C}_0} \tag{80}$$

yields the finite formulation of *macroscopic mass conservation*

$$\varrho(\mathbf{x},t)\det\big(\mathbf{F}(\mathbf{X},t)\big) \underset{\mathrm{m}}{\cong} \frac{M_{\mathscr{C}_\chi}(t_0)}{V_{\mathscr{C}_0}} = \varrho(\mathbf{X},t_0) =: \varrho_0(\mathbf{X}), \tag{81}$$

with $\varrho_0$ the *referential mass density*. Taking the time derivative of (81)—with due attention to (53): cf. footnote 14—and using Jacobi's formula for the derivative of the determinant, one finally obtains the differential formulation of macroscopic mass conservation in terms of the pair of spatial fields $(\mathbf{x},t) \mapsto (\varrho(\mathbf{x},t), \mathbf{v}(\mathbf{x},t))$:[18]

$$\dot{\varrho} + \operatorname{div}(\varrho\mathbf{v}) \underset{\mathrm{m}}{\cong} 0. \tag{82}$$

Because of this (approximate) equality, the fields $\varrho$ and $\mathbf{v}$, though separately defined in (31), are not altogether independent of each other: if we knew $\mathbf{v}$ over space-time and just the instantaneous space field $\varrho(\cdot,t_0) \equiv \varrho_0$ at any time $t_0$, we could reconstruct $\varrho$ (at least locally) by integrating (82) over time, without any further recourse to microscopic information. This opportunity stems crucially from the (standard) choice of defining mesoscopic averages by weighting particle velocities by the corresponding masses, as in (8). Weighting them differently would destroy property (76) and introduce a nontrivial source term on the right side of (82), whose identification would require fresh microscopic information.

A second fundamental consequence of the smoothness requirement (34a)–(34b) on the mesoscopic properties of mass distribution concerns the mesoscopic cell momentum (13). Thanks to (25), (30), (33) and (71),

$$\begin{aligned}\mathbf{m}_{\mathscr{C}\times\mathscr{I}} &\underset{\mathrm{m}}{\cong} \frac{1}{T_{\mathrm{m}}}\int_{\mathscr{C}\times\mathscr{I}} \sum_i m_i\big(\mathbf{v}(\mathbf{x},t) + \mathbf{L}(\mathbf{x},t)\,\overline{\mathbf{r}}_i(\tau)\big)\delta_{\mathbf{q}_i(\tau)} \times d\tau \\ &\underset{\mathrm{m}}{\cong} \varrho_{\mathscr{C}\times\mathscr{I}} \int_{\mathscr{E}} \mathbf{1}_{\mathscr{C}}(\mathbf{z})\big(\mathbf{v}(\mathbf{x},t) + \mathbf{L}(\mathbf{x},t)(\mathbf{z} - \overline{\mathbf{c}}_{\mathscr{C}})\big)H^3 = M_{\mathscr{C}\times\mathscr{I}}\mathbf{v}(\mathbf{x},t),\end{aligned} \tag{83}$$

Paired with $(19)_1$, (83) implies that, for all mesoscopic space-time cell $\mathscr{C}' \times \mathscr{I}'$ centred at $(\mathbf{x},t)$,

$$\varrho_{\mathscr{C}'\times\mathscr{I}'} \underset{\mathrm{m}}{\cong} \varrho_{\mathscr{C}\times\mathscr{I}} \quad \Rightarrow \quad \mathbf{v}_{\mathscr{C}'\times\mathscr{I}'} \underset{\mathrm{m}}{\cong} \mathbf{v}_{\mathscr{C}\times\mathscr{I}}, \tag{84}$$

[18] This approximate equality is customarily written with a plain equal sign, and rightly so: distinguishing between $\underset{\mathrm{m}}{\cong}$ and $=$ is pointless in equalities between macroscopic quantities.

thus proving the consistency of $(31)_2$—the consistency of $(31)_1$ being a tautological paraphrase of the smoothness assumption itself. Other two fundamental consequences of (34a)–(34b) concern the mesoscopic cell central Euler tensor (21a) and moment of momentum (21b). In analogy with (83), we have that

$$\mathbf{E}_{\mathscr{C}_\chi}(t) := \mathbf{E}_{\mathscr{C}\times\mathscr{I}} \underset{\mathsf{m}}{\cong} \varrho_{\mathscr{C}\times\mathscr{I}} \int_{\mathscr{E}} \mathbf{1}_{\mathscr{C}}(\mathbf{z})(\mathbf{z}-\bar{\mathbf{c}}_{\mathscr{C}}) \otimes (\mathbf{z}-\bar{\mathbf{c}}_{\mathscr{C}}) H^3$$
$$\underset{\mathsf{m}}{\cong} \mathbf{F}(\mathbf{X},t)\mathbf{E}_{\mathscr{C}_\chi}(t_0)\mathbf{F}(\mathbf{X},t)^\top, \tag{85}$$

where

$$\mathbf{E}_{\mathscr{C}_\chi}(t_0) := \varrho_0(\mathbf{X}) \int_{\mathscr{E}} \mathbf{1}_{\mathscr{C}_0}(\mathbf{z})(\mathbf{z}-\bar{\mathbf{c}}_{\mathscr{C}_0}) \otimes (\mathbf{z}-\bar{\mathbf{c}}_{\mathscr{C}_0}) H^3 \tag{86}$$

is the referential central Euler tensor, with $\mathbf{1}_{\mathscr{C}_0}$ the indicator function of $\mathscr{C}_0$. Note, in particular, that if the reference cell $\mathscr{C}_0$ has cubic symmetry, then $\mathbf{E}_{\mathscr{C}_0}(t_0)$ is spherical. Combining (85) with (21a) and (59) yields

$$\frac{1}{T_\mathsf{m}} \int_{\mathscr{C}\times\mathscr{I}} \sum_i m_i \mathbf{s}_i(\tau) \otimes \mathbf{s}_i(\tau) \delta_{\mathbf{q}_i(\tau)} \times d\tau \underset{\mathsf{m}}{\cong} \mathbf{E}_{\mathscr{C}_\chi}(t_0), \tag{87}$$

thus showing that (despite appearances) the quantity on the left side of (87) is mesoscopically stationary. Finally, thanks to (26), (30), (33) and (71),

$$\mathbf{M}^{\mathsf{mom}}_{\mathscr{C}\times\mathscr{I}} \underset{\mathsf{m}}{\cong} \frac{1}{T_\mathsf{m}} \int_{\mathscr{C}\times\mathscr{I}} \sum_i m_i \bar{\mathbf{r}}_i(\tau) \otimes \big(\mathbf{v}(\mathbf{x},\tau) + \mathbf{L}(\mathbf{x},\tau)\bar{\mathbf{r}}_i(\tau)\big) \delta_{\mathbf{q}_i(\tau)} \times d\tau$$
$$\underset{\mathsf{m}}{\cong} \varrho_{\mathscr{C}\times\mathscr{I}} \int_{\mathscr{E}} \mathbf{1}_{\mathscr{C}}(\mathbf{z})(\mathbf{z}-\bar{\mathbf{c}}_{\mathscr{C}}) \otimes \big(\mathbf{v}(\mathbf{x},t) + \mathbf{L}(\mathbf{x},t)(\mathbf{z}-\bar{\mathbf{c}}_{\mathscr{C}})\big) H^3$$
$$\underset{\mathsf{m}}{\cong} \mathbf{L}(\mathbf{x},t)\mathbf{E}_{\mathscr{C}\times\mathscr{I}}. \tag{88}$$

Paired with (85), (88) implies that, for all mesoscopic space-time cell $\mathscr{C}' \times \mathscr{I}'$ centred at $(\mathbf{x}, t)$,

$$\varrho_{\mathscr{C}'\times\mathscr{I}'} \underset{\mathsf{m}}{\cong} \varrho_{\mathscr{C}\times\mathscr{I}} \quad \Rightarrow \quad \mathbf{L}_{\mathscr{C}'\times\mathscr{I}'} \underset{\mathsf{m}}{\cong} \mathbf{L}_{\mathscr{C}\times\mathscr{I}}, \tag{89}$$

thus proving the consistency of $(31)_3$.

## 3 Split/Concerted Dynamics

Let us now focus on the evolution of the macroscopic configuration of the material system $\mathscr{M}$ over a time interval of macroscopic duration $T$ (typically, a large multiple of $T_\mathsf{M}$), leaving particles in the background. Due to the smoothness of the macroscopic placement $\chi(\cdot, t)$ for all time $t$ in the said interval (cf. Sect. 2.5), we may assume that a finite set of (suitably chosen) macroscopic points, labelled with their reference positions

$$\mathbf{X}^1, \mathbf{X}^2, \ldots, \mathbf{X}^M, \tag{90}$$

suffices to sample $\chi(\cdot, t)$ for all $t$. So we are left with $3M$ macroscopic degrees of freedom:

$$\chi(\mathbf{X}^1, \cdot), \chi(\mathbf{X}^2, \cdot), \dots, \chi(\mathbf{X}^M, \cdot), \tag{91}$$

assumedly evolving on the $T_{\mathrm{M}}$ time scale. With the general representative point $\mathbf{X}^J$ I associate a co-moving cell of size $L_{\mathrm{m}}$, $\mathscr{C}_\chi^J$, populated by a cloud of $m^J$ representative particles, whose thermal motion is parameterised by the thermal degrees of freedom

$$\mathbf{s}_1^J(\cdot), \mathbf{s}_2^J(\cdot), \dots, \mathbf{s}_{m^J}^J(\cdot) \tag{92}$$

(cf. (60)). Overall, we have

$$3\sum_{J=1}^{M} m^J \ll 3N (= 3\#\mathscr{M}) \tag{93}$$

microscopic degrees of freedom, evolving on the $T_\mu$ time scale. The equations of motion governing the above macroscopic (91) and microscopic (92) degrees of freedom are constructed in Sects. 3.1–3.6 and presented in Sect. 3.7.

One peculiar issue to be addressed is the lack of a strict parameterisation of thermal velocities. In fact, while in each cell streaming velocities (24a) are *explicitly* parameterised in terms of a few macroscopic parameters (amounting to $3 + 3 \times 3 = 12$ scalar degrees of freedom), thermal velocities (24b) are only *implicitly* characterised by the property of having zero mesoscopic momentum (25) and moment of momentum (26). In the a posteriori decomposition of (supposedly known) microscopic velocities operated in Sect. 2, this property is a direct by-product of the definition of the local streaming velocity field as the best spatially affine, constant-in-time approximation to particle velocities in the sense of (8). In Sect. 3.5, equalities (25), (26) will be enforced as *nonholonomic*[19] constraints on the time rates of the reference radius vectors (92), via the approximate equality (64).

In the following, I consider the general representative point $\mathbf{X}^J$ and the co-moving cell $\mathscr{C}_\chi^J$ associated with it, dropping the superscript $J$ whenever unnecessary. Before entering into details, I offer here a simplified preview of the contents of Sect. 3.7, to help the reader catch the structure and functioning of the equations governing the concerted macroscopic/microscopic dynamics.

The evolution of the 3 macroscopic degrees of freedom $\chi(\mathbf{X}, \cdot)$ is governed by equation (143) which, in the absence of external fields, simplifies to

$$\varrho_0(\mathbf{X})\ddot{\chi}(\mathbf{X}, t) = \operatorname{div} \mathbf{S}|_{(\mathbf{X},t)}, \tag{94}$$

where $\varrho_0(\mathbf{X})$ is the referential mass density defined in (81), and the right side is the divergence of the current Piola stress field $\mathbf{S}(\cdot, t)$, to be determined from microscopic information: see (96) below. The evolution of the 3 microscopic degrees of freedom $\mathbf{s}_i(\cdot)$ is governed by

[19] A scholarly overview of nonholonomic constraints *linear* in the velocities—such as (130a), (130b)—is given by Block, Marsden and Zenkov in [5]. The trickier treatment of nonholonomic constraints nonlinear in the velocities and/or involving higher time rates has long been controversial: see the essay [13] by Flannery. Also the two papers [29, 30] by Ray provide interesting information.

equation (153) which, after neglecting minor terms, simplifies to

$$\ddot{\mathbf{s}}_i(\tau) = \mathbf{F}(\mathbf{X},t)^{-1}\big(m_i^{-1}\mathbf{f}_i^{\mathsf{shr}}(\tau) - 2\dot{\mathbf{F}}(\mathbf{X},t)\dot{\mathbf{s}}_i(\tau)\big), \tag{95}$$

where $\mathbf{F}$ is the deformation gradient (51), $m_i$ is the mass of the $i$-th particle in the $J$-th cloud, and $\mathbf{f}_i^{\mathsf{shr}}(\tau)$ is the current short-range force exerted on it. The stress is obtained from

$$\mathbf{S}(\mathbf{X},t) = -\frac{1}{V_{\mathscr{C}_0}}\mathbf{M}^{\mathsf{vir}}_{\mathscr{C}_\chi}(t)\mathbf{F}(\mathbf{X},t)^{-\top}, \tag{96}$$

where $V_{\mathscr{C}_0}$ is the reference volume of $\mathscr{C}_\chi$ (cf. (80)) and

$$\begin{aligned}\mathbf{M}^{\mathsf{vir}}_{\mathscr{C}_\chi}(t) = {}& \frac{1}{T_{\mathsf{m}}}\int_{\mathscr{I}} d\tau \int_{\mathscr{C}_\chi(t)} \sum_i m_i\big(\mathbf{F}(\mathbf{X},t)\dot{\mathbf{s}}_i(\tau)\big)\otimes\big(\mathbf{F}(\mathbf{X},t)\dot{\mathbf{s}}_i(\tau)\big)\delta_{\mathbf{q}_i(\tau)} \\ & + \frac{1}{T_{\mathsf{m}}}\int_{\mathscr{I}} d\tau \int_{\mathscr{C}_\chi(t)} \sum_i \mathbf{F}(\mathbf{X},t)\mathbf{s}_i(\tau)\otimes\mathbf{f}_i^{\mathsf{shr}\mathscr{S}}(\tau)\delta_{\mathbf{q}_i(\tau)}\end{aligned} \tag{97}$$

is the mesoscopic cell virial (135).[20]

The time integration algorithm proceeds basically as follows:

(1) The macroscopic state $(\chi(\cdot,t), \dot{\chi}(\cdot,t))$ being known, and the microscopic state of the particle clouds having been suitably prepared, equations (95) are integrated over the mesoscopic time interval $\mathscr{I}$ with microscopic time steps $\approx T_\mu$. Note that $\mathbf{F}$ and $\dot{\mathbf{F}}$ are evaluated once and for all at time $t$. Particles in different clouds do not interact, so that this computation is naturally parallel.
(2) The time history of the microscopic degrees of freedom (92) is used to compute the mesoscopic cell virial (97), in order to construct the current stress field $\mathbf{S}(\cdot,t)$ via (96) and to compute its divergence.
(3) The macroscopic acceleration $\ddot{\chi}(\cdot,t)$ is computed from (94), and the macroscopic state is updated to time $t + T_{\mathsf{M}}$.

## 3.1 Two-Level Test Velocities and Lagrangian Forces

As in Appendix B.10, and independently of whether forces entering the Newtonian equation (236) are conservative or not, the Lagrangian forces dual to the Lagrangian degrees of freedom $\chi(\mathbf{X},\cdot)$, $\mathbf{F}(\mathbf{X},\cdot)$, and $\mathbf{s}_i(\cdot)$ introduced in (60) are obtained straightforwardly by computing the work integral

$$\int_{\tau_-}^{\tau_+} d\tau \int_{\mathscr{C}_\chi(\tau)} \sum_i \mathbf{f}_i^{\mathsf{tot}}(\tau)\cdot\mathbf{v}_i(\tau)\delta_{\mathbf{q}_i(\tau)}, \tag{98}$$

on *test* velocities obtained by combining linearly the variations

$$\widetilde{\chi}(\mathbf{X},\tau), \qquad \widetilde{\mathbf{F}}(\mathbf{X},\tau) := \nabla\widetilde{\chi}|_{(\mathbf{X},\tau)}; \quad \widetilde{\mathbf{s}}_i(\tau) \tag{99}$$

[20] $\mathbf{f}_i^{\mathsf{shr}\mathscr{S}}(\tau)$ is the short-range force exerted on the $i$-particle of the $J$-th cloud by the other particles currently inside $\mathscr{C}_\chi^J$ (cf. Sects. 3.4 and 4.1).

in the same way as the time derivatives of the corresponding degrees of freedom are combined in (65a), i.e.,

$$\mathbf{v}_i(\tau) = \widetilde{\chi}(\mathbf{X}, \tau) + \widetilde{\mathbf{F}}(\mathbf{X}, \tau)\mathbf{s}_i(\tau) + \mathbf{F}(\mathbf{X}, \tau)\widetilde{\mathbf{s}}_i(\tau). \tag{100}$$

The pair $(\widetilde{\chi}(\mathbf{X}, \cdot), \widetilde{\mathbf{F}}(\mathbf{X}, \cdot))$ parameterises the referential streaming test field

$$\mathscr{C}_0 \times [\tau_-, \tau_+] \ni (\mathbf{X} + \mathbf{u}, \tau) \mapsto \widetilde{\chi}(\mathbf{X}, \tau) + \widetilde{\mathbf{F}}(\mathbf{X}, \tau)\mathbf{u}. \tag{101}$$

In analogy with (63),

$$\widetilde{\chi}(\mathbf{X}, \tau) \underset{\mathsf{m}}{\cong} \widetilde{\chi}(\mathbf{X}, t), \qquad \widetilde{\mathbf{F}}(\mathbf{X}, \tau) \underset{\mathsf{m}}{\cong} \widetilde{\mathbf{F}}(\mathbf{X}, t), \quad \text{whenever } |\tau - t| \leq T_{\mathsf{m}}; \tag{102}$$

under the same condition, $\mathscr{C}_\chi(\tau) \underset{\mathsf{m}}{\cong} \mathscr{C}_\chi(t)$. The total force $\mathbf{f}_i^{\mathsf{tot}}$ is defined as in Appendix B.7, with the proviso that all interaction forces are short-ranged, as postulated in (2):

$$\mathbf{f}_i^{\mathsf{tot}}(\tau) = -m_i \ddot{\mathbf{q}}_i(\tau) + \mathbf{f}_i^{\mathsf{ext}}(\tau) + \mathbf{f}_i^{\mathsf{shr}}(\tau) + \mathbf{f}_i^{\mathsf{rct}}(\tau). \tag{103}$$

For integral (98) to make sense, the time interval $[\tau_-, \tau_+]$ should be at least of mesoscopic size:

$$\tau_+ - \tau_- \geq T_{\mathsf{m}} \tag{104}$$

(in practice, it is typically a large multiple of $T_{\mathsf{M}}$). Substituting (100) into (98) yields

$$\begin{aligned}
&\int_{\tau_-}^{\tau_+} d\tau \int_{\mathscr{C}_\chi(\tau)} \sum_i \mathbf{f}_i^{\mathsf{tot}}(\tau) \cdot \mathbf{v}_i(\tau) \delta_{\mathbf{q}_i(\tau)} \\
&\underset{\mathsf{m}}{\cong} \int_{\tau_-}^{\tau_+} d\tau \Big(\mathbf{f}^{\mathsf{tot}}_{\mathscr{C}_\chi}(\tau) \cdot \widetilde{\chi}(\mathbf{X}, \tau) + \big(\mathbf{M}^{\mathsf{tot}}_{\mathscr{C}_\chi}(\tau) \mathbf{F}(\mathbf{X}, \tau)^{-\top}\big) \cdot \widetilde{\mathbf{F}}(\mathbf{X}, \tau)\Big) \\
&\quad + \int_{\tau_-}^{\tau_+} d\tau \int_{\mathscr{C}_\chi(\tau)} \sum_i \big(\mathbf{F}(\mathbf{X}, \tau)^{\top} \mathbf{f}_i^{\mathsf{tot}}(\tau)\big) \cdot \widetilde{\mathbf{s}}_i(\tau) \delta_{\mathbf{q}_i(\tau)},
\end{aligned} \tag{105}$$

where

$$\mathbf{f}^{\mathsf{tot}}_{\mathscr{C}_\chi}(\tau) := \frac{1}{T_{\mathsf{m}}} \int_{\mathscr{I}_\tau} d\sigma \int_{\mathscr{C}_\chi(\tau)} \sum_i \mathbf{f}_i^{\mathsf{tot}}(\sigma) \delta_{\mathbf{q}_i(\sigma)}, \tag{106a}$$

$$\mathbf{M}^{\mathsf{tot}}_{\mathscr{C}_\chi}(\tau) := \frac{1}{T_{\mathsf{m}}} \int_{\mathscr{I}_\tau} d\sigma \int_{\mathscr{C}_\chi(\tau)} \sum_i \mathbf{r}_i(\sigma) \otimes \mathbf{f}_i^{\mathsf{tot}}(\sigma) \delta_{\mathbf{q}_i(\sigma)} \tag{106b}$$

are, respectively, the total mesoscopic force and the total mesoscopic moment on (the particles in) the cell. The proof of (105) goes as follows. Consider an integral such as

$$\int_{\tau_-}^{\tau_+} d\tau \int_{\mathscr{C}_\chi(\tau)} \sum_i \mathtt{s}(\tau) \diamond \mathtt{f}_i(\tau) \delta_{\mathbf{q}_i(\tau)}, \tag{107}$$

whose integrand is a sum over $i$ of bilinear products of two functions of time, (at least) one of which, namely $\mathtt{s}$, is *slowly-varying* on the $T_{\mathsf{m}}$ time scale and independent of $i$. Hypothesis (104) allows the time interval $[\tau_-, \tau_+]$ to be subdivided into internally disjoint intervals

$\{\mathscr{I}_k\}$, each of which is of mesoscopic size: $T_k := |\mathscr{I}_k| \approx T_{\rm m}$. If $\tau_k$ is the central time of $\mathscr{I}_k$, then

$$\begin{aligned}
&\int_{\tau_-}^{\tau_+} d\tau \int_{\mathscr{C}_\chi(\tau)} \sum_i \mathtt{s}(\tau) \diamond \mathtt{f}_i(\tau)\delta_{\mathbf{q}_i(\tau)} \\
&\underset{\rm m}{\cong} \sum_k \left( \mathtt{s}(\tau_k) \diamond \int_{\mathscr{I}_k} d\sigma \int_{\mathscr{C}_\chi(\tau_k)} \sum_i \mathtt{f}_i(\sigma)\delta_{\mathbf{q}_i(\sigma)} \right) \\
&\underset{\rm m}{\cong} \int_{\tau_-}^{\tau_+} d\tau \left( \mathtt{s}(\tau) \diamond \frac{1}{T_{\rm m}} \int_{\mathscr{I}_\tau} d\sigma \int_{\mathscr{C}_\chi(\tau)} \sum_i \mathtt{f}_i(\sigma)\delta_{\mathbf{q}_i(\sigma)} \right).
\end{aligned} \tag{108}$$

Statements analogous to (105) in Sects. 3.2–3.6 and related appendices are proved along the same lines. Then, enforcing the condition that the work integral (98) be zero for all variations $(\widetilde{\chi}, \widetilde{\mathbf{F}}, \widetilde{\mathbf{s}}_i)$ yields the Lagrangian equations

$$\mathbf{f}^{\rm tot}_{\mathscr{C}_\chi}(\tau) \underset{\rm m}{\cong} \mathbf{0}, \tag{109a}$$

$$\mathbf{M}^{\rm tot}_{\mathscr{C}_\chi}(\tau)\mathbf{F}(\mathbf{X},\tau)^{-\top} \underset{\rm m}{\cong} \mathbf{0}, \tag{109b}$$

$$\mathbf{F}(\mathbf{X},\tau)^{\top}\mathbf{f}^{\rm tot}_i(\tau) \underset{\rm m}{\cong} \mathbf{0}. \tag{109c}$$

*Prima facie*, equation (109c) appears to be a mesoscopically accurate replica of (218);[21] equations (109a) and (109b) are scarcely enlightening. To make them more informative, the contributions of each of the four components of the total force (103) to $\mathbf{f}^{\rm tot}_{\mathscr{C}_\chi}$ and $\mathbf{M}^{\rm tot}_{\mathscr{C}_\chi}$ should be brought to the fore and separately analysed, as is done in the following. While the treatment of external forces (Sect. 3.3) is straightforward and the parameterisation of reactive forces in terms of Lagrangian multipliers (Sect. 3.5) is standard, handling of inertial forces (Sect. 3.2 and Appendix C.1) and (short-ranged) interaction forces requires special attention: interaction forces, in particular, need to be carefully distinguished between those exerted from *within* and from *without* the cell (Sect. 3.4 and Appendix C.2).

### 3.2 (Nominal) Force of Inertia

The time derivative of the referential representation of particle velocities (65a) reads

$$\ddot{\mathbf{q}}_i(\tau) \underset{\rm m}{\cong} \ddot{\chi}(\mathbf{X},\tau) + \ddot{\mathbf{F}}(\mathbf{X},\tau)\mathbf{s}_i(\tau) + 2\dot{\mathbf{F}}(\mathbf{X},\tau)\dot{\mathbf{s}}_i(\tau) + \mathbf{F}(\mathbf{X},\tau)\ddot{\mathbf{s}}_i(\tau), \tag{110}$$

whence the work done by the nominal[22] force of inertia on the test velocities (100):

$$\begin{aligned}
&-\int_{\tau_-}^{\tau_+} d\tau \int_{\mathscr{C}_\chi(\tau)} \sum_i m_i \ddot{\mathbf{q}}_i(\tau) \cdot \mathbf{v}_i(\tau)\delta_{\mathbf{q}_i(\tau)} \\
&\underset{\rm m}{\cong} -\int_{\tau_-}^{\tau_+} M_{\mathscr{C}_\chi}(\tau)\ddot{\chi}(\mathbf{X},\tau) \cdot \widetilde{\chi}(\mathbf{X},\tau) d\tau
\end{aligned}$$

[21] But see the discussion in Sect. 3.4.

[22] In the terminology introduced in Appendix B.5. Fictitious forces are incorporated in the external force.

$$- \int_{\tau_-}^{\tau_+} d\tau \big(\ddot{\mathbf{F}}(\mathbf{X},\tau)\mathbf{E}_{\mathscr{C}_\chi}(t_0) - \mathbf{M}^{\mathsf{kin}}_{\mathscr{C}_\chi}(\tau)\,\mathbf{F}(\mathbf{X},\tau)^{-\top}\big)\cdot\widetilde{\mathbf{F}}(\mathbf{X},\tau)$$

$$- \int_{\tau_-}^{\tau_+} d\tau \int_{\mathscr{C}_\chi(\tau)} \sum_i m_i \big(\mathbf{F}(\mathbf{X},\tau)^{\top}\ddot{\mathbf{q}}_i(\tau)\big)\cdot\widetilde{\mathbf{s}}_i(\tau)\delta_{\mathbf{q}_i(\tau)}, \tag{111}$$

where $M_{\mathscr{C}_\chi}(\tau)$ is the (mesoscopically $\tau$-independent) $\tau$-centred mesoscopic mass content (cf. (78), (79)), $\mathbf{E}_{\mathscr{C}_\chi}(t_0)$ is the referential central Euler tensor (cf. (86), (87)), the mesoscopic *cell kinetic moment* $\mathbf{M}^{\mathsf{kin}}_{\mathscr{C}_\chi}$ is defined as

$$\mathbf{M}^{\mathsf{kin}}_{\mathscr{C}_\chi}(\tau) := \frac{1}{T_{\mathsf{m}}}\int_{\mathscr{I}_\tau} d\sigma \int_{\mathscr{C}_\chi(\tau)} \sum_i m_i \mathbf{w}_i(\sigma)\otimes\mathbf{w}_i(\sigma)\delta_{\mathbf{q}_i(\sigma)}$$

$$\overset{\simeq}{\scriptstyle \mathsf{m}} \frac{1}{T_{\mathsf{m}}}\int_{\mathscr{I}_\tau} d\sigma \int_{\mathscr{C}_\chi(\tau)} \sum_i m_i \big(\mathbf{F}(\mathbf{X},\sigma)\dot{\mathbf{s}}_i(\sigma)\big)\otimes\big(\mathbf{F}(\mathbf{X},\sigma)\dot{\mathbf{s}}_i(\sigma)\big)\delta_{\mathbf{q}_i(\sigma)}, \tag{112}$$

and $\ddot{\mathbf{q}}_i(\tau)$ is shorthand for the right side of (110). Proofs are given in Appendix C.1. For the sake of comparison, the customary Lagrangian treatment in terms of kinetic energy via the work-energy theorem (215) is presented in Appendix C.3, where it is shown to be simpler but *non*-equivalent to the direct treatment given here and expounded in Appendix C.1.

## 3.3 External Force

Changing labels to (105) and (106a)–(106b) in Sect. 3.1 suffices to obtain the work done by the external force:

$$\int_{\tau_-}^{\tau_+} d\tau \int_{\mathscr{C}_\chi(\tau)} \sum_i \mathbf{f}^{\mathsf{ext}}_i(\tau)\cdot\mathbf{v}_i(\tau)\delta_{\mathbf{q}_i(\tau)}$$

$$\overset{\simeq}{\scriptstyle \mathsf{m}} \int_{\tau_-}^{\tau_+} d\tau \big(\mathbf{f}^{\mathsf{ext}}_{\mathscr{C}_\chi}(\tau)\cdot\widetilde{\chi}(\mathbf{X},\tau) + \big(\mathbf{M}^{\mathsf{ext}}_{\mathscr{C}_\chi}(\tau)\,\mathbf{F}(\mathbf{X},\tau)^{-\top}\big)\cdot\widetilde{\mathbf{F}}(\mathbf{X},\tau)\big)$$

$$+ \int_{\tau_-}^{\tau_+} d\tau \int_{\mathscr{C}_\chi(\tau)} \sum_i \big(\mathbf{F}(\mathbf{X},\tau)^{\top}\mathbf{f}^{\mathsf{ext}}_i(\tau)\big)\cdot\widetilde{\mathbf{s}}_i(\tau)\delta_{\mathbf{q}_i(\tau)}, \tag{113}$$

where

$$\mathbf{f}^{\mathsf{ext}}_{\mathscr{C}_\chi}(\tau) := \frac{1}{T_{\mathsf{m}}}\int_{\mathscr{I}_\tau} d\sigma \int_{\mathscr{C}_\chi(\tau)} \sum_i \mathbf{f}^{\mathsf{ext}}_i(\sigma)\delta_{\mathbf{q}_i(\sigma)}, \tag{114a}$$

$$\mathbf{M}^{\mathsf{ext}}_{\mathscr{C}_\chi}(\tau) := \frac{1}{T_{\mathsf{m}}}\int_{\mathscr{I}_\tau} d\sigma \int_{\mathscr{C}_\chi(\tau)} \sum_i \mathbf{r}_i(\sigma)\otimes\mathbf{f}^{\mathsf{ext}}_i(\sigma)\delta_{\mathbf{q}_i(\sigma)} \tag{114b}$$

are, respectively, the mesoscopic external force and the mesoscopic external moment on (the particles in) the cell.

## 3.4 Short-Range Interaction Force

As far as short-ranged interaction forces are concerned, it is crucial to split the force $\mathbf{f}^{\mathsf{shr}}_i(\tau)$ applied to the general $i$-th particle in $\mathscr{C}_\chi(\tau)$ at time $\tau$ into the sum of the force $\mathbf{f}^{\mathsf{shr}\mathscr{S}}_i(\tau)$

exerted by the other particles in $\mathscr{C}_\chi(\tau)$ and the force $\mathbf{f}_i^{\mathsf{shr}\mathscr{S}\mathsf{C}}(\tau)$ exerted by the particles outside of $\mathscr{C}_\chi(\tau)$:

$$\mathbf{f}_i^{\mathsf{shr}}(\tau) = \mathbf{f}_i^{\mathsf{shr}\mathscr{S}}(\tau) + \mathbf{f}_i^{\mathsf{shr}\mathscr{S}\mathsf{C}}(\tau). \tag{115}$$

Borrowing the notations introduced in Appendix B.10, I denote by $\mathscr{S}_{\mathscr{C}_\chi(\tau)}$ the material subsystem comprised of all particles in $\mathscr{C}_\chi(\tau)$ at time $\tau$, and by

$$\mathscr{S}^{\mathsf{C}}_{\mathscr{C}_\chi(\tau)} := \mathscr{M} \setminus \mathscr{S}_{\mathscr{C}_\chi(\tau)} \tag{116}$$

its complement in the macroscopic material system $\mathscr{M}$. To avoid notational clutter, in superscripts I write $\mathscr{S}$ for $\mathscr{S}_{\mathscr{C}_\chi(\tau)}$. The work done by the *in-cell* interaction force reads

$$\begin{aligned}
&\int_{\tau_-}^{\tau_+} d\tau \int_{\mathscr{C}_\chi(\tau)} \sum_i \mathbf{f}_i^{\mathsf{shr}\mathscr{S}}(\tau)\cdot \mathbf{v}_i(\tau)\delta_{\mathbf{q}_i(\tau)} \\
&\underset{\mathsf{m}}{\cong} \int_{\tau_-}^{\tau_+} d\tau \big(\mathbf{f}_{\mathscr{C}_\chi}^{\mathsf{shr}\mathscr{S}}(\tau)\cdot \widetilde{\chi}(\mathbf{X},\tau) + \big(\mathbf{M}_{\mathscr{C}_\chi}^{\mathsf{shr}\mathscr{S}}(\tau)\,\mathbf{F}(\mathbf{X},\tau)^{-\top}\big)\cdot \widetilde{\mathbf{F}}(\mathbf{X},\tau)\big) \\
&\quad + \int_{\tau_-}^{\tau_+} d\tau \int_{\mathscr{C}_\chi(\tau)} \sum_i \big(\mathbf{F}(\mathbf{X},\tau)^{\top}\mathbf{f}_i^{\mathsf{shr}\mathscr{S}}(\tau)\big)\cdot \widetilde{\mathbf{s}}_i(\tau)\delta_{\mathbf{q}_i(\tau)},
\end{aligned} \tag{117}$$

where

$$\mathbf{f}_{\mathscr{C}_\chi}^{\mathsf{shr}\mathscr{S}}(\tau) := \frac{1}{T_{\mathsf{m}}}\int_{\mathscr{I}_\tau} d\sigma \int_{\mathscr{C}_\chi(\tau)} \sum_i \mathbf{f}_i^{\mathsf{shr}\mathscr{S}}(\sigma)\delta_{\mathbf{q}_i(\sigma)}, \tag{118a}$$

$$\mathbf{M}_{\mathscr{C}_\chi}^{\mathsf{shr}\mathscr{S}}(\tau) := \frac{1}{T_{\mathsf{m}}}\int_{\mathscr{I}_\tau} d\sigma \int_{\mathscr{C}_\chi(\tau)} \sum_i \mathbf{r}_i(\sigma)\otimes \mathbf{f}_i^{\mathsf{shr}\mathscr{S}}(\sigma)\delta_{\mathbf{q}_i(\sigma)}. \tag{118b}$$

Due to the invariance requirement (222), the mesoscopic mutual force $\mathbf{f}_{\mathscr{C}_\chi}^{\mathsf{shr}\mathscr{S}}$ is *null* and the mesoscopic mutual moment $\mathbf{M}_{\mathscr{C}_\chi}^{\mathsf{shr}\mathscr{S}}$ is *symmetric* (cf. Appendix B.8):

$$\mathbf{f}_{\mathscr{C}_\chi}^{\mathsf{shr}\mathscr{S}}(\tau) = \mathbf{0}, \qquad \mathbf{M}_{\mathscr{C}_\chi}^{\mathsf{shr}\mathscr{S}}(\tau) \in \mathrm{Sym}. \tag{119}$$

At any time $\tau$, the general $i$-th particle in $\mathscr{C}_\chi(\tau)$ interacts effectively with all the particles in the microscopically sized ball $\mathscr{B}_R(\mathbf{q}_i(\tau))$ (cf. (1)). Particles in $\mathscr{B}_R(\mathbf{q}_i(\tau)) \setminus \mathscr{C}_\chi(\tau)$ contribute to $\mathbf{f}_i^{\mathsf{shr}\mathscr{S}\mathsf{C}}(\tau)$, but their contribution has to be *estimated*, since we ignore the exact position and velocity of particles outside of $\mathscr{C}_\chi(\tau)$. The assumed robustness against microscopic variations of the sampling procedure expounded in Sect. 2 leads us to expect that the mesoscopically significant properties of the particle distribution within the microscopically thin shell[23]

$$\big(\textstyle\bigcup_{\mathbf{q}_i(\tau)\in\mathscr{C}_\chi(\tau)} \mathscr{B}_R\big(\mathbf{q}_i(\tau)\big)\big) \setminus \mathscr{C}_\chi(\tau) \underset{\mathsf{m}}{\cong} \mathscr{F}_R\big(\partial\mathscr{C}_\chi(\tau)\big) \setminus \mathscr{C}_\chi(\tau) \tag{120}$$

can be legitimately approximated by extrapolating the particle distribution prevailing within $\mathscr{C}_\chi(\tau)$. The standard technique is to imagine the parallelepipedal cell $\mathscr{C}_\chi(\tau)$ surrounded by periodic copies of the molecular pattern inside it. Since cell edges have mesoscopic lengths $\approx L_{\mathsf{m}} \gg L_\mu = R$, the general $j$-th particle such that $\mathbf{q}_j(\tau) \in \mathscr{C}_\chi(\tau) \setminus \mathscr{B}_R(\mathbf{q}_i(\tau))$ has at

[23] $\mathscr{F}_R(\partial\mathscr{C}_\chi(\tau)) := \bigcup_{\mathbf{z}\in\partial\mathscr{C}_\chi(\tau)} \mathscr{B}_R(\mathbf{z})$ is the *R-fattening* of the boundary of the cell $\mathscr{C}_\chi(\tau)$.

most one replica such that $\mathbf{q}_j^{\mathsf{rep}}(\tau) \in \mathscr{B}_R(\mathbf{q}_i(\tau))$,[24] thus contributing to the constructible approximation $\mathbf{f}_i^{\mathsf{rep}\mathscr{S}}(\tau)$ of the unattainable quantity $\mathbf{f}_i^{\mathsf{shr}\mathscr{S}^\complement}(\tau)$. It is hoped that

$$\int_{\tau_-}^{\tau_+} d\tau \int_{\mathscr{C}_\chi(\tau)} \sum_i \big(\mathbf{F}(\mathbf{X},\tau)^\top \mathbf{f}_i^{\mathsf{rep}\mathscr{S}}(\tau)\big) \cdot \widetilde{\mathbf{s}}_i(\tau)\delta_{\mathbf{q}_i(\tau)} \underset{\mathsf{m}}{\cong} \int_{\tau_-}^{\tau_+} d\tau \int_{\mathscr{C}_\chi(\tau)} \sum_i \big(\mathbf{F}(\mathbf{X},\tau)^\top \mathbf{f}_i^{\mathsf{shr}\mathscr{S}^\complement}(\tau)\big) \cdot \widetilde{\mathbf{s}}_i(\tau)\delta_{\mathbf{q}_i(\tau)} \tag{121}$$

for all test velocities $\widetilde{\mathbf{s}}_i$ compatible with the thermal constraints (130a)–(130b) (cf. Sect. 3.5). Periodicity restricts the microscopically instantaneous net force exterted by the replicas on (the particles in) the cell to be null:

$$\check{\mathbf{f}}_{\mathscr{C}_\chi}^{\mathsf{rep}\mathscr{S}}(\tau) := \int_{\mathscr{C}_\chi(\tau)} \sum_i \mathbf{f}_i^{\mathsf{rep}\mathscr{S}}(\tau)\delta_{\mathbf{q}_i(\tau)} = \mathbf{0} \quad \Longrightarrow \quad \int_{\mathscr{I}_\tau} \check{\mathbf{f}}_{\mathscr{C}_\chi}^{\mathsf{rep}\mathscr{S}}(\sigma)d\sigma = \mathbf{0}, \tag{122a}$$

while the microscopically instantaneous net moment

$$\check{\mathbf{M}}_{\mathscr{C}_\chi}^{\mathsf{rep}\mathscr{S}}(\tau) := \int_{\mathscr{C}_\chi(\tau)} \sum_i \mathbf{r}_i(\tau) \otimes \mathbf{f}_i^{\mathsf{rep}\mathscr{S}}(\tau)\delta_{\mathbf{q}_i(\tau)} \tag{122b}$$

exterted by them remains unrestricted. Bona fide dynamical quantities $\mathbf{f}_{\mathscr{C}_\chi}^{\mathsf{shr}\mathscr{S}^\complement}$ and $\mathbf{M}_{\mathscr{C}_\chi}^{\mathsf{shr}\mathscr{S}^\complement}$, respectively dual to the streaming degrees of freedom $\chi(\mathbf{X},\cdot)$ and $\mathbf{F}(\mathbf{X},\cdot)$, have to be estimated via completely different means, altogether independent of the particle-wise approximation (121) and the consequent cell quantities (122a)–(122b). Due to the short-range assumption (2), the interaction between the multitudinous particles in $\mathscr{S}^\complement_{\mathscr{C}_\chi(\tau)}$ (cf. (116)) and the relatively few particles in $\mathscr{S}_{\mathscr{C}_\chi(\tau)}$ is mediated through the short-range forces exerted by the even fewer particles currently occupying the microscopically thin shell (120). Relying on the foundational concepts of continuum dynamics,[25] I postulate the existence of two slowly varying *spatial* fields: the vector-valued *traction field* $\mathbf{t}(\cdot,\tau)$ on the cell boundary $\partial\mathscr{C}_\chi(\tau)$ and the tensor-valued *stress field* $\mathbf{T}(\cdot,\tau)$ on a neighbourhood of $\mathscr{C}_\chi(\tau)$, such that

$$\int_{\tau_-}^{\tau_+} d\tau \int_{\partial\mathscr{C}_\chi(\tau)} \mathbf{t}(\cdot,\tau)\cdot \underset{\sim}{\mathbf{y}}(\cdot,\tau) H^2 \llcorner \partial\mathscr{C}_\chi(\tau) \underset{\mathsf{m}}{\cong} \int_{\tau_-}^{\tau_+} d\tau \int_{\mathscr{C}_\chi(\tau)} \sum_i \mathbf{f}_i^{\mathsf{shr}\mathscr{S}^\complement}(\tau)\cdot\big(\widetilde{\chi}(\mathbf{X},\tau) + \widetilde{\mathbf{F}}(\mathbf{X},\tau)\mathbf{s}_i(\tau)\big)\delta_{\mathbf{q}_i(\tau)}$$

[24] This circumstance is customarily referred to as the "minimum-image convention."

[25] When introducing the concept of stress still central to present-day continuum mechanics, A.L. Cauchy [9, p. 42] wrote: « Les géomètres qui ont recherché les équations d'équilibre ou de mouvement des lames ou des surfaces élastiques ou non élastiques, ont distingué deux espèces de forces produites les unes par la dilatation ou la contraction, les autres par la flexion de ces mêmes surfaces. De plus, ils ont généralement supposé, dans leurs calculs, que les forces de la première espèce, nommées tensions, restent perpendiculaires aux lignes contre lesquelles elles s'exercent. Il m'a semblé ques ces deux espèces de forces pouvaient être réduites à une seule, qui doit constamment s'appeler *tension* ou *pression*, qui agit sur chaque élément d'une section faite à volonté, non seulement dans une surface flexible, mais encore dans un solide élastique ou non élastique, et qui est de la même nature que la pression hydrostatique exercée par un fluide en repos contre la surface extérieure d'un corps. Seulement la nouvelle pression ne demeure pas toujours perpendiculaire aux faces qui lui sont soumises, ni la même dans tous les sens en un point donné. En développant cette idée, je suis parvenu à reconnaitre que la pression ou tension exercée contre un plan quelconque en un point donné d'un corps solide se déduit très-aisément, tant en grandeur qu'en direction, des pressions ou tensions exercées contre trois plans perpendiculaires menés par le même point. »

$$\underset{\overline{\text{m}}}{\cong} \int_{\tau_-}^{\tau_+} d\tau \big( \mathbf{f}_{\mathscr{C}_\chi}^{\mathsf{shr},\mathscr{S}\mathsf{C}}(\tau) \cdot \widetilde{\chi}(\mathbf{X},\tau) + \big( \mathbf{M}_{\mathscr{C}_\chi}^{\mathsf{shr},\mathscr{S}\mathsf{C}}(\tau)\, \mathbf{F}(\mathbf{X},\tau)^{-\top} \big) \cdot \widetilde{\mathbf{F}}(\mathbf{X},\tau) \big), \tag{123}$$

where $H^2 \llcorner \partial \mathscr{C}_\chi(\tau)$ is the area measure restricted to the current cell boundary, and

$$\underset{\sim}{\mathbf{y}}(\cdot,\tau) := \widetilde{\chi}\big(\chi^{-1}(\cdot,\tau),\tau\big) \tag{124}$$

is the *spatial* streaming test field (cf. (54) for notation),

$$\mathbf{f}_{\mathscr{C}_\chi}^{\mathsf{shr},\mathscr{S}\mathsf{C}}(\tau) := \frac{1}{T_{\mathsf{m}}} \int_{\mathscr{I}_\tau} d\sigma \int_{\mathscr{C}_\chi(\tau)} \sum_i \mathbf{f}_i^{\mathsf{shr},\mathscr{S}\mathsf{C}}(\sigma) \delta_{\mathbf{q}_i(\sigma)}, \tag{125a}$$

$$\mathbf{M}_{\mathscr{C}_\chi}^{\mathsf{shr},\mathscr{S}\mathsf{C}}(\tau) := \frac{1}{T_{\mathsf{m}}} \int_{\mathscr{I}_\tau} d\sigma \int_{\mathscr{C}_\chi(\tau)} \sum_i \mathbf{r}_i(\sigma) \otimes \mathbf{f}_i^{\mathsf{shr},\mathscr{S}\mathsf{C}}(\sigma) \delta_{\mathbf{q}_i(\sigma)}, \tag{125b}$$

and, moreover,

$$\mathbf{M}_{\mathscr{C}_\chi}^{\mathsf{shr},\mathscr{S}\mathsf{C}}(\tau) \underset{\overline{\text{m}}}{\cong} \int_{\mathscr{C}_\chi(\tau)} \mathbf{T}(\cdot,\tau) H^3 \underset{\overline{\text{m}}}{\cong} V_{\mathscr{C}_\chi}(\tau) \mathbf{T}(\boldsymbol{\xi},\tau), \tag{126}$$

where

$$V_{\mathscr{C}_\chi}(\tau) := H^3\big(\mathscr{C}_\chi(\tau)\big). \tag{127}$$

Then, Cauchy's fundamental theorem establishes that[26]

$$\mathbf{t}(\cdot,\tau) = \mathbf{T}(\cdot,\tau)\mathbf{n}_{\partial\mathscr{C}_\chi(\tau)}, \tag{128}$$

with $\mathbf{n}_{\partial\mathscr{C}_\chi(\tau)}$ the unit outer normal field to $\partial\mathscr{C}_\chi(\tau)$. It follows that[27]

$$\mathbf{f}_{\mathscr{C}_\chi}^{\mathsf{shr},\mathscr{S}\mathsf{C}}(\tau) \underset{\overline{\text{m}}}{\cong} \int_{\partial\mathscr{C}_\chi(\tau)} \mathbf{t}(\cdot,\tau) H^2 \llcorner \partial\mathscr{C}_\chi(\tau) \underset{\overline{\text{m}}}{\cong} V_{\mathscr{C}_\chi}(\tau)\, \operatorname{div} \mathbf{T}|_{(\boldsymbol{\xi},\tau)}. \tag{129}$$

Proofs are given in Appendix C.2. The approximate equality (129) expresses the *macroscopic dynamical consistency* between the net short-range force (125a) and the stress field $\mathbf{T}(\cdot,\tau)$. This consistency condition—dual to the *macroscopic kinematic compatibility* (37)—is peculiar to short-range interactions.

### 3.5 Reactive Force

Via the approximate equality (64), equalities (25) and (26) characterising thermal velocities translate into the nonholonomic constraints

$$\int_{\mathscr{I}_\tau} d\sigma \int_{\mathscr{C}_\chi(\tau)} \sum_i m_i \mathbf{F}(\mathbf{X},\sigma)\dot{\mathbf{s}}_i(\sigma)\delta_{\mathbf{q}_i(\sigma)} \underset{\overline{\text{m}}}{\cong} \mathbf{0}, \tag{130a}$$

$$\int_{\mathscr{I}_\tau} d\sigma \int_{\mathscr{C}_\chi(\tau)} \sum_i m_i \big(\mathbf{F}(\mathbf{X},\sigma)\mathbf{s}_i(\sigma)\big) \otimes \big(\mathbf{F}(\mathbf{X},\sigma)\dot{\mathbf{s}}_i(\sigma)\big)\delta_{\mathbf{q}_i(\sigma)} \underset{\overline{\text{m}}}{\cong} \mathbf{0}, \tag{130b}$$

[26] The assumption, implied by (123) and (126), that the stress expends power only on the *first* gradient of the streaming test field is *equivalent* to the original hypothesis postulated by Cauchy that the value at a point of the traction exerted across a surface depends solely on the oriented unit normal to the surface at that point.

[27] Compare and contrast with (122a).

The reaction to these constraints is required to do null work on all test velocities compatible with them, i.e., to satisfy

$$\int_{\tau_-}^{\tau_+} d\tau \int_{\mathscr{C}_\chi(\tau)} \sum_i \big(\mathbf{F}(\mathbf{X},\tau)^\top \mathbf{f}_i^{\mathsf{rct}}(\tau)\big)\cdot \widetilde{\mathbf{s}}_i(\tau)\delta_{\mathbf{q}_i(\tau)} \underset{\overline{m}}{\cong} 0 \tag{131}$$

for all $\widetilde{\mathbf{s}}_i$ such that, for each $\tau \in [\tau_-, \tau_+]$,

$$\int_{\mathscr{I}_\tau} d\sigma \int_{\mathscr{C}_\chi(\tau)} \sum_i m_i \mathbf{F}(\mathbf{X},\sigma)\widetilde{\mathbf{s}}_i(\sigma)\delta_{\mathbf{q}_i(\sigma)} \underset{\overline{m}}{\cong} \mathbf{0} \quad \text{and} \tag{132a}$$

$$\int_{\mathscr{I}_\tau} d\sigma \int_{\mathscr{C}_\chi(\tau)} \sum_i m_i \big(\mathbf{F}(\mathbf{X},\sigma)\mathbf{s}_i(\sigma)\big)\otimes\big(\mathbf{F}(\mathbf{X},\sigma)\widetilde{\mathbf{s}}_i(\sigma)\big)\delta_{\mathbf{q}_i(\sigma)} \underset{\overline{m}}{\cong} \mathbf{0}. \tag{132b}$$

As a consequence, the reactive force $\mathbf{f}_i^{\mathsf{rct}}$ is parameterised by two slowly varying functions of time, a vector-valued multiplier $\lambda_{\mathscr{C}_\chi}$ (having physical dimension acceleration) and a tensor-valued multiplier $\Lambda_{\mathscr{C}_\chi}$ (having physical dimension angular acceleration), such that

$$\mathbf{f}_i^{\mathsf{rct}}(\tau) = m_i\big(\lambda_{\mathscr{C}_\chi}(\tau) + \Lambda_{\mathscr{C}_\chi}(\tau)\mathbf{F}(\mathbf{X},\tau)\mathbf{s}_i(\tau)\big). \tag{133}$$

Lagrange multipliers $\lambda_{\mathscr{C}_\chi}$, $\Lambda_{\mathscr{C}_\chi}$ are then determined by the condition that the Lagrangian equations of motion, to be written up in Sect. 3.6, do generate an evolution complying with constraints (130a) and (130b) at all time $\tau$.

## 3.6 Lagrangian Equations of Motion

Collecting results from Sects. 3.2–3.5, the Lagrangian equations (109a)–(109c) are given the more explicit form:

$$M_{\mathscr{C}_\chi}(\tau)\ddot{\chi}(\mathbf{X},\tau) \underset{\overline{m}}{\cong} \mathbf{f}_{\mathscr{C}_\chi}^{\mathsf{ext}}(\tau) + \mathbf{f}_{\mathscr{C}_\chi}^{\mathsf{shr}\,\mathscr{S}\mathsf{C}}(\tau), \tag{134a}$$

$$\ddot{\mathbf{F}}(\mathbf{X},\tau)\mathbf{E}_{\mathscr{C}_\chi}(t_0) \underset{\overline{m}}{\cong} \big(\mathbf{M}_{\mathscr{C}_\chi}^{\mathsf{ext}}(\tau) + \mathbf{M}_{\mathscr{C}_\chi}^{\mathsf{vir}}(\tau) + \mathbf{M}_{\mathscr{C}_\chi}^{\mathsf{shr}\,\mathscr{S}\mathsf{C}}\big)\mathbf{F}(\mathbf{X},\tau)^{-\top}, \tag{134b}$$

$$\begin{aligned} m_i\mathbf{C}(\mathbf{X},\tau)\ddot{\mathbf{s}}_i(\tau) \underset{\overline{m}}{\cong}\ & \mathbf{F}(\mathbf{X},\tau)^\top\big(\mathbf{f}_i^{\mathsf{ext}}(\tau) + \mathbf{f}_i^{\mathsf{shr}\,\mathscr{S}}(\tau) + \mathbf{f}_i^{\mathsf{rep}\,\mathscr{S}}(\tau) \\ & - m_i\big(2\dot{\mathbf{F}}(\mathbf{X},\tau)\dot{\mathbf{s}}_i(\tau) + \big(\ddot{\mathbf{F}}(\mathbf{X},\tau) - \Lambda_{\mathscr{C}_\chi}(\tau)\mathbf{F}(\mathbf{X},\tau)\big)\mathbf{s}_i(\tau) \\ & + \ddot{\chi}(\mathbf{X},\tau) - \lambda_{\mathscr{C}_\chi}(\tau)\big)\big), \end{aligned} \tag{134c}$$

where the net mesoscopic short-range force $\mathbf{f}_{\mathscr{C}_\chi}^{\mathsf{shr}\,\mathscr{S}\mathsf{C}}$ and the net mesoscopic short-range moment $\mathbf{M}_{\mathscr{C}_\chi}^{\mathsf{shr}\,\mathscr{S}\mathsf{C}}$ depend on the Cauchy stress field $\mathbf{T}$ via (129) and (126), respectively, the *mesoscopic cell virial* $\mathbf{M}_{\mathscr{C}_\chi}^{\mathsf{vir}}$ is the sum of the cell kinetic moment (112) and the mesoscopic mutual moment (118b):

$$\mathbf{M}_{\mathscr{C}_\chi}^{\mathsf{vir}}(\tau) := \mathbf{M}_{\mathscr{C}_\chi}^{\mathsf{kin}}(\tau) + \mathbf{M}_{\mathscr{C}_\chi}^{\mathsf{shr}\,\mathscr{S}}(\tau), \tag{135}$$

and $\mathbf{C}$ is the change-of-metric tensor field (58). Equations (134a)–(134c) are supplemented by the time derivative of constraints (130a), (130b), evaluated on states compatible with the constraints (with due account of (61), (63)):

$$\int_{\mathscr{I}_\tau} d\sigma \int_{\mathscr{C}_\chi(\tau)} \sum_i m_i \mathbf{F}(\mathbf{X},\sigma)\ddot{\mathbf{s}}_i(\sigma)\delta_{\mathbf{q}_i(\sigma)} \underset{\overline{m}}{\cong} \mathbf{0}, \tag{136a}$$

$$\int_{\mathscr{I}_\tau} d\sigma \int_{\mathscr{C}_X(\tau)} \sum_i m_i \big(\mathbf{F}(\mathbf{X},\sigma)\dot{\mathbf{s}}_i(\sigma)\big) \otimes \big(\mathbf{F}(\mathbf{X},\sigma)\dot{\mathbf{s}}_i(\sigma)\big)\delta_{\mathbf{q}_i(\sigma)}$$

$$+\int_{\mathscr{I}_\tau} d\sigma \int_{\mathscr{C}_X(\tau)} \sum_i m_i \big(\mathbf{F}(\mathbf{X},\sigma)\mathbf{s}_i(\sigma)\big) \otimes \big(\mathbf{F}(\mathbf{X},\sigma)\ddot{\mathbf{s}}_i(\sigma)\big)\delta_{\mathbf{q}_i(\sigma)} \underset{\mathsf{m}}{\cong} \mathbf{0}. \qquad (136b)$$

Substituting (134c) into (136a), (136b) and accounting for (134a), (134b) and (122a), (122b), it is found[28] that

$$\lambda_{\mathscr{C}_X}(\tau) \underset{\mathsf{m}}{\cong} M_{\mathscr{C}_X}(\tau)^{-1}\mathbf{f}^{\mathsf{shr}\mathscr{S}\mathsf{C}}_{\mathscr{C}_X}(\tau), \qquad (137a)$$

$$\Lambda_{\mathscr{C}_X}(\tau) \underset{\mathsf{m}}{\cong} \big(\mathbf{M}^{\mathsf{shr}\mathscr{S}\mathsf{C}}_{\mathscr{C}_X}(\tau) - \mathbf{M}^{\mathsf{rep}\mathscr{S}}_{\mathscr{C}_X}(\tau)\big)\mathbf{E}_{\mathscr{C}_X}(\tau)^{-1}, \qquad (137b)$$

where $M_{\mathscr{C}_X}(\tau)$ is the (mesoscopically $\tau$-independent: cf. (111)) cell mass, $\mathbf{E}_{\mathscr{C}_X}(\tau)$ is the current cell central Euler tensor, and

$$\mathbf{M}^{\mathsf{rep}\mathscr{S}}_{\mathscr{C}_X}(\tau) := \frac{1}{T_{\mathsf{m}}}\int_{\mathscr{I}_\tau} \check{\mathbf{M}}^{\mathsf{rep}\mathscr{S}}_{\mathscr{C}_X}(\sigma)d\sigma. \qquad (138)$$

Substituting (137a), (137b) into (133) and accounting for (85) gives

$$\mathbf{f}^{\mathsf{rct}}_i(\tau) \underset{\mathsf{m}}{\cong} \frac{m_i}{M_{\mathscr{C}_X}(\tau)}\mathbf{f}^{\mathsf{shr}\mathscr{S}\mathsf{C}}_{\mathscr{C}_X}(\tau)$$
$$+ m_i\big(\mathbf{M}^{\mathsf{shr}\mathscr{S}\mathsf{C}}_{\mathscr{C}_X}(\tau) - \mathbf{M}^{\mathsf{rep}\mathscr{S}}_{\mathscr{C}_X}(\tau)\big)\mathbf{F}(\mathbf{X},\tau)^{-\top}\mathbf{E}_{\mathscr{C}_X}(t_0)^{-1}\mathbf{s}_i(\tau), \qquad (139)$$

which, substituted back into (134c), yields

$$m_i\big(\mathbf{F}(\mathbf{X},\tau)\ddot{\mathbf{s}}_i(\tau) + 2\dot{\mathbf{F}}(\mathbf{X},\tau)\dot{\mathbf{s}}_i(\tau) + \ddot{\mathbf{F}}(\mathbf{X},\tau)\mathbf{s}_i(\tau) + \ddot{\chi}(\mathbf{X},\tau)\big)$$
$$\underset{\mathsf{m}}{\cong} \mathbf{f}^{\mathsf{ext}}_i(\tau) + \mathbf{f}^{\mathsf{shr}\mathscr{S}}_i(\tau) + \widetilde{\mathbf{f}}^{\mathsf{shr}\mathscr{S}\mathsf{C}}_i(\tau), \qquad (140)$$

where

$$\widetilde{\mathbf{f}}^{\mathsf{shr}\mathscr{S}\mathsf{C}}_i(\tau) := \mathbf{f}^{\mathsf{rep}\mathscr{S}}_i(\tau) + \mathbf{f}^{\mathsf{rct}}_i(\tau) \qquad (141)$$

should be considered an improved estimate of the unavailable quantity $\mathbf{f}^{\mathsf{shr}\mathscr{S}\mathsf{C}}_i(\tau)$. In fact, $\mathbf{f}^{\mathsf{rct}}_i(\tau)$ satisfies (131) by construction, so that, with respect to criterion (121), $\widetilde{\mathbf{f}}^{\mathsf{shr}\mathscr{S}\mathsf{C}}_i(\tau)$ is as good as $\mathbf{f}^{\mathsf{rep}\mathscr{S}}_i(\tau)$ as an approximation of $\mathbf{f}^{\mathsf{shr}\mathscr{S}\mathsf{C}}_i(\tau)$. However,

$$\widetilde{\mathbf{f}}^{\mathsf{shr}\mathscr{S}\mathsf{C}}_{\mathscr{C}_X}(\tau) := \frac{1}{T_{\mathsf{m}}}\int_{\mathscr{I}_\tau} d\sigma \int_{\mathscr{C}_X(\tau)} \sum_i \widetilde{\mathbf{f}}^{\mathsf{shr}\mathscr{S}\mathsf{C}}_i(\sigma)\delta_{\mathbf{q}_i(\sigma)} \underset{\mathsf{m}}{\cong} \mathbf{f}^{\mathsf{shr}\mathscr{S}\mathsf{C}}_{\mathscr{C}_X}(\tau), \qquad (142a)$$

$$\widetilde{\mathbf{M}}^{\mathsf{shr}\mathscr{S}\mathsf{C}}_{\mathscr{C}_X}(\tau) := \frac{1}{T_{\mathsf{m}}}\int_{\mathscr{I}_\tau} d\sigma \int_{\mathscr{C}_X(\tau)} \sum_i \mathbf{r}_i(\sigma) \otimes \widetilde{\mathbf{f}}^{\mathsf{shr}\mathscr{S}\mathsf{C}}_i(\sigma)\delta_{\mathbf{q}_i(\sigma)} \underset{\mathsf{m}}{\cong} \mathbf{M}^{\mathsf{shr}\mathscr{S}\mathsf{C}}_{\mathscr{C}_X}(\tau), \qquad (142b)$$

with the right sides defined in (129) and (126), respectively. A comparison between (142a), (142b) and (122a), (122b) shows that the reaction to the thermal constraints (130a), (130b) repairs the inconsistency between the particle-wise and cell-wise estimates of short-range interactions with the exterior of the cell.

[28] Intermediate steps are essentially identical to the ones expounded in Appendix C.1.

### 3.7 Concerted Macro- and Micro-Dynamics

For each pair $(\mathbf{X}, t_0) \in \mathscr{D}_{\mathsf{M}}$ (cf. Sect. 2.5), the *mesoscopic force balance equation* (134a) associated with any mesoscopic co-moving cell $\mathscr{C}_\chi$ centred at $\mathbf{X}$ at time $t_0$ results in a unique—to within a mesoscopically negligible error—evolution equation for $\chi(\mathbf{X}, \cdot)$ in terms of referential force densities, i.e., current forces per unit reference volume:[29]

$$\varrho_0(\mathbf{X})\ddot{\chi}(\mathbf{X}, t) = \mathbf{b}_0^{\mathsf{ext}}(\mathbf{X}, t) + \operatorname{div} \mathbf{S}|_{(\mathbf{X},t)}, \tag{143}$$

where $\varrho_0$ is the referential mass density defined in (81), the external body force per unit reference volume $\mathbf{b}_0^{\mathsf{ext}}$ is defined as

$$\mathbf{b}_0^{\mathsf{ext}}(\mathbf{X}, t) := \frac{1}{V_{\mathscr{C}_0}}\mathbf{f}_{\mathscr{C}_\chi}^{\mathsf{ext}}(t), \tag{144}$$

and the Piola stress $\mathbf{S}$ is defined as the Piola transform of the Cauchy stress:

$$\mathbf{S}(\mathbf{X}, t) := \det\big(\mathbf{F}(\mathbf{X}, t)\big)\mathbf{T}(\mathbf{x}, t)\mathbf{F}(\mathbf{X}, t)^{-\top}, \tag{145}$$

in such a way that, on account of (80) and (129),

$$\operatorname{div} \mathbf{S}|_{(\mathbf{X},t)} = \det\big(\mathbf{F}(\mathbf{X}, t)\big) \operatorname{div} \mathbf{T}|_{(\mathbf{x},t)} = \frac{1}{V_{\mathscr{C}_0}}\mathbf{f}_{\mathscr{C}_\chi}^{\mathsf{shr},\mathscr{S}\mathsf{C}}(t). \tag{146}$$

In terms of spatial fields, equation (143) translates into[30]

$$\varrho(\mathbf{x}, t)\,\mathbf{a}(\mathbf{x}, t) = \mathbf{b}^{\mathsf{ext}}(\mathbf{x}, t) + \operatorname{div} \mathbf{T}|_{(\mathbf{x},t)}, \tag{147}$$

where $\varrho$ is the mass per unit current volume $(31)_1$, the spatial acceleration field $\mathbf{a}$ is defined as

$$\mathbf{a}(\mathbf{x}, t) := \ddot{\chi}\big(\chi^{-1}(\mathbf{x}, t), t\big) = \dot{\mathbf{v}}(\mathbf{x}, t) + \nabla\mathbf{v}|_{(\mathbf{x},t)}\mathbf{v}(\mathbf{x}, t) \tag{148}$$

(cf. (50a)), and

$$\mathbf{b}^{\mathsf{ext}}(\mathbf{x}, t) := \frac{1}{V_{\mathscr{C}_\chi}(t)}\mathbf{f}_{\mathscr{C}_\chi}^{\mathsf{ext}}(t) \tag{149}$$

is the external body force per unit current volume. On the left side of (143), the (time independent) referential mass density subsumes m-local (i.e., mesoscopically local) information on particle mass via (12) and (81). On its right side, the two terms making up the net applied force per unit volume come from radically different sources. The external body force per unit reference volume collects m-local information through (114a), the external fields acting on particles being assumedly known a priori (cf. Appendix B.7). Contrariwise, the net short-range force per unit reference volume is obtained by applying a M-local (i.e., macroscopically local) space differential operator to the slowly varying stress field $\mathbf{S}(\cdot, t)$; as such, this term extracts essential dynamical information from a M-neighbourhood of $\mathbf{X}$. If, at any

[29] From here onwards, the distinction between $\underset{\mathsf{m}}{\cong}$ and the equal sign will be dismissed (cf. footnote 18).

[30] On account of mass conservation in differential form (82), equation (147) has an equivalent formulation in terms of the time rate of the spatial field $\varrho\mathbf{v}$ of macroscopic momentum density: $(\varrho\mathbf{v})^{\cdot} = \mathbf{b}^{\mathsf{ext}} + \operatorname{div}(\mathbf{T} - \varrho\mathbf{v} \otimes \mathbf{v})$.

time $t$, the macroscopic state of the material system $(\chi(\cdot,t), \dot{\chi}(\cdot,t))$ and the net applied force density $\mathbf{b}_0^{\mathsf{ext}}(\cdot,t) + \operatorname{div}\mathbf{S}|_{(\cdot,t)}$ are known, (143) may be used to evolve the state to time $t + T_{\mathsf{M}}$.

On account of (51), (126) and (145), the *mesoscopic moment balance equation* (134b) may be rewritten in the form

$$\mathbf{S}(\mathbf{X},t) = -\frac{1}{V_{\mathscr{C}_0}}\big(\mathbf{M}^{\mathsf{ext}}_{\mathscr{C}_\chi}(t) + \mathbf{M}^{\mathsf{vir}}_{\mathscr{C}_\chi}(t)\big)\nabla\chi|^{-\mathsf{T}}_{(\mathbf{X},t)} + \mathcal{O}\bigg(\frac{L_{\mathsf{m}}}{L_{\mathsf{M}}}\bigg)^2, \tag{150}$$

and used to compute the Piola stress—a couple per unit reference volume. In fact, due to (143), the contribution to (150) of the term on the left side of (134b) equals

$$\begin{aligned}&\nabla\ddot{\chi}|_{(\mathbf{X},t)}\mathbf{E}_{\mathscr{C}_\chi}(t_0)\\ &\quad= \big(\nabla\mathbf{b}_0^{\mathsf{ext}} + \Delta\mathbf{S} - (\nabla\ln\varrho_0)\otimes\big(\mathbf{b}_0^{\mathsf{ext}} + \operatorname{div}\mathbf{S}\big)\big)\Big|_{(\mathbf{X},t)}\frac{\mathbf{E}_{\mathscr{C}_\chi}(t_0)}{M_{\mathscr{C}_\chi}(t)} \approx \mathcal{O}\bigg(\frac{L_{\mathsf{m}}}{L_{\mathsf{M}}}\bigg)^2\end{aligned} \tag{151}$$

(with $\Delta := \nabla\operatorname{div}$), under the assumption that the external body force is at most of the order of the net contact force, i.e., if

$$\|\mathbf{b}_0^{\mathsf{ext}}(\mathbf{X},t)\| \lessapprox \|\operatorname{div}\mathbf{S}|_{(\mathbf{X},t)}\|. \tag{152}$$

The quantity (151) can safely be neglected.[31] In (150), both the external moment $\mathbf{M}^{\mathsf{ext}}_{\mathscr{C}_\chi}(t)$ and the cell virial $\mathbf{M}^{\mathsf{vir}}_{\mathscr{C}_\chi}(t)$ collect m-local information through (114b) and through (112), (118b) and (135), respectively. The placement gradient $\nabla\chi|_{(\mathbf{X},t)}$, on the contrary, extracts essential kinematic information from all cells in a M-neighbourhood of $\mathscr{C}_\chi$.

Finally, the *microscopic force balance equation* (140), rewritten in the form

$$\begin{aligned}\ddot{\mathbf{s}}_i(\tau) = \nabla\chi|^{-1}_{(\mathbf{X},t)}\big(&m_i^{-1}\big(\mathbf{f}_i^{\mathsf{ext}}(\tau) + \mathbf{f}_i^{\mathsf{shr}\mathscr{S}}(\tau) + \widetilde{\mathbf{f}}_i^{\mathsf{shr}\mathscr{S}\mathsf{C}}(\tau)\big)\\ &- \big(2\nabla\dot{\chi}|_{(\mathbf{X},t)}\dot{\mathbf{s}}_i(\tau) + \nabla\ddot{\chi}|_{(\mathbf{X},t)}\mathbf{s}_i(\tau) + \ddot{\chi}(\mathbf{X},t)\big)\big)\end{aligned} \tag{153}$$

governs the thermal dynamics of the general $i$-th particle moving within $\mathscr{C}_\chi$ during the mesoscopic time interval $\mathscr{I}$, centred at $t$ (cf. (5)). Note that all the slowly varying macroscopic fields involved in (153) are frozen at the central time $t$, while $\mathbf{s}_i$ is let evolve on the $T_\mu$ timescale. At each time $\tau \in \mathscr{I}$, after the state of the particles in $\mathscr{C}_\chi$ has been reconstructed by joining macroscopic and microscopic information via (60) and (65a), forces $\mathbf{f}_i^{\mathsf{ext}}(\tau)$, $\mathbf{f}_i^{\mathsf{shr}\mathscr{S}}(\tau)$ and $\widetilde{\mathbf{f}}_i^{\mathsf{shr}\mathscr{S}\mathsf{C}}(\tau)$ are computed and (153) is used to evolve $(\mathbf{s}_i, \dot{\mathbf{s}}_i)$ to time $\tau + T_\mu$. Since the reference radius vectors $\mathbf{s}_i$ fluctuate on the microscopic time scale $T_\mu$, while the placement $\chi$ and its space gradient $\nabla\chi$ vary significantly only at the macroscopic time scale $T_{\mathsf{M}}$, the four acceleration terms appearing in (153) are of different order in the small dimensionless parameter $T_\mu/T_{\mathsf{M}}$, namely

$$\begin{aligned}&\ddot{\mathbf{s}}_i(\tau) \approx \mathcal{O}\bigg(\frac{T_\mu}{T_{\mathsf{M}}}\bigg)^0, \qquad \nabla\dot{\chi}|_{(\mathbf{X},t)}\dot{\mathbf{s}}_i(\tau) \approx \mathcal{O}\bigg(\frac{T_\mu}{T_{\mathsf{M}}}\bigg)^1,\\ &\nabla\ddot{\chi}|_{(\mathbf{X},t)}\mathbf{s}_i(\tau) \approx \ddot{\chi}(\mathbf{X},t) \approx \mathcal{O}\bigg(\frac{T_\mu}{T_{\mathsf{M}}}\bigg)^2.\end{aligned} \tag{154}$$

[31] But, if desired, it can be taken into account by cycling between (143) and (150). Typically, $L_{\mathsf{m}} \approx 10\text{nm}$, while $L_{\mathsf{M}} \gtrapprox 1\ \mu\text{m}$, so that $(L_{\mathsf{m}}/L_{\mathsf{M}})^2 \lessapprox 10^{-4}$.

Typically, $T_\mu \approx 1\,\text{fs}$, $T_{\mathsf{M}} \gtrsim 1\ \mu\text{s}$, so that $T_\mu/T_{\mathsf{M}} \lesssim 10^{-9}$, and at least the last two terms on the right side of (153) can be neglected. The third-to-last term, though small, is the only one that couples directly the streaming and the thermal velocity. In fact, due to (51), (52a) and (64),

$$\nabla\dot{\chi}\,|_{(\mathbf{x},t)}\dot{\mathbf{s}}_i(\tau) = \mathbf{G}(\mathbf{x},t)\mathbf{w}_i(\tau). \tag{155}$$

# 4 Remarks and Reflections

## 4.1 A Critical Comparison with Parrinello-Rahman Dynamics

It is appropriate and instructive to compare equations (134b) and (134c)—and the rewrite (153) of the latter—with those governing the original Parrinello-Rahman dynamics.[32] Equations (2) and (3) of [25], translated into my notations, read

$$\ddot{\mathbf{s}}_i = m_i^{-1}\mathbf{F}^{-1}\mathbf{f}_i - \mathbf{C}^{-1}\dot{\mathbf{C}}\dot{\mathbf{s}}_i, \tag{156a}$$

$$\ddot{\mathbf{F}} = W^{-1}\big(\mathbf{M}_{\mathsf{PR}}^{\mathsf{vir}} - p\det(\mathbf{F})V_{\mathscr{C}_0}\mathbf{I}\big)\mathbf{F}^{-\top}. \tag{156b}$$

No macroscopic *fields* are contemplated and, accordingly, the cell deformation $\mathbf{F}$ is not thought of as the value taken by the (slowly varying) placement gradient over the cell, as in (51). Since

$$-\mathbf{C}^{-1}\dot{\mathbf{C}}\dot{\mathbf{s}}_i = -2\mathbf{F}^{-1}\mathrm{sym}\big(\dot{\mathbf{F}}\mathbf{F}^{-1}\big)\mathbf{F}\dot{\mathbf{s}}_i, \tag{157}$$

the second term on the right side of (156a) is a near match to the second line of (153), which may be rewritten in the form

$$-2\mathbf{F}^{-1}\big(\dot{\mathbf{F}}\mathbf{F}^{-1}\big)\mathbf{F}\dot{\mathbf{s}}_i - \mathbf{F}^{-1}\ddot{\mathbf{F}}\mathbf{s}_i - \mathbf{F}^{-1}\ddot{\chi}. \tag{158}$$

The mismatch between (157) and (158) is exactly the same as that occurring between the results I obtain in Sect. 3.2 and those in Appendix C.3, where a Lagrangian formalism is used, similarly to what is done in [25]. The applied force $\mathbf{f}_i$ entering the right side of (156a) is given by

$$\mathbf{f}_i = \sum_{j\neq i}\mathbf{f}_{ij} \quad \text{with} \quad \mathbf{f}_{ij} = -f(d_{ij})(\mathbf{r}_i - \mathbf{r}_j) \quad \text{and} \quad f(d) := \varphi'(d)/d, \tag{159}$$

where $d \mapsto \varphi(d)$ is the (assumedly binary) interaction potential, a prime denotes differentiation, and $d_{ij} := \|\mathbf{q}_i - \mathbf{q}_j\| = \|\mathbf{r}_i - \mathbf{r}_j\|$ is the mutual distance between the $i$-th and $j$-th particles. Since $\mathbf{r}_i$ is *a priori defined* by Parrinello and Rahman as

$$\mathbf{r}_i := \mathbf{F}\mathbf{s}_i \tag{160}$$

(note the difference with my (59), which goes the other way round), the force term in (156a) is actually written as follows:

$$\mathbf{F}^{-1}\mathbf{f}_i = \sum_{j\neq i} f(d_{ij})(\mathbf{s}_i - \mathbf{s}_j). \tag{161}$$

[32] In the Parrinello-Rahman method, as in any other single-cell application of molecular dynamics, there is no analogue of (134a), since the net force applied to the cell is tacitly assumed to be zero, and force balance is trivially satisfied by a null acceleration of its centre of mass ($\ddot{\chi} = \mathbf{0}$).

The sum over $j$ extends over the periodic replicas of the particles in the cell,

$$\mathbf{f}_i = \mathbf{f}_i^{\mathsf{shr}\mathscr{S}} + \mathbf{f}_i^{\mathsf{rep}\mathscr{S}}, \tag{162}$$

the external force $\mathbf{f}_i^{\mathsf{ext}}$ being assumed to be zero.

The scalar quantity $W > 0$ appearing in (156b) is a fudge factor, left at users' disposal. If

$$\mathbf{E}_{\mathscr{C}_\chi}(t_0) = W\mathbf{I}, \tag{163}$$

then the product $W\ddot{\mathbf{F}}$ matches the left side of (134b). As stated at the end of Sect. 2.7, this cannot be, unless the reference cell has cubic symmetry—hence the criticisms reported in footnote 15. However, (87) proves that the authors' guess was less wild than it appears to be at first sight.[33] If equality (163) is taken for granted and, moreover, $\mathbf{M}_{\mathsf{PR}}^{\mathsf{vir}}$ is identified with the mesoscopic cell virial (135), then (156b) matches (134b), for $\mathbf{M}_{\mathscr{C}_\chi}^{\mathsf{ext}} = \mathbf{0}$ and a spherical applied Cauchy stress

$$\mathbf{T} = -p\,\mathbf{I}, \tag{164}$$

with $p$ a given constant pressure—the only case considered by the authors in [25].[34] In my notations, Parrinello and Rahman's cell virial $\mathbf{M}_{\mathsf{PR}}^{\mathsf{vir}}$ reads (equation (4) of [25]):

$$\mathbf{M}_{\mathsf{PR}}^{\mathsf{vir}} = \sum_i m_i \mathbf{w}_i \otimes \mathbf{w}_i + \sum_i \sum_{j>i} f(d_{ij})(\mathbf{r}_i - \mathbf{r}_j) \otimes (\mathbf{r}_i - \mathbf{r}_j), \tag{165}$$

the velocity $\mathbf{w}_i$ being *defined* as

$$\mathbf{w}_i := \mathbf{F}\dot{\mathbf{s}}_i, \tag{166}$$

Definition (166) should be compared and contrasted with the approximate equality (64), which establishes a relation between *independently* defined quantities. Note also that in (165) the sum over $j$ extends over the periodic replicas of the particles in the cell, as in (161), so that $\mathbf{M}_{\mathsf{PR}}^{\mathsf{vir}}$ includes a spurious contribution from *outside* the cell, due to $\mathbf{f}_i^{\mathsf{rep}\mathscr{S}}$ (in my notations). As a consequence,

$$\mathbf{M}_{\mathsf{PR}}^{\mathsf{vir}} \neq \mathbf{M}_{\mathscr{C}_\chi}^{\mathsf{vir}}. \tag{167}$$

The fallacy in definition (165) is inherent to the way in which Parrinello and Rahman—and Andersen before them—devised their Lagrangians, by rescaling the particle degrees of freedom in the pristine Newtonian Lagrangian and *independently adding* to it terms meant to associate a kinetic and a potential energy with the cell degrees of freedom introduced—the cell volume for Andersen [4], the cell deformation $\mathbf{F}$ for Parrinello and Rahman. As a matter of fact, this criticism applies to all the molecular dynamics formulations I cite in this paper.

All of the above being said, the general concordance between (156a), (156b) and (134b), (134c) is more striking than their discrepancies, considering how much standpoint and target of Andersen, Parrinello and Rahman differed from mine. All of them knew very little, if any, of continuum mechanics, and did not care about it. Andersen allowed the cell to dilate, giving its volume the status of an extra degree of freedom, aiming at molecular dynamics

[33] The precursory results obtained by Podio-Guidugli [28] in this direction are reported and discussed in the second part of Sect. 4.4.

[34] The natural extension to a general applied stress is considered in [26], though with needless variational contortions and some blunders. The issue was set straight by Ray and Rahman [31].

simulations of fluids under prescribed pressure conditions. Parrinello and Rahman modified Andersen's method so as to allow for changes in volume and shape. My point here is that, in their quest for a molecular dynamics method suited for simulating solid-to-solid phase transitions, these authors were led, by an intrinsic necessity, to allow the computational cell—a mesoscopic space cell, in my own terms—to deform in exactly the same way as an infinitesimally small piece of a continuous medium is assumed to do in continuum mechanics.

## 4.2 Ribarsky-Landman Dynamics

In the originally meant—and still largely prevalent—application of the Parrinello-Rahman dynamics for barostatting a single cell, (the analogue of) equation (134b) is solved in terms of $\ddot{\mathbf{F}}$, the Cauchy stress $\mathbf{T}$ being assigned.[35] In the absence of a co-evolving continuum, hence of the regulatory effects of its slow dynamics, there is no reason why the cell deformation $\mathbf{F}$ should not exhibit mesoscopically fast transients—even though tending to an equilibrium value, hopefully. An alternative simulation procedure, sort of dual to the previous one, was put forward and applied in a notable paper by Ribarsky and Landman [32], where the computational cell is let to deform "in a prescribed way at a prescribed constant rate," as if controlled by a hard loading device. In this regime (mislabelled CSR—for constant strain rate—by the authors)

$$\ddot{\mathbf{F}} = \mathbf{0}, \tag{168}$$

so that the $\mathcal{O}(L_{\mathsf{m}}/L_{\mathsf{M}})^2$ term on the right side of (150) is exactly zero, and the stress is obtained from (134b) as the reaction to the holonomic, time-dependent constraint (168). It should be emphasised that the stress response computed in this way may be either the equilibrium response under an assigned deformation, if $\dot{\mathbf{F}} = \mathbf{0}$, or else the response under stationary non-equilibrium conditions. The difference between equilibrium and non-equilibrium is determined by the second term on the right side of (156a), coupling the cell stretching $\mathrm{sym}(\dot{\mathbf{F}}\mathbf{F}^{-1})$ and the thermal velocity (166) in the Parrinello-Rahman dynamics (cf. (157)).[36] Of course, the control exerted on mesoscopic cells by the co-evolving continuum (cf. Sect. 3.7) is softer than the CRS constraint and nonlocal, but equally effective in slowing down the time evolution of $\chi$ and $\mathbf{F} = \nabla\chi$ at each cell.

## 4.3 Space-Time Versus Ensemble Averages

In Sect. 1—and, more extensively, in [11]—I argued that space-time averages make more physical sense than ensemble averages, in line with Irving and Kirkwood's physical interpretation of these different kinds of averages [14]. Without doubt, the idea of introducing statistical ensembles has had an immense heuristic value, and it is fortunate that—less infrequently than one should expect—ensemble and space-time averages seem to agree. However, whenever they differ, it is the first ones that should be taken as (often convenient) approximations to the second ones, not vice versa.

Space-time averages are more laborious to work with than ensemble averages. As a compensation for this, the inner workings of the method are more evident. An example to the

[35] Or the Piola stress $\mathbf{S}$, or the Cosserat stress $\mathbf{Z} := \mathbf{F}^{-1}\mathbf{S}$, as in [31], where $-\mathbf{Z}$ is dubbed "thermodynamic tension."

[36] Or, in the split/concerted dynamics put forth in Sect. 3, by the analogous term in (153), coupling the full streaming velocity gradient $\mathbf{G} = \dot{\mathbf{F}}\mathbf{F}^{-1}$ with the thermal velocity (64).

point is macroscopic mass conservation (cf. Sect. 2.7). As stated explicitly by Noll, proving the "continuity equation" is extremely easy: it takes less than four lines in [24], a few more lines in [14]. In fact, through ensemble averaging this physically nontrivial result is apparently obtained as a trivial consequence of the principle of conservation of probability. To prove the same theorem via space-time averaging, I had to use explicitly the notion of thermal velocity—which does not appear in [14], nor in [24]—and the fact that its moment of momentum is mesoscopically negligible, in order to show that its net mass flux across the boundary of any mesoscopic co-moving cell is mesoscopically negligible. Moreover, my proof hinges on the hypothesis that the macroscopic field of mass density varies slowly on the mesoscopic space scale. However, I feel I understood *why* (79)—and consequently (81) and (82)—hold only after proving it in this way.

As regards the construction of the stress tensor in terms of molecular quantities, the results I obtain here from space-time averages are stronger, more general and more straightforward than those obtained by Irving and Kirkwood, and also by Noll, from ensemble averages. The stress is identified directly, not through its divergence. No artificial localisation of the interaction potential energy is used—hence, the stress is uniquely determined for short-ranged $n$-ary interactions of any order $n$.[37] However, this outcome is not due to space-time averaging per se, but rather to its use to extract a cellwise *affine* streaming velocity field from particle velocities, instead of its cellwise average. In fact, nothing similar is to be found in Murdoch's work—from which I learned much—despite his extensive use of space averaging techniques (much less of space-*time* averages, which, on the contrary, I find essential). On the other hand, in 2010 Rossi and Testa [33] succeeded in proving that the stress tensor is uniquely determined, using standard tools from statistical mechanics.[38] Taking for granted the symmetry of the stress tensor—which in general does not hold—they computed the variational derivative of the partition function with respect to the infinitesimal strain $\mathbf{E}$, *without* enforcing a priori the compatibility condition $\mathbf{E} = \mathrm{sym}\nabla\mathbf{u}$ for some infinitesimal displacement field $\mathbf{u}$.[39] The Cauchy stress emerges then as the (symmetric) tensor-valued Lagrange multiplier needed to enforce the compatibility constraint. Hence, as the authors state, "there is no freedom to add to [it] any arbitrary divergenceless, symmetric rank-two tensor [field.]"

### 4.4 On Scale Separation

Dealing properly and efficiently with the three separate space and time scales involved in the present theory is vital to its successful application. In particular, exploiting the gap between the macroscopic space scale $L_\mathsf{M}$ and the mesoscopic space scale $L_\mathsf{m}$ is key to an effective transfer of molecular information to truly macroscopic extents. Consider, for instance, a three-dimensional body whose characteristic linear size is $L = \lambda L_\mathsf{M}$, with $\lambda$ larger—typically, much larger—than 1. A fully molecular model of it would comprise about

$$(L/d)^3 = (\lambda\mu L_\mathsf{M}/L_\mu)^3 \tag{169}$$

[37] This issue is discussed at length in [2, 34, 37]. To cure the non-uniqueness of binary representations of the cluster expansion (238) for $n > 4$, Torres-Sánchez, Vanegas and Arroyo [37] introduce a connection on the shape manifold, and compute the corresponding covariant derivative of the interaction potential, thus skirting its non-unique extension to an ambient linear space. It would be interesting to compare their unique result with the one obtained here in a direct, straightforward way.

[38] Also the proof by Rossi and Testa (completely different from mine) rests on the key assumption that the interaction potential between the body elementary constituents—for the rest arbitrary—is short-ranged.

[39] Isomorphic to $\mathbf{D} = \mathrm{sym}\nabla\mathbf{v}$ for some velocity field $\mathbf{v}$, with $\mathbf{D} := \mathrm{sym}\mathbf{L}$ (cf. (37)).

particles, assuming that the mean inter-particle distance $d$ equals a fraction of the microscopic length scale (1): $d = L_\mu/\mu$, with $\mu$ larger—but, typically, not much larger—than 1. The multiscale method based on the present theory would sample the molecular level only at scattered locations, the typical distance between them being $L_\mathsf{M}$. Each mesoscopic space cell would contain about $(L_\mathsf{m}/d)^3$ particles, and there would be $(L/L_\mathsf{M})^3 = \lambda^3$ cells. With respect to a fully molecular model, the number of particles *explicitly* handled by this method would therefore be reduced by a factor $(L_\mathsf{M}/L_\mathsf{m})^3$, i.e., by many orders of magnitude. This pivotal boost factor is missing from the atomistic-to-continuum method developed by Li and coworkers [17, 18, 36, 39], where the whole continuum domain is *tessellated* by Parrinello-Rahman-like cells and hence, by necessity, stays of mesoscopic size.

The gain in the time domain is even more substantial. To evolve the fully molecular model over a time interval of duration $T = \nu T_\mathsf{M}$, with $\nu$ larger—typically, much larger—than 1, would require as many as

$$T/T_\mu = \nu T_\mathsf{M}/T_\mu \tag{170}$$

time steps for each of the particles counted in (169). With the present method, the microscopic evolution equation (153) is integrated only on intervals of mesoscopic duration $T_\mathsf{m}$, spaced by pauses of macroscopic duration $T_\mathsf{M}$, corresponding to the time-stepping intervals at which the macroscopic dynamics is evolved. In conclusion, considering also the reduction in the number of particles, the overall computational burden is cut down by a factor $(T_\mathsf{M}/T_\mathsf{m})(L_\mathsf{M}/L_\mathsf{m})^3$.

When dealing simultaneously with widely disparate scales, one has to acknowledge that even the equality relation between homogeneous physical quantities is scale-dependent—hence the necessity of mesoscopic tolerances, formally introduced in this paper. Before, the idea was already in the air, but the lack of mathematical explicitness blurred the issues at stake, causing difficulties and misinterpretations. In [28] Podio-Guidugli examined under which circumstances the kinetic energy concocted by Parrinello and Rahman in [25, 26] matches *exactly* the Newtonian kinetic energy of the population of particles occupying the cell. He proved the following sufficient conditions (rephrased here in the terminology and notations of the present paper): (i) the microscopically instantaneous "scaled" cell Euler tensor[40]

$$\int_{\mathscr{C}_\chi(\tau)} \sum_i m_i \mathbf{s}_i(\tau) \otimes \mathbf{s}_i(\tau) \delta_{\mathbf{q}_i(\tau)} \tag{171}$$

should be stationary; (ii) the cell spin $\operatorname{skw}(\dot{\mathbf{F}}\mathbf{F}^{-1})$ should be zero; (iii) the microscopically instantaneous "scaled" cell Euler tensor (171) should be spherical. In particular, (i) and (ii) together imply that the second and third terms of the sum on the second line of (242) in Appendix C.3 are *instantaneously* orthogonal (the first being absent in [25, 26]), so that the mixed kinetic energy term vanishes exactly. A few years later, in [38] Ulz checked whether these conditions were satisfied in a replica of the groundbreaking simulation of the fcc-to-hcp phase transformation in a nickel crystal that Parrinello and Rahman had presented in [26]. He concluded that condition (i) could be considered approximately satisfied, since the amplitude of the fluctuations of the "scaled" cell Euler tensor (171) were reasonably smaller than its average, their ratio being $\sim 2 \cdot 10^{-3}$. He did not even check condition (ii), having observed that the fluctuations of the mixed kinetic energy were even larger that those of the Parrinello-Rahman kinetic energy. In doing so, he paid no attention to the fact that only the

[40] Compare with (87), whose left side is the mesoscopic time average of (171).

first, but not the second, was fluctuating rapidly around zero, as one can see in Fig. 11 of [38]. Now, it should be clear that the property of being mesoscopically stationary, proved in (87) for the referential cell Euler tensor $\mathbf{E}_{\mathscr{C}_\chi}(t_0)$, implies nothing on the amplitude of its microscopic fluctuations. Analogously, the fact that the three terms of the sum on the second line of (242) are *mesoscopically* mutually orthogonal gives no information about the amplitude of small-scale fluctuations of their pairwise products. The reader should have no difficulty in mapping Podio-Guidugli's conditions onto the (more or less) corresponding statements in Sects. 2.3, 2.7 and 4.1, and to appreciate similarities and differences.

## 4.5 Local Streaming Velocity Fields of Higher-Order?

Formally, there is no reason not to raise the order of the vector-valued polynomial entering (8), defining the local streaming velocity field as the second-order polynomial

$$\pi : \mathbf{x} \mapsto \pi(\mathbf{o}) + \mathbf{L}(\mathbf{x} - \mathbf{o}) + \frac{1}{2}\mathbb{L}\big((\mathbf{x} - \mathbf{o}) \otimes (\mathbf{x} - \mathbf{o})\big) \tag{172}$$

(where $\mathbb{L} : \mathrm{Sym} \to \mathsf{V}\mathscr{E}$ linearly) that minimises the quadratic functional

$$\pi \mapsto \int_{\mathscr{C} \times \mathscr{I}} \sum_i m_i \left\| \dot{\mathbf{q}}_i(\tau) - \pi\big(\mathbf{q}_i(\tau)\big) \right\|^2 \delta_{\mathbf{q}_i(\tau)} \times d\tau.$$

However, there is a snag. Under assumption (34a)–(34b), the kinematic compatibility (37) extends to the Hessian of the macroscopic velocity field $\mathbf{v}$:

$$\mathbf{L} \underset{\mathrm{m}}{\cong} \nabla\mathbf{v}, \qquad \mathbb{L} \underset{\mathrm{m}}{\cong} \nabla\nabla\mathbf{v}. \tag{173}$$

Now, the streaming velocity field (172) affects the time rate of the mutual distance between two particles, respectively at $\mathbf{x}$ and $\mathbf{o}$, through the scalar quantity

$$\big((\nabla\mathbf{v})\mathbf{r}\big) \cdot \frac{\mathbf{r}}{\|\mathbf{r}\|} + \frac{1}{2}\big((\nabla\nabla\mathbf{v})(\mathbf{r} \otimes \mathbf{r})\big) \cdot \frac{\mathbf{r}}{\|\mathbf{r}\|}, \tag{174}$$

where $\mathbf{r} := \mathbf{x} - \mathbf{o}$. Hence, in order for the second-order correction to have a significant effect within a space cell of size $L_{\mathrm{m}}$, the field $\mathbf{v}$ should vary substantially on distances possibly larger than $L_{\mathrm{m}}$, but certainly not larger in order of magnitude, as required by assumption (34a). This negative result attests to the difficulty of giving a direct atomistic foundation to higher-gradient continuum theories. In a sense, affine streaming velocity fields are privileged, being the only ones for which the relative velocity between particle pairs is *nontrivial* and *scale-invariant*:

$$\alpha(\mathbf{o} + \ell\mathbf{r}) - \alpha(\mathbf{o}) = \mathbf{L}(\ell\mathbf{r}) = \ell(\mathbf{L}\mathbf{r}) = \ell\big(\alpha(\mathbf{o} + \mathbf{r}) - \alpha(\mathbf{o})\big). \tag{175}$$

Flexible beams and thin plates are prototypes of second-gradient one-dimensional and two-dimensional continua, respectively. Interestingly, a very specialised second-gradient variant of the Parrinello-Rahman scheme was developed by Passerone et al. [27] and Tartaglino et al. [35] to study the Au(111) surface reconstruction of two slabs of crystalline gold—only 16 or 32 layers thick—caused by bending. It would be interesting to reconsider that problem from the multiscale perspective developed in this paper.

## 4.6 Limitations of This Work

The continuum notion of stress formalises satisfactorily the predominant macroscopic effects of short-range interactions between microscopic particles. It is much less clear whether and how *long*-range interactions could be properly coarse-grained. An even more severe limitation is that I am presently unable to obtain continuum *thermo*mechanics as a coarse-grained picture of molecular dynamics, as would be desirable. In fact, decoupling impulse flux from heat flux, while formally feasible (and possibly useful) in macrophysics, is intrinsically impossible in microphysics. The difficulty here is in identifying and managing a bona fide mesoscopic representative of the heat flux vector.

**Acknowledgements** This paper is the provisional completion of ruminations and lucubrations that started about a decade ago and intensified in the last two-three years. I wish to acknowledge my debt towards my former students and postdocs, M. Ribezzi-Crivellari, M. Paoluzzi and M. Minozzi, who collaborated with me on this project in the now remote years of its long incubation. Without the ongoing interactions—sometimes harsh, always instructive—with S. Bonella, G. Ciccotti, M. Ferrario, and P. Podio-Guidugli, my ideas would have never developed to the present stage—much less would they have been written down.



# Appendix A: List of Notations

| | |
|---|---|
| $\approx$ | $a \approx b$: $a$ and $b$ have the same order of magnitude |
| $\gtrapprox$ | $a \gtrapprox b$: $a$ is at least of the order of magnitude of $b$ |
| $\lessapprox$ | $a \lessapprox b$: $a$ is at most of the order of magnitude of $b$ |
| $\underset{\mathrm{m}}{\cong}$ | $a \underset{\mathrm{m}}{\cong} b$: $a$ and $b$ are mesoscopically indistinguishable |
| $\underset{\mathrm{m}}{>}$ | $a \underset{\mathrm{m}}{>} b$: $a$ is mesoscopically larger than $b$ |
| $\#$ | cardinality sign |
| $\times$ | Cartesian product, product of measures, Gibbs cross product |
| $\otimes$ | tensor product |
| $\llcorner$ | measure restriction sign |
| $\diamond$ | generic bilinear product |
| $\mathbf{1}_{\mathscr{C}}$ | indicator function of the space cell $\mathscr{C}$ |
| $\alpha$ | generic affine vector field on $\mathscr{E}$ |
| $\alpha_{\mathscr{C}\times\mathscr{I}}$ | local streaming velocity field on the space-time cell $\mathscr{C}\times\mathscr{I}$ |
| $\delta_{\mathbf{x}}$ | point Dirac measure of mass 1 at $\mathbf{x}\in\mathscr{E}$ |
| $\lambda_{\mathscr{C}_\chi}$ | vector-valued Lagrange multiplier |
| $\Lambda_{\mathscr{C}_\chi}$ | tensor-valued Lagrange multiplier |
| $\varrho$ | mass density |
| $\varrho_0$ | referential mass density |
| $\varrho_{\mathscr{C}\times\mathscr{I}}$ | mesoscopic mass density on the space-time cell $\mathscr{C}\times\mathscr{I}$ |
| $\chi$ | macroscopic motion |
| $\mathbf{a}$ | spatial acceleration field |
| $\mathcal{A}^{\mathrm{kin}}_{\mathscr{C}_\chi}$ | kinetic action functional |
| $\mathbf{b}^{\mathrm{ext}}$ | external body force per unit current volume |
| $\mathbf{b}^{\mathrm{ext}}_0$ | external body force per unit reference volume |
| $\mathscr{B}_R(\mathbf{x})$ | space ball of radius $R$ centred at $\mathbf{x}$ |
| $\mathbf{c}_{\mathscr{C}\times\mathscr{I}}$ | time-averaged centre of mass of the space-time cell $\mathscr{C}\times\mathscr{I}$ |

| | |
|---|---|
| $\mathbf{c}_{\mathscr{S}}$ | centre of mass of the material subsystem $\mathscr{S}$ |
| $\bar{\mathbf{c}}_{\mathscr{C}}$ | centroid of the space cell $\mathscr{C}$ |
| $\mathscr{C}$ | mesoscopic space cell |
| $\mathscr{C} \times \mathscr{I}$ | mesoscopic space-time cell |
| $\mathscr{C}_{\chi}$ | co-moving space cell |
| $\mathscr{C}_0$ | reference space cell |
| $\mathbf{C}$ | change-of-metric tensor |
| $D_{\mathsf{in}}$ | inner diameter |
| $D_{\mathsf{out}}$ | outer diameter |
| $\mathscr{D}_{\mathsf{M}}$ | domain of definition of macroscopic fields |
| $\mathscr{E}$ | three-dimensional Euclidean ambient space |
| $\mathbf{E}_{\mathscr{C}\times\mathscr{I}}$ | time-averaged central Euler tensor of the space-time cell $\mathscr{C} \times \mathscr{I}$ |
| $\mathbf{E}_{\mathscr{C}_{\chi}}$ | mesoscopic central Euler tensor of the co-moving space cell $\mathscr{C}_{\chi}$ |
| $\mathbf{E}_{\mathscr{C}_0}$ | referential central Euler tensor |
| $\mathbf{E}_{\mathscr{S}}$ | central Euler tensor of the material subsystem $\mathscr{S}$ |
| $\mathbf{f}^{\mathsf{ext}}$ | external force |
| $\mathbf{f}^{\mathsf{int}}$ | interaction force |
| $\mathbf{f}^{\mathsf{shr}}$ | short-ranged interaction force |
| $\mathbf{f}^{\mathsf{nrt}}$ | (nominal) force of inertia |
| $\mathbf{f}^{\mathsf{fct}}$ | fictitious force |
| $\mathbf{f}^{\mathsf{rct}}$ | reactive force |
| $\mathbf{f}^{\mathsf{ext}}_{\mathscr{C}_{\chi}}$ | mesoscopic external force |
| $\mathbf{f}^{\mathsf{shr}\mathscr{S}}_{\mathscr{C}_{\chi}}$ | mesoscopic mutual force |
| $\mathbf{f}^{\mathsf{shr}\mathscr{S}\complement}_{\mathscr{C}_{\chi}}$ | net mesoscopic short-range force |
| $\mathbf{f}^{\mathsf{shr}\mathscr{S}}_{i}$ | short-range force exerted on the $i$-th particle by the particles in $\mathscr{C}_{\chi}$ |
| $\mathbf{f}^{\mathsf{rep}\mathscr{S}}_{i}$ | short-range force exterted on the $i$-th particle by replica particles |
| $\mathbf{f}^{\mathsf{shr}\mathscr{S}\complement}_{i}$ | short-range force exerted on the $i$-th particle by the particles out of $\mathscr{C}_{\chi}$ |
| $\widetilde{\mathbf{f}}^{\mathsf{shr}\mathscr{S}\complement}_{i}$ | improved estimate of $\mathbf{f}^{\mathsf{shr}\mathscr{S}\complement}_{i}$ |
| $\mathbf{F}$ | deformation gradient |
| $\mathscr{F}_R$ | $R$-fattening |
| $\mathbf{G}$ | spatial velocity gradient |
| $\mathbf{I}$ | identity tensor |
| $\mathscr{I}$ | mesoscopic time interval (centred at time $t$) |
| $\mathscr{I}_{\tau}$ | mesoscopic time interval centred at time $\tau$ |
| $\mathbf{J}_{\mathscr{S}}$ | moment of inertia tensor of the material subsystem $\mathscr{S}$ |
| $\mathcal{K}^{\mathscr{S}}$ | kinetic energy function of the material subsystem $\mathscr{S}$ |
| $\mathcal{K}_{\mathscr{C}_{\chi}}$ | mesoscopic cell kinetic energy |
| $\mathcal{L}$ | Lagrangian function |
| $\mathbf{L}_{\mathscr{C}\times\mathscr{I}}$ | linear part of the affine streaming velocity field $\alpha_{\mathscr{C}\times\mathscr{I}}$ |
| $L_{\mu}$ | microscopic length scale |
| $L_{\mathsf{m}}$ | mesoscopic length scale |
| $L_{\mathsf{M}}$ | macroscopic length scale |
| $\mathbf{m}_{\mathscr{C}\times\mathscr{I}}$ | time-averaged momentum of the space-time cell $\mathscr{C} \times \mathscr{I}$ |
| $\mathscr{M}$ | macroscopic material system, comprised of $N$ particles |
| $M_{\mathscr{C}\times\mathscr{I}}$ | time-averaged mass of the space-time cell $\mathscr{C} \times \mathscr{I}$ |
| $M_{\mathscr{C}_{\chi}}$ | mesoscopic mass content of the co-moving cell $\mathscr{C}_{\chi}$ |
| $M_{\mathscr{S}}$ | mass of the material subsystem $\mathscr{S}$ |

| | |
|---|---|
| $\mathbf{M}_{\mathscr{S}}^{\mathsf{mom}}$ | central moment of momentum of the material subsystem $\mathscr{S}$ |
| $\mathbf{M}_{\mathscr{S}}^{\wedge}$ | rotational momentum of the material subsystem $\mathscr{S}$ |
| $\mathbf{M}_{\mathscr{C}_\chi}^{\mathsf{ext}}$ | mesoscopic external moment |
| $\mathbf{M}_{\mathscr{C}_\chi}^{\mathsf{shr}\mathscr{S}}$ | mesoscopic moment of mutual forces between the particles in $\mathscr{C}_\chi$ |
| $\mathbf{M}_{\mathscr{C}_\chi}^{\mathsf{shr}\mathscr{S}\complement}$ | mesoscopic moment of forces exterted by the particles out of $\mathscr{C}_\chi$ |
| $\mathbf{M}_{\mathscr{C}_\chi}^{\mathsf{rep}\mathscr{S}}$ | mesoscopic moment of forces exterted by replica particles |
| $\mathbf{M}_{\mathscr{C}_\chi}^{\mathsf{kin}}$ | cell kinetic moment |
| $\mathbf{M}_{\mathscr{C}\times\mathscr{I}}^{\mathsf{mom}}$ | time-averaged central moment of momentum of $\mathscr{C}\times\mathscr{I}$ |
| $\mathbf{M}_{\mathscr{C}_\chi}^{\mathsf{vir}}$ | mesoscopic cell virial |
| $\mathbf{M}_{\mathsf{PR}}^{\mathsf{vir}}$ | Parrinello and Rahman's cell virial |
| $\mathbf{n}_{\partial\mathscr{C}}$ | unit outer normal field to $\partial\mathscr{C}$ |
| $\mathscr{O}(\mathsf{V}\mathscr{E})$ | group of orthogonal transformations of $\mathsf{V}\mathscr{E}$ |
| $\mathbf{p}_i$ | momentum of the $i$-th particle |
| $\mathbf{q}_i$ | position of the $i$-th particle |
| $\mathscr{Q}$ | configuration manifold |
| $\mathbf{r}_i$ | radius vector of the $i$-th particle with respect to the centre of mass |
| $\bar{\mathbf{r}}_i$ | radius vector of the $i$-th particle with respect to the centroid |
| $\mathbf{s}_i$ | reference radius vector of the $i$-th particle |
| skw | skew-symmetrisation operator on $\mathsf{V}\mathscr{E}\otimes\mathsf{V}\mathscr{E}$ |
| sym | symmetrisation operator on $\mathsf{V}\mathscr{E}\otimes\mathsf{V}\mathscr{E}$ |
| $\mathbf{S}$ | Piola stress tensor |
| $\mathscr{S}$ | subsystem of the material system $\mathscr{M}$ |
| $\mathsf{SL}(\mathsf{V}\mathscr{E})$ | special linear group on $\mathsf{V}\mathscr{E}$ |
| Sym | subspace of symmetric double tensors |
| $\mathbf{t}$ | traction vector |
| tr | trace operator |
| $\mathbf{T}$ | Cauchy stress tensor |
| $T_\mu$ | microscopic time scale |
| $T_{\mathsf{m}}$ | mesoscopic time scale |
| $T_{\mathsf{M}}$ | macroscopic time scale |
| $\mathsf{T}\mathscr{Q}$ | state manifold |
| $\mathbf{v}$ | macroscopic velocity field |
| $\underset{\sim}{\mathbf{v}}$ | spatial streaming test field |
| $\mathbf{v}_{\mathscr{C}\times\mathscr{I}}$ | mesoscopic central velocity of the space-time cell $\mathscr{C}\times\mathscr{I}$ |
| $V_{\mathscr{C}}$ | volume of the space cell $\mathscr{C}$ |
| $V_{\mathscr{C}_\chi}$ | current volume of the co-moving space cell $\mathscr{C}_\chi$ |
| $V_{\mathscr{C}_0}$ | reference volume of the co-moving space cell $\mathscr{C}_\chi$ |
| $\mathcal{V}$ | interaction potential |
| $\mathsf{V}\mathscr{E}$ | translation space of $\mathscr{E}$ |
| $\mathsf{V}\mathscr{E}\otimes\mathsf{V}\mathscr{E}$ | space of double tensors on $\mathsf{V}\mathscr{E}$ (linear maps of $\mathsf{V}\mathscr{E}$ into itself) |
| $\mathbf{w}_i$ | thermal velocity of the $i$-th particle |
| $\mathbf{X}$ | reference position labelling a macroscopic point |

## Appendix B: Newtonian Mechanics of Interacting Point Particles

Here I assume as basic the standard microscopic picture of a macroscopic material system $\mathscr{M}$ as composed of a finite—but monstrously large—number of identifiable point particles, labelled for convenience with the first $N$ natural numbers.

### B.1 Configuration, State, Motion

Let $\mathbf{q}_i \in \mathscr{E}$ denote the instantaneous position in the three-dimensional Euclidean ambient space $\mathscr{E}$ of the $i$-th point particle, and $\mathbf{v}_i \in \mathsf{V}\mathscr{E}$ denote its instantaneous velocity, a vector in the translation space of $\mathscr{E}$. The ordered $N$-tuples

$$\mathfrak{q} := (\mathbf{q}_1, \ldots, \mathbf{q}_i, \ldots, \mathbf{q}_N) \in \mathscr{E}^N, \qquad \mathfrak{v} := (\mathbf{v}_1, \ldots, \mathbf{v}_i, \ldots, \mathbf{v}_N) \in \mathsf{V}\mathscr{E}^N \tag{176}$$

parameterise, respectively, the instantaneous configuration and velocity of the system. Since distinct particles are not allowed to simultaneously sit in the same place, the $3N$-dimensional configuration manifold $\mathscr{Q}$ of the system is the Cartesian product $\mathscr{E}^N$ minus the union of the 3-codimensional pairwise diagonal slices where at least two particles share the same position, namely

$$\mathscr{Q} := \mathscr{E}^N \setminus \bigcup_{i<j} \big\{(\mathbf{x}_1, \ldots, \mathbf{x}_i, \ldots, \mathbf{x}_j, \ldots, \mathbf{x}_N) \in \mathscr{E}^N \,|\, \mathbf{x}_i = \mathbf{x}_j\big\}, \tag{177}$$

endowed with the differentiable structure inherited from $\mathscr{E}$. A pair $(\mathfrak{q}, \mathfrak{v})$, i.e., the vector $\mathfrak{v} \in \mathsf{V}\mathscr{E}^N$ based at $\mathfrak{q} \in \mathscr{Q}$, characterises an instantaneous state of the system. The structured collection of such pairs—the state manifold of the system—constitutes the (trivial) tangent bundle $\mathsf{T}\mathscr{Q}$ to $\mathscr{Q}$. A motion of the system is a curve in $\mathsf{T}\mathscr{Q}$ smoothly parameterised by a time interval, such that

$$\dot{\mathfrak{q}}(\tau) = \mathfrak{v}(\tau) \tag{178}$$

for all time $\tau$, as measured by a clock ticking with a precision sufficient to resolve the microscopic details of particle motion, a precision characterised by the microscopic time scale $T_\mu$. Here and elsewhere, a superposed dot denotes time derivative.

### B.2 Impulse Balance

Along all motion, the balance equation

$$\dot{\mathfrak{p}}(\tau) = \mathfrak{f}(\tau), \tag{179}$$

where

$$\mathfrak{p} := (\mathbf{p}_1, \ldots, \mathbf{p}_i, \ldots, \mathbf{p}_N) \in \mathsf{V}\mathscr{E}^N, \qquad \mathfrak{f} := (\mathbf{f}_1, \ldots, \mathbf{f}_i, \ldots, \mathbf{f}_N) \in \mathsf{V}\mathscr{E}^N, \tag{180}$$

has to be satisfied at all time $\tau$. The covectors (identified with vectors via the Euclidean structure of $\mathsf{V}\mathscr{E}$) $\mathbf{p}_i(\tau)$ and $\mathbf{f}_i(\tau)$ are, respectively, the momentum of the $i$-th particle and the applied force (i.e., impulse supply per unit time) exerted on it at time $\tau$. Once $\mathfrak{p}(\tau)$ and $\mathfrak{f}(\tau)$ are related to the current state $(\mathfrak{q}(\tau), \mathfrak{v}(\tau))$ through the (smooth) prescriptions

$$\mathfrak{p}(\tau) = \widehat{\mathfrak{p}}\big(\mathfrak{q}(\tau), \mathfrak{v}(\tau); \tau\big), \qquad \mathfrak{f}(\tau) = \widehat{\mathfrak{f}}\big(\mathfrak{q}(\tau), \mathfrak{v}(\tau); \tau\big), \tag{181}$$

Eqs. (181), (179) and (178) uniquely determine the motion passing through any given initial state $(\mathfrak{q}_0, \mathfrak{v}_0)$ at time $\tau_0$. Impulse balance admits of an integral formulation, equivalent to the differential formulation (179) under the stated smoothness assumption. The (real-valued) linear impulse functional

$$\mathfrak{v} \mapsto \int_{\tau_-}^{\tau_+} \big(\mathfrak{f}(\tau) \cdot \mathfrak{v}(\tau) + \mathfrak{p}(\tau) \cdot \dot{\mathfrak{v}}(\tau)\big) d\tau - \big[\mathfrak{p}_{\mathsf{cut}}(\tau) \cdot \mathfrak{v}(\tau)\big]_{\tau_-}^{\tau_+} \tag{182}$$

(where $\mathfrak{p}_{\mathsf{cut}}$ is the boundary impulse, defined at the two ends of the time interval) should be identically zero, for all test velocity[41] $\mathfrak{v}$ on all time interval $]\tau_-, \tau_+[$. Integration by parts yields (179) at all $\tau \in ]\tau_-, \tau_+[$, plus the boundary conditions

$$\mathfrak{p}_{\mathsf{cut}}(\tau_-) = \lim_{\tau \downarrow \tau_-} \mathfrak{p}(\tau), \qquad \mathfrak{p}_{\mathsf{cut}}(\tau_+) = \lim_{\tau \uparrow \tau_+} \mathfrak{p}(\tau), \tag{183}$$

which ensure momentum continuity across time cuts when joining consecutive time intervals. Both in (179) and (182), it is intended that $\mathfrak{p}(\tau)$ and $\mathfrak{f}(\tau)$ are evaluated along the putative motion $\mathfrak{q}$, i.e., after substituting (178) into (181):

$$\mathfrak{p}(\tau) = \widehat{\mathfrak{p}}\big(\mathfrak{q}(\tau), \dot{\mathfrak{q}}(\tau); \tau\big), \qquad \mathfrak{f}(\tau) = \widehat{\mathfrak{f}}\big(\mathfrak{q}(\tau), \dot{\mathfrak{q}}(\tau); \tau\big). \tag{184}$$

## B.3 Mass, Centre of Mass, Central Moment of Momentum, Euler Tensor

Motion is defined—possibly only implicitly, as in Appendix B.1—with respect to an observer. A central tenet of Newtonian mechanics is that there exists (at least) one observer, called *inertial*, for which $(181)_1$ specialises to

$$\mathbf{p}_i(\tau) = m_i \mathbf{v}_i(\tau), \tag{185}$$

with $m_i > 0$ the time-independent mass of the $i$-th particle. For an inertial observer, (179) simplifies to

$$m_i \mathbf{a}_i(\tau) = \mathbf{f}_i(\tau), \tag{186}$$

where

$$\mathbf{a}_i(\tau) := \dot{\mathbf{v}}_i(\tau) = \ddot{\mathbf{q}}_i(\tau). \tag{187}$$

Several mass-related quantities can be usefully associated with each subsystem $\mathscr{S}$ of the material system $\mathscr{M}$. The total mass of $\mathscr{S}$ is obviously defined as the sum

$$M_{\mathscr{S}} := \sum_{i \in \mathscr{S}} m_i, \tag{188}$$

positive (if $\mathscr{S}$ is nonempty, as admitted in what follows) and time-independent. At each time $\tau$, its centre of mass $\mathbf{c}_{\mathscr{S}}(\tau)$ is the unique minimiser of the quadratic functional

$$\mathscr{E} \ni \mathbf{x} \mapsto \sum_{i \in \mathscr{S}} m_i \|\mathbf{q}_i(\tau) - \mathbf{x}\|^2, \tag{189}$$

[41] As a quantifier in (182), $\mathfrak{v}$ is *not* constrained by (178), i.e., $\mathfrak{v}(\tau)$ is *any* vector in $\mathsf{T}_{\mathfrak{q}(\tau)}\mathscr{Q} = \mathsf{V}\mathscr{E}^N$. In other terms, $\tau \mapsto \mathfrak{v}(\tau)$ is an arbitrary variation of the putative motion $\tau \mapsto \mathfrak{q}(\tau)$.

namely

$$\mathbf{c}_{\mathscr{S}}(\tau) = \mathbf{o} + \frac{1}{M_{\mathscr{S}}} \sum_{i \in \mathscr{S}} m_i \big(\mathbf{q}_i(\tau) - \mathbf{o}\big) \qquad (\forall\, \mathbf{o} \in \mathscr{E}). \tag{190}$$

Since all particle masses are positive, the current centre of mass of $\mathscr{S}$ is in the convex hull of the current positions of all particles in $\mathscr{S}$:

$$\mathrm{Conv}\big(\{\mathbf{q}_i(\tau) \mid i \in \mathscr{S}\}\big) := \mathbf{o} + \sum_{i \in \mathscr{S}} \ell_i \big(\mathbf{q}_i(\tau) - \mathbf{o}\big) \;\Big|\; \ell_j > 0 \;\&\; \sum_{j \in \mathscr{S}} \ell_j = 1. \tag{191}$$

It is readily checked that the minimum value attained by functional (189) is null. The linear momentum of $\mathscr{S}$ is defined in analogy with its mass:

$$\mathbf{m}_{\mathscr{S}}(\tau) := \sum_{i \in \mathscr{S}} m_i \dot{\mathbf{q}}_i(\tau). \tag{192}$$

Of course, it is a time-dependent vector quantity. The central moment of momentum of $\mathscr{S}$ is a tensor quantity that depends on time through both particle positions and velocities:

$$\mathbf{M}^{\mathsf{mom}}_{\mathscr{S}}(\tau) := \sum_{i \in \mathscr{S}} m_i \mathbf{r}_i(\tau) \otimes \dot{\mathbf{q}}_i(\tau), \tag{193}$$

where

$$\mathbf{r}_i(\tau) := \mathbf{q}_i(\tau) - \mathbf{c}_{\mathscr{S}}(\tau) \tag{194}$$

is the current central radius vector of the $i$-th particle. The central Euler tensor of $\mathscr{S}$

$$\mathbf{E}_{\mathscr{S}}(\tau) := \sum_{i \in \mathscr{S}} m_i \mathbf{r}_i(\tau) \otimes \mathbf{r}_i(\tau) \tag{195}$$

is symmetric positive semi-definite by construction. It is positive definite if and only if the convex hull (191) is full dimensional. The moment of inertia tensor $\mathbf{J}_{\mathscr{S}}$, more commonly used in rigid-body mechanics, is related to the Euler tensor via[42]

$$\mathbf{J}_{\mathscr{S}} := (\mathrm{tr}\mathbf{E}_{\mathscr{S}})\mathbf{I} - \mathbf{E}_{\mathscr{S}}. \tag{196}$$

It is readily checked that $\mathbf{J}_{\mathscr{S}}$ shares the structural properties of $\mathbf{E}_{\mathscr{S}}$. The rotational momentum $\mathbf{M}^{\wedge}_{\mathscr{S}}$ is twice the skew part of $\mathbf{M}^{\mathsf{mom}}_{\mathscr{S}}$:

$$\mathbf{M}^{\wedge}_{\mathscr{S}}(\tau) := \sum_{i \in \mathscr{S}} m_i \mathbf{r}_i(\tau) \wedge \dot{\mathbf{q}}_i(\tau), \tag{197}$$

customarily represented by its dual axial vector

$$\mathbf{m}^{\wedge}_{\mathscr{S}}(\tau) := \sum_{i \in \mathscr{S}} m_i \mathbf{r}_i(\tau) \times \dot{\mathbf{q}}_i(\tau), \tag{198}$$

[42]The trace $\mathrm{tr} : \mathsf{V}\mathscr{E} \otimes \mathsf{V}\mathscr{E} \to \mathbb{R}$ is linear and acts on dyads as follows: $\mathrm{tr}(\mathbf{a} \otimes \mathbf{b}) = \mathbf{a} \cdot \mathbf{b}$. By linearity it extends to all tensors.

where $\times$ denotes the Gibbs cross product. The axial vector $\omega$ dual of the angular velocity $\mathbf{W}$ is such that, for all vector $\mathbf{r}$,

$$\omega \times \mathbf{r} = \mathbf{W}\mathbf{r}. \tag{199}$$

Then, the axial vector $\mathbf{J}_{\mathscr{S}}\omega$ is dual to the skew tensor $\mathbf{W}\mathbf{E}_{\mathscr{S}} + \mathbf{E}_{\mathscr{S}}\mathbf{W} = 2\,\mathrm{skw}(\mathbf{W}\mathbf{E}_{\mathscr{S}})$.

## B.4 Change of Observer

The group of space-time transformations characterising Newtonian mechanics is the semidirect product[43] of the group of time translations and the group of smoothly time-dependent Euclidean isometries of space. In turn, the group of Euclidean isometries is the semidirect product of the orthogonal group $\mathscr{O}(\mathsf{V}\mathscr{E})$ and the additive group underlying $\mathsf{V}\mathscr{E}$. The evolution of the system that one observer records as the motion $\tau \mapsto \mathbf{q}(\tau)$ is perceived by another, equally legitimate, observer as the motion $\overset{*}{\tau} \mapsto \overset{*}{\mathbf{q}}(\overset{*}{\tau})$, with

$$\overset{*}{\tau} = \tau + s, \qquad \overset{*}{\mathbf{q}}_i(\overset{*}{\tau}) = \overset{*}{\mathbf{q}}_1(\overset{*}{\tau}) + \mathbf{Q}(\tau)\big(\mathbf{q}_i(\tau) - \mathbf{q}_1(\tau)\big), \tag{200}$$

where $s$ is a time shift and $\mathbf{Q}$ is a smooth map from the time line to $\mathscr{O}(\mathsf{V}\mathscr{E})$. The orthogonal tensor $\mathbf{Q}(\tau)$ is the attitude taken by the "unstarred" observer with respect to the "starred" observer at the current time $\tau$. Hence, in a change of observer, at any time $\tau$ the vector difference between the positions of any two particles—say, the $i$-th and $j$-th ones—is rotated (and possibly reflected) by $\mathbf{Q}(\tau)$ and time-shifted by $s$:

$$\overset{*}{\mathbf{q}}_i(\overset{*}{\tau}) - \overset{*}{\mathbf{q}}_j(\overset{*}{\tau}) = \mathbf{Q}(\tau)\big(\mathbf{q}_i(\tau) - \mathbf{q}_j(\tau)\big). \tag{201}$$

Vector quantities that transform as position differences under all change of observer will be called covariant. Since (200) holds identically in time, particle velocities transform as follows:

$$\overset{*}{\mathbf{v}}_i(\overset{*}{\tau}) = \mathbf{Q}(\tau)\mathbf{v}_i(\tau) + \overset{*}{\mathbf{v}}_1(\overset{*}{\tau}) - \mathbf{Q}(\tau)\mathbf{v}_1(\tau) + \mathbf{A}(\tau)\big(\overset{*}{\mathbf{q}}_i(\overset{*}{\tau}) - \overset{*}{\mathbf{q}}_1(\overset{*}{\tau})\big), \tag{202}$$

where the skew tensor

$$\mathbf{A}(\tau) := \dot{\mathbf{Q}}(\tau)\mathbf{Q}(\tau)^{\top} \tag{203}$$

is the current angular velocity of the unstarred observer with respect to the starred one. It appears that the covariance defect of velocity is the same for all particles (whatever their positions) if and only if the $\mathscr{O}(\mathsf{V}\mathscr{E})$-valued map $\mathbf{Q}$ is stationary, i.e., if and only if there is an orthogonal tensor $\overline{\mathbf{Q}}$ such that, for all time $\tau$,

$$\mathbf{Q}(\tau) = \overline{\mathbf{Q}}, \tag{204}$$

and hence the skew-valued map $\mathbf{A}$ is null. Condition (204) entails the existence of a $\mathsf{V}\mathscr{E}$-valued map $\varpi$ such that, for all $i$,

$$\overset{*}{\mathbf{v}}_i(\overset{*}{\tau}) - \overline{\mathbf{Q}}\mathbf{v}_i(\tau) = \varpi(\tau), \tag{205}$$

the vector $\varpi(\tau)$ being the current translational velocity of the unstarred observer with respect to the starred one. Since (202) holds identically in time, particle accelerations (187)

[43] For an entry-level introduction to the notion of semidirect group product, see [1].

transform as follows:

$$\overset{*}{\mathbf{a}}_i(\overset{*}{\tau}) = \mathbf{Q}(\tau)\mathbf{a}_i(\tau) + \overset{*}{\mathbf{a}}_1(\overset{*}{\tau}) - \mathbf{Q}(\tau)\mathbf{a}_1(\tau)$$
$$+ \big(\dot{\mathbf{A}}(\tau) + \mathbf{A}(\tau)^2\big)\mathbf{Q}(\tau)\big(\mathbf{q}_i(\tau) - \mathbf{q}_1(\tau)\big) + 2\mathbf{A}(\tau)\mathbf{Q}(\tau)\big(\mathbf{v}_i(\tau) - \mathbf{v}_1(\tau)\big) \tag{206}$$

If (and only if) condition (204) prevails, then the covariance defect of acceleration equals the time derivative of $\varpi$ for all particles:

$$\overset{*}{\mathbf{a}}_i(\overset{*}{\tau}) - \overline{\mathbf{Q}}\mathbf{a}_i(\tau) = \dot{\varpi}(\tau). \tag{207}$$

Therefore, particle acceleration is covariant if and only if condition (204) is satisfied and, moreover, there is a vector $\overline{\varpi}$ such that, for all time $\tau$,

$$\varpi(\tau) = \overline{\varpi}, \tag{208}$$

i.e., the translational velocity of the unstarred observer with respect to the starred one is time independent. Conditions (204) and (208) characterise the small, finite-dimensional subgroup of Galilean space-time transformations

$$\overset{*}{\tau} = \tau + s, \qquad \overset{*}{\mathbf{q}}_i(\overset{*}{\tau}) = \overset{*}{\mathbf{q}}_1(\overset{*}{\tau}_0) + \overline{\varpi}(\tau - \tau_0) + \overline{\mathbf{Q}}\big(\mathbf{q}_i(\tau) - \mathbf{q}_1(\tau)\big). \tag{209}$$

### B.5 Force of Inertia

A fundamental dynamical postulate is that all forces are covariant, i.e., they transform as position differences under a general change in observer (200). The time derivative of momentum is a force—namely, the opposite of the force of inertia:

$$\mathbf{f}_i^{\mathsf{nrt}} := -\dot{\mathbf{p}}_i. \tag{210}$$

As such, it transforms according to

$$\overset{*}{\mathbf{f}}{}_i^{\mathsf{nrt}}(\overset{*}{\tau}) = \mathbf{Q}(\tau)\mathbf{f}_i^{\mathsf{nrt}}(\tau). \tag{211}$$

If the starred observer is inertial, then (185) holds for $\overset{*}{\mathbf{p}}_i$, so that

$$\overset{*}{\mathbf{f}}{}_i^{\mathsf{nrt}}(\overset{*}{\tau}) = -m_i\overset{*}{\mathbf{a}}_i(\overset{*}{\tau}) \quad\Longleftrightarrow\quad \mathbf{f}_i^{\mathsf{nrt}}(\tau) = -m_i\mathbf{Q}(\tau)^{\top}\overset{*}{\mathbf{a}}_i(\overset{*}{\tau}), \tag{212}$$

under the tacit but essential assumption that mass is invariant under all change of observer. Substituting (206) into $(212)_2$ yields

$$\mathbf{f}_i^{\mathsf{nrt}}(\tau) = -m_i\mathbf{a}_i(\tau) + \mathbf{f}_i^{\mathsf{fct}}(\tau), \tag{213}$$

where the so-called fictitious force $\mathbf{f}_i^{\mathsf{fct}}$ is the sum of the linear acceleration force (on the first line of (214)), the angular acceleration force (second line), the centrifugal force (third line), and the Coriolis force (fourth line):

$$\mathbf{f}_i^{\mathsf{fct}}(\tau) = -m_i\big(\mathbf{Q}(\tau)^{\top}\overset{*}{\mathbf{a}}_1(\overset{*}{\tau}) - \mathbf{a}_1(\tau) + 2\mathbf{Q}(\tau)^{\top}\mathbf{A}(\tau)\mathbf{Q}(\tau)\mathbf{v}_1(\tau)\big)$$
$$- m_i\mathbf{Q}(\tau)^{\top}\dot{\mathbf{A}}(\tau)\mathbf{Q}(\tau)\big(\mathbf{q}_i(\tau) - \mathbf{q}_1(\tau)\big)$$

$$
\begin{aligned}
&- m_i \mathbf{Q}(\tau)^\top \mathbf{A}(\tau)^2 \mathbf{Q}(\tau)\big(\mathbf{q}_i(\tau) - \mathbf{q}_1(\tau)\big) \\
&- 2m_i \mathbf{Q}(\tau)^\top \mathbf{A}(\tau)\mathbf{Q}(\tau)\mathbf{v}_i(\tau).
\end{aligned}
\tag{214}
$$

According to (213), the force of inertia $\mathbf{f}_i^{\mathrm{nrt}}$ experienced by the starred observer is the sum of the nominal[44] force of inertia $-m_i\mathbf{a}_i$ and the fictitious force $\mathbf{f}_i^{\mathrm{fct}}$. The fictitious force vanishes if and only if the unstarred observer is inertial—in particular, if the starred observer is inertial and the change of observer (200) is Galilean.

## B.6 Kinetic Energy; Work-Energy Theorem

The work done by a force along a motion $]\tau_-, \tau_+[\ \ni \tau \mapsto \mathbf{q}(\tau)$ restricted to any subsystem $\mathscr{S}$ is, by definition, the integral over time of the power expended by that force on the instantaneous velocity $\dot{\mathbf{q}}(\tau)$ restricted to $\mathscr{S}$. In particular, the work done by the force of inertia (213) is

$$
\begin{aligned}
\int_{\tau_-}^{\tau_+} \sum_{i\in\mathscr{S}} \mathbf{f}_i^{\mathrm{nrt}}(\tau)\cdot\dot{\mathbf{q}}_i(\tau)d\tau &= \int_{\tau_-}^{\tau_+} \sum_{i\in\mathscr{S}} \big(-m_i\ddot{\mathbf{q}}_i(\tau) + \mathbf{f}_i^{\mathrm{fct}}(\tau)\big)\cdot\dot{\mathbf{q}}_i(\tau)d\tau \\
&= \mathcal{K}^{\mathscr{S}}\big(\dot{\mathbf{q}}(\tau_-)\big) - \mathcal{K}^{\mathscr{S}}\big(\dot{\mathbf{q}}(\tau_+)\big) \\
&\quad + \int_{\tau_-}^{\tau_+} \sum_{i\in\mathscr{S}} \mathbf{f}_i^{\mathrm{fct}}(\tau)\cdot\dot{\mathbf{q}}_i(\tau)d\tau,
\end{aligned}
\tag{215}
$$

where the kinetic energy function $\mathcal{K}^{\mathscr{S}}$ is defined as

$$
\mathcal{K}^{\mathscr{S}}(\mathfrak{v}) := \sum_{i\in\mathscr{S}} \mathcal{K}_i(\mathbf{v}_i), \qquad \mathcal{K}_i(\mathbf{v}_i) := \frac{1}{2}m_i\|\mathbf{v}_i\|^2.
\tag{216}
$$

If the motion is taken with respect to an inertial observer, $\mathfrak{f}^{\mathrm{fct}} = \mathbf{0}$. Then, the work done by the force of inertia equals the difference between the kinetic energy in the initial state and that in the final state, and is therefore path independent. If, on the contrary, the motion is taken with respect to a non-inertial observer, the contributions to the work (215) of the linear acceleration force, the angular acceleration force, and the centrifugal force (as defined in (214)) are, in general, path dependent. The Coriolis force is powerless, being orthogonal to the current velocity.

## B.7 External, Interactive, Reactive, and Total Force

The applied force $\mathbf{f}_i$ is the sum of the external force $\mathbf{f}_i^{\mathrm{ext}}$ that (supposedly known) external fields exert on the $i$-th particle, the interaction force $\mathbf{f}_i^{\mathrm{int}}$ exerted on it by the other $N-1$ particles, and the reaction $\mathbf{f}_i^{\mathrm{rct}}$ to constraints (if any):

$$
\mathbf{f}_i = \mathbf{f}_i^{\mathrm{ext}} + \mathbf{f}_i^{\mathrm{int}} + \mathbf{f}_i^{\mathrm{rct}}.
\tag{217}
$$

On account of (187), (210) and (213), balance (179) may be rewritten in the synthetic form

$$
\mathbf{f}_i^{\mathrm{tot}}(\tau) := -m_i\ddot{\mathbf{q}}_i(\tau) + \mathbf{f}_i^{\mathrm{ext}}(\tau) + \mathbf{f}_i^{\mathrm{int}}(\tau) + \mathbf{f}_i^{\mathrm{rct}}(\tau) = \mathbf{0},
\tag{218}
$$

[44] My own terminology.

where the total force $\mathbf{f}_i^{\text{tot}}$ is defined as the sum of the force of inertia $\mathbf{f}_i^{\text{nrt}}$ and the applied force $\mathbf{f}_i$, the fictitious force $\mathbf{f}_i^{\text{fct}}$ being incorporated in the external force $\mathbf{f}_i^{\text{ext}}$. For inertial observers, $\mathbf{f}_i^{\text{fct}} = \mathbf{0}$, and the nominal force of inertia $-m_i\ddot{\mathbf{q}}_i$ is minus the time derivative of the true particle momentum $\mathbf{p}_i$, as delivered by (185). Formally, (185) may be kept also for non-inertial observers, provided that the fictitious force delivered by (214) is added to the external force, as implied in (218).

### B.8 Interaction Force

In principle, each particle interacts with all the other $N-1$ particles in the system, but the intensity of interparticle interactions decays to zero—faster or slower—for increasing interparticle distance. It is sensible to organise the contribution to $\mathbf{f}_i^{\text{int}}$ of single particles, particle pairs, ..., particle $n$-tuples (with $n \le N-1$) in a hierarchical way [2, 12, 34, 37]:

$$\mathbf{f}_i^{\text{int}} = \sum_{j\neq i} \mathbf{f}_{ij} + \sum_{j<k}^{\neq i} \mathbf{f}_{ijk} + \sum_{j<k<l}^{\neq i} \mathbf{f}_{ijkl} + \cdots + \sum_{j_1<\cdots<j_n}^{\neq i} \mathbf{f}_{ij_1\ldots j_n}, \tag{219}$$

where the notation $\sum^{\neq i}$ means that the sum extends to indices different from $i$. The binary contribution $\mathbf{f}_{ij}$ is the force that would be applied by the $j$-th particle to the $i$-th particle if all other particles were removed to infinity; the ternary contribution $\mathbf{f}_{ijk}$ is the *extra* force applied to the $i$-th particle by the $j$-th and $k$-th particles, additional to the sum $\mathbf{f}_{ij} + \mathbf{f}_{ik}$ of their binary contributions, all other particles being removed to infinity; and so on, up to the $n$-ary contributions (typically, $n \ll N$). Moreover, to keep the system evolving within the (non-compact) configuration manifold $\mathcal{Q}$ for whatever initial state and external fields, any two particles should repel each other with diverging intensity whenever their distance

$$d_{ij}(\tau) := \|\mathbf{q}_i(\tau) - \mathbf{q}_j(\tau)\| > 0 \tag{220}$$

tends to zero:

$$d_{ij}(\tau) \downarrow 0 \qquad \Longrightarrow \qquad \mathbf{f}_{ij}(\tau) \cdot \frac{\mathbf{q}_i(\tau) - \mathbf{q}_j(\tau)}{d_{ij}(\tau)} \to +\infty. \tag{221}$$

An essential requirement, peculiar to interaction forces, is the invariance of the instantaneous power they expend on any test velocity under a general change of observer (200). For any subsystem $\mathcal{S} \subset \mathcal{M}$ comprised of more than one particle,

$$\sum_{i\in\mathcal{S}} \big(\overset{*}{\mathbf{f}}{}_i^{\text{int}_\mathcal{S}}(\overset{*}{\tau}) \cdot \overset{*}{\mathbf{v}}_i(\overset{*}{\tau}) - \mathbf{f}_i^{\text{int}_\mathcal{S}}(\tau) \cdot \mathbf{v}_i(\tau)\big) = 0, \tag{222}$$

where $\mathbf{f}_i^{\text{int}_\mathcal{S}}$ is the force applied to the $i$-th particle by *all and only* the other particles in $\mathcal{S}$. Since forces are covariant (cf. the general statement preceding (211)),

$$\overset{*}{\mathbf{f}}{}_i^{\text{int}_\mathcal{S}}(\overset{*}{\tau}) = \mathbf{Q}(\tau)\mathbf{f}_i^{\text{int}_\mathcal{S}}(\tau), \tag{223}$$

and since test velocities transform according to (202), equality (222) holds for all changes of observer if and only if

$$\sum_{i\in\mathcal{S}} \mathbf{f}_i^{\text{int}_\mathcal{S}}(\tau) = \mathbf{0}, \qquad \operatorname{skw} \sum_{i\in\mathcal{S}} \mathbf{r}_i(\tau) \otimes \mathbf{f}_i^{\text{int}_\mathcal{S}}(\tau) = \mathbf{0}, \tag{224}$$

where $\mathbf{r}_i(\tau) := \mathbf{q}_i(\tau) - \mathbf{q}_1(\tau)$, i.e., if and only if the mutual interaction forces within $\mathscr{S}$ sum up to zero and their resultant torque with respect to $\mathbf{q}_1(\tau)$ (and hence to any place in $\mathscr{E}$, thanks to $(224)_1$) is null. In particular, for the binary contribution $\mathbf{f}_{ij}$, equalities (224) are equivalent to

$$\mathbf{f}_{ij}(\tau) = -\mathbf{f}_{ji}(\tau), \qquad \mathbf{f}_{ij}(\tau) = \frac{f_{ij}(\tau)}{d_{ij}(\tau)}\big(\mathbf{q}_i(\tau) - \mathbf{q}_j(\tau)\big), \tag{225}$$

with $f_{ij}(\tau) \uparrow +\infty$ for $d_{ij}(\tau) \downarrow 0$ (cf. (221)).

In most, if not all, physical regimes where Newtonian mechanics of point particles applies, $\mathfrak{f}^{\mathsf{int}}$ is not detectably affected by the physical agencies determining $\mathfrak{f}^{\mathsf{ext}}$ and $\mathfrak{f}^{\mathsf{fct}}$. This motivates postulating that interaction forces depend on time only through the current state of the system along the motion $\tau \mapsto \mathfrak{q}(\tau)$, i.e.,

$$\mathfrak{f}^{\mathsf{int}}(\tau) = \widehat{\mathfrak{f}}^{\mathsf{int}}\big(\mathfrak{q}(\tau), \dot{\mathfrak{q}}(\tau)\big), \tag{226}$$

and, moreover, that $\widehat{\mathfrak{f}}^{\mathsf{int}}$ should be *invariant* under all change of observer (200):[45]

$$\overset{*}{\widehat{\mathfrak{f}}}{}^{\mathsf{int}} = \widehat{\mathfrak{f}}^{\mathsf{int}}. \tag{227}$$

Combined with (223), equality (227) significantly restricts the map $\widehat{\mathfrak{f}}^{\mathsf{int}}$, which has to satisfy identically the condition that, along all motions and for all change of observer,

$$\begin{aligned}&\widehat{\mathbf{f}}_i^{\mathsf{int}}\Big(\big(\mathbf{Q}(\tau)\big(\mathbf{q}_i(\tau) - \mathbf{q}_1(\tau)\big), \ldots\big), \big(\dot{\mathbf{q}}_i(\tau) + \mathbf{d}_1(\tau) + \mathbf{A}(\tau)\mathbf{Q}(\tau)\big(\mathbf{q}_i(\tau) - \mathbf{q}_1(\tau)\big), \ldots\big)\Big)\\ &\quad = \mathbf{Q}(\tau)\widehat{\mathbf{f}}_i^{\mathsf{int}}\big(\mathfrak{q}(\tau), \dot{\mathfrak{q}}(\tau)\big),\end{aligned} \tag{228}$$

where $\mathbf{d}_1(\tau) := \overset{*}{\mathbf{v}}_1(\overset{*}{\tau}) - \mathbf{Q}(\tau)\mathbf{v}_1(\tau)$. Condition (228) implies that interaction forces can depend on the current state of the system only through pairwise interparticle distances $d_{ij}(\tau)$ and their time derivatives $\dot{d}_{ij}(\tau)$, both invariant under a change of observer. Most often, $\mathfrak{f}^{\mathsf{int}}$ is supposed to be conservative, by postulating the existence of a (smooth) real-valued function, the interaction potential $\mathcal{V} : \mathcal{Q} \to \mathbb{R}$ such that

$$\mathbf{f}_i^{\mathsf{int}}(\tau) = -\partial_i \mathcal{V}|_{\mathfrak{q}(\tau)}, \tag{229}$$

where $\partial_i$ denotes differentiation with respect to the current position of the $i$-th particle. The force delivered by (229) is covariant if and only if the potential $\mathcal{V}$ is invariant under change of observer. In turn, condition (228) is satisfied if and only if $\mathcal{V}$ depends on the current configuration only through pairwise interparticle distances.

## B.9 Interaction Potential Energy

Unlike kinetic energy, the energy of a conservative interaction force cannot be defined particle by particle, independently of the state of all other particles, and then summed, as in (216). For any subsystem $\mathscr{S} \subset \mathscr{M}$ comprised of more than one particle, it is apposite to consider the work integral, analogous to (215),

$$\int_{\tau_-}^{\tau_+} \sum_{i \in \mathscr{S}} \mathbf{f}_i^{\mathsf{int}\mathscr{S}}(\tau) \cdot \dot{\mathbf{q}}_i(\tau)\, d\tau \tag{230a}$$

[45] In my opinion, the modest proviso preceding (226) settles the issue behind the confusing controversy about the validity of the invariance postulate (227), reported by Murdoch in [23].

where, as in (222), $\mathbf{f}_i^{\text{int}\mathscr{S}}$ is the force applied to the $i$-th particle by all and only the other particles in $\mathscr{S}$. If $\mathbf{f}_i^{\text{int}\mathscr{S}}$ admits a potential $\mathcal{V}^{\mathscr{S}} : \mathscr{Q}^{\mathscr{S}} \to \mathbb{R}$, i.e., if

$$\mathbf{f}_i^{\text{int}\mathscr{S}}(\tau) = -\partial_i \mathcal{V}^{\mathscr{S}}|_{\mathbf{q}|_{\mathscr{S}}(\tau)}, \tag{230b}$$

with $\mathscr{Q}^{\mathscr{S}}$ the submanifold of $\mathscr{Q}$ traced out by the motions of particles in $\mathscr{S}$, then

$$\begin{aligned}\int_{\tau_-}^{\tau_+} \sum_{i\in\mathscr{S}} \mathbf{f}_i^{\text{int}\mathscr{S}}(\tau)\cdot\dot{\mathbf{q}}_i(\tau)d\tau &= -\int_{\tau_-}^{\tau_+} \big(\mathcal{V}^{\mathscr{S}} \circ \mathbf{q}|_{\mathscr{S}}\big)^{\dot{}}(\tau)d\tau \\ &= \mathcal{V}^{\mathscr{S}}\big(\mathbf{q}|_{\mathscr{S}}(\tau_-)\big) - \mathcal{V}^{\mathscr{S}}\big(\mathbf{q}|_{\mathscr{S}}(\tau_+)\big),\end{aligned} \tag{231}$$

so that the work done by $\mathfrak{f}^{\text{int}\mathscr{S}}$ is path independent, being equal to the difference between the interaction potential energy in the initial configuration and that in the final configuration of the subsystem $\mathscr{S}$.

## B.10 Lagrangian Formulation

Under the hypothesis that (216) and (231) hold, it is useful to consider the action functional

$$\begin{aligned}&\mathcal{A}^{\mathscr{S}}\big(]\tau_-,\tau_+[\ni\tau\mapsto\mathbf{q}|_{\mathscr{S}}(\tau)\big) \\ &:= \int_{\tau_-}^{\tau_+} \mathcal{L}^{\mathscr{S}}\big(\mathbf{q}|_{\mathscr{S}}(\tau),\dot{\mathbf{q}}|_{\mathscr{S}}(\tau)\big)d\tau - \big[\mathbf{p}_{\text{cut}}|_{\mathscr{S}}(\tau)\cdot\big(\mathbf{q}|_{\mathscr{S}}(\tau)-\mathbf{o}\big)\big]_{\tau_-}^{\tau_+}\end{aligned} \tag{232}$$

where the Lagrangian function $\mathcal{L}^{\mathscr{S}}$ is defined as the difference between the kinetic energy (216) and the potential energy (230b):

$$\mathcal{L}^{\mathscr{S}}\big(\mathbf{q}|_{\mathscr{S}}(\tau),\dot{\mathbf{q}}|_{\mathscr{S}}(\tau)\big) := \mathcal{K}^{\mathscr{S}}\big(\dot{\mathbf{q}}|_{\mathscr{S}}(\tau)\big) - \mathcal{V}^{\mathscr{S}}\big(\mathbf{q}|_{\mathscr{S}}(\tau)\big), \tag{233}$$

and $\mathbf{o} := (\mathbf{o},\dots,\mathbf{o}) \in \mathscr{E}^N$. The variation of $\mathcal{A}^{\mathscr{S}}$ along $\mathbf{v}|_{\mathscr{S}}$ equals

$$\sum_{i\in\mathscr{S}}\Big(\int_{\tau_-}^{\tau_+}\big(-m_i\ddot{\mathbf{q}}_i(\tau)+\mathbf{f}_i^{\text{int}\mathscr{S}}(\tau)\big)\cdot\mathbf{v}_i(\tau)d\tau + \big[\big(m_i\dot{\mathbf{q}}_i(\tau)-\mathbf{p}_{i\text{cut}}(\tau)\big)\cdot\mathbf{v}_i(\tau)\big]_{\tau_-}^{\tau_+}\Big), \tag{234}$$

so that enforcing the stationarity of the action functional would yield

$$m_i\ddot{\mathbf{q}}_i(\tau) = \mathbf{f}_i^{\text{int}\mathscr{S}}(\tau),$$

plus momentum continuity conditions through time cuts (cf. (183)). The equation governing the evolution of the subsystem $\mathscr{S}$ is obtained by adding the work done on $\mathbf{v}|_{\mathscr{S}}$ by the forces not accounted for by the Lagrangian (233), namely

$$\sum_{i\in\mathscr{S}}\int_{\tau_-}^{\tau_+}\big(\mathbf{f}_i^{\text{int}\mathscr{S}\mathsf{C}}(\tau)+\mathbf{f}_i^{\text{ext}}(\tau)+\mathbf{f}_i^{\text{rct}}(\tau)\big)\cdot\mathbf{v}_i(\tau)d\tau, \tag{235}$$

to (234) and enforcing the condition that the sum be zero for all $\mathbf{v}|_{\mathscr{S}}$.[46] Equation (218) is recovered:

$$m_i\ddot{\mathbf{q}}_i(\tau) = \mathbf{f}_i(\tau) := \mathbf{f}_i^{\text{int}\mathscr{S}}(\tau) + \mathbf{f}_i^{\text{int}\mathscr{S}\mathsf{C}}(\tau) + \mathbf{f}_i^{\text{ext}}(\tau) + \mathbf{f}_i^{\text{rct}}(\tau), \tag{236}$$

[46] With this addition, the Lagrangian formulation is equivalent to the integral formulation of impulse balance (182) augmented with the work-energy theorem (215).

where the interaction force (219) on the $i$-th particle in $\mathscr{S}$ is split into the sum of the force exerted on it by the other particles in $\mathscr{S}$ and that exerted by the particles *not* in $\mathscr{S}$, i.e., in $\mathscr{S}^{\mathsf{C}} := \mathscr{M} \setminus \mathscr{S}$:

$$\mathbf{f}_i^{\mathsf{int}} = \mathbf{f}_i^{\mathsf{int}\mathscr{S}} + \mathbf{f}_i^{\mathsf{int}\mathscr{S}^{\mathsf{C}}}. \tag{237}$$

According to (219), $\mathbf{f}_i^{\mathsf{int}\mathscr{S}}$ is computed as follows:

$$\mathbf{f}_i^{\mathsf{int}\mathscr{S}} = \sum_{\mathscr{S}\ni j\neq i} \mathbf{f}_{ij} + \sum_{j<k\in\mathscr{S}}^{\neq i} \mathbf{f}_{ijk} + \sum_{j<k<l\in\mathscr{S}}^{\neq i} \mathbf{f}_{ijkl} + \cdots + \sum_{j_1<\cdots<j_n\in\mathscr{S}}^{\neq i} \mathbf{f}_{i j_1\ldots j_n}, \tag{238}$$

where the provisos $j_1 < \cdots < j_m \in \mathscr{S}$ $(\forall m \leq n)$ mean that the $j_1$-th, ..., $j_m$-th particles are all in $\mathscr{S}$. The force $\mathbf{f}_i^{\mathsf{int}\mathscr{S}^{\mathsf{C}}}$ is computed analogously, with the complementary provisos $j_1 < \cdots < j_m \in \mathscr{S}^{\mathsf{C}}$ $(\forall m \leq n)$.

## Appendix C: Complements to Sect. 3

### C.1 Inertial Forces

In order to spare space and to put the line of reasoning in better evidence, abridged notations are used in this appendix.

**Work Done on *Constant* Streaming Test Velocities** Let us first compute term-by-term the work done by (the opposite of) the nominal force of inertia on test velocities $\mathbf{v}_i = \widetilde{\chi}$ (cf. (100)):

$$\int_{\tau_-}^{\tau_+} d\tau \int_{\mathscr{C}_\chi(\tau)} \sum_i m_i \ddot{\chi} \cdot \widetilde{\chi} \delta_{\mathbf{q}_i(\tau)} \underset{\mathrm{m}}{\cong} \int_{\tau_-}^{\tau_+} M_{\mathscr{C}_\chi} \ddot{\chi} \cdot \widetilde{\chi}\, d\tau, \tag{239a}$$

$$\int_{\tau_-}^{\tau_+} d\tau \int_{\mathscr{C}_\chi(\tau)} \sum_i (m_i \ddot{\mathbf{F}} \mathbf{s}_i) \cdot \widetilde{\chi}\, \delta_{\mathbf{q}_i(\tau)} \underset{\mathrm{m}}{\cong} \int_{\tau_-}^{\tau_+} \left( \ddot{\mathbf{F}}\mathbf{F}^{-1} \int_{\mathscr{C}_\chi(\tau)} \sum_i m_i \mathbf{r}_i \delta_{\mathbf{q}_i(\tau)} \right) \cdot \widetilde{\chi}\, d\tau = 0, \tag{239b}$$

$$\int_{\tau_-}^{\tau_+} d\tau \int_{\mathscr{C}_\chi(\tau)} \sum_i (m_i \dot{\mathbf{F}} \dot{\mathbf{s}}_i) \cdot \widetilde{\chi}\, \delta_{\mathbf{q}_i(\tau)} \underset{\mathrm{m}}{\cong} \int_{\tau_-}^{\tau_+} \left( \dot{\mathbf{F}}\mathbf{F}^{-1} \int_{\mathscr{C}_\chi(\tau)} \sum_i m_i \mathbf{w}_i \delta_{\mathbf{q}_i(\tau)} \right) \cdot \widetilde{\chi}\, d\tau = 0, \tag{239c}$$

$$\begin{aligned} &\int_{\tau_-}^{\tau_+} d\tau \int_{\mathscr{C}_\chi(\tau)} \sum_i (m_i \mathbf{F} \ddot{\mathbf{s}}_i) \cdot \widetilde{\chi}\, \delta_{\mathbf{q}_i(\tau)} \\ &\underset{\mathrm{m}}{\cong} \int_{\tau_-}^{\tau_+} \left( \int_{\mathscr{C}_\chi(\tau)} \sum_i m_i (\dot{\mathbf{w}}_i - \dot{\mathbf{F}} \dot{\mathbf{s}}_i) \delta_{\mathbf{q}_i(\tau)} \right) \cdot \widetilde{\chi}\, d\tau \\ &\underset{\mathrm{m}}{\cong} \int_{\tau_-}^{\tau_+} \left( \int_{\mathscr{C}_\chi(\tau)} \sum_i m_i \mathbf{w}_i \delta_{\mathbf{q}_i(\tau)} \right)^{\boldsymbol{\cdot}} d\tau \underset{\mathrm{m}}{\cong} 0. \end{aligned} \tag{239d}$$

Equation (239a) makes use of (12) and (78); equalities (18) and (59) have been used in (239b); equality (25) and the approximate equality (64) have been used in (239c); this result and the fact that (25) and (64) hold at all time have been used in (239d).

**Work Done on *Linear* Streaming Test Velocities** Computing the work done on test velocities $\mathbf{v}_i = \widetilde{\mathbf{F}}\,\mathbf{s}_i$ is more challenging:

$$\int_{\tau_-}^{\tau_+} d\tau \int_{\mathcal{C}_\chi(\tau)} \sum_i m_i \ddot{\chi} \cdot (\widetilde{\mathbf{F}}\mathbf{s}_i)\delta_{\mathbf{q}_i(\tau)} \underset{\mathrm{m}}{\cong} \int_{\tau_-}^{\tau_+} \Big(\ddot{\mathbf{F}}\mathbf{F}^{-1} \int_{\mathcal{C}_\chi(\tau)} \sum_i m_i \mathbf{r}_i \delta_{\mathbf{q}_i(\tau)}\Big) \cdot \widetilde{\mathbf{F}} d\tau = 0, \tag{240a}$$

$$\begin{aligned} \int_{\tau_-}^{\tau_+} d\tau \int_{\mathcal{C}_\chi(\tau)} \sum_i (m_i \ddot{\mathbf{F}}\mathbf{s}_i) \cdot (\widetilde{\mathbf{F}}\mathbf{s}_i)\delta_{\mathbf{q}_i(\tau)} &\underset{\mathrm{m}}{\cong} \int_{\tau_-}^{\tau_+} \Big(\ddot{\mathbf{F}} \int_{\mathcal{C}_\chi(\tau)} \sum_i m_i \mathbf{s}_i \otimes \mathbf{s}_i \delta_{\mathbf{q}_i(\tau)}\Big) \cdot \widetilde{\mathbf{F}} d\tau \\ &\underset{\mathrm{m}}{\cong} \int_{\tau_-}^{\tau_+} (\ddot{\mathbf{F}}\mathbf{E}_{\mathcal{C}_\chi}(t_0)) \cdot \widetilde{\mathbf{F}} d\tau, \end{aligned} \tag{240b}$$

$$\begin{aligned} &\int_{\tau_-}^{\tau_+} d\tau \int_{\mathcal{C}_\chi(\tau)} \sum_i (m_i \dot{\mathbf{F}}\dot{\mathbf{s}}_i) \cdot (\widetilde{\mathbf{F}}\mathbf{s}_i)\delta_{\mathbf{q}_i(\tau)} \\ &\quad \underset{\mathrm{m}}{\cong} \int_{\tau_-}^{\tau_+} \Big(\int_{\mathcal{C}_\chi(\tau)} \sum_i m_i \mathbf{s}_i \otimes (\dot{\mathbf{F}}\mathbf{F}^{-1}\mathbf{w}_i)\delta_{\mathbf{q}_i(\tau)}\Big) \cdot \widetilde{\mathbf{F}} d\tau \\ &\quad \underset{\mathrm{m}}{\cong} \int_{\tau_-}^{\tau_+} \Big(\dot{\mathbf{F}}\mathbf{F}^{-1}\Big(\int_{\mathcal{C}_\chi(\tau)} \sum_i m_i \mathbf{r}_i \otimes \mathbf{w}_i \delta_{\mathbf{q}_i(\tau)}\Big)\mathbf{F}^{-\top}\Big) \cdot \widetilde{\mathbf{F}} d\tau = 0, \end{aligned} \tag{240c}$$

$$\begin{aligned} &\int_{\tau_-}^{\tau_+} d\tau \int_{\mathcal{C}_\chi(\tau)} \sum_i (m_i \mathbf{F}\ddot{\mathbf{s}}_i) \cdot (\widetilde{\mathbf{F}}\mathbf{s}_i)\delta_{\mathbf{q}_i(\tau)} \\ &\quad \underset{\mathrm{m}}{\cong} \int_{\tau_-}^{\tau_+} \Big(\int_{\mathcal{C}_\chi(\tau)} \sum_i m_i \mathbf{r}_i \otimes (\dot{\mathbf{w}}_i - \dot{\mathbf{F}}\dot{\mathbf{s}}_i)\mathbf{F}^{-\top}\delta_{\mathbf{q}_i(\tau)}\Big) \cdot \widetilde{\mathbf{F}} d\tau \\ &\quad \underset{\mathrm{m}}{\cong} -\int_{\tau_-}^{\tau_+} \Big(\int_{\mathcal{C}_\chi(\tau)} \sum_i (m_i \dot{\mathbf{r}}_i \otimes \mathbf{w}_i)\mathbf{F}^{-\top}\delta_{\mathbf{q}_i(\tau)}\Big) \cdot \widetilde{\mathbf{F}} d\tau \\ &\quad \underset{\mathrm{m}}{\cong} -\int_{\tau_-}^{\tau_+} \Big(\int_{\mathcal{C}_\chi(\tau)} \sum_i (m_i (\dot{\mathbf{F}}\mathbf{F}^{-1}\mathbf{r}_i + \mathbf{w}_i) \otimes \mathbf{w}_i)\mathbf{F}^{-\top}\delta_{\mathbf{q}_i(\tau)}\Big) \cdot \widetilde{\mathbf{F}} d\tau \\ &\quad \underset{\mathrm{m}}{\cong} -\int_{\tau_-}^{\tau_+} \Big(\int_{\mathcal{C}_\chi(\tau)} \Big(\textstyle\sum_i m_i \mathbf{w}_i \otimes \mathbf{w}_i\Big)\mathbf{F}^{-\top}\delta_{\mathbf{q}_i(\tau)}\Big) \cdot \widetilde{\mathbf{F}} d\tau \\ &\quad = -\int_{\tau_-}^{\tau_+} (\mathbf{M}^{\mathsf{kin}}_{\mathcal{C}_\chi}\mathbf{F}^{-\top}) \cdot \widetilde{\mathbf{F}} d\tau. \end{aligned} \tag{240d}$$

Equation (240a) uses (18) and (59); equations (240b) and (240c) use, respectively, (87) and (26), (64); equation (240d) uses a result obtained within (240c), the fact that (26), (59) and (64) hold at all time, and definition (112).

**Work Done on *Thermal* Test Velocities** Computing the work done on test velocities $\mathbf{v}_i = \mathbf{F}\widetilde{\mathbf{s}}_i$ is straightforward.

### C.2 Interaction Forces

Let us compute step by step the boundary integral on the first line of (123):

$$\begin{aligned} &\int_{\partial\mathcal{C}_\chi(\tau)} \mathbf{t}(\cdot,\tau) \cdot \underset{\sim}{\mathbf{v}}(\cdot,\tau) H^2 \llcorner \partial\mathcal{C}_\chi(\tau) \\ &\quad = \int_{\partial\mathcal{C}_\chi(\tau)} (\mathbf{T}(\cdot,\tau)\mathbf{n}_{\partial\mathcal{C}_\chi(\tau)}) \cdot \underset{\sim}{\mathbf{v}}(\cdot,\tau) H^2 \llcorner \partial\mathcal{C} \end{aligned}$$

$$
\begin{aligned}
&= \int_{\partial\mathscr{C}_\chi(\tau)} \big(\mathbf{T}(\cdot,\tau)^\top \underset{\sim}{\mathbf{y}}(\cdot,\tau)\big)\cdot \mathbf{n}_{\partial\mathscr{C}_\chi(\tau)}\, H^2 \llcorner \partial\mathscr{C} = \int_{\mathscr{C}_\chi(\tau)} \operatorname{div}\big(\mathbf{T}(\cdot,\tau)^\top \underset{\sim}{\mathbf{y}}(\cdot,\tau)\big) H^3 \\
&= \int_{\mathscr{C}_\chi(\tau)} \big(\operatorname{div}\mathbf{T}|_{(\cdot,\tau)}\cdot \underset{\sim}{\mathbf{y}}(\cdot,\tau) + \mathbf{T}(\cdot,\tau)\cdot \nabla\underset{\sim}{\mathbf{y}}|_{(\cdot,\tau)}\big) H^3 \\
&\underset{\mathrm{m}}{\cong} V_{\mathscr{C}}(\tau)\big(\operatorname{div}\mathbf{T}|_{(\boldsymbol{\xi},\tau)}\cdot \underset{\sim}{\mathbf{y}}(\boldsymbol{\xi},\tau) + \mathbf{T}(\boldsymbol{\xi},\tau)\cdot \nabla\underset{\sim}{\mathbf{y}}|_{(\boldsymbol{\xi},\tau)}\big) \\
&\underset{\mathrm{m}}{\cong} \mathbf{f}^{\mathsf{shr}\mathscr{S}\mathsf{C}}_{\mathscr{C}_\chi}(\tau)\cdot \widetilde{\chi}(\mathbf{X},\tau) + \mathbf{M}^{\mathsf{shr}\mathscr{S}\mathsf{C}}_{\mathscr{C}_\chi}(\tau)\cdot \big(\widetilde{\mathbf{F}}(\mathbf{X},\tau)\mathbf{F}(\mathbf{X},\tau)^{-1}\big).
\end{aligned}
\tag{241}
$$

Cauchy's theorem (128) is used first, then the divergence theorem and the definition of the divergence of a tensor field. The equality $\nabla\underset{\sim}{\mathbf{y}}|_{(\boldsymbol{\xi},\tau)} = \widetilde{\mathbf{F}}(\mathbf{X},\tau)\mathbf{F}(\mathbf{X},\tau)^{-1}$ stems from definitions $(99)_2$ and (124).

## C.3 Nominal Force of Inertia: Lagrangian Formalism

A representation of the *mesoscopic cell kinetic energy* $\mathcal{K}_{\mathscr{C}_\chi}$ in terms of the independent referential descriptors $\chi(\mathbf{X},\cdot)$, $\mathbf{F}(\mathbf{X},\cdot)$ and $\mathbf{s}_i(\cdot)$ is based on (65a):

$$
\begin{aligned}
\mathcal{K}_{\mathscr{C}_\chi}(\tau) &:= \frac{1}{2T_{\mathrm{m}}}\int_{\mathscr{I}_\tau} d\sigma \int_{\mathscr{C}_\chi(\sigma)} \sum_i m_i \|\dot{\mathbf{q}}_i(\sigma)\|^2 \delta_{\mathbf{q}_i(\sigma)} \\
&\underset{\mathrm{m}}{\cong} \frac{1}{2T_{\mathrm{m}}}\int_{\mathscr{I}_\tau} d\sigma \int_{\mathscr{C}_\chi(\sigma)} \sum_i m_i \|\dot{\chi}(\mathbf{X},\sigma) + \dot{\mathbf{F}}(\mathbf{X},\sigma)\mathbf{s}_i(\sigma) + \mathbf{F}(\mathbf{X},\sigma)\dot{\mathbf{s}}_i(\sigma)\|^2 \delta_{\mathbf{q}_i(\sigma)} \\
&\underset{\mathrm{m}}{\cong} \frac{1}{2} M_{\mathscr{C}_\chi}(\tau)\dot{\chi}(\mathbf{X},\tau)\cdot\dot{\chi}(\mathbf{X},\tau) + \tfrac{1}{2}\big(\dot{\mathbf{F}}(\mathbf{X},\tau)\mathbf{E}_{\mathscr{C}_\chi}(t_0)\big)\cdot\dot{\mathbf{F}}(\mathbf{X},\tau) \\
&\quad + \frac{1}{2T_{\mathrm{m}}}\int_{\mathscr{I}_\tau} d\sigma \int_{\mathscr{C}_\chi(\sigma)} \sum_i m_i \big(\mathbf{C}(\mathbf{X},\sigma)\dot{\mathbf{s}}_i(\sigma)\big)\cdot \dot{\mathbf{s}}_i(\sigma)\delta_{\mathbf{q}_i(\sigma)},
\end{aligned}
\tag{242}
$$

with $M_{\mathscr{C}_\chi}(\tau)$ the mesoscopic cell mass (78), $\mathbf{E}_{\mathscr{C}_\chi}(t_0)$ the referential cell Euler tensor (87), and $\mathbf{C}(\mathbf{X},\tau)$ the change-of-metric tensor (58). The second approximate equality relies on the fact that the three terms of the sum on the second line of (242) are (approximately) mutually orthogonal (cf. Sects. 2.2 and 2.6). The quantity on the fourth line—the time-averaged *cell thermal energy*—equals one half of the trace of the mesoscopic cell kinetic moment (cf. (112)).

The kinetic action functional[47]

$$
\begin{aligned}
&\mathcal{A}^{\mathsf{kin}}_{\mathscr{C}_\chi}\big(]\tau_-,\tau_+[\ni\tau\mapsto\big(\chi(\mathbf{X},\tau),\mathbf{F}(\mathbf{X},\tau),\mathbf{s}_i(\tau)\big)\big) \\
&:= \frac{1}{2}\int_{\tau_-}^{\tau_+} \big(M_{\mathscr{C}_\chi}(\tau)\dot{\chi}(\mathbf{X},\tau)\cdot\dot{\chi}(\mathbf{X},\tau) + \big(\dot{\mathbf{F}}(\mathbf{X},\tau)\mathbf{E}_{\mathscr{C}_\chi}(t_0)\big)\cdot\dot{\mathbf{F}}(\mathbf{X},\tau)\big)d\tau \\
&\quad + \frac{1}{2}\int_{\tau_-}^{\tau_+} d\tau \int_{\mathscr{C}_\chi(\tau)} \sum_i m_i\big(\mathbf{C}(\mathbf{X},\tau)\dot{\mathbf{s}}_i(\tau)\big)\cdot\dot{\mathbf{s}}_i(\tau)\delta_{\mathbf{q}_i(\tau)} + \text{boundary terms}
\end{aligned}
\tag{243}
$$

is mesoscopically equivalent to the Newtonian kinetic action, i.e., to the time integral of the instantaneous kinetic energy, provided that the time interval $]\tau_-,\tau_+[$ is at least of mesoscopic

[47] The boundary terms in (243), whose function is to ensure continuity of Lagrangian momenta across time cuts, are straightforward to obtain by analogy with (232), but too cumbersome to write down.

size: $\tau_+ - \tau_- \geq T_\mathsf{m}$. The variational derivatives of $\mathcal{A}^{\mathsf{kin}}_{\mathscr{C}_\chi}$ with respect to $\chi(\mathbf{X},\cdot)$, $\mathbf{F}(\mathbf{X},\cdot)$, and $\mathbf{s}_i(\cdot)$ are, respectively,

$$-M_{\mathscr{C}_\chi}(\tau)\ddot{\chi}(\mathbf{X},\tau), \tag{244a}$$

$$-\ddot{\mathbf{F}}(\mathbf{X},\tau)\mathbf{E}_{\mathscr{C}_\chi}(t_0) + \mathbf{M}^{\mathsf{kin}}_{\mathscr{C}_\chi}(\tau)\mathbf{F}(\mathbf{X},\tau)^{-\top}, \tag{244b}$$

$$-m_i\big(\mathbf{C}(\mathbf{X},\tau)\ddot{\mathbf{s}}_i(\tau) + \dot{\mathbf{C}}(\mathbf{X},\tau)\dot{\mathbf{s}}_i(\tau)\big). \tag{244c}$$

The contribution of the cell kinetic moment $\mathbf{M}^{\mathsf{kin}}_{\mathscr{C}_\chi}$ to the variational derivative of $\mathcal{A}^{\mathsf{kin}}_{\mathscr{C}_\chi}$ with respect to $\mathbf{F}(\mathbf{X},\cdot)$ is a direct consequence of the fact that, contrary to the tangent bundle $\mathsf{T}\mathscr{Q}_{\mathscr{C}\times\mathscr{I}}$, the streaming subbundle and the thermal subbundle into which $\mathsf{T}\mathscr{Q}_{\mathscr{C}\times\mathscr{I}}$ gets decomposed by (65a) are nontrivial. Comparing (244a)–(244c) with (134a)–(134c) shows that (244a) and (244b) match with (134a) and (134b), respectively; (244c), on the contrary, differs from (134c) by not including the terms

$$-m_i\big(2\mathrm{skw}\big(\mathbf{F}(\mathbf{X},\tau)^\top\dot{\mathbf{F}}(\mathbf{X},\tau)\big)\dot{\mathbf{s}}_i(\tau) + \mathbf{F}(\mathbf{X},\tau)^\top\big(\ddot{\mathbf{F}}(\mathbf{X},\tau)\mathbf{s}_i(\tau) + \ddot{\chi}(\mathbf{X},\tau)\big)\big). \tag{245}$$

The first term in (245) is absent from (134c) because, by construction, it is orthogonal to $\dot{\mathbf{s}}_i(\tau)$; as such, it is powerless and does not contribute to the kinetic energy. The other two terms, while present in (153), are regarded as negligible in Sect. 3.7 (cf. (154)); in the present variational treatment, they are filtered out a priori by the slowness assumptions on $\dot{\chi}(\mathbf{X},\cdot)$ and $\dot{\mathbf{F}}(\mathbf{X},\cdot)$, implicit in the definition of the mesoscopic cell kinetic energy (242). In conclusion, the only substantial difference between (134c) and (244c) is the first term of (245). Whether it can produce significant consequences is a question deserving investigation.

## References


1. Abhyankar, S.S., Christensen, C.: Semidirect products: $x \mapsto ax - b$ as a first example. Math. Mag. **75**(4), 284–289 (2002). https://doi.org/10.1080/0025570X.2002.11953146
2. Admal, N.C., Tadmor, E.B.: A unified interpretation of stress in molecular systems. J. Elast. **100**(1), 63–143 (2010). https://doi.org/10.1007/s10659-010-9249-6
3. Alder, B.J.: Berni J. Alder, interview. In: Mac Kernan, D. (ed.) SIMU Challenges in Molecular Simulations: Bridging the Length- and Timescales Gap, vol. 4, pp. 15–58. Centre Européen de Calcul Atomique et Moléculaire, Écublens (2002). https://doi.org/10.13140/2.1.2562.7843
4. Andersen, H.C.: Molecular dynamics simulations at constant pressure and/or temperature. J. Chem. Phys. **72**(4), 2384–2393 (1980). https://doi.org/10.1063/1.439486
5. Bloch, A.M., Marsden, J.E., Zenkov, D.V.: Nonholonomic dynamics. Not. Am. Math. Soc. **52**(3), 324–333 (2005)
6. Bouchet, F., Gupta, S., Mukamel, D.: Thermodynamics and dynamics of systems with long-range interactions. Physica A **389**(20), 4389–4405 (2010). https://doi.org/10.1016/j.physa.2010.02.024
7. Capriz, G.: On ephemeral continua. Phys. Mesomech. **11**(5–6), 285–298 (2008)
8. Capriz, G., Giovine, P.: Classes of ephemeral continua. Math. Methods Appl. Sci. **41**(3), 1175–1196 (2018). https://doi.org/10.1002/mma.4658
9. Cauchy, A.L.: Exercises de Mathématiques—Seconde Année. Chez de Bure Frères, Libraires du Roi et de la Bibliothèque du Roi, Paris, Rue Serpente, 7 (1827)
10. Cleveland, C.L.: New equations of motion for molecular dynamics systems that change shape. J. Chem. Phys. **89**(8), 4987–4993 (1988). https://doi.org/10.1063/1.455642
11. DiCarlo, A.: A major serendipitous contribution to continuum mechanics. Mech. Res. Commun. **93**, 41–46 (2018). https://doi.org/10.1016/j.mechrescom.2017.10.002
12. Fisher, M.E.: The free energy of a macroscopic system. Arch. Ration. Mech. Anal. **17**(5), 377–410 (1964). https://doi.org/10.1007/BF00250473
13. Flannery, M.R.: d'Alembert-Lagrange analytical dynamics for nonholonomic systems. J. Math. Phys. **52**(3), 032,705 (2011). https://doi.org/10.1063/1.3559128

14. Irving, J.H., Kirkwood, J.G.: The statistical mechanical theory of transport processes. IV. The equations of hydrodynamics. J. Chem. Phys. **18**(6), 817–829 (1950). https://doi.org/10.1063/1.1747782
15. Lamé, G.: Leçons sur la Théorie Mathématique de l'Elasticité des Corps Solides, 1 edn. Bachelier, Imprimeur-Libraire du Bureau des Longitudes et de l'École Polytechnique, Paris, Quai des Augustins, 55 (1852)
16. Lehoucq, R.B., Von Lilienfeld-Toal, A.: Translation of Walter Noll's "Derivation of the fundamental equations of continuum thermodynamics from statistical mechanics". J. Elast. **100**(1), 5–24 (2010). https://doi.org/10.1007/s10659-010-9246-9
17. Li, S., Tong, Q.: A concurrent multiscale micromorphic molecular dynamics. J. Appl. Phys. **117**(15), 154,303 (2015). https://doi.org/10.1063/1.4916702
18. Li, S., Urata, S.: An atomistic-to-continuum molecular dynamics: theory, algorithm, and applications. Comput. Methods Appl. Mech. Eng. **306**, 452–478 (2016). https://doi.org/10.1016/j.cma.2016.03.048
19. Murdoch, A.I.: A corpuscular approach to continuum mechanics: basic considerations. Arch. Ration. Mech. Anal. **88**(4), 291–321 (1985). https://doi.org/10.1007/BF00250868
20. Murdoch, A.I.: A critique of atomistic definitions of the stress tensor. J. Elast. **88**(2), 113–140 (2007). https://doi.org/10.1007/s10659-007-9121-5
21. Murdoch, A.I.: On molecular modelling and continuum concepts. J. Elast. **100**(1), 33–61 (2010). https://doi.org/10.1007/s10659-010-9248-7
22. Murdoch, A.I.: Physical Foundations of Continuum Mechanics. Cambridge University Press, Cambridge (2012)
23. Murdoch, A.I.: A personal appreciation of Walter Noll. J. Elast. (2019, this issue). https://doi.org/10.1007/s10659-018-9704-3
24. Noll, W.: Die Herleitung der Grundgleichungen der Thermomechanik der Kontinua aus der Statistischen Mechanik. J. Ration. Mech. Anal. **4**, 627–646 (1955). https://doi.org/10.1512/iumj.1955.4.54022
25. Parrinello, M., Rahman, A.: Crystal structure and pair potentials: a molecular-dynamics study. Phys. Rev. Lett. **45**, 1196–1199 (1980). https://doi.org/10.1103/PhysRevLett.45.1196
26. Parrinello, M., Rahman, A.: Polymorphic transitions in single crystals: a new molecular dynamics method. J. Appl. Phys. **52**(12), 7182–7190 (1981). https://doi.org/10.1063/1.328693
27. Passerone, D., Tosatti, E., Chiarotti, G.L., Ercolessi, F.: Variable-curvature-slab molecular dynamics as a method to determine surface stress. Phys. Rev. B **59**, 7687–7696 (1999). https://doi.org/10.1103/PhysRevB.59.7687
28. Podio-Guidugli, P.: On (Andersen-)Parrinello-Rahman molecular dynamics, the related metadynamics, and the use of the Cauchy-Born rule. J. Elast. **100**(1), 145–153 (2010). https://doi.org/10.1007/s10659-010-9250-0
29. Ray, J.R.: Nonholonomic constraints. Am. J. Phys. **34**(12), 1202–1203 (1966). https://doi.org/10.1119/1.1972703. Erratum: Am. J. Phys. **34**, 406 (1966)
30. Ray, J.R.: Nonholonomic constraints and Gauss' principle of least constraint. Am. J. Phys. **40**(1), 179–188 (1972). https://doi.org/10.1119/1.1986465
31. Ray, J.R., Rahman, A.: Statistical ensembles and molecular dynamics studies of anisotropic solids. J. Chem. Phys. **80**(9), 4423–4428 (1984). https://doi.org/10.1063/1.447221
32. Ribarsky, M.W., Landman, U.: Dynamical simulations of stress, strain, and finite deformations. Phys. Rev. B **38**, 9522–9537 (1988). https://doi.org/10.1103/PhysRevB.38.9522
33. Rossi, G.C., Testa, M.: The stress tensor in thermodynamics and statistical mechanics. J. Chem. Phys. **132**(7), 074902 (2010). https://doi.org/10.1063/1.3316134
34. Tadmor, E.B., Miller, R.E.: Modeling Materials: Continuum, Atomistic and Multiscale Techniques. Cambridge University Press, Cambridge (2011)
35. Tartaglino, U., Tosatti, E., Passerone, D., Ercolessi, F.: Bending strain-driven modification of surface reconstructions: Au(111). Phys. Rev. B **65**, 241406 (2002). https://doi.org/10.1103/PhysRevB.65.241406
36. Tong, Q., Li, S.: From molecular systems to continuum solids: a multiscale structure and dynamics. J. Chem. Phys. **143**(6), 064101 (2015). https://doi.org/10.1063/1.4927656
37. Torres-Sánchez, A., Vanegas, J.M., Arroyo, M.: Geometric derivation of the microscopic stress: a covariant central force decomposition. J. Mech. Phys. Solids **93**, 224–239 (2016). https://doi.org/10.1016/j.jmps.2016.03.006
38. Ulz, M.H.: Comments on a continuum-related Parrinello-Rahman molecular dynamics formulation. J. Elast. **113**(1), 93–112 (2013). https://doi.org/10.1007/s10659-012-9412-3
39. Urata, S., Li, S.: A multiscale model for amorphous materials. Comput. Mater. Sci. **135**, 64–77 (2017). https://doi.org/10.1016/j.commatsci.2017.03.029
40. Wentzcovitch, R.M.: Invariant molecular-dynamics approach to structural phase transitions. Phys. Rev. B **44**, 2358–2361 (1991). https://doi.org/10.1103/PhysRevB.44.2358

J Elast (2019) 135:237–260
https://doi.org/10.1007/s10659-018-9693-2

# Material Geometry

**Marcelo Epstein**[1] · **Víctor Manuel Jiménez**[2] ·
**Manuel de León**[2]

Received: 15 April 2018 / Published online: 30 October 2018


**Abstract** Walter Noll's trailblazing constitutive theory of material defects in smoothly uniform bodies is recast in the language of Lie groupoids and their associated Lie algebroids. From this vantage point the theory is extended to non-uniform bodies by introducing the notion of singular material distributions and the physically cognate idea of graded uniformity and homogeneity.

**Keywords** Uniformity · Homogeneity · Groupoids · Lie algebroids · Singular distributions · Stefan-Sussman theorem

**Mathematics Subject Classification** 74A20 · 22A22

## 1 Introduction

The intimate kinship between Continuum Mechanics and Differential Geometry is apparent from the very fact that a continuum is, by definition, a differentiable manifold. Moreover, just as in classical Analytical Mechanics the configuration space of a system of particles is regarded as a finite-dimensional differentiable manifold, so too the configuration space of a continuum can be regarded as an infinite dimensional manifold of maps [7, 18]. Thus,

---

Walter Noll's scientific legacy stand as a paradigm of Apollonian clarity, but it is also able to inspire the ecstasy of aesthetic inebriation. The first acquaintance with his work has been a decisive turning point in the intellectual lives of many, who are thereby forever grateful. This work is dedicated to his memory.

✉ M. Epstein
mepstein@ucalgary.ca

V.M. Jiménez
victor.jimenez@icmat.es

M. de León
mdeleon@icmat.es

1 University of Calgary, Calgary, Canada

2 Instituto de Ciencias Matemáticas (CSIC-UAM-UC3M-UCM), Madrid, Spain

the kinematic aspects of the discipline are completely embraceable within a differential geometric context. One of the seminal contributions of Walter Noll [17] was the realization that the specification of a constitutive response of a materially uniform body automatically endows the body with an additional differential geometric structure that we call *material geometry*. This point of view is consonant with the rich tradition of the theory of continuous distribution of defects, pioneered earlier by Kondo [13], Kroener [14], Eshelby [8], Bilby [1], Frank [9], and others. Noll's original work was soon significantly extended by Wang [19] and Bloom [2]. Furthermore, the formalism of $G$-structures [3] and groupoids [4] was shown to be the natural tool to express the geometry associated with the constitutive structure of uniform bodies of simple and structured media.

Our objective in this article is to present a unified treatment of material geometry from the vantage point of the geometric theory of groupoids. What is gained in so doing is the awareness that to any constitutive response of a body, whether uniform or not, one can uniquely associate a corresponding *material groupoid*. If the body happens to be smoothly uniform, an associated *material Lie algebroid* becomes available and plays a fundamental role in the determination of homogeneity conditions. If, on the other hand, the constitutive response is smooth but the body is not uniform, we show that a unique singular *material distribution* can be defined that divides the body into parts of various dimensions, each of which is smoothly uniform. Some of these results have been separately introduced in recent publications [10, 11].

## 2 Constitutive Response and Material Isomorphisms

The simplest example of a geometric material structure is provided by (first-grade) elastic materials. The *constitutive response* at a point $X$ of a *body manifold* $\mathcal{B}$ is defined as follows. Let $\mathcal{L}(U, V)$ denote the space of linear isomorphisms between two vector spaces $U, V$ of the same finite dimension. A constitutive response at $X \in \mathcal{B}$ is a (smooth) map

$$\psi_X : \mathcal{L}\big(T_X\mathcal{B}, \mathbb{R}^3\big) \to \mathcal{R}, \tag{1}$$

where $T_X\mathcal{B}$ is the tangent space of $\mathcal{B}$ at $X$ and where $\mathcal{R}$ is a vector space of values. The *Cauchy stress*, for example, takes values in the space of contravariant symmetric tensors on $\mathbb{R}^3$. The constitutive response of the body is the collection of the constitutive responses of its points. A *configuration* $\kappa$ of the body is an embedding

$$\kappa : \mathcal{B} \to \mathbb{E}^3, \tag{2}$$

where $\mathbb{E}^3$ is the standard Euclidean space, whose tangent space at each point is precisely $\mathbb{R}^3$. For any configuration $\kappa$ we can evaluate the field $\psi$ (such as the Cauchy stress field) in that configuration as the map $\psi_\kappa : \mathcal{B} \to \mathcal{R}$ given by

$$\psi_\kappa(X) = \psi_X\big(\kappa_*(X)\big), \tag{3}$$

where $\kappa_*(X)$ is the derivative of $\kappa$ at $X$. Notice that $\kappa_*(X)$ belongs indeed to $\mathcal{L}(T_X\mathcal{B}, \mathbb{R}^3)$, as implied by Eq. (2) and the notion of derivative of a map between manifolds.

These definitions do not make use of any specific reference configuration. If such a reference configuration, say $\kappa_0$, is specified, then the constitutive response with respect to that configuration can be deduced by composition of maps. The use of a reference configuration is justified in practical applications by the need to have an explicit analytic expression for the constitutive response, for which the terminology *constitutive equation* is usually applied. Notice that the 'master' constitutive response $\psi$ defined on the body manifold guarantees the usual consistency rules between the subsidiary constitutive equations referred to different reference configurations.

Two body points, $X, Y \in \mathcal{B}$, are *materially isomorphic* if there exists a map $\mathbf{P} \in \mathcal{L}(T_X\mathcal{B}, T_Y\mathcal{B})$ such that

$$\psi_Y(\mathbf{F}) = \psi_X(\mathbf{FP}) \quad \forall \mathbf{F} \in \mathcal{L}\big(T_Y\mathcal{B}, \mathbb{R}^3\big). \tag{4}$$

The physical meaning of the existence of a material isomorphism $\mathbf{P}$ between $X$ and $Y$ is that these two points are made of the same material. A body all of whose points are mutually materially isomorphic is said to be *materially uniform*. A body is *smoothly uniform* if it can be covered with materially uniform open sets such that in each of them the material isomorphisms can be chosen smoothly.

A material automorphism at a point $X \in \mathcal{B}$ is nothing but a *material symmetry* of $X$. Under a material symmetry, the constitutive response remains *invariant*, as implied by Eq. (4) when setting $Y = X$. The collection of all material symmetries at $X$ forms a group $\mathcal{G}_X$, known as the *symmetry group* at $X$. A group is essentially an algebraic structure. If it happens to be a *Lie group*, $\mathcal{G}_X$ also has the differential geometric structure of a smooth manifold. Thus, every point of a material body endowed with a constitutive response carries its own algebraic and/or geometric structure. In a celebrated imagery [20], we may think of the body as a bathroom floor made of infinitesimal tiles, each of which has some symmetries (rotations and reflections). These are *local symmetries*. On the other hand, different tiles may also be compared with each other. If two tiles are congruent, they can be mutually transplanted so as to occupy each other's former place. Under this action, the floor remains invariant. It seems justifiable to call these congruences *distant symmetries* and to ask the question as to whether there might exist an algebraic/geometric structure that can encompass *both local and distant symmetries* under a single umbrella enjoying properties similar, if not identical, to those of a group.

# 3 The Geometry Induced by the Material Isomorphisms

## 3.1 An Intuitive View

Each material isomorphism $\mathbf{P} : T_X\mathcal{B} \to T_Y\mathcal{B}$ can be imagined as an arrow with its tail at $X$ and its tip at $Y$. Let $\mathcal{P}$ denote the set of all these arrows corresponding to a given constitutive response. This set is never empty. Indeed, it contains at the very least the identity automorphisms $id_X : T_X\mathcal{B} \to T_X\mathcal{B}$ for all $X \in \mathcal{B}$. In the arrow description, these are loop-shaped arrows, one at each body point. Moreover, if a point $X$ has additional non-trivial material symmetries, we obtain a collection of loop-shaped arrows at $X$ which embody the local symmetry group $\mathcal{G}_X$.

Consider two different points $X, Y$ that happen to be materially isomorphic. In that case, we include an arrow $\mathbf{P}(X, Y)$ from $X$ to $Y$. Since material isomorphism is an equivalence relation, we must also include the reverse arrow $\mathbf{P}(Y, X) = \mathbf{P}^{-1}(X, Y)$. Moreover, if $\mathbf{P}(Y, Z)$ is an additional material isomorphism between $Y$ and $Z$, by transitivity we must also include the composite arrow $\mathbf{P}(X, Z) = \mathbf{P}(Y, Z)\mathbf{P}(X, Y)$, as suggested in Fig. 1. In other words, arrows are composed in the standard tip-to-tail fashion, and in no other way. Two arrows cannot be composed unless the tip of the first coincides with the tail of the second. We will

Fig. 1 Material isomorphisms as groupoid arrows and their composition

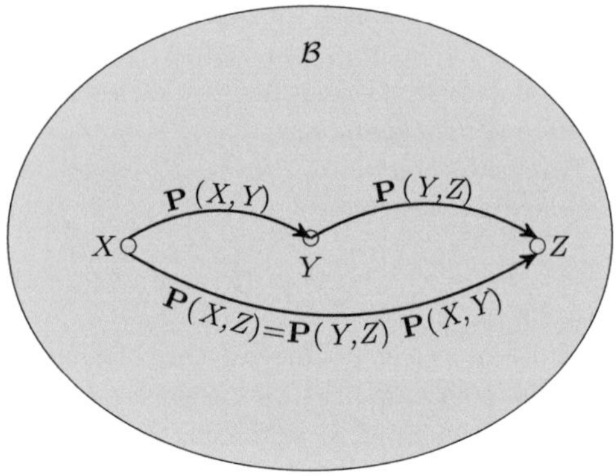


call the collection $\mathcal{P}$ of all these arrows the *material groupoid* associated with the given constitutive response, a terminology that will be presently justified. At this stage we can state that

> *To any given material response $\psi_X(\cdot)$, whether or not continuous with respect to X, we can associate a uniquely defined material groupoid.*

### 3.2 Groupoids: Definition and Terminology

Following [15], a *groupoid* $\mathcal{Z} \rightrightarrows \mathcal{M}$ consists of

- A *total set* $\mathcal{Z}$, whose elements are called *arrows*;
- A *base set* $\mathcal{M}$, whose elements are called *objects*;
- Two surjective projection maps, $\alpha : \mathcal{Z} \to \mathcal{M}$ and $\beta : \mathcal{Z} \to \mathcal{M}$, called the *source map* and the *target map*, respectively;
- An injective map $i : \mathcal{M} \to \mathcal{Z}$, called the *object inclusion map*;
- An associative binary internal operation in $\mathcal{Z}$ called *composition* or *product* defined for, and only for, all pairs $u, v \in \mathcal{Z}$ such that $\beta(u) = \alpha(v)$. The result of the composition is denoted as $vu$.

The product, moreover, satisfies

- $\alpha(vu) = \alpha(u)$ and $\beta(vu) = \beta(v)$;
- $\alpha \circ i(X) = \beta \circ i(X) = X \ \forall X \in \mathcal{M}$;
- $u(i \circ \alpha(u)) = u$ and $(i \circ \beta(u))u = u \ \forall u \in \mathcal{Z}$;
- For each $u \in \mathcal{Z}$ there is an *inverse* $u^{-1} \in \mathcal{Z}$ such that $\alpha(u^{-1}) = \beta(u)$, $\beta(u^{-1}) = \alpha(u)$ and $u^{-1}u = i \circ \alpha(u)$, $uu^{-1} = i \circ \beta(u)$.

Although the overly formal statement of these properties is somewhat intimidating, they simply state that the source and target of a product are provided by the source of one factor and the target of the other, and that the object inclusion map acts as an assignation of a *unit element* (identity arrow) to each point of the base set $\mathcal{M}$. Finally, for each arrow $u$ with source at $X = \alpha(u)$ and target at $Y = \beta(u)$, there is an opposite arrow $u^{-1}$ with source at $Y$ and target at $X$. Their compositions, $uu^{-1}$ and $u^{-1}u$, produce the respective units at the target and the source of $u$. It is not difficult to prove that the units and inverses are unique.

For each object $X \in \mathcal{M}$, the subsets $\mathcal{Z}_X = \alpha^{-1}(X) \subset \mathcal{Z}$ and $\mathcal{Z}^X = \beta^{-1}(X) \subset \mathcal{Z}$ are called the *$\alpha$-fibre* and the *$\beta$-fibre* over $X$, respectively. Accordingly, the subset $\mathcal{Z}_X^Y = \mathcal{Z}_X \cap \mathcal{Z}^Y$ is

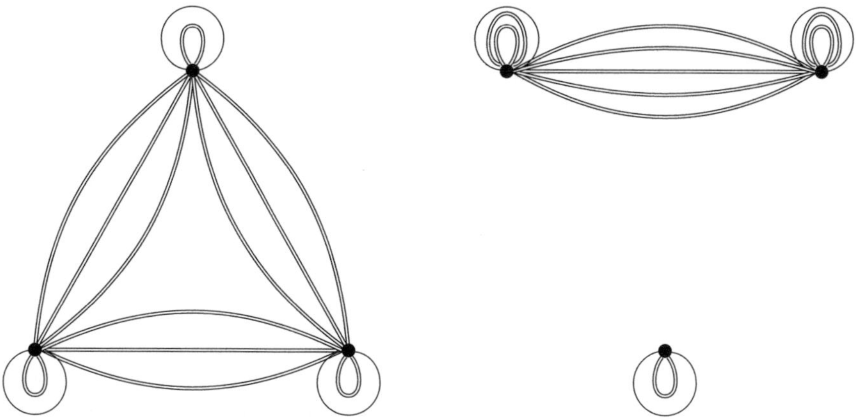

**Fig. 2** A groupoid on a discrete base set. The transitivity components are clearly visible. A double stem represents an arrow and its inverse. Single-stemmed arrows represent the units

the collection of all arrows that start at $X$ and end at $Y$. Thus, $\mathcal{Z}_X^X$ is the collection of all loop-shaped arrows at $X$. It is not difficult to conclude, from the rules for the product, that each of these subsets $\mathcal{Z}_X^X$ is a group, known as the *vertex group* at $X$.

A groupoid is *transitive* if $\mathcal{Z}_X^Y \neq \emptyset$ for all pairs $X, Y \in \mathcal{M}$. In a transitive groupoid all the vertex groups are conjugate of each other in the sense that

$$\mathcal{Z}_Y^Y = u\mathcal{Z}_X^X u^{-1}, \tag{5}$$

where $u \in \mathcal{Z}_X^Y$. Any one of these conjugate groups can be declared to be the *structure group* of the transitive groupoid.

A groupoid is *totally intransitive* if, for all $X, Y \in \mathcal{M}$, it happens that $X \neq Y \Rightarrow \mathcal{Z}_X^Y = \emptyset$. Put differently, no two distinct objects are connected by an arrow.

We can always define an equivalence relation $\sim$ on $\mathcal{M}$ by stipulating that $X \sim Y$ if, and only if, $\mathcal{Z}_X^Y \neq \emptyset$. The corresponding equivalence classes are denoted by $\mathcal{M}_X$. The restriction of $\mathcal{Z}$ to an equivalence class $\mathcal{M}_X$ is obviously a transitive groupoid, called the *transitivity component* of $\mathcal{Z}$ that contains $X$. In short, every groupoid can be seen as the disjoint union of its transitivity components. In particular, a totally intransitive groupoid is precisely the disjoint union of its vertex groups. Some of these concepts are illustrated, albeit naively, in Fig. 2.

Just as in the case of a group, a groupoid is defined as a purely algebraic structure. Topological and geometrical aspects are introduced whenever the sets of arrows and objects, the maps relating them and the composition and inversion operations enjoy further topological and differential properties. Thus, a *topological groupoid* $\mathcal{Z} \rightrightarrows \mathcal{M}$ is a groupoid for which $\mathcal{Z}$ and $\mathcal{M}$ are topological spaces and the projections $\alpha, \beta$, the inclusion map $i$ and the operations of composition and inversion are continuous in the respective topologies. A *Lie groupoid* $\mathcal{Z} \rightrightarrows \mathcal{M}$ is a groupoid where $\mathcal{Z}$ and $\mathcal{M}$ are smooth manifolds, the projections $\alpha, \beta$ are surjective submersions, the object inclusion map $i$ is smooth and so are the operations of composition and inversion.[1]

[1]The assumption of smoothness of the inversion is not necessary, since it is implied by the other properties. Mackenzie [15] distinguishes between Lie groupoids and differentiable groupoids, the latter not being necessarily locally trivial.

### 3.3 Correspondences with Continuum Mechanics

Combining the intuitive considerations developed in Sect. 3.1 with the formal definition of a groupoid introduced in Sect. 3.2, we have at our disposal the material groupoid $\mathcal{P} \rightrightarrows \mathcal{B}$ associated with the constitutive response $\psi_X(\cdot)$, and we can assign to the maps $\alpha, \beta$ and $i$ and to the operations of composition and inversion the obvious physical meanings implied in the very definition of material isomorphisms. The vertex group $\mathcal{P}_X^X$ is precisely the material symmetry group $\mathcal{G}_X$.

Comparing the concept of a materially uniform body with the definition of a transitive groupoid we conclude that

*A body $\mathcal{B}$ is materially uniform if, and only if, the material groupoid $\mathcal{P}$ is transitive.*

At the other extreme, a totally intransitive material groupoid describes a situation in which no two material points of the body are made of the same material. Between these two extreme cases, we can state that

*A constitutive response partitions the body into uniform components, each one corresponding to a transitivity component of the material groupoid.*

Assume that the constitutive response $\psi_X(\cdot)$ is a smooth function of its arguments and that the body $\mathcal{B}$ is considered as a smooth manifold. It may appear that these properties imply that the resulting material groupoid is a Lie groupoid, but this is not the case in general, as can be shown by counterexamples. When the body is smoothly uniform, however, it can be shown [11] that the corresponding transitive material groupoid is a Lie groupoid.

### 3.4 Subgroupoids

A map between two groupoids, $\mathcal{Z}_1 \rightrightarrows \mathcal{M}_1$ and $\mathcal{Z}_2 \rightrightarrows \mathcal{M}_2$ is a *groupoid morphism* if it is consistent with the projections and the compositions. More precisely, a morphism consists of two maps, $\Phi : \mathcal{Z}_1 \to \mathcal{Z}_2$ and $\phi : \mathcal{M}_1 \to \mathcal{M}_2$, such that

$$\alpha_2 \circ \Phi = \phi \circ \alpha_1 \qquad \beta_2 \circ \Phi = \phi \circ \beta_1, \tag{6}$$

and

$$\Phi(uv) = \Phi(u)\Phi(v) \quad \forall u, v \in \mathcal{Z}_1. \tag{7}$$

In Eq. (6), $\alpha_1, \beta_1$ and $\alpha_2, \beta_2$ denote the source and target maps of $\mathcal{Z}_1 \rightrightarrows \mathcal{M}_1$ and $\mathcal{Z}_2 \rightrightarrows \mathcal{M}_2$, respectively. It should be noted that, as a direct consequence of (7), a morphism automatically preserves the identities. Moreover, as a consequence of (6), the map $\phi$ is implied by the map $\Phi$. A *groupoid isomorphism* is an invertible groupoid morphism whose inverse is also a groupoid morphism. A *Lie groupoid morphism* is a differentiable morphism between two Lie groupoids.

A *subgroupoid* $\mathcal{Z}' \rightrightarrows \mathcal{M}'$ of a groupoid $\mathcal{Z} \rightrightarrows \mathcal{M}$ is a groupoid such that $\mathcal{Z}' \subset \mathcal{Z}$ and $\mathcal{M}' \subset \mathcal{M}$ and such that the inclusion map is a groupoid morphism. A *Lie subgroupoid* is defined similarly, except that $\mathcal{Z}'$ and $\mathcal{M}'$ are required to be submanifolds of $\mathcal{Z}$ and $\mathcal{M}$, respectively, and the inclusion map is required to be a Lie groupoid morphism.

A transitive Lie subgroupoid $\mathcal{Z}'$ of a transitive Lie groupoid $\mathcal{Z}$ is said to be a *reduction*, if the base manifolds coincide. This terminology arises from considering that the only difference between the two groupoids is that the fibres have been curtailed. In other words, the structure group of the reduced subgroupoid is precisely a subgroup of the structure group of the original groupoid.

*Remark 1* Given a groupoid $\mathcal{Z} \rightrightarrows \mathcal{M}$, and given a subset $\mathcal{U} \subset \mathcal{M}$, the set $\mathcal{Z}_{\mathcal{U}}^{\mathcal{U}} = \{z_X^Y \mid X, Y \in \mathcal{U}\}$ is a subgroupoid of $\mathcal{Z}$ with base $\mathcal{U}$. It is called the *restriction* of $\mathcal{Z}$ to $\mathcal{U}$, since it consists of only all the arrows that start and end in $\mathcal{U}$, the remaining arrows being discarded. Notice that if $\mathcal{Z}$ is transitive, so is $\mathcal{Z}_{\mathcal{U}}^{\mathcal{U}}$. In particular, for the material groupoid $\mathcal{P} \rightrightarrows \mathcal{B}$, the restriction $\mathcal{P}_{\mathcal{U}}^{\mathcal{U}}$ of $\mathcal{P}$ to a sub-body $\mathcal{U}$ is precisely the material groupoid of $\mathcal{U}$.

## 3.5 The Frame Groupoid and Homogeneity

An important example of a transitive Lie groupoid defined on a manifold $\mathcal{M}$ is the *frame groupoid* $\mathcal{F}(\mathcal{M})$. Starting from the tangent bundle $T\mathcal{M}$, we consider each linear isomorphism $z : T_X\mathcal{M} \to T_Y\mathcal{M}$ between the tangent spaces at $X, Y \in \mathcal{M}$ as an arrow in the groupoid. The groupoid operation is the ordinary composition of maps. The importance of the frame groupoid in the context of Continuum Mechanics is that

*Every material groupoid is a subgroupoid of the frame groupoid of $\mathcal{B}$. Moreover, if the body is smoothly uniform, the material groupoid is a reduction of the frame groupoid of the body.*

The frame groupoid $\mathcal{F}(\mathcal{U})$ of an open subset $\mathcal{U} \subset \mathbb{R}^3$ plays an important role in the definition of the concept of *material homogeneity*. We note that, by the existence of a trivial translation, there is a natural Lie groupoid isomorphism between $\mathcal{F}(\mathbb{R}^3)$ and the Cartesian product $\mathbb{R}^3 \times \mathbb{R}^3 \times GL(3; \mathbb{R})$. Indeed, any arrow of $\mathcal{F}(\mathbb{R}^3)$ consists of a pair of points of $\mathbb{R}^3$ and a non-singular $3 \times 3$ matrix. For any (Lie) subgroup $G$ of the general linear group $GL(3; \mathbb{R})$, we can accordingly generate a reduction of $\mathcal{F}(\mathcal{U})$ by means of its natural isomorphism with $\mathcal{U} \times \mathcal{U} \times G$. This reduction, which is clearly a transitive Lie groupoid, is called the *standard flat reduction* of $\mathcal{F}(\mathcal{U})$ to $G$.

To introduce the notion of homogeneity, we start by noting that the material groupoid $\mathcal{P}$ defined on the body manifold $\mathcal{B}$ is isomorphic (as a groupoid) to its representation in each configuration $\kappa : \mathcal{B} \to \mathbb{E}^3$. A uniform body is said to be *materially homogeneous* if, for some configuration $\kappa$, its material groupoid coincides with a standard flat reduction of $\mathcal{F}(\kappa(\mathcal{B}))$. In other words, the canonical translations in that configuration are material isomorphisms! A uniform body is *locally homogeneous* if it can be covered with homogeneous subbodies.

*Remark 2* It is important to note that, since the material isomorphisms must of necessity transform according to the rules of change of configuration, the groupoid isomorphisms brought about by any changes of configurations are *lifts* of the respective deformations $\phi$ of the body, as shown schematically in Fig. 3. By the term 'lift' we mean that the map $\Phi$ between the groupoid arrows is obtained from the derivative of the map $\phi$ between the set of objects. This is possible only because we are restricting attention to subgroupoids of the frame groupoid. Let $\mathbf{P} : T_X(\kappa_0(\mathcal{B})) \to T_Y(\kappa_0(\mathcal{B}))$ be a material isomorphism between points $X$ and $Y$ in the configuration $\kappa_0$. Let $\phi = \kappa_1 \circ \kappa_0^{-1}$ be the deformation between $\kappa_0$ and another configuration $\kappa_1$. The material isomorphism $\mathbf{P}'$ between the image points $X' = \phi(X)$ and $Y' = \phi(Y)$ corresponding to $\mathbf{P}$ is given by

$$\mathbf{P}' = \phi_*(Y) \circ \mathbf{P} \circ \phi_*^{-1}(X'). \tag{8}$$

To summarize,

*Homogeneity of a smoothly uniform body $\mathcal{B}$ is tantamount to the existence of a configuration $\kappa$ such that the induced material groupoid coincides with a standard flat*

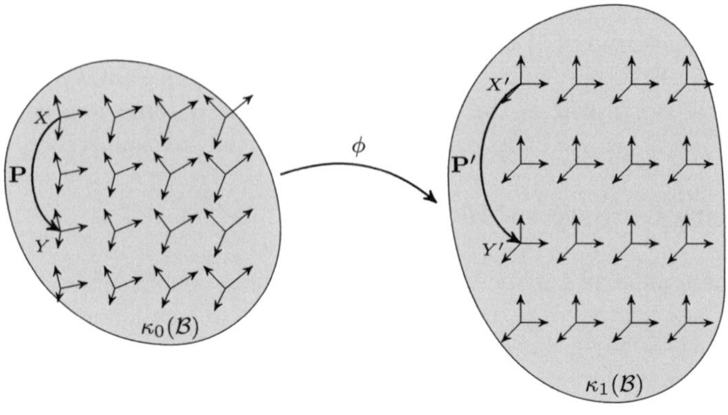


**Fig. 3** Schematic depiction of homogeneity

*reduction of the frame groupoid of $\kappa(\mathcal{B}) \subset \mathbb{E}^3$. Accordingly, homogeneity implies the existence of a configuration (unique modulo a congruence) whereby the trivial Euclidean translations are material isomorphisms. Intuitively, a uniform body is homogeneous if it can be 'straightened' so that its constitutive equation is independent of the (Cartesian) coordinates of its points. Local homogeneity requires only that this property be attainable chunk-wise.*

## 4 Lie Algebroids

### 4.1 Introduction

As an algebraic structure, a groupoid can be colorfully described as a group 'on steroids'. Indeed, a group can be regarded as the particular case of a groupoid whose set of objects $\mathcal{M}$ is a singleton. All the elements of a group can be mutually composed (multiplied) and there is a single, uniquely defined, unit element $e$. In a general groupoid $\mathcal{Z} \rightrightarrows \mathcal{M}$, however, $\mathcal{M}$ is an arbitrary set, but, except for the totally intransitive case, $\mathcal{Z}$ is not just the union of its individual vertex groups. The elements of $\mathcal{Z}$ are 'arrows' that may have different sources and targets. Arrows can be composed only if they satisfy the extra condition of being in tandem, tip-to-tail fashion. Moreover, rather than a single unit element, each element $X$ of the set of objects $\mathcal{M}$ carries its own (unique) unit $i(X)$, an arrow in the form of a loop, so to speak.

Recall that a *Lie algebra* is a vector space endowed with an antisymmetric binary operation called a *Lie bracket*. Lie algebras are defined independently from groups, but the fundamental work of Sophus Lie (1842–1899) demonstrated the intimate connection that exists between Lie algebras and Lie groups, that is, groups that are also manifolds in which the operations of multiplication and inversion are smooth.

The Lie algebra of a Lie group represents an infinitesimal version of the latter in a precise sense. Its underlying vector space can be identified with the tangent space of the Lie group at the unit element. The vehicle to this identification is provided by the notion of left- (or right-) invariant vector fields on the Lie group. Similarly, the concept of a *Lie algebroid* can be introduced independently and eventually related to the notion of Lie groupoid. As an infinitesimal version of the latter, however, it involves certain tangent spaces to the groupoid

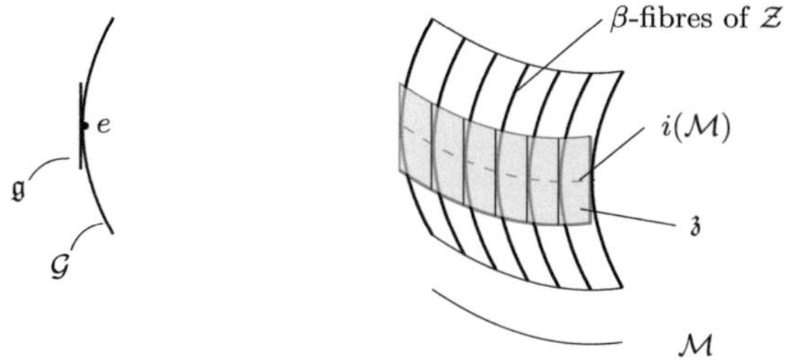


**Fig. 4** A preliminary mental picture of the Lie algebroid $\mathfrak{z}$ of a Lie groupoid $\mathcal{Z}$ (right) as compared with the Lie algebra $\mathfrak{g}$ of a Lie group $\mathcal{G}$ (left). The $\beta$-fibre at $X \in \mathcal{M}$ is the collection of all the arrows arriving at $X$

$\mathcal{Z}$ at each of its unit elements. Again, these notions are intermediated via left- (or right-) invariant vector fields on the groupoid. As everything else pertaining to groupoids, these notions acquire a further degree of sophistication as compared with their group counterparts. Although certainly premature for this introduction, we take the liberty of depicting, in Fig. 4, a schematic drawing that may serve as an intuitive basis for a mental representation of the concepts that will be advanced below in a more precise fashion.

## 4.2 Definition

Let $\pi : A \to \mathcal{M}$ denote a *vector bundle* over a *base manifold* $\mathcal{M}$, and let $\Gamma(A)$ denote the space of its smooth global sections $\sigma : \mathcal{M} \to A$. A *Lie algebroid* structure on this vector bundle is obtained by specifying a bilinear *bracket* operation $[\cdot,\cdot] : \Gamma(A) \times \Gamma(A) \to \Gamma(A)$ and a vector-bundle morphism $\sharp : A \to T\mathcal{M}$, called the *anchor map*. These maps must satisfy the following properties;

1. Skew-symmetry:

$$[\rho, \sigma] = -[\sigma, \rho] \quad \forall \rho, \sigma \in \Gamma(A). \tag{9}$$

2. Jacobi identity:

$$\big[\rho, [\sigma, \tau]\big] + \big[\tau, [\rho, \sigma]\big] + \big[\sigma, [\tau, \rho]\big] = \mathbf{0} \quad \forall \rho, \sigma, \tau \in \Gamma(A). \tag{10}$$

3. Consistency:

$$[\rho, \sigma]^\sharp = \big[\rho^\sharp, \sigma^\sharp\big] \quad \forall \rho, \sigma \in \Gamma(A). \tag{11}$$

4. Leibniz rule:

$$[\rho, f\sigma] = f[\rho, \sigma] + \rho^\sharp(f)\sigma \quad \forall \rho, \sigma \in \Gamma(A), f \in C^\infty(\mathcal{M}). \tag{12}$$

*Remark 3* The first two properties are self-explanatory. The third property can be shown to be a consequence of the other ones. For compactness of notation, we have indicated by $\rho^\sharp$ the image $\sharp(\rho) \in \Gamma(T\mathcal{M})$ of the section $\rho \in \Gamma(A)$. Moreover, the bracket appearing on the right-hand side of (11) is the ordinary *Lie bracket* of vector fields in $T\mathcal{M}$. The fourth property requires some further clarification, as it displays the reason behind the need for an anchor map. In an arbitrary vector bundle, there is in principle no canonical action of the vectors in the bundle on a smooth real-valued function $f \in C^\infty(\mathcal{M})$ defined on the base manifold $\mathcal{M}$. It is only in the *tangent bundle* $T\mathcal{M}$ that such an action exists, providing us with the *directional derivative* $\mathbf{v}(f)$ of $f$ in the direction of $\mathbf{v} \in T\mathcal{M}$.

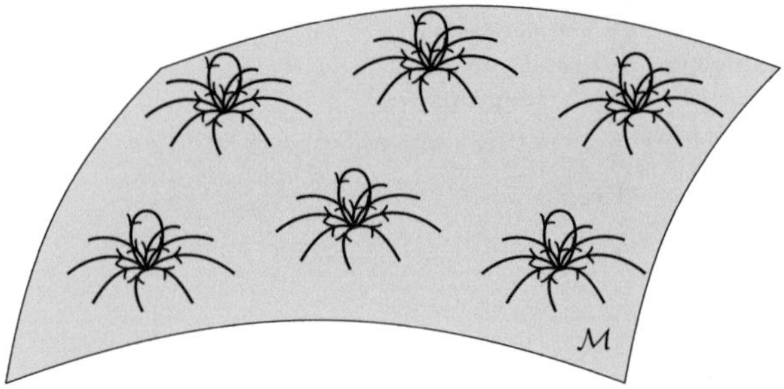


**Fig. 5** The $\beta$-bundle $\mathcal{Z}^{\mathcal{M}}$ as a spider colony

A Lie algebroid is *transitive* if the anchor map $\sharp$ is surjective. It is *totally intransitive* if the anchor is the zero map (assigning to each vector in $A$ the zero tangent vector at the corresponding point of the base manifold). The reason for this terminology will become apparent later.

## 4.3 The Lie Algebroid of a Lie Groupoid

### *4.3.1 The $\beta$-Bundle*

Consider the disjoint union $\mathcal{Z}^{\mathcal{M}}$ of all the $\beta$-fibres of a Lie groupoid $\mathcal{Z}$, that is,

$$\mathcal{Z}^{\mathcal{M}} = \bigcup_{X \in \mathcal{M}} \mathcal{Z}^X. \tag{13}$$

This set, which we call the $\beta$-bundle, can be regarded as a fibre bundle over the base manifold $\mathcal{M}$ with projection $\beta$. In terms of arrows, $\mathcal{Z}^{\mathcal{M}}$ looks like a spider colony, each fibre $\mathcal{Z}^X$ being a spider with legs arriving at $X$ and issuing from some point $Y \in \mathcal{M}$, as shown schematically in Fig. 5 [6]. Notice that the total set of this fibre bundle is the same as the total set of the original transitive groupoid $\mathcal{Z}$. They both consist of the set of all arrows.

### *4.3.2 Left-Invariant Vector Fields on a Lie Groupoid*

In any groupoid $\mathcal{Z} \rightrightarrows \mathcal{M}$ we can define the concept of *left translation*. Let $g \in \mathcal{Z}$ and let $z \in \mathcal{Z}$ be such that $\beta(z) = \alpha(g)$. The left translation of $z$ by $g$ is given by

$$L_g(z) = gz. \tag{14}$$

It follows from this definition that $L_g$ maps $\beta$-fibres into $\beta$-fibres, that is, the spider at $\alpha(g)$ gets mapped to the fibre at $\beta(g)$. Specifically,

$$L_g : \mathcal{Z}^{\beta(z)} = \mathcal{Z}^{\alpha(g)} \to \mathcal{Z}^{\beta(g)}. \tag{15}$$

If the groupoid happens to be a Lie groupoid, the left action is differentiable and we can consider the tangent map $(L_g)_* : T\mathcal{Z}^{\alpha(g)} \to T\mathcal{Z}^{\beta(g)}$. A vector field $V : \mathcal{Z} \to T\mathcal{Z}$ on $\mathcal{Z}$ is *left-invariant* if

$$(L_g)_*\big(V(z)\big) = V\big(L_g(z)\big) \quad \forall g, z \in \mathcal{Z}. \tag{16}$$

Of necessity, a left-invariant vector field must be $\beta$-*vertical*, that is, it must dwell on the tangent spaces of the $\beta$-fibres of $\mathcal{Z} \rightrightarrows \mathcal{M}$.

A left-invariant vector field on a Lie groupoid is completely determined by its values at the unit elements $i(X)$, for all $X \in \mathcal{M}$. Indeed, setting $z = i \circ \alpha(g)$ in Eq. (16), we obtain

$$(L_g)_*\big(V\big(i \circ \alpha(g)\big)\big) = V\big(L_g\big(i \circ \alpha(g)\big)\big) = V\big(g\big(i \circ \alpha(g)\big)\big) = V(g). \tag{17}$$

Recall that, given any smooth vector field $V$ on a manifold $\mathcal{M}$, the fundamental theorem of the theory of ODEs guarantees the existence and uniqueness of maximal smooth *integral curves* defined at each point of $\mathcal{M}$. If $\gamma_x = \gamma_x(t)$ is the integral curve containing the point $x \in \mathcal{M}$, the curve parameter can be adjusted by a mere translation such that $\gamma_x(0) = x$. By definition of integral curve, we have

$$V(x) = \left.\frac{d\gamma_x(t)}{dt}\right|_{t=0}. \tag{18}$$

Moreover, every smooth vector field acts as the *infinitesimal generator* of a *local flow* $\phi_t^V$. For each $t$ in a certain interval of $\mathbb{R}$ containing the origin, $\phi_t^V$ is a local diffeomorphism of $\mathcal{M}$ defined by the prescription

$$\phi_t^V(x) = \gamma_x(t). \tag{19}$$

Clearly, $\phi_0^V = id_{\mathcal{M}}$ and $\phi_{-t}^V = (\phi_t^V)^{-1}$. Applying these concepts to Eq. (16) we obtain the result of the following lemma.

**Lemma 1** *A ($\beta$-vertical) vector field $V$ on a Lie groupoid is left invariant if, and only if, its local flow commutes with left translations, that is,*

$$L_g \circ \phi_t^V(z) = \phi_t^V \circ L_g(z), \tag{20}$$

*for all $g, z$ such that $\alpha(g) = \beta(z)$.*

### *4.3.3 The Associated Lie Algebroid*

After the foregoing properties of left-invariant vector fields on a Lie groupoid $\mathcal{Z} \rightrightarrows \mathcal{M}$ have been established, we introduce the vector bundle $\pi : A\mathcal{Z} \to \mathcal{M}$ whose fibre at each $X \in \mathcal{M}$ is the tangent space to the $\beta$-fibre of $\mathcal{Z}$ at the identity $i(X)$. It is clear that a section of this vector bundle corresponds exactly to a left-invariant vector field on $\mathcal{Z}$. Since each map $L_g$ is a diffeomorphism between two $\beta$-fibres, and since the left-invariant vector fields are tangent to these fibres, it follows from Lemma 1 that the ordinary Lie bracket between two left-invariant vector fields is again left-invariant. Therefore, given two sections of $A\mathcal{Z}$, we can define a Lie-algebroid bracket operation by considering the Lie bracket of the corresponding vector fields in $\mathcal{Z}$ and then considering its value at the unit section.

To complete the determination of the Lie algebroid associated with the Lie groupoid $\mathcal{Z}$, we declare the anchor map to be given by the restriction of $\alpha_* : T\mathcal{Z} \to T\mathcal{M}$ to $A\mathcal{Z}$. An intuitive idea of the anchor map (and of the meaning of the Lie algebroid) can be gathered from Fig. 6. Starting from the identity loop-like arrow at a point $X \in \mathcal{M}$, we explore its vicinity in $\mathcal{Z}$ by keeping the tip of the arrow fixed at $X$, so as to stay always in the same $\beta$-fibre $\beta^{-1}(X)$. If we keep the tail of the arrow also at $X$ (that is, if we explore just the loop-like neighbors), we are clearly moving within the vertex group at $X$. As a result, we recover the Lie algebra of this vertex group. In the case of the material groupoid $\mathcal{P}$ introduced above, we obtain the Lie algebra of the *material symmetry group* at $X$.

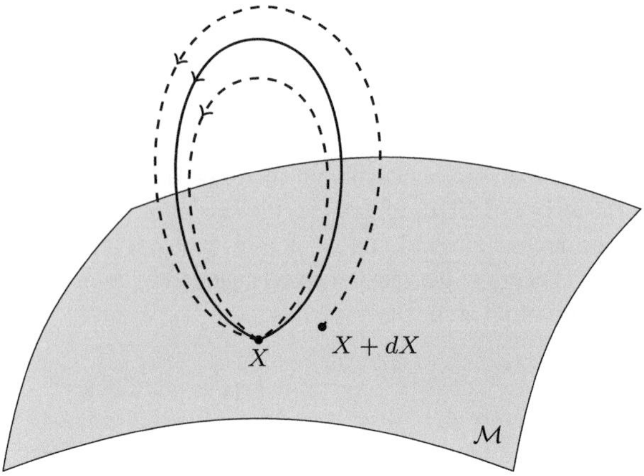


**Fig. 6** Intuiting the Lie algebroid of a Lie groupoid. The identity $i(X)$ is drawn as a solid arrow, while elements in its vicinity are drawn as dashed arrows. The anchor map assigns to each dashed arrow the opening between its tail and $X$

Let us further explore the vicinity of the unit $i(X)$ by considering an arrow $z$ with its tip at $X$, but with its tail elsewhere, at say $X + dX$. The differential projection $\alpha(z) - \alpha(i(X))$ is precisely $dX$. Thus, intuitively enough, we see how the map $\alpha_*$ acts as the anchor of the algebroid. We see, moreover, that the Lie algebra of the vertex group at $X$ is precisely the kernel of the anchor at $X$. Finally, if $\alpha_*$ is a surjective map, there are arrows between $X$ and every point in an open neighborhood of $X$ in $\mathcal{M}$. This picture perfectly justifies the terminology introduced above for transitive and totally intransitive algebroids. In the case of the material groupoid $\mathcal{P}$ we conclude that a smoothly uniform body has a transitive *material Lie algebroid*.

## 5 The Material Lie Algebroid and Homogeneity

### 5.1 Lie-Algebroid Morphisms and Lie Subalgebroids

A morphism $\phi : A \to A'$ of two Lie algebroids $\pi : A \to \mathcal{M}$ and $\pi' = A' \to \mathcal{M}$ over a common base manifold $\mathcal{M}$ is defined as a vector-bundle morphism that is consistent with the respective anchors and brackets. That is, in an obvious notation,

$$\sharp = \sharp' \circ \phi, \tag{21}$$

and

$$\phi[\rho, \sigma] = [\phi \circ \rho, \phi \circ \sigma] \quad \forall \rho, \sigma \in \Gamma(A). \tag{22}$$

Let $A'$ be a vector subbundle of $A$, and let the inclusion map $incl : A' \hookrightarrow A$ be a Lie-algebroid morphism. We say that $A'$ is a *Lie subalgebroid* of $A$. When both $A$ and $A'$ are transitive Lie algebroids, $A'$ is called a *reduction* of $A$.

Consider the frame groupoid $\mathcal{F}(\mathbb{R}^3) \approx \mathbb{R}^3 \times \mathbb{R}^3 \times GL(3; \mathbb{R})$, which we have already encountered. The $\beta$-fibre at $X \in \mathbb{R}^3$ consists of elements of the form $(\{Y\}, \{X\}, [A])$, where $\{Y\}$ is a triple containing the coordinates of the variable tail and $\{X\}$ are the coordinates of the common tip $X$, and where $[A]$ is a $3 \times 3$ non-singular matrix. A vector tangent to this fibre will, therefore, have the component form $(\{v\}, \{X\}, [B])$, where $\{v\}$ are the 3 components of a tangent vector to $\mathbb{R}^3$ at $X$ and where $[B]$ is an arbitrary $3 \times 3$ matrix. The same reasoning applies to the standard flat reduction of $\mathcal{F}(\mathbb{R}^3)$ to a subgroup $G \subset GL(3;\mathbb{R})$. In other words, the Lie algebroid associated to this reduction is $T\mathbb{R}^3 \oplus (\mathbb{R}^3 \times \mathfrak{g})$, where $\mathfrak{g}$ denotes the Lie algebra of $G$. In the imagery suggested by Fig. 6, the first component corresponds to the opening of arrows with tip at $X$, while the last component is spanned by all the loop-like arrows at $X$, namely, the Lie algebra of the symmetry group.

*Remark 4* Let $\mathcal{M}$ be a manifold and $\mathfrak{g}$ a Lie algebra. We can construct a Lie algebroid structure over the vector bundle $A = T\mathcal{M} \oplus (\mathcal{M} \times \mathfrak{g}) \to \mathcal{M}$ such that

(i) The anchor $\sharp : T\mathcal{M} \oplus (\mathcal{M} \times \mathfrak{g}) \to T\mathcal{M}$ is the projection over the first component of the Whitney sum.
(ii) The Lie algebra structure over the space of sections is given by

$$[\mathbf{u} \oplus g, \mathbf{v} \oplus h] = [\mathbf{u}, \mathbf{v}] \oplus \{\mathbf{u}(h) - \mathbf{v}(g) + [g, h]\},$$

for all $\mathbf{u} \oplus g, \mathbf{v} \oplus h \in \Gamma(A)$.

This Lie algebroid is called the *trivial Lie algebroid on* $\mathcal{M}$ *with structure algebra* $\mathfrak{g}$. Note that $g$ and $h$ are Lie-algebra valued functions over $\mathcal{M}$, so that $\mathbf{u}(h)$ and $\mathbf{v}(g)$ are directional derivatives of these functions.

## 5.2 Homogeneity and Material Connections

Following the lead of the groupoid treatment counterpart, we assert that a smoothly uniform body $\mathcal{B}$ is homogeneous if its (transitive) material Lie algebroid $A\mathcal{P}$ in some configuration $\kappa$ coincides with the Lie algebroid of a standard flat reduction of the frame groupoid of (an open subset of) $\mathbb{R}^3$. Local homogeneity is defined accordingly.

Our intention in this section is to demonstrate the relation between this definition of homogeneity and the notion of *material connections* introduced by Noll [17] and Wang [19]. As a first step in that direction, consider a transitive Lie groupoid $\mathcal{Z} \rightrightarrows \mathcal{M}$ and arbitrarily fix a *reference or archetypal point* $X_0 \in \mathcal{M}$. It is not difficult to verify that the disjoint union

$$\mathcal{Z}_0 = \bigcup_{X \in \mathcal{M}} \mathcal{Z}_{X_0}^{X} \tag{23}$$

is a principal fibre bundle with structure group $\mathcal{G}$ equal to the vertex group at $X_0$ and with bundle projection $\beta$. Recall [12] that a *connection* in a principal bundle is a *horizontal distribution*, namely, the point-wise specification of a non-vertical subspace $\mathcal{H}$ such that (i) the tangent space is the direct sum of $\mathcal{H}$ and the (vertical) tangent space to the fibre; (ii) the horizontal distribution is invariant under the right action of the structure group.

Let $\Lambda : T\mathcal{M} \to A\mathcal{Z}$ be a vector-bundle morphism such that $\sharp \circ \Lambda = id_{T\mathcal{M}}$. One can directly verify that there is a one-to-one correspondence between such vector-bundle morphisms and connections in the principal bundle $\mathcal{Z}_0$. In the context of Noll's formulation, local homogeneity is equivalent to the existence of a (locally defined, say) material connection with vanishing torsion. We conclude, therefore, that local homogeneity is guaranteed by the existence of a (local) vector-bundle morphism from the tangent bundle of a (sub-)body to its (restricted) material Lie algebroid such that the associated connection is torsionless.

## 6 The Material Distribution

### 6.1 General Considerations

The material groupoid is essentially an algebraic entity based on the establishment of an equivalence relation between body points made of the same material. This algebraic entity has also a definite topological and geometric flavor arising from the fact that the elements of a groupoid can be visualized as arrows connecting points. A differential geometric aspect is added when the material groupoid is a Lie groupoid. In this last case, the study of its associated *Lie algebroid* is also of great relevance, particularly in the study of local homogeneity conditions.

As already mentioned in Sect. 3.3, even when the constitutive response is a perfectly smooth function, the resulting material groupoid is not necessarily a Lie groupoid. This somewhat surprising fact leads us to introduce and investigate a new differential geometric construct, called the *material distribution*, that answers the question as to what is the most general differentiable structure that can most faithfully reflect the consequences of having a smooth constitutive response even in the absence of material uniformity.

A heuristic clue is gathered by thinking of the constitutive law $\psi$ as a function defined over the product $\kappa_0(\mathcal{B}) \times GL(3, \mathbb{R})$ of the body manifold in a reference configuration $\kappa_0$ times the general linear group (of non-singular matrices) in 3 dimensions. Appropriately, the function $\psi$ assigns to each material point and each deformation gradient a value of the constitutive quantity. Consider now a parametric curve $\gamma(t)$ in $\mathcal{B}$ such that all the points along this curve are smoothly materially isomorphic, so that there exists a (not necessarily unique) smooth field $\mathbf{P}(t)$ of material isomorphisms from the point $\gamma(0)$ to the points $\gamma(t)$, such that $\mathbf{P}(0) = \mathbf{I}$. This means that the function $\psi$, when restricted to the curve, can be expressed as $\psi(\gamma(t), \mathbf{F}) = \psi(\gamma(0), \mathbf{FP}(t))$. For each fixed value of the deformation gradient $\mathbf{F}$ we can *lift* the curve $\gamma(t)$ to a curve $\hat{\gamma}(t)$ in $\kappa_0(\mathcal{B}) \times GL(3, \mathbb{R})$ by using the value of $\mathbf{FP}(t)$ as the coordinate along $GL(3, \mathbb{R})$. By the assumed material isomorphism, the value of $\psi$ will remain constant along $\hat{\gamma}(t)$. Moreover, for two different values of $\mathbf{F}$, the curves will be "parallel" in the sense that they differ only by multiplication of their matrix argument to the left by a different fixed factor $\mathbf{F}$, while $\mathbf{P}(t)$ remains unchanged.

Since a small piece of a curve is, in the limit, a tangent vector, the description just outlined leads us to consider the collection of all vector fields in $\kappa_0(\mathcal{B}) \times GL(3, \mathbb{R})$ that are left invariant under the action of $GL(3, \mathbb{R})$ and, at the same time, annihilate the differential of the constitutive function $\psi$. This approach was adopted in [10]. Here, we generalize this construction by working directly on the frame groupoid $\mathcal{F}(\mathcal{B})$, rather than on the product $\kappa_0(\mathcal{B}) \times GL(3, \mathbb{R})$. The local vector fields enjoying the properties outlined above can be regarded as the generators of a unique *singular distribution* in $\mathcal{F}(\mathcal{B})$ completely characterizing the smoothness of the constitutive law of departure and its physical consequences in terms of a unique partition of the body into disjoint smoothly uniform components.

### 6.2 Constitutive Invariance

In a reference configuration $\kappa_0$, a constitutive response such as (1) is expressed as a function

$$\psi = \psi\left(X^J, F^i_I\right). \tag{24}$$

It is convenient, however, to reinterpret it as a function $\tilde{\psi}$ defined over the frame groupoid $\mathcal{F}(\mathcal{B})$. This interpretation is possible because, having chosen a reference configuration $\kappa_0$, the maps between tangent spaces can be regarded as deformation gradients **F** if so desired. At the element $(X^L, Y^I, A^J_K)$ of the $\beta$-fibre at $Y$, the constitutive response is given by

$$\tilde{\psi}(X^L, Y^I, A^J_K) = \psi(X^L, A^J_K). \tag{25}$$

We have exploited the fact that the constitutive response at a point does not depend on the target point, since it is invariant under translations in $\mathbb{E}^3$. The differential of the function $\tilde{\psi}$ is

$$d\tilde{\psi} = \frac{\partial \psi}{\partial X^L} dX^L + \frac{\partial \psi}{\partial A^J_K} dA^J_K. \tag{26}$$

We are interested in finding the collection $\mathcal{V}$ of all the smooth local vector fields $V$ on $\mathcal{F}(\mathcal{B})$ that satisfy three conditions: (a) they are everywhere tangent to the $\beta$-fibres; (b) they annihilate the differential of the constitutive response; and (c) they are left-invariant. Condition (a) is implied by (c). We shall call these vector fields *constitutively admissible*, or just *admissible*. It is certainly possible to proceed analytically so as to identify exactly what these vector fields represent from the point of view of continuum mechanics. Instead, at the risk of some loss of rigor, we will advance a more geometric argument.

We commence by noticing that condition (b) implies that these vector fields are, by construction, everywhere tangent to the level sets of the function $\tilde{\psi}$. As a consequence of this simple observation, we conclude that the integral curves of the admissible vector fields are entirely contained in these level sets. By condition (a), the desired integral curves belong to the intersection of the level sets of $\tilde{\psi}$ with the $\beta$-fibres. Moreover, because of the already noticed independence of $\tilde{\psi}$ on the target point, we observe that the picture of the integral-curve-portraits in each and every $\beta$-fibre are all identical. Finally, the left invariance condition (c) means that, under the left action of an arbitrary element $z$ of the frame groupoid, integral curves are mapped into integral curves.

If $z(t)$ is the parametric representation of an integral curve of $\mathcal{V}$, by condition (a) we have necessarily $\beta(z(t)) = \text{constant}$. On the other hand, the projected curve $\alpha(z(t))$ consists entirely of mutually materially isomorphic points in $\mathcal{B}$. Indeed, by conditions (b) and (c) above, they respond in "parallel" ways to all possible left translations in the frame groupoid. The main conclusion of this observation is that the set of integral curves of $\mathcal{V}$ and the material groupoid are intimately related according to the following statement.

*If the material groupoid has a point (that is, an arrow $z$) in common with an integral curve of $\mathcal{V}$, then the whole integral curve belongs to the material groupoid.*

## 6.3 Singular Distributions

A *distribution* $\mathcal{D}$ over a manifold $M$ is obtained by assigning to each point $x \in M$ a vector subspace $D_x$ of the tangent space $T_x M$. The distribution is defined as the disjoint union

$$\mathcal{D} = \bigcup_{x \in M} D_x. \tag{27}$$

The distribution $\mathcal{D}$ is said to be *regular* or of *constant rank* if the dimension of $D_x$ is the same for all $x \in M$. Otherwise, the distribution is called *singular* [16]. Consider the collection $\mathcal{V}_\mathcal{D}$ of all local smooth vector fields that belong to the distribution. The distribution $\mathcal{D}$ is *smooth* if it is spanned by $\mathcal{V}_\mathcal{D}$. Equivalently, for every $x \in M$, $D_x$ coincides with the set of all linear combinations of all vectors $V(x)$ of all local vector fields $V \in \mathcal{V}_\mathcal{D}$ defined at $x$.

An *integral manifold* of a distribution $\mathcal{D}$ in $M$ is an immersed submanifold $N \subset M$ such that, at each point $x \in N$, $T_x N = D_x$.[2] A *maximal integral manifold* of a distribution is an integral manifold that is not contained in any strictly larger integral manifold. Given a distribution on $M$, it can be shown that each point $x \in M$ that is contained in some integral manifold is contained in a unique maximal integral manifold.

A smooth distribution $\mathcal{D}$ on a manifold $M$ is said to be *integrable* if every point $x \in M$ is contained in an integral manifold of $\mathcal{D}$ and, hence, in a unique maximal integral manifold. An integrable distribution induces a partition of the manifold $M$ into its maximal integral manifolds. This partition is, in fact, a *foliation* of $M$, with each maximal submanifold referred to as a *leaf* of the foliation. Notice that in a strictly singular integrable distribution there will be leaves with different dimensions. To emphasize this fact, we sometimes refer to the partition induced by a singular integrable distribution as a *singular* foliation. The following theorem establishes necessary and sufficient conditions for integrability of a singular distribution.

**Theorem 1** (Stefan-Sussman) *Let $\mathcal{D}$ be a smooth singular distribution on a smooth manifold $M$. Then the following three conditions are equivalent*:

1. *$\mathcal{D}$ is integrable*;
2. *$\mathcal{D}$ is spanned by a family of local vector fields, with respect to which it is invariant*;
3. *$\mathcal{D}$ is the tangent distribution of a smooth singular foliation.*

Condition 2 generalizes to singular distributions the condition of involutivity in the classical Frobenius theorem, which is valid for regular distributions. For singular distributions, involutivity is insufficient to guarantee integrability. What may fail is not involutivity but *stability*, which is the meaning of condition 2 above. A distribution is stable if it is invariant under the flow of any of its generating vector fields.

*Example 1* Consider the distribution on $\mathbb{R}^3$ (with coordinates $x, y, z$) spanned by the vector fields

$$\mathbf{u} = \frac{\partial}{\partial y} \qquad \mathbf{v} = f(y)\frac{\partial}{\partial x}, \tag{28}$$

where $f \in C^\infty(\mathbb{R}^3)$ is the function

$$f(y) = \begin{cases} 0 & y \leq 0 \\ e^{-1/y} & y > 0 \end{cases} \tag{29}$$

This distribution is not integrable. What fails here is the condition of stability. Indeed, the image of the vector field $\mathbf{v}$ by the flow $\phi_t^u$ of $\mathbf{u}$ for some $t < 0$ would translate the non-zero vectors $\mathbf{v}(x, y, z)$ with $0 < y < t$ to places with $y < 0$, where $\mathbf{v} = 0$. If, however, we were to restrict the domain of the distribution to $\mathbb{R}^3 \backslash \{(x, y, z) \mid y = 0\}$, the distribution would become integrable. The integral manifolds are shown schematically in Fig. 7.

[2] In the case of a regular one-dimensional distribution, every point is contained in an integral manifold. A regular distribution of higher dimension need not have any integral manifolds.

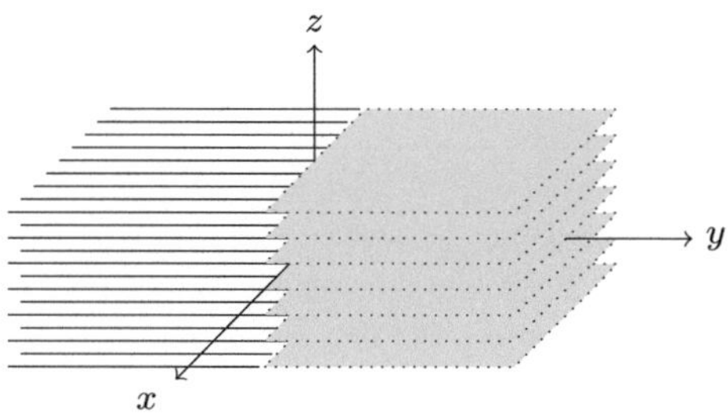


**Fig. 7** Integral manifolds of the singular distribution generated by the vector fields (28)

## 6.4 The Material Distribution and Its Body Projection

The collection $\mathcal{V}$ of admissible vector fields introduced in Sect. 6.2 generates, by considering all point-wise linear combinations, a smooth singular distribution, denoted by $A\mathcal{P}$, called the *material distribution*. The following proposition is crucial to obtain relevant physical results.

**Proposition 1** *The material distribution $A\mathcal{P}$ is integrable.*

*Proof* According to Theorem 1, we only need to prove that the image of any admissible vector field by the flow of any other admissible vector field is again an admissible vector field. Let $V$ and $W$ be two admissible vector fields and denote by $W'$ the image at time $t$ of $W$ by the flow of $V$. By definition of derivative, the field $W'$ is everywhere tangent to the image at time $t$ of the integral curves of $W$ by the flow of $V$. At a point $z \in \mathcal{F}(\mathcal{B})$, the mapped integral curve is, accordingly, given by

$$\phi_s^{W'}(z) = \phi_t^V \circ \phi_s^W \circ \phi_{-t}^V(z). \tag{30}$$

Since the vector fields of departure, $V$ and $W$, are left invariant, their respective flows commute with the left translation and, therefore, so does their composition on the right-hand side of Eq. (30). Consequently, the flow of $W'$ also commutes with left translations. By Lemma 1, we conclude that $W'$ is left-invariant. Moreover, since the function $\tilde{\psi}$ is constant on each integral curve, it is also constant on the integral curves of $W'$, so that the field $W'$ is admissible. □

Having just proved that the material distribution is always integrable, we can invoke the third part of Theorem 1 to conclude that there exists a smooth singular foliation $\mathcal{S}$ in $\mathcal{F}(\mathcal{B})$ such that $A\mathcal{P}$ is its tangent distribution. This fact has important ramifications, stemming from the fact that it can be shown that each leaf of the foliation is entirely contained in a $\beta$-fibre of $\mathcal{F}(\mathcal{B})$. Moreover, if $z \in \mathcal{P}$ then the leaf containing $z$ is entirely contained in a $\beta$-fibre of $\mathcal{P}$. We call the collection of all the leaves of $\mathcal{S}$ that are contained in $\mathcal{P}$ the *material foliation*, denoted by $\mathcal{S}(\mathcal{P})$.[3]

The material distribution $A\mathcal{P}$ associated with any given smooth constitutive response is a geometrical object defined over the frame groupoid $\mathcal{F}(\mathcal{B})$ that can be restricted to the material groupoid $\mathcal{P}$. The derivative of the source projection $\alpha$ is a map $\alpha_* : T\mathcal{P} \to T\mathcal{B}$. Accordingly, in purposeful analogy with the definition of the anchor of an associated Lie algebroid, we can use $\alpha_*$ evaluated at the groupoid units, to define a "projected" distribution on

[3]This terminology is slightly misleading, since $\mathcal{S}(\mathcal{P})$ is technically not a foliation of $\mathcal{P}$, which may not even be a manifold.

the body itself. The resulting object $A\mathcal{P}^\sharp$ is called the *body distribution* associated with the given smooth constitutive response. This distribution can be shown to be always integrable. Hence, we obtain uniquely a corresponding singular foliation $\mathcal{H}$. The unique leaf containing a point $X \in \mathcal{B}$ is denoted as $\mathcal{H}(X)$. Each leaf of the foliation is a smoothly uniform *material submanifold.* A material submanifold of dimension equal to 3 is also a material sub-body. For lower dimensions we obtain material surfaces, lines and points. Their uniformity, nevertheless, is to be understood in terms of fully-fledged three-dimensional material isomorphisms (as opposed to shells and beams and point-like particles). Summarizing,

*Every material body with a specified smooth material response can be uniquely partitioned as a foliation of smoothly uniform connected material submanifolds of various dimensions. Smoothness and connectedness make this partition in general different from the partition induced by the transitivity components of the material groupoid.*

## 6.5 A Worked-out Example

### 6.5.1 Formulation

Let

$$\bar{\psi} = \mathbb{D}(\mathbf{E} \otimes \mathbf{E}) = D^{IJKL} E_{IJ} E_{KL} \tag{31}$$

be an archetypal constitutive law for a material point. In this equation, $\mathbb{D}$ is a fourth-order tensor of material constants and

$$\mathbf{E} = \frac{1}{2}(\mathbf{F}^T \mathbf{F} - \mathbf{I}) \tag{32}$$

is the *Lagrangian strain tensor*. Let the body $\mathcal{B}$ in a given reference configuration occupy the open ball with center at the origin of $\mathbb{R}^3$ and with radius $R = 2$. Define the constitutive response of $\mathcal{B}$ as

$$\psi = \big(1 - f(r-1)\big)\bar{\psi}, \tag{33}$$

where $f$ is the function (29) and $r$ is the radial coordinate in $\mathbb{R}^3$.

### 6.5.2 Transitivity Components

At the outset, by construction, it should be clear that the transitivity components of the material groupoid $\mathcal{P} \rightrightarrows \mathcal{B}$ coincide with the level sets of the function $f(r-1)$. These level sets are: (i) the *closed* ball defined by $0 \le r \le 1$, and (ii) each of the 2-spheres $r = k$, with $1 < k < 2$. This partition is unrelated to the differentiability properties of the non-decreasing function $f$. The Euclidean translations within each component are material isomorphisms. Additional material isomorphisms are obtained by composition with non-trivial elements (if any) of the symmetry group of the archetypal function $\bar{\psi}$. A schematic representation of the partition into transitivity components is shown in Fig. 8, where, for the sake of clarity, only a hemisphere has been drawn. The body resembles a planet with a uniform core surrounded by uniform concentric layers with progressively weaker material properties.

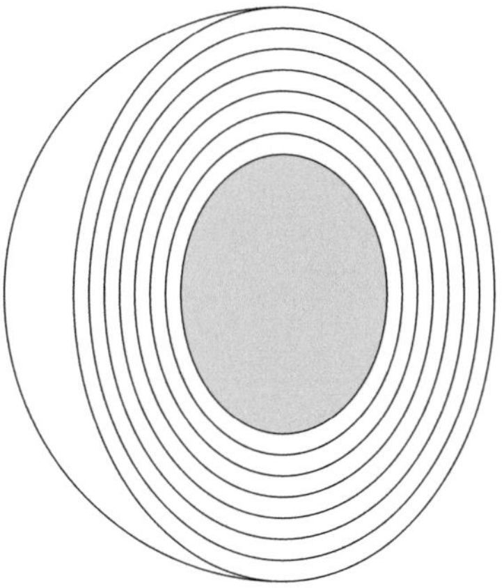

**Fig. 8** Transitivity components of $\mathcal{B}$ for the constitutive law (33)

### 6.5.3 Material Distribution: Annihilation Condition

To independently construct the material distribution $A\mathcal{P}$, we recall that the (admissible) vector fields that generate it must satisfy two conditions. The first condition stipulates that they must annihilate the differential of the constitutive law, namely,

$$d\psi = -\frac{df(r-1)}{dr}\bar{\psi}\,dr + \big(1 - f(r-1)\big)\frac{\partial\bar{\psi}}{\partial F_I^J}dF_I^J. \tag{34}$$

A general vector field **V** tangent to the $\beta$-fibres of the frame groupoid $\mathcal{F}(\mathcal{B})$ can be written as

$$\mathbf{V} = v^L\frac{\partial}{\partial X^L} + V_I^J\frac{\partial}{\partial F_I^J}, \tag{35}$$

or, more precisely, at a point $(X^L, Y^I, A_K^J)$,

$$\mathbf{V} = \big(X^L, Y^I, A_K^J, v^L, 0, V_K^J\big). \tag{36}$$

Accordingly, the components of any admissible vector field must satisfy the identity

$$-v^L\frac{df(r-1)}{dr}\frac{\partial r}{\partial X^L}\bar{\psi} + V_I^J\frac{\partial\bar{\psi}}{\partial F_I^J}\big(1 - f(r-1)\big) = 0. \tag{37}$$

### 6.5.4 Material Distribution: Left-Invariance Condition

The second condition demands that the admissible vector fields be left-invariant under the action of the elements of the frame groupoid. An element $a$ in the $\beta$-fibre of $\mathcal{F}(\mathcal{B})$ at a point $Y \in \mathcal{B}$ has coordinates $(X^L, Y^I, A_K^J)$, where $X^L$ are the coordinates of $X = \alpha(a) \in \mathcal{B}$. The unit element of this fibre has coordinates $(Y^L, Y^I, \delta_K^J)$. Let $b \in \mathcal{F}(\mathcal{M})$ be an element such that $\alpha(b) = \beta(a)$, and let its coordinates, as an element of the $\beta$-fibre at $Z = \beta(b)$, be $(Y^L, Z^I, B_K^J)$. The left action by $b$, as an element of the $\beta$-fibre at $\beta(b)$, has the coordinate representation

$$L_b(a) = \big(X^L, Z^I, B_M^J A_K^M\big). \tag{38}$$

The tangent vector **V** in Eq. (36) is mapped to the vector

$$(L_b)_*\big(\mathbf{V}(X^L, Y^I, A^J_K)\big) = \big(X^L, Z^I, B^J_S A^S_K, v^L, 0, B^J_S V^S_K\big). \tag{39}$$

For a left-invariant vector field this vector should be the same as the value of the vector field at the point $(X^L, Z^I, B^J_S A^S_K)$. Let the vector field attain values at the units given in coordinates by

$$\mathbf{V}(Y^L, Y^I, \delta^J_K) = \big(Y^L, Y^I, \delta^J_K, w^L(Y), 0, W^J_K(Y)\big). \tag{40}$$

Left-invariance demands that

$$\mathbf{V}(Y^L, Z^I, B^J_K) = \big(Y^L, Z^I, B^J_K, w^L(Y), 0, B^J_S W^S_K(Y)\big). \tag{41}$$

### 6.5.5 Generating Vector Fields

As a consequence of the left-invariance condition we obtained that the components $v^L$ of the admissible vector fields depend only on the coordinates of the base manifold. Accordingly, the identity (37) implies that

$$v^L \frac{df(r-1)}{dr}\frac{\partial r}{\partial X^L} = 0 \tag{42}$$

and

$$V^J_I \frac{\partial \bar{\psi}}{\partial F^J_I} = 0. \tag{43}$$

Equation (42) implies that

$$\mathbf{v} = v^L \frac{\partial}{\partial X^L} = \begin{cases} \text{arbitrary} & 0 \le r \le 1 \\ \text{tangent to } r = \text{constant} & r > 1 \end{cases} \tag{44}$$

Since, however, the admissible vector fields must also be locally smooth, we must modify this equation slightly to read

$$\mathbf{v} = v^L \frac{\partial}{\partial X^L} = \begin{cases} \text{arbitrary} & 0 \le r < 1 \\ \text{tangent to } r = \text{constant} & r \ge 1 \end{cases} \tag{45}$$

As far as Eq. (43) is concerned, recalling that a symmetry **G** of $\bar{\psi}$ satisfies the identity $\bar{\psi}(\mathbf{FG}) = \bar{\psi}(\mathbf{F})$, we conclude that

$$V^J_I = F^J_K \mathfrak{g}^K_I, \tag{46}$$

where $\mathfrak{g}^K_I$ is any element of the Lie algebra of the symmetry group of $\bar{\psi}$.

### 6.5.6 Body Distribution

According to the previous results, the generators of the material distribution are all the vector fields of the form (41), where $w^L(Y)$ and $W^J_K(Y)$ (specified at the unit section) are locally smooth and satisfy Eqs. (45) and (46). These admissible vector fields are defined over the whole frame groupoid $\mathcal{F}(\mathcal{B})$. As explained in Sect. 6.2, however, the admissible vector fields can be restricted to the material groupoid $\mathcal{P}$ itself.

To obtain the body distribution, we project the admissible vector fields by means of $\alpha_*$ and obtain corresponding vector fields over $\mathcal{B}$. As a result, we obtain the vector fields with components $w^L(Y)$ satisfying Eq. (45). Thus, the leaves of the body distribution in this case coincide with the transitivity components, except for the 2-sphere $r = 1$, which in the body distribution acts as an independent 2-dimensional material manifold, just as all the other layers with $r > 1$, while the core component is an open ball of unit radius.

We note, finally, that the construction of the body distribution can, in principle, be carried out independently of the explicit determination of the material groupoid or its transitivity components.

## 7 *p*-Uniformity

### 7.1 Introduction

The material distribution is a well-defined geometrical object uniquely associated with a given smooth constitutive response. It plays a role akin to that of the Lie algebroid associated to a Lie groupoid of a uniform body. Thus, from the point of view of Continuum Mechanics, it offers a tool apt to provide a general classification of smoothly non-uniform bodies and opens the possibility to distinguish various degrees of homogeneity and their associated physical counterparts, such as continuous distributions of defects. Moreover, extending the domain of definition of the material groupoid to the body-time realm, processes of material remodeling, growth, aging and morphogenesis can be subsumed under the same theoretical umbrella.

A particular instance of the rigorous analysis of non-uniform bodies will be presented in the remainder of this section. Roughly speaking, a (connected) body $\mathcal{B}$ is said to be *p-uniform* if it is the disjoint union of material submanifolds of the same dimension $p$. The case $p = 3$ corresponds to standard uniformity. It is important to remark that, by not having included manifolds with boundary in the definition of material submanifolds, we cannot possibly obtain a partition of $\mathcal{B}$ into two or more uniform subbodies (of dimension 3), since there would have to be at least one extra material submanifold of dimension 2 between them. Uniformity of degree 2 corresponds to the important practical application of *laminated bodies*, obtained as an assembly of uniform layers whose material properties vary continuously from layer to layer. The case $p = 1$ represents *filamented bundles*, such as is the case in many biological applications. Finally, the case $p = 0$ can be considered as that of a general FGM (functionally graded material).

### 7.2 *p*-Uniformity

**Definition 1** Let $\mathcal{B}$ be a body with material groupoid $\mathcal{P}$ and body distribution $A\mathcal{P}^\sharp$. $\mathcal{B}$ is said to be *p-uniform at* $X \in \mathcal{B}$ if $(A\mathcal{P}^\sharp)_X$ has dimension $p$. $\mathcal{B}$ is *p-uniform* if it is $p$-uniform at all its points.

Smooth uniformity is a particular case of $p$-uniformity. In fact, $\mathcal{B}$ is smoothly uniform if, and only if, $\mathcal{B}$ is 3-uniform. Equivalently, $\mathcal{B}$ is 3-uniform if, and only if, $(A\mathcal{P}^\sharp)_X$ has dimension 3 for all $X \in \mathcal{B}$, i.e., there exists just one leaf of the material foliation equal to $\mathcal{B}$. Hence, the material groupoid is the Lie groupoid whose $\beta$-fibres integrate the material distribution.

**Corollary 1** *Let $\mathcal{B}$ be a body and let $X \in \mathcal{B}$. $\mathcal{B}$ is $p$-uniform at $X$ if, and only if, the uniform body submanifold at $X$ (that is, the leaf $\mathcal{H}(X)$ of the body foliation) has dimension $p$.*

**Corollary 2** *Let $\mathcal{B}$ be a body. $\mathcal{B}$ is $p$-uniform if, and only if, the body foliation $\mathcal{H}$ is regular of rank $p$.*

Note that the body-material foliation has certain condition of maximality. In fact, suppose that there exists another foliation $\mathcal{H}'$ of $\mathcal{B}$ by smoothly uniform material submanifolds. Then, for all $X \in \mathcal{B}$ we must have

$$\mathcal{H}'(X) \subseteq \mathcal{H}(X), \ \forall X \in \mathcal{B}.$$

As a result, we obtain

**Corollary 3** *Let $\mathcal{B}$ be a body and let $X \in \mathcal{B}$. $\mathcal{B}$ is $q$-uniform with $q \geq p$ at $X$ if, and only if, there exists a foliation $\{\mathcal{H}'(X)\}$ of $\mathcal{B}$ such that the leaf $\mathcal{H}'(X)$ has dimension greater or equal to $p$.*

**Corollary 4** *Let $\mathcal{B}$ be a body. $\mathcal{B}$ is $q$-uniform with $q \geq p$ if, and only if, the body can be foliated by smoothly uniform material submanifolds of dimension $p$.*

### 7.3 Standard Flat Structures in $\mathbb{R}^n$

For a given $p$ (with $0 \leq p \leq n$), $\mathbb{R}^n$ can be trivially foliated as the product $\mathbb{R}^p \times \mathbb{R}^{n-p}$, each leaf $\mathbb{S}_c$ being defined as

$$\mathbb{S}_c = \{(x, c) \mid x \in \mathbb{R}^p\}, \tag{47}$$

where $c$ is a fixed element of $\mathbb{R}^{n-p}$. The set $\mathcal{F}^p(\mathbb{R}^n)$ defined as

$$\mathcal{F}^p(\mathbb{R}^n) = \bigcup_{c \in \mathbb{R}^{n-p}} \mathbb{S}_c \times \mathbb{S}_c \times GL(n, \mathbb{R}) \tag{48}$$

will be called the *p-frame bundle of* $\mathbb{R}^n$. This set has the structure of a Lie groupoid.

Let $G$ be a (Lie) subgroup of the general linear group $GL(n, \mathbb{R})$. The Lie subgroupoid

$$\mathcal{F}^p_G(\mathbb{R}^n) = \bigcup_{c \in \mathbb{R}^{n-p}} \mathbb{S}_c \times \mathbb{S}_c \times G \tag{49}$$

is a reduction of $\mathcal{F}^p(\mathbb{R}^n)$ that will be called the *standard flat reduction of* $\mathcal{F}^p(\mathbb{R}^n)$ *to* $G$.

*Remark 5* A more general definition can be envisioned. Let $Conj(G)$ denote the set of all the Lie subgroups of $GL(n, \mathbb{R})$ that are conjugate to $G$, and let $\hat{G} : \mathbb{R}^{n-p} \to Conj(G)$ be a (smooth) function. Then the set

$$\mathcal{F}^p_{\hat{G}}(\mathbb{R}^n) = \bigcup_{c \in \mathbb{R}^{n-p}} \mathbb{S}_c \times \mathbb{S}_c \times \hat{G}(c) \tag{50}$$

is a (Lie) subgroupoid of $\mathcal{F}^p(\mathbb{R}^n)$.

A (vector-valued) function $\tilde{\psi}$ over the frame groupoid $\mathcal{F}(\mathbb{R}^n)$ is of the *constitutive type* if it is of the form of Eq. (25), namely,

$$\tilde{\psi}(X, Y, A) = \psi(X, A) \quad X, Y \in \mathbb{R}^n, \quad A \in GL(n, \mathbb{R}). \tag{51}$$

Consider the particular case

$$\psi(X, A) = \psi\big((x, c), A\big) = \hat{\psi}(c, A) \quad c \in \mathbb{R}^{n-p} \quad A \in GL(n, \mathbb{R}), \tag{52}$$

namely, the function depends on the leaf $\mathbb{S}_c$, but is independent of the location within the leaf. Moreover, let the derivative with respect to $c$ not vanish identically on any submanifold $c =$ constant. Finally, let the symmetry group $G$ of $\hat{\psi}$ be independent of $c$. The material distribution $A^p_G(\mathbb{R}^n)$ associated with such a function will be called *the p-standard flat distribution with respect to G*.

**Proposition 2** *The maximal integral manifolds of $A^p_G(\mathbb{R}^n)$ are the $\beta$-fibres of the standard flat reduction $\mathcal{F}^p_G(\mathbb{R}^n)$.*

Clearly, all the structures introduced in this section can be restricted to any open subset of $\mathbb{R}^n$.

### 7.4 Homogeneity of $p$-Uniform Bodies

As originally introduced by Noll [17], a precondition for a body to be homogeneous is that it must be materially uniform. In the putative physical picture, however, if homogeneity is to be interpreted as the absence of defects, it is conceivable that in some sense a non-uniform body may be defect-free and that, if it were not, a measure of defect density may be obtainable. A generalization of this kind for certain classes of FGMs has been presented in [5], where the notion of material isomorphism was greatly relaxed by declaring points to be isomorphic if they just have conjugate symmetry groups. For $p$-uniform bodies with $p = 1, 2$, which are indeed FGMs, stricter criteria of homogeneity can be established without thus relaxing the definition of material isomorphism.

**Definition 2** A $p$-uniform body $\mathcal{B}$ is *$p$-homogeneous* if its material groupoid in some configuration $\kappa$ coincides with the standard flat reduction $\mathcal{F}^p_G(\kappa(\mathcal{B}))$, for some subgroup $G \subset GL(3, \mathbb{R})$. It is *locally $p$-homogeneous* if it can be covered with $p$-homogeneous sub-bodies.

**Corollary 5** *A $p$-uniform body $\mathcal{B}$ is $p$-homogeneous if, and only if, its material distribution in some reference configuration $\kappa$ coincides with the standard flat distribution $A^p_G(\kappa(\mathcal{B}))$ for some subgroup $G \subset GL(3, \mathbb{R})$.*

*Remark 6* A more general concept of $p$-homogeneity can be achieved by considering the standard Lie subgroupoids suggested in Remark 5.

## References


1. Bilby, B.A.: Continuous distributions of dislocations. In: Sneddon, I.N., Hill, R. (eds.) Progress in Solid Mechanics I, pp. 329–398. North-Holland, Amsterdam (1960)
2. Bloom, F.: Modern Differential Geometric Techniques in the Theory of Continuous Distributions of Dislocations. Lecture Notes in Mathematics, vol. 733. Springer, Berlin (1979)
3. Elżanowski, M., Epstein, M., Śniatycki, J.: G-structures and material homogeneity. J. Elast. **23**, 167–180 (1990)
4. Epstein, M., de León, M.: Geometrical theory of uniform Cosserat media. J. Geom. Phys. **26**, 127–170 (1998)
5. Epstein, M., de León, M.: Homogeneity without uniformity: toward a mathematical theory of functionally graded materials. Int. J. Solids Struct. **37**, 7577–7591 (2000)
6. Epstein, M., de León, M.: Material groupoids and algebroids. Math. Mech. Solids (2018). https://doi.org/10.1177/1081286518755229
7. Epstein, M., Segev, R.: Differentiable manifolds and the principle of virtual work in continuum mechanics. J. Math. Phys. **21**, 1243–1245 (1980)
8. Eshelby, J.D.: The continuum theory of lattice defects. In: Seitz, F., Turnbull, D. (eds.) Solid State Physics 3, pp. 79–144. Elsevier, Amsterdam (1956)
9. Frank, F.C.: Crystal dislocations—elementary concepts and definitions. Philos. Mag. **42**, 809–819 (1951)
10. Jiménez, V.M., de León, M., Epstein, M.: Material distributions. Math. Mech. Solids (2018). https://doi.org/10.1177/1081286517736922
11. de Jiménez, V.M., León, M., Epstein, M.: Characteristic distribution: an application to material bodies. J. Geom. Phys. **127**, 19–31 (2018)
12. Kobayashi, S., Nomizu, K.: Foundations of Differential Geometry. Wiley, New York (1963)
13. Kondo, K.: Geometry of Elastic Deformation and incompatibility. RAAG Memoirs of the Unifying Study of Basic Problems in Engineering and Physical Sciences by means of Geometry, Tokyo Gakujutsu Benken Fukyu-Kai 1/C, 361–373 (1955)
14. Kroener, E.: Allgemeine Kontinuumstheorie der Versetzungen und Eigenspannungen. Arch. Ration. Mech. Anal. **4**, 273–334 (1959)
15. Mackenzie, K.: Lie Groupoids and Lie Algebroids in Differential Geometry. London Math. Soc. Lecture Notes Series, vol. 124. Cambridge University Press, Cambridge (1987)
16. Michor, P.W.: Topics in Differential Geometry. Am. Math. Soc., Providence (2008)
17. Noll, W.: Materially uniform simple bodies with inhomogeneities. Arch. Ration. Mech. Anal. **27**, 1–32 (1967)
18. Segev, R.: Forces and the existence of stresses in invariant continuum mechanics. J. Math. Phys. **27**, 163–170 (1986)
19. Wang, C-C.: On the geometric structures of simple bodies, a mathematical foundation for the theory of continuous distributions of dislocations. Arch. Ration. Mech. Anal. **27**, 33–94 (1967)
20. Weinstein, A.: Groupoids: unifying internal and external symmetry. Not. Am. Math. Soc. **43**(7), 744–752 (1996)

J Elast (2019) 135:261–293
https://doi.org/10.1007/s10659-018-09718-4

# A Causality Setting for Elasticity Theory

**Roger Fosdick**[1]

*Dedicated to the memory of Walter Noll with affection and a deep sense of gratitude for his clarity of style and expression of concept*

Received: 2 June 2018 / Published online: 3 January 2019


**Abstract** We give a causality approach to nonlinear elasticity theory within a pure mechanics setting. Based on the notion of a fixed frame in $\mathbb{E}^3$ and Euclidean invariance considerations, we show how a broadly general and novel evolutionary hypothesis concerning 'cause' and 'effect' must reduce to the classical statement of the balance of energy, and we obtain all of the classical balance laws of continuum mechanics. The concept of mass and its balance is derived within this theory. The mass density naturally emerges from the theory without preconception as an inertial scalar field for the body which is associated with the speed of its material particles and a measure of their kinetic behavior. Aside from the causality hypothesis and its invariance, the fundamental notions of body, motion, force and internal power are primitive.

**Keywords** Causality · Continuum mechanics · Elasticity · Invariance · Objectivity

**Mathematics Subject Classification (2010)** 74A99

## Contents



✉ R. Fosdick
fosdick@aem.umn.edu

[1] Department of Aerospace Engineering and Mechanics, University of Minnesota, Minneapolis, MN 55455-0153, USA

## 1 Introduction

It has long been a curiosity why the classical balance of energy law for a material body is assumed to be additive in the sense that the *sum* of the time rate of change of the internal plus the kinetic energy is equal to the *sum* of the internal power supply plus the rate of heat supply. Accepting this set-up in a fixed frame as a basic principle, Green and Rivlin [7, 8], in 1964, proposed that the law be form invariant under time-dependent Euclidean changes of frame,[1] and that certain appropriate physical fields used in expressing the law, such as specific internal energy, contact force, heat flux, radiant heating and mass density, be objective. Their aim was to derive the classical balances of linear and angular momentum and they noticed as an afterthought that the balance of mass also was derivable. One year earlier, Noll [14] proposed a principle of frame indifference of the mechanical power in which he regarded the 'inertial force' as a constitutive postulate expressing the interaction of the body and the space in which it is emersed, and he derived the classical balances of linear and angular momentum. In the above referenced works, the body force field was given special attention essentially because of its long history of being considered dependent upon the particular frame in which the balance of energy and the definition of internal power is expressed. Newton assumed the existence of a fixed frame in absolute Euclidean space in formulating his first law of motion and he observed that there was a whole class of frames, related by Galilean transformations, under which his first law was invariant. Thus, in most modern developments this observation has been taken to mean that the body force field is invariant under translational Galilean changes of frame from a given fixed frame; otherwise, it is allowed to be frame-dependent.

The classical balance of energy approach, taken together with invariance as forming a unifying basis for the balance equations of mechanics, has been improved over the years and made more fundamental at various levels. At a primitive level, Newton's introduction of a fixed frame, related to the existence of an absolute space, might be replaced by a more modern relativistic idea of space, together with an appropriate non-relativistic approximation. Such a major revision has not yet been accomplished though it may help to explain

[1] Green and Rivlin preferred to state this invariance in terms of superposed rigid body motions relative to a fixed frame. This somewhat imprecise 'change of frame' interpretation of their notion is not meant to infer equivalence of the two concepts. It is used here solely for the sake of convenience and simplicity in drawing a parallel to other works.

why the body force field in the non-relativistic limit contains the strange Coriolis effect in moving frames—strange mainly because of its explicit dependence on the velocity of the material particles relative to the moving frame. In the present work, we accept Newton's introduction of a fixed frame, and within a non-relativistic setting we address other pertinent questions which are at the root of the classical balance laws of mechanics. In order to exhibit most transparently the core elements of our causality approach within a continuum setting, we shall concentrate on the pure mechanics of elastic bodies. We admit motion, force and internal power as the fundamental concepts from mechanics and relative to a fixed frame we propose an evolutionary law for arbitrary parts of a body which expresses a cause-effect relationship[2] between these fundamental variables. The law of causality we introduce, which may be considered to be a broad generalization of the balance of energy, is restricted to be form invariant under Euclidean changes of frame and certain minimal objective properties of force are assumed.

A primary goal of this work is to address two major questions based upon the hypothesis of a fairly general law of causality within the mechanics of elastic bodies—a law which is supposed to express the evolutionary behavior of the internal shape and particle speeds of a physical body: (1) Does the classical form of the balance of energy and, in particular, the splitting of the energy into the sum of an internal and kinetic part, follow from our law of causality as a result of invariance considerations? (2) Is the origin of mass and its balance derivable from such a causality law? Such questions have been investigated previously[3] within a setting different from causality, the contribution of Šilhavý [17] being of greatest depth, scope and generality. See also, Serrin [16], Podio-Guidugli [15], Favata, Podio-Guidugli and Tomassetti [3], and Fosdick and Serrin [5]. A brief comment on the overall distinction between these various works in their approach to the energy-splitting issue as compared to the current investigation is given in Sect. 2 following this introduction. Recently, Fosdick [4] has proposed and investigated a causality approach to particle dynamics for systems. The present paper is a generalization of the foundations and strategy of that approach within the area of continuum mechanics, especially suitable for describing the mechanics of elastic bodies.

We begin Sect. 3 with a short description of frames in Euclidean 3-space and time-dependent proper changes of frame. Then, in Sects. 3.2, 3.3, and 3.4, we introduce the fundamental concepts of motion and force, and the definition of internal power as a linear function of particle velocity—concepts and definitions that are essential to this work. Body force and contact surface traction are distinguished, and the surface traction on an oriented surface in the deformed state of a body is assumed to depend upon the unit normal field

[2] In early days, causality, as a form of determinism, broadly influenced the origin of scientific thought and the attainment of rational knowledge. In Bunge's [2] 1959 philosophical essay, *Causality*, he examines critically the causal principle and the meaning of the law of causation. In Part IV, he inquires into the place held by the causal principle in modern science, with the special interest of investigating the function of a philosophical principle in scientific research. He traces the relation between cause and reason, a subject of the $17^{\text{th}}$ century rationalists, and he gives an evolutionary account of the causal principle in scientific thought and law, its meaning and interpretation. He draws the following conclusion [2, p. 335]: "The causal principle is one of the various valuable guides of scientific research and, like most of them, it enjoys an approximate validity in limited ranges; it is a general hypothesis with a high heuristic value–a fact suggesting that, in certain domains, it does correspond rather closely to reality."

[3] Being inspired by the clarity and depth of Walter Noll's publications on the foundations of continuum mechanics, J.E. Dunn and I began working on the question of splitting in the late 1960's, and in December, 1969 R.F. presented invited colloquium talks at the University of Illinois and the University of Kentucky titled "A Causality Approach to Continuum Mechanics". In these talks, a strategy for splitting and the existence of mass and its balance was presented, albeit the work was not yet completed.

of the surface and, to characterize a constitutive elastic behavior, the deformation gradient field.

In Sect. 4, we introduce and characterize our causality hypothesis for arbitrary parts of a body relative to a fixed frame F as an evolutionary ordinary differential equation which contains a causal functional and a response functional, both fundamental elements to be prescribed as part of the theory. The law of causality which we adopt states that for arbitrary parts of a body the present time rate of change of the response functional is equal to the present value of the causal functional, and, because of the generality of our hypotheses regarding these two functionals, our latter analysis, while based on invariance considerations, is not a part of any of the earlier referenced works within continuum mechanics that address the energy-splitting issue. The current work, though, is restricted to elastic bodies. The response functional for an arbitrary part of the body at each time is assumed to depend rather generally upon the particle speeds and deformation gradient field throughout that part, and its time derivative is supposed to track the evolutionary behavior of these fields during a process. The causal functional for an arbitrary part of the body at each time also is broadly defined and assumed to depend upon the internal power supplied to that part and the deformation gradient field throughout that part. The two entries in this functional are identified as elements that are supposed to cause the process to evolve, modulated and measured through the form of the causal functional. Smoothness and other mild hypotheses for the causal and response functionals are given and a main hypothesis which characterizes mechanical processes $\mathbb{P}_t$ relative to the fixed frame F is introduced at the end of this section.

In Sect. 5, we obtain the general consequences of the first of our two axioms which concerns the form invariance of the causality law under translational Galilean changes of frame relative to a fixed frame: The major results of this paper rest upon the consequences of this axiom. We obtain the splitting of the response functional for an arbitrary part of the body into the sum of a static (denoted as an internal energy functional) and an inertial (denoted as a kinetic energy functional) contribution and show that the causal functional for an arbitrary part of the body reduces to a linear function of its internal power with positive multiplying constant. This split of the response functional and the coincidental reduction of the causal functional requires a somewhat intricate invariance argument. The result is of critical importance as it leads to the existence of a mass measure through the identification of a crude form of the balance of linear momentum for arbitrary parts of a body relative to a fixed frame. This crude form involves the inertial contribution of the response functional and it supports the conclusion that the inertial contribution is additive on separate parts of the body and absolutely continuous with respect to volume measure. This gives rise to the existence of a mass measure and its balance, as we show in Sect. 5.1.

We begin Sect. 6 with our second axiom which asserts the existence of a frame-dependent body force field and proposes a related invariance property of the causality law for mechanical processes $\mathbb{P}_t^*$ under proper time-dependent Euclidean changes of frame $\mathsf{F} \to \mathsf{F}^*(t)$. This axiom allows us to express the causality law under any time-dependent proper Euclidean change of frame as a form-invariant copy of the so-called balance of energy reached in Sect. 5. We then obtain, in Sect. 6.1, the classical form of the balance of linear momentum for arbitrary parts of a body relative to the fixed frame F. Consequently, the classical Cauchy stress theorem holds and, because of the earlier constitutive hypothesis on the surface traction, the related Cauchy stress tensor becomes expressed as a constitutive function of the deformation gradient field. Then, in Sect. 6.2 we develop two fundamental identities, using the corresponding balance of energy expressions for the mechanical processes $\mathbb{P}_t$ and $\mathbb{P}_t^*$ under the proper Euclidean change of frame $\mathsf{F} \to \mathsf{F}^*(t)$.

In Sect. 6.3, the first of the identities from Sect. 6.2 is used to show that the Cauchy stress tensor is symmetric and, correspondingly, the balance of angular momentum is satisfied for the process $\mathbb{P}_t$ relative to the fixed frame $\mathsf{F}$. In addition, the first identity is used to show that the static part of the response functional for an arbitrary part of the body is represented as a volume integral of a frame-indifferent specific stored energy function which depends upon the deformation gradient field. It is noted that this specific stored energy function serves as a potential for calculating the Cauchy stress, as is normal in elasticity theory.

In Sect. 6.4, the second of the identities from Sect. 6.2 is used to determine the form of the frame-dependent body force relative to the moving frame $\mathsf{F}^*(t)$. It is significant that this frame-dependent body force field does not contain the so-called Coriolis force, a consequence which stems from the fundamental hypothesis in Sect. 3.4 that power is a linear function of the particle velocity and body force is supposed to be independent of particle velocity. Of course, at this point the balance laws of linear and angular momentum have been determined only relative to the fixed frame $\mathsf{F}$.

In Sect. 7, we first determine the balance of linear momentum for mechanical processes $\mathbb{P}_t^*$ relative to a moving frame $\mathsf{F}^*(t)$. This balance equation is shown to contain an effective body force that is the sum of the frame-dependent body force found in Sect. 6.4 and the classical Coriolis term, i.e., the term which accounts for an interaction between the velocity of the particles of the body and the rotation of the moving frame. It is noteworthy that the Coriolis force for a mechanical process $\mathbb{P}_t^*$ relative to $\mathsf{F}^*(t)$ is orthogonal to the corresponding particle velocity and does not contribute to the expression for power. We end Sect. 7 with reference to the symmetry of the Cauchy stress for mechanical processes $\mathbb{P}_t^*$ and its consequence of balance of angular momentum relative to the moving frame $\mathsf{F}^*(t)$.

The Appendix contains some elementary lemmas that are used in the body of this work.

## 2 Comments on Previous Works and Present Development

The contributions of Šilhavý [17], Serrin [16], Podio-Guidugli [15], Favata, Podio-Guidugli and Tomassetti [3], and Fosdick and Serrin [5] that address the splitting of the local density of intrinsic energy of a body into the sum of a frame indifferent internal energy plus kinetic energy are all based on invariance arguments related to 'frame indifference'. However, the specific starting hypotheses of what is taken to be fundamentally invariant and other working hypotheses of lesser significance differ between these works, and they all are different from the set-up and detailed invariance requirements surrounding the causality law that is considered in the present paper. In all of these works, the splitting issue and the existence (and balance) of mass, are resolved by applying only the special requirement of translational Galilean invariance. For obtaining the remaining balance laws and determining various field invariance conditions, the full invariance notion associated with frame indifference is required.

Šilhavý [17], Serrin [16] and Fosdick and Serrin [5] start with a balance of energy hypothesis for arbitrary parts of a body in which the local (reference volume) density of intrinsic energy is taken as fundamental and not considered, as classically, to be the sum of internal plus kinetic energy. In these works, the time rate of change of the total intrinsic energy for a part is balanced against the power input plus the heat input to that part. With additional technical hypotheses, including the notion of frame indifference, Šilhavý [17] obtains the balance of linear and angular momentum, the existence (and balance) of a non-negative mass density field, and the transformation law for the body force density. The so-called Coriolis force is built into Šilhavý's frame indifference body force hypothesis. Šilhavý defines the density of internal energy as the density of intrinsic energy minus the density of kinetic energy, and then shows that the density of internal energy is invariant under changes of frame. Thus, he establishes the split of intrinsic energy, as it is classically assumed, into the additive form. Serrin [16] and Fosdick and Serrin [5] take a different approach than that of Šilhavý in formulating their relation between force and power, and this effects the way in which the Coriolis force emerges from the development. The meaning of 'state', constitutive assumptions, and related detailed hypotheses of importance, but of lesser significance, differ between these three works, but an overall requirement of invariance is common and the classical additive splitting of intrinsic energy is a result of translational Galilean invariance.

Podio-Guidugli [15] and Favata, Podio-Guidugli and Tomassetti [3] assume a referential setting for their development of the splitting of the density of intrinsic energy into the sum of internal plus kinetic energy. To resolve the splitting issue, they introduce as fundamental the density of internal (i.e., intrinsic) power as the time derivative of the density of intrinsic energy plus the density of inertial power, the latter of which is defined as the inner product of the inertial body force and the particle velocity. They take the intrinsic power to be translational Galilean invariant, and in their work the body force is assumed to be composed of a noninertial part, which plays no role in the inertial power, plus an inertial part. In addition, since the inertial power is assumed to be linear in the velocity, the inertial body force consists of the sum of a velocity-independent component, which is responsible for all of the inertial power, plus a vector component which depends on particle velocity but is orthogonal to it. The latter component is 'powerless' and is termed the Coriolis part, following an idea expressed by Serrin[16]. The notions and detailed hypotheses concerning body force, 'state', and constitutive assumption are important elements in the two above-mentioned works. In the second, Favata, Podio-Guidugli and Tomassetti [3] generalize the constitutive structure considered by Podio-Guidugli [15] to include fading memory materials. In both works, the splitting of intrinsic energy into the additive form as it is classically presented, is achieved and the existence of a conservative mass density tensor field is proved.

The approach of the present work is based on an evolutionary cause-effect notion which is expressed in general terms as an equivalence between the time rate of change of a response functional and a causal functional for arbitrary parts of a body. While the response and causal functionals are broadly defined and generalizations seem possible, the developments here are restricted to the mechanics of elastic bodies. As with all other investigations of the splitting issue, invariance hypotheses related to frame indifference are an integral part of the causality law of this work. In particular, with translational Galilean invariance, the response functional, being defined as a (regular) spatial nonlinear functional of the deformation gradient field, the square magnitude of the particle velocity field and the particle, over an arbitrary part of a body, is shown to split into the sum of a static (internal energy) spatial functional of the deformation gradient field and particle plus a inertial (kinetic energy) spatial functional of the square magnitude of the particle velocity field and particle. In addition, both of these functionals are found to be absolutely continuous with respect to reference volume, and therefore are expressible as volume integrals with corresponding densities of internal energy depending on deformation gradient and particle, and kinetic energy depending quadratically on the particle speed and particle. The causal functional corresponding to an arbitrary part of the body is defined as a (regular) nonlinear functional of the power supplied to that part, and the deformation gradient field and the particle over that part. This functional is shown to reduce considerably to a linear function (with positive multiplying constant) of the power supplied to the part. A positive, conserved mass density measure emerges naturally from, and is defined in terms of, the inertial spatial functional, and so has an interpretation as a 'capacity' for kinetic energy. The remaining fundamental balance equations of linear and angular momentum, the stress constitutive equation for hyperelasticity, the transformation of body force under change of frame together with the existence and form of the Coriolis body force in moving frames of reference, and the frame indifference of the density of internal energy are all dependent upon the causality law being invariant under arbitrary changes of frame.

## 3 Setup and Fundamental Fields

Frames, motion, force and internal power are fundamental to this work and all have been described elsewhere in the continuum mechanics literature. This short section is included so that the latter developments given herein are reasonably self-contained.

### 3.1 Frames

The elements $\boldsymbol{x} := (x_1, x_2, x_3) \in \mathbb{E}^3$ of a three dimensional Euclidean point space $\mathbb{E}^3$ are triples of all real numbers, and with $\mathbb{E}^3$ we associate a translation vector space $\mathbb{V}^3$. A frame $\mathsf{F}$ is defined as a basis $\{\boldsymbol{e}_1, \boldsymbol{e}_2, \boldsymbol{e}_3\} \in \mathbb{V}^3$, together with a specified element $\boldsymbol{o} \in \mathbb{E}^3$, called the origin. Without loss of generality, we let $\boldsymbol{o} := (0, 0, 0)$ and refer to $\mathsf{F}$ as fixed. Thus, all elements $\boldsymbol{x} \in \mathbb{E}^3$ are equally characterized in terms of the frame $\mathsf{F}$ by

$$\boldsymbol{x} = x^i \boldsymbol{e}_i + \boldsymbol{o} \quad \forall \left(x^1, x^2, x^3\right) \in \mathbb{R}^3, \tag{3.1}$$

where $(x^1, x^2, x^3) \in \mathbb{R}^3$ is unique for each $\boldsymbol{x} \in \mathbb{E}^3$.

A second moving frame $\mathsf{F}^*(t)$, where $t$ denotes time, is defined by a $t$-dependent basis $\{\boldsymbol{e}_1^*, \boldsymbol{e}_2^*, \boldsymbol{e}_3^*\}(t) \in \mathbb{V}^3$, together with a specified element $\boldsymbol{c}(t) \in \mathbb{E}^3$, again called the origin. The two frames $\mathsf{F}$ and $\mathsf{F}^*(t)$ preserve orientation and are isometrically equivalent if and only if

$$\boldsymbol{e}_i^*(t) = \boldsymbol{Q}(t)\boldsymbol{e}_i \quad i = 1, 2, 3, \qquad \boldsymbol{Q}(t) \in \mathrm{Orth}^+ := \{\boldsymbol{Q} \in \mathrm{Orth} \,|\, \det \boldsymbol{Q} = +1\}, \tag{3.2}$$

where Orth denotes the set of all orthogonal linear transformations of $\mathbb{V}^3 \to \mathbb{V}^3$. This means that length, angle and relative orientation are preserved for the corresponding basis elements. Also, with the introduction of $\mathsf{F}^*(t)$ and the unique identification of $(x^1, x^2, x^3)$ with $\boldsymbol{x}$ for all $\boldsymbol{x} \in \mathbb{E}^3$, every element $\boldsymbol{x}^* \in \mathbb{E}^3$ may be characterized at time $t$ by

$$\boldsymbol{x}^* = x^i \boldsymbol{e}_i^*(t) + \boldsymbol{c}(t) \quad \forall \left(x^1, x^2, x^3\right) \in \mathbb{R}^3. \tag{3.3}$$

Thus, it follows that

$$\boldsymbol{x}^* = \boldsymbol{Q}(t)(\boldsymbol{x} - \boldsymbol{o}) + \boldsymbol{c}(t) \quad \forall \boldsymbol{x} \in \mathbb{E}^3. \tag{3.4}$$

The transformation (3.4) is an isometric, orientation preserving transformation of $\mathbb{E}^3 \mapsto \mathbb{E}^3$ and it is referred to as a proper Euclidean re-framing of $\mathbb{E}^3$, or more briefly as a proper change of frame $\mathsf{F} \to \mathsf{F}^*(t)$. This time-dependent change of frame simply represents two isometrically equivalent ways of characterizing the points of $\mathbb{E}^3$ and it is common to refer to this as a proper Euclidean change of frame. By definition, we refer to the frame $\mathsf{F}$ as a fixed frame, independent of time. A Galilean change of frame corresponds to the special case where $\boldsymbol{Q}(t) = \boldsymbol{Q}$ and $\boldsymbol{c}(t) = \boldsymbol{c}t + \boldsymbol{a}$, where $\boldsymbol{Q} \in \mathrm{Orth}^+$, $\boldsymbol{c} \in \mathbb{V}^3$ and $\boldsymbol{a} \in \mathbb{E}^3$ are constant, independent of time $t$.

### 3.2 Motion

It suffices for our purposes to consider a body $\mathcal{B}$ to be a bounded, open and connected set of points $X \in \mathbb{E}^3$ with piecewise smooth boundary $\partial\mathcal{B}$ relative to the fixed frame F. The points $X$ are called the particles of $\mathcal{B}$ and we will write $X \in \mathcal{B}$. We refer to $\mathcal{B}$ as a reference configuration.

A motion of $\mathcal{B}$ is a smooth time-parameter sequence of invertible placements of $\mathcal{B}$ in $\mathbb{E}^3$ during an interval of time $t \in \mathcal{I} \subset \mathbb{R}$; we assume that the initial instant $t = 0 \in \mathcal{I}$ and denote a motion as

$$\chi_t(\cdot) : \mathcal{B} \longrightarrow \mathbb{E}^3, \quad \forall t \in \mathcal{I}, \tag{3.5}$$

differentiable on $\mathcal{B} \times \mathcal{I}$. The particle velocity and the deformation gradient are denoted, respectively, as $\boldsymbol{v}_t(X) := \dot{\chi}_t(X)$ and $\boldsymbol{F}_t(X) := \nabla \chi_t(X)$, and we assume that $\det \boldsymbol{F}_t(X) > 0$ for all $X \in \mathcal{B}$, i.e., $\boldsymbol{F}_t(X) \in \mathrm{Lin}^+$. Here, and throughout, the superposed 'dot' notation $\dot{\overline{(\cdot)}}$ means partial time differentiation at fixed $X \in \mathcal{B}$. It is convenient to associate $\mathcal{B}$ with the initial configuration $\chi_0(\mathcal{B})$, so that $X = \chi_0(X)$ and $\boldsymbol{F}_0(X) = \boldsymbol{1}$, and, thus, refer to $\mathcal{B}$ as undistorted. We denote the configuration of $\mathcal{B}$ at any time $t$ as $\mathcal{B}_t := \chi_t(\mathcal{B})$. Of course, $\partial\mathcal{B} := \chi_0(\partial\mathcal{B})$ and $\partial\mathcal{B}_t := \chi_t(\partial\mathcal{B})$. An arbitrary part $\mathcal{P} \subseteq \mathcal{B}$ is open and connected, with piecewise smooth boundary $\partial\mathcal{P}$. Naturally, we write $\mathcal{P}_t := \chi_t(\mathcal{P})$ and $\partial\mathcal{B}_t := \chi_t(\partial\mathcal{B})$.

Under a proper change of frame $\mathsf{F} \to \mathsf{F}^*(t)$, according to (3.4) the motion (3.5) is represented as

$$\chi_t^*(\cdot) = \boldsymbol{Q}(t)\big(\chi_t(\cdot) - \boldsymbol{o}\big) + \boldsymbol{c}(t) : \mathcal{B} \longrightarrow \mathbb{E}^3, \quad \forall t \in \mathcal{I},$$

and the particle velocity for $X \in \mathcal{B}$ satisfies the transformation rule

$$\dot{\chi}_t^*(X) = \boldsymbol{Q}(t)\dot{\chi}_t(X) + \dot{\boldsymbol{Q}}(t)\,\chi_t(X) + \dot{\boldsymbol{c}}(t).$$

Throughout this work, we shall assume that motions of $\mathcal{B}$ are sufficiently regular. To express this more precisely, we borrow the notion of 'tameness' as introduced by Gurtin [9].

**Definition 3.1** A motion $\chi_t$ is tame on $\mathcal{B}$ for all $t \in \mathcal{I}$ if the particle velocity field $\boldsymbol{v}_t := \dot{\chi}_t$ and the deformation gradient field $\boldsymbol{F}_t := \nabla \chi_t \in \mathrm{Lin}^+$ satisfy

(1) $\boldsymbol{v}_t$ and $\boldsymbol{F}_t$ belong to the space $C(\mathcal{B}) \times C^1(\mathcal{I})$ of continuous functions on $\mathcal{B}$ and differentiable functions on $\mathcal{I}$;
(2) $\dot{\boldsymbol{v}}_t$ and $\dot{\boldsymbol{F}}_t$ belong to the space $C(\mathcal{B})$ of continuous functions on $\mathcal{B}$ for all $t \in \mathcal{I}$;
(3) Given a norm $\|\cdot\|$ on $C(\mathcal{B})$,

$$\lim_{\tau \to t} \left\| \frac{\boldsymbol{v}_\tau - \boldsymbol{v}_t}{\tau - t} - \dot{\boldsymbol{v}}_t \right\| = 0 \tag{3.6a}$$

and

$$\lim_{\tau \to t} \left\| \frac{\boldsymbol{F}_\tau - \boldsymbol{F}_t}{\tau - t} - \dot{\boldsymbol{F}}_t \right\| = 0, \tag{3.6b}$$

for every $t \in \mathcal{I}$.

The following lemma shows that one can arbitrarily specify three functions $\chi = \chi(\cdot)$, $\boldsymbol{v} = \boldsymbol{v}(\cdot)$ and $\boldsymbol{a} = \boldsymbol{a}(\cdot)$ as differentiable fields on $\mathcal{B}$, with $\nabla \chi \in \mathrm{Lin}^+$, and be assured that there is an interval $\mathcal{I}$ and a tame motion $\chi_t$ on $\mathcal{B}$ for all $t \in \mathcal{I}$ such that $\chi_{t_0} = \chi$, $\dot{\chi}_{t_0} = \boldsymbol{v}$ and $\ddot{\chi}_{t_0} = \boldsymbol{a}$ at a fixed time $t_0 \in \mathcal{I}$. This will be an important property of tame motions used later in this work. Formally, we have

**Lemma 3.1** *Let $t_0 \in \mathbb{R}$, and suppose $\chi$ maps $\mathcal{B} \to \mathbb{E}^3$, with $\nabla \chi \in \mathrm{Lin}^+$, and $\boldsymbol{v}$ and $\boldsymbol{a}$ are differentiable fields mapping $\mathcal{B} \to \mathbb{V}^3$. Then, there exists an interval $\mathcal{I} \subset \mathbb{R}$ with $t_0 \in \mathcal{I}$, and a tame motion $\chi_t$ on $\mathcal{B}$ for all $t \in \mathcal{I}$, which has the properties*

$$\chi_{t_0} = \chi, \qquad \dot{\chi}_{t_0} = \boldsymbol{v}, \qquad \ddot{\chi}_{t_0} = \boldsymbol{a}. \tag{3.7}$$

*Proof* Consider the motion defined by the field

$$\chi_t \equiv \chi + (t - t_0)\boldsymbol{v} + \frac{1}{2}(t - t_0)^2 \boldsymbol{a}, \quad \forall t \in \mathcal{I}, \tag{3.8}$$

where $t_0 \in \mathcal{I}$, and where the interval $\mathcal{I}$ is sufficiently small so that $\nabla \chi_t \in \mathrm{Lin}^+$ for all $t \in \mathcal{I}$. Clearly, this motion is tame on $\mathcal{B}$ for all $t \in \mathcal{I}$. ■

*Remark 3.1* The motion (3.8) is not the most general tame motion on $\mathcal{B}$ for $t \in \mathcal{I}$, but it will suffice for later developments. Note that for this motion, $\boldsymbol{v}_t$ and $\dot{\boldsymbol{v}}_t$ belong to the space $C^1(\mathcal{B})$ of differentiable functions on $\mathcal{B}$ for all $t \in \mathcal{I}$. □

### 3.3 Force

We suppose that in the configuration $\mathcal{B}_t$ there is acting at each $\boldsymbol{X} \in \mathcal{B}$ a body force $\boldsymbol{b}_t(\boldsymbol{X}) \in \mathbb{V}^3$, measured per unit volume of $\mathcal{B}_t$, and a contact force (traction) $\boldsymbol{\sigma}_t(\boldsymbol{X}; \boldsymbol{n}) \in \mathbb{V}^3$, defined for each vector $\boldsymbol{n} \in \mathrm{Unit} := \{\boldsymbol{n} \in \mathbb{V}^3 \,|\, |\boldsymbol{n}| = 1\}$ and measured per unit area of an oriented surface through $\chi_t(\boldsymbol{X})$ with unit normal $\boldsymbol{n}$. We assume that $\boldsymbol{b}_t(\cdot)$ is a continuous function and that $\boldsymbol{\sigma}_t(\cdot; \boldsymbol{n}) \in C^1(\mathcal{B})$ for each $\boldsymbol{n} \in \mathrm{Unit}$ while $\boldsymbol{\sigma}_t(\boldsymbol{X}; \cdot)$ is continuous for each $\boldsymbol{X} \in \mathcal{B}$. For an elastic body $\mathcal{B}$, the contact force is given in terms of a (sufficiently smooth) constitutive function $\bar{\boldsymbol{\sigma}}(\boldsymbol{X}, \cdot; \boldsymbol{n}) : \mathrm{Lin}^+ \to \mathbb{V}^3$ for every $\boldsymbol{X} \in \mathcal{B}$ and for all $\boldsymbol{n} \in \mathrm{Unit}$, and we shall assume that the undistorted reference configuration $\mathcal{B}$ is traction-free. Thus, for the motion $\chi_t$ of an elastic body we have

$$\boldsymbol{\sigma}_t(\boldsymbol{X}; \boldsymbol{n}) = \bar{\boldsymbol{\sigma}}\big(\boldsymbol{X}, \boldsymbol{F}_t(\boldsymbol{X}); \boldsymbol{n}\big), \quad \boldsymbol{F}_t(\boldsymbol{X}) = \nabla \chi_t(\boldsymbol{X}) \tag{3.9a}$$

at each particle $\boldsymbol{X} \in \mathcal{B}$ and $t \in \mathcal{I}$, and

$$\bar{\boldsymbol{\sigma}}(\boldsymbol{X}, \boldsymbol{1}; \boldsymbol{n}) = \boldsymbol{0}. \tag{3.9b}$$

The total external force acting on a part $\mathcal{P} \subseteq \mathcal{B}$ in the configuration $\mathcal{P}_t = \chi_t(\mathcal{P})$ is given by

$$\boldsymbol{f}_t(\mathcal{P}) := \int_{\partial \mathcal{P}_t} \boldsymbol{\sigma}_t \, \mathrm{d}a + \int_{\mathcal{P}_t} \boldsymbol{b}_t \, \mathrm{d}v, \tag{3.10}$$

wherein the unit normal $\boldsymbol{n}$ in $\boldsymbol{\sigma}_t$ is understood to be the outer unit normal field to $\partial \mathcal{P}_t$. Otherwise, in order to keep the notation less complicated here and elsewhere in this work, the domains of surface and volume integration will be used to identify the domains in which the integrands must be evaluated. Thus, in (3.10) $\boldsymbol{\sigma}_t$ is actually $\boldsymbol{\sigma}_t \circ \chi_t^{-1}$ and $\boldsymbol{b}_t$ is actually $\boldsymbol{b}_t \circ \chi_t^{-1}$. This abuse of notation is natural to the continuum mechanics community and should cause no problem.

### 3.4 Internal Power

The internal power $w_t(\mathcal{P})$ of $\mathcal{P} \subseteq \mathcal{B}$ in the configuration $\mathcal{P}_t = \chi_t(\mathcal{P})$ represents the work per unit time *entering* (if $w_t(\mathcal{P}) > 0$), or *exiting* (if $w_t(\mathcal{P}) < 0$) $\mathcal{P}$ at time $t \in \mathcal{I}$, due to the forces that are acting on $\mathcal{P}_t \cup \partial\mathcal{P}_t$. The internal power $w_t(\mathcal{P})$ is given by the following linear function of the particle velocity:

$$w_t(\mathcal{P}) := \int_{\partial\mathcal{P}_t} \boldsymbol{\sigma}_t \cdot \boldsymbol{v}_t \,\mathrm{d}a + \int_{\mathcal{P}_t} \boldsymbol{b}_t \cdot \boldsymbol{v}_t \,\mathrm{d}v, \tag{3.11}$$

where the notation described at the end of the last paragraph is in effect. Note that because $w_t(\mathcal{P})$ is considered to be a linear function of the particle velocity, the contact force $\boldsymbol{\sigma}_t(\boldsymbol{X})$ and the body force $\boldsymbol{b}_t(\boldsymbol{X})$ on the particle $\boldsymbol{X}$ at time $t$ are supposed to be independent of $\boldsymbol{v}_t(\boldsymbol{X})$. Of course, for an elastic body the constitutive relation (3.9a), (3.9b) already confirms this for $\boldsymbol{\sigma}_t$. It is a fundamental hypothesis in the case of the body force $\boldsymbol{b}_t$.

## 4 The Causality Principle

To state the causality principle, we need to introduce appropriate response and causal functionals. The idea is based on a balance between 'cause' and 'effect', and it requires a quantitative identification of these two notions. Within pure mechanics, the classical balance of energy for a material body $\mathcal{B}$ requires that the internal power $w_t(\mathcal{P})$ of an arbitrary part $\mathcal{P} \subseteq \mathcal{B}$ be in balance at all times $t \in \mathcal{I}$ with the time rate of change of the total intrinsic energy of $\mathcal{P}$. In the classical setting, the total intrinsic energy is defined as the sum of the internal and the kinetic energies of $\mathcal{P}$ at time $t$, and formulae are given for the forms of these two energies. The resulting balance is, at base, an evolution law which describes how the internal power distributed throughout $\mathcal{B}$ at time $t$ causes a corresponding response of the body by effecting its internal energy and kinetic energy throughout.

Here, we consider a broad generalization of this classical evolution law which, as we shall show, when coupled with certain invariance considerations, reduces naturally to the classical balance of energy, and produces the balances of linear and angular momentum as well as the existence of mass and its balance. Our governing 'law of causality', to be stated more formally below, is of the form

$$\frac{\mathrm{d}}{\mathrm{d}t}\mathsf{R}_t(\mathcal{P}) = \mathsf{C}_t(\mathcal{P}) \quad \forall \mathcal{P} \subseteq \mathcal{B}, \quad \forall t \in \mathcal{I}. \tag{4.1}$$

The value of the 'causal function' $\mathsf{C}_t(\mathcal{P})$ is determined by a functional relation which depends on both the instantaneous internal power of $\mathcal{P}$ and the complete field of deformation gradient throughout $\mathcal{P}$ at time $t$. Thus, the instantaneous internal power and the present distortion of $\mathcal{P}$ are identified as the 'cause' and this cause is modulated through a causal functional to determine $\mathsf{C}_t(\mathcal{P})$. We take the point of view that the 'response function' $\mathsf{R}_t(\mathcal{P})$ is determined by a functional relation which depends on the complete fields of particle speed and deformation gradient throughout $\mathcal{P}$ at time $t$. Thus, the 'effect' is identified as the evolution at time $t$ of the fields of particle speeds and distortion in $\mathcal{P}$. The response functional modulates the effect, and because of the time derivative of $\mathsf{R}_t(\mathcal{P})$ in the causality law it gives an evolutionary character to the process of response. Of course, the causal and response functionals will be assumed to be smooth in their dependencies, but generally nonlinear and not additive.

It is intentional that there is no mention of mass in stating the law of causality (4.1). Motion, force and internal power are fundamental to the statement. The notion of mass as an inertial constant for all parts of the body and attributed to each particle through a referential mass density field in $\mathcal{B}$ will be *derived*.

## 4.1 Response Functional

A response functional for any $\mathcal{P} \subseteq \mathcal{B}$ satisfies the following

**Definition 4.1** Let $C(\mathcal{B})$ denote the set of all continuous functions on $\mathcal{B}$ and define $C^+(\mathcal{B}) := \{\varphi : C(\mathcal{B}) \to \mathbb{R} \mid \varphi \geq 0\}$ and $C^{\mathrm{Lin}^+}(\mathcal{B}) := \{\boldsymbol{A} : C(\mathcal{B}) \to \mathrm{Lin}^+\}$, with suitable norms $\|\cdot\|$. A response functional for $\mathcal{B}$ is the assignment of a single continuous functional

$$\mathsf{R}_{\mathcal{P}}[\varphi, \boldsymbol{A}] := \underset{\boldsymbol{X} \in \mathcal{P}}{\bar{\mathsf{R}}\big(\varphi(\boldsymbol{X}), \boldsymbol{A}(\boldsymbol{X}); \boldsymbol{X}\big)} \quad \forall \mathcal{P} \subseteq \mathcal{B}, \tag{4.2}$$

mapping $C^+(\mathcal{B}) \times C^{\mathrm{Lin}^+}(\mathcal{B}) \to \mathbb{R}$, with the following regularity properties:

(1) $\mathsf{R}_{\mathcal{P}}[\cdot, \cdot]$ is F-differentiable in the sense that for every $\varphi \in C^+(\mathcal{B})$ and $\boldsymbol{A} \in C^{\mathrm{Lin}^+}(\mathcal{B})$ there exist continuous, linear functionals (bounded)

$$\delta_1 \mathsf{R}_{\mathcal{P}}[\varphi, \boldsymbol{A} \mid \cdot] : C(\mathcal{P}) \to \mathbb{R}, \qquad \delta_2 \mathsf{R}_{\mathcal{P}}[\varphi, \boldsymbol{A} \mid \cdot] : C^{\mathrm{Lin}}(\mathcal{P}) \to \mathbb{R}$$

such that

$$\mathsf{R}_{\mathcal{P}}[\varphi + u, \boldsymbol{A}] = \mathsf{R}_{\mathcal{P}}[\varphi, \boldsymbol{A}] + \delta_1 \mathsf{R}_{\mathcal{P}}[\varphi, \boldsymbol{A} \mid u] + o\big(\|u\|\big) \tag{4.3a}$$

for all $u \in C(\mathcal{B})$ with $\varphi + u \in C^+(\mathcal{B})$, and

$$\mathsf{R}_{\mathcal{P}}[\varphi, \boldsymbol{A} + \boldsymbol{H}] = \mathsf{R}_{\mathcal{P}}[\varphi, \boldsymbol{A}] + \delta_2 \mathsf{R}_{\mathcal{P}}[\varphi, \boldsymbol{A} \mid \boldsymbol{H}] + o\big(\|\boldsymbol{H}\|\big) \tag{4.3b}$$

for all $\boldsymbol{H} \in C^{\mathrm{Lin}}(\mathcal{B})$ with $\boldsymbol{A} + \boldsymbol{H} \in C^{\mathrm{Lin}^+}(\mathcal{B})$.

(2) For every $\varphi \in C^+(\mathcal{B})$ and $\boldsymbol{A} \in C^{\mathrm{Lin}^+}(\mathcal{B})$, and for $\alpha \in \mathbb{R}$,

$$\lim_{\alpha \to \infty} \delta_1 \mathsf{R}_{\mathcal{P}}[\varphi + \alpha, \boldsymbol{A} \mid \cdot] = \lim_{\alpha \to \infty} \delta_1 \mathsf{R}_{\mathcal{P}}[\alpha, \boldsymbol{A} \mid \cdot]. \tag{4.4}$$

(3) For any given $\boldsymbol{A} \in C^{\mathrm{Lin}^+}(\mathcal{B})$ and $\mathcal{P} \subseteq \mathcal{B}$, $\mathsf{R}_{\mathcal{P}}[\cdot, \boldsymbol{A}]$ is minimized over $C^+(\mathcal{B})$ at $\varphi \equiv 0$ in $\mathcal{P}$, i.e.,

$$\mathsf{R}_{\mathcal{P}}[\varphi, \boldsymbol{A}] > \mathsf{R}_{\mathcal{P}}[0, \boldsymbol{A}] \tag{4.5}$$

for all $\varphi \in C^+(\mathcal{B})$ not identically zero in $\mathcal{P}$.

According to the Definition 4.1, for a given tame motion $\chi_t$ on $\mathcal{B}$ for all $t \in \mathcal{I}$, the response functional $\mathsf{R}_{\mathcal{P}}[|\dot{\chi}_t|^2, \nabla \chi_t]$ assigns a real value to the fields of particle speed and deformation gradient that reside within the part $\mathcal{P} \subseteq \mathcal{B}$ at time $t$. Moreover, for such motions the functional is sufficiently regular to allow chain rule and mean value theorems to hold, as is noted in the Appendix; these properties will be of considerable importance later in this work. In addition, (4.5) requires that the value of the response functional for a part $\mathcal{P} \subseteq \mathcal{B}$ is greater when its particles are moving than when they are at rest. Later, after we formally define the class of mechanical processes considered in this work and state our governing law of causality, we shall interpret the response function $\mathsf{R}_t(\mathcal{P})$ in (4.1) as

$$\mathsf{R}_t(\mathcal{P}) := \mathsf{R}_{\mathcal{P}}\big[|\dot{\chi}_t|^2, \nabla \chi_t\big] = \underset{\boldsymbol{X} \in \mathcal{P}}{\bar{\mathsf{R}}\big(|\dot{\chi}_t|^2(\boldsymbol{X}), \nabla \chi_t(\boldsymbol{X}); \boldsymbol{X}\big)} \quad \forall \mathcal{P} \subseteq \mathcal{B}. \tag{4.6}$$

Thus, the evolution of particle speed and distortion of the body will be identified as the appropriate measures of response, of course modulated by the form of the response functional $\mathsf{R}_{\mathcal{P}}[\cdot,\cdot]$. With this interpretation, the inequality restriction (4.5) is reasonable.

*Remark 4.1* Whenever a part $\mathcal{P} \subseteq \mathcal{B}$ is in a state of rest and undistorted, the response functional has the value $\mathsf{R}_{\mathcal{P}}[0, \boldsymbol{I}]$, and it is reasonable that this value is fixed independent of the particular part $\mathcal{P}$ that is in such a state. Thus, without loss of generality we set

$$\mathsf{R}_{\mathcal{P}}[0, \boldsymbol{I}] = 0, \quad \forall \mathcal{P} \subseteq \mathcal{B}. \tag{4.7}$$

□

### 4.2 Causal Functional

A causal functional for any $\mathcal{P} \subseteq \mathcal{B}$ satisfies the following

**Definition 4.2** A causal functional for $\mathcal{B}$ is the assignment of a single continuous functional

$$\mathsf{C}_{\mathcal{P}}[\alpha, \boldsymbol{A}] := \bar{\mathsf{C}}\big(\alpha, \underset{\boldsymbol{X}\in\mathcal{P}}{\boldsymbol{A}(\boldsymbol{X}); \boldsymbol{X}}\big) \quad \forall \mathcal{P} \subseteq \mathcal{B}, \tag{4.8}$$

mapping $\mathbb{R} \times C^{\mathrm{Lin}^+}(\mathcal{B}) \to \mathbb{R}$, with the following property:

(1) $\mathsf{C}_{\mathcal{P}}[\cdot, \boldsymbol{A}]$ is differentiable and

$$\frac{\partial}{\partial \alpha} \mathsf{C}_{\mathcal{P}}[\alpha, \boldsymbol{A}] > 0 \tag{4.9}$$

for all $\boldsymbol{A} \in C^{\mathrm{Lin}^+}(\mathcal{B})$.

According to the Definition 4.2, for a given tame motion $\chi_t$ on $\mathcal{B}$ for all $t \in \mathcal{I}$, the causal functional $\mathsf{C}_{\mathcal{P}}[w_t(\mathcal{P}), \nabla \chi_t]$ assigns a real value to the internal power and the field of deformation gradient that reside within the part $\mathcal{P} \subseteq \mathcal{B}$ at time $t$. In addition, (4.9) requires that for such motions the functional value, which is a measure of the causal effect, is monotone increasing as a function of internal power. Later, after we formally define the class of mechanical processes considered in this work and state our governing law of causality, we shall interpret the causal function $\mathsf{C}_t(\mathcal{P})$ in (4.1) as

$$\mathsf{C}_t(\mathcal{P}) := \mathsf{C}_{\mathcal{P}}\big[w_t(\mathcal{P}), \nabla \chi_t\big] = \bar{\mathsf{C}}\big(w_t(\mathcal{P}), \underset{\boldsymbol{X}\in\mathcal{P}}{\nabla \chi_t(\boldsymbol{X}); \boldsymbol{X}}\big) \quad \forall \mathcal{P} \subseteq \mathcal{B}. \tag{4.10}$$

Thus, the internal power and distortion of the body will be identified as the appropriate causal measures, of course modulated by the form of the causal functional $\mathsf{C}_{\mathcal{P}}[\cdot,\cdot]$. Because of this, the inequality (4.9) is a natural fundamental requirement.

### 4.3 Processes and the Causality Law

Given a fixed frame $\mathsf{F}$, the ordered set

$$\mathbb{P}_t(\cdot) := \big\{\chi_t(\cdot), \boldsymbol{b}_t(\cdot), \boldsymbol{\sigma}_t(\cdot; \boldsymbol{n})\big\}, \tag{4.11}$$

prescribed for all $t \in \mathcal{I}$ and for all $\boldsymbol{n} \in$ Unit, is a *mechanical process for the body $\mathcal{B}$ during an interval $\mathcal{I}$* if the motion $\chi_t$ is tame on $\mathcal{B}$ for all $t \in \mathcal{I}$, if the body force satisfies $\boldsymbol{b}_t(\cdot) \in C(\mathcal{B})$

for all $t \in \mathcal{I}$, if the traction satisfies $\boldsymbol{\sigma}_t(\cdot;\boldsymbol{n}) \in C^1(\mathcal{B})$ for each $\boldsymbol{n} \in \text{Unit}$ and $\boldsymbol{\sigma}_t(\boldsymbol{X};\cdot) \in C(\text{Unit})$ for each $\boldsymbol{X} \in \mathcal{B}$ for all $t \in \mathcal{I}$, and if for every part $\mathcal{P} \subseteq \mathcal{B}$ and for all $t \in \mathcal{I}$ the following causality law holds:

$$\frac{\mathrm{d}}{\mathrm{d}t}\mathsf{R}_{\mathcal{P}}\big[|\dot{\chi}_t|^2, \nabla\chi_t\big] = \mathsf{C}_{\mathcal{P}}\big[w_t(\mathcal{P}), \nabla\chi_t\big], \tag{4.12}$$

where $\mathsf{R}_{\mathcal{P}}[\cdot,\cdot]$ and $\mathsf{C}_{\mathcal{P}}[\cdot,\cdot]$ are restricted as in Sects. 4.1 and 4.2, respectively.

Throughout this work we apply the following hypothesis regarding the existence of mechanical processes:

**Hypothesis 4.1** (Existence of a mechanical process) Given an interval $\mathcal{I} \subset \mathbb{R}$, any tame motion $\chi_t$ on $\mathcal{B}$ during $\mathcal{I}$ and any traction field $\boldsymbol{\sigma}_t(\cdot;\boldsymbol{n}) \in C^1(\mathcal{B})$ for each $\boldsymbol{n} \in \text{Unit}$ and $\boldsymbol{\sigma}_t(\boldsymbol{X};\cdot) \in C(\text{Unit})$ for each $\boldsymbol{X} \in \mathcal{B}$ for all $t \in \mathcal{I}$, as noted in (3.9a), (3.9b). Then, there exists a body force field $\boldsymbol{b}_t(\cdot) : C(\mathcal{B}) \to \mathbb{V}^3$ such that

$$\mathbb{P}_t(\cdot) := \big\{\chi_t(\cdot), \boldsymbol{b}_t(\cdot), \boldsymbol{\sigma}_t(\cdot;\boldsymbol{n})\big\}$$

is a mechanical process for $\mathcal{B}$ during $\mathcal{I}$.

*Remark 4.2* The internal power $w_t(\mathcal{P})$, being introduced in Sect. 3.4 as a linear function of the particle velocity field $\boldsymbol{v}_t$ in this work, admits only contact and body force fields that are independent of the particle velocity; thus, for any mechanical process $\mathbb{P}_t$ for $\mathcal{B}$ during $\mathcal{I}$, the fields $\boldsymbol{\sigma}_t$ and $\boldsymbol{b}_t$ are independent of $\boldsymbol{v}_t$. While (3.9a), (3.9b) automatically verifies that such is the case for $\boldsymbol{\sigma}_t$, the Hypothesis 4.1 regarding the existence of mechanical processes implicitly ensures that such is the case for $\boldsymbol{b}_t$. In this sense, the hypothesis is expected to place a restriction on the form of the causality law (4.12); a restriction which we begin to investigate in Sect. 5. □

## 5 First Invariance Axiom and Its Consequences

A translational Galilean change of frame plays an important role in this work. As we shall see, many fundamental conclusions are a result of the following first axiom of invariance:

**Axiom 5.1** (First invariance axiom) Let $\mathsf{F}$ be a fixed frame and suppose $\mathbb{P}_t$ of (4.11) is a mechanical process for $\mathcal{B}$ during a time interval $\mathcal{I}$; thus, (4.12) holds. Let $\mathsf{F} \mapsto \mathsf{F}^*(t)$ be a translational Galilean change of frame for which $\boldsymbol{Q}(t) = \boldsymbol{1}$ and $\boldsymbol{c}(t) = t\boldsymbol{c} + \boldsymbol{o}$ in (3.4). Then,

$$\mathbb{P}_t^{(c)}(\cdot) := \big\{\chi_t^{(c)}(\cdot), \boldsymbol{b}_t^{(c)}(\cdot), \boldsymbol{\sigma}_t^{(c)}(\cdot;\boldsymbol{n})\big\}$$

is a mechanical process for $\mathcal{B}$ during the time interval $\mathcal{I}$ for all $\boldsymbol{c} \in \mathbb{V}^3$, where

$$\chi_t^{(c)}(\cdot) := \chi_t(\cdot) + t\boldsymbol{c}, \qquad \boldsymbol{b}_t^{(c)}(\cdot) := \boldsymbol{b}_t(\cdot), \qquad \boldsymbol{\sigma}_t^{(c)}(\cdot;\boldsymbol{n}) := \boldsymbol{\sigma}_t(\cdot;\boldsymbol{n}),$$

the latter for all $\boldsymbol{n} \in \text{Unit}$.

*Remark 5.1* This axiom requires $\mathbb{P}_t^{(c)}$ to satisfy

$$\frac{\mathrm{d}}{\mathrm{d}t}\mathsf{R}_{\mathcal{P}}\big[\big|\dot{\chi}_t^{(c)}\big|^2, \nabla\chi_t^{(c)}\big] = \mathsf{C}_{\mathcal{P}}\big[w_t^{(c)}(\mathcal{P}), \nabla\chi_t^{(c)}\big]$$

for all $\mathcal{P} \subseteq \mathcal{B}$ and for all $t \in \mathcal{I}$, where

$$w_t^{(c)}(\mathcal{P}) := \int_{\partial \mathcal{P}_t} \boldsymbol{\sigma}_t \cdot (\boldsymbol{v}_t + \boldsymbol{c})\, \mathrm{d}a + \int_{\mathcal{P}_t} \boldsymbol{b}_t \cdot (\boldsymbol{v}_t + \boldsymbol{c})\, \mathrm{d}v = w_t(\mathcal{P}) + \boldsymbol{c} \cdot \boldsymbol{f}_t(\mathcal{P}).$$

Thus, in addition to (4.12), for every part $\mathcal{P} \subseteq \mathcal{B}$ and for all $t \in \mathcal{I}$ the following must hold for every tame motion $\chi_t$ on $\mathcal{B}$ for all $t \in \mathcal{I}$:

$$\frac{\mathrm{d}}{\mathrm{d}t} \mathsf{R}_{\mathcal{P}}\big[|\dot{\chi}_t + \boldsymbol{c}|^2, \nabla \chi_t\big] = \mathsf{C}_{\mathcal{P}}\big[w_t(\mathcal{P}) + \boldsymbol{c} \cdot \boldsymbol{f}_t(\mathcal{P}), \nabla \chi_t\big] \quad \forall \boldsymbol{c} \in \mathbb{V}^3. \tag{5.1}$$

□

We are now in a position to show that for any mechanical process of $\mathcal{B}$ during a time-interval $\mathcal{I}$, the response and causal functionals reduce significantly to simpler expressions. The results of this first theorem will then be used to show the existence of a mass density field in $\mathcal{B}$ and to obtain the classical forms of the balance laws of energy, mass and momenta. The generality of the primitive forms of the causal and response functionals in the causality law (4.12) will require special attention in the proof of this theorem.

**Theorem 5.1** *Let* $\mathbb{P}_t$ *of* (4.11) *be a mechanical process for* $\mathcal{B}$ *during a time interval* $\mathcal{I}$, *for which* (4.12) *holds relative to a fixed frame* $\mathsf{F}$. *Suppose that the motion* $\chi_t$ *of* $\mathbb{P}_t$, *in addition to being tame, is such that* $\boldsymbol{v}_t := \dot{\chi}_t \in C^1(\mathcal{B})$ *for all* $t \in \mathcal{I}$. *Then*:

(1) *There exist functionals*

$$\mathsf{E}_{\mathcal{P}}[\nabla \chi_t] := \underset{\boldsymbol{X} \in \mathcal{P}}{\bar{\mathsf{E}}}\big(\nabla \chi_t(\boldsymbol{X}), \boldsymbol{X}\big), \quad \mathsf{K}_{\mathcal{P}}\big[|\dot{\chi}_t|^2\big] := \underset{\boldsymbol{X} \in \mathcal{P}}{\bar{\mathsf{K}}}\big(|\dot{\chi}_t|^2(\boldsymbol{X}), \boldsymbol{X}\big) \quad \forall \mathcal{P} \subseteq \mathcal{B},$$

*each having the regularity of the response functional* $\mathsf{R}_{\mathcal{P}}[\cdot,\cdot]$ *and such that* $\mathsf{K}_{\mathcal{P}}[\cdot] : C(\mathcal{P}) \longrightarrow \mathbb{R}$ *is* linear *with* $\mathsf{K}_{\mathcal{P}}[\varphi] > 0$ *for all* $\varphi \geq 0$ (*not identically* 0) *everywhere in* $\mathcal{P}$, $\mathsf{E}_{\mathcal{P}}[\boldsymbol{I}] = 0$ *and*

$$\mathsf{R}_{\mathcal{P}}\big[|\dot{\chi}_t|^2, \nabla \chi_t\big] = \underset{\boldsymbol{X} \in \mathcal{P}}{\bar{\mathsf{R}}}\big(|\dot{\chi}_t|^2(\boldsymbol{X}), \nabla \chi_t(\boldsymbol{X}); \boldsymbol{X}\big) = \mathsf{E}_{\mathcal{P}}[\nabla \chi_t] + \mathsf{K}_{\mathcal{P}}\big[|\dot{\chi}_t|^2\big] \tag{5.2}$$

*for all* $\mathcal{P} \subseteq \mathcal{B}$ *and all* $t \in \mathcal{I}$.

(2) *There exists a positive constant* $C > 0$ *such that*

$$\mathsf{C}_{\mathcal{P}}\big[w_t(\mathcal{P}), \nabla \chi_t)\big] = \underset{\boldsymbol{X} \in \mathcal{P}}{\bar{\mathsf{C}}}\big(w_t(\mathcal{P}), \nabla \chi_t(\boldsymbol{X}); \boldsymbol{X}\big) = C\, w_t(\mathcal{P}) \tag{5.3}$$

*for all* $\mathcal{P} \subseteq \mathcal{B}$ *and all* $t \in \mathcal{I}$.

*Proof* Let $t_0 \in \mathcal{I}$ be arbitrary and suppose that the motion $\chi_t$ of the mechanical process $\mathbb{P}_t$ and the interval $\mathcal{I}$ are such that (3.8) and Lemma 3.1 hold. Then, with the aid of Axiom 5.1 and the chain rule Lemma A.1, we may write (4.12) and (5.1) at time $t_0$ as

$$\delta_1 \mathsf{R}_{\mathcal{P}}\big[|\boldsymbol{v}|^2, \nabla \chi \,|\, 2\boldsymbol{v} \cdot \boldsymbol{a}\big] + \delta_2 \mathsf{R}_{\mathcal{P}}\big[|\boldsymbol{v}|^2, \nabla \chi \,|\, \nabla \boldsymbol{v}\big] = \mathsf{C}_{\mathcal{P}}\big[w_{t_0}(\mathcal{P}), \nabla \chi\big] \tag{5.4a}$$

and

$$\begin{aligned} &\delta_1 \mathsf{R}_{\mathcal{P}}\big[|\boldsymbol{v} + \boldsymbol{c}|^2, \nabla \chi \,|\, 2(\boldsymbol{v} + \boldsymbol{c}) \cdot \boldsymbol{a}\big] + \delta_2 \mathsf{R}_{\mathcal{P}}\big[|\boldsymbol{v} + \boldsymbol{c}|^2, \nabla \chi \,|\, \nabla \boldsymbol{v}\big] \\ &\quad = \mathsf{C}_{\mathcal{P}}\big[w_{t_0}(\mathcal{P}) + \boldsymbol{c} \cdot \boldsymbol{f}_{t_0}(\mathcal{P}), \nabla \chi\big], \end{aligned} \tag{5.4b}$$

both of which are to hold for all smooth functions $\chi$, $\boldsymbol{v}$ and $\boldsymbol{a}$ with $\nabla\chi \in \mathrm{Lin}^+$, and the latter of which is to hold for all $\boldsymbol{c} \in \mathbb{V}^3$. We note that $\boldsymbol{f}_{t0}(\mathcal{P})$ and $w_{t0}(\mathcal{P})$ have the forms

$$\boldsymbol{f}_{t0}(\mathcal{P}) = \int_{\partial\mathcal{P}_{t0}} \boldsymbol{\sigma}_{t0}\,\mathrm{d}a + \int_{\mathcal{P}_{t0}} \boldsymbol{b}_{t0}\,\mathrm{d}v$$

and

$$w_{t0}(\mathcal{P}) = \int_{\partial\mathcal{P}_{t0}} \boldsymbol{\sigma}_{t0}\cdot\boldsymbol{v}\,\mathrm{d}a + \int_{\mathcal{P}_{t0}} \boldsymbol{b}_{t0}\cdot\boldsymbol{v}\,\mathrm{d}v,$$

where, according to (3.9a) and (3.9b),

$$\boldsymbol{\sigma}_{t0}(\cdot;\boldsymbol{n}) = \bar{\boldsymbol{\sigma}}\big(\cdot, \boldsymbol{F}(\cdot);\boldsymbol{n}\big), \quad \boldsymbol{F}(\cdot) := \nabla\chi(\cdot),$$

and, because $w_{t0}(\mathcal{P})$ is a linear function of the particle velocity field $\boldsymbol{v}_{t0} = \boldsymbol{v}$, the contact force $\boldsymbol{\sigma}_{t0}$ and the body force $\boldsymbol{b}_{t0}$ are independent of $\boldsymbol{v}$ (see Remark 4.2). Note that the Hypothesis 4.1 guarantees the existence of such a body force field $\boldsymbol{b}_t$ so that $\mathbb{P}_t$ is a mechanical process for $\mathcal{B}$ during $\mathcal{I}$.

As a first step, let us choose the field $\boldsymbol{v} \equiv \lambda$ in $\mathcal{B}$, where $\lambda \in \mathbb{V}^3$ is an arbitrary constant vector. Note that $\boldsymbol{f}_{t0}(\mathcal{P})$, being independent of the particle velocity field, is independent of $\lambda$ and that we may write $w_{t0}(\mathcal{P}) = \lambda\cdot\boldsymbol{f}_{t0}(\mathcal{P})$. Consequently, (5.4a) yields

$$\boldsymbol{a} \equiv \boldsymbol{0} \;\text{ in } \mathcal{B} \implies \mathsf{C}_{\mathcal{P}}\big[\lambda\cdot\boldsymbol{f}_{t0}(\mathcal{P}), \nabla\chi\big] = 0 \quad \forall\mathcal{P}\subseteq\mathcal{B}.$$

But, since $\lambda \in \mathbb{V}^3$ is arbitrary, we may use (4.9) of Definition 4.2 to conclude that

$$\boldsymbol{a} \equiv \boldsymbol{0} \;\text{ in } \mathcal{B} \implies \boldsymbol{f}_{t0}(\mathcal{P}) = 0 \quad \forall\mathcal{P}\subseteq\mathcal{B}. \tag{5.6}$$

Now, we return to (5.4a), (5.4b) with (5.6) and set $\boldsymbol{a} \equiv \boldsymbol{0}$ to reach

$$\delta_2\mathsf{R}_{\mathcal{P}}\big[|\boldsymbol{v}|^2, \nabla\chi \mid \nabla\boldsymbol{v}\big] = \delta_2\mathsf{R}_{\mathcal{P}}\big[|\boldsymbol{v}+\boldsymbol{c}|^2, \nabla\chi \mid \nabla\boldsymbol{v}\big] \quad \forall\mathcal{P}\subseteq\mathcal{B}, \tag{5.7}$$

which is to hold for all smooth functions $\chi$ and $\boldsymbol{v}$, and for all $\boldsymbol{c} \in \mathbb{V}^3$. Consequently, returning to (5.4a), (5.4b) and subtracting the two equations, we find that

$$\delta_1\mathsf{R}_{\mathcal{P}}\big[|\boldsymbol{v}|^2, \nabla\chi \mid 2\boldsymbol{v}\cdot\boldsymbol{a}\big] = \delta_1\mathsf{R}_{\mathcal{P}}\big[|\boldsymbol{v}+\boldsymbol{c}|^2, \nabla\chi \mid 2(\boldsymbol{v}+\boldsymbol{c})\cdot\boldsymbol{a}\big] \quad \forall\mathcal{P}\subseteq\mathcal{B}, \tag{5.8}$$

which is supposed to hold for all smooth functions $\chi$, $\boldsymbol{v}$ and $\boldsymbol{a}$, and for all $\boldsymbol{c} \in \mathbb{V}^3$, now such that $\boldsymbol{c}\cdot\boldsymbol{f}_{t0}(\mathcal{P}) = 0$. It is worth emphasizing that for given fields $\chi$, $\boldsymbol{v}$ and $\boldsymbol{a}$, according to (3.8) and Hypothesis 4.1 there exists an interval $\mathcal{I}$ and a mechanical process $\mathbb{P}_t$. Moreover, $\boldsymbol{f}_{t0}(\mathcal{P})$ is independent of the field $\boldsymbol{v}$ (see Remark 4.2). A major consequence of this is that in applying (5.8), while $\boldsymbol{c}$ must be orthogonal to $\boldsymbol{f}_{t0}(\mathcal{P})$, the choice of $\boldsymbol{c}$ is not effected by the choice of the field $\boldsymbol{v}$.

Let us consider the consequences of (5.8) and begin by arbitrarily fixing the differentiable function $\chi$. Suppose $\boldsymbol{e} \in \mathrm{Unit}$ is fixed and let $\alpha(\cdot)$ and $\beta(\cdot)$ be two fixed differentiable scalar functions in $\mathcal{B}$. Set $\boldsymbol{a}(\cdot) = \alpha(\cdot)\boldsymbol{e}$ and $\boldsymbol{v}(\cdot) = \beta(\cdot)\boldsymbol{e}$, and take $\boldsymbol{c} = \gamma\boldsymbol{m}$, where $\boldsymbol{m} \in \mathrm{Unit}$ with $\boldsymbol{m}\cdot\boldsymbol{e} = 0$ and $\gamma$ is an arbitrary scalar constant. Note that according to Hypothesis 4.1, the quantities $\boldsymbol{f}_{t0}(\mathcal{P})$ and $w_{t0}(\mathcal{P})$ are now fixed, and observe that $\boldsymbol{f}_{t0}(\mathcal{P})$ is independent of the field $\beta$. Adjust $\boldsymbol{m}$ so that it is not only orthogonal to $\boldsymbol{e}$, but also such that $\boldsymbol{m}\cdot\boldsymbol{f}_{t0}(\mathcal{P}) = 0$. Thus, (5.8) yields

$$\delta_1\mathsf{R}_{\mathcal{P}}\big[\beta^2, \nabla\chi \mid 2\alpha\beta\big] = \delta_1\mathsf{R}_{\mathcal{P}}\big[\beta^2+\gamma^2, \nabla\chi \mid 2\alpha\beta\big]. \tag{5.9}$$

Setting $\beta = c$, where $c$ is an arbitrary constant, we readily find that

$$\delta_1 \mathsf{R}_{\mathcal{P}}\big[c^2, \nabla\chi \,|\, \alpha\big] = \delta_1 \mathsf{R}_{\mathcal{P}}\big[c^2 + \gamma^2, \nabla\chi \,|\, \alpha\big],$$

which yields

$$\delta_1 \mathsf{R}_{\mathcal{P}}\big[0, \nabla\chi \,|\, \alpha\big] = \delta_1 \mathsf{R}_{\mathcal{P}}\big[\gamma^2, \nabla\chi \,|\, \alpha\big].$$

Then, with this and the aid of the limit property (2) of Definition 4.1, we observe that the right hand side of (5.9) satisfies

$$\lim_{\gamma\to\infty} \delta_1 \mathsf{R}_{\mathcal{P}}\big[\beta^2 + \gamma^2, \nabla\chi \,|\, \cdot\big] = \lim_{\gamma\to\infty} \delta_1 \mathsf{R}_{\mathcal{P}}\big[\gamma^2, \nabla\chi \,|\, \cdot\big] = \delta_1 \mathsf{R}_{\mathcal{P}}[0, \nabla\chi \,|\, \cdot],$$

which, with (5.9) and the definition $\psi := \alpha\beta$, yields

$$\delta_1 \mathsf{R}_{\mathcal{P}}\big[\beta^2, \nabla\chi \,|\, \psi\big] = \delta_1 \mathsf{R}_{\mathcal{P}}[0, \nabla\chi \,|\, \psi].$$

Because $\beta$, $\psi$ and $\chi$ are arbitrary and smooth with $\nabla\chi \in \mathrm{Lin}^+$, we may use Lemma A.3 to conclude that

$$\mathsf{R}_{\mathcal{P}}\big[\beta^2, \nabla\chi\big] = \mathsf{R}_{\mathcal{P}}[0, \nabla\chi] + \mathsf{L}_{\mathcal{P}}\big[\nabla\chi \,|\, \beta^2\big],$$

where

$$\mathsf{L}_{\mathcal{P}}\big[\nabla\chi \,|\, \beta^2\big] := \delta_1 \mathsf{R}_{\mathcal{P}}\big[0, \nabla\chi \,|\, \beta^2\big]$$

has the properties of Lemma A.3 and, thus, is F-differentiable and is linear in its second place, i.e., for every $\boldsymbol{A} \in C^{\mathrm{Lin}^+}(\mathcal{B})$ there exists $\delta\mathsf{L}_{\mathcal{P}}[\cdot| \cdot |\cdot]$, linear in its second and third places, such that

$$\mathsf{L}_{\mathcal{P}}[\boldsymbol{A} + \boldsymbol{H} | \cdot] = \mathsf{L}_{\mathcal{P}}[\boldsymbol{A} | \cdot] + \delta\mathsf{L}_{\mathcal{P}}[\boldsymbol{A} | \cdot |\boldsymbol{H}] + o\big(\|\boldsymbol{H}\|\big) \tag{5.10}$$

for all $\boldsymbol{H} \in C^{\mathrm{Lin}}(\mathcal{B})$ with $\boldsymbol{A} + \boldsymbol{H} \in C^{\mathrm{Lin}^+}(\mathcal{B})$.

Clearly, we may replace $\beta$ with $|\boldsymbol{v}|$, define $\mathsf{E}_{\mathcal{P}}[\nabla\chi] := \mathsf{R}_{\mathcal{P}}[0, \nabla\chi]$, and at this point arrive at the conclusion

$$\mathsf{R}_{\mathcal{P}}\big[|\boldsymbol{v}|^2, \nabla\chi\big] = \mathsf{E}_{\mathcal{P}}[\nabla\chi] + \mathsf{L}_{\mathcal{P}}\big[\nabla\chi \,|\, |\boldsymbol{v}|^2\big], \tag{5.11}$$

which is to hold for all smooth fields $\chi$ and $\boldsymbol{v}$ with $\nabla\chi \in \mathrm{Lin}^+$. Note, from (4.7) in Remark 4.1, that $\mathsf{E}_{\mathcal{P}}[\boldsymbol{I}] = 0$. In addition, because of the inequality (4.5) of Definition 4.1, we observe that

$$\mathsf{L}_{\mathcal{P}}\big[\nabla\chi \,|\, |\boldsymbol{v}|^2\big] > 0 \quad \forall \boldsymbol{v} \neq \boldsymbol{0} \text{ identically in } \mathcal{P}. \tag{5.12}$$

To go further, we now use the F-differentiability property noted in (5.10), substitute (5.11) into (5.7) and observe, using the fixed orthonormal basis $\{\boldsymbol{e}_1, \boldsymbol{e}_2, \boldsymbol{e}_3\} \in \mathbb{V}^3$ and the arbitrariness of $\boldsymbol{c}$ in (5.7), that

$$\delta\mathsf{L}_{\mathcal{P}}\big[\nabla\chi \,|\, v_i | \nabla\boldsymbol{v}\big]\boldsymbol{e}_i = \boldsymbol{0}, \tag{5.13a}$$

where $v_i := \boldsymbol{v} \cdot \boldsymbol{e}_i$, and

$$\delta\mathsf{L}_{\mathcal{P}}\big[\nabla\chi \,| 1 | \nabla\boldsymbol{v}\big] = 0, \tag{5.13b}$$

which, again, are to hold for all smooth functions $\chi$ and $\boldsymbol{v}$ with $\nabla\chi \in \mathrm{Lin}^+$.

*Remark 5.2* Returning to (5.8) with (5.11) and, again, introducing the fixed orthonormal basis $\{\boldsymbol{e}_1, \boldsymbol{e}_2, \boldsymbol{e}_3\} \in \mathbb{V}^3$, we readily find

$$\boldsymbol{c} \cdot \mathsf{L}_{\mathcal{P}}[\nabla \chi \,|\, a_i]\boldsymbol{e}_i = 0 \quad \forall \mathcal{P} \subseteq \mathcal{B},$$

where $a_i := \boldsymbol{a} \cdot \boldsymbol{e}_i$, which is supposed to hold for all smooth functions $\chi$ and $\boldsymbol{a}$, and for all $\boldsymbol{c} \in \mathbb{V}^3$ such that $\boldsymbol{c} \cdot \boldsymbol{f}_{t_0}(\mathcal{P}) = 0$. Of course, for given fields $\chi$, $\boldsymbol{v}$ and $\boldsymbol{a}$, according to (3.8) and Hypothesis 4.1 there exists an interval $\mathcal{I}$ and a mechanical process $\mathbb{P}_t$, which means that the causality law (4.12) holds, simplified by the use of (5.11). Moreover, while $\boldsymbol{c}$ must be orthogonal to $\boldsymbol{f}_{t_0}(\mathcal{P})$, it should be recalled that the choice of $\boldsymbol{c}$ is not effected by the choice of the field $\boldsymbol{v}$ because $\boldsymbol{f}_{t_0}(\mathcal{P})$ is independent of $\boldsymbol{v}$. Thus, here $\boldsymbol{c}$ is any vector in the plane perpendicular to $\boldsymbol{f}_{t_0}(\mathcal{P})$, and we conclude that

$$\mathsf{L}_{\mathcal{P}}[\nabla \chi \,|\, a_i]\boldsymbol{e}_i \parallel \boldsymbol{f}_{t_0}(\mathcal{P}).$$

Of course, knowing that $\nabla \chi = \nabla \chi_{t_0}$ and $\boldsymbol{a} = \dot{\boldsymbol{v}}_{t_0}$, and that $t_0$ is arbitrary, we may infer that for any mechanical process for $\mathcal{B}$ during $\mathcal{I}$,

$$\mathsf{L}_{\mathcal{P}}[\nabla \chi_t \,|\, \dot{\boldsymbol{v}}_t \cdot \boldsymbol{e}_i]\boldsymbol{e}_i \parallel \boldsymbol{f}_t(\mathcal{P}) \quad \forall \mathcal{P} \subseteq \mathcal{B}\, \forall t \in \mathcal{I}.$$

While we shall not need to use this intermediate conclusion in subsequent arguments, nevertheless, it is of passing interest because it expresses a precursor to the balance of linear momentum and the association of acceleration and force. □

We now show that (5.13a) requires $\mathsf{L}_{\mathcal{P}}[\nabla \chi \,|\, |\boldsymbol{v}|^2]$ of (5.11) to be independent of $\nabla \chi$. Thus, with (5.12), we will conclude that there is a linear functional $\mathsf{K}_{\mathcal{P}}[\cdot] : C(\mathcal{P}) \longrightarrow \mathbb{R}$ such that

$$\mathsf{L}_{\mathcal{P}}\big[\nabla \chi \,|\, |\boldsymbol{v}|^2\big] = \mathsf{K}_{\mathcal{P}}\big[|\boldsymbol{v}|^2\big] > 0 \quad \forall \boldsymbol{v} \neq \boldsymbol{0} \text{ identically in } \mathcal{P}. \tag{5.14}$$

As a consequence, recalling (5.11) and that $\chi = \chi_{t_0}$ and $\boldsymbol{v} = \dot{\chi}_{t_0}$, where $t_0 \in \mathcal{I}$ is arbitrary, we will have proven part (1) of Theorem 5.1.

The strategy for reaching (5.14) is to first show that (5.13a) implies

$$\delta \mathsf{L}_{\mathcal{P}}\big[\nabla \chi \,|\varphi|\nabla \boldsymbol{v}\big] = 0 \tag{5.15}$$

for all $\varphi \in C^1(\mathcal{B})$ and all smooth $\chi$ and $\boldsymbol{v}$ in $\mathcal{B}$ with $\nabla \chi \in \mathrm{Lin}^+$, and then to appeal to Lemma A.4. Toward this end, as a first step consider the inner product of $\boldsymbol{e}_2$ with (5.13a) to obtain

$$\delta \mathsf{L}_{\mathcal{P}}\big[\nabla \chi \,|\, v_2 |\, \boldsymbol{e}_i \otimes \nabla v_i\big] = 0. \tag{5.16}$$

Since this is to hold for all smooth $\boldsymbol{v}$, we may choose $v_1 = v_3 \equiv 0$ to see that

$$\delta \mathsf{L}_{\mathcal{P}}\big[\nabla \chi \,|\, v_2 |\, \boldsymbol{e}_2 \otimes \nabla v_2\big] = 0 \quad \forall v_2,$$

and then return to (5.16) with this result, with the choice $v_1 \equiv 0$, and with the linearity of $\delta \mathsf{L}_{\mathcal{P}}[\nabla \chi \,|\, v_2 |\cdot]$ to reach

$$\delta \mathsf{L}_{\mathcal{P}}\big[\nabla \chi \,|\, v_2 |\, \boldsymbol{e}_3 \otimes \nabla v_3\big] = 0 \quad \forall v_2, v_3. \tag{5.17}$$

Using the same argument, by choosing $v_3 = 0$ we find

$$\delta \mathsf{L}_{\mathcal{P}}\big[\nabla \chi \,|\, v_2 |\, \boldsymbol{e}_1 \otimes \nabla v_1\big] = 0 \quad \forall v_2, v_1. \tag{5.18}$$

Now, as a second step, consider the inner product of $\boldsymbol{e}_1$ with (5.13a) to obtain

$$\delta\mathsf{L}_{\mathcal{P}}\big[\nabla\chi \mid v_1 \mid \boldsymbol{e}_i \otimes \nabla v_i\big] = 0.$$

Then, using an argument similar to the above and adjusting the components $v_2$ and $v_3$ relative to 0, we easily arrive at

$$\delta\mathsf{L}_{\mathcal{P}}\big[\nabla\chi \mid v_1 \mid \boldsymbol{e}_2 \otimes \nabla v_2\big] = 0 \quad \forall v_1, v_2. \tag{5.19}$$

As a final step, we simply relabel $v_2$ in (5.17) and (5.18) as $\varphi$, relabel $v_1$ in (5.19) as $\varphi$, and add the three equations, using the linearity of $\delta\mathsf{L}_{\mathcal{P}}[\nabla\chi \mid \varphi \mid \cdot]$ to obtain

$$\delta\mathsf{L}_{\mathcal{P}}\big[\nabla\chi \,|\varphi|\boldsymbol{e}_i \otimes \nabla v_i\big] = \delta\mathsf{L}_{\mathcal{P}}\big[\nabla\chi \,|\varphi|\nabla\boldsymbol{v}\big] = 0,$$

which confirms (5.15). Then, with Lemma A.4 we verify (5.14) and complete the proof of part (1) of Theorem 5.1.

We now consider part (2) of Theorem 5.1 and show that (5.3) holds for some positive constant $C > 0$. To begin, let us return to the two fundamental equations of (5.4a), (5.4b) and write them for the specific form of $\mathsf{R}_{\mathcal{P}}[\cdot,\cdot]$ found in part (1) of the theorem, i.e., (5.2). Using the linearity of the functional $\mathsf{K}_{\mathcal{P}}[\cdot]$, we readily find

$$\mathsf{K}_{\mathcal{P}}[2\boldsymbol{v}\cdot\boldsymbol{a}] + \delta\mathsf{E}_{\mathcal{P}}[\nabla\chi \mid \nabla\boldsymbol{v}] = \mathsf{C}_{\mathcal{P}}\big[w_{t_0}(\mathcal{P}), \nabla\chi\big] \tag{5.20a}$$

and

$$\mathsf{K}_{\mathcal{P}}\big[2(\boldsymbol{v}+\boldsymbol{c})\cdot\boldsymbol{a}\big] + \delta\mathsf{E}_{\mathcal{P}}[\nabla\chi \mid \nabla\boldsymbol{v}] = \mathsf{C}_{\mathcal{P}}\big[w_{t_0}(\mathcal{P}) + \boldsymbol{c}\cdot\boldsymbol{f}_{t_0}(\mathcal{P}), \nabla\chi\big], \tag{5.20b}$$

both of which are to hold for all smooth functions $\chi$, $\boldsymbol{v}$ and $\boldsymbol{a}$ with $\nabla\chi \in \mathrm{Lin}^+$, and the latter of which is to hold for all $\boldsymbol{c} \in \mathbb{V}^3$. Thus, by subtraction we find

$$\mathsf{K}_{\mathcal{P}}[2\boldsymbol{c}\cdot\boldsymbol{a}] = \mathsf{C}_{\mathcal{P}}\big[w_{t_0}(\mathcal{P}) + \boldsymbol{c}\cdot\boldsymbol{f}_{t_0}(\mathcal{P}), \nabla\chi\big] - \mathsf{C}_{\mathcal{P}}\big[w_{t_0}(\mathcal{P}), \nabla\chi\big] \quad \forall\mathcal{P} \subseteq \mathcal{B}, \tag{5.21}$$

which must hold for all $\boldsymbol{c} \in \mathbb{V}^3$.

Of course, this last holds true if we replace $\boldsymbol{c}$ by $\boldsymbol{h}$ or if we replace $\boldsymbol{c}$ by $\boldsymbol{c}-\boldsymbol{h}$, where $\boldsymbol{h}$ is arbitrary. If, after replacing $\boldsymbol{c}$ in (5.21) by $\boldsymbol{h}$, we subtract the resulting equation from (5.21), and then use the equation that results from replacing $\boldsymbol{c}$ in (5.21) by $\boldsymbol{c}-\boldsymbol{h}$, we readily find that

$$\begin{aligned}
&\mathsf{C}_{\mathcal{P}}\big[w_{t_0}(\mathcal{P}) + \boldsymbol{c}\cdot\boldsymbol{f}_{t_0}(\mathcal{P}) - \boldsymbol{h}\cdot\boldsymbol{f}_{t_0}(\mathcal{P}), \nabla\chi\big] \\
&\quad = \mathsf{C}_{\mathcal{P}}\big[w_{t_0}(\mathcal{P}) + \boldsymbol{c}\cdot\boldsymbol{f}_{t_0}(\mathcal{P}), \nabla\chi\big] + \mathsf{C}_{\mathcal{P}}\big[w_{t_0}(\mathcal{P}), \nabla\chi\big] - \mathsf{C}_{\mathcal{P}}\big[w_{t_0}(\mathcal{P}) + \boldsymbol{c}\cdot\boldsymbol{f}_{t_0}(\mathcal{P}), \nabla\chi\big].
\end{aligned}$$

For convenience, let us replace $\boldsymbol{c}$ by $\boldsymbol{d} := \boldsymbol{c}-\boldsymbol{h}$, where, now, $\boldsymbol{d}$ and $\boldsymbol{h}$ are arbitrary constant vectors. Then, for $\boldsymbol{f}_{t_0}(\mathcal{P}) \neq 0$, this last equation may be written as the equality of difference quotients

$$\begin{aligned}
&\frac{\mathsf{C}_{\mathcal{P}}[w_{t_0}(\mathcal{P}) + \boldsymbol{d}\cdot\boldsymbol{f}_{t_0}(\mathcal{P}), \nabla\chi] - \mathsf{C}_{\mathcal{P}}[w_{t_0}(\mathcal{P}), \nabla\chi]}{\boldsymbol{d}\cdot\boldsymbol{f}_{t_0}(\mathcal{P})} \\
&\quad = \frac{\mathsf{C}_{\mathcal{P}}[w_{t_0}(\mathcal{P}) + \boldsymbol{h}\cdot\boldsymbol{f}_{t_0}(\mathcal{P}) + \boldsymbol{d}\cdot\boldsymbol{f}_{t_0}(\mathcal{P}), \nabla\chi] - \mathsf{C}_{\mathcal{P}}[w_{t_0}(\mathcal{P}) + \boldsymbol{h}\cdot\boldsymbol{f}_{t_0}(\mathcal{P}), \nabla\chi]}{\boldsymbol{d}\cdot\boldsymbol{f}_{t_0}(\mathcal{P})},
\end{aligned}$$

and, recalling that $\boldsymbol{d} \in \mathbb{V}^3$ is arbitrary, we may take the limit as $\boldsymbol{d} \cdot \boldsymbol{f}_{t_0}(\mathcal{P}) \to 0$ to reach

$$\frac{\partial}{\partial w}\mathsf{C}_{\mathcal{P}}[w, \nabla \chi]|_{w=w_{t_0}(\mathcal{P})} = \frac{\partial}{\partial w}\mathsf{C}_{\mathcal{P}}[w, \nabla \chi]|_{w=w_{t_0}(\mathcal{P})+\boldsymbol{h}\cdot \boldsymbol{f}_{t_0}(\mathcal{P})},$$

which is to hold for all $\boldsymbol{h} \in \mathbb{V}^3$. Thus, by choosing $\boldsymbol{h}$ such that $\boldsymbol{h} \cdot \boldsymbol{f}_{t_0}(\mathcal{P}) = -w_{t_0}(\mathcal{P})$, we see that the function $\mathsf{C}_{\mathcal{P}}[\cdot, \nabla \chi]$ has a constant slope $\frac{\partial}{\partial w}\mathsf{C}_{\mathcal{P}}[w, \nabla \chi]|_{w=0}$. With this, and the monotonicity condition (4.9) in property (1) of Definition 4.2, we conclude that there exists a *positive-valued* continuous functional

$$\mathsf{G}_{\mathcal{P}}[\nabla \chi] := \bar{\mathsf{G}}\big(\nabla \chi(\boldsymbol{X}), \boldsymbol{X}\big)_{\boldsymbol{X}\in\mathcal{P}} > 0 \quad \forall \mathcal{P} \subseteq \mathcal{B},$$

mapping $C^{\mathrm{Lin}^+}(\mathcal{B}) \to \mathbb{R}$, such that

$$\mathsf{C}_{\mathcal{P}}\big[w_{t_0}(\mathcal{P}), \nabla \chi\big] = \mathsf{G}_{\mathcal{P}}[\nabla \chi] w_{t_0}(\mathcal{P}). \tag{5.22}$$

Of course, $\mathsf{G}_{\mathcal{P}}[\nabla \chi] = \frac{\partial}{\partial w}\mathsf{C}_{\mathcal{P}}[w, \nabla \chi]|_{w=0}$.

Finally, to complete the proof of part (2) of Theorem 5.1 we must show that $\mathsf{G}_{\mathcal{P}}[\nabla \chi]$ reduces to a positive constant $C > 0$. To this aim, first note that because of the linearity of $\mathsf{K}_{\mathcal{P}}[\cdot]$ and the arbitrariness of $\boldsymbol{c} \in \mathbb{V}^3$ we may, again, introduce the fixed orthonormal basis $\{\boldsymbol{e}_1, \boldsymbol{e}_2, \boldsymbol{e}_3\} \in \mathbb{V}^3$ and reduce (5.21) to

$$2\mathsf{K}_{\mathcal{P}}[a_i]\boldsymbol{e}_i = \mathsf{G}_{\mathcal{P}}[\nabla \chi]\boldsymbol{f}_{t_0}(\mathcal{P}), \quad \mathsf{G}_{\mathcal{P}}[\nabla \chi] > 0, \ \forall \mathcal{P} \subseteq \mathcal{B}. \tag{5.23}$$

Now, suppose $\mathcal{P}_1$ and $\mathcal{P}_2$ are two subparts of $\mathcal{B}$ with $\mathcal{P}_1 \cap \mathcal{P}_2 = \emptyset$, and that the field $\boldsymbol{a}(\cdot) : \mathcal{B} \to \mathbb{E}^3$ is such that $\boldsymbol{a} \equiv \boldsymbol{0}$ in $\mathcal{P}_1$ and $\mathsf{K}_{\mathcal{P}_2}[a_i]\boldsymbol{e}_i \neq \boldsymbol{0}$. Then,

$$2\mathsf{K}_{\mathcal{P}_1\cup\mathcal{P}_2}[a_i]\boldsymbol{e}_i = 2\mathsf{K}_{\mathcal{P}_2}[a_i]\boldsymbol{e}_i = \mathsf{G}_{\mathcal{P}_1\cup\mathcal{P}_2}[\nabla \chi]\boldsymbol{f}_{t_0}(\mathcal{P}_1 \cup \mathcal{P}_2).$$

But, also,

$$2\mathsf{K}_{\mathcal{P}_2}[a_i]\boldsymbol{e}_i = \mathsf{G}_{\mathcal{P}_2}[\nabla \chi]\boldsymbol{f}_{t_0}(\mathcal{P}_2) \neq 0, \quad \mathsf{G}_{\mathcal{P}_2}[\nabla \chi] > 0,$$

which implies that $\boldsymbol{f}_{t_0}(\mathcal{P}_2) \neq \boldsymbol{0}$. On the other hand, we know that

$$2\mathsf{K}_{\mathcal{P}_1}[a_i]\boldsymbol{e}_i = \mathsf{G}_{\mathcal{P}_1}[\nabla \chi]\boldsymbol{f}_{t_0}(\mathcal{P}_1) = 0, \quad \mathsf{G}_{\mathcal{P}_1}[\nabla \chi] > 0,$$

which implies that $\boldsymbol{f}_{t_0}(\mathcal{P}_1) = \boldsymbol{0}$. Thus, because $\boldsymbol{f}_{t_0}(\mathcal{P}_1 \cup \mathcal{P}_2) = \boldsymbol{f}_{t_0}(\mathcal{P}_1) + \boldsymbol{f}_{t_0}(\mathcal{P}_2)$, we see that

$$\mathsf{G}_{\mathcal{P}_1\cup\mathcal{P}_2}[\nabla \chi]\boldsymbol{f}_{t_0}(\mathcal{P}_2) = \mathsf{G}_{\mathcal{P}_2}[\nabla \chi]\boldsymbol{f}_{t_0}(\mathcal{P}_2) \neq 0,$$

and we find that $\mathsf{G}_{\mathcal{P}_1\cup\mathcal{P}_2}[\nabla \chi] = \mathsf{G}_{\mathcal{P}_2}[\nabla \chi]$.

Using a similar argument, we readily find that $\mathsf{G}_{\mathcal{P}_1\cup\mathcal{P}_2}[\nabla \chi] = \mathsf{G}_{\mathcal{P}_1}[\nabla \chi]$, which allows the conclusion that

$$\mathsf{G}_{\mathcal{P}_1}[\nabla \chi] = \mathsf{G}_{\mathcal{P}_2}[\nabla \chi] \quad \forall \mathcal{P}_1, \mathcal{P}_2 \text{ with } \mathcal{P}_1 \cap \mathcal{P}_2 = \emptyset.$$

Thus, if $\mathcal{P}_3$ is any other subset of $\mathcal{B}$ such that $\mathcal{P}_1 \cap \mathcal{P}_3 = \emptyset$, we may draw the same conclusion with $\mathcal{P}_2$ replaced by $\mathcal{P}_3$, which then yields

$$\mathsf{G}_{\mathcal{P}_2}[\nabla \chi] = \mathsf{G}_{\mathcal{P}_3}[\nabla \chi] \quad \forall \mathcal{P}_2, \mathcal{P}_3.$$

This has the consequence that $\mathsf{G}_{\mathcal{P}}[\nabla\chi] > 0$ is independent of $\mathcal{P} \subset \mathcal{B}$, and equal to a positive constant. Recalling (5.22) and that $\chi = \chi_{t_0}$, where $t_0 \in \mathcal{I}$ is arbitrary, we have proven part (2) of Theorem 5.1 and thereby completed the proof of the theorem. ■

*Remark 5.3* Given a fixed frame F, we have shown that the causality law (4.12) that governs a mechanical process $\mathbb{P}_t$ of (4.11) has the reduced form

$$\frac{\mathrm{d}}{\mathrm{d}t}\mathsf{E}_{\mathcal{P}}[\nabla\chi_t] + \frac{\mathrm{d}}{\mathrm{d}t}\mathsf{K}_{\mathcal{P}}\big[|\dot{\chi}_t|^2\big] = C w_t(\mathcal{P}) \quad \forall \mathcal{P} \subseteq \mathcal{B},\ \forall t \in \mathcal{I}, \tag{5.24}$$

where $\mathsf{K}_{\mathcal{P}}[\cdot] : C(\mathcal{P}) \longrightarrow \mathbb{R}$ is linear and such that $\mathsf{K}_{\mathcal{P}}[\varphi] > 0$ for all $\varphi \geq 0$ (not identically 0) everywhere in $\mathcal{P}$, $\mathsf{E}_{\mathcal{P}}[\boldsymbol{I}] = 0$ and $C > 0$. Also, recalling (5.23) and that $\chi = \chi_{t_0}$ and $\boldsymbol{a} = \ddot{\chi}_{t_0} = \dot{\boldsymbol{v}}_{t_0}$, where $t_0 \in \mathcal{I}$ is arbitrary, we have found that relative to a fixed frame F

$$2\mathsf{K}_{\mathcal{P}}[\dot{\boldsymbol{v}}_t \cdot \boldsymbol{e}_i]\boldsymbol{e}_i = C\boldsymbol{f}_t(\mathcal{P}) \quad \forall \mathcal{P} \subseteq \mathcal{B}, \quad \forall t \in \mathcal{I}, \tag{5.25}$$

which is a residual result of Axiom 5.1.

The constant $C > 0$ is universal and allows us to arrange for the dimensional interpretation of $\mathsf{E}_{\mathcal{P}}[\cdot]/C$ and $\mathsf{K}_{\mathcal{P}}[\cdot]/C$ in (5.24) as the total internal and kinetic energy functionals, respectively, and to interpret this causality law as a statement of the fundamental balance of energy. In that case, $2\mathsf{K}_{\mathcal{P}}[\dot{\boldsymbol{v}}_t \cdot \boldsymbol{e}_i]\boldsymbol{e}_i/C$ in (5.25) will then turn out to have the dimension of force, as it should. □

## 5.1 Mass

Consider the special translational motion $\chi_t = \boldsymbol{X} + \boldsymbol{p}(t)$, $\boldsymbol{p}(0) = \boldsymbol{0}$, for which $\nabla\chi_t = \boldsymbol{I}$. Then, using the linearity of $\mathsf{K}_{\mathcal{P}}[\cdot]$ we see from (5.25) and (3.9a), (3.9b) that

$$2\ddot{\boldsymbol{p}}(t)\mathsf{K}_{\mathcal{P}}[1] = C\int_{\mathcal{P}_t} \boldsymbol{b}_t\,\mathrm{d}v = C\int_{\mathcal{P}} \boldsymbol{b}_t\,\mathrm{d}v,$$

the latter of which is due to a natural mapping of the domain of integration from $\mathcal{P}_t$ to $\mathcal{P}$ under the translational motion $\chi_t = \boldsymbol{X} + \boldsymbol{p}(t)$. Clearly, this shows that $\mathsf{K}_{\mathcal{P}}[1]$, which we recall from (5.14) is positive, is additive on disjoint subsets of $\mathcal{B}$, i.e.,

$$\mathsf{K}_{\mathcal{P}_1\cup\mathcal{P}_2}[1] = \mathsf{K}_{\mathcal{P}_1}[1] + \mathsf{K}_{\mathcal{P}_2}[1], \quad \mathcal{P}_1 \cap \mathcal{P}_2 = \emptyset,$$

and that $\mathsf{K}_{\mathcal{P}}[1]$ is absolutely continuous with respect to volume measure in $\mathcal{B}$, i.e.,

$$\mathsf{K}_{\mathcal{P}}[1] \leq k\,\mathrm{vol}(\mathcal{P}) \quad \forall \mathcal{P} \subseteq \mathcal{B},\ k > 0.$$

Since it is well-known[4] that $\mathsf{K}_{\mathcal{P}}[1]$ with the above properties can be extended to a measure on the Borel sets of $\mathcal{B}$, we can apply the Raydon-Nikodym Theorem to arrive at the conclusion that

$$\mathsf{K}_{\mathcal{P}}[1] = \int_{\mathcal{P}} \kappa\,\mathrm{d}v, \tag{5.26}$$

where $\kappa = \kappa(\boldsymbol{X}) > 0$ is essentially bounded (herein assumed continuous) for all $\boldsymbol{X} \in \mathcal{B}$. We now define

$$\mathsf{M}_{\mathcal{P}} := \frac{2}{C}\mathsf{K}_{\mathcal{P}}[1] = \int_{\mathcal{P}} \rho_0\,\mathrm{d}v, \quad \rho_0 := \frac{2\kappa}{C}, \ \forall \mathcal{P} \subseteq \mathcal{B},$$

[4] See, e.g., Gurtin and Williams [11].

and denote $\mathsf{M}_{\mathcal{P}}$ as the *mass*[5] of $\mathcal{P}$, and $\rho_0(\boldsymbol{X})$ as the *referential mass density* of the particle $\boldsymbol{X} \in \mathcal{B}$, measured per unit volume of $\mathcal{B}$. Observe that $\mathsf{M}_{\mathcal{P}}$ is positive, independent of time, and that by transformation of the domain of integration from $\mathcal{P}$ to $\mathcal{P}_t$ for an arbitrary mechanical process $\mathbb{P}_t$, we have, equivalently,

$$\mathsf{M}_{\mathcal{P}} = \int_{\mathcal{P}_t} \rho_t \, \mathrm{d}v, \qquad \rho_t := \frac{\rho_0}{\det \nabla \chi_t}. \tag{5.27}$$

We denote $\rho_t(\boldsymbol{X})$ as the *present mass density* of the particle $\boldsymbol{X} \in \mathcal{B}$ which is located at $\chi_t(\boldsymbol{X}) \in \mathcal{B}_t$, measured per unit volume of $\mathcal{B}_t$. While the mass density for the particle $\boldsymbol{X} \in \mathcal{B}$ changes in time, the mass of any part $\mathcal{P} \subseteq \mathcal{B}$ is conserved. The relationship, here, between $\rho_0$ and $\rho_t$ is commonly referred to as the Lagrangian form of the balance of mass.

## 6 Second Invariance Axiom and its Consequences

A proper Euclidean change of frame $\mathsf{F} \to \mathsf{F}^*$ as characterized in (3.4) also plays a role in this work. The conclusions of Theorem 5.1, summarized in Remark 5.3, were based upon the invariance requirements of Axiom 5.1. We now introduce an additional second axiom of invariance and consider its consequences.

**Axiom 6.1** (Second invariance axiom) Let $\mathsf{F}$ be a fixed frame and suppose $\mathbb{P}_t$ of (4.11) is a mechanical process for $\mathcal{B}$ during a time interval $\mathcal{I}$; thus, (4.12) holds and, because of Theorem 5.1, the conclusions (5.24) and (5.25) hold. Let $\mathsf{F} \mapsto \mathsf{F}^*(t)$ be a proper Euclidean change of frame (3.4) for which $\boldsymbol{c}(t) \in \mathbb{E}^3$ and $\boldsymbol{Q}(t) \in \mathrm{Orth}^+$. Then, there is a frame-dependent body force field $\boldsymbol{b}_t^* = \boldsymbol{b}_{\mathsf{F}^*(t)}(\boldsymbol{b}_t)$ such that

$$\mathbb{P}_t^*(\cdot) := \{\chi_t^*(\cdot), \boldsymbol{b}_t^*(\cdot), \boldsymbol{\sigma}_t^*(\cdot; \boldsymbol{n}^*)\}, \tag{6.1a}$$

with

$$\chi_t^*(\cdot) := \boldsymbol{Q}(t)\big(\chi_t(\cdot) - \boldsymbol{o}\big) + \boldsymbol{c}(t), \qquad \boldsymbol{\sigma}_t^*(\cdot; \boldsymbol{n}^*) := \boldsymbol{Q}(t)\boldsymbol{\sigma}_t(\cdot; \boldsymbol{n}), \qquad \boldsymbol{n}^* := \boldsymbol{Q}(t)\boldsymbol{n} \tag{6.1b}$$

for all $\boldsymbol{n} \in \mathrm{Unit}$, is a mechanical process for $\mathcal{B}$ during the time interval $\mathcal{I}$ with the *same* response and causal functionals as shown in (4.12) for the mechanical process $\mathbb{P}_t$.

*Remark 6.1* The significance of Axiom 6.1 is that it requires $\mathbb{P}_t$ and $\mathbb{P}_t^*$ to be mechanical processes for $\mathcal{B}$ during $\mathcal{I}$ with the *same* response and causal functionals. Equivalently, the causality law (4.12), and therefore its reduced form (5.24), is supposed to be *form invariant* under any proper Euclidean change of frame in the sense that $\mathbb{P}_t^*$ satisfies

$$\frac{\mathrm{d}}{\mathrm{d}t}\mathsf{E}_{\mathcal{P}}[\nabla \chi_t^*] + \frac{\mathrm{d}}{\mathrm{d}t}\mathsf{K}_{\mathcal{P}}[|\dot{\chi}_t^*|^2] = C w_t^*(\mathcal{P}) \quad \forall \mathcal{P} \subseteq \mathcal{B},\ \forall t \in \mathcal{I}, \tag{6.2}$$

[5] Later, in (6.8), we define the kinetic energy functional for $\mathcal{P}$ as the ratio $\mathsf{K}_{\mathcal{P}}[\cdot]/C$. It is interesting that in early 1900 Wilhelm Ostwald, according to Jammer [12, p. 108], entered the long-standing debate about the concept of mass and supported the fundamental idea that 'mass' should be defined in terms of energy. He conceived 'mass' as merely a capacity for kinetic energy and the definition given here, which is connected to the identification of kinetic energy, supports that point of view.

where $\mathsf{K}_{\mathcal{P}}[\cdot]: C(\mathcal{P}) \longrightarrow I\!R$ is linear and such that $\mathsf{K}_{\mathcal{P}}[\varphi] > 0$ for all $\varphi \geq 0$ (not identically 0) everywhere in $\mathcal{P}$, where $C > 0$, and where

$$w_t^*(\mathcal{P}) = \int_{\partial \mathcal{P}_t^*} \boldsymbol{\sigma}_t^* \cdot \boldsymbol{v}_t^* \, \mathrm{d}a + \int_{\mathcal{P}_t^*} \boldsymbol{b}_t^* \cdot \boldsymbol{v}_t^* \, \mathrm{d}v \tag{6.3a}$$

is a linear function of $\boldsymbol{v}_t^*$ with $\mathcal{P}_t^* := \chi_t^*(\mathcal{P})$, $\partial \mathcal{P}_t^* := \chi_t^*(\partial \mathcal{P})$, $\boldsymbol{v}_t^* := \dot{\chi}_t^*$ and, according to (3.9a) and (3.9b),

$$\boldsymbol{\sigma}_t^*(\cdot; \boldsymbol{n}^*) = \bar{\boldsymbol{\sigma}}\big(\cdot, \nabla \chi_t^*(\cdot); \boldsymbol{n}^*\big). \tag{6.3b}$$

Thus, applying (3.9a), (3.9b) once again, together with (6.1b), we obtain the following traction invariance condition for the constitutive function $\bar{\boldsymbol{\sigma}}(\boldsymbol{X}, \cdot; \boldsymbol{n}) : \mathrm{Lin}^+ \rightarrow I\!V^3$:

$$\bar{\boldsymbol{\sigma}}\big(\cdot, \boldsymbol{Q}(t)\nabla \chi_t(\cdot); \boldsymbol{Q}(t)\boldsymbol{n}\big) = \boldsymbol{Q}(t)\bar{\boldsymbol{\sigma}}\big(\cdot, \nabla \chi_t(\cdot); \boldsymbol{n}\big). \tag{6.4}$$

Of course, the total external force acting on $\mathcal{P}$ at time $t$ during the mechanical process $I\!P_t^*$ is

$$\boldsymbol{f}_t^*(\mathcal{P}) := \int_{\partial \mathcal{P}_t^*} \boldsymbol{\sigma}_t^* \, \mathrm{d}a + \int_{\mathcal{P}_t^*} \boldsymbol{b}_t^* \, \mathrm{d}v, \tag{6.5}$$

and, in the above,

$$\boldsymbol{v}_t^* = \dot{\chi}_t^* = \boldsymbol{Q}(t)\dot{\chi}_t + \dot{\boldsymbol{Q}}(t)(\chi_t - \boldsymbol{o}) + \dot{\boldsymbol{c}} = \boldsymbol{Q}(t)\boldsymbol{v}_t + \dot{\boldsymbol{Q}}(t)(\chi_t - \boldsymbol{o}) + \dot{\boldsymbol{c}}. \tag{6.6}$$

The unit vector field $\boldsymbol{n}^*$ that occurs in $\boldsymbol{\sigma}_t^*$ in the surface integrals above is the outer unit normal to $\partial \mathcal{P}_t^*$ defined by $\boldsymbol{n}^* := \boldsymbol{Q}(t)\boldsymbol{n}$, where $\boldsymbol{n}$ is the corresponding outer unit normal to $\partial \mathcal{P}_t$.

It is worth emphasizing that because $w_t^*(\mathcal{P})$ fundamentally is a linear function of the particle velocity field $\boldsymbol{v}_t^*$, the contact force $\boldsymbol{\sigma}_t^*$ and the body force $\boldsymbol{b}_t^*$ are independent of $\boldsymbol{v}_t^*$ (see Remark 4.2). The constitutive equation (6.3b) confirms this for $\boldsymbol{\sigma}_t^*$. The Axiom 6.1 states that $I\!P_t^*$ is a mechanical process for $\mathcal{B}$ during $\mathcal{I}$, and the Hypothesis 4.1 concerning mechanical processes confirms the existence of a $\boldsymbol{v}_t^*$-independent body force field $\boldsymbol{b}_t^*$. Since the body force $\boldsymbol{b}_t^*$ is assumed to depend on $\boldsymbol{b}_t$ and the frame $\mathsf{F}^*(t)$, and we know that $\boldsymbol{b}_t$ is independent of $\boldsymbol{v}_t$, the independence of $\boldsymbol{b}_t^*$ and $\boldsymbol{v}_t^*$ is naturally consistent. □

## 6.1 Balance of Energy and Linear Momentum Relative to the Frame F

As a consequence of (5.26) and the linearity of $\mathsf{K}_{\mathcal{P}}[\cdot]: C(\mathcal{P}) \longrightarrow I\!R$, it is an exercise in integration theory, using simple functions and the Lebesgue Dominated Convergence Theorem (see, e.g., Bartle [1]) to show the following integral representation:

$$\mathsf{K}_{\mathcal{P}}[\varphi] = \int_{\mathcal{P}} \varphi \kappa \, \mathrm{d}v = \frac{C}{2} \int_{\mathcal{P}} \varphi \rho_0 \, \mathrm{d}v \quad \forall \mathcal{P} \subseteq \mathcal{B}, \ \forall \varphi \in C(\mathcal{P}). \tag{6.7}$$

We now formally define

$$E_{\mathcal{P}}[\cdot] := \frac{1}{C}\mathsf{E}_{\mathcal{P}}[\cdot], \qquad K_{\mathcal{P}}[\cdot] := \frac{1}{C}\mathsf{K}_{\mathcal{P}}[\cdot] \tag{6.8}$$

as the total internal and total kinetic energy functionals, respectively, for $\mathcal{P} \subseteq \mathcal{B}$, and write (5.24) as

$$\frac{\mathrm{d}}{\mathrm{d}t} E_{\mathcal{P}}[\nabla \chi_t] + \frac{\mathrm{d}}{\mathrm{d}t} K_{\mathcal{P}}\big[|\dot{\chi}_t|^2\big] = w_t(\mathcal{P}) \quad \forall \mathcal{P} \subseteq \mathcal{B}, \ \forall t \in \mathcal{I}, \tag{6.9}$$

which represents the classical balance of energy for a mechanical process $\mathbb{P}_t$ of (4.11) for the body $\mathcal{B}$ during $\mathcal{I}$, where $E_{\mathcal{P}}[\boldsymbol{1}] = 0$ and, according to (6.7),

$$K_{\mathcal{P}}\big[|\dot{\chi}_t|^2\big] = \frac{1}{2}\int_{\mathcal{P}} |\dot{\chi}_t|^2 \rho_0 \, \mathrm{d}v. \tag{6.10}$$

Clearly, we may also write (5.25) as

$$\begin{aligned} \frac{2}{C}\mathsf{K}_{\mathcal{P}}[\dot{\boldsymbol{v}}_t \cdot \boldsymbol{e}_i]\boldsymbol{e}_i &= 2K_{\mathcal{P}}[\dot{\boldsymbol{v}}_t \cdot \boldsymbol{e}_i]\boldsymbol{e}_i \\ &= \int_{\mathcal{P}} \dot{\boldsymbol{v}}_t \rho_0 \, \mathrm{d}v = \frac{\mathrm{d}}{\mathrm{d}t}\int_{\mathcal{P}_t} \boldsymbol{v}_t \rho_t \, \mathrm{d}v \\ &= \boldsymbol{f}_t(\mathcal{P}) \quad \forall \mathcal{P} \subseteq \mathcal{B}, \ \forall t \in \mathcal{I}, \end{aligned} \tag{6.11}$$

which represents the classical balance of linear momentum for a mechanical process $\mathbb{P}_t$ of (4.11) for the body $\mathcal{B}$ during $\mathcal{I}$ relative to the fixed frame $\mathsf{F}$. We remark that the natural mapping of $\mathcal{P}$ to $\mathcal{P}_t$ via the motion $\chi_t$ and (5.27) were used in establishing this result.

*Remark 6.2* It is well-known that (3.10), (5.27) and (6.11) imply both[6] the Cauchy Stress Theorem

$$\boldsymbol{\sigma}_t(\boldsymbol{X}; \boldsymbol{n}) = \boldsymbol{T}_t^{\top}(\boldsymbol{X})\boldsymbol{n}, \tag{6.12}$$

where $\boldsymbol{T}_t(\cdot) \in C^1(\mathcal{B})$ is the Cauchy stress tensor field associated with the mechanical process $\mathbb{P}_t$ at time $t$, and the Eulerian form of the local balance of linear momentum

$$\mathrm{div}\boldsymbol{T}_t^{\top} + \boldsymbol{b}_t = \rho_t \dot{\boldsymbol{v}}_t, \tag{6.13}$$

which is to hold in $\mathcal{B}_t$ for all $t \in \mathcal{I}$. Note that (6.12) and the constitutive assumption (3.9a), (3.9b) implies that there is a (sufficiently smooth) constitutive function $\mathsf{T}(\boldsymbol{X}, \cdot) : \mathrm{Lin}^+ \to \mathrm{Lin}$ for all $\boldsymbol{X} \in \mathcal{B}$ such that

$$\bar{\sigma}\big(\cdot, \boldsymbol{F}_t(\cdot); \boldsymbol{n}\big) = \mathsf{T}^{\top}\big(\cdot, \boldsymbol{F}_t(\cdot)\big)\boldsymbol{n}, \tag{6.14}$$

with

$$\boldsymbol{T}_t(\cdot) = \mathsf{T}\big(\cdot, \boldsymbol{F}_t(\cdot)\big), \qquad \boldsymbol{F}_t(\cdot) = \nabla \chi_t(\cdot), \qquad \mathsf{T}(\cdot, \boldsymbol{1}) = \boldsymbol{0} \tag{6.15}$$

in $\mathcal{B}$ for $t \in \mathcal{I}$. Considering the traction invariance condition (6.4), and introducing the convenient, but slightly abusive, abbreviated notation

$$\boldsymbol{T}_t = \mathsf{T}(\boldsymbol{F}_t) = \mathsf{T}\big(\cdot, \boldsymbol{F}_t(\cdot)\big) \tag{6.16}$$

for the field in $\mathcal{B}$, we see, using (6.14), that

$$\mathsf{T}\big(\boldsymbol{Q}(t)\boldsymbol{F}_t\big)\boldsymbol{Q}(t)\boldsymbol{n} = \boldsymbol{Q}(t)\mathsf{T}(\boldsymbol{F}_t)\boldsymbol{n},$$

for all $\boldsymbol{n} \in \mathrm{Unit}$ and all $\boldsymbol{Q}(t) \in \mathrm{Orth}^+$, which readily implies the stress invariance condition

$$\mathsf{T}\big(\boldsymbol{Q}(t)\boldsymbol{F}_t\big) = \boldsymbol{Q}(t)\mathsf{T}(\boldsymbol{F}_t)\boldsymbol{Q}^{\top} \tag{6.17}$$

for all $\boldsymbol{Q}(t) \in \mathrm{Orth}^+$.

[6] Recall that for any mechanical process we have $\boldsymbol{\sigma}_t(\cdot; \boldsymbol{n}) \in C^1(\mathcal{B})$ for each $\boldsymbol{n} \in \mathrm{Unit}$ and $\boldsymbol{\sigma}_t(\boldsymbol{X}; \cdot) \in C(\mathrm{Unit})$ for each $\boldsymbol{X} \in \mathcal{B}$, as was first noted in Sect. 3.3.

Now, with the aid of (3.10), (5.27), (6.10), (6.13) and the notational change $\boldsymbol{F}_t = \nabla \chi_t$, the balance of energy (6.9) for any mechanical process $\mathbb{P}_t$ readily may be reduced to the form

$$\frac{\mathrm{d}}{\mathrm{d}t} E_{\mathcal{P}}[\boldsymbol{F}_t] = \int_{\mathcal{P}_t} \boldsymbol{T}_t^{\top} \cdot \operatorname{grad} \boldsymbol{v}_t \, \mathrm{d}v = \int_{\mathcal{P}} (\det \boldsymbol{F}_t) \boldsymbol{T}_t^{\top} \boldsymbol{F}_t^{-\top} \cdot \dot{\boldsymbol{F}}_t \, \mathrm{d}v$$
$$= \int_{\mathcal{P}} (\det \boldsymbol{F}_t) \mathsf{T}^{\top}(\boldsymbol{F}_t) \boldsymbol{F}_t^{-\top} \cdot \dot{\boldsymbol{F}}_t \, \mathrm{d}v, \tag{6.18}$$

where we have used (6.16), the identity $\dot{\boldsymbol{F}}_t = (\operatorname{grad} \boldsymbol{v}_t)\boldsymbol{F}_t$ and the natural change of domain of integration from $\mathcal{P}_t$ to $\mathcal{P}$ via the motion $\chi_t$. □

## 6.2 Two Fundamental Identities

We now use Axiom 6.1 in the broadest sense and consider the mechanical process $\mathbb{P}_t^*$ which, accordingly, satisfies (6.2), (6.3a), (6.5) and (6.6). Our aim is to obtain the frame indifference and localization properties of the internal energy functional $E_{\mathcal{P}}[\cdot]$, to show that the balance of angular momentum for mechanical processes $\mathbb{P}_t$ relative to the fixed frame $\mathsf{F}$ holds, and to obtain the form of the frame-dependent body force field $\boldsymbol{b}_t^* = \boldsymbol{b}_{\mathsf{F}^*(t)}(\boldsymbol{b}_t)$ relative to the moving frame $\mathsf{F}^*(t)$. To facilitate the discussion, we first record two relevant elementary identities concerning the mechanical processes $\mathbb{P}_t$, $\mathbb{P}_t^*$ and the proper Euclidean change of frame $\mathsf{F} \mapsto \mathsf{F}^*(t)$.

Starting with (6.3a), we readily see that

$$w_t^*(\mathcal{P}) = \int_{\partial \mathcal{P}_t^*} \sigma_t^* \cdot \boldsymbol{v}_t^* \, \mathrm{d}a + \int_{\mathcal{P}_t^*} \boldsymbol{b}_t^* \cdot \boldsymbol{v}_t^* \, \mathrm{d}v$$
$$= w_t(\mathcal{P}) + \dot{\boldsymbol{Q}}^{\top}(t) \boldsymbol{Q}(t) \cdot \left( \int_{\partial \mathcal{P}_t} (\chi_t - \boldsymbol{o}) \otimes \boldsymbol{\sigma}_t \, \mathrm{d}a + \int_{\mathcal{P}_t} (\chi_t - \boldsymbol{o}) \otimes \boldsymbol{b}_t \, \mathrm{d}v \right)$$
$$+ \boldsymbol{Q}^{\top}(t) \dot{\boldsymbol{c}}(t) \cdot \left( \int_{\partial \mathcal{P}_t} \boldsymbol{\sigma}_t \, \mathrm{d}a + \int_{\mathcal{P}_t} \boldsymbol{b}_t \, \mathrm{d}v \right) + \int_{\mathcal{P}_t} \left( \boldsymbol{b}_{\mathsf{F}^*(t)}(\boldsymbol{b}_t) - \boldsymbol{Q}(t) \boldsymbol{b}_t \right) \cdot \bar{\boldsymbol{v}}_t^* \, \mathrm{d}v,$$

where in the last two lines we have transformed the integrals over $\partial \mathcal{P}_t^*$ and $\mathcal{P}_t^*$ to $\partial \mathcal{P}_t$ and $\mathcal{P}_t$ using the proper Euclidean change of frame $\mathsf{F} \mapsto \mathsf{F}^*(t)$ of (3.4). Also, with the aid of (6.6) and (6.10) it is straightforward to show that

$$\frac{\mathrm{d}}{\mathrm{d}t} K_{\mathcal{P}}\big[|\dot{\chi}_t^*|^2\big] = \int_{\mathcal{P}} \dot{\boldsymbol{v}}_t^* \cdot \boldsymbol{v}_t^* \rho_0 \, \mathrm{d}v$$
$$= \int_{\mathcal{P}} \left( \ddot{\boldsymbol{Q}}(t)(\chi_t - \boldsymbol{o}) + 2\dot{\boldsymbol{Q}}(t)\boldsymbol{v}_t + \boldsymbol{Q}(t)\dot{\boldsymbol{v}}_t + \ddot{\boldsymbol{c}}(t) \right) \cdot \boldsymbol{v}_t^* \rho_0 \, \mathrm{d}v$$
$$= \int_{\mathcal{P}} \Big( \left( \ddot{\boldsymbol{Q}}(t)(\chi_t - \boldsymbol{o}) + 2\dot{\boldsymbol{Q}}(t)\boldsymbol{v}_t + \ddot{\boldsymbol{c}}(t) \right) \cdot \boldsymbol{v}_t^*$$
$$+ \boldsymbol{Q}(t)\dot{\boldsymbol{v}}_t \cdot \left( \dot{\boldsymbol{Q}}(t)(\chi_t - \boldsymbol{o}) + \boldsymbol{Q}(t)\boldsymbol{v}_t + \dot{\boldsymbol{c}}(t) \right) \Big) \rho_0 \, \mathrm{d}v$$
$$= \frac{\mathrm{d}}{\mathrm{d}t} K_{\mathcal{P}}\big(|\dot{\chi}_t|^2\big) + \int_{\mathcal{P}} \left( \ddot{\boldsymbol{Q}}(t)(\chi_t - \boldsymbol{o}) + 2\dot{\boldsymbol{Q}}(t)\boldsymbol{v}_t + \ddot{\boldsymbol{c}}(t) \right) \cdot \boldsymbol{v}_t^* \rho_0 \, \mathrm{d}v$$
$$+ \dot{\boldsymbol{Q}}^{\top}(t) \boldsymbol{Q}(t) \cdot \frac{\mathrm{d}}{\mathrm{d}t} \int_{\mathcal{P}} (\chi_t - \boldsymbol{o}) \otimes \boldsymbol{v}_t \rho_0 \, \mathrm{d}v + \boldsymbol{Q}^{\top}(t) \dot{\boldsymbol{c}}(t) \cdot \frac{\mathrm{d}}{\mathrm{d}t} \int_{\mathcal{P}} \boldsymbol{v}_t \rho_0 \, \mathrm{d}v.$$

Here, we also have used $\boldsymbol{Q}(t) \in \text{Orth}^{+} \Rightarrow \dot{\boldsymbol{Q}}^{\top}(t)\boldsymbol{Q}(t) \in \text{Skew}$. Now, we use (5.27) and put these last equations together to reach

$$\frac{\mathrm{d}}{\mathrm{d}t} K_{\mathcal{P}}[|\dot{\chi}_t^*|^2] - w_t^*(\mathcal{P}) = \frac{\mathrm{d}}{\mathrm{d}t} K_{\mathcal{P}}[|\dot{\chi}_t|^2] - w_t(\mathcal{P})$$

$$+ \dot{\boldsymbol{Q}}^{\top}(t)\boldsymbol{Q}(t) \cdot \left( \frac{\mathrm{d}}{\mathrm{d}t} \int_{\mathcal{P}_t} (\chi_t - \boldsymbol{o}) \otimes \boldsymbol{v}_t \rho_t \, \mathrm{d}v \right.$$

$$\left. - \left( \int_{\partial \mathcal{P}_t} (\chi_t - \boldsymbol{o}) \otimes \boldsymbol{\sigma}_t \, \mathrm{d}a + \int_{\mathcal{P}_t} (\chi_t - \boldsymbol{o}) \otimes \boldsymbol{b}_t \, \mathrm{d}v \right) \right)$$

$$+ \boldsymbol{Q}^{\top}(t)\dot{\boldsymbol{c}}(t) \cdot \left( \frac{\mathrm{d}}{\mathrm{d}t} \int_{\mathcal{P}} \boldsymbol{v}_t \rho_0 \, \mathrm{d}v - \left( \int_{\partial \mathcal{P}_t} \boldsymbol{\sigma}_t \, \mathrm{d}a + \int_{\mathcal{P}_t} \boldsymbol{b}_t \, \mathrm{d}v \right) \right)$$

$$- \int_{\mathcal{P}_t} \left( \boldsymbol{b}_{\mathsf{F}^*(t)}(\boldsymbol{b}_t) - \left( \boldsymbol{Q}(t)\boldsymbol{b}_t + \left( \ddot{\boldsymbol{Q}}(t)\, \chi_t + 2\dot{\boldsymbol{Q}}(t)\boldsymbol{v}_t + \ddot{\boldsymbol{c}}(t) \right) \rho_t \right) \right) \cdot \boldsymbol{v}_t^* \, \mathrm{d}v.$$

Observe that the fourth line above vanishes because of the balance of linear momentum (6.11) and (3.10). Also, note that by mapping the domain $\mathcal{P}_t$ to $\mathcal{P}$ and using the balance of mass (5.27), we readily have the identity

$$\dot{\boldsymbol{Q}}^{\top}(t)\boldsymbol{Q}(t) \cdot \frac{\mathrm{d}}{\mathrm{d}t} \int_{\mathcal{P}_t} (\chi_t - \boldsymbol{o}) \otimes \boldsymbol{v}_t \rho_t \, \mathrm{d}v = \dot{\boldsymbol{Q}}^{\top}(t)\boldsymbol{Q}(t) \cdot \int_{\mathcal{P}_t} (\chi_t - \boldsymbol{o}) \otimes \dot{\boldsymbol{v}}_t \rho_t \, \mathrm{d}v,$$

so, with the Cauchy Stress theorem (6.12), the Eulerian form of the local balance of linear momentum (6.13), and the divergence theorem, the second and third lines of the above major equation before this last identity may be replaced by

$$-\dot{\boldsymbol{Q}}^{\top}(t)\boldsymbol{Q}(t) \cdot \int_{\mathcal{P}_t} \boldsymbol{T}_t \, \mathrm{d}v.$$

Thus, with the aid of (6.2), (6.8), (6.9) and the notation $\boldsymbol{F}_t = \nabla \chi_t$, we conclude that

$$\frac{\mathrm{d}}{\mathrm{d}t} E_{\mathcal{P}}[\boldsymbol{Q}(t)\boldsymbol{F}_t] - \frac{\mathrm{d}}{\mathrm{d}t} E_{\mathcal{P}}[\boldsymbol{F}_t] - \dot{\boldsymbol{Q}}^{\top}(t)\boldsymbol{Q}(t) \cdot \int_{\mathcal{P}_t} \boldsymbol{T}_t \, \mathrm{d}v$$

$$- \int_{\mathcal{P}_t} \left( \boldsymbol{b}_{\mathsf{F}^*(t)}(\boldsymbol{b}_t) - \left( \boldsymbol{Q}(t)\boldsymbol{b}_t + \left( \ddot{\boldsymbol{Q}}(t)(\chi_t - \boldsymbol{o}) + 2\dot{\boldsymbol{Q}}(t)\boldsymbol{v}_t + \ddot{\boldsymbol{c}}(t) \right) \rho_t \right) \right) \cdot \boldsymbol{v}_t^* \, \mathrm{d}v = 0, \quad (6.19)$$

where $\boldsymbol{v}_t^*$ is defined in (6.6).

Finally, with the replacement of $\boldsymbol{F}_t$ by $\boldsymbol{Q}(t)\boldsymbol{F}_t$ in (6.18), we first note that

$$\frac{\mathrm{d}}{\mathrm{d}t} E_{\mathcal{P}}[\boldsymbol{Q}(t)\boldsymbol{F}_t] = \int_{\mathcal{P}} \left( \det(\boldsymbol{Q}(t)\boldsymbol{F}_t) \right) \mathbf{T}^{\top}(\boldsymbol{Q}(t)\boldsymbol{F}_t) \left( \boldsymbol{Q}(t)\boldsymbol{F}_t \right)^{-\top} \cdot \overline{\left( \boldsymbol{Q}(t)\boldsymbol{F}_t \right)}^{\boldsymbol{\cdot}} \, \mathrm{d}v. \quad (6.20)$$

Thus, with the aid of the stress invariance condition (6.17), $\boldsymbol{Q}(t) \in \text{Orth}^{+}$, and a return to (6.18), (6.20) readily simplifies to the form

$$\frac{\mathrm{d}}{\mathrm{d}t} E_{\mathcal{P}}[\boldsymbol{Q}(t)\boldsymbol{F}_t] - \frac{\mathrm{d}}{\mathrm{d}t} E_{\mathcal{P}}[\boldsymbol{F}_t] - \dot{\boldsymbol{Q}}^{\top}(t)\boldsymbol{Q}(t) \cdot \int_{\mathcal{P}_t} \boldsymbol{T}_t \, \mathrm{d}v = 0, \quad (6.21)$$

which, with (6.19), yields

$$\int_{\mathcal{P}_t} \left( \boldsymbol{b}_{\mathsf{F}^*(t)}(\boldsymbol{b}_t) - \left( \boldsymbol{Q}(t)\boldsymbol{b}_t + \left( \ddot{\boldsymbol{Q}}(t)(\chi_t - \boldsymbol{o}) + 2\dot{\boldsymbol{Q}}(t)\boldsymbol{v}_t + \ddot{\boldsymbol{c}}(t) \right) \rho_t \right) \right) \cdot \boldsymbol{v}_t^* \, \mathrm{d}v = 0. \quad (6.22)$$

These last two fundamental identities, (6.21) and (6.22), are supposed to hold for all mechanical processes $\mathbb{P}_t$ of $\mathcal{B}$ during a time interval $\mathcal{I}$, for all $\boldsymbol{Q}(t) \in \text{Orth}^+$, all $\boldsymbol{c}(t) \in \mathbb{E}^3$, and for all $\mathcal{P} \subseteq \mathcal{B}$.

There are major consequences that result from the identities (6.21) and (6.22), and we develop these in the following two subsections.

## 6.3 Internal Energy and Balance of Angular Momentum Relative to the Frame F: Frame Indifference and Local Properties

Consider now (6.21), which is to hold for all mechanical processes $\mathbb{P}_t$ of $\mathcal{B}$ during a time interval $\mathcal{I}$, for all $\boldsymbol{Q}(t) \in \text{Orth}^+$, and for all $\mathcal{P} \subseteq \mathcal{B}$.

Suppose $\boldsymbol{W} \in \text{Skew}$ is constant with $|\boldsymbol{W}| = \sqrt{2}$ and let $\alpha > 0$ be a positive constant, and define

$$\boldsymbol{Q}(t) := \exp \frac{2\pi}{\alpha} t \boldsymbol{W} \in \text{Orth}^+ .$$

Then, with this choice of $\boldsymbol{Q}(t)$ entered into (6.21), after integration over the interval $t \in (0, \alpha) \subset \mathcal{I}$ and observing that $\boldsymbol{Q}(0) = \boldsymbol{Q}(\alpha) = \boldsymbol{1}$ and $\dot{\boldsymbol{Q}}^{\top}(t)\, \boldsymbol{Q}(t) = \frac{2\pi}{\alpha} \boldsymbol{W}^{\top}$, we find

$$\frac{2\pi}{\alpha} \boldsymbol{W}^{\top} \cdot \int_0^{\alpha} \left( \int_{\mathcal{P}_t} \boldsymbol{T}_t \, \mathrm{d}v \right) \mathrm{d}t = 0.$$

As a consequence of the limit $\alpha \to 0$, we see that

$$\boldsymbol{W}^{\top} \cdot \int_{\mathcal{P}_t} \boldsymbol{T}_t \, \mathrm{d}v = 0,$$

which is to hold for all $\boldsymbol{W} \in \text{Skew}$ and for all parts $\mathcal{P} \subset \mathcal{B}$. Thus, by localization the Cauchy stress tensor field is symmetric,

$$\boldsymbol{T}_t(\boldsymbol{X}) = \boldsymbol{T}_t^{\top}(\boldsymbol{X}) \quad \forall \boldsymbol{X} \in \mathcal{B},\ \forall t \in \mathcal{I}, \tag{6.23}$$

which is well-known to express the balance of angular momentum relative to the frame $\mathsf{F}$ for the process $\mathbb{P}_t$.

Now, returning to (6.21) and noting that $\dot{\boldsymbol{Q}}^{\top}(t)\, \boldsymbol{Q}(t) \in \text{Skew}$, we have

$$\frac{\mathrm{d}}{\mathrm{d}t} E_{\mathcal{P}}\big[\boldsymbol{Q}(t)\boldsymbol{F}_t\big] - \frac{\mathrm{d}}{\mathrm{d}t} E_{\mathcal{P}}[\boldsymbol{F}_t] = 0 \quad \forall \boldsymbol{Q}(t) \in \text{Orth}^+, \quad \forall \mathcal{P} \subseteq \mathcal{B},$$

which, when integrated over an interval including $t = 0$ with $\boldsymbol{Q}(0) = \boldsymbol{1}$, yields

$$E_{\mathcal{P}}\big[\boldsymbol{Q}(t)\boldsymbol{F}_t\big] = E_{\mathcal{P}}[\boldsymbol{F}_t] \quad \forall \boldsymbol{Q}(t) \in \text{Orth}^+, \quad \forall \mathcal{P} \subseteq \mathcal{B}, \tag{6.24}$$

which states that the total internal energy functional $E_{\mathcal{P}}[\cdot]$ is frame indifferent under a proper orthogonal change of frame $\mathsf{F} \mapsto \mathsf{F}^*(t)$.

To go farther, suppose we integrate (6.18) and switch the order of integrations to reach

$$E_{\mathcal{P}}[\boldsymbol{F}_t] - E_{\mathcal{P}}[\boldsymbol{1}] = \int_{\mathcal{P}} \left( \int_0^t (\det \boldsymbol{F}_t) \boldsymbol{T}_t^{\top} (\boldsymbol{F}_t)^{-\top} \cdot \dot{\boldsymbol{F}}_t \, \mathrm{d}t \right) \mathrm{d}v, \tag{6.25}$$

wherein we have used $\chi_0(\boldsymbol{X}) = \boldsymbol{X}$ with the consequence that $\boldsymbol{F}_0 = \boldsymbol{1}$. Note that because of (4.7) we may set $E_{\mathcal{P}}[\boldsymbol{1}] = 0$ and conclude, from (6.25), that $E_{\mathcal{P}}[\cdot]$ is additive on separate parts $\mathcal{P} \subseteq \mathcal{B}$ and is absolutely continuous with respect to volume measure in $\mathcal{B}$. Moreover, since (6.25) holds for any mechanical process $\mathbb{P}_t$ and we know that (6.15) holds, then we may assert, using the theorem of Gurtin [10, p. 186], that there exists a stored energy function $W(\boldsymbol{X}, \cdot) : \mathrm{Lin}^+ \to \mathbb{R}$ for all $\boldsymbol{X} \in \mathcal{B}$, measured per unit volume of $\mathcal{B}$, such that

$$(\det \boldsymbol{F}_t)\mathsf{T}^\top(\boldsymbol{X}, \boldsymbol{F}_t)(\boldsymbol{F}_t)^{-\top} = \mathsf{D}_{\boldsymbol{F}} W(\boldsymbol{X}, \boldsymbol{F}_t), \tag{6.26}$$

where $\mathsf{D}_{\boldsymbol{F}} W(\boldsymbol{X}, \cdot)$ denotes the gradient of $W(\boldsymbol{X}, \cdot)$ in the set $\mathrm{Lin}^+$. Without loss of generality, we may set $W(\boldsymbol{X}, \boldsymbol{I}) = 0$ and conclude, using (6.25), that

$$E_{\mathcal{P}}[\boldsymbol{F}_t] = \int_{\mathcal{P}} W(\boldsymbol{X}, \boldsymbol{F}_t)\, \mathrm{d}v, \quad \forall \mathcal{P} \subseteq \mathcal{B}. \tag{6.27}$$

Moreover, because of (6.24) and (6.27), a localization shows that the stored energy function $W(\boldsymbol{X}, \cdot)$ is frame indifferent under a proper orthogonal change of frame $\mathsf{F} \mapsto \mathsf{F}^*(t)$, i.e., $W(\boldsymbol{X}, \boldsymbol{Q}(t)\boldsymbol{F}_t) = W(\boldsymbol{X}, \boldsymbol{F}_t)$ for all $\boldsymbol{Q}(t) \in \mathrm{Orth}^+$. Also, because of (6.15), we note that $\mathsf{D}_{\boldsymbol{F}} W(\boldsymbol{X}, \boldsymbol{I}) = \boldsymbol{0}$.[7]

*Remark 6.3* The Piola-Kirchhoff stress $\boldsymbol{S}_t = \boldsymbol{S}_t(\boldsymbol{X})$ commonly is defined as

$$\boldsymbol{S}_t := (\det \boldsymbol{F}_t)\boldsymbol{F}_t\boldsymbol{T}_t,$$

in which case its constitutive equation is given by the first of

$$\boldsymbol{S}_t(\boldsymbol{X}) = \mathsf{S}(\boldsymbol{X}, \boldsymbol{F}_t) := \mathsf{D}_{\boldsymbol{F}} W^\top(\boldsymbol{X}, \boldsymbol{F}_t), \quad \mathrm{Div}\, \boldsymbol{S}_t^\top + \boldsymbol{b}_t = \rho_0 \boldsymbol{v}_t,$$

the latter of which expresses a form of the balance of linear momentum alternative to (6.13), and which is to hold in $\mathcal{B}$ for all $t \in \mathcal{I}$. □

## 6.4 Body Force Relative to the Frame $\mathsf{F}^*(t)$

We consider here the consequence of the identity (6.22), which is supposed to hold for all mechanical processes $\mathbb{P}_t$ of $\mathcal{B}$ during a time interval $\mathcal{I}$, for all $\boldsymbol{Q}(t) \in \mathrm{Orth}^+$, all $\boldsymbol{c}(t) \in \mathbb{E}^3$, and for all $\mathcal{P} \subseteq \mathcal{B}$. Of course, by localization (and using smoothness requirements) the integrand must vanish everywhere in $\mathcal{B}_t$, but a more insightful expression of this conclusion is obtained if we first map the domain of integration from $\mathcal{P}_t$ to $\mathcal{P}_t^*$ via the change of frame $\mathsf{F} \mapsto \mathsf{F}^*(t)$, i.e., $\boldsymbol{x}^* = \boldsymbol{Q}(t)(\boldsymbol{x} - \boldsymbol{o}) + \boldsymbol{c}(t)$ of (3.4). Thus, using $\boldsymbol{x} := \chi_t(\boldsymbol{X}, t) \in \mathcal{B}_t$, $\boldsymbol{x}^* := \chi_t^*(\boldsymbol{X}, t) \in \mathcal{B}_t^*$, $\rho_t^* := \rho_0 / \det \nabla \chi_t^* (= \rho_t)$ analogous to (5.27), and (6.6), we may write (6.22) equivalently as

$$\begin{aligned} \int_{\mathcal{P}_t^*} \Big( & \boldsymbol{b}_{\mathsf{F}^*(t)}(\boldsymbol{b}_t) - \boldsymbol{Q}(t)\boldsymbol{b}_t - \rho_t^*\big(\dot{\boldsymbol{\omega}}(t) \times \big(\boldsymbol{x}^* - \boldsymbol{c}(t)\big) - \boldsymbol{\omega}(t) \times \big(\boldsymbol{\omega}(t) \times \big(\boldsymbol{x}^* - \boldsymbol{c}(t)\big)\big) \\ & + 2\boldsymbol{\omega}(t) \times \big(\boldsymbol{v}_t^* - \dot{\boldsymbol{c}}(t)\big) + \ddot{\boldsymbol{c}}(t)\big)\Big) \cdot \boldsymbol{v}_t^*\, \mathrm{d}v \\ = 0, & \end{aligned}$$

[7] Noll [13, p. 44] showed in his thesis that with the constitutive relation implicit in (6.26), the frame indifference of the stored energy function $W(\boldsymbol{X}, \cdot)$ and the symmetry of the Cauchy stress (6.23) are equivalent. Remarkably, three quarters of a century earlier Gibbs [6, p.190] argued, within nonlinear elasto-statics, for the symmetry of the stress based on invariance considerations, and he emphasized that the symmetry would persist in dynamical situations. It is interesting that Gibbs did not consider this symmetry to be connected to any balance law and that he made no mention of moment equilibrium in his work.

where $\boldsymbol{\omega}(t)$, the axial vector associated with $\dot{\boldsymbol{Q}}(t)\boldsymbol{Q}^\top(t) \in \text{Skew}$, represents the angular velocity of the frame $\mathsf{F}^*(t)$.

Now, since this must hold for all $\mathcal{P} \subseteq \mathcal{B}$ we may conclude that the integrand must vanish throughout $\mathcal{B}_t^*$. Clearly, as a consequence, there exists a vector field $\boldsymbol{\Omega}_t : \mathcal{B}_t^* \to \mathbb{V}^3$ such that

$$\begin{aligned} &\boldsymbol{b}_{\mathsf{F}^*(t)}(\boldsymbol{b}_t) - \boldsymbol{Q}(t)\boldsymbol{b}_t - \rho_t^*\big(\dot{\boldsymbol{\omega}}(t) \times \big(\boldsymbol{x}^* - \boldsymbol{c}(t)\big) - \boldsymbol{\omega}(t) \times \big(\boldsymbol{\omega}(t) \times \big(\boldsymbol{x}^* - \boldsymbol{c}(t)\big)\big) \\ &\quad + 2\boldsymbol{\omega}(t) \times \big(\boldsymbol{v}_t^* - \dot{\boldsymbol{c}}(t)\big) + \ddot{\boldsymbol{c}}(t)\big) = \boldsymbol{\Omega}_t \times \boldsymbol{v}_t^* \end{aligned}$$

throughout $\mathcal{B}_t^*$. Finally, since $\boldsymbol{b}_t^* = \boldsymbol{b}_{\mathsf{F}^*(t)}(\boldsymbol{b}_t)$ is independent of $\boldsymbol{v}_t^*$, as noted in Remark 6.1, we find

$$\begin{aligned} \boldsymbol{b}_{\mathsf{F}^*(t)}(\boldsymbol{b}_t) &= \boldsymbol{Q}(t)\boldsymbol{b}_t + \rho_t^*\big(\dot{\boldsymbol{\omega}}(t) \times \big(\boldsymbol{x}^* - \boldsymbol{c}(t)\big) \\ &\quad - \boldsymbol{\omega}(t) \times \big(\boldsymbol{\omega}(t) \times \big(\boldsymbol{x}^* - \boldsymbol{c}(t)\big)\big) - 2\boldsymbol{\omega}(t) \times \dot{\boldsymbol{c}}(t) + \ddot{\boldsymbol{c}}(t)\big) \end{aligned} \tag{6.28}$$

and

$$\boldsymbol{\Omega}_t = -2\rho_t^*\boldsymbol{\omega}(t), \tag{6.29}$$

both of which hold throughout $\mathcal{B}_t^*$ The former of these, i.e., (6.28), explicitly characterizes how $\boldsymbol{b}_{\mathsf{F}^*(t)}(\boldsymbol{b}_t)$ depends on $\boldsymbol{b}_t$ and the frame $\mathsf{F}^*(t)$. Note, in particular, that there is no 'Coriolis force' term contained in this expression, and there should not be such a term because $\boldsymbol{b}_t^*$ is supposed[8] to be independent of $\boldsymbol{v}_t^*$.

As a cautionary observation, it is important to emphasize that at this point the balance laws of linear and angular momentum have been obtained only relative to a *fixed* frame $\mathsf{F}$. One may expect that corresponding balance laws also hold relative to a moving frame $\mathsf{F}^*(t)$, but their form and structure has yet to be determined. In the final short section we explore this question.

## 7 Balance of Linear and Angular Momentum Relative to the Frame $\mathsf{F}^*(t)$

The form of the balance of linear momentum relative to the moving frame $\mathsf{F}^*(t)$ cannot be inferred directly from its expression (6.11) relative to the fixed frame $\mathsf{F}$ because Axiom 5.1, the fundamental basis of (6.11), is predicated on mechanical processes $\mathbb{P}_t$ relative to a *fixed* frame. Note, however, that based upon what we introduced and have concluded in Sect. 6 we may write

$$\begin{aligned} &\frac{\mathrm{d}}{\mathrm{d}t}\int_{\mathcal{P}_t^*} \rho_t^*\boldsymbol{v}_t^*\,\mathrm{d}v - \left(\int_{\partial\mathcal{P}_t^*} \boldsymbol{\sigma}_t^*\,\mathrm{d}a + \int_{\mathcal{P}_t^*} \boldsymbol{b}_t^*\,\mathrm{d}v\right) \\ &\quad = \frac{\mathrm{d}}{\mathrm{d}t}\int_{\mathcal{P}_t} \rho_t\big(\dot{\boldsymbol{Q}}(t)(\chi_t - \boldsymbol{o}) + \boldsymbol{Q}(t)\boldsymbol{v}_t + \dot{\boldsymbol{c}}(t)\big)\,\mathrm{d}v \\ &\qquad - \boldsymbol{Q}(t)\int_{\partial\mathcal{P}_t} \boldsymbol{\sigma}_t\,\mathrm{d}a - \int_{\mathcal{P}_t} \boldsymbol{b}_{\mathsf{F}^*(t)}(\boldsymbol{b}_t)\,\mathrm{d}v \\ &\quad = \frac{\mathrm{d}}{\mathrm{d}t}\int_{\mathcal{P}_t} \rho_t\big(\dot{\boldsymbol{Q}}(t)(\chi_t - \boldsymbol{o}) + \boldsymbol{Q}(t)\boldsymbol{v}_t + \dot{\boldsymbol{c}}(t)\big)\,\mathrm{d}v - \boldsymbol{Q}(t)\frac{\mathrm{d}}{\mathrm{d}t}\int_{\mathcal{P}_t} \rho_t\boldsymbol{v}_t\,\mathrm{d}v \end{aligned}$$

[8] See Remark 6.1.

$$
\begin{aligned}
& - \int_{\mathcal{P}_t} \left( \boldsymbol{b}_{\mathsf{F}^*(t)}(\boldsymbol{b}_t) - \boldsymbol{Q}(t)\boldsymbol{b}_t \right) \mathrm{d}v \\
&= - \int_{\mathcal{P}_t} \left( \boldsymbol{b}_{\mathsf{F}^*(t)}(\boldsymbol{b}_t) - \left( \boldsymbol{Q}(t)\boldsymbol{b}_t + \rho_t \left( \ddot{\boldsymbol{Q}}(t)(\chi_t - \boldsymbol{o}) + 2\dot{\boldsymbol{Q}}(t)\boldsymbol{v}_t + \ddot{\boldsymbol{c}}(t) \right) \right) \right) \mathrm{d}v \\
&= - \int_{\mathcal{P}_t*} \left( \boldsymbol{b}_t^* - \boldsymbol{Q}(t)\boldsymbol{b}_t - \rho_t^* \left( \dot{\boldsymbol{\omega}}(t) \times \left( \boldsymbol{x}^* - \boldsymbol{c}(t) \right) \right. \right. \\
& \left. \left. - \boldsymbol{\omega}(t) \times \left( \boldsymbol{\omega}(t) \times \left( \boldsymbol{x}^* - \boldsymbol{c}(t) \right) \right) + 2\boldsymbol{\omega}(t) \times \left( \boldsymbol{v}_t^* - \dot{\boldsymbol{c}}(t) \right) + \ddot{\boldsymbol{c}}(t) \right) \right) \mathrm{d}v \\
&= - \int_{\mathcal{P}_t*} \boldsymbol{\Omega}_t \times \boldsymbol{v}_t^* \, \mathrm{d}v \\
&= \int_{\mathcal{P}_t*} 2\rho_t^* \boldsymbol{\omega}(t) \times \boldsymbol{v}_t^* \, \mathrm{d}v \quad \forall \mathcal{P} \in \mathcal{B}, \ \forall t \in \mathcal{I}.
\end{aligned}
$$

Thus, defining

$$
\begin{aligned}
\bar{\boldsymbol{b}}_t^* &:= \boldsymbol{b}_t^* + 2\rho_t^* \boldsymbol{\omega}(t) \times \boldsymbol{v}_t^* \\
&= \boldsymbol{Q}(t)\boldsymbol{b}_t + \rho_t^* (\dot{\boldsymbol{\omega}}(t) \times (\boldsymbol{x}^* - \boldsymbol{c}(t)) - \boldsymbol{\omega}(t) \times (\boldsymbol{\omega}(t) \times (\boldsymbol{x}^* - \boldsymbol{c}(t))) \\
&\quad + 2\boldsymbol{\omega}(t) \times (\boldsymbol{v}_t^* - \dot{\boldsymbol{c}}(t)) + \ddot{\boldsymbol{c}}(t)),
\end{aligned} \tag{7.1}
$$

we see that

$$
\frac{\mathrm{d}}{\mathrm{d}t} \int_{\mathcal{P}_t^*} \rho_t^* \boldsymbol{v}_t^* \, \mathrm{d}v = \int_{\partial \mathcal{P}_t^*} \boldsymbol{\sigma}_t^* \, \mathrm{d}a + \int_{\mathcal{P}_t^*} \bar{\boldsymbol{b}}_t^* \, \mathrm{d}v \quad \forall \mathcal{P} \in \mathcal{B}, \ \forall t \in \mathcal{I}, \tag{7.2}
$$

which expresses the balance of linear momentum relative to the moving frame $\mathsf{F}^*(t)$. Clearly, the effective body force field $\bar{\boldsymbol{b}}_t^*$ that naturally enters this balance is the sum of the frame-dependent body force field $\boldsymbol{b}_t^* = \boldsymbol{b}_{\mathsf{F}^*(t)}(\boldsymbol{b}_t)$ which is fundamental to the expression of internal power as a linear function of particle velocity, and the velocity-dependent field $2\rho_t^* \boldsymbol{\omega}(t) \times \boldsymbol{v}_t^*$ which measures an additional body force that acts on a material particle of the body when that particle is moving relative to the rotating frame $\mathsf{F}^*(t)$. This latter force is known as the Coriolis force.

*Remark 7.1* Note that the body force $\boldsymbol{b}_t^*$ in the expression (6.3a) of the internal power $w_t^*(\mathcal{P})$ relative to the moving frame may be replaced formally by $\bar{\boldsymbol{b}}_t^*$ since the difference, i.e., the Coriolis force, automatically cancels out of the calculation—the Coriolis force is said to be 'powerless'. However, there is a difference in these forces at a fundamental level, and the independence of $\boldsymbol{b}_t^*$ and $\boldsymbol{v}_t^*$ has been instrumental in the preceding developments. □

Finally, recalling from Axiom 6.1 that $\boldsymbol{\sigma}_t^* = \boldsymbol{Q}(t)\boldsymbol{\sigma}_t$, from (6.12) that $\boldsymbol{\sigma}_t(\boldsymbol{X}; \boldsymbol{n}) = \boldsymbol{T}_t^\top(\boldsymbol{X})\boldsymbol{n}$, from (6.23) that $\boldsymbol{T}_t \in \mathrm{Sym}$, and that $\boldsymbol{n}^* = \boldsymbol{Q}(t)\boldsymbol{n}$ for all $\boldsymbol{n} \in \mathrm{Unit}$, we readily find that $\sigma_t^*(\boldsymbol{X}) = \boldsymbol{T}_t^{*\top}(\boldsymbol{X})\boldsymbol{n}^*$, where, according to (6.3b) and (6.14), $\boldsymbol{T}_t^*(\cdot) = \mathsf{T}(\cdot, \boldsymbol{F}_t^*(\cdot))$, and where $\boldsymbol{T}_t^* = \boldsymbol{Q}(t)\boldsymbol{T}_t\boldsymbol{Q}^\top(t)$, so that $\boldsymbol{T}_t^* \in \mathrm{Sym}$. This essentially expresses the balance of angular momentum relative to the moving frame $\mathsf{F}^*(t)$.

## Appendix

Here, we summarize several special properties of F-differentiable functionals, introduced in Sect. 4.1 and used throughout Sects. 4 and 5.

**Lemma A.1** (*Chain-rule*) *Let* $\chi_t$ *be a tame motion according to Definition* 3.1 *on* $\mathcal{B}$ *for all* $t \in \mathcal{I}$. *Let* $\mathsf{R}_{\mathcal{P}}[\cdot,\cdot]$ *be a F-differentiable functional for* $\mathcal{B}$ *according to property* (1) *of Definition* 4.1. *Then, the following chain-rule holds*:

$$\frac{\mathrm{d}}{\mathrm{d}t}\mathsf{R}_{\mathcal{P}}\big[|\dot{\chi}_t|^2, \nabla\chi_t\big] = \delta_1\mathsf{R}_{\mathcal{P}}\big[|\dot{\chi}_t|^2, \nabla\chi_t \,|\, 2\dot{\chi}_t\cdot\ddot{\chi}_t\big] + \delta_2\mathsf{R}_{\mathcal{P}}\big[|\dot{\chi}_t|^2, \nabla\chi_t \,|\, \dot{\overline{\nabla\chi_t}}\big] \tag{A.1}$$

*for all* $t \in \mathcal{I}$.

*Proof* Note that for a tame motion $\chi_t$, it follows from Definition 3.1 that $\varphi_t := |\dot{\chi}_t|^2 \in C^+(\mathcal{B})\times C^1(\mathcal{I})$ and $\boldsymbol{A}_t := \nabla\chi_t \in C^{\mathrm{Lin}^+}(\mathcal{B})\times C^1(\mathcal{I})$ and that $\dot{\varphi}_t$ and $\dot{\boldsymbol{A}}_t$ are in $C(\mathcal{B})$ for all $t \in \mathcal{I}$. Then, since $\mathsf{R}_{\mathcal{P}}[\cdot,\cdot]$ is F-differentiable, we know that for all $\tau, t \in \mathcal{I}$

$$\begin{aligned}\mathsf{R}_{\mathcal{P}}[\varphi_\tau, \boldsymbol{A}_\tau] &= \mathsf{R}_{\mathcal{P}}[\varphi_t + \varphi_\tau - \varphi_t, \boldsymbol{A}_t + \boldsymbol{A}_\tau - \boldsymbol{A}_t] \\ &= \mathsf{R}_{\mathcal{P}}[\varphi_t, \boldsymbol{A}_t] + \delta_1\mathsf{R}_{\mathcal{P}}[\varphi_t, \boldsymbol{A}_t |\, \varphi_\tau - \varphi_t] + o\big(\|\varphi_\tau - \varphi_t\|\big) \\ &\quad + \delta_2\mathsf{R}_{\mathcal{P}}[\varphi_t, \boldsymbol{A}_t |\, \boldsymbol{A}_\tau - \boldsymbol{A}_t] + o\big(\|\boldsymbol{A}_\tau - \boldsymbol{A}_t\|\big).\end{aligned}$$

Thus, by computing

$$\frac{\mathrm{d}}{\mathrm{d}t}\mathsf{R}_{\mathcal{P}}[\varphi_t, \boldsymbol{A}_t] = \lim_{\tau\to t}\frac{\mathsf{R}_{\mathcal{P}}[\varphi_t, \boldsymbol{A}_t] - \mathsf{R}_{\mathcal{P}}[\varphi_\tau, \boldsymbol{A}_\tau]}{t-\tau},$$

we readily find

$$\frac{\mathrm{d}}{\mathrm{d}t}\mathsf{R}_{\mathcal{P}}[\varphi_t, \boldsymbol{A}_t] = \delta_1\mathsf{R}_{\mathcal{P}}[\varphi_t, \boldsymbol{A}_t |\, \dot{\varphi}_t] + \delta_2\mathsf{R}_{\mathcal{P}}[\varphi_t, \boldsymbol{A}_t |\, \dot{\boldsymbol{A}}_t], \tag{A.2}$$

and by using the definitions of $\varphi_t$ and $\boldsymbol{A}_t$, we arrive at (A.1). ■

**Lemma A.2** (*Mean-value theorem*)

(1) *Let* $\{\varphi^{(1)}, \varphi^{(2)}\} \in C^+(\mathcal{B})$ *and* $\boldsymbol{A} \in C^{\mathrm{Lin}^+}(\mathcal{B})$. *Let* $\mathsf{R}_{\mathcal{P}}[\cdot, \boldsymbol{A}]$ *be a F-differentiable functional for* $\mathcal{B}$ *according to property* (1) *of Definition* 4.1. *Then, there exists* $\bar{\varphi} \in C^+(\mathcal{B})$ *such that*

$$\mathsf{R}_{\mathcal{P}}\big[\varphi^{(1)}, \boldsymbol{A}\big] - \mathsf{R}_{\mathcal{P}}\big[\varphi^{(2)}, \boldsymbol{A}\big] = \delta_1\mathsf{R}_{\mathcal{P}}\big[\bar{\varphi}, \boldsymbol{A} |\, \varphi^{(1)} - \varphi^{(2)}\big]. \tag{A.3}$$

(2) *Let* $\{\boldsymbol{A}^{(1)}, \boldsymbol{A}^{(2)}\} \in C^{\mathrm{Lin}^+}(\mathcal{B})$ *be such that* $\boldsymbol{A}^{(2)} + t(\boldsymbol{A}^{(1)} - \boldsymbol{A}^{(2)}) \in C^{\mathrm{Lin}^+}(\mathcal{B})$ *for all* $t \in [0,1]$, *and let* $\varphi \in C^+(\mathcal{B})$. *Let* $\mathsf{R}_{\mathcal{P}}[\varphi, \cdot]$ *be a F-differentiable functional for* $\mathcal{B}$ *according to property* (1) *of Definition* 4.1. *Then, there exists* $\bar{\boldsymbol{A}} \in C^{\mathrm{Lin}^+}(\mathcal{B})$ *such that*

$$\mathsf{R}_{\mathcal{P}}\big[\varphi, \boldsymbol{A}^{(1)}\big] - \mathsf{R}_{\mathcal{P}}\big[\varphi, \boldsymbol{A}^{(2)}\big] = \delta_2\mathsf{R}_{\mathcal{P}}\big[\varphi, \bar{\boldsymbol{A}} |\, \boldsymbol{A}^{(1)} - \boldsymbol{A}^{(2)}\big]. \tag{A.4}$$

*Proof* To prove part (1), first note that $\varphi_t := \varphi^{(2)} + t(\varphi^{(1)} - \varphi^{(2)}) \in C^+(\mathcal{B})$ for all $t \in [0,1]$. Now, define

$$\omega(t) := \mathsf{R}_{\mathcal{P}}[\varphi_t, \boldsymbol{A}] \quad \forall t \in [0,1].$$

Then, using the F-differentiability of $\mathsf{R}_{\mathcal{P}}[\cdot, \boldsymbol{A}]$ in (4.3a), and (A.2) in the proof of Lemma A.1, we reach

$$\dot{\omega}(t) = \delta_1 \mathsf{R}_{\mathcal{P}}\big[\varphi_t, \boldsymbol{A} \,|\, \varphi^{(1)} - \varphi^{(2)}\big],$$

and the elementary mean value theorem for real-valued functions implies that

$$\omega(1) - \omega(0) = \dot{\omega}(\bar{t})$$

for some $\bar{t} \in [0, 1]$, to complete the proof of (A.3).

To prove part (2), we first note that $\boldsymbol{A}_t := \boldsymbol{A}^{(2)} + t(\boldsymbol{A}^{(1)} - \boldsymbol{A}^{(2)}) \in C^{\mathrm{Lin}^+}(\mathcal{B})$ for all $t \in [0, 1]$. Then, defining

$$\omega(t) := \mathsf{R}_{\mathcal{P}}[\varphi, \boldsymbol{A}_t] \quad \forall t \in [0, 1],$$

and using the F-differentiability of $\mathsf{R}_{\mathcal{P}}[\varphi, \cdot]$ in (4.3b) we may follow an argument similar to that used above to reach (A.4). ■

*Remark A.1* Note that if $\{\varphi^{(1)}, \varphi^{(2)}\} \in C^+(\mathcal{B}) \cap C^1(\mathcal{B})$ in part (1) of Lemma A.2, then a similar argument shows that there exists a $\bar{\varphi} \in C^+(\mathcal{B}) \cap C^1(\mathcal{B})$ for which (A.3) holds.

Note, also, that if $\boldsymbol{A}^{(1)}$ and $\boldsymbol{A}^{(2)}$ in part (2) of Lemma A.2 are gradients of vector-valued functions, then $\boldsymbol{A}_t$ is a gradient. This means that $\bar{\boldsymbol{A}}$ in (A.4), being equal to $\boldsymbol{A}_{\bar{t}}$ for some $\bar{t} \in [0, 1]$, also is a gradient. □

**Lemma A.3** *Let $\varphi \in C^+(\mathcal{B}) \cap C^1(\mathcal{B})$ and $\boldsymbol{A} \in C^{\mathrm{Lin}^+}(\mathcal{B})$. Suppose that $\mathsf{R}_{\mathcal{P}}[\cdot, \cdot]$ is a F-differentiable functional for $\mathcal{B}$ according to Definition* 4.1, *and let*

$$\delta_1 \mathsf{R}_{\mathcal{P}}[\varphi, \boldsymbol{A} \,|\, \psi] = \delta_1 \mathsf{R}_{\mathcal{P}}[0, \boldsymbol{A} \,|\, \psi] \tag{A.5}$$

*for all such $\varphi$ and $\boldsymbol{A}$, and for all $\psi \in C^1(\mathcal{B})$. Then,*

$$\mathsf{R}_{\mathcal{P}}[\varphi, \boldsymbol{A}] - \mathsf{R}_{\mathcal{P}}[0, \boldsymbol{A}] = \mathsf{L}_{\mathcal{P}}[\boldsymbol{A} \,|\, \varphi],$$

*where $\mathsf{L}_{\mathcal{P}}[\cdot|\cdot] : C^{\mathrm{Lin}^+}(\mathcal{P}) \times C(\mathcal{P}) \to \mathbb{R}$ is continuous and F-differentiable, and $\mathsf{L}_{\mathcal{P}}[\boldsymbol{A}|\cdot]$ is linear for every $\boldsymbol{A} \in C^{\mathrm{Lin}^+}(\mathcal{B})$.*

*Proof* In part (1) of Lemma A.2, set

$$\varphi^{(1)} = \varphi, \qquad \varphi^{(2)} = 0.$$

Then, because of Remark A.1 following Lemma A.2, we know that there exists a $\bar{\varphi} \in C^+(\mathcal{B}) \cap C^1(\mathcal{B})$ such that

$$\mathsf{R}_{\mathcal{P}}[\varphi, \boldsymbol{A}] - \mathsf{R}_{\mathcal{P}}[0, \boldsymbol{A}] = \delta_1 \mathsf{R}_{\mathcal{P}}[\bar{\varphi}, \boldsymbol{A} \,|\, \varphi].$$

This, together with the hypothesis (A.5), allows the conclusion

$$\mathsf{R}_{\mathcal{P}}[\varphi, \boldsymbol{A}] - \mathsf{R}_{\mathcal{P}}[0, \boldsymbol{A}] = \delta_1 \mathsf{R}_{\mathcal{P}}[0, \boldsymbol{A} \,|\, \varphi],$$

and by defining

$$\mathsf{L}_{\mathcal{P}}[\boldsymbol{A} \,|\, \cdot] := \delta_1 \mathsf{R}_{\mathcal{P}}[0, \boldsymbol{A} \,|\, \cdot]$$

for every $\boldsymbol{A} \in C^{\mathrm{Lin}^+}(\mathcal{B})$, and recalling the property (1) of F-differentiability in Definition 4.1, we complete this proof. ■

**Lemma A.4** *Let $\varphi \in C^{+}(\mathcal{B})$ and let $\chi$ and $\boldsymbol{v}$ be smooth fields over $\mathcal{B}$ with $\nabla\chi \in \mathrm{Lin}^{+}$. Suppose $\mathsf{L}_{\mathcal{P}}[\cdot|\psi] : C^{\mathrm{Lin}^{+}}(\mathcal{P}) \to \mathbb{R}$ is a continuous and F-differentiable functional for $\mathcal{B}$ according to Definition* 4.1, *so that for every $\boldsymbol{A} \in C^{\mathrm{Lin}^{+}}(\mathcal{B})$ there exists $\delta\mathsf{L}_{\mathcal{P}}[\cdot|\cdot|\cdot]$, linear in its second and third places, such that*

$$\mathsf{L}_{\mathcal{P}}[\boldsymbol{A}+\boldsymbol{H}|\cdot] = \mathsf{L}_{\mathcal{P}}[\boldsymbol{A}|\cdot] + \delta\mathsf{L}_{\mathcal{P}}\big[\boldsymbol{A}|\cdot|\boldsymbol{H}\big] + o\big(\|\boldsymbol{H}\|\big) \tag{A.6}$$

*for all $\boldsymbol{H} \in C^{\mathrm{Lin}}(\mathcal{B})$ with $\boldsymbol{A}+\boldsymbol{H} \in C^{\mathrm{Lin}^{+}}(\mathcal{B})$. Finally, suppose*

$$\delta\mathsf{L}_{\mathcal{P}}[\nabla\chi|\varphi|\nabla\boldsymbol{v}] = 0 \tag{A.7}$$

*for all such $\chi$ and $\boldsymbol{v}$. Then, there exists a linear functional $\mathsf{K}_{\mathcal{P}}[\cdot] : C(\mathcal{B}) \longrightarrow \mathbb{R}$ such that*

$$\mathsf{L}_{\mathcal{P}}[\nabla\chi|\varphi] = \mathsf{K}_{\mathcal{P}}[\varphi].$$

*Proof* In part (2) of Lemma A.2, let us first identify $\mathsf{R}_{\mathcal{P}}[\varphi, \boldsymbol{A}]$ with $\mathsf{L}_{\mathcal{P}}[\boldsymbol{A}|\varphi]$ and set

$$\boldsymbol{A}^{(1)} = \nabla\chi + \nabla\boldsymbol{v}, \qquad \boldsymbol{A}^{(2)} = \nabla\chi\,.$$

Now, choose $\chi(\boldsymbol{X}) = \boldsymbol{X}$ and $\boldsymbol{v}(\boldsymbol{X}) = (\lambda-1)\boldsymbol{X}$, with $\lambda > 0$. In this case, we have $\boldsymbol{A}^{(1)} = \lambda\boldsymbol{1}$ and $\boldsymbol{A}^{(2)} = \boldsymbol{1}$, which implies that $\boldsymbol{A}_t = (1+t(\lambda-1))\boldsymbol{1}$. Thus, $\boldsymbol{A}_t \in \mathrm{Lin}^{+}$ is the gradient of a vector-valued function for all $t \in [0, 1]$, and we may use (A.4) (see also Remark A.1) to conclude that there exists $\bar{\boldsymbol{A}} \in C^{\mathrm{Lin}^{+}}(\mathcal{B})$ which is a gradient and such that

$$\mathsf{L}_{\mathcal{P}}[\lambda\boldsymbol{1}|\varphi] - \mathsf{L}_{\mathcal{P}}[\boldsymbol{1}|\varphi] = \delta\mathsf{L}_{\mathcal{P}}\big[\bar{\boldsymbol{A}}|\varphi|(\lambda-1)\boldsymbol{1}\big].$$

But, because of (A.7), the right hand side here vanishes, which implies

$$\mathsf{L}_{\mathcal{P}}[\lambda\boldsymbol{1}|\varphi] = \mathsf{L}_{\mathcal{P}}[\boldsymbol{1}|\varphi].$$

As a second choice, suppose $\chi$ is left unspecified with $\nabla\chi \in \mathrm{Lin}^{+}$, and $\boldsymbol{v}(\boldsymbol{X}) = \lambda\boldsymbol{X} - \chi(\boldsymbol{X})$, with $\lambda > 0$. In this case, we have $\boldsymbol{A}^{(1)} = \lambda\boldsymbol{1}$ and $\boldsymbol{A}^{(2)} = \nabla\chi$, which implies that $\boldsymbol{A}_t = (1-t)\nabla\chi + t\lambda\boldsymbol{1}$. Thus, for sufficiently small $\lambda > 0$, we see that $\boldsymbol{A}_t \in \mathrm{Lin}^{+}$, and is the gradient of a vector-valued function for all $t \in [0, 1]$. Again, we may use (A.4) (see also Remark A.1) to conclude that there exists $\bar{\boldsymbol{A}} \in C^{\mathrm{Lin}^{+}}(\mathcal{B})$, the gradient of a vector-valued function, such that

$$\mathsf{L}_{\mathcal{P}}[\lambda\boldsymbol{1}|\varphi] - \mathsf{L}_{\mathcal{P}}[\nabla\chi|\varphi] = \delta\mathsf{L}_{\mathcal{P}}\big[\bar{\boldsymbol{A}}|\varphi|\lambda\boldsymbol{1} - \nabla\chi\big].$$

But, again, the right hand side here vanishes because of (A.7), and we see that

$$\mathsf{L}_{\mathcal{P}}[\lambda\boldsymbol{1}|\varphi] = \mathsf{L}_{\mathcal{P}}[\nabla\chi|\varphi].$$

From the above two choices, we take $\lambda > 0$, sufficiently small, and conclude that

$$\mathsf{L}_{\mathcal{P}}[\nabla\chi|\varphi] = \mathsf{L}_{\mathcal{P}}[\boldsymbol{1}|\varphi] =: \mathsf{K}_{\mathcal{P}}[\varphi],$$

to complete this proof. ■

## References


1. Bartle, R.G.: The Elements of Integration Theory. Wiley, New York (1966)
2. Bunge, M.: Causality, The Place of the Causal Principle in Modern Science. Harvard University Press, Cambridge (1959)
3. Favata, A., Podio-Guidugli, P., Tomassetti, G.: Energy splitting theorems for materials with memory. J. Elast. **101**, 59–67 (2010)
4. Fosdick, R.: A causality approach to particle dynamics for systems. Arch. Ration. Mech. Anal. **207**(1), 247–302 (2013). https://doi.org/10.1007/s00205-012-0567-7
5. Fosdick, R., Serrin, J.: The splitting of intrinsic energy and the origin of mass density in continuum mechanics. Contin. Mech. Thermodyn. **26**, 287–302 (2014). https://doi.org/10.1007/s00161-013-0301-1
6. Gibbs, J.W.: On the equilibrium of heterogeneous substances. In: Trans. Conn. Acad. 1875–1878. The Scientific Papers, Vol. I, pp. 55–353. Yale University Press, New Haven (1907)
7. Green, A.E., Rivlin, R.S.: On Cauchy's equations of motion. Z. Angew. Math. Phys. **15**, 290–292 (1964)
8. Green, A.E., Rivlin, R.S.: Multipolar continuum mechanics. Arch. Ration. Mech. Anal. **17**, 113–147 (1964)
9. Gurtin, M.E.: Thermodynamics and the possibility of spatial interaction in rigid heat conductors. Arch. Ration. Mech. Anal. **18**, 335–342 (1965)
10. Gurtin, M.E.: An Introduction to Continuum Mechanics. Academic Press, San Diego (1981)
11. Gurtin, M.E., Williams, W.O.: An axiomatic foundation for continuum thermodynamics. Arch. Ration. Mech. Anal. **26**, 83–117 (1967)
12. Jammer, M.: Concepts of Mass in Classical and Modern Physics. Harvard University Press, Cambridge (1961)
13. Noll, W.: On the continuity of the solid and fluid states. J. Ration. Mech. Anal. **4**, 3–81 (1955)
14. Noll, W.: La mécanique classique, basée sur un axiome d'objectivité. La méthode axiomatique dans les mécaniques classiques et nouvelles, pp. 47–63, Paris (1963)
15. Podio-Guidugli, P.: Inertia and invariance. Ann. Mat. Pura Appl. (4) **CLXXII**, 103–124 (1997)
16. Serrin, J.: The equations of continuum mechanics as a consequence of group invariance. In: Ferrarese, G., Editrice, P. (eds.) Advances in Modern Continuum Mechanics, Bologna, pp. 217–225 (1992)
17. Šilhavý, M.: Mass, internal energy, and Cauchy's equations in frame-indifferent thermodynamics. Arch. Ration. Mech. Anal. **107**, 1–22 (1989)

J Elast (2019) 135:295–350
https://doi.org/10.1007/s10659-018-09722-8

# The Symmetries of Octupolar Tensors

**Giuseppe Gaeta**[1] · **Epifanio G. Virga**[2]

Received: 10 July 2018 / Published online: 14 January 2019


**Abstract** Octupolar tensors are third order, completely symmetric and traceless tensors. Whereas in 2D an octupolar tensor has the same symmetries as an equilateral triangle and can ultimately be identified with a vector in the plane, the symmetries that it enjoys in 3D are quite different, and only exceptionally reduce to those of a regular tetrahedron. By use of the octupolar potential, that is, the cubic form associated on the unit sphere with an octupolar tensor, we shall classify all inequivalent octupolar symmetries. This is a mathematical study which also reviews and incorporates some previous, less systematic attempts.

**Keywords** Order tensors · Phase transitions · Octupolar tensors · Generalized (nonlinear) eigenvalues and eigenvectors

**Mathematics Subject Classification** 76A15 · 15A69

## 1 Introduction

It is well known that the Landau theory of phase transitions [21–23, 31, 43] describes the states of matter in the vicinity of a critical point in terms of an *order parameter*; in the simplest cases this is a scalar quantity, but it can be a vector, or more generally a tensor of any order.

In fact, in the case of liquid crystals it is rather common to describe their state in terms of a second order tensor [15, 16, 47]. More recently, it became apparent that certain materials displaying *tetrahedral* nematic phases [11, 12] are better described in terms of a *third*

✉ E.G. Virga
eg.virga@unipv.it

G. Gaeta
giuseppe.gaeta@unimi.it

1 Dipartimento di Matematica, Università degli Studi di Milano, via Saldini 50, 20133 Milano, Italy

2 Dipartimento di Matematica, Università di Pavia, via Ferrata 5, 27100 Pavia, Italy

order tensor **A**—more precisely, a fully symmetric and completely traceless one (see below for definitions). We stress that it is conceivable, even probable, that order parameters described by still higher order tensors will be needed in considering generalized liquid crystals [25–27].

This is precisely what is meant in this paper by an *octupolar tensor*: a third order, fully symmetric and completely traceless tensor. Although, as also shown below, octupolar tensors feature in many branches of physics, we shall systematically use the paradigm of the Landau theory of phase transitions to illustrate the physical significance of our study. The reader is advised from the start that this is but one of many incarnations of our mathematical theory, which is concerned with the symmetries of the most general octupolar tensor in three space dimensions.

One of us provided a complete description of the physics of a material represented by an octupolar tensor in the *two dimensional* case [48]. In this case, one obtains a remarkably simple description, and the physical state is basically identified by the orientation of an equilateral triangle in the order parameter space.

Unfortunately, such a simple description is peculiar to the 2D case, and as soon as we pass to consider a *three-dimensional* situation things become much more involved. In a recent contribution [14], we have studied this situation, providing a representation of the physics described by an octupolar tensor in three dimensions; this study also showed an unexpected feature awaiting experimental confirmation, i.e., the existence—together with higher symmetric special phases—of *two* different generic phases; the interface between these has been investigated in detail in a subsequent work [6].

In our efforts to classify all symmetries of an octupolar tensor **A** a prominent role is played by the appropriate notion of eigenvalues and eigenvectors applicable for **A**. In multilinear algebra, such a notion is not as univocally defined as one might naively think. For real-valued tensors that bear a physical meaning, as the ones we are interested in, the issue arises as to whether complex eigenvalues, which would perfectly be allowed according to certain definitions, should be admitted or not. The definition of eigenvalues (and associated eigenvectors) that we adopt in this paper is essentially the one put forward in [32–34] (see also the recent book [35], especially Chap. 7, which is specifically concerned with octupolar tensors and their mechanical applications). However, we give an equivalent characterization of this notion in terms of the critical points of a cubic polynomial defined over the unit sphere $S^2$. This is indeed the natural extension of what one learns from the lucid (and now rare) book of Noll [30]. In Sect. 84, the maximum and minimum of the spectrum of a symmetric, second order tensor **S** in a $n$-dimensional space are characterized as the corresponding extrema of the quadratic form associated with **S** on the unit sphere $S^{n-1}$. Noll's book was published in 1987, but its contents were available many years back.[1]

In our previous work [14] we focused on the physics of the problem, and in particular of its generic phases—discovering an unexpected phenomenon, i.e., the existence of two generic (octupolar) phases, hence the possibility of an intra-octupolar phase transition (see also [6] for details on this)—providing little mathematical detail; the purpose of the present paper is twofold:

[1] In the Introduction to [30] (p. IX), we read:

> About 25 years ago I started to write notes for a course for seniors and beginning graduate students at Carnegie Institute of Technology (renamed Carnegie-Mellon University in 1968). At first, the course was entitled "Tensor Analysis". [...] The notes were rewritten several times. They were widely distributed and they served as the basis for appendices to the books [9] and [44].

(a) On the one hand, we want to give a full account of the mathematical details needed for the study of such a problem. We trust that—beside the interest *per se*—this will also be relevant to systems described by higher order tensors [25–27].
(b) On the other hand, also thanks to this higher mathematical detail provided here, we want to discuss in more detail the non-generic phases (and the transitions between these), thus completing the physical description provided in previous work [6, 14].

The plan of the paper is as follows. In Sect. 2 we present the physical motivation of our work; in Sect. 3 we discuss the general features of a prototypical Landau potential, which will be referred to as the *octupolar potential*, for short. As this potential is based on octupolar tensors in three spatial dimensions, the subsequent Sect. 4 is devoted to study these objects and their eigenvectors. We can then pass, in Sect. 5, to study the general octupolar potential; this depends a priori on seven parameters, but by a suitable choice of reference frame and of the potential scale—as discussed in Sect. 5—we can reduce to study a problem depending only on three parameters; the allowed parameters are described by a cylinder in parameter space. It turns out that this potential and its critical points bear some relation to the tetrahedral group; Sect. 6 is thus devoted to recalling some basic facts about this. We can then pass to study the extremals of the Landau potential; these depend of course on the parameters and different regions in the allowed cylinder in parameter space correspond actually to different symmetries (phases) for the potential and its critical set, as discussed in detail in Sect. 7 and its subsections, each devoted to one of these phases. Finally in Sect. 8 we review and summarize our findings, with special emphasis on the distribution in parameter space of the critical points of the octupolar potential. In the final Sect. 9 we draw our conclusions. The paper is completed by an appendix with details on the tetrahedral group beyond the brief discussion of Sect. 6.

Whenever indices are needed, summation over repeated pairs will be routinely understood, unless explicitly stated otherwise.

## 2 Physical Motivation

Octupolar tensors have many manifestations in physics. In this section, we shall review a few of them, ranging from the most classical to the most innovative ones.

Among the former is Buckingham's formula [4] for the probability density $\varrho$ of the distribution of a molecular director $\boldsymbol{\ell}$ over the unit sphere $S^2$, which can be written as

$$\varrho(\boldsymbol{\ell}) = \frac{1}{4\pi}\left(1 + \sum_{k=1}^{\infty} \frac{(2k+1)!!}{k!} \langle \overline{\boldsymbol{\ell}^{\otimes k}} \rangle_{\varrho} \cdot \boldsymbol{\ell}^{\otimes k}\right), \tag{2.1}$$

where, for a generic vector $\mathbf{x}$, $\mathbf{x}^{\otimes k}$ is the $k$th order tensor defined as in [45] by

$$\mathbf{x}^{\otimes k} = \mathbf{x} \otimes \cdots \otimes \mathbf{x} \tag{2.2}$$

(with $k$ factors, of course), $\cdot$ denotes tensor contraction, and $\langle \overline{\mathbf{P}} \rangle_{\varrho}$ is the multipole average,

$$\langle \overline{\mathbf{P}} \rangle_{\varrho} := \frac{1}{4\pi} \int_{S^2} \langle \overline{\mathbf{P}} \rangle \varrho(\boldsymbol{\ell}) \mathrm{d}\boldsymbol{\ell}, \tag{2.3}$$

of the symmetric, traceless part $\overline{\mathbf{P}}$ of a tensor $\mathbf{P}$.

With this notation, our octupolar order tensor $\mathbf{A}$ is identified with

$$\mathbf{A} = \langle \overline{\boldsymbol{\ell} \otimes \boldsymbol{\ell} \otimes \boldsymbol{\ell}} \rangle_{\varrho}. \tag{2.4}$$

Passing to spherical coordinates, and writing $\boldsymbol{\ell}$ as

$$\boldsymbol{\ell} = \sin\vartheta \cos\varphi \mathbf{e}_1 + \sin\vartheta \sin\varphi \mathbf{e}_2 + \cos\vartheta \mathbf{e}_3, \tag{2.5}$$

we easily express the scalar parameters that represent $\mathbf{A}$ in the Cartesian frame $(\mathbf{e}_1, \mathbf{e}_2, \mathbf{e}_3)$ in terms of multipole averages, which also reveal the bounds they are subject to [14].

In nonlinear optics (see, for example Sect. 1.5 of [3], and also [54] and [18]), higher order susceptibilities tensors, often called *hypersusceptibilities* (and also *hyperpolarizabilities*), are introduced to decompose the electromagnetic energy density in multipoles. In particular, the cubic term has the general form

$$U^{(2)} = \chi^{(2)}_{ijk} F_i F_j F_k, \tag{2.6}$$

where $F_i$ are the components of an external field and $\chi^{(2)}_{ijk}$ are the components of the first hypersusceptibility tensor $\boldsymbol{\chi}^{(2)}$, which is a third order tensor.[2]

When the frequencies of the applied fields are much smaller than the resonance frequency, $\boldsymbol{\chi}^{(2)}$ can safely be assumed to be independent of frequencies and fully symmetric in all its indices. Though this symmetry, which is often called Kleinman's symmetry after the name of the author who first introduced it [20], has been widely criticized [10] and also found in disagreement with some computational schemes [53], it is still often accepted as an approximation for its simplicity. Assuming Kleinman's symmetry to be valid, we can extract an octupolar tensor $\mathbf{A} = \overline{\boldsymbol{\chi}^{(2)}}$ out of $\boldsymbol{\chi}^{(2)}$ and rewrite the latter in the equivalent form

$$\chi^{(2)}_{ijk} = A_{ijk} + \frac{1}{5}\big(\chi^{(2)}_i \delta_{jk} + \chi^{(2)}_j \delta_{ki} + \chi^{(2)}_k \delta_{ij}\big), \tag{2.7}$$

where $\chi^{(2)}_k := \chi^{(2)}_{iik} = \chi^{(2)}_{iki} = \chi^{(2)}_{kii}$. Using (2.7) in (2.6), we also write

$$U^{(2)} = A_{ijk} F_i F_j F_k + \frac{3}{5} F^2 \chi^{(2)}_i F_i, \tag{2.8}$$

where $F$ is the strength of the applied field. Normalizing $F$ to unity, $U^{(2)}$ becomes the sum of an octupolar and a dipolar potential, the latter of which contributes to a lower multipole.

A third order tensor very similar to $\boldsymbol{\chi}^{(2)}$ was introduced in [28] to describe the ordering of bent-core molecules that possess liquid crystal phases. As also recalled in [6], the theory of bent-core liquid crystal phases features a mesoscopic third order tensor derived from $\overline{\boldsymbol{\alpha}^{(3)}}$; here $\boldsymbol{\alpha}^{(3)}$ is the molecular structural tensor defined by

$$\boldsymbol{\alpha}^{(3)} := \sum_{\mu=1}^{N} m_\mu \mathbf{r}_\mu \otimes \mathbf{r}_\mu \otimes \mathbf{r}_\mu, \tag{2.9}$$

[2] The superscript $^{(2)}$ reminds us that this tensor expresses the field induced by polarization as a *quadratic* function of the external field, whereas the ordinary susceptibility establishes a linear relationship between the two fields.

where the sum is extended to all the atoms in a constituent molecule, $m_\mu$ is the mass of each individual atom and $\mathbf{r}_\mu$ is its position vector relative to the molecule's center of mass.

The *ensemble* average $\mathbf{A} = \langle\, \overline{\boldsymbol{\alpha}^{(3)}}\, \rangle$ is an octupolar tensor that plays an important role in classifying all possible new phases that bent-core liquid crystals are allowed to exhibit. They have been collectedly called *tetrahedral* by a symmetry that $\mathbf{A}$ can indeed enjoy, but our study has shown to be only too special, rather than generic.

Lately, octupolar tensors have also become popular with the classical elastic theory of nematic liquid crystals (both passive and active). The nematic director field $\mathbf{n}$ represents at the macroscopic scale the average orientation of the elongated molecules that constitute the medium. Elastic distortions of $\mathbf{n}$ are measured locally by its spatial gradient $\nabla\mathbf{n}$, which may become singular at certain points in space in response to external distorting stimuli. These are the *defects* of $\mathbf{n}$, where molecular order is degraded; they can be classified into distinct topological classes, associated in 2D with the *winding* number $m$ of $\mathbf{n}$ around a point defect ($m$, which is often referred to as the *topological charge*, is half an integer; more information about defects in liquid crystals and their topological charges can be obtained from the monographs [40] and [47]).

It has recently been proposed [49] that a vector be associated with a point defect in 2D to represent the average direction along which the field $\mathbf{n}$ is fluted away from the defect. It is believed that such a vector could play a role in describing the interaction of defects, as if they were *oriented* particles interacting like *vessels* in a viscous *sea*. If this image is indeed suggestive for the charge $m = \frac{1}{2}$, when the integral lines of $\mathbf{n}$ around the defect resemble a flame, associating a single direction with a defect with charge $m = -\frac{1}{2}$ seems somehow troublesome at first sight, as in that case the integral lines of $\mathbf{n}$ escape along three directions, ideally separating the plane in three equal sectors. To overcome this difficulty, it has been proposed in [42] to describe the orientation of a $-\frac{1}{2}$ defect through the octupolar tensor

$$\mathbf{A} = \langle\, \overline{\nabla\mathbf{n} \otimes \mathbf{n}}\, \rangle, \tag{2.10}$$

where the average $\langle\cdots\rangle$ is now meant to be computed on a sufficiently small neighborhood surrounding the defect. As shown in [48], any octupolar tensor $\mathbf{A}$ in 2D possesses the symmetries of an equilateral triangle, and its most general representation is given by

$$\mathbf{A} = \alpha\, \overline{\mathbf{e} \otimes \mathbf{e} \otimes \mathbf{e}}\,, \tag{2.11}$$

where $\alpha$ is a scalar and $\mathbf{e}$ is a unit vector in the plane. Thus, in 2D, $\mathbf{A}$ effectively reduces to a vector and both approaches in [49] and [42] are equivalent. However, as shown in [14], in 3D $\mathbf{A}$ cannot in general be represented in terms of a single vector as in (2.11) and (2.10) becomes a more versatile tool to describe the directions along which the integral lines of $\mathbf{n}$ are fluted around a point defect. In particular, it might be interesting to compute $\mathbf{A}$ in (2.10) for the *combed* defects described in [38] and their distortions possibly due to the interaction with other point defects nearby (see also [39]). The complete classification of the symmetries enjoyed by $\mathbf{A}$, which we give in this paper, may supplement the topological classification of defects by providing extra synthetic information on the qualitative features of the director field surrounding the defect.

## 3 Octupolar Potential

We will work in three-dimensional space, with standard coordinates $(x_1, x_2, x_3)$ for $\mathbf{x}$ in a Cartesian frame $(\mathbf{e}_1, \mathbf{e}_2, \mathbf{e}_3)$. In the theory we are interested in, the octupolar potential $\Phi(\mathbf{x})$

is described by a three-dimensional third order tensor[3] **A** via

$$\Phi = A_{ijk} x_i x_j x_k; \tag{3.1}$$

the physical states will be described by minima of this function. In view of the homogeneity of $\Phi$, it is not restrictive to look for extrema of $\Phi$ constrained to the unit sphere $S^2 \subset \mathbf{R}^3$.

The third order tensor **A** has some additional properties:

1. **A** is completely symmetric; in terms of the components of **A**, this means $A_{ijk} = A_{\pi(ijk)}$, with $\pi$ any permutation;
2. **A** is completely traceless, i.e., $A_{iik} = A_{iki} = A_{kii} = 0$ for any $k$.

Since $\Phi$ is homogeneous of odd degree, we always have

$$\Phi(-\mathbf{x}) = -\Phi(\mathbf{x}). \tag{3.2}$$

This also implies that if $\mathbf{x}_0$ is a minimum of $\Phi$, then $-\mathbf{x}_0$ is a maximum, and conversely; on the other hand, if $\mathbf{x}_0$ is saddle point with $p$ unstable directions, then $-\mathbf{x}_0$ is again a saddle point, albeit with $p$ *stable* directions (and hence $\widetilde{p} = 3 - p$ unstable ones).

This means that we can equivalently describe the system in terms of *maxima* of $\Phi$ (this amounts to changing a global sign); this is more convenient in graphical terms, and we will thus adhere to this convention.

## 3.1 Extrema, Eigenvectors, Ray Solutions

We thus have to maximize (or minimize) $\Phi(\mathbf{x})$ given by (3.1) subject to the constraint $|\mathbf{x}| = 1$. This can be obtained in two ways:

(a) by augmenting $\Phi$ to a function

$$\Phi_\lambda(\mathbf{x}) := \Phi(\mathbf{x}) - \frac{1}{2}\lambda\big(|\mathbf{x}|^2 - 1\big), \tag{3.3}$$

where $\lambda$ is a Lagrange multiplier;

(b) or passing to spherical coordinates $(r, \varphi, \vartheta)$ and setting $r = 1$, thus obtaining a reduced potential[4] $\Psi(\varphi, \vartheta) : S^2 \to \mathbf{R}$.

We will mainly use the latter approach, but where convenient we also employ the former.[5]

The condition of constrained extremum results in requiring that $(\nabla\Phi)(\mathbf{x})$ is collinear to $\mathbf{x}$, i.e.,

$$(\nabla\Phi)(\mathbf{x}) = k\lambda\mathbf{x}. \tag{3.4}$$

[3] More generally, we might consider potentials with contributions *up to* third order; thus we would have the sum of a scalar part, a vector one, another part described by a second order tensor, and finally the one described by the third order one. Here we focus on this last contribution, as the study of theories with scalar, vector, or second order tensor order parameters is standard (in principle; obviously concrete applications can present endless complications).

[4] As a general convention, we will denote the potentials in Cartesian coordinates by $\Phi$ (with several suffixes) and those in spherical coordinates—which we always consider only for $r = 1$—by $\Psi$ (again with corresponding suffixes).

[5] It may be worth mentioning that (in particular, if we are satisfied with studying $\Phi$ on one hemisphere, which is justified by (3.2)) a third option is present, i.e., setting $z = \pm\sqrt{1 - x^2 - y^2}$ and considering $\Phi$ as a function of $x$ and $y$; these take value in the unit disk. This will be used in Sect. 5.3.

We will look at this problem in terms of *eigenvectors* (and eigenvalues) of a higher order tensor [5, 29, 32–34], see Sect. 3.3 below.

It should be noted that the same problem can be seen in a slightly different way. That is, we can consider the associated dynamical system

$$\frac{d\mathbf{x}}{dt} = (\nabla\Phi)(\mathbf{x}), \tag{3.5}$$

and search for *ray solutions*, i.e., solutions of the form

$$\mathbf{x}(t) = \alpha(t)\mathbf{x}(0). \tag{3.6}$$

This problem has been considered in the literature, and a number of results (in particular, concerning the number of such solutions) are available [36, 37, 50]. These coincide with the results obtained in terms of eigenvectors of higher order tensors [5, 29, 32–34].

In the following Sects. 3.2 and 3.3 we recall both kind of results. We stress that we just want to report results present in the literature, but these should not be seen in terms of priority.[6]

Unfortunately, as we will see, both approaches provide a complete answer in terms of *complex* numbers, while we need results in the field of *real* numbers. The theorems to be reported below do not specify how many of these ray solutions or equivalently eigenvectors, and the associated eigenvalues, will be real (more or less in the same way as the fundamental theorem of algebra does not say how many of the roots of a polynomial are real).

## 3.2 Ray Solutions of Homogeneous Dynamical Systems

We start by reporting the results obtained by Röhrl [37] for ray solutions of dynamical systems in $\mathbf{C}^N$.

**Proposition 1** *Consider the dynamical system*

$$\dot{x}^i = B^i(x), \quad i = 1, \ldots, q, \tag{3.7}$$

*with $B^i$ homogeneous polynomials of degree $p$ in $x$. If the coefficients of the polynomials $B^i$ are algebraically independent over the field of the rationals, then* (3.7) *fails to have a critical point in the origin and has precisely*

$$N_R = p^q - 1 \tag{3.8}$$

*ray solutions.*

In the case of interest here we deal with three-dimensional systems, thus $q = 3$, and homogeneous polynomials of degree two, as these result from considering the gradient of $\Phi(\mathbf{x})$, thus $p = 2$. This yields $2^3 - 1 = 7$ ray solutions. Each of these intersects the unit sphere in two (antipodal) points, hence we get 14 critical points for our constrained variational problem.

It should also be stressed that, strictly speaking, Röhrl's theorem holds under the condition of algebraic independence of the coefficients $B^i_{jk}$ in $B^i = B^i_{jk}x^j x^k$; this is a generic

[6] In fact, as pointed out by Walcher [51], this kind of results follow ultimately from the work of Bezout on intersection theory dating back to the 18th century. See his paper [52] for details.

property for general polynomials, but fails for symmetric ones. However one should note that only the symmetrized sums $B^i_{jk} + B^i_{kj}$ will play a role, hence one should understand that the condition of algebraic independence refers to these;[7] moreover, one should note that algebraic independence is only a sufficient—but not necessary—condition in Röhrl's theorem.

The reader is referred to the original contribution by Röhrl [37], or to [36, 50], for more detail.

*Remark 1* Note also that the Equivariant Branching Lemma [7, 8, 13, 46] gives information about the existence and stability of solutions along such critical rays when we consider theories depending on external parameters.

### 3.3 Eigenvalues and Eigenvectors for Higher Order Tensors

Determining eigenvalues and eigenvectors for a second order, symmetric tensor $\mathbf{S}$ in an $n$-dimensional vector space $V$ is a very classical problem in linear algebra and amounts to solving the linear problem

$$\mathbf{S}\mathbf{v} = \lambda \mathbf{v}; \tag{3.9}$$

one can without loss of generality restrict to unit vectors,[8] i.e., complement the problem with the side condition

$$(\mathbf{v}, \mathbf{v}) = |\mathbf{v}|^2 = 1. \tag{3.10}$$

The same problem is obtained if one looks for minimizers of a function $\Phi : \mathbf{R}^n \to \mathbf{R}$ which is a quadratic form defined by $\mathbf{S}$, i.e., $\Phi = S_{ij} x_i x_j$, with $S_{ij} = S_{ji}$, restricted to the unit sphere $S^{n-1}$. In fact, in this case one introduces the Lagrange multiplier $\lambda$ and considers the extended (constrained) potential

$$\Phi_\lambda = \Phi - \frac{1}{2} \lambda x^i x^i, \tag{3.11}$$

obtaining the same equation (3.9) as the critical point equation for the potential $\Phi_\lambda : \mathbf{R}^n \times \mathbf{R} \to \mathbf{R}$. It is well known (see, for example, Sect. 84 of [30]) that all eigenvalues of $\mathbf{S}$ are *real* and bear a physical meaning.

It seems in many ways natural to consider the same problem in *multilinear algebra*, i.e., for higher order tensors. Quite surprisingly, not only very little is known in this respect, but moreover the available results are rather recent [5, 29, 32–34]—albeit, as mentioned above, older results dealing with algebraic analysis of differential equations and which can be interpreted in this direction are available in the literature [36, 37, 50].

In this respect, it turns out that the problem we are interested in (that is, third order symmetric traceless tensors in three dimensions) is precisely the simplest nontrivial and non-degenerate class of tensors (the same problem in dimension two turns out to display a rather special and in many ways degenerate behavior [48]), so we believe our results are also of general interest, as they show the kind of—rather counterintuitive—behavior one can meet in studying the eigenvalues problem in multilinear algebra.

---

[7] It should be noted that the "disappearance" of real critical points—w.r.t. the generic situation described by Röhrl's theorem—is related, at least in our model, to the appearance of a "monkey saddle" [14], i.e., of a critical point with a non-generic index; see below for detail.

[8] In fact, if $\mathbf{v}$ is an eigenvector of $M$ with eigenvalue $\lambda$, then for any number $\alpha \neq 0$ also $\mathbf{w} = \alpha \mathbf{v}$ is an eigenvector with the same eigenvalue $\lambda$.

As well known, a tensor $T$ of order $k$ on $V$ is a $k$-linear map,

$$T : V \times \cdots \times V = V^k \to \mathbf{R}; \tag{3.12}$$

it is also well known that the algebra of completely symmetric tensors on $\mathbf{R}^n$ is isomorphic to the algebra of polynomials in $\mathbf{R}^n$.

By duality (3.12) also defines a $(k-1)$-linear map, which we denote by $\widehat{T}$ (by a standard abuse of notation, in the following we will also denote this by $T$),

$$\widehat{T} : V^{k-1} \to V; \tag{3.13}$$

for second order tensors this is just the standard description of a matrix as a linear operator in $V$. For third order tensors, this associates to $T$ a quadratic map $\widehat{T} : V \times V \to V$.

We say that $\mathbf{v} \in V$ is an *eigenvector* of $T$ (with *eigenvalue* $\lambda \in \mathbf{C}$) if

$$\widehat{T}(\mathbf{v}, \ldots, \mathbf{v}) = \lambda \mathbf{v}. \tag{3.14}$$

For second order tensors, this coincides with the standard definition of eigenvectors and eigenvalues.

*Remark 2* It should be stressed that for tensors of order $k \neq 2$, eigenvectors come in linear spaces, but these do not share a common eigenvalue—that is, eigenvectors are well defined but eigenvalues are not. In fact, if we consider multiples of the eigenvector $\mathbf{v}$, i.e., $\mathbf{w} = \alpha \mathbf{v}$ (with $\alpha \neq 0$), we have

$$\begin{aligned} \widehat{T}(\mathbf{w}, \ldots \mathbf{w}) &= \widehat{T}(\alpha \mathbf{v}, \ldots, \alpha \mathbf{v}) = \alpha^{k-1} \widehat{T}(\mathbf{v}, \ldots, \mathbf{v}) = \alpha^{k-1} \lambda \mathbf{v} \\ &= \alpha^{k-2} \lambda (\alpha \mathbf{v}) = \alpha^{k-2} \lambda \mathbf{w}. \end{aligned} \tag{3.15}$$

The situation is of course different if we require the eigenvectors to be of unit length, as this requirement also uniquely determines the eigenvalue (up to a sign in case $k$ is odd).

We (obviously) reach the same equation (3.14) if we consider the minimization of the homogeneous function $\Phi_T$ of degree $k$ associated to the tensor $T$, i.e., of

$$\Phi_T := T_{i_1,\ldots,i_k} x^{i_1} \cdots x^{i_k}, \tag{3.16}$$

constrained to the unit sphere $|\mathbf{x}| = 1$. In fact, in this case one introduces the Lagrange multiplier $\lambda$ and the constraint term

$$-\frac{k}{2} \lambda x^i x^i,$$

and minimization of

$$\Psi_T = \Phi_T - \frac{k}{2} \lambda x^i x^i \tag{3.17}$$

leads precisely to (3.14). It should be noted that, again, in this context one is only interested in *real* solutions.

As well known, when one is properly taking into account multiplicities, a square matrix in $\mathbf{R}^n$ always admits $n$ algebraic eigenvalues. This is generalized by Proposition 2 below [5, 34] for arbitrary tensors of rank $m$ over $\mathbf{C}^n$. Note that for an eigenpair $(\lambda, \mathbf{v})$ we have an (equivalent) eigenpair $(t^{m-2}\lambda, t\mathbf{v})$ for any $t$ with $|t| = 1$ (if we are in $\mathbf{R}^n$, only $t = \pm 1$ are admissible); thus we should speak of equivalence classes of eigenpairs. The following result is given in Cartwright & Sturmfels [5]; see also Qi [29, 32–34].

**Proposition 2** *If a tensor $A$ of rank $m \geq 3$ over $\mathbf{C}^n$ admits a finite number of equivalence classes of eigenpairs, their number counted with multiplicity is*

$$E(m,n) = \frac{(m-1)^n - 1}{m-2} = \sum_{j=0}^{n-1} (m-1)^j. \tag{3.18}$$

For the case of interest here, i.e., $m = 3$, formula (3.18) provides

$$E(3,n) = 2^n - 1, \tag{3.19}$$

in agreement with Röhrl formula (3.8); in particular, we get

$$E(3,2) = 3, \qquad E(3,3) = 7. \tag{3.20}$$

Note also that if we work in $\mathbf{R}^n$, the number of eigenpairs is obtained simply by multiplying the number of equivalence classes by two; on the other hand, some of the eigenpairs could be complex rather than real, so Proposition 2 only provides an upper bound on the number of *real* critical points. Thus the maximal number of critical points for the constrained potential associated to a tensor of order three in two dimensions is six, i.e., three pairs (as in [48]), in dimension three this number is fourteen (as in [14]), and in dimension four it is thirty.

More detail about eigenvectors of higher order tensors is given in a paper by Walcher [52] (whom we thank for discussing a preliminary version of this with us).

## 4 Octupolar Tensors in Dimension 3 and Their Eigenvalues

In the following we will be especially interested in *completely symmetric* tensors. In this case, it is convenient to extract the trace terms. More precisely, one considers tensors all of whose partial traces are zero.

We thus want to consider a fully symmetric and completely traceless tensor $\mathbf{A}$ with components $A_{ijk}$ (all partial traces being zero). Taking into account the full symmetry and the condition of zero partial traces, the only independent components are

$$\begin{aligned} &A_{123} = \alpha_0; \\ &A_{111} = \alpha_1, \qquad A_{222} = \alpha_2, \qquad A_{333} = \alpha_3; \\ &A_{122} = \beta_1, \qquad A_{233} = \beta_2, \qquad A_{311} = \beta_3. \end{aligned} \tag{4.1}$$

The partial trace condition gives

$$A_{133} = -(\alpha_1 + \beta_1), \qquad A_{211} = -(\alpha_2 + \beta_2), \qquad A_{322} = -(\alpha_3 + \beta_3). \tag{4.2}$$

The other components are immediately recovered by the tensor symmetry. The equations $A(\mathbf{x}, \mathbf{x}) = \lambda \mathbf{x}$ for eigenvectors and eigenvalues are then written as

$$\begin{aligned} &2\alpha_0 x_2 x_3 + \alpha_1(x_1^2 - x_3^2) - 2\alpha_2 x_1 x_2 + \beta_1(x_2^2 - x_3^2) - 2\beta_2 x_1 x_2 + 2\beta_3 x_1 x_3 = \lambda x_1; \\ &2\alpha_0 x_3 x_1 + \alpha_2(x_2^2 - x_1^2) - 2\alpha_3 x_2 x_3 + \beta_2(x_3^2 - x_1^2) - 2\beta_3 x_2 x_3 + 2\beta_1 x_1 x_2 = \lambda x_2; \\ &2\alpha_0 x_1 x_2 + \alpha_3(x_3^2 - x_2^2) - 2\alpha_1 x_3 x_1 + \beta_3(x_1^2 - x_2^2) - 2\beta_1 x_3 x_1 + 2\beta_2 x_3 x_2 = \lambda x_3. \end{aligned} \tag{4.3}$$

*Remark 3* A general completely symmetric tensor would depend on *ten* parameters. Correspondingly, the potential we will consider in Sect. 5 will depend on *seven* parameters (these will then be reduced, see Sect. 5.2), while a generic potential homogeneous of order three in three spatial dimension would depend on ten parameters.

*Remark 4* By introducing the vectors

$$\mathbf{a} = \begin{pmatrix} \alpha_1 \\ \alpha_2 \\ \alpha_3 \end{pmatrix}, \qquad \mathbf{b} = \begin{pmatrix} \beta_1 \\ \beta_2 \\ \beta_3 \end{pmatrix}, \qquad \mathbf{u} = \begin{pmatrix} x_1 \\ x_2 \\ x_3 \end{pmatrix}, \qquad \mathbf{v} \begin{pmatrix} x_2 x_3 \\ x_1 x_3 \\ x_1 x_2 \end{pmatrix} \tag{4.4}$$

and the matrices

$$\begin{aligned} L &= \begin{pmatrix} (x_3^2 - x_1^2) & 2x_1x_2 & 0 \\ 0 & (x_1^2 - x_2^2) & 2x_2x_3 \\ 2x_1x_3 & 0 & (x_2^2 - x_3^2) \end{pmatrix}, \\ M &= \begin{pmatrix} (x_2^2 - x_3^2) & -2x_1x_2 & 2x_1x_3 \\ 2x_1x_2 & (x_3^2 - x_1^2) & -2x_2x_3 \\ -2x_1x_3 & 2x_2x_3 & (x_1^2 - x_2^2) \end{pmatrix}, \end{aligned} \tag{4.5}$$

Eqs. (4.3) are compactly rewritten as

$$M\mathbf{b} = L\mathbf{a} + \lambda\mathbf{u} - 2\alpha_0\mathbf{v}. \tag{4.6}$$

Provided $\det(L) \neq 0$ and/or $\det(M) \neq 0$ we can solve for **a** and/or for **b**; e.g., assuming $\det(M) \neq 0$ we have

$$\mathbf{b} = M^{-1}[L\mathbf{a} + \lambda\mathbf{u} - 2\alpha_0\mathbf{v}]. \tag{4.7}$$

It may be noted that there are special points on the unit sphere where *both* $L$ and $M$ have zero determinant; these are the two poles $(0, 0, \pm 1)$ together with $(\pm 1, 0, 0)$ and $(0, \pm 1, 0)$; and the four points $(0, \pm 1/\sqrt{2}, \pm 1/\sqrt{2})$ together with $(\pm 1/\sqrt{2}, 0, \pm 1/\sqrt{2})$ and $(\pm 1/\sqrt{2}, \pm 1/\sqrt{2}, 0)$.

This approach provides an expression for the value of the parameters (belonging to one subset) which make a certain point $\mathbf{x} \in S^2$ a critical one for given values of the other parameters; unfortunately, we are interested in the inverse—and natural—problem, i.e., determining the critical points for a given (full) set of parameters. These relations can however be used to check the correctness of computations and results.

# 5 Critical Points

## 5.1 General Potential and the Critical Point Equations

It follows from our discussion on the tensor **A** that the general (unconstrained) potential (3.1) is thus written, using coordinates $(x, y, z)$ rather than $(x_1, x_2, x_3)$ as we are from now on working in three dimensions, as

$$\begin{aligned} \Phi = {} & 6\alpha_0 xyz + \alpha_1 x(x^2 - 3z^2) + \alpha_2 y(y^2 - 3x^2) + \alpha_3 z(z^2 - 3y^2) \\ & + 3[\beta_1 x(y^2 - z^2) + \beta_2 y(z^2 - x^2) + \beta_3 z(x^2 - y^2)]. \end{aligned} \tag{5.1}$$

*Remark 5* This is obviously covariant under inversion (this implies that the potential can be constant only if it is identically zero), i.e.,

$$\Phi(-x,-y,-z) = -\Phi(x,y,z). \tag{5.2}$$

Actually this is just the consequence of having a homogeneous (of odd degree) potential. This potential is also covariant under inversion of the parameters (which is a consequence of its being homogeneous of degree one in these); including the dependence on parameters in our notation we have

$$\Phi(-\mathbf{x};\mathbf{p}) = \Phi(\mathbf{x};-\mathbf{p}) = -\Phi(\mathbf{x};\mathbf{p}), \tag{5.3}$$

where $\mathbf{x} = (x,y,z)$ and $\mathbf{p} = (\alpha_0;\alpha_1,\alpha_2,\alpha_3;\beta_1,\beta_2,\beta_3)$ is the vector of parameters.

*Remark 6* It follows from (5.2) that the average of $\Phi$ over the unit sphere is zero. This also implies that (unless the potential is identically zero) the absolute maximum—or maxima in case of degenerate ones—is necessarily positive.

*Remark 7* The potential $\Phi$ is invariant under simultaneous identical permutations in the triples $(x,y,z)$, $(\alpha_1,\alpha_2,\alpha_3)$ and $(\beta_1,\beta_2,\beta_3)$. This invariance does *not* extend to simultaneous general rotations in the same three-dimensional spaces.

We are actually interested in critical points for the potential constrained to the unit sphere $S^2$ (this constraint does not interfere with the symmetries mentioned in Remark 5 above). The constrained potential

$$\Phi_\lambda := \Phi + \Phi_c \tag{5.4}$$

is obtained by adding to $\Phi$ the constraint term

$$\Phi_c := -\frac{3}{2}\lambda\left(x^2+y^2+z^2-1\right); \tag{5.5}$$

this breaks the covariance under spatial inversion, i.e., in general

$$\Phi_\lambda(-x,-y,-z;\lambda) \neq -\Phi_\lambda(x,y,z;\lambda). \tag{5.6}$$

On the other hand, it is clear that if we also take into account the possibility of reversing $\lambda$, we get

$$\Phi_\lambda(-x,-y,-z;-\lambda) = -\Phi_\lambda(x,y,z;\lambda). \tag{5.7}$$

Since we shall only consider critical points of the potential, any constant term can be omitted, and the constraint term can be written simply as

$$\Phi_c = -\frac{3}{2}\lambda\left(x^2+y^2+z^2\right). \tag{5.8}$$

It is then immediate to obtain the equation for critical points of the constrained potential $\Phi_\lambda$, which are

$$\begin{aligned}
2\alpha_0 yz + \alpha_1\left(x^2-z^2\right) - 2\alpha_2 xy + \beta_1\left(y^2-z^2\right) - 2\beta_2 xy + 2\beta_3 xz &= \lambda x,\\
2\alpha_0 zx + \alpha_2\left(y^2-x^2\right) - 2\alpha_3 yz + \beta_2\left(z^2-x^2\right) - 2\beta_3 yz + 2\beta_1 yx &= \lambda y,\\
2\alpha_0 xy + \alpha_3\left(z^2-y^2\right) - 2\alpha_1 zx + \beta_3\left(x^2-y^2\right) - 2\beta_1 zx + 2\beta_2 zy &= \lambda z.
\end{aligned} \tag{5.9}$$

*Remark 8* These equations are invariant—as it should be in view of (5.7)—under the transformation

$$(x, y, z; \lambda) \to (-x, -y, -z; -\lambda). \tag{5.10}$$

If we also consider inversion in the parameters, denoting the critical point equations (5.9) as $E(\mathbf{x}, \lambda, \mathbf{p}) = 0$ (with the same notation used in Remark 5 above), we have

$$\begin{aligned} E(-\mathbf{x}; -\lambda; \mathbf{p}) &= E(\mathbf{x}; \lambda; \mathbf{p}), \\ E(\mathbf{x}; -\lambda; -\mathbf{p}) &= E(\mathbf{x}; \lambda; \mathbf{p}), \\ E(-\mathbf{x}; -\lambda; -\mathbf{p}) &= -E(\mathbf{x}; \lambda; \mathbf{p}). \end{aligned} \tag{5.11}$$

(Obviously the map $E \to -E$ leaves the equations $E = 0$ invariant as well.) Moreover they are still covariant under the three simultaneous permutations mentioned in Remark 7 above.

The property (5.10) together with (5.2) guarantee that to each critical point $\mathbf{x}$ (with $k$ stable directions) is associated another critical point $-\mathbf{x}$ (with $k$ unstable directions, and hence $\widetilde{k} = n - k$ stable ones, with $n$ the space dimensionality; in our case $n = 3$). This also means that it would suffice to consider one of the hemispheres.

*Remark 9* As remarked above, the constrained potential reads

$$\Phi_\lambda = \Phi - \frac{3}{2}\lambda|\mathbf{x}|^2; \tag{5.12}$$

the equations for critical points are

$$\nabla\Phi_\lambda := \nabla\Phi - 3\lambda\mathbf{x} = 0. \tag{5.13}$$

If we perform scalar product of (both members of) this equality with $\mathbf{x}$, we get

$$(\mathbf{x} \cdot \nabla\Phi) = 3\lambda(\mathbf{x} \cdot \mathbf{x}). \tag{5.14}$$

Recalling now that $\Phi$ is homogeneous of degree three and that critical points of $\Phi_\lambda$ are located, by construction, on the unit sphere $|\mathbf{x}|^2 = 1$, we obtain in the end that at critical points we always have

$$\lambda = \Phi. \tag{5.15}$$

That is, there is a simple relation—actually, an identity—between eigenvalues and the value of the potential at the critical point identified by the corresponding eigenvector.

## 5.2 Oriented Potential

The potential $\Phi$ in (5.1) depends on seven parameters; however, it is possible to simplify it by choosing an adapted reference frame in $\mathbf{R}^3$, as we are now going to discuss.

If the potential $\Phi$ is not constant on the sphere (the potential is constant on $S^2$ only if all parameters are zero; we are obviously not interested in this case), it will have at least a critical point and actually at least a maximum—and by symmetry, a minimum.

We will then choose the $z$ axis so that one of the critical points is in $(0, 0, 1)$; by trivial computations, this requires to have

$$\beta_1 = -\alpha_1, \quad \beta_2 = 0. \tag{5.16}$$

*Remark 10* We can actually choose $(0, 0, 1)$ to be a maximum; this will set further constraints, as discussed in a moment (in Sect. 5.3).

Moreover, we can still choose an orientation in the $(x, y)$ plane; in fact (5.2) implies that there is at least a point on the circle $S^1 \subset S^2$ identified by $z = 0$ in which $\Phi$ vanishes. Thus we choose an orientation in the $(x, y)$ plane by requiring that $\Phi(1, 0, 0) = 0$. Again by trivial computations this implies, recalling also the expression for $\beta_1$ in (5.16),

$$\alpha_1 = 0, \qquad \beta_1 = 0. \tag{5.17}$$

In this way we obtain an "oriented" potential which depends only on the four parameters $\alpha_0, \alpha_2, \alpha_3, \beta_3$; this reads

$$\Phi_{or} = 6\alpha_0 xyz + \alpha_2(y^2 - 3x^2)y + \alpha_3(z^2 - 3y^2)z + 3\beta_3(x^2 - y^2)z. \tag{5.18}$$

From now on we will only deal with this oriented potential.

*Remark 11* By orienting the potential we have disposed of the invariance under simultaneous permutations discussed in Remark 7 (see also below). On the other hand, the inversion symmetries discussed in Remark 8 are obviously inherited by the oriented potential.

*Remark 12* When we have several critical points, each of them can be chosen as the main orienting one; this trivial remark may be used to simplify the discussion avoiding redundance in some classification. In particular, in [14] (see in particular Sect. 5) an alternative parametrization based on this remark is used and allows one to reduce the parameter space from the cylinder to a part of it.

As we are working in the unit sphere, it is often convenient to use angular coordinates (with $r = 1$); we will set

$$\begin{aligned} x &= \cos\vartheta_1 \cos\vartheta_2, \\ y &= \cos\vartheta_1 \sin\vartheta_2, \\ z &= \sin\vartheta_1; \end{aligned} \tag{5.19}$$

with this choice of angular coordinates,

$$\vartheta_1 \in [-\pi/2, \pi/2], \qquad \vartheta_2 \in [-\pi, \pi], \tag{5.20}$$

and the volume element becomes

$$\mathrm{d}x \wedge \mathrm{d}y \wedge \mathrm{d}z = r^2 \cos\vartheta_1 \mathrm{d}r \wedge \mathrm{d}\vartheta_2 \wedge \mathrm{d}\vartheta_1. \tag{5.21}$$

*Remark 13* It should be noted that this potential has generically (i.e., for generic[9] values of the parameters appearing in it) no invariance under subgroups of $O(3)$; it retains the covariance under inversion, which is described by

$$\vartheta_1 \to -\vartheta_1, \qquad \vartheta_2 \to -\vartheta_2$$

[9] In this paper, the adjective "generic" is given the meaning common in algebraic geometry, that is, it designates a property valid away from the roots of a polynomial in parameter space [5].

in polar coordinates. Special invariance properties are possible, and will be studied, for special values of the parameters

The critical point equations (5.9) (CPE) read now

$$\begin{aligned} 2(\alpha_0 yz - \alpha_2 xy + \beta_3 xz) &= \lambda x, \\ 2\alpha_0 xz - \alpha_2(x^2 - y^2) - 2(\alpha_3 + \beta_3) yz &= \lambda y, \\ 2\alpha_0 xy - \alpha_3(y^2 - z^2) + \beta_3(x^2 - y^2) &= \lambda z. \end{aligned} \tag{5.22}$$

In $(0, 0, 1)$ the first two are identically satisfied, while the third one yields

$$\lambda_N = \alpha_3, \tag{5.23}$$

where $\lambda_N$ is the eigenvalue corresponding to the eigenvector $(0, 0, 1)$ identifying the North Pole.[10] Thus normalizing the potential by setting

$$\alpha_3 = 1, \tag{5.24}$$

as we do in the following, is equivalent to setting $\lambda_N = 1$, i.e., to normalizing the "orienting" eigenvalue. This will also correspond to normalizing the potential in the maximum located at the North Pole. The oriented potential is then written as

$$\Psi_{or} = \sin^3\vartheta_1 - \alpha_2 \sin 3\vartheta_2 \cos^3\vartheta_1 + \frac{3}{2}\sin\vartheta_1\big[(2\beta_3 + 1)\cos 2\vartheta_2 + 2\alpha_0 \sin 2\vartheta_2 - 1\big]\cos^2\vartheta_1. \tag{5.25}$$

Needless to say, the CPE are also obtained by considering the gradient of $\Psi_{or}$ w.r.t. the angular coordinates:[11]

$$\begin{aligned} &\frac{3}{2}\cos\vartheta_1\big[\big(-1 + (1 + 2\beta_3)\cos 2\vartheta_2 + 2\alpha_0 \sin 2\vartheta_2\big)\cos^2\vartheta_1 \\ &\qquad - 2\sin^2\vartheta_1\big(-2 + (1 + 2\beta_3)\cos 2\vartheta_2 + 2\alpha_0 \sin 2\vartheta_2\big) + \alpha_2 \sin 2\vartheta_1 \sin 3\vartheta_2\big] = 0, \\ &\frac{3}{2}\cos^2\vartheta_1\big[\sin\vartheta_1\big(4\alpha_0\cos 2\vartheta_2 - 2(1 + 2\beta_3)\sin 2\vartheta_2\big) - 2\alpha_2\cos\vartheta_1\cos 3\vartheta_2\big] = 0. \end{aligned} \tag{5.26}$$

We obtain, in the end, a potential which depends on *three* parameters; this is still a substantial problem, but much simpler than the initial one, depending on *seven* parameters.

## 5.3 The Maximum Condition

In the previous Sect. 5.2 we have implemented the requirement to have a critical point in the North Pole $(0, 0, 1)$, setting some conditions on the potential parameters. Our discussion,

---

[10] This means that we can rule out the possibility to have $\alpha_3 = 0$. In fact, even in the case this is a local maximum at height zero, we can always—see Remark 6—choose the North Pole to be an absolute maximum, and this is necessarily positive.

[11] In order to know the value for the corresponding $\lambda$, one needs to express the solution in Cartesian coordinates and go back to (5.22); this is due to the fact that our change of coordinates was performed imposing $r = 1$ and thus the constraint term, which represents the dynamical origin of $\lambda$, is absent in the angular coordinates.

however, did not enter into the nature of this critical point. This is what we will presently do.

To distinguish between maxima, minima and saddles, we just have to compute the Hessian of the constrained potential at its critical points. We need not consider the full potential (5.1), but we can work directly on the oriented potential (5.18); moreover, we are mainly interested in the critical point at the North Pole.

If we work in the northern hemisphere, the constraint condition can be implemented simply by setting $z=\sqrt{1-x^2-y^2}$. In this way the oriented potential is rewritten as

$$\Phi_N=\left[6\alpha_0 xy+3\beta_3(x^2-y^2)+(1-x^2-4y^2)\right]\sqrt{1-x^2-y^2}+\alpha_2(y^2-3x^2)y. \quad (5.27)$$

The Hessian at the North Pole is simply computed as[12]

$$H_N=\begin{pmatrix}\partial^2\Phi_N/\partial x^2 & \partial^2\Phi_N/\partial x\partial y\\ \partial^2\Phi_N/\partial x\partial y & \partial^2\Phi_N/\partial y^2\end{pmatrix}_{(0,0)}; \quad (5.28)$$

with trivial computations we get

$$H_N=3\begin{pmatrix}2\beta_3-1 & 2\alpha_0\\ 2\alpha_0 & -(3+2\beta_3)\end{pmatrix}. \quad (5.29)$$

With the reparametrization[13]

$$\alpha_0=\frac{\rho}{2}\cos\chi,\qquad \beta_3=-\frac{1}{2}+\frac{\rho}{2}\sin\chi,\qquad \alpha_2=K, \quad (5.30)$$

where

$$\rho\geq 0,\qquad \chi\in[-\pi,\pi],\qquad K\in\mathbf{R}, \quad (5.31)$$

we have

$$H_N=3\begin{pmatrix}\rho\sin\chi-2 & \rho\cos\chi\\ \rho\cos\chi & -\rho\sin\chi-2\end{pmatrix}, \quad (5.32)$$

and its eigenvalues are

$$\sigma_\pm=-3(2\pm\rho); \quad (5.33)$$

thus we have a maximum in the North Pole if and only if

$$0\leq\rho\leq 2. \quad (5.34)$$

With (5.30), the oriented (and reparametrized) potential in (5.25) now reads as

$$\Psi_{or}=-K\sin 3\vartheta_2\cos^3\vartheta_1+\frac{3}{2}\sin\vartheta_1\cos^2\vartheta_1\left[\rho\sin(\chi+2\vartheta_2)-1\right]+\sin^3\vartheta_1, \quad (5.35)$$

[12]To compare the expressions worked out in this paper for the Hessian matrix of the octupolar potential with those featuring in [14], the reader should heed that these differ by a scaling factor: the Hessian matrix here is three times the Hessian matrix there.

[13]It should be noted that in our previous work [14] we have used a slightly different reparametrization, with $\rho$ instead of $\rho/2$. This accounts for the differences in many of the forthcoming formulas.

**Table 1** Combinations of extrema and saddle points allowed by the Hopf-Poincaré index constraints for scalar potentials on a sphere $S^2 \subset \mathbf{R}^3$, allowing only non-degenerate critical points. The last column indicates if this is realized in our model; note that at bifurcations degenerate critical points can be present

| | $M$ | $S$ | $2N$ | |
|---|---|---|---|---|
| $(a)$ | 1 | 0 | 2 | no |
| $(b)$ | 2 | 1 | 6 | no |
| $(c)$ | 3 | 2 | 10 | yes |
| $(d)$ | 4 | 3 | 14 | yes |

and its analogue in Cartesian coordinates as

$$\Phi_{or} = z^3 - 3y^2 z + 3\rho\cos\chi\, xyz + \frac{3}{2}(\rho\sin\chi - 1)\big(x^2 - y^2\big)z + Ky\big(y^2 - 3x^2\big). \tag{5.36}$$

The oriented octupolar potential, in either of its representations (5.35) or (5.36), will be the founding stone of our future development.

*Remark 14* It is immediately apparent from (5.35) that for $K = 0$ the oriented potential is invariant under rotation by an angle $\pi$ in $\vartheta_2$, i.e., for $\vartheta_2 \to \vartheta_2 \pm \pi$; similarly, for $\rho = 0$ the oriented potential is invariant under rotation by an angle $2\pi/3$ in $\vartheta_2$, i.e., for $\vartheta_2 \to \vartheta_2 \pm 2\pi/3$. More generally, the potential is always invariant under a rotation by an angle $(2/3)n\pi$ in $\vartheta_2$ accompanied by a rotation by an angle $(2m + 4n/3)\pi$ in the parameter $\chi$.

## 5.4 Critical Points and Index

The general results mentioned above (Propositions 1 and 2) show that generically we have 14 critical points (that is, 7 pairs of parity-conjugated ones); some of these could be complex, thus not acceptable in the present context.

As we are mainly interested in maxima, it would be convenient to have further information about the nature of these critical points. Such information can be obtained through the use of Poincaré-Hopf index. To discuss this, we pass to consider the vector field

$$\mathbf{v} = \frac{\nabla\Phi}{|\nabla\Phi|}. \tag{5.37}$$

Obviously critical points of $\Phi$ correspond to singularities for $\mathbf{v}$, and to these singular points we apply Poincaré-Hopf theory [41, pp. 239–247]. In fact, each (isolated) non-degenerate singular point $p_k$ has an index $\iota_k$, which takes the value $\iota_k = +1$ if $p_k$ is a (local) maximum or minimum, and $\iota_k = -1$ if $p_k$ is a saddle. The sum of the indices for all critical points must equal the Euler characteristic of the two-sphere, i.e.,

$$\sum_k \iota_k = 2. \tag{5.38}$$

In other words, if all critical points are non-degenerate and $2N$ is their number, $2M$ is the number of maxima and minima (together) and $2S$ is the number of saddles, we must have $M + S = N$ and $M - S = 1$.

The different combinations allowed by these constraints are reported in Table 1.

**Table 2** Combinations of extrema and saddle points allowed by the Hopf-Poincaré index constraints for scalar potentials on a sphere $S^2 \subset \mathbf{R}^3$, allowing regular critical points as well as finitely degenerate saddles of index $\iota = -2, -3$. The last column indicates if this is realized in our model; cases $(c_2)$ and $(d_3)$ will only be realized at bifurcation points. We shall indeed encounter instances where the octupolar potential possesses 12 critical points, but these do not fall under case $(d_2)$ (see Remark 16)

| | $M$ | $S_1$ | $S_2$ | $S_3$ | $2N$ | |
|---|---|---|---|---|---|---|
| $(a)$ | 1 | 0 | 0 | 0 | 2 | no |
| $(b)$ | 2 | 1 | 0 | 0 | 6 | no |
| $(c_1)$ | 3 | 2 | 0 | 0 | 10 | yes |
| $(c_2)$ | 3 | 0 | 1 | 0 | 8 | yes |
| $(d_1)$ | 4 | 3 | 0 | 0 | 14 | yes |
| $(d_2)$ | 4 | 1 | 1 | 0 | 12 | no |
| $(d_3)$ | 4 | 0 | 0 | 1 | 10 | yes |

Thus, if all eigenvalues are real (hence $2N = 14$) and all critical points are isolated and non-degenerate (which is not always the case, as we will see in our discussion), we will have 4 maxima, 4 minima, and 6 saddles; parity conjugation relates maxima to minima, and saddles to saddles.

In the following, we will meet this "expected" situation; but we will also meet the case where there are only ten real eigenvalues (thus $2N = 10$); in this case we will have 3 maxima, 3 minima and 4 saddles. Both these situations will be shown to be generic, but cases $(a)$ and $(b)$ will never be met.

*Remark 15* It should be mentioned that the transition between the generic cases $(d)$ and $(c)$ is made possible by the occurrence of degenerate critical points (see also below); more precisely, of "monkey saddles", see [14]; these have index $\iota = -2$, to be compared with the index $\iota = -1$ for ordinary saddles. Thus it appears possible that the non-appearance of the "more degenerate" situations $(a)$ and $(b)$ is related to the fact that, due to the small degree of the potential and to its symmetries (which, e.g., forbid $\Phi(x, y, z) = z^3$, leading to case $(a)$ above), there is no possibility for the appearance of "more degenerate" critical points, with greater (in absolute value) index, which would be needed for the associate bifurcation to take place.

In the above discussion, we have supposed that only non-degenerate critical points are present. The index approach cannot give results in the case of infinitely-degenerate critical points (e.g., if we have a line of critical points, as we will find in Sect. 7.3), but critical points with a finite degeneration can also be present and can be taken into account. In our present context, this concerns in particular the "monkey saddles" we will meet in Sect. 7.4, which have index $\iota = -2$. In principle, degenerate saddles with index $\iota = -3$ could also be relevant to our classification. If we denote the number of ordinary saddles by $2S_1$, that of monkey saddles by $2S_2$, and the number of saddles with index $-3$ by $2S_3$, while still denoting by $2N = 2(M + S_1 + S_2 + S_3)$ the number of critical points, the different possible situations—all having total index $\iota = 2$—are summarized in Table 2. It is perhaps worth noting that if we classify the different cases just by the number of maxima, this gives the same classification as before, albeit the same number of maxima can correspond to different numbers of saddles and hence of critical points.

*Remark 16* Both Tables 1 and 2 describe possible scenarios with increasing, but moderate degrees of complexity. We have abstained from considering a type of critical points that may be added freely, as they do not affect the global constraint (5.38). These are critical points with index $\iota = 0$. The vector field $\mathbf{v}$ in (5.37) is still singular at these points, but it can be continuously altered all around each of them so as to be made locally equivalent to a uniform field. All such singularities of $\mathbf{v}$ then turn out to be *removable*. As will be shown below (see Sect. 7.6), the octupolar potential can indeed exhibit critical points of this type, albeit in rather special circumstances, which are not included in either of the above tables.

### 5.5 Symmetry of the Reduced Potential

As already mentioned, orienting the potential destroys the invariance under rotations (in the $\mathbf{x}$ space and in the parameter space). However, the high degree of symmetry of the tensor $\mathbf{A}$ makes that even the oriented potential has some remnant of this in the form of discrete symmetries. These are better described by passing to the representation of $\Phi_{or}$ in polar coordinates and with the reparametrization (5.30), i.e., in terms of $\Psi_{or}$, see (5.35).

It is easily checked that this is invariant under several discrete maps; as expected these do not involve the $\vartheta_1$ angle, which is fixed by the requirement to have the North Pole as a maximum. They do act on the other coordinate $\vartheta_2$ and on the parameters[14] $K, \chi$ (the action given below on the angles $\vartheta_2$ and $\chi$ are of course to be meant *modulo* $2\pi$):

$$\begin{aligned}
\gamma_1 &: (\vartheta_2; K, \rho, \chi) \mapsto (\vartheta_2 + \pi, -K, \rho, \chi), \\
\gamma_2 &: (\vartheta_2; K, \rho, \chi) \mapsto (-\vartheta_2, -K, \rho, -\chi - \pi), \\
\gamma_3 &: (\vartheta_2; K, \rho, \chi) \mapsto (\vartheta_2 + 2\pi/3, K, \rho, \chi + 2\pi/3).
\end{aligned} \tag{5.39}$$

The invariance under $\gamma_1$ means that we can limit our study to the region with $K \geq 0$, which we will do. When taking into account the invariance under the three maps, we can limit our study to the parameter region

$$K \geq 0, \qquad \chi_0 \leq \chi \leq \chi_0 + \frac{\pi}{3}, \tag{5.40}$$

for any given $\chi_0$. This can be handy in the study of the more complex situations.

## 6 The Tetrahedral Group

A special role in our discussion will be played by the tetrahedral group. Some notions about it—and explicit features of its fundamental representation—are collected in the Appendix. (The reader is referred to classical books [1, 17, 19, 24] for further detail.)

In this section we collect the main facts needed for our following discussion; in fact, the potential—and in particular its maxima—are organized according to various subgroups of the tetrahedral group, depending on the values assumed by the parameters.

We will denote by $(g\Phi)(\mathbf{x}) = \Phi(g\mathbf{x})$ the value of the potential computed at a transformed point; here $g$ is an element of the group, and this is identified with its representation (we always use the defining representation discussed in detail in the Appendix; this agrees with the orientation we have considered and singles out accordingly a reference tetrahedron).

[14] There are also maps acting on $\rho$ by changing its sign and leaving the potential invariant; these are not admitted as we have required $\rho \in [0, 2]$.

The full tetrahedron group consists of ternary rotations around the main tetrahedral axes (identified by vertices of the tetrahedron), of binary rotations around axes joining the middle points of opposite tetrahedron edges, and by reflections through the planes identified by two rotation axes.

Note also that we have considered an *oriented* potential. Thus we expect that in general, only the subgroup of $T_d$ leaving the North Pole fixed (we will refer to these as *oriented subgroups*) will be relevant. This is given by ternary rotations around the $z$ axis, and by reflections through the planes passing through the $z$ axis; in the notations of the Appendix, these are the elements $\{M_1, M_2, M_3, M_{13}, M_{14}, M_{15}\}$, where $M_1$ is the identity $I$.

The nontrivial subgroups formed by these are the rotation group $\{M_1, M_2, M_3\}$ (no reflections); and the reflection groups $\{M_1, M_{13}\}$, $\{M_1, M_{14}\}$, $\{M_1, M_{15}\}$.

Actually, there are special values of the parameters such that $g\Phi = \Phi$ for all elements of the tetrahedron group. These are $\rho = 0$, $K = \pm\sqrt{2}/2$.

It will turn out that for each of the oriented subgroups mentioned above there are regions in the parameter space where the potential admits these as symmetry groups; this will be discussed in Sect. 7 below.

## 7 Classification of Symmetries

As mentioned above, we can always rescale the potential and adopt (5.24), thus setting $\lambda = 1$ for the critical point at the North Pole. In such a way the potential scale is set by the "principal eigenvalue", the one which sets orientation, but is not necessarily the largest.

### 7.1 The Cylinder in Parameter Space

We have discussed above the limits to be set on the parameters in order to accommodate our condition to have a maximum at the North Pole. These limits mean we have to investigate a limited subset of the full parameter space. Having reduced the latter to $(\alpha_0, \alpha_2, \beta_3)$, by (5.30), (5.31), and (5.34), we readily see that we can effectively confine our study to an infinite cylinder $\mathsf{C}$ of unit radius and axis along the line $\alpha_0 = 0$, $\beta_3 = -1/2$. In the equivalent parameters $(K, \rho, \chi)$, which we shall hereafter turn to, $\mathsf{C}$ is identified by the combination of (5.31) and (5.34), and its axis is the line $\rho = 0$. The discussion of Sect. 5.5 shows that we do not actually have to study the full cylinder $\mathsf{C}$; in particular, it would suffice to study the half with $K \geq 0$, which we denote as $\mathsf{C}_+$.

In Table 3 we distinguish several subsets in the unit cylinder $\mathsf{C}$, i.e. its center $\mathcal{C}$ ($K = 0$, $\rho = 0$), its axis $\mathcal{A}$ ($\rho = 0$), the disk $\mathcal{D}$ at $K = 0$, and the special points $\mathcal{T}$ on the axis at height $K = 1/\sqrt{2}$, together with generic points in the bulk $\mathcal{B}$ of the cylinder. To each of these set

**Table 3** Different subsets in the cylinder $\mathsf{C}$

| | $G$ | parameters $(\alpha_0, \alpha_2, \beta_3)$ | parameters $(K, \rho)$ | subset in $\mathsf{C}$ |
|---|---|---|---|---|
| (0) | $\{e\}$ | | | bulk $\mathcal{B}$ |
| (1) | $D_{\infty h}$ | $\alpha_0 = \alpha_2 = 0$; $\beta_3 = -1/2$ | $K = \rho = 0$ | center $\mathcal{C}$ |
| (2) | $D_{2h}$ | $\alpha_2 = 0$ | $K = 0$ | disk $\mathcal{D}$ |
| (3) | $D_{3h}$ | $\alpha_0 = 0, \beta_3 = -1/2$ | $\rho = 0$ | axis $\mathcal{A}$ |
| (4) | $T_d$ | $\alpha_0 = 0, \beta_3 = -1/2$; $\alpha_2 = \pm 1/\sqrt{2}$ | $K = \pm 1/\sqrt{2}, \rho = 0$ | points $\mathcal{T} \in \mathcal{A}$ |

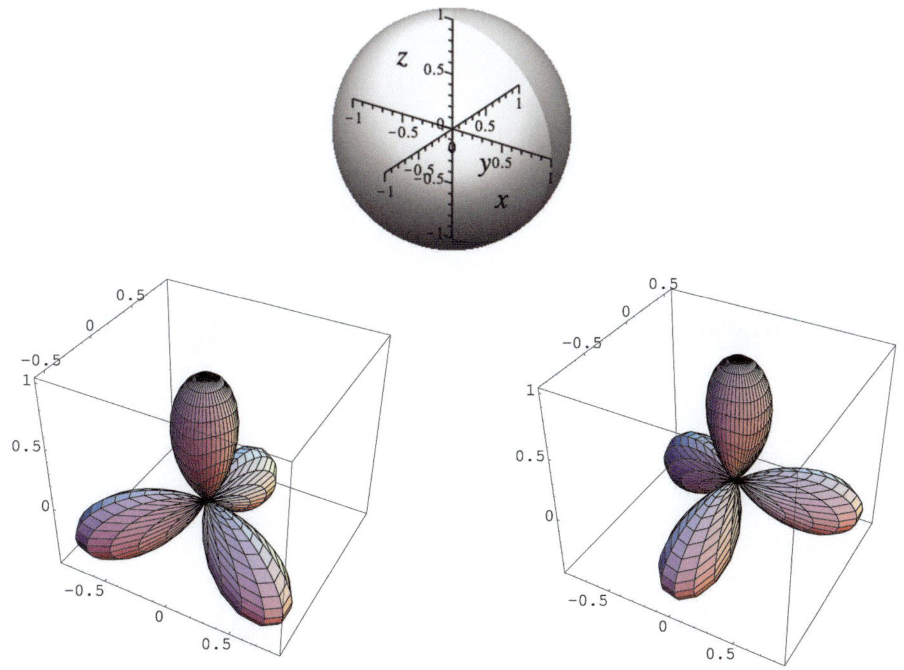


**Fig. 1** The potential $\Phi_T^{(\pm)}$ on the unit sphere. Left: $\Phi_T^{(+)}$ (i.e. $K = 1/\sqrt{2}$); right: $\Phi_T^{(-)}$ (i.e. $K = -1/\sqrt{2}$). The two cases are equivalent via a rotation by an angle $(2n+1)\pi/3$. The "spherical compass" on top of these panels shows the orientation of the Cartesian frame shared by the polar plots of the octupolar potential in all remaining panels below

will correspond a given invariance group for the potential. Further splitting of these regions is also possible, as shown below. In particular, we will find that there are special planes $\mathcal{P}$ in the bulk $\mathcal{B}$ with special symmetry properties, and that $\mathcal{B}$ has a richer structure than one could think of at first.

We will now consider in detail the different cases, starting from the more symmetric ones.

### 7.2 Tetrahedral Symmetry: Special Points $\mathcal{T}$

For the points $\mathcal{T}$, i.e., for $\rho = 0$ and $K = \pm 1/\sqrt{2}$, the potential in (5.36) reduces to

$$\Phi_T^{(\pm)} = z^3 - \frac{3}{2}(x^2 + y^2)z \pm \frac{1}{\sqrt{2}}(y^2 - 3x^2)y. \tag{7.1}$$

This is invariant under the full tetrahedron group (referred to the "standard" regular tetrahedron; by this we mean the one complying with our orientation choices; see the Appendix); note that for generic points on $\mathcal{A}$ (see the next subsection) we do not have tetrahedral invariance, but only that under the $D_{3h}$ group.

Now the maxima (and hence the minima) are all non-degenerate, take the same value, and their position are just at the vertices of the "standard" regular tetrahedron. This is depicted in Fig. 1. All polar plots such as those in Fig. 1 are three-dimensional renderings of the

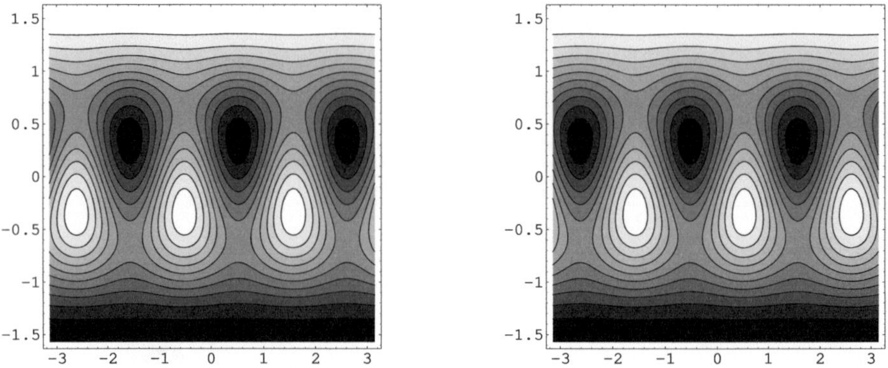


**Fig. 2** Contour plots of the potentials $\Psi_T^{(+)}$ (left) and $\Psi_T^{(-)}$ (right) on the plane $(\vartheta_2, \vartheta_1)$; see (7.2). In these plots, as in the others that follow, the brightest points represent the maxima of the potential, whereas the darkest points represent the minima. One easily checks that there are indeed 14 critical points (two of them at the poles). The symmetry properties of $\Psi_T^{(\pm)}$ are evident from this picture. The upper and lower border, at $\vartheta_1 = \pm\pi/2$, correspond to the North and South Poles, respectively. The two plots are coincident up to a shift in $\vartheta_2$, in accord with the predicted invariance under $\gamma_1$ in (5.39)

surface covered by the vector $\Phi(\mathbf{e}_r)\mathbf{e}_r$ as $\mathbf{e}_r = \mathbf{x}/|\mathbf{x}|$ spans the unit sphere $S^2$. Since $\Phi(\mathbf{e}_r) = -\Phi(-\mathbf{e}_r)$, in all these plots the minima of $\Phi$ are invaginated underneath its maxima.

In angular coordinates, from (5.35) we have

$$\Psi_T^{(\pm)} = \sin^3\vartheta_1 - \frac{3}{2}\sin\vartheta_1\cos^2\vartheta_1 \mp \frac{1}{\sqrt{2}}\sin 3\vartheta_1\cos^3\vartheta_1. \tag{7.2}$$

See Fig. 2 for a contour plot of this function. All contour plots such as those in Fig. 2 are on the plane $(\vartheta_2, \vartheta_1)$, which develops the unit sphere $S^2$ so that the upper side $\vartheta_1 = \pi/2$ corresponds to the North Pole and the lower side $\vartheta_1 = -\pi/2$ corresponds to the South Pole.

The potential is obviously invariant under shifts by $(2/3)\pi$ in $\vartheta_2$, and the parameter inversion $K \to -K$ is equivalent to an inversion or to a shift by $\pi/3$ in $\vartheta_2$; i.e.,

$$\Psi_T^{(\pm)}(\vartheta_1, \vartheta_2 + 2\pi/3) = \Psi_T^{(\pm)}(\vartheta_1, \vartheta_2), \tag{7.3}$$

$$\Psi_T^{(-)}(\vartheta_1, \vartheta_2) = \Psi_T^{(+)}(\vartheta_1, -\vartheta_2) = \Psi_T^{(-)}(\vartheta_1, \vartheta_2 + \pi/3). \tag{7.4}$$

These properties imply that it suffices to study one of the two cases, say $\Psi_T^{(+)}$; from now on we will just consider this, and refer to it simply as $\Psi_T$ (and correspondingly for $\Phi_T^{(+)}$ and $\Phi_T$).

Eigenvalues and the corresponding critical points (i.e., normalized eigenvectors) for $\Phi_T$ are listed in Table 4 (recall that $\lambda = \Phi$ at critical points).

### 7.3 Symmetry $D_{\infty h}$: Center $\mathcal{C}$

At the center $\mathcal{C}$ of the cylinder C, i.e. for $K = \rho = 0$, the potential in (5.36) is just

$$\Phi_\infty = z^3 - \frac{3}{2}(x^2 + y^2)z; \tag{7.5}$$

**Table 4** Critical points for tetrahedral symmetry; here $\nu_m = \arcsin(1/3)$, $\nu_s = \arcsin(1/\sqrt{3})$

| $n$ | $\lambda$ | $\vartheta_1$ | $\vartheta_2$ | $\Psi$ | type |
|---|---|---|---|---|---|
| 1 | $-1$ | $-\pi/2$ | – | $-1$ | min |
| 2 | $-1$ | $-\nu_m$ | $-\pi/2$ | $-1$ | min |
| 3 | $-1$ | $-\nu_m$ | $5\pi/6$ | $-1$ | min |
| 4 | $-1$ | $-\nu_m$ | $\pi/6$ | $-1$ | min |
| 5 | 0 | $-\nu_s$ | $-\pi/2$ | 0 | saddle |
| 6 | 0 | $-\nu_s$ | $5\pi/6$ | 0 | saddle |
| 7 | 0 | $-\nu_s$ | $\pi/6$ | 0 | saddle |
| 8 | 0 | $\nu_s$ | $\pi/2$ | 0 | saddle |
| 9 | 0 | $\nu_s$ | $-5\pi/6$ | 0 | saddle |
| 10 | 0 | $\nu_s$ | $-\pi/6$ | 0 | saddle |
| 11 | 1 | $\nu_m$ | $\pi/2$ | 1 | max |
| 12 | 1 | $\nu_m$ | $5\pi/6$ | 1 | max |
| 13 | 1 | $\nu_m$ | $-\pi/6$ | 1 | max |
| 14 | 1 | $\pi/2$ | – | 1 | max |

correspondingly, its variant in spherical coordinates (5.35) is

$$\Psi_\infty = \frac{1}{8}(3\sin\vartheta_1 - 5\sin 3\vartheta_1). \tag{7.6}$$

The symmetry under rotations (about the $z$ axis) is immediately apparent, as well as the symmetry under reflection in any vertical plane. We thus have a $D_{\infty h}$ symmetry. The potential is also covariant under inversion in $z$,

$$\Phi_\infty(x, y, -z) = -\Phi_\infty(x, y, z). \tag{7.7}$$

The critical point equations (5.9) are now

$$\begin{aligned} -3xz &= \lambda x, \\ -3yz &= \lambda y, \\ 3z^2 - \frac{3}{2}(x^2 + y^2) &= \lambda z. \end{aligned} \tag{7.8}$$

It follows immediately from the first two equations that the critical points not lying at the poles of the sphere have $\lambda = -3z$. More precisely, inserting this into the third equation—and recalling that $x^2 + y^2 = 1 - z^2$—it turns out that for these we have

$$z = \pm 1/\sqrt{5}, \qquad \lambda = \mp 3/\sqrt{5}. \tag{7.9}$$

*All* points on this circle are (obviously, degenerate) critical points; the only non-degenerate critical points are the maximum and the minimum in the North and South Poles.

We get of course the same result working in angular coordinates. From (7.6), we have that the critical points are located in the North and South Poles ($\vartheta_1 = \pm\pi/2$, value $\pm 1$); and on the parallels with $\vartheta_1 = \pm\arccos(2/\sqrt{5}) \doteq \pm 0.46$. In particular, we have a circle

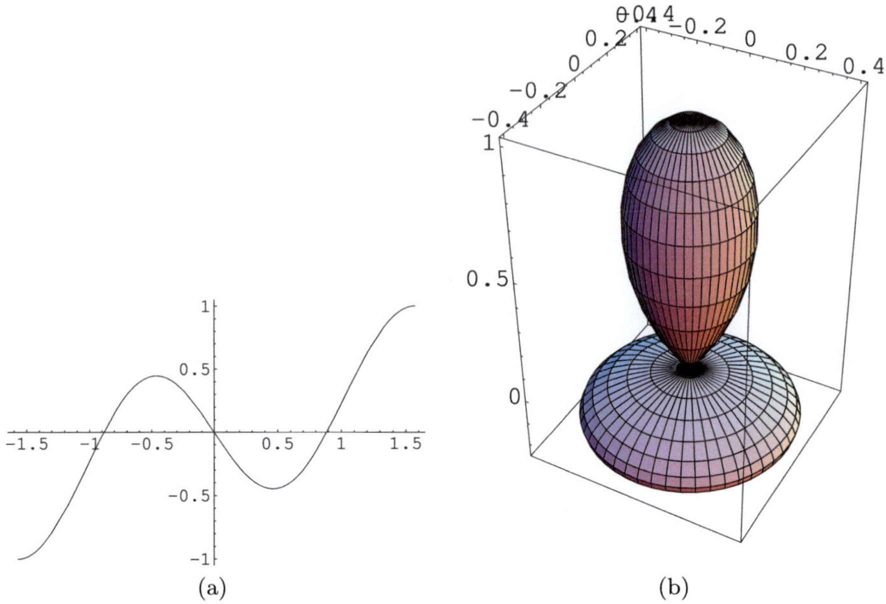

**Fig. 3** The potential $\Psi_\infty$ depends only on $\vartheta_1$. (*a*) Plot of $\Psi_\infty$ as a function of $\vartheta_1$: we observe the (circle of) maxima for $\vartheta_1 = \arcsin(-1/\sqrt{5}) \doteq -0.46$, and the (circle of) minima for $\vartheta_1 = \arcsin(1/\sqrt{5})$. (*b*) The potential $\Phi_\infty$ as a "polar plot": again we observe the (invaginated) circle of maxima at $z = -1/\sqrt{5}$, and the circle of minima for $z = 1/\sqrt{5}$

of degenerate minima (value $-1/\sqrt{5}$) for $\vartheta_1 = \arccos(2/\sqrt{5})$, and a circle of degenerate maxima (value $1/\sqrt{5}$) for $\vartheta_1 = -\arccos(2/\sqrt{5})$.

The critical points are all non-degenerate *modulo* the degeneration enforced by the SO(2) symmetry (this also requires one of the eigenvalues of the Hessian to be zero). Critical points at the poles have an Hessian with nontrivial eigenvalue $\sigma_0 = \pm 6$; those on the circles have an Hessian with nontrivial eigenvalue $\sigma_1 = \pm 12/\sqrt{5}$. This case is depicted in Fig. 3.

### 7.4 Symmetry $D_{3h}$: The Axis $\mathcal{A}$

On the axis $\mathcal{A}$ of the cylinder (i.e. for $\rho = 0$), the potential in (5.36) reads

$$\Phi_3 = Ky\left(y^2 - 3x^2\right) + z\left(z^2 - \frac{3}{2}\left(x^2 + y^2\right)\right), \tag{7.10}$$

and the potential expressed in (5.35) in angular coordinates is

$$\Psi_3 = \sin^3\vartheta_1 - K\cos^3\vartheta_1 \sin 3\vartheta_2 - \frac{3}{2}\cos^2\vartheta_1 \sin\vartheta_1. \tag{7.11}$$

Of course, we need only consider points different from the special points $\mathcal{C}$ and $\mathcal{T}$ considered before. Here $K \geq 0$ is a free parameter. Graphical illustrations of both $\Phi_3$ and $\Psi_3$ are shown in Figs. 4 and 5.

The potential is invariant under the inversion $x \to -x$ and under rotations by $2\pi/3$ around the $z$ axis (i.e. shift by $2\pi/3$ in $\vartheta_2$); thus it would suffice to study the potential

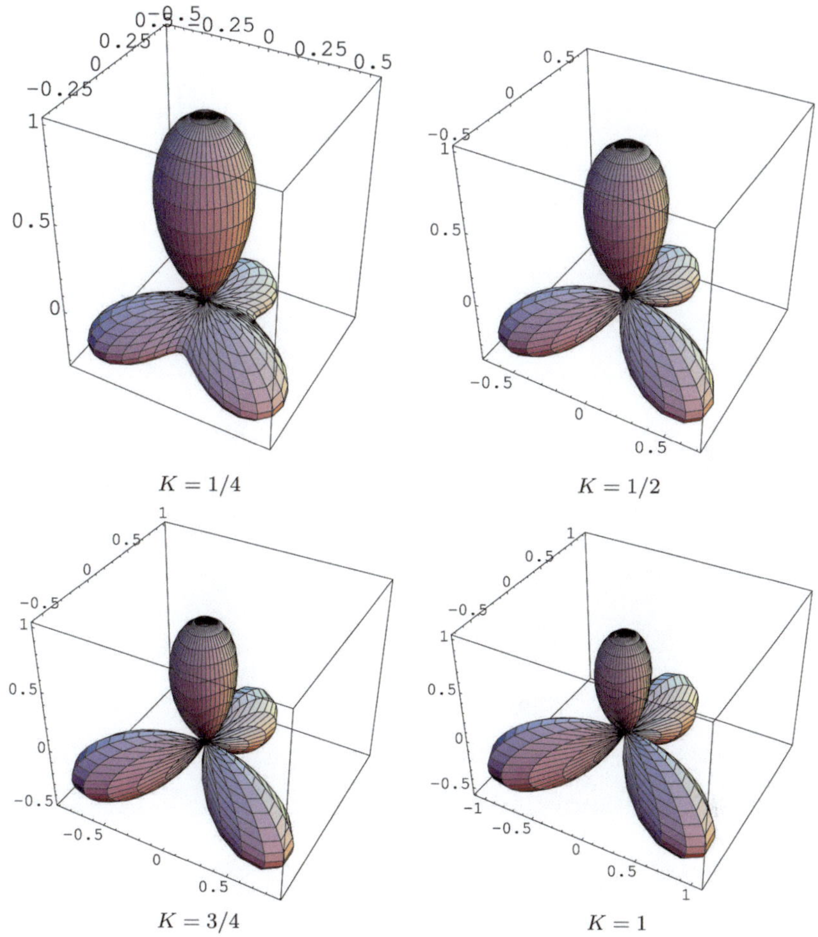


**Fig. 4** Polar plots of the potential $\Phi_3$ for different values of the parameter $K$. For low $K$ the maximum in the North Pole is the highest, while for higher $K$ it becomes smaller than the three symmetric maxima in the southern hemisphere

in the sector $0 \leq \vartheta_2 \leq 2\pi/3$; that is, its symmetry group is fully described by the matrices

$$\begin{pmatrix} 1 & 0 & 0 \\ 0 & 1 & 0 \\ 0 & 0 & 1 \end{pmatrix}, \quad \begin{pmatrix} -\frac{1}{2} & -\frac{\sqrt{3}}{2} & 0 \\ \frac{\sqrt{3}}{2} & -\frac{1}{2} & 0 \\ 0 & 0 & 1 \end{pmatrix}, \quad \begin{pmatrix} -\frac{1}{2} & \frac{\sqrt{3}}{2} & 0 \\ -\frac{\sqrt{3}}{2} & -\frac{1}{2} & 0 \\ 0 & 0 & 1 \end{pmatrix}; \\ \begin{pmatrix} -1 & 0 & 0 \\ 0 & 1 & 0 \\ 0 & 0 & 1 \end{pmatrix}, \quad \begin{pmatrix} \frac{1}{2} & -\frac{\sqrt{3}}{2} & 0 \\ -\frac{\sqrt{3}}{2} & -\frac{1}{2} & 0 \\ 0 & 0 & 1 \end{pmatrix}, \quad \begin{pmatrix} \frac{1}{2} & \frac{\sqrt{3}}{2} & 0 \\ \frac{\sqrt{3}}{2} & -\frac{1}{2} & 0 \\ 0 & 0 & 1 \end{pmatrix}. \tag{7.12}$$

The potential is of course also covariant under inversion,

$$\Phi_3(-\vartheta_1, -\vartheta_2) = -\Phi_3(\vartheta_1, \vartheta_2), \tag{7.13}$$

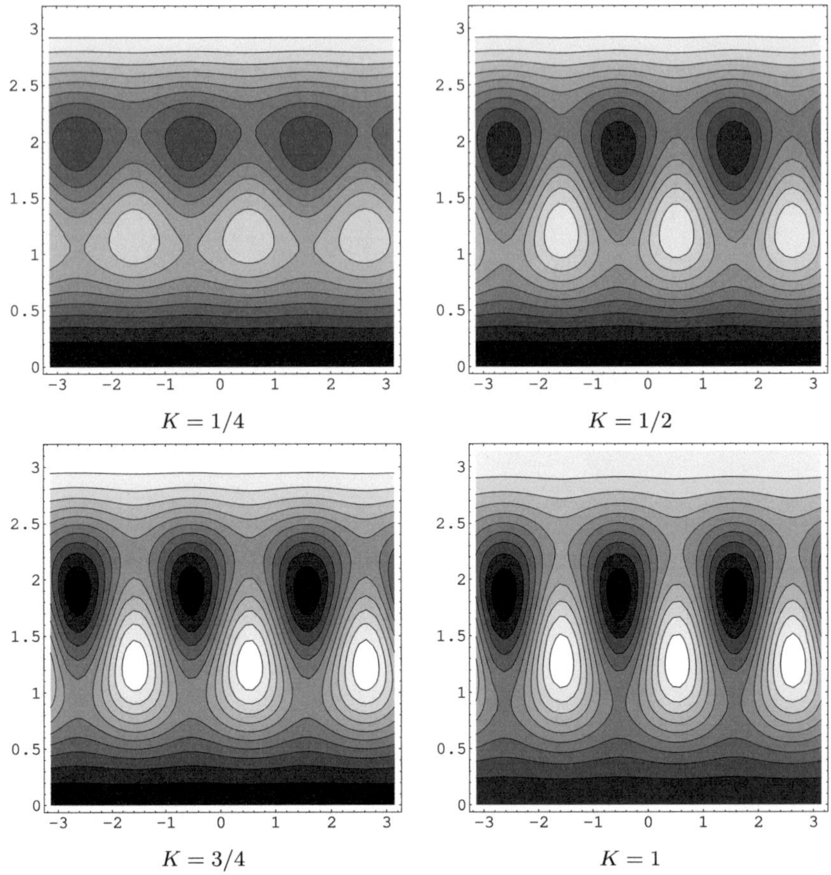


**Fig. 5** The same as in Fig. 4 but with contour plots. Note again that for low $K$ the maximum in the North Pole is the highest, while for higher $K$ it becomes smaller than the three symmetric maxima in the southern hemisphere

which corresponds to $(x, y, z) \to (-x, -y, -z)$; thus it suffices to consider $\vartheta_1 > 0$ (i.e., the northern hemisphere, as we already know from general discussion).

The critical point equations are now

$$\cos\vartheta_1\left(\cos^2\vartheta_1 - 4\sin^2\vartheta_1 - K\sin 2\vartheta_1 \sin 3\vartheta_2\right) = 0,$$
$$K\cos^3\vartheta_1\cos 3\vartheta_2 = 0. \tag{7.14}$$

The second of these implies that critical points are either at the poles, with $\vartheta_1 = \pm\pi/2$ (there is of course a maximum, value $\Phi_3 = 1$, at the North Pole; and a minimum, value $\Phi_3 = -1$, at the South Pole), or on the meridians identified by

$$\vartheta_2 = \pm\frac{\pi}{6}, \pm\frac{\pi}{2}, \pm\frac{5\pi}{6}. \tag{7.15}$$

As $\vartheta_2$ enters the equations only through terms $3\vartheta_2$, it is obvious that cases with $\vartheta_2$ differing by $2\pi/3$ are equivalent (so the cases $\vartheta_2 = -\pi/2, +\pi/6, +5\pi/6$ are equivalent to each other, and so are the cases $\vartheta_2 = -5\pi/6, -\pi/6, \pi/2)$; it thus suffices to consider $\vartheta_2 = \pm\pi/6$, which yields $\sin 3\vartheta_2 = \pm 1$. With this choice for $\vartheta_2$, the first of (7.14) reads

$$\cos\vartheta_1(3 - 5\cos 2\vartheta_1 \pm 2K\sin 2\vartheta_1) = 0. \tag{7.16}$$

Assuming $\vartheta_1 \neq \pm\pi/2$, this further reduces to

$$\cos^2\vartheta_1 - 4\sin^2\vartheta_1 \mp 2K\sin\vartheta_1\cos\vartheta_1 = 0. \tag{7.17}$$

Writing $\eta = \tan\vartheta_1$, and taking off an overall factor $\cos^2\vartheta_1 \neq 0$, this reads

$$1 - 4\eta^2 \mp 2K\eta = 0. \tag{7.18}$$

Thus for $\vartheta_2 = -5\pi/6, -\pi/6, \pi/2$ we get

$$\eta_\pm^- = -\frac{K}{4}\left(1 \pm \sqrt{1 + 4/K^2}\right), \tag{7.19}$$

while for $\vartheta_2 = -\pi/2, \pi/6, 5\pi/6$ we get

$$\eta_\pm^+ = \frac{K}{4}\left(1 \pm \sqrt{1 + 4/K^2}\right). \tag{7.20}$$

This means that we have in total twelve nontrivial critical points, i.e., fourteen counting also those at the poles. We know (see Sect. 5.4) that four of these will be maxima, other four will be minima, and the remaining six will be saddles.

Recalling that $-\pi/2 < \vartheta_1 < \pi/2$, one then easily shows that critical points can be represented as $\vartheta_1 = \xi_\pm^\pm$, where

$$\xi_\pm^+ = \arcsin\left(\eta_\pm^+/\sqrt{1 + (\eta_\pm^+)^2}\right), \qquad \xi_\pm^- = \arcsin\left(\eta_\pm^-/\sqrt{1 + (\eta_\pm^-)^2}\right). \tag{7.21}$$

To characterize the nature of critical points, we can either consider the potential on the meridians (7.15) and on the parallels identified by (7.21) (with the aid of (7.19) and (7.20)); or study the eigenvalues of the Hessian at these critical points. In fact, the eigenvalues of the Hessian on $\vartheta_2 = \pi/6$ (and equivalent meridians) are

$$\Lambda^+ = \left\{9K\cos^3\vartheta_1, \frac{3}{8}(2K\cos\vartheta_1 + 6K\cos 3\vartheta_1 - \sin\vartheta_1 + 15\sin 3\vartheta_1)\right\}, \tag{7.22}$$

while those on $\vartheta_2 = -\pi/6$ (and equivalent meridians) are

$$\Lambda^- = \left\{-9K\cos^3\vartheta_1, -\frac{3}{8}(2K\cos\vartheta_1 + 6K\cos 3\vartheta_1 + \sin\vartheta_1 - 15\sin 3\vartheta_1)\right\}. \tag{7.23}$$

Recalling that $\cos\vartheta_1 > 0$ and $K > 0$, which fixes the sign of the first eigenvalue in both cases, we see that on $\vartheta_2 = \pi/6$ (and equivalent meridians) we can only have saddles or minima, and on $\vartheta_2 = -\pi/6$ (and equivalent meridians) we can only have maxima or saddles.

The above formulas (7.22), (7.23) also allow us to compute explicitly the eigenvalues at critical points, and hence the nature of these. The results are summarized in Table 5 (which also includes the critical points at the poles, considered as trivial in the previous discussion); we stress that there is no change of stability as $K$ is varied, as follows from (7.22), (7.23).

**Table 5** Critical points for the case $D_{3h}$. The relation prescribed by Eq. (5.15) are satisfied, hence $\lambda$ is not displayed. The shorthand notation (7.24) is used

| $n$ | $\vartheta_1$ | $\vartheta_2$ | $\Psi$ | type |
|---|---|---|---|---|
| 1 | $-\pi/2$ | – | $-1$ | min |
| 2 | $-\tau_-$ | $-\pi/2$ | $-\zeta_-$ | min |
| 3 | $-\tau_-$ | $\pi/6$ | $-\zeta_-$ | min |
| 4 | $-\tau_-$ | $5\pi/6$ | $-\zeta_-$ | min |
| 5 | $-\tau_+$ | $-\pi/2$ | $-\zeta_+$ | saddle |
| 6 | $-\tau_+$ | $\pi/6$ | $-\zeta_+$ | saddle |
| 7 | $-\tau_+$ | $5\pi/6$ | $-\zeta_+$ | saddle |
| 8 | $\tau_+$ | $-5\pi/6$ | $\zeta_+$ | saddle |
| 9 | $\tau_+$ | $-\pi/6$ | $\zeta_+$ | saddle |
| 10 | $\tau_+$ | $\pi/2$ | $\zeta_+$ | saddle |
| 11 | $\tau_-$ | $-5\pi/6$ | $\zeta_-$ | max |
| 12 | $\tau_-$ | $-\pi/6$ | $\zeta_-$ | max |
| 13 | $\tau_-$ | $\pi/2$ | $\zeta_-$ | max |
| 14 | $\pi/2$ | – | 1 | max |

To improve readability of Table 5, we have used the shorthand notations

$$
\begin{aligned}
q_\pm &:= \left[10 + K^2 \pm K\sqrt{4+K^2}\right]^{-1},\\
\tau_\pm &:= \arcsin\left(\frac{K_\pm q_\pm}{2}\right),\\
\zeta_\pm &:= \sqrt{2q_\pm^3}\left[K^3 \pm K^2\sqrt{4+K^2} + 13K \mp 5\sqrt{4+K^2}\right].
\end{aligned}
\tag{7.24}
$$

*Remark 17* We note that the value of the potential at the non-orienting maxima—i.e., at the critical points 11, 12, and 13 in Table 5, is in explicit terms

$$
\zeta_- = \frac{\sqrt{2}(K^3 - \sqrt{K^2+4}K^2 + 13K + 5\sqrt{K^2+4})}{(K^2 - \sqrt{K^2+4}K + 10)^{3/2}};
\tag{7.25}
$$

it is a simple matter to check that this is always (positive and) increasing with $K$, and that $\zeta_- = 1$ for $K = K_0 := \sqrt{2}/2 \doteq 0.71$. Thus, these are secondary maxima for $K < K_0$, and become absolute maxima for $K > K_0$. We thus have a "global bifurcation" between the phases $D_{3h}^+$ and $D_{3h}^-$ taking place at $K = K_0$. For this value of $K$, the value taken by the potential at the saddle points 5–10 in Table 5 is $\zeta_+ = 0$. It should also be noted that $\zeta_+$ is always increasing with $K$.

*Remark 18* Let us focus on the maxima. In those other than the North Pole, we have $\Phi = \Phi_M$ and $\lambda = \lambda_M$. These are smaller than the corresponding quantities for the maximum in the North Pole for $K < K_0$, and greater than those for $K > K_0$. In other words, albeit at $K = K_0$ there is no local bifurcation, we have a rearrangement of the maxima and saddles in terms of their ordering according to value of the potential. When choosing the orienting maximum, we could also choose it in such a way that it should always be the largest; this means that the case with $K > K_0$ necessarily corresponds to a case present with other values of the parameters, in which the orienting maximum is the largest. Such a choice would break the $D_{3h}$ symmetry around the orienting axis, and we would be left with a $D_{2h}$ symmetry around the new orienting axis.

*Remark 19* Inverting the relation between $K$ and $\lambda$, we can express $K$ in terms of $\lambda$; similarly, we can invert the relation between $K$ and $\vartheta_1$. Eliminating $K$ from these two equations according to the two variants in (7.21), they provide the corresponding (not too simple) relations linking $\lambda_S$ and $\lambda_M$ with $\vartheta_1$, for saddles and maxima, respectively; more precisely, we have

$$\lambda_S = \frac{1}{2}\frac{2 - 3\cos^2\vartheta_1}{\sin\vartheta_1}, \qquad \lambda_M = \frac{8 - 9\cos^2\vartheta_1}{\cos\vartheta_1\sqrt{16 - 15\cos^2\vartheta_1}}. \tag{7.26}$$

### 7.4.1 *Symmetry Breaking: From $D_{\infty h}$ to $D_{3h}$*

It is interesting to consider the situation in the $D_{3h}$ phase for values of $K$ near zero; in other words, to consider the symmetry breaking from $D_{\infty h}$ (case $K = 0$, the center of the disk $\mathcal{D}$) to $D_{3h}$.

We will write $K = \varepsilon$ and work at first order in $\varepsilon$. With this, we obtain easily

$$\cos\vartheta_1 = \frac{10 + \varepsilon}{5\sqrt{5}}; \tag{7.27}$$

in the same way we also get

$$\lambda_M = \frac{15 + 24\varepsilon}{5\sqrt{5}}. \tag{7.28}$$

In particular, denoting by a "0" the limit for $K \to 0$, we have

$$\frac{\lambda_M - \lambda_M^0}{\cos\vartheta_1 - \cos\vartheta_1^0} = 24 + O(\varepsilon). \tag{7.29}$$

### 7.4.2 *Symmetry Breaking: From $T_d$ to $D_{3h}$*

We can also consider the symmetry breaking from the tetrahedral phase to $D_{3h}$. In this case we will write $K = K_0 + \varepsilon$ and work again at first order in $\varepsilon$. Now we get

$$\cos\vartheta_1 = \frac{2\sqrt{2}}{3} + \frac{4}{81}\varepsilon, \qquad \lambda_M = 3 + \frac{16\sqrt{2}}{9}\varepsilon. \tag{7.30}$$

Thus in this case, denoting now by a superscript $^T$ the tetrahedral limit, we have

$$\frac{\lambda_M - \lambda^T}{\cos\vartheta_1 - \cos\vartheta_1^T} = 36\sqrt{2} + O(\varepsilon). \tag{7.31}$$

## 7.5 Symmetry $D_{2h}$: The Disk $\mathcal{D}$

On the disk $\mathcal{D}$ identified by $K = 0$ (with $\rho \neq 0$, or we would be at the center $\mathcal{C}$ and in case $D_{\infty h}$ considered above), we have a symmetry $D_{2h}$. In fact, the potential in (5.36) reduces to

$$\Phi_2 = z^3 - 3y^2 z + 3\rho\cos\chi\, xyz + \frac{3}{2}(\rho\sin\chi - 1)\left(x^2 - y^2\right)z, \tag{7.32}$$

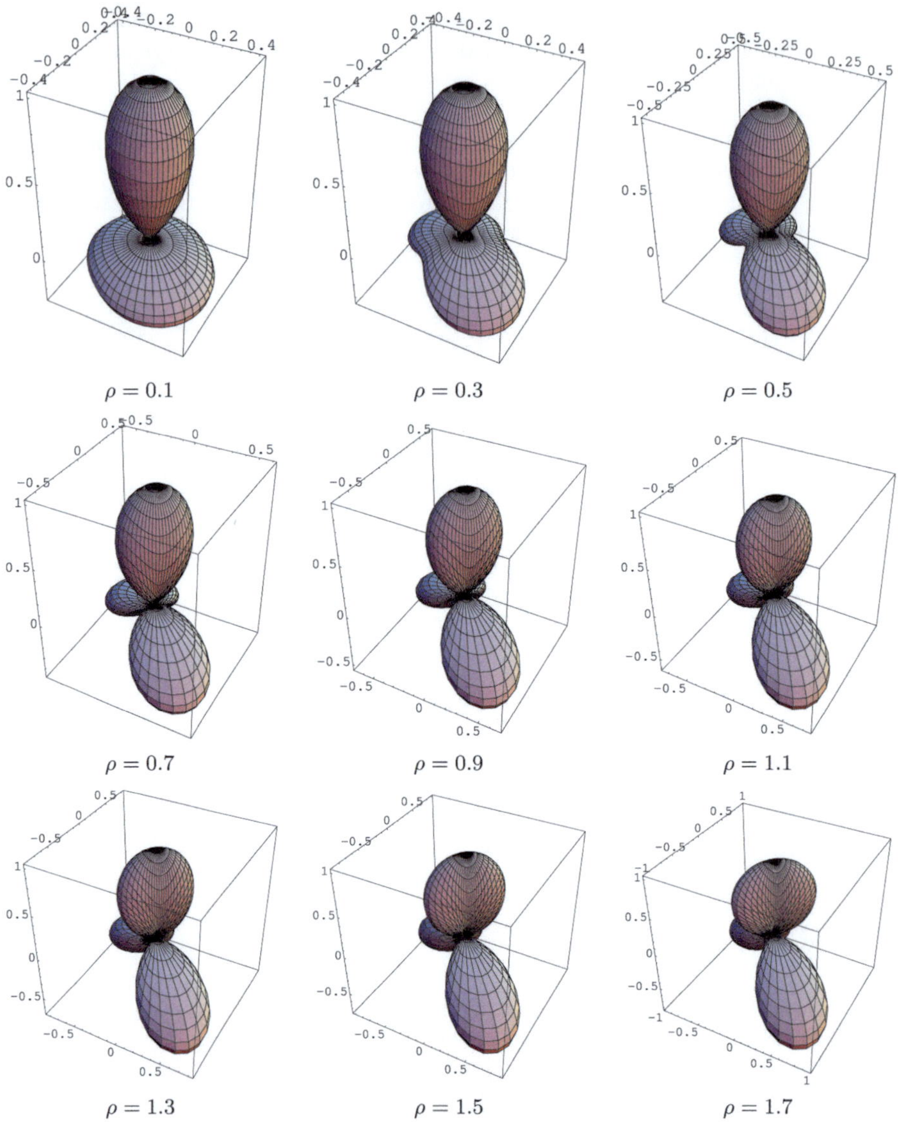


**Fig. 6** The potential $\Phi_2$ for $\chi = 0$ and different values of the parameter $\rho$

and similarly its counterpart in angular coordinates is readily obtained from (5.35) as

$$\Psi_2 = \sin\vartheta_1 \left[ \sin^2\vartheta_1 + \frac{3}{2}\cos^2\vartheta_1 \big(\rho\sin(\chi + 2\vartheta_2) - 1\big) \right]. \tag{7.33}$$

Graphical illustrations for both $\Phi_2$ and $\Psi_2$ in the styles introduced above are provided in Figs. 6, 7, and 8.

*Remark 20* It is clear from this expression that a shift in $\chi$ corresponds to a shift (of half the amplitude) in $\vartheta_2$; see also Fig. 9 for a visual demonstration.

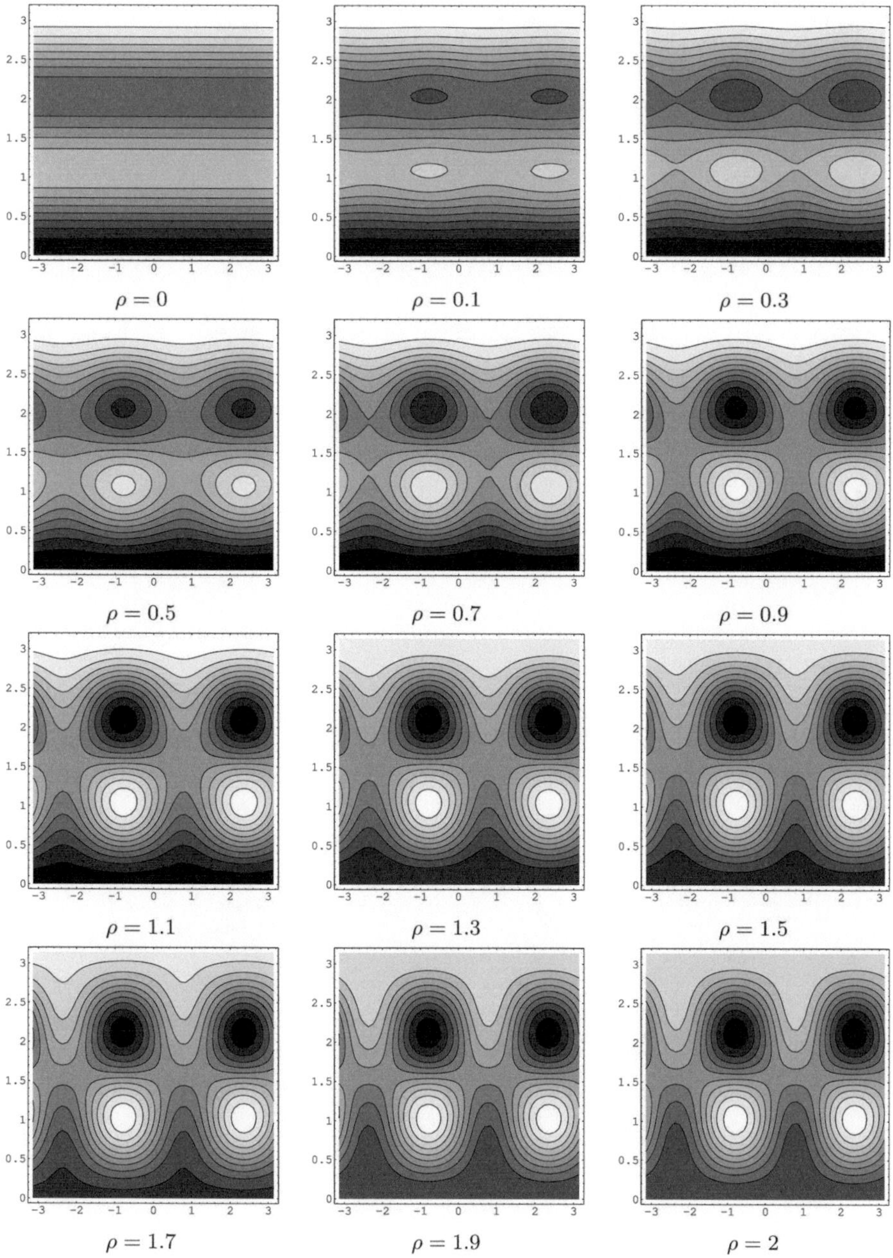


**Fig. 7** Contour plot of the potential $\Psi_2$ for $\chi = 0$ and for various values of $\rho > 0$. There always are three maxima, three minima and four saddles; for $\rho = 1.0$, pairs of saddles originally one on top of the other collide and re-emerge side to side. See Fig. 8 for more details

*Remark 21* It is also apparent from (7.33) that $\Psi_2$ is identically zero on the Equator $\vartheta_1 = 0$ (i.e., $\Phi_2$ is identically zero for $z = 0$); this implies that any critical point lying on the Equator will necessarily be non-hyperbolic, hence degenerate.

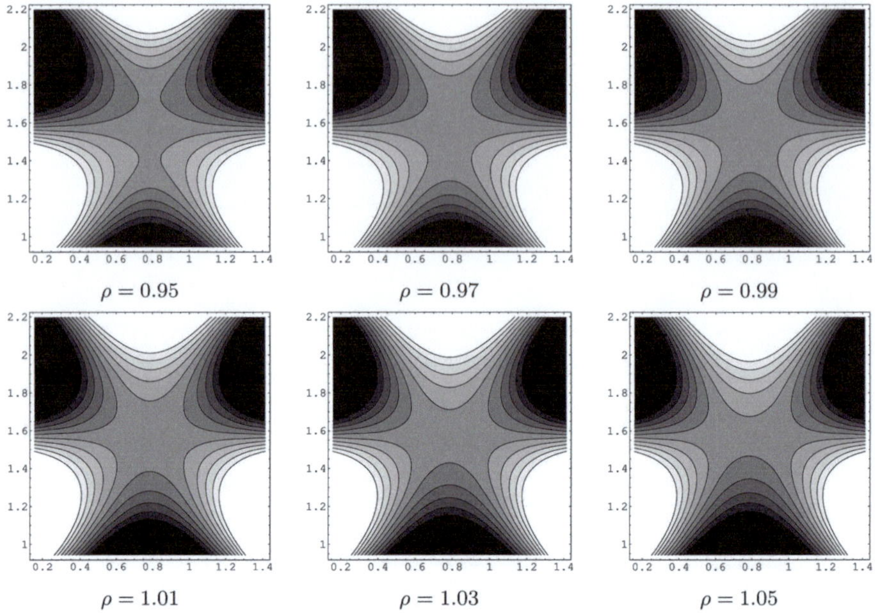


**Fig. 8** Contour plot of the potential $\Psi_2$ in the region where the saddle/saddle bifurcation takes place (here $\vartheta_1 \in [-\pi/5, \pi/5]$, $\vartheta_2 \in [\pi/4 - \pi/5, \pi/4 + \pi/5]$) for $\chi = 0$ and for various values of $\rho$. We observe the saddle/saddle bifurcation taking place; in this a pair of saddles originally one on top of the other collide (at $\rho = 1.0$) and re-emerge side to side. At $\rho = 1.0$ there is a monkey saddle

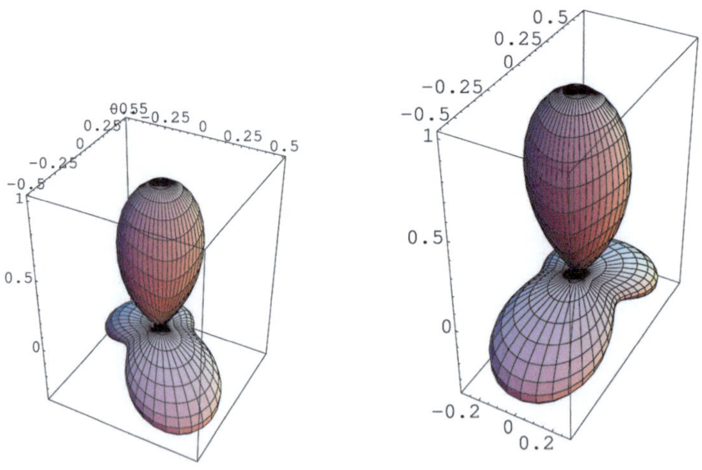

**Fig. 9** Three-dimensional polar plot of the potential $\Phi_2$, see (7.33), for $\rho = 0.4$. Here we have chosen $\chi = 0$ (left) and $\chi = \pi/2$ (right). It is clear that the rotations by $\pi/2$ in $\chi$ corresponds to a rotation by $\pi/4$ in $\vartheta_2$

*Remark 22* The invariance under rotations by $\pi$ in $\vartheta_2$ is evident. We also have invariance under rotations by any $\delta$ in $\vartheta_2$ accompanied by a rotation by $2(\pi - \delta)$ in $\chi$. In particular, we have invariance under a rotation by $\pi/2$ in $\vartheta_2$ accompanied by a rotation by $\pi$ in $\chi$; thus, we can just consider $-\pi/2 \leq \chi \leq \pi/2$.

We could also restrict the domain in which we study the potential by noting that $\Psi_2 \to -\Psi_2$ under $\vartheta_1 \to -\vartheta_1$; this also follows by the usual skew-symmetry of the potential under the antipodal map combined with invariance under $\vartheta_2 \to \vartheta_2 + \pi$.

### 7.5.1 Symmetry of the Potential

The potential is invariant under a subgroup $O(2) \subset T_d$; this is explicitly given by the matrices

$$M_0 = \begin{pmatrix} 1 & 0 & 0 \\ 0 & 1 & 0 \\ 0 & 0 & 1 \end{pmatrix}, \qquad M_\pi = \begin{pmatrix} -1 & 0 & 0 \\ 0 & -1 & 0 \\ 0 & 0 & 1 \end{pmatrix}; \tag{7.34}$$

$$R_1 = \begin{pmatrix} -\sin\chi & -\cos\chi & 0 \\ -\cos\chi & \sin\chi & 0 \\ 0 & 0 & 1 \end{pmatrix}, \qquad R_2 = \begin{pmatrix} \sin\chi & \cos\chi & 0 \\ \cos\chi & -\sin\chi & 0 \\ 0 & 0 & 1 \end{pmatrix}. \tag{7.35}$$

Both $M_0$ and $M_\pi$ have determinant $+1$, while both $R_1$ and $R_2$ have determinant $-1$.

The matrices $R_i$ describe a reflection in a vertical plane. More precisely, recalling that the matrix describing a reflection in the vertical plane $y = mx$ is given by

$$R_m = \begin{pmatrix} a & b & 0 \\ b & -a & 0 \\ 0 & 0 & 1 \end{pmatrix}, \quad a = \frac{1-m^2}{m^2+1}, \quad b = \frac{2m}{m^2+1}, \tag{7.36}$$

and that conversely a matrix $R_m$ as in (7.36) with $a^2 + b^2 = 1$ represents a reflection through the plane $y = mx$ with

$$m = \frac{1-a}{b}, \tag{7.37}$$

the planes of reflection for the matrices $R_i$ (orthogonal to one another) have equations $y = m_i x$, where

$$m_1 = -\frac{1+\sin\chi}{\cos\chi}, \qquad m_2 = \frac{1-\sin\chi}{\cos\chi}, \tag{7.38}$$

which clearly satisfy $m_1 m_2 = -1$.

We have special cases for $\chi = \pm\pi/2$ or $\chi = 0, \pi$ (note that when both these conditions are met, we are back to the case $\rho = 0$ considered above; in fact, we have characterized $D_{2h}$ by also requiring $\rho \neq 0$).

*Remark 23* To appreciate better our previous Remark 20, consider the case $\chi = 0$. Now $R_i$ read as

$$R_1 = \begin{pmatrix} 0 & -1 & 0 \\ -1 & 0 & 0 \\ 0 & 0 & 1 \end{pmatrix}, \qquad R_2 = \begin{pmatrix} 0 & 1 & 0 \\ 1 & 0 & 0 \\ 0 & 0 & 1 \end{pmatrix}. \tag{7.39}$$

These represent reflections in the vertical planes $y = x$ and $y = -x$, as illustrated in the left panel of Fig. 9.

### 7.5.2 Critical Points

The critical points are identified as solutions to the equations

$$\begin{aligned} \frac{3}{4}\cos\vartheta_1\big[-5\cos 2\vartheta_1+\rho(3\cos 2\vartheta_1-1)\sin(\chi+2\vartheta_2)+3\big]=0,\\ 3\rho\cos^2\vartheta_1\cos(\chi+2\vartheta_2)\sin\vartheta_1=0. \end{aligned} \tag{7.40}$$

*Remark 24* It is immediately apparent that any change in $\chi$ can be compensated by a change in $\vartheta_2$, and conversely; in other words, the relevant angle is $\Theta=(\vartheta_2+\chi/2)$. Thus it suffices to study the problem with a given value of $\chi$, e.g., $\chi=\pi/2$ (hence $\alpha_0=\rho$, $\gamma=0$) or $\chi=0$ (hence $\alpha_0=0$, $\gamma=\rho$); the general case (i.e., the case of general $\chi$) will be obtained via a suitable rotation in $\vartheta_2$.

Solving Eqs. (7.40) in general is a matter of standard algebra and trigonometry (and some patience). The results are reported in Table 6; it should be stressed that some of the solutions exist only for certain ranges of $\rho$.

Let us first focus on the second of (7.40); discarding as usual the "trivial" (in this context) solutions for $\vartheta_1=\pm\pi/2$, which corresponds to the poles, we need either $\cos(\chi+2\vartheta_2)=\pm\pi/2$; or $\vartheta_1=0$. But inserting $\vartheta_1=0$ in the first of (7.40), we obtain

$$-\frac{3}{2}\big[1-\rho\sin(\chi+2\vartheta_2)\big]=0. \tag{7.41}$$

This admits solutions *only* for $\rho\geq 1$; in view of our general restriction on $\rho$, see (5.34), this means $\rho\in[1,2]$.

On the other hand, if $\cos(\chi+2\vartheta_2)=0$, i.e., $\chi+2\vartheta_2=\pm\pi/2$, and hence $\sin(\chi+2\vartheta_2)=\pm 1$, the first of (7.40) reads (after taking away the inessential factor $(3/4)\cos\vartheta_1$)

$$3-5\cos 2\vartheta_1\mp\rho(1-3\cos 2\vartheta_1)=0; \tag{7.42}$$

this means

$$\cos 2\vartheta_1=\frac{3\mp\rho}{5\mp 3\rho}. \tag{7.43}$$

While the solution with the plus sign does exist for all values of $\rho$, the solution with the minus sign exists only for $\rho=2$ and $\rho\leq 1$.

In other words, we have some family of solutions existing through the whole range $\rho\in[0,2]$, while for other families we will have to consider separately the subranges $\rho\in[0,1]$ and $\rho\in[1,2]$. It has to be expected (and it will indeed result) that a multiple bifurcation takes place at $\rho=1$. It should be noted that at the bifurcation point the saddles are degenerate, so carrying a different index.

The situation is rather clear if we think of a fixed value of $\chi$, say $\chi=\pi/2$, and focus only on families of solutions not existing for the whole range of $\rho$ (it turns out that all of these are saddles). For $\rho\in[0,1)$ there are solutions on the meridians identified by $\chi+2\vartheta_2=(k+1/2)\pi$ and approaching the Equator as $\rho\to 1$, but none on the equator; on the other hand, for $\rho\in(1,2]$ there are no solutions on those meridians, but we have instead families of solutions on the Equator, drifting away from the meridians identified by $\chi+2\vartheta_2=(k+1/2)\pi$ as $\rho$ grows away from $\rho=1$. This is clearly illustrated in Fig. 10.

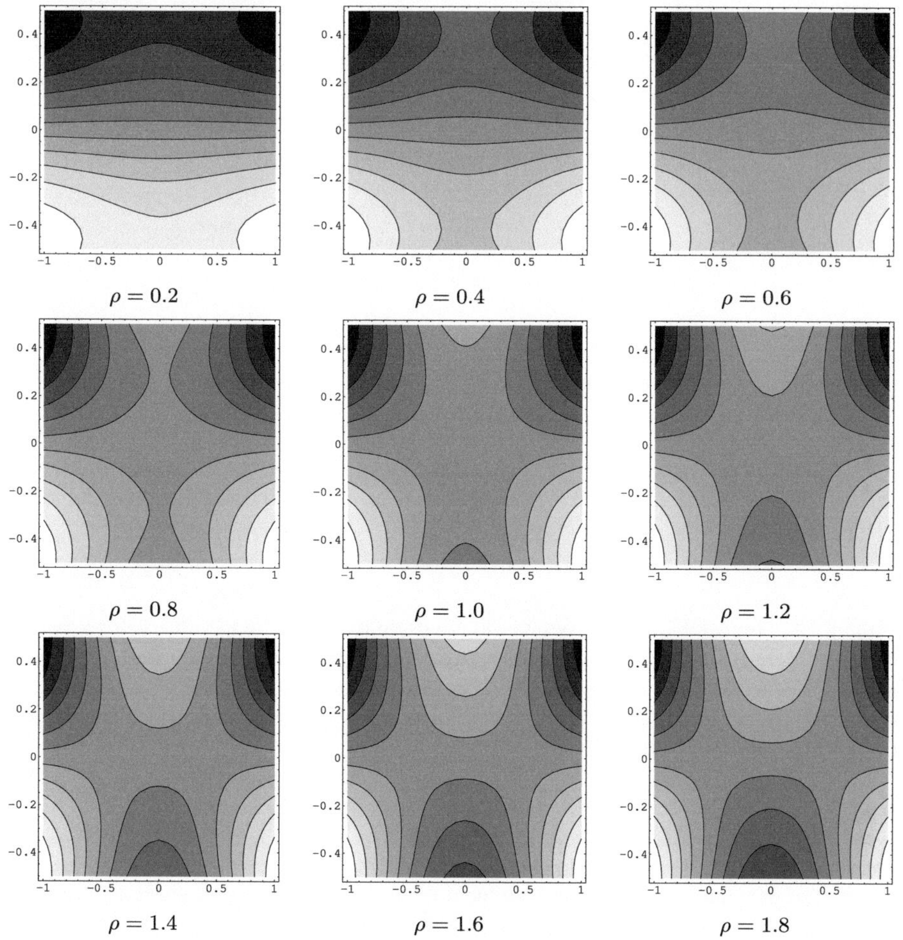


**Fig. 10** The saddle-saddle bifurcation in the $D_{2h}$ case taking place at $\rho = 1$. We have chosen $\chi = \pi/2$ and plotted a neighborhood of the point $(\vartheta_1, \vartheta_2) = (0, 0)$—more precisely the region $|\vartheta_1| \leq 1/2$, $|\vartheta_2| \leq 1$—as $\rho$ is varied. The two saddles located (symmetrically w.r.t. the Equator) on the meridian $\vartheta_2 = 0$ collide at $\rho = 1$, giving raise to a new pair of saddles located on the Equator, symmetrically w.r.t. the meridian $\vartheta_2 = 0$

We now return to the solutions to the critical point equations (7.40). The nature of these critical points is easily ascertained by considering the eigenvalues of the Hessian at them; again the result of this analysis is reported in Table 6. Here again we resort to some shorthand notation to make the table more readable; in that we have defined

$$\begin{aligned} \rho_\pm &:= \frac{\rho \pm 3}{3\rho \pm 5}, \qquad & r_\pm &:= \arccos(\rho_\pm), \qquad & r_0 &:= \arccos(1/\rho), \\ \mu_\pm &:= \frac{\chi}{2} \pm \frac{\pi}{4}, \qquad & \nu_\pm &:= \frac{\chi}{2} \pm \frac{3\pi}{4}, \qquad & \omega_\pm &:= \frac{\chi \pm r_0}{2}. \end{aligned} \tag{7.44}$$

*Remark 25* Looking at Table 6, we note that maxima and minima belong to families running through the whole range of admitted values for $\rho$, while the saddles undergo bifurcations.

**Table 6** Critical points for the case $D_{2h}$. The shorthand notation (7.44) is used here

| $n$ | $\vartheta_1$ | $\vartheta_2$ | $\Psi$ | range | type |
|---|---|---|---|---|---|
| 1 | $-\pi/2$ | – | $-1$ | always | min |
| 2 | $r_+$ | $-\mu_+$ | $-(\rho+1)\sin r_+$ | always | min |
| 3 | $r_+$ | $-\nu_-$ | $-(\rho+1)\sin r_+$ | always | min |
| 4 | $r_-$ | $-\mu_-$ | $(\rho-1)\sin r_-$ | $\rho \le 1$ | saddle |
| 5 | $-r_-$ | $-\mu_-$ | $-(\rho-1)\sin r_-$ | $\rho \le 1$ | saddle |
| 6 | $-r_-$ | $-\nu_+$ | $-(\rho-1)\sin r_-$ | $\rho \le 1$ | saddle |
| 7 | $r_-$ | $-\nu_+$ | $(\rho-1)\sin r_-$ | $\rho \le 1$ | saddle |
| 8 | 0 | $-\omega_-$ | 0 | $\rho \ge 1$ | saddle |
| 9 | 0 | $\pi/2-\omega_+$ | 0 | $\rho \ge 1$ | saddle |
| 10 | 0 | $\pi-\omega_-$ | 0 | $\rho \ge 1$ | saddle |
| 11 | 0 | $-\omega_+-\pi/2$ | 0 | $\rho \ge 1$ | saddle |
| 12 | $-r_+$ | $-\mu_+$ | $(\rho+1)\sin r_+$ | always | max |
| 13 | $-r_+$ | $-\nu_-$ | $(\rho+1)\sin r_+$ | always | max |
| 14 | $\pi/2$ | – | 1 | always | max |

For $\rho \le 1$ the four saddles are at symmetric points on two opposite meridians (for $\chi = \pi/2$ these are identified by $y = 0$) and drift towards the Equator as $\rho$ approaches the critical value $\rho = 1$, while for $\rho > 1$ the four saddles are at symmetric points on the Equator and drift away from the previously mentioned meridians as $\rho$ increases. This means that there is a (saddle/saddle) *local* bifurcation.

*Remark 26* Note also that a *global* change takes place at the same value $\rho = 1$. That is, for $\rho < 1$ the orienting local maximum in the North Pole is also the absolute maximum, the other two being (degenerate and) lower than this; for $\rho > 1$, on the other hand, the other two maxima are (degenerate and) higher than the orienting one. Similarly to what we have done for the $D_{3h}$ phase, we will distinguish these as $D_{2h}^+$ and $D_{2h}^-$ phases.

*Remark 27* We could have defined the orientation requiring that the North Pole is not only a local maximum, but actually the absolute maximum. In this case the parameter range would be further restricted from the cylinder $\mathsf{C}$ to a subset $\mathsf{C}_0$; and Remark 26 shows that the intersection of $\mathsf{C}_0$ with the disk $\mathcal{D}$ (of radius $\rho = 2$) would just be the disk of radius $\rho = 1$. This would however introduce rather complex mappings involving both the physical and the parameter space [6, 14], and we prefer not to discuss it here; the reader is referred to [6] for a detailed discussion.

*Remark 28* If we look at the potential for $\rho = 1$, say for the "reference case" $\chi = \pi/2$, it turns out this is invariant under a subgroup of the group $O(2)$ acting in the $(y, z)$ plane; this is generated by the matrices

$$M_1 = \begin{pmatrix} 1 & 0 \\ 0 & 1 \end{pmatrix}, \quad M_2 = \begin{pmatrix} -\frac{1}{2} & -\frac{\sqrt{3}}{2} \\ \frac{\sqrt{3}}{2} & -\frac{1}{2} \end{pmatrix}, \quad M_3 = \begin{pmatrix} -\frac{1}{2} & \frac{\sqrt{3}}{2} \\ -\frac{\sqrt{3}}{2} & -\frac{1}{2} \end{pmatrix};$$
$$M_4 = \begin{pmatrix} -1 & 0 \\ 0 & 1 \end{pmatrix}, \quad M_5 = \begin{pmatrix} \frac{1}{2} & \frac{\sqrt{3}}{2} \\ \frac{\sqrt{3}}{2} & -\frac{1}{2} \end{pmatrix}, \quad M_6 = \begin{pmatrix} \frac{1}{2} & -\frac{\sqrt{3}}{2} \\ -\frac{\sqrt{3}}{2} & -\frac{1}{2} \end{pmatrix}. \tag{7.45}$$

These satisfy

$$\begin{aligned} &M_2^3 = M_3^3 = M_4^2 = M_1 = I; \\ &M_2^2 = M_3, \qquad M_5 = M_4 M_2, \qquad M_6 = M_4 M_3; \\ &M_5^2 = M_6^2 = I. \end{aligned} \tag{7.46}$$

The $SO(2)$ matrices $\{M_1, M_2, M_3\}$ span the group $S_3$ of rotations through multiples of $2\pi/3$, while $\{M_4, M_5, M_6\}$ (having determinant $-1$) are reflections in $y$—that is, in three-dimensional terms, through the plane $(x, z)$—and through planes obtained from this by rotations of $2\pi/3$ about the $x$ axis. Thus the six matrices provide a representation of the group $D_3$. The potential is also obviously invariant under reflections in $x$, i.e., through the $(y, z)$ plane, hence *at the bifurcation point* we have a $D_{3h}$ symmetry (as on the axis $\mathcal{A}$).

*Remark 29* Note also that the group $S_3$ acts mapping maxima to maxima and minima to minima; the reflections map maxima into minima and minima into maxima. Saddle points are obviously invariant, as the transformations we are considering do not act on the $x$ coordinate.

*Remark 30* Our discussion in the last two Remarks has been conducted in the "reference case" $\chi = \pi/2$. For different values of $\chi$, we have the same situation but with an overall rotation of the whole picture (see Remark 20).

### 7.5.3 Bifurcations

We are again interested in relations between the eigenvalues and the direction of eigenvectors, in particular near the bifurcation points. In this case, $\Phi_+ = (1+\rho)\sin\xi_+$, where $\Phi_+$ is the height of the "secondary" maxima (which for $\rho > 1$ are actually higher than the orienting one) and $(\xi_+ + \pi/2)$ is the angle between these maxima and the orienting one ("secondary maxima" can be recognized by the fact they are always degenerate). We express $\rho$ in terms of $\xi_+$ through the equation

$$\rho = -\frac{3 - 5\cos 2\xi_+}{1 - 3\cos 2\xi_+}. \tag{7.47}$$

At the bifurcation point $\rho = 1$, we have $\xi_+ = \pi/6$ and $\Phi_+ = 1$. Thus, using (7.47) and with some trivial algebra, we see that

$$\Phi_+ = 4\frac{\sin^3 \xi_+}{3\cos 2\xi_+ - 1}. \tag{7.48}$$

By series expansion at the bifurcation point, i.e., for $\xi_+ = \pi/6 + \varepsilon$, we get

$$\Phi_+(\varepsilon) - \Phi_+(0) = 9\sqrt{3}\varepsilon + O(\varepsilon^2). \tag{7.49}$$

Let us also consider the bifurcation from the $D_{\infty h}$ to the $D_{2h}$ phase, taking place at $\rho = 0$. Using again (7.47), and writing $\xi_+ = \xi_+^{(0)} + \varepsilon$ where

$$\xi_+^{(0)} = \frac{1}{2}\arccos\left(\frac{3}{5}\right) \doteq 0.46 \tag{7.50}$$

is the value taken by $\xi_+$ for $\rho = 0$, we get

$$\Phi_+(\varepsilon) - \Phi_+(0) = \left(15\cos\xi_+^{(0)}\sin^2\xi_+^{(0)} + 30\sin^3\xi_+^{(0)}\right)\varepsilon + O\left(\varepsilon^2\right) \doteq 5.37\varepsilon + O\left(\varepsilon^2\right). \quad (7.51)$$

## 7.6 Reflection Symmetry: Special Planes in the Bulk $\mathcal{B}$

As suggested by the classification of $T_d$ subgroups, see the Appendix, we expect that there are specific values of the parameters such that the potential is invariant under a reflection in a vertical plane, i.e. under a $Z_2 = D_h$ group.

This is indeed the case for $\chi = \pm\pi/2$, $\chi = \pm\pi/6$, $\chi = \pm 5\pi/6$. Let us just consider the first case (the others are obtained from this by a $2\pi/3$ rotation, see below).

For $\chi = \pm\pi/2$, the potential (5.36) reduces to

$$\Phi_{or} = z^3 + Ky\left(y^2 - 3x^2\right) - \frac{3}{2}\left(x^2 + y^2\right)z \pm \frac{3}{2}\rho\left(x^2 - y^2\right)z; \quad (7.52)$$

this is manifestly invariant under the reflection in the $(y, z)$ plane, i.e., under $x \to -x$. Equivalently, we have invariance under the subgroup of $T_d$ generated by $M_{13}$ (see the Appendix for the matrices $M_i$).

Similar considerations apply for the subgroups generated by $M_{14}$, with invariance subject to the condition $\chi = -\pi/6$ or $\chi = 5\pi/6$; in this case the reflection is through the plane $x = -\sqrt{3}y$. In complete analogy with this is the subgroup generated by $M_{15}$, where now one has to require $\chi = -5\pi/6$ or $\chi = \pi/6$; the reflection plane is then $x = \sqrt{3}y$.

We will refer to these planes (collectively) as $\mathcal{P}$; when we want to be more specific (see Table 7 at the end of this section) we will call them, respectively, $\mathcal{P}_0$, $\mathcal{P}_-$, $\mathcal{P}_+$. These reflection symmetries had evaded our previous studies [6, 14] and have proved quite significant in the present one.

For the special $D_h$ phases considered here, as for all others, the transition from the octupolar potential having four maxima to that having only three takes place for parameters chosen on the *separatrix* identified in our previous work [6, 14] and also recalled in the following Sect. 7.7. This is illustrated in Figs. 11 and 12. The separatrix is a surface in parameter space that marks the border between the subregions $\mathcal{B}_3$ and $\mathcal{B}_4$ of $\mathcal{B}$, where $\Phi_{or}$ has three or four maxima, respectively.

To discuss in more detail the critical points in these phases, we find it more convenient to express the octupolar potential in angular coordinates as in (5.35).

*Remark 31* Since, by $\gamma_3$ in (5.39), the potential in (5.35) is invariant under the simultaneous shifts

$$\chi \to \chi + \frac{2m}{3}\pi, \qquad \vartheta_2 \to \vartheta + \frac{2m}{3}\pi,$$

by studying the case $\chi = \pi/2$ we also obtain information about the cases $\chi = -5\pi/6$ and $\chi = -\pi/6$; and by studying the case $\chi = -\pi/2$ we also obtain information about the cases $\chi = \pi/6$ and $\chi = 5\pi/6$.

We will only consider the cases $\chi = \pm\pi/2$.

### 7.6.1 The Case $\chi = \pi/2$

By setting $\chi = \pi/2$, the potential (5.35) reduces to

$$\Psi_{or} = \sin^3\vartheta_1 - K\cos^3\vartheta_1\sin 3\vartheta_2 - \frac{3}{2}\sin\vartheta_1\cos^2\vartheta_1(1 - \rho\cos 2\vartheta_2). \quad (7.53)$$

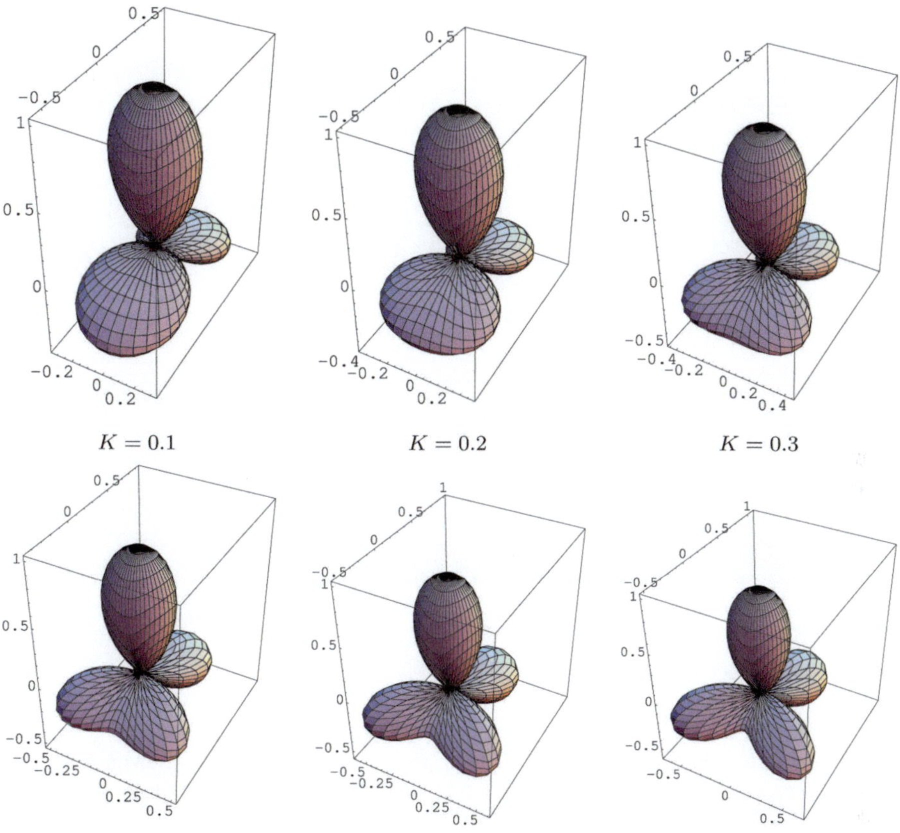


**Fig. 11** The potential for $\chi = \pi/2$ and $\rho = 1/2$, for different values of $K > 0$. We observe the transition from a phase with three maxima at low $K$ to a phase with four maxima at higher $K$. The octupolar potential is always reflection invariant through a vertical plane spanned by the axes $y$ and $z$

The conditions for a critical point are then

$$\frac{\partial \Psi_{or}}{\partial \vartheta_1} = 0, \qquad \frac{\partial \Psi_{or}}{\partial \vartheta_2} = 0; \tag{7.54}$$

both equations have a power of $\cos\vartheta_1$ as an overall factor, which of course vanishes only at the poles; quotienting this factor (and constants) out, we remain with

$$\begin{aligned} K\sin 2\vartheta_1 \sin 3\vartheta_2 - \cos^2\vartheta_1(1-\rho\cos 2\vartheta_2) + 2\sin^2\vartheta_1(2-\rho\cos 2\vartheta_2) = 0, \\ K\cos\vartheta_1\cos 3\vartheta_2 + \rho\sin\vartheta_1\sin 2\vartheta_2 = 0. \end{aligned} \tag{7.55}$$

The second equation (7.55) is solved for $\vartheta_2 = \pm\pi/2$ and moreover (assuming $\vartheta_2 \neq \pm\pi/2$) by

$$\tan\vartheta_1 = -\frac{K}{\rho}\frac{\cos 3\vartheta_2}{\sin 2\vartheta_2}, \tag{7.56}$$

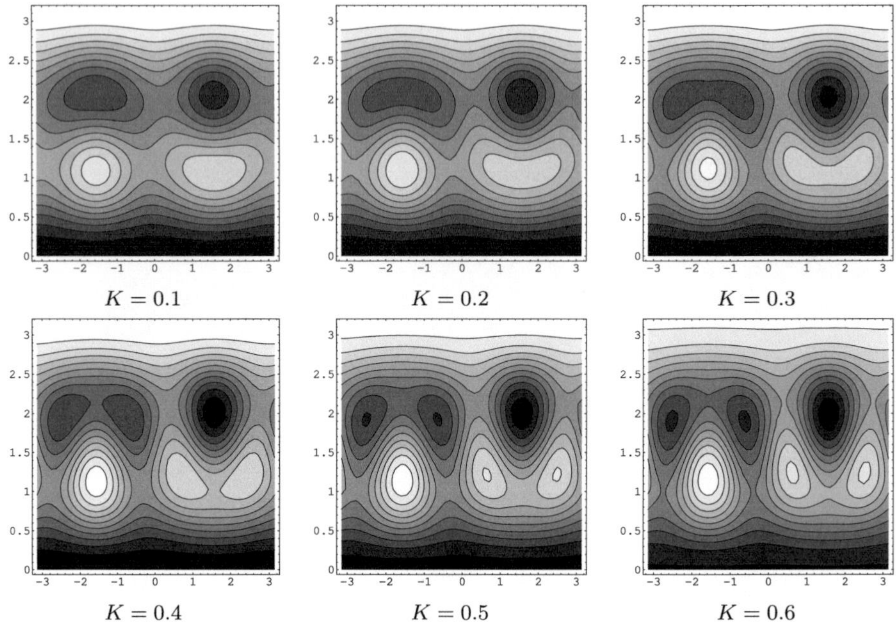

Fig. 12 Same as Fig. 11 but with contour plots

which determines uniquely $\vartheta_1 \in (-\pi/2, \pi/2)$ once $\vartheta_2$ is given. In the following it will be useful to express this, and more specifically $X = \sin\vartheta_1$, in terms of $Y = \sin\vartheta_2$. With some trigonometry, it turns out that

$$X = -K \frac{1 - 4Y^2}{2\rho Y \sqrt{1 + \frac{K^2(1-4Y^2)^2}{4\rho^2 Y^2}}}. \tag{7.57}$$

The right hand of (7.57) is an odd function of $Y$; this equation delivers the same result for $X$, and hence for $\vartheta_1 = \arcsin X \in [-\pi/2, \pi/2]$, for the two determinations of $\vartheta_2 = \arcsin Y \in [-\pi, \pi]$.

**The Solutions $\boldsymbol{\vartheta_2 = \pm\pi/2}$** Let us consider first the solutions $\vartheta_2 = \pm\pi/2$. Inserting this into the first equation (7.55), we obtain

$$3 + \rho(5 + 3\rho)\cos 2\vartheta_1 \mp 2K \sin 2\vartheta_1 = 0. \tag{7.58}$$

This is better written in terms of $X$ as

$$10X^2 - 2 + \rho\big(6X^2 - 2\big) = \pm 4KX\sqrt{1 - X^2}. \tag{7.59}$$

The solutions to (7.59) for the case $\vartheta_2 = \pi/2$ are

$$X_1 = \sqrt{\frac{\alpha + \beta}{\gamma}}, \qquad X_2 = -\sqrt{\frac{\alpha - \beta}{\gamma}}; \tag{7.60}$$

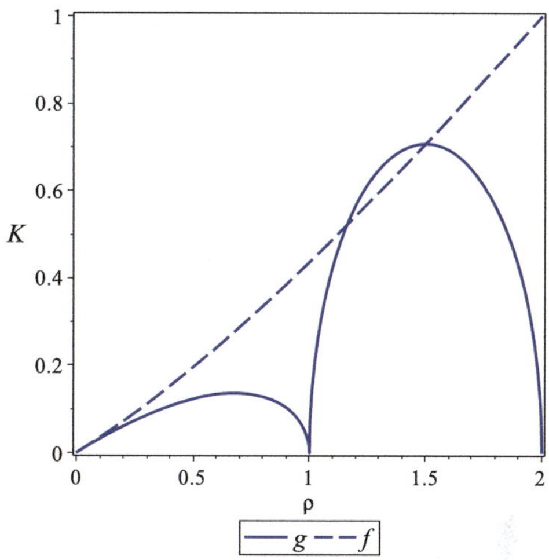


**Fig. 13** The graph of the function $g$ defined by (7.70) is superimposed to the graph of the function $f$ defined by (7.63)

in the case $\vartheta_2 = -\pi/2$ we have instead

$$X_3 = -X_2, \qquad X_4 = -X_1. \tag{7.61}$$

In the formulas (7.60) and (7.61) we have set

$$\begin{aligned} \alpha &= 5 + 2K^2 + 8\rho + 3\rho^2, \\ \beta &= 2K\sqrt{4 + K^2 + 6\rho + 2\rho^2}, \\ \gamma &= 4K^2 + (5 + 3\rho)^2. \end{aligned} \tag{7.62}$$

It is obvious that the argument of the square root is always positive (hence $\beta$ is always real), and the same applies for $\alpha$ and $\gamma$; moreover $\alpha \geq \beta$ in our range $0 \leq \rho \leq 2$. Thus the four solutions are all real. It is easy to check, even numerically, that the solutions (7.60) and (7.61) also satisfy (for $0 \leq \rho \leq 2$) the condition $|X| \leq 1$, necessary to be in accord with the definition of $X$ as $\sin \vartheta_1$.

By looking at the Hessian of the potential computed in these critical points, we can ascertain their nature.

For $\vartheta_2 = \pi/2$, it turns out that $X_2$ is a maximum for all values of $K$ and $\rho$ in the considered range, while $X_1$ undergoes a bifurcation along a certain curve $K = f(\rho)$, and is a minimum (for $K < f(\rho)$) or a saddle (for $K > f(\rho)$) depending on the values of the parameters.

Similarly, for $\vartheta_2 = -\pi/2$, it turns out that $X_3$ is a minimum for all allowed values of $K$ and $\rho$, while $X_4$ undergoes a bifurcation along the same curve $K = f(\rho)$, and is a maximum (for $K < f(\rho)$) or a saddle (for $K > f(\rho)$) depending on the values of the parameters.

The function $f(\rho)$ is given explicitly by

$$f(\rho) = \sqrt{\frac{2\rho^2(1+\rho)}{3(6+\rho)}} \tag{7.63}$$

and plotted in Fig. 13.

**The Other Solutions ($\vartheta_2 \neq \pm\pi/2$)** The other solutions, i.e., those with $\vartheta_2 \neq \pm\pi/2$, are characterized by (7.56) as solution to the second equation (7.55). Plugging this into the first equation (7.55), recalling again that $\vartheta_1 \in [-\pi/2, \pi/2]$, applying some standard trigonometry and writing for ease of notation $Y = \sin\vartheta_2$, we obtain

$$K^2(\rho-2) + \left[4K^2(4-\rho) + 2\rho^2(1-\rho)\right]Y^2 + 4\left(\rho^3 - 8\right)Y^4 = 0. \tag{7.64}$$

The solutions to this equation are

$$Y = \pm\sqrt{\frac{A \pm B}{C}}, \tag{7.65}$$

where we have written

$$\begin{aligned} A &= 2K^2(4-\rho) + \rho^2(1-\rho), \\ B &= \rho\sqrt{4K^4\rho^2 + 4K^2\rho^2(4-3\rho) + \rho^4(1-\rho)^2} := \sqrt{b}; \\ C &= 4\left(8K^2 - \rho^3\right). \end{aligned} \tag{7.66}$$

For these to be real we need that $b \geq 0$; and moreover, if this condition is satisfied, that $(A \pm B)/C \geq 0$. One easily checks, e.g., numerically, that (for $0 \leq \rho \leq 2$) indeed $b \geq 0$, hence $B$ is real (and non-negative, as we take the positive determination of the root).

Actually, we have

$$\begin{aligned} b &= A^2 + 4K^2(2-\rho)\left(\rho^3 - 8K^2\right); \\ B &= |A|\sqrt{1 + \frac{4K^2(2-\rho)(\rho^3 - 8K^2)}{A^2}}. \end{aligned} \tag{7.67}$$

Thus we can have $A \pm B = 0$ only for $K = 0$, for $\rho = 2$, and on the curve $K^2 = (\rho/2)^3$. Note that $A \geq 0$ for

$$K^2 \geq \frac{\rho^2(\rho-1)}{2(4-\rho)}; \tag{7.68}$$

this implies in particular that $A$ is always positive for $\rho < 1$.

In order to study if the solutions

$$Y_a = \pm\sqrt{\frac{A-B}{C}}, \qquad Y_b = \pm\sqrt{\frac{A+B}{C}} \tag{7.69}$$

are real, we consider the signs of $A \pm B$ and of $C$. It turns out that $A + B \geq 0$ for all values of $K$ and for all $0 \leq \rho \leq 2$. As for $A - B$, it follows from (7.67) that it has the same sign as $C = 8K^2 - \rho^3$. Finally, it is obvious that $C > 0$ for $K^2 > (\rho/2)^3$.

This means that, by requiring $Y$ to be real, for $K^2 < (\rho/2)^3$ we only have the solutions $Y_a$, while for $K^2 > (\rho/2)^3$ we have both $Y_a$ and $Y_b$.

This is not enough: in fact, $Y = \sin\vartheta_2$ requires also $|Y| \leq 1$. This condition is always satisfied by $Y_a$, while for $Y_b$ it requires $K \geq f(\rho)$. It follows from (7.63) that, in the relevant range for $\rho$, $f$ satisfies $f(\rho) \geq (\rho/2)^{3/2}$.

Finally we note that each solution for $Y = \sin\vartheta_2$ corresponds to two solutions for $\vartheta_2 \in [-\pi, \pi]$; and, as mentioned above, once $\vartheta_2$ is given, $\vartheta_1$ is uniquely determined.

Thus we conclude that:

1. For $K > f(\rho)$ we have four solutions for $Y$, two of type $(a)$ and two of type $(b)$ as in (7.69), and hence eight critical points beside the two at the poles and the four with $\vartheta_2 = \pm\pi/2$, for a total of *fourteen* critical points;
2. For $K < f(\rho)$ we have two solutions for $Y$, of type $(a)$, and hence four critical points beside those at the Poles and for $\vartheta_2 = \pm\pi/2$, for a total of *ten* critical points;
3. On the curve $K = f(\rho)$ there is a bifurcation, in which the two solutions of type $(b)$ disappear as $K$ is reduced. On the curve $K = f(\rho)$ the solutions $Y_b$ have $\vartheta_2 = \pm\pi/2$, hence merge with those studied before.

*Remark 32* Clearly, the simple expression $K = f(\rho)$ for the bifurcation curve was possible only because we have fixed the value of $\chi$. This curve is the section of the separatrix with the plane $\chi = \pi/2$. The separatrix in the full three-dimensional $(K, \rho, \chi)$ parameter space has been studied in [14] and [6], but it has an awkward analytic expression, which duly reduces to (7.63) for $\chi = \pi/2$ (*modulo* the different scaling of $\rho$, as shown by (25) of [6]).

### 7.6.2 The Case $\chi = -\pi/2$

The case $\chi = -\pi/2$ is analyzed in the same way, though it entails a somewhat dissimilar outcome, on which se shall particularly concentrate.

Paralleling $f$ in (7.63), there is a continuous function $g$ defined for $0 \leq \rho \leq 2$ by

$$g(\rho) = \begin{cases} \sqrt{\frac{2\rho^2(1-\rho)}{3(6-\rho)}} & \text{for } 0 \leq \rho \leq 1 \\ \sqrt{2(2-\rho)(\rho-1)} & \text{for } 1 \leq \rho \leq 2, \end{cases} \tag{7.70}$$

whose graph is reproduced in Fig. 13 for the reader's ease.

1. For $K > g(\rho)$ we have eight generic critical points beside the two at the poles and four on the special meridians with $\vartheta_2 = \pm\pi/2$, for a total of *fourteen* critical points. Four are maxima, four minima, and the remaining six are saddles.
2. For $K < g(\rho)$, we have a total of *ten* critical points, of which three are maxima, three minima, and the remaining four are saddles.
3. For $K = g(\rho)$, two different scenarios present themselves, according to whether $0 < \rho < 1$ or $1 < \rho < 2$. In the former case, the critical points are ten, whereas in the latter case they are *twelve*. In both cases, the total number of maxima is three, as many as the minima; only the number of saddles differs: there are four for $0 < \rho < 1$ and six for $1 < \rho < 2$. In the former case, two saddles are degenerate, but all four have index $\iota = -1$. In the latter case, two out of the six saddles are degenerate and have index $\iota = 0$ (see Remark 16), while the remaining four are not degenerate and have the usual index $\iota = -1$.
4. A special note is deserved by the limiting values $\rho = 1$ and $\rho = 2$. For the former value, the total number of critical points is *eight*, whereas it is ten for the latter. For $\rho = 1$, three maxima and three minima are accompanied by two degenerate saddles, each with index $\iota = -2$. For $\rho = 2$, the same number of maxima and minima is accompanied by four degenerate saddles, each with index $\iota = -1$, for a total of ten critical points. The total number of critical points for both these limiting cases were predicted by our taxonomic analysis in Sect. 5.4: the case $\rho = 1$ falls under row $(c_2)$ in Table 2, while the case $\rho = 2$ falls under row $(c)$ in Table 1.

We now describe in more detail how the different components of this varied landscape of critical points are combined together. The critical points for $K < g(\rho)$ are related to those for

$K > g(\rho)$ in two different ways, corresponding to the two branches of $g$ in (7.70), according to whether $0 < \rho < 1$ or $1 < \rho < 2$.

For $0 < \rho < 1$, the four critical points on the special meridians $\vartheta_2 = \pm\pi/2$ survive as $K$ decreases below $g(\rho)$; two saddles, one for each meridian, stay saddles, whereas a maximum on one meridian ($\vartheta_2 = \pi/2$) becomes a saddle, as does a minimum on the other meridian ($\vartheta_2 = -\pi/2$). On the other hand, always for $0 < \rho < 1$, two generic critical points, which are both saddles, approach each meridian $\vartheta_2 = \pm\pi/2$ as $K$ approaches $g(\rho)$ from above; they coalesce for $K = g(\rho)$ on the extremal that on the targeted meridian will transform into a saddle, and disappear as $K < g(\rho)$. For $K = g(\rho)$, on each meridian $\vartheta_2 = \pm\pi/2$ the coalesced critical point is a degenerate saddle with index $\iota = -1$.

For $1 < \rho < 2$, the evolution of critical points is somewhat different, though the final outcome is identical. As $K$ decreases below $g(\rho)$, the four critical points on meridians $\vartheta_2 = \pm\pi/2$ cease to exist, but they do not mingle with the generic critical points, which instead survive. They rather annihilate in pairs on each meridian for $K = g(\rho)$. The superposition of a maximum with a saddle (for $\vartheta_2 = \pi/2$) and that of a minimum with a saddle (for $\vartheta_2 = -\pi/2$) give rise to a critical point with index $\iota = 0$, so that for $K = g(\rho)$ and $1 < \rho < 2$ the total number of critical points is *twelve*: three maxima, three minima, and six saddles.

To determine the nature of the latter critical points, we expanded $\Psi_{or}$ in their vicinity. For $\vartheta_2 = -\pi/2$ and $K = g(\rho)$, we found a degenerate saddle located at

$$\vartheta_1 = \vartheta_S = -\arcsin\sqrt{\frac{\rho-1}{3-\rho}} \tag{7.71}$$

and we computed

$$\begin{aligned}\Psi_{or}(\vartheta_1, \vartheta_2) = &-\sqrt{(\rho-1)(3-\rho)} + 12(2-\rho)\sqrt{\frac{\rho-1}{3-\rho}}\left(\vartheta_2 + \frac{\pi}{2}\right)^2 \\ &+ \frac{1}{2}\sqrt{2(2-\rho)(3-\rho)}(\vartheta_1 - \vartheta_S)^3 \\ &+ 3\sqrt{2}(5\rho-6)\sqrt{\frac{2-\rho}{3-\rho}}\left(\vartheta_2 + \frac{\pi}{2}\right)^2(\vartheta_1 - \vartheta_S) \\ &+ O(4).\end{aligned} \tag{7.72}$$

A similar formula applies to the degenerate saddle on the meridian $\vartheta_2 = \pi/2$. It is then easy to conclude that both critical points are degenerate saddles with index $\iota = 0$ (see, for example [2]). They migrate towards the poles as $\rho$ approaches 2 along the line $K = g(\rho)$ and towards the Equator as $\rho$ approaches 1. Correspondingly, the North Pole becomes a degenerate maximum (while the South Pole becomes a degenerate minimum) and the Equator hosts two symmetric "monkey saddles".

## 7.7 Trivial Symmetry $\{e\}$: The Bulk $\mathcal{B}$

We have so far discussed the critical strata that correspond to nontrivial symmetries in the cylinder $\mathsf{C}$ representing the parameter space of our theory.

The situation in the bulk $\mathcal{B}$ of the cylinder, i.e., for the generic case, has been discussed in detail in our previous work [14]; there we have shown that—quite surprisingly—there are *two* generic octupolar phases, characterized by a different number (10 and 14) of critical

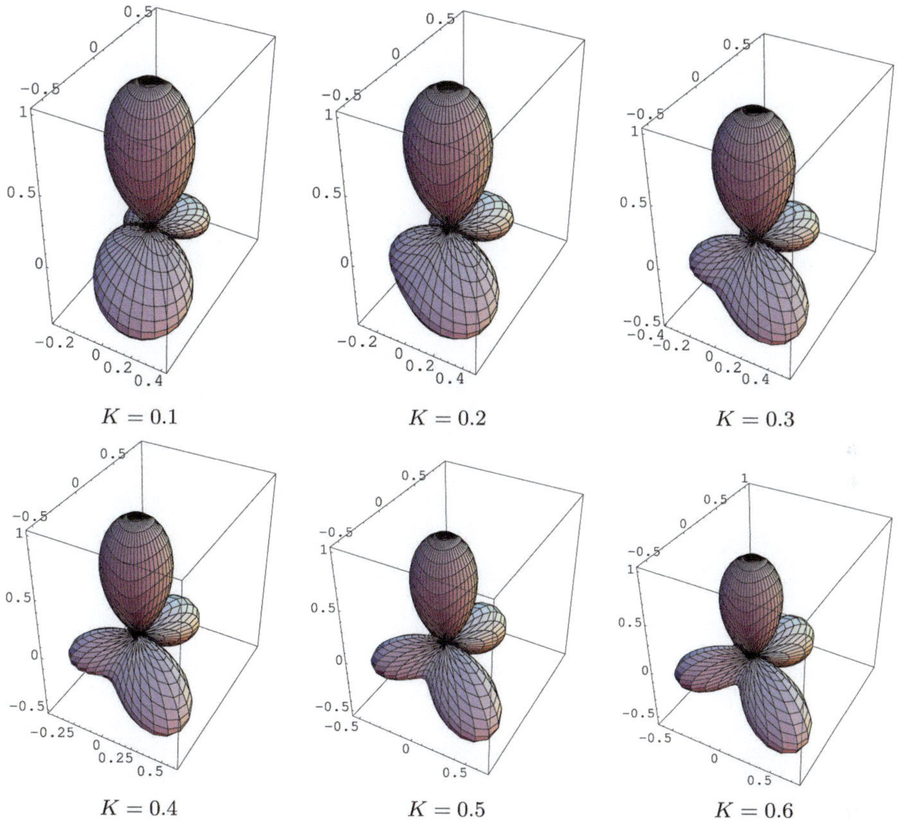


**Fig. 14** The potential for $\chi = \pi/3$ and $\rho = 1/2$, for different values of $K > 0$. We observe the transition from a phase with three maxima at low $K$ to a phase with four maxima at higher $K$. Generically, the potential is never reflection invariant through a vertical plane

points, and separated in parameter space by a *separatrix*. Moreover, one could also distinguish the cases where the maximum in the North Pole is the absolute one, and that where it is a local maximum but not the absolute one; these two regions are separated by a *dome*, having its vertex in one of the tetrahedral points $\mathcal{T}$ and meeting the disk $\mathcal{D}$ on the circle of radius $\rho = 1$. This dome has been further investigated, providing more detailed information, in [6]. The generic incarnations of the octupolar potential below the dome in parameter space are illustrated in Figs. 14 and 15.

One easily distinguishes two different regions where the potential can be seen as topologically equivalent to the highly symmetric incarnations met on the disk $\mathcal{D}$ and at the tetrahedral points $\mathcal{T}$. In fact, the study of the generic configurations in the bulk, and of the separatrix between the two octupolar phases, was based on continuation techniques starting from these singular strata. Much of this study has already been presented in our previous work [6, 14].

Here, the more detailed insight gained through the analysis of the highly symmetric cases $\chi = \pm\pi/2$ allows us to refine our understanding of the separatrix by establishing the existence of an extension where the total number of critical points for the octupolar potential is either 8 or 12, instead of the 10 that had already been found in [6, 14].

Figure 16 shows the outcome a standard numerical continuation technique applied to the sector $-\pi/2 \leq \chi \leq -\pi/6$ in parameter space $(K, \rho, \chi)$, which in view of the symmetries

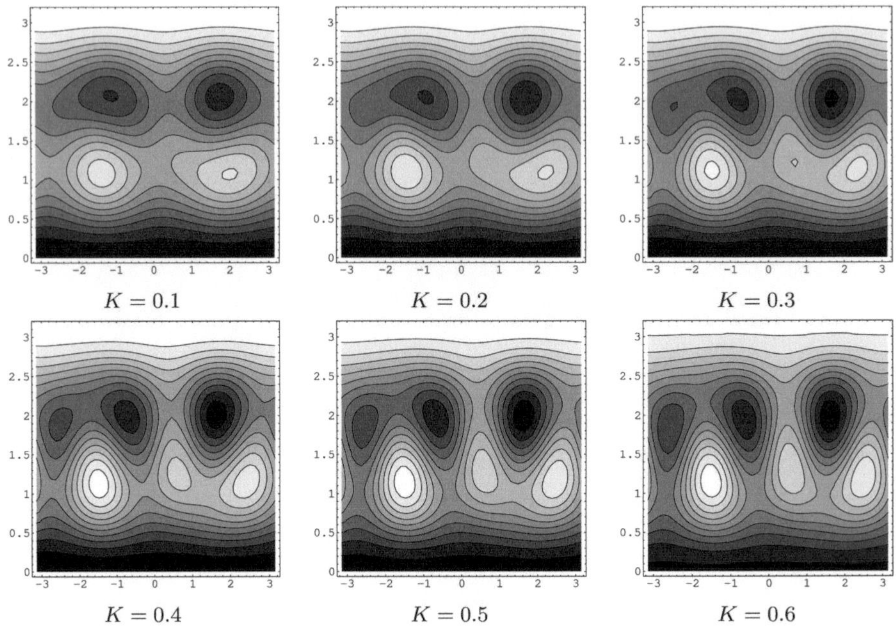

Fig. 15 Same as Fig. 14 but with contour plots

described above is the only one bearing essential information. The educated eye will discern the graphs of functions $g$ and $f$, plotted against $\rho$ upon the sections $\chi = -\pi/2$ and $\chi = -\pi/6$, respectively. The remaining curves outline the whole extended separatrix.

A line of cusps connects the point $K = 0$, $\rho = 1$, $\chi = -\pi/2$ with the point $K = 1$, $\rho = 2$, $\chi = -\pi/6$: it bears 8 critical points for the octupolar potential. The line that connects the point $K = 0$, $\rho = 2$, $\chi = -\pi/2$ with the point $K = 1$, $\rho = 2$, $\chi = -\pi/6$ consists of an arc of circle of radius 2 in the plane $K = 0$ and a straight segment orthogonal to that plane: it bears instead 10 critical points. The bump delimited by these degenerate lines bears 12 critical points. Always, above the separatrix the critical points are 14, whereas they are 10 below it. Our previous studies were confined to the cylinder $0 \leq \rho \leq 1$, and so they were blind to the outer component of the separatrix with 12 critical points for the octupolar potential.

## 7.8 Special Transformations and Potential Invariance

Our choice of orientation and scaling allowed to simplify the potential, passing from seven to three parameters. This also reduced the allowed transformations. In fact, except at the special tetrahedral point $\rho = 0$, $K = K_0$ (see above) we can only consider transformations which preserve the unit sphere *and* leave invariant the $z$ axis; we refer to these as *oriented orthogonal transformations*. They are just maps in $O(2)$ acting on the $(x, y)$ variables, and hence correspond to matrices of the form

$$M = \begin{pmatrix} \cos\gamma & -s\sin\gamma \\ \sin\gamma & s\cos\gamma \end{pmatrix}, \tag{7.73}$$

where $s$ is just a sign, $s = \pm 1$ and corresponds to the determinant of $M$.

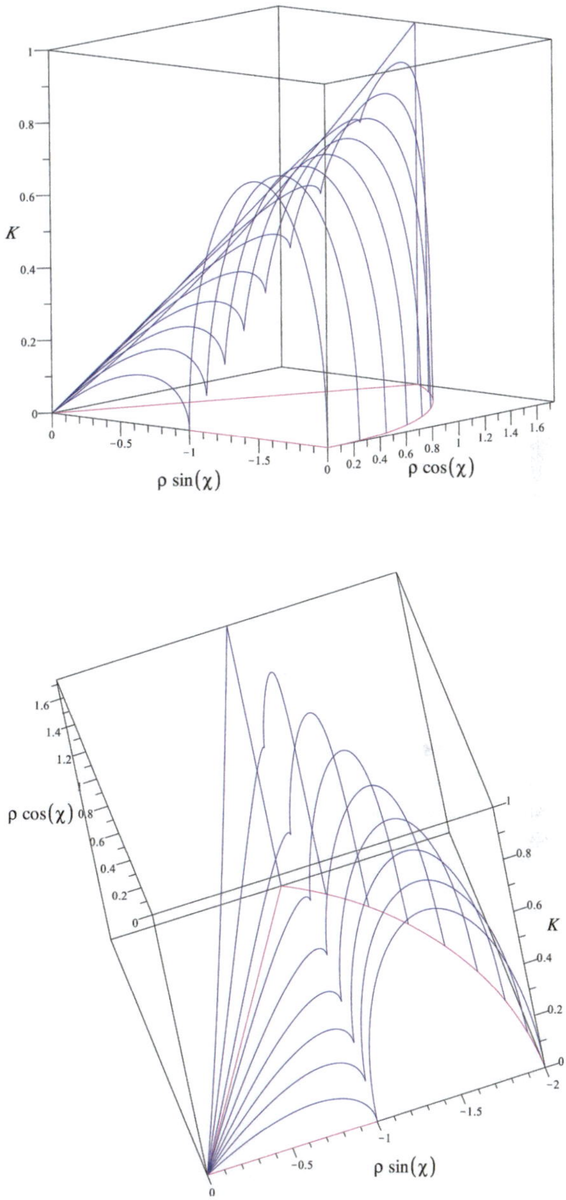


**Fig. 16** Two views of the whole separatrix, which is outlined by curves at equally spaced values of $\chi$ in the interval $[-\pi/2, -\pi/6]$. A line of cusps, which bears 8 critical points for the octupolar potential, separates the inner component of the separatrix with 10 critical points from the outer component with 12 critical points

When we require invariance of $\Phi$ under such transformations, we obtain—by standard algebra—five classes of nontrivial solutions; each of these can exist only for certain values of the parameters, i.e., for certain regions $R$ in the parameter space. The results are summarized in Table 7. This table confirms that our classification of different phases is complete. In fact, the first class gives $G = O(2)$ in $\mathcal{C}$; the second one provides $G = D_{2h}$ in $\mathcal{D}$; the third one provides $G = D_{3h}$ in $\mathcal{A}$, while the fourth points out at the fact that a symmetry $C_3$ is also present in $\mathcal{A}$; finally the fifth class shows that on the special planes $\mathcal{P}$ there is a reflection symmetry.

**Table 7** Different oriented transformations as symmetries of the oriented potential in different regions $R$ of the parameter space. The symbol – means that any value of the corresponding parameter is allowed

| $K$ | $\rho$ | $\chi$ | $R$ | $s$ | $\gamma$ |
|---|---|---|---|---|---|
| 0 | 0 | – | $\mathcal{C}$ | $\pm 1$ | – |
| 0 | – | – | $\mathcal{D}$ | $-1$ | $\pm\arccos(\pm\sin\chi)$ |
| 0 | – | – | $\mathcal{D}$ | $+1$ | $\pm\pi$ |
| – | 0 | – | $\mathcal{A}$ | $-1$ | $-5\pi/3$ |
| – | 0 | – | $\mathcal{A}$ | $-1$ | $-\pi/2$ |
| – | 0 | – | $\mathcal{A}$ | $-1$ | $-\pi/3$ |
| – | 0 | – | $\mathcal{A}$ | $-1$ | $+\pi/3$ |
| – | 0 | – | $\mathcal{A}$ | $-1$ | $+\pi/2$ |
| – | 0 | – | $\mathcal{A}$ | $-1$ | $+5\pi/3$ |
| – | 0 | – | $\mathcal{A}$ | $+1$ | $-\pi$ |
| – | 0 | – | $\mathcal{A}$ | $+1$ | $-4\pi/3$ |
| – | 0 | – | $\mathcal{A}$ | $+1$ | $-2\pi/3$ |
| – | 0 | – | $\mathcal{A}$ | $+1$ | $+2\pi/3$ |
| – | 0 | – | $\mathcal{A}$ | $+1$ | $+4\pi/3$ |
| – | 0 | – | $\mathcal{A}$ | $+1$ | $+\pi$ |
| – | – | $-5\pi/6$ | $\mathcal{P}_+$ | $\pm 1$ | $\pi/3$ |
| – | – | $-\pi/2$ | $\mathcal{P}_0$ | $\pm 1$ | $\pi$ |
| – | – | $-\pi/6$ | $\mathcal{P}_-$ | $\pm 1$ | $-\pi/3$ |
| – | – | $+\pi/6$ | $\mathcal{P}_+$ | $\pm 1$ | $\pi/3$ |
| – | – | $+\pi/2$ | $\mathcal{P}_0$ | $\pm 1$ | $\pi$ |
| – | – | $+5\pi/6$ | $\mathcal{P}_-$ | $\pm 1$ | $-\pi/3$ |

**Table 8** Summary of the totality of critical points in parameter space with index $\iota \neq 0$, irrespective of all their possible additional features and characterization

| $G$ | set | crit. pts. |
|---|---|---|
| $\{e\}$ | $\mathcal{B}$ | 10 or 14 |
| $D_{\infty h}$ | $\mathcal{C}$ | $2+\infty$ |
| $D_{2h}$ | $\mathcal{D}$ | 10 |
| $D_{3h}$ | $\mathcal{A}$ | 14 |
| $T_d$ | $\mathcal{T}$ | 14 |

## 8 Summary of Critical Points

In Table 8, we summarize our findings concerning critical points of the octupolar potential $\Phi$ in different regions of the admissible semi-definite cylinder $\mathsf{C}_+$ in parameter space. We preliminary leave out the critical points with index $\iota = 0$, which we found on some special bifurcation loci. A more refined description of the landscape of critical points is outlined at the end of the section.

The relations between geometric (and symmetry) strata of the cylinder are summarized in this diagram

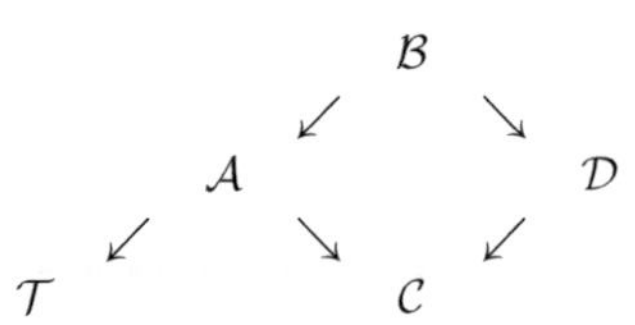

It should be noted that $\mathcal{C}$ and $\mathcal{T}$ can be reached from the bulk $\mathcal{B}$ also directly, i.e., without passing through $\mathcal{A}$ or $\mathcal{D}$.

Similarly, the relation between isotropy subgroups are summarized in the next diagram

$$\begin{array}{ccccccc} D_{\infty h} & & & T_d & & & D_{\infty h} \\ \downarrow & & \swarrow & & \searrow & & \downarrow \\ D_{3h} & & & & & & D_{2h} \\ & & \searrow & & \swarrow & & \\ & & & \{e\} & & & \end{array}$$

Here we do not distinguish in $\mathcal{B}$ between the region $\mathcal{B}_4$ with four maxima of $\Phi$ from the region $\mathcal{B}_3$ with three maxima, as these correspond to different phases which do not differ in terms of symmetry. However, if we wish to classify the regions in $\mathsf{C}_+$ according to the cardinality of the real spectrum of the octupolar tensor $\mathbf{A}$ corresponding to $\Phi$ (which is the number of critical points of $\Phi$ on the unit sphere $S^2$), we need to be more refined. Then even the distinction between $\mathcal{B}_3$ and $\mathcal{B}_4$ is too gross.

We have shown that $\mathcal{B}_3$ and $\mathcal{B}_4$ are separated by a separatrix $\mathcal{S}$, a surface in parameter space which consists of two folds, $\mathcal{S}_1$ and $\mathcal{S}_2$, where $\Phi$ has 10 and 12 critical points, respectively. These subsurfaces are divided by a line of cusps $\mathcal{L}_1$ where $\Phi$ has 8 critical points. Moreover, $\mathcal{S}_2$ is bordered by another line, $\mathcal{L}_2$, where $\Phi$ has 10 critical points.

## 9 Conclusions

An octupolar tensor $\mathbf{A}$ is any third order, completely symmetric and completely traceless tensor. In 2D, $\mathbf{A}$ has the symmetries of an equilateral triangle and it can be effectively represented by a vector in the plane. In 3D, the symmetries enjoyed by $\mathbf{A}$ outline a much richer landscape. This paper has classified all of them by introducing the octupolar potential $\Phi$ associated with $\mathbf{A}$, that is, the cubic form of $\mathbf{A}$ restricted to the unit sphere $S^2$. The maxima (and antipodal minima) of $\Phi$ and their locations on the unit sphere describe the whole variety of octupolar tensors and allow for a visual representation of their symmetries.

We showed that a semi-indefinite cylinder $\mathsf{C}_+$ in a three-dimensional parameter space suffices to represent all possible incarnations of $\mathbf{A}$ in a three-dimensional physical space. Such a reduction (of the originally seven-dimensional parameter space) is obtained by fixing a maximum of $\Phi$ on the North Pole of $S^2$ and scaling its value to unity. We identified in $\mathsf{C}_+$ a number of special regions characterized by different symmetry groups for $\Phi$ and ordering of its maxima relative to the *orienting* maximum at the North Pole. For sake of illustration, each of these regions could be further divided in two subregions, which we distinguish by a $^+$ or a $^-$ superscript indicating where the maxima of $\Phi$ that supplement the orienting maximum have a larger or a smaller value than the latter.

The axis $\mathcal{A}$ of the cylinder $\mathsf{C}_+$ is characterized by a potential $\Phi$ enjoying the $D_{3h}$ symmetry, with 4 maxima, 3 of which are equal and either exceed the fourth, orienting maximum (in $\mathcal{A}^+$) or fall short of it (in $\mathcal{A}^-$). A special point of $\mathcal{A}$ separates $\mathcal{A}^+$ and $\mathcal{A}^-$: this is the only point with tetrahedral symmetry $T_d$, which, despite its singularity, has given its name to a whole class of bent-core liquid crystal phases.[15]

On a disk $\mathcal{D}$, which is the base of $\mathsf{C}_+$, $\Phi$ has the $D_{2h}$ symmetry and possesses 3 maxima. $\mathcal{D}$ can be separated into an inner disk $\mathcal{D}_1$ and an outer annulus $\mathcal{D}_2$. The center $\mathcal{C}$ of $\mathcal{D}$ has a special nature: there the potential $\Phi$ is axially symmetric. The symmetry group is $D_{\infty h}$

[15] Which thus are presumably more complicated than anticipated.

and the (primary) maximum at the North Pole is accompanied by a full circle of (secondary) maxima in the southern hemisphere of $S^2$.

Away from all these special loci is the generic bulk $\mathcal{B}$, where $\Phi$ has either 4 (generically unequal) maxima (in $\mathcal{B}_4$) or 3 (unequal) maxima (in $\mathcal{B}_3$), separated by a surface $\mathcal{S}$ in parameter space, called the separatrix. Both $\mathcal{B}_3$ and $\mathcal{B}_4$ (where, if necessary, we could distinguish the variants $\mathcal{B}_3^\pm$ and $\mathcal{B}_4^\pm$) enjoy three planes of symmetry, collectively denoted as $\mathcal{P}$. Correspondingly, $\Phi$ has on $S^2$ a reflection symmetry across a plane through the poles. Therefore, when the parameters fall in particular in $\mathcal{B}_4 \cap \mathcal{P}$, two secondary maxima of $\Phi$ are equal.

Much like the decomposition of $\mathcal{D}$ into the union of $\mathcal{D}_1$ and $\mathcal{D}_2$, the separatrix $\mathcal{S}$ can be further decomposed into the union of two surfaces, an inner $\mathcal{S}_1$ and an outer $\mathcal{S}_2$, where $\Phi$ has 10 and 12 critical points, respectively. The boundary $\mathcal{L}_1$ of $\mathcal{S}_1$ is a line of cusps in parameter space where $\Phi$ has 8 critical points, whereas the outer boundary $\mathcal{L}_2$ of $\mathcal{S}_2$ is a line where $\Phi$ has 10 critical points.

All secondary maxima of $\Phi$ have a remarkable, universal feature, irrespective of the choice of parameters: they fall in the southern hemisphere of $S^2$, when the North Pole is marked by the primary, orienting maximum.

We trust that all the qualitative features of the octupolar potential described in this paper would prompt a better understanding of the many physical instances where an octupolar tensor is at play.

**Acknowledgements** E.G. Virga acknowledges the kind hospitality of the Oxford Centre for Nonlinear PDE, where part of this work was done while he was visiting the Mathematical Institute at the University of Oxford.



## Appendix: The Tetrahedron Group

In this Appendix we give further detail—beyond those mentioned in Sect. 6—on the tetrahedron group. We will work on a concrete realization of it in three-dimensional space; the points identifying the tetrahedron will be

$$\left(0, \frac{2\sqrt{2}}{3}, -\frac{1}{3}\right), \left(\sqrt{\frac{2}{3}}, -\frac{\sqrt{2}}{3}, -\frac{1}{3}\right), \left(-\sqrt{\frac{2}{3}}, -\frac{\sqrt{2}}{3}, -\frac{1}{3}\right), (0, 0, 1).$$

In angular coordinates $(\vartheta_1, \vartheta_2)$, these are

$$\left(\theta_0, -\frac{\pi}{2}\right), \left(\theta_0, -\frac{\pi}{6}\right), \left(\theta_0, \frac{5\pi}{6}\right), \left(\frac{\pi}{2}, *\right),$$

where the symbol $*$ means that in this case $\vartheta_2$ is not determined, and

$$\theta_0 = -\arcsin\left(\frac{1}{3}\right) \doteq -0.34.$$

The tetrahedron group $T_d \subset O(3)$ is made of 12 matrices of determinant $+1$, associated to rotations of an angle $2\pi/3$ and multiples around each of the four axes of the tetrahedron, denoted as $T_d^+ \subset SO(3)$; and other 12 matrices of determinant $-1$, collectively denoted as $T_d^-$, among which are those associated to reflections through planes containing axes of the tetrahedron.

We now give the twelve matrices in $T_d^+$; these are:

$$M_1 = \begin{pmatrix} 1 & 0 & 0 \\ 0 & 1 & 0 \\ 0 & 0 & 1 \end{pmatrix}, \qquad M_2 = \begin{pmatrix} -\frac{1}{2} & \frac{\sqrt{3}}{2} & 0 \\ -\frac{\sqrt{3}}{2} & -\frac{1}{2} & 0 \\ 0 & 0 & 1 \end{pmatrix},$$

$$M_3 = \begin{pmatrix} -\frac{1}{2} & -\frac{\sqrt{3}}{2} & 0 \\ \frac{\sqrt{3}}{2} & -\frac{1}{2} & 0 \\ 0 & 0 & 1 \end{pmatrix}, \qquad M_4 = \begin{pmatrix} \frac{1}{2} & \frac{\sqrt{3}}{2} & 0 \\ \frac{1}{2\sqrt{3}} & -\frac{1}{6} & \frac{2\sqrt{2}}{3} \\ \sqrt{\frac{2}{3}} & -\frac{\sqrt{2}}{3} & -\frac{1}{3} \end{pmatrix},$$

$$M_5 = \begin{pmatrix} \frac{1}{2} & \frac{1}{2\sqrt{3}} & \sqrt{\frac{2}{3}} \\ \frac{\sqrt{3}}{2} & -\frac{1}{6} & -\frac{\sqrt{2}}{3} \\ 0 & \frac{2\sqrt{2}}{3} & -\frac{1}{3} \end{pmatrix}, \qquad M_6 = \begin{pmatrix} \frac{1}{2} & -\frac{\sqrt{3}}{2} & 0 \\ -\frac{1}{2\sqrt{3}} & -\frac{1}{6} & \frac{2\sqrt{2}}{3} \\ -\sqrt{\frac{2}{3}} & -\frac{\sqrt{2}}{3} & -\frac{1}{3} \end{pmatrix},$$

$$M_7 = \begin{pmatrix} \frac{1}{2} & -\frac{1}{2\sqrt{3}} & -\sqrt{\frac{2}{3}} \\ -\frac{\sqrt{3}}{2} & -\frac{1}{6} & -\frac{\sqrt{2}}{3} \\ 0 & \frac{2\sqrt{2}}{3} & -\frac{1}{3} \end{pmatrix}, \qquad M_8 = \begin{pmatrix} -\frac{1}{2} & \frac{1}{2\sqrt{3}} & \sqrt{\frac{2}{3}} \\ -\frac{1}{2\sqrt{3}} & \frac{5}{6} & -\frac{\sqrt{2}}{3} \\ -\sqrt{\frac{2}{3}} & -\frac{\sqrt{2}}{3} & -\frac{1}{3} \end{pmatrix},$$

$$M_9 = \begin{pmatrix} -\frac{1}{2} & -\frac{1}{2\sqrt{3}} & -\sqrt{\frac{2}{3}} \\ \frac{1}{2\sqrt{3}} & \frac{5}{6} & -\frac{\sqrt{2}}{3} \\ \sqrt{\frac{2}{3}} & -\frac{\sqrt{2}}{3} & -\frac{1}{3} \end{pmatrix}, \qquad M_{10} = \begin{pmatrix} 0 & \frac{1}{\sqrt{3}} & -\sqrt{\frac{2}{3}} \\ \frac{1}{\sqrt{3}} & -\frac{2}{3} & -\frac{\sqrt{2}}{3} \\ -\sqrt{\frac{2}{3}} & -\frac{\sqrt{2}}{3} & -\frac{1}{3} \end{pmatrix},$$

$$M_{11} = \begin{pmatrix} 0 & -\frac{1}{\sqrt{3}} & \sqrt{\frac{2}{3}} \\ -\frac{1}{\sqrt{3}} & -\frac{2}{3} & -\frac{\sqrt{2}}{3} \\ \sqrt{\frac{2}{3}} & -\frac{\sqrt{2}}{3} & -\frac{1}{3} \end{pmatrix}, \qquad M_{12} = \begin{pmatrix} -1 & 0 & 0 \\ 0 & \frac{1}{3} & \frac{2\sqrt{2}}{3} \\ 0 & \frac{2\sqrt{2}}{3} & -\frac{1}{3} \end{pmatrix}.$$

The multiplication table for these matrices is the following:

$$P_{11} = \begin{pmatrix} 1 & 2 & 3 & 4 & 5 & 6 & 7 & 8 & 9 & 10 & 11 & 12 \\ 2 & 3 & 1 & 11 & 7 & 8 & 12 & 10 & 4 & 6 & 9 & 5 \\ 3 & 1 & 2 & 9 & 12 & 10 & 5 & 6 & 11 & 8 & 4 & 7 \\ 4 & 12 & 6 & 5 & 1 & 11 & 9 & 2 & 10 & 7 & 3 & 8 \\ 5 & 8 & 11 & 1 & 4 & 3 & 10 & 12 & 7 & 9 & 6 & 2 \\ 6 & 4 & 12 & 10 & 8 & 7 & 1 & 11 & 3 & 2 & 5 & 9 \\ 7 & 10 & 9 & 2 & 11 & 1 & 6 & 5 & 12 & 4 & 8 & 3 \\ 8 & 11 & 5 & 6 & 10 & 12 & 2 & 9 & 1 & 3 & 7 & 4 \\ 9 & 7 & 10 & 12 & 3 & 4 & 11 & 1 & 8 & 5 & 2 & 6 \\ 10 & 9 & 7 & 8 & 6 & 5 & 3 & 4 & 2 & 1 & 12 & 11 \\ 11 & 5 & 8 & 7 & 2 & 9 & 4 & 3 & 6 & 12 & 1 & 10 \\ 12 & 6 & 4 & 3 & 9 & 2 & 8 & 7 & 5 & 11 & 10 & 1 \end{pmatrix}.$$

These generate several subgroups; in particular—apart from the trivial ones consisting of $M_1$ alone and of the full group $T_d^+$—we have four subgroups of order three,

$$G_1 = \{M_1, M_2, M_3\}, \qquad G_2 = \{M_1, M_4, M_5\}, \qquad G_3 = \{M_1, M_6, M_7\}, \\ G_4 = \{M_1, M_8, M_9\};$$

three groups of order two,

$$G_5 = \{M_1, M_{10}\}, \qquad G_6 = \{M_1, M_{11}\}, \qquad G_7 = \{M_1, M_{12}\};$$

and one group of order four,

$$G_8 = \{M_1, M_{10}, M_{11}, M_{12}\}.$$

The latter is the only nontrivial normal subgroup, and also the only one acting freely.

We can also easily determine the subspaces $F_k$ admitting each of these $G_k$ as symmetry subgroups; in particular,

$$F_1 = (0, 0, z), \qquad F_2 = (\sqrt{6}z, \sqrt{2}z, z), \qquad F_3 = (-\sqrt{6}z, \sqrt{2}z, z), \\ F_4 = (0, -2\sqrt{2}z, z);$$

the subgroups $G_k$, $k = 1, 2, 3, 4$ act as rotations (by an angle $2\pi/3$) around these axes, which are just the axes of the tetrahedron. Moreover,

$$F_5 = \left(-\sqrt{\frac{3}{2}}z, -\sqrt{\frac{1}{2}}z, z\right), \qquad F_6 = \left(\sqrt{\frac{3}{2}}z, -\sqrt{\frac{1}{2}}z, z\right), \qquad F_7 = (0, \sqrt{2}z, z);$$

these subgroups $G_k$, $k = 5, 6, 7$, represent rotations by $\pi$ around the given axes $F_k$. Note that $F_8 = \{(0, 0, 0)\}$, and correspondingly $G_8$ represents combined $\pi$ rotations around the $F_5, F_6, F_7$ axes.

We can give as well the twelve matrices in $T_d^-$; these are:

$$M_1^- = \begin{pmatrix} -1 & 0 & 0 \\ 0 & 1 & 0 \\ 0 & 0 & 1 \end{pmatrix}, \qquad M_2^- = \begin{pmatrix} \frac{1}{2} & -\frac{\sqrt{3}}{2} & 0 \\ -\frac{\sqrt{3}}{2} & -\frac{1}{2}, 0 & \\ 0 & 0 & 1 \end{pmatrix},$$

$$M_3^- = \begin{pmatrix} \frac{1}{2} & \frac{\sqrt{3}}{2} & 0 \\ \frac{\sqrt{3}}{2} & -\frac{1}{2}, 0 & \\ 0 & 0 & 1 \end{pmatrix}, \qquad M_4^- = \begin{pmatrix} -\frac{1}{2} & -\frac{\sqrt{3}}{2} & 0 \\ \frac{1}{2\sqrt{3}} & -\frac{1}{6} & \frac{2\sqrt{2}}{3} \\ \sqrt{\frac{2}{3}} & -\frac{\sqrt{2}}{3} & -\frac{1}{3} \end{pmatrix},$$

$$M_5^- = \begin{pmatrix} -\frac{1}{2} & -\frac{1}{2\sqrt{3}} & -\sqrt{\frac{2}{3}} \\ \frac{\sqrt{3}}{2} & -\frac{1}{6} & -\frac{\sqrt{2}}{3} \\ 0 & \frac{2\sqrt{2}}{3} & -\frac{1}{3} \end{pmatrix}, \qquad M_6^- = \begin{pmatrix} -\frac{1}{2} & \frac{\sqrt{3}}{2} & 0 \\ -\frac{1}{2\sqrt{3}} & -\frac{1}{6} & \frac{2\sqrt{2}}{3} \\ -\sqrt{\frac{2}{3}} & -\frac{\sqrt{2}}{3} & -\frac{1}{3} \end{pmatrix},$$

$$M_7^- = \begin{pmatrix} -\frac{1}{2} & \frac{1}{2\sqrt{3}} & \sqrt{\frac{2}{3}} \\ -\frac{\sqrt{3}}{2} & -\frac{1}{6} & -\frac{\sqrt{2}}{3} \\ 0, \frac{2\sqrt{2}}{3} & -\frac{1}{3} & \end{pmatrix}, \qquad M_8^- = \begin{pmatrix} \frac{1}{2} & -\frac{1}{2\sqrt{3}} & -\sqrt{\frac{2}{3}} \\ -\frac{1}{2\sqrt{3}} & \frac{5}{6} & -\frac{\sqrt{2}}{3} \\ -\sqrt{\frac{2}{3}} & -\frac{\sqrt{2}}{3} & -\frac{1}{3} \end{pmatrix},$$

$$M_9^- = \begin{pmatrix} \frac{1}{2} & \frac{1}{2\sqrt{3}} & \sqrt{\frac{2}{3}} \\ \frac{1}{2\sqrt{3}} & \frac{5}{6} & -\frac{\sqrt{2}}{3} \\ \sqrt{\frac{2}{3}} & -\frac{\sqrt{2}}{3} & -\frac{1}{3} \end{pmatrix}, \qquad M_{10}^- = \begin{pmatrix} 0 & -\frac{1}{\sqrt{3}} & \sqrt{\frac{2}{3}} \\ \frac{1}{\sqrt{3}} & -\frac{2}{3} & -\frac{\sqrt{2}}{3} \\ -\sqrt{\frac{2}{3}} & -\frac{\sqrt{2}}{3} & -\frac{1}{3} \end{pmatrix},$$

$$M_{11}^- = \begin{pmatrix} 0 & \frac{1}{\sqrt{3}} & -\sqrt{\frac{2}{3}} \\ -\frac{1}{\sqrt{3}} & -\frac{2}{3} & -\frac{\sqrt{2}}{3} \\ \sqrt{\frac{2}{3}} & -\frac{\sqrt{2}}{3} & -\frac{1}{3} \end{pmatrix}, \qquad M_{12}^- = \begin{pmatrix} 1 & 0 & 0 \\ 0 & \frac{1}{3} & \frac{2\sqrt{2}}{3} \\ 0 & \frac{2\sqrt{2}}{3} & -\frac{1}{3} \end{pmatrix}.$$

Once $M_1^-$ has been defined, they are built by

$$M_k^- = M_1^- M_k.$$

It is obvious that $M_1^-$ represents a reflection (across the $(y, z)$ plane), so that the matrices $M_k^-$ represent the combination of rotations and reflections.

If we write $M_{k+12} = M_k^-$, the full multiplication table is given in block form by

$$P = \begin{pmatrix} P_{11} & P_{12} \\ P_{21} & P_{22} \end{pmatrix},$$

where $P_{11}$ has been given above, and the other blocks are:

$$P_{12} = \begin{pmatrix} 13 & 14 & 15 & 16 & 17 & 18 & 19 & 20 & 21 & 22 & 23 & 24 \\ 15 & 13 & 14 & 21 & 24 & 22 & 17 & 18 & 23 & 20 & 16 & 19 \\ 14 & 15 & 13 & 23 & 19 & 20 & 24 & 22 & 16 & 18 & 21 & 17 \\ 18 & 16 & 24 & 22 & 20 & 19 & 13 & 23 & 15 & 14 & 17 & 21 \\ 19 & 22 & 21 & 14 & 23 & 13 & 18 & 17 & 24 & 16 & 20 & 15 \\ 16 & 24 & 18 & 17 & 13 & 23 & 21 & 14 & 22 & 19 & 15 & 20 \\ 17 & 20 & 23 & 13 & 16 & 15 & 22 & 24 & 19 & 21 & 18 & 14 \\ 21 & 19 & 22 & 24 & 15 & 16 & 23 & 13 & 20 & 17 & 14 & 18 \\ 20 & 23 & 17 & 18 & 22 & 24 & 14 & 21 & 13 & 15 & 19 & 16 \\ 23 & 17 & 20 & 19 & 14 & 21 & 16 & 15 & 18 & 24 & 13 & 22 \\ 22 & 21 & 19 & 20 & 18 & 17 & 15 & 16 & 14 & 13 & 24 & 23 \\ 24 & 18 & 16 & 15 & 21 & 14 & 20 & 19 & 17 & 23 & 22 & 13 \end{pmatrix},$$

$$P_{21} = \begin{pmatrix} 13 & 14 & 15 & 16 & 17 & 18 & 19 & 20 & 21 & 22 & 23 & 24 \\ 14 & 15 & 13 & 23 & 19 & 20 & 24 & 22 & 16 & 18 & 21 & 17 \\ 15 & 13 & 14 & 21 & 24 & 22 & 17 & 18 & 23 & 20 & 16 & 19 \\ 16 & 24 & 18 & 17 & 13 & 23 & 21 & 14 & 22 & 19 & 15 & 20 \\ 17 & 20 & 23 & 13 & 16 & 15 & 22 & 24 & 19 & 21 & 18 & 14 \\ 18 & 16 & 24 & 22 & 20 & 19 & 13 & 23 & 15 & 14 & 17 & 21 \\ 19 & 22 & 21 & 14 & 23 & 13 & 18 & 17 & 24 & 16 & 20 & 15 \\ 20 & 23 & 17 & 18 & 22 & 24 & 14 & 21 & 13 & 15 & 19 & 16 \\ 21 & 19 & 22 & 24 & 15 & 16 & 23 & 13 & 20 & 17 & 14 & 18 \\ 22 & 21 & 19 & 20 & 18 & 17 & 15 & 16 & 14 & 13 & 24 & 23 \\ 23 & 17 & 20 & 19 & 14 & 21 & 16 & 15 & 18 & 24 & 13 & 22 \\ 24 & 18 & 16 & 15 & 21 & 14 & 20 & 19 & 17 & 23 & 22 & 13 \end{pmatrix},$$

$$P_{22} = \begin{pmatrix} 1 & 2 & 3 & 4 & 5 & 6 & 7 & 8 & 9 & 10 & 11 & 12 \\ 3 & 1 & 2 & 9 & 12 & 10 & 5 & 6 & 11 & 8 & 4 & 7 \\ 2 & 3 & 1 & 11 & 7 & 8 & 12 & 10 & 4 & 6 & 9 & 5 \\ 6 & 4 & 12 & 10 & 8 & 7 & 1 & 11 & 3 & 2 & 5 & 9 \\ 7 & 10 & 9 & 2 & 11 & 1 & 6 & 5 & 12 & 4 & 8 & 3 \\ 4 & 12 & 6 & 5 & 1 & 11 & 9 & 2 & 10 & 7 & 3 & 8 \\ 5 & 8 & 11 & 1 & 4 & 3 & 10 & 12 & 7 & 9 & 6 & 2 \\ 9 & 7 & 10 & 12 & 3 & 4 & 11 & 1 & 8 & 5 & 2 & 6 \\ 8 & 11 & 5 & 6 & 10 & 12 & 2 & 9 & 1 & 3 & 7 & 4 \\ 11 & 5 & 8 & 7 & 2 & 9 & 4 & 3 & 6 & 12 & 1 & 10 \\ 10 & 9 & 7 & 8 & 6 & 5 & 3 & 4 & 2 & 1 & 12 & 11 \\ 12 & 6 & 4 & 3 & 9 & 2 & 8 & 7 & 5 & 11 & 10 & 1 \end{pmatrix}.$$

We now have subgroups involving elements of both $T_d^+$ and $T_d^-$; in particular, using again the 1 through 24 numeration and denoting $M_k$ directly by $k$, we have the following subgroups; those of types $\mathcal{G}_a$ through $\mathcal{G}_c$ extending those seen above, while those of type $\mathcal{G}_d$ involve no other element of $T_d^+$ but the identity:

$$\mathcal{G}_a : (1,2,3,13,14,15), (1,2,3,16,21,23), (1,2,3,17,19,24), (1,2,3,18,20,22);$$
$$(1,4,5,13,18,19), (1,4,5,14,16,22), (1,4,5,15,21,24), (1,4,5,17,20,23);$$
$$(1,8,9,13,20,21), (1,8,9,14,19,23), (1,8,9,15,17,22), (1,8,9,16,18,24);$$
$$\mathcal{G}_b : (1,10,13,23), (1,10,14,17), (1,10,15,20), (1,10,16,19), (1,10,18,21),$$
$$(1,10,22,24);$$
$$(1,11,13,22), (1,11,14,21), (1,11,15,19), (1,11,16,20), (1,11,17,18),$$
$$(1,11,23,24);$$
$$(1,12,13,24), (1,12,14,18), (1,12,15,16), (1,12,17,21), (1,12,19,20),$$
$$(1,12,22,23);$$
$$\mathcal{G}_c : (1,10,11,12,13,22,23,24), (1,10,11,12,14,17,18,21),$$
$$(1,10,11,12,15,16,19,20);$$
$$\mathcal{G}_d : (1,13), (1,14), (1,15), (1,20), (1,21), (1,24).$$

Groups of type $\mathcal{G}_a$ contain rotations by $2\pi/3$ around an axis and reflections through a plane containing that axis, hence they are of type $D_{3h}$, and only the first one leaves the distinguished point at the North Pole untouched; those of type $\mathcal{G}_b$ contain rotations by $\pi$ around an axis and reflections through a plane containing that axis, hence are of type $D_{2h}$, but none of these leaves the distinguished point at the North Pole untouched; those of type $\mathcal{G}_c$ combine rotations and reflections through different axes, and none of them preserves the North Pole; and those of type $\mathcal{G}_d$ consist just of reflections in a plane, hence are of type $D_h$; the first three preserve the North Pole.

## References


1. Ashcroft, N.W., Mermin, D.N.: Solid State Physics. Academic Press/Saunders, San Diego/Philadelphia (1976)

2. Bolis, T.S.: Degenerate critical points. Math. Mag. **53**(5), 294–299 (1980)
3. Boyd, R.W.: Nonlinear Optics, 3rd edn. Academic Press, Burlington (2008)
4. Buckingham, A.D.: Angular correlation in liquids. Discuss. Faraday Soc. **43**, 205–211 (1967)
5. Cartwright, D., Sturmfels, B.: The number of eigenvalues of a tensor. Linear Algebra Appl. **438**(2), 942–952 (2013)
6. Chen, Y., Qi, L., Virga, E.G.: Octupolar tensors for liquid crystals. J. Phys. A, Math. Theor. **51**(2), 025206 (2018)
7. Cicogna, G.: Symmetry breakdown from bifurcation. Lett. Nuovo Cimento **31**, 600–602 (1981)
8. Cicogna, G.: A nonlinear version of the equivalent bifurcation lemma. J. Phys. A **23**, L1339–L1343 (1990)
9. Coleman, B.D., Markovitz, H., Noll, W.: Viscometric Flows of Non-Newtonian Fluids, Theory and Experiment. Springer Tracts in Natural Philosophy, vol. 5. Springer, Berlin (1966)
10. Dailey, C.A., Burke, B.J., Simpson, G.J.: The general failure of Kleinman symmetry in practical nonlinear optical applications. Chem. Phys. Lett. **390**(1), 8–13 (2004)
11. Fel, L.G.: Symmetry of the Fréedericksz transition in nonchiral nematic liquid crystals. Phys. Rev. E **52**, 2692–2701 (1995)
12. Fel, L.G.: Tetrahedral symmetry in nematic liquid crystals. Phys. Rev. E **52**, 702–717 (1995)
13. Gaeta, G.: Reduction and equivariant branching lemma: dynamical systems, evolution PDEs, and gauge theories. Acta Appl. Math. **28**, 43–68 (1992)
14. Gaeta, G., Virga, E.G.: Octupolar order in three dimensions. Eur. Phys. J. E **39**, 113 (2016)
15. de Gennes, P., Prost, J.: The Physics of Liquid Crystals. Oxford University Press, Oxford (1993)
16. Gramsbergen, E.F., Longa, L., de Jeu, W.H.: Landau theory of the nematic-isotropic phase transition. Phys. Rep. **135**(4), 195–257 (1986)
17. Hamermesh, M.: Group Theory and Its Application to Physical Problems p. 11501. Dover, Mineola (1989)
18. Kanis, D.R., Ratner, M.A., Marks, T.J.: Design and construction of molecular assemblies with large second-order optical nonlinearities. Quantum chemical aspects. Chem. Rev. **94**, 195–242 (1994)
19. Kittel, C.: Introduction to Solid State Physics. Wiley, New York (1996)
20. Kleinman, D.A.: Nonlinear dielectric polarization in optical media. Phys. Rev. **126**, 1977–1979 (1962)
21. Landau, L.D.: The theory of phase transitions. Nature **138**, 840–841 (1936)
22. Landau, L.D.: On the theory of phase transitions I. Zh. Èksp. Teor. Fiz. **7**, 19–32 (1937)
23. Landau, L.D., Lifshitz, E.M.: Statistical Physics. Pergamon Press, Oxford (1958)
24. Landau, L.D., Lifshitz, E.M.: Quantum Mechanics. Pergamon Press, Oxford (1965)
25. Liu, K., Greitemann, J., Pollet, L.: Generic first-order phase transitions between isotropic and orientational phases with polyhedral symmetries. Phys. Rev. E **97**, 012706 (2018)
26. Liu, K., Nissinen, J., Nussinov, Z., Slager, R.J., Wu, K., Zaanen, J.: Classification of nematic order in 2+1 dimensions: dislocation melting and $o(2)/Z_N$ lattice gauge theory. Phys. Rev. B **91**, 075103 (2015)
27. Liu, K., Nissinen, J., Slager, R.J., Wu, K., Zaanen, J.: Generalized liquid crystals: giant fluctuations and the vestigial chiral order of $i$, $o$, and $t$ matter. Phys. Rev. X **6**, 041025 (2016)
28. Lubensky, T.C., Radzihovsky, L.: Theory of bent-core liquid-crystal phases and phase transitions. Phys. Rev. E **66**, 031704 (2002)
29. Ni, G., Qi, L., Wang, F., Wang, Y.: The degree of the e-characteristic polynomial of an even order tensor. J. Math. Anal. Appl. **329**(2), 1218–1229 (2007)
30. Noll, W.: Finite-Dimensional Spaces: Algebra, Geometry, and Analysis. Mechanics: Analysis, vol. 10. Martinus Nijhoff, Dordrecht (1987)
31. Peliti, L.: Statistical Mechanics in a Nutshell. Princeton University Press, Princeton (2011)
32. Qi, L.: Eigenvalues of a real supersymmetric tensor. J. Symb. Comput. **40**(6), 1302–1324 (2005)
33. Qi, L.: Rank and eigenvalues of a supersymmetric tensor, the multivariate homogeneous polynomial and the algebraic hypersurface it defines. J. Symb. Comput. **41**(12), 1309–1327 (2006)
34. Qi, L.: Eigenvalues and invariants of tensors. J. Math. Anal. Appl. **325**(2), 1363–1377 (2007)
35. Qi, L., Chen, H., Chen, Y.: Tensor Eigenvalues and Their Applications. Advances in Mechanics and Mathematics, vol. 39. Springer, Singapore (2018)
36. Röhrl, H.: Algebras and differential equations. Nagoya Math. J. **68**, 59–122 (1977)
37. Röhrl, H.: A theorem on non-associative algebras and its application to differential equations. Manuscr. Math. **21**(2), 181–187 (1977)
38. Saupe, A.: Disclinations and properties of the directorfield in nematic and cholesteric liquid crystals. Mol. Cryst. Liq. Cryst. **21**, 211–238 (1973)
39. Sonnet, A.M., Virga, E.G.: Reorientational dynamics of conjugated nematic point defects. Liq. Cryst. **37**, 785–797 (2010)
40. Stewart, I.W.: The Static and Dynamic Continuum Theory of Liquid Crystals. Taylor & Francis, London (2004)

41. Stoker, J.J.: Differential Geometry. Pure and Applied Mathematics, vol. XX. Wiley-Interscience, New York (1969)
42. Tang, X., Selinger, J.V.: Orientation of topological defects in 2D nematic liquid crystals. Soft Matter **13**, 5481–5490 (2017)
43. Tolédano, P., Dmitriev, V.: Reconstructive Phase Transitions in Crystals and Quasicrystals. World Scientific, Singapore (1996)
44. Truesdell, C.A.: A First Course in Rational Continuum Mechanics. Academic Press, Boston (1977). A second edition, corrected, revised, and augmented, was published by the same Publisher in 1991
45. Turzi, S.S.: On the Cartesian definition of orientational order parameters. J. Math. Phys. **52**(5), 053517 (2011)
46. Vanderbauwhede, A.: Local Bifurcation and Symmetry. Pitman, Boston (1982)
47. Virga, E.G.: Variational Theories for Liquid Crystals. Chapman & Hall, London (1994)
48. Virga, E.G.: Octupolar order in two dimensions. Eur. Phys. J. E **38**(6), 1–7 (2015)
49. Vromans, A.J., Giomi, L.: Orientational properties of nematic disclinations. Soft Matter **12**, 6490–6495 (2016)
50. Walcher, S.: Algebras and Differential Equations. Hadronic Press, Palm Harbor (1991)
51. Walcher, S.: Eigenvectors of tensors (2017). Talk given at GSD2017 Conference
52. Walcher, S.: Eigenvectors of tensors—A primer. Acta Appl. Math. (2019, in press). https://link.springer.com/article/10.1007%2Fs10440-018-0225-7
53. Wergifosse, M., Liégeois, V., Champagne, B.: Evaluation of the molecular static and dynamic first hyperpolarizabilities. Int. J. Quant. Chem. **114**(14), 900–910 (2014)
54. Zyss, J., Ledoux, I.: Nonlinear optics in multipolar media: theory and experiments. Chem. Rev. **94**(1), 77–105 (1994)

J Elast (2019) 135:351–373
https://doi.org/10.1007/s10659-018-09720-w

# Explicit Relaxation of a Two-Well Hadamard Energy

**Yury Grabovsky**[1] · **Lev Truskinovsky**[2]

*The paper is dedicated to the memory of W. Noll*

Received: 4 October 2018 / Published online: 10 January 2019


**Abstract** We compute an explicit quasiconvex envelope for a subclass of double-well Hadamard energies which model materials undergoing isotropic-to-isotropic elastic phase transitions. The construction becomes possible because of stability of the entire jump set, representing points that can coexist at phase boundaries. To prove stability we apply a recently developed technique for establishing polyconvexity of points on the jump set.

**Keywords** Quasiconvexity · Polyconvexity · Rank-one convexity · Nonlinear elasticity · Hadamard material · Jump set · Elastic stability · Binodal

**Mathematics Subject Classification** 74A50 · 74G65 · 49K40 · 49S05

## 1 Introduction

The object of study of this paper is the behavior of Hadamard material [21], a term coined by F. John [22] for the nonlinear elasticity problem with the energy density

$$W(\boldsymbol{F}) = \frac{\mu}{2}|\boldsymbol{F}|^2 + h(\det \boldsymbol{F}), \qquad \det \boldsymbol{F} > 0. \tag{1.1}$$

Specifically, we'll be interested in the case when $h(d)$ is a $C^2(0, +\infty)$ function with a double-well shape shown in Fig. 1; the precise technical definition will be given later. Such energies can be used to model materials undergoing isotropic-to-isotropic martensitic phase transitions which are the simplest transitions of this type [14, 26]. Even though the energy wells are described by a function of a single variable, $h(d)$, the results are nontrivial due to the incompatibility and nonlinearity of the energy wells [5].

The fundamental challenge in the theory of martensitic phase transitions is to construct the quasiconvex envelope of the elastic energy, and in this paper we solve this problem for

✉ Y. Grabovsky
yury@temple.edu

1 Temple University, Philadelphia, PA, USA

2 ESPCI, Paris, France

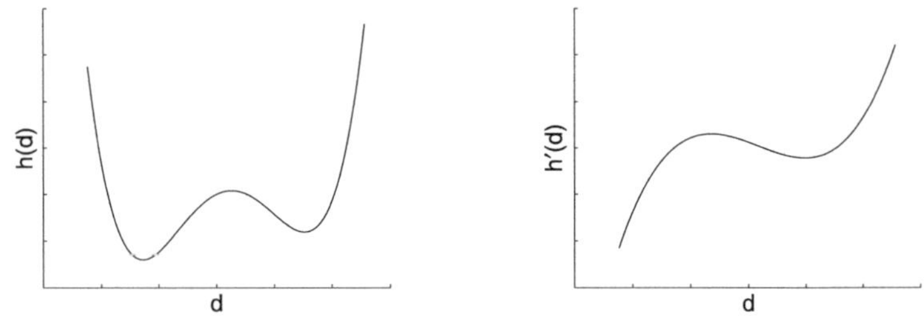


**Fig. 1** Double-well structure of the energy density $h$

the Hadamard material (1.1), when the shear modulus $\mu$ is sufficiently large. It is well-known that $W(\boldsymbol{F})$, given by (1.1) is quasiconvex if and only if $h(d)$ is convex [6]. This statement may be thought of as a bit surprising, since it asserts failure of quasiconvexity of $W(\boldsymbol{F})$ for *any* non-convex $h(d)$, regardless of how large the coefficient $\mu$ in front of the quadratic term is. The reason is a highly anisotropic behavior of the determinant at infinity, where it can remain bounded, while $|\boldsymbol{F}|$ grows, and where small perturbations of $\boldsymbol{F}$, can lead to substantial variations of the determinant. This allows for variations that lower the energy due to the non-convexity of $h(d)$, while the increment of $\mu|\boldsymbol{F}|^2$ remains relatively small.

The relaxation of the Hadamard material with non-convex $h(d)$ is presently known only in the case $\mu = 0$, where $QW(\boldsymbol{F}) = h^{**}(\det \boldsymbol{F})$, [11]. This result may be taken as a hint that the quasiconvex envelope of (1.1) is

$$U(\boldsymbol{F}) = \frac{\mu}{2}|\boldsymbol{F}|^2 + h^{**}(\det \boldsymbol{F}), \tag{1.2}$$

where $h^{**}$ denotes the convex hull of $h(d)$. However, this guess is wrong not just as a general statement but for every single non-convex function $h(d)$ of class $C^1$ with superlinear growth at infinity and/or blow-up at $d = 0$. To make our presentation self-contained, we supply the proof of this result in Appendix A even though this result is probably known to experts. Curiously, a version of formula (1.2) for linearized kinematics is, in fact, correct [20]. The effect of nonlinear kinematics is in the nontrivial coupling between $|\boldsymbol{F}|$ and $\det \boldsymbol{F}$, which is responsible for quasiconvexity of $W(\boldsymbol{F})$ at some $\boldsymbol{F}$ for which $h(\det \boldsymbol{F}) > h^{**}(\det \boldsymbol{F})$. By linearizing kinematics we effectively decouple the deviatoric part of $\boldsymbol{F}$ from its trace which can be then relaxed separately, see [20] for more details.

In this paper we present an explicit formula for $QW(\boldsymbol{F})$ in the case when $\mu > 0$ is sufficiently large and the quadratic term dominates. The formula couples $|\boldsymbol{F}|$ and $\det \boldsymbol{F}$, and $QW(\boldsymbol{F})$ is sandwiched between $W(\boldsymbol{F})$ above and $U(\boldsymbol{F})$, given by (1.2). Our constraint on $\mu$ is not a technical limitation, and as $\mu$ decreases our formula for $QW(\boldsymbol{F})$ ceases to be valid in some subsets of the binodal region. A preliminary study shows that in the range of small $\mu$ the relaxation of $W(\boldsymbol{F})$ goes through a complex chain of structural transitions whose nature will be revealed in a separate publication.

Our approach is rooted in the observation that in the course of a hard device loading program, the homogeneous deformation can become unstable in strong topology if the applied affine deformation crosses into the binodal region, where the quasi-convexity of the energy is lost. The instability manifests itself through a *strong bifurcation* [17], whereby a heterogeneous configuration with the same energy as the homogeneous one becomes available. In the case of materials supporting phase transitions, such inhomogeneous configurations will feature smooth surfaces of jump discontinuity of the deformation gradient and the disjoint regions of the phase space between which the gradient jumps are regarded as phases. The set of such "broken extremals" is substantially larger than the set of smooth ones, and finding the location of jump discontinuities is usually a complex free-boundary problem, e.g., [1, 2, 8, 15, 27].

As in the case of quasiconvexification, there is no general algorithm for finding the boundary of the binodal region, known as the *binodal*. Nevertheless, in our prior work we have developed a general method for identifying its subset supporting the laminate type energy-minimizing configurations [16, 19, 20]. Behind this method is the study of stability of the jump set—a codimension one variety in the phase space that has a dual nature. On the one hand it determines the set of pairs $\boldsymbol{F}_{\pm}$ that could be the traces of the deformation gradient at the phase boundary in a stable configuration. On the other, the jump set must consist of points that are at most marginally stable in the sense that their every neighborhood contains points where quasiconvexity fails. Therefore, if one can prove quasiconvexity at a point on the jump set, then this point must lie on the binodal. Checking this is an algebraic problem posed in [20] for general $W(\boldsymbol{F})$ in any number of dimensions. In this paper we solve this problem for the Hadamard energy (1.1) in two space dimensions and with sufficiently large $\mu$. We show that in this limit, the full jump set becomes stable which ensures that the method delivers the *entire* binodal. A natural consequence of this result is that all nontrivial energy-minimizing configurations are simple laminates.

## 2 Preliminaries

We begin by recalling the standard definitions.

**Definition 2.1**

(i) We say that $W(\boldsymbol{F})$ is quasiconvex at $\boldsymbol{F} \in \mathbb{M} = \mathbb{R}^{m\times n}$, if

$$\int_{\mathbb{R}^n} \{W(\boldsymbol{F} + \nabla\boldsymbol{\phi}(z)) - W(\boldsymbol{F})\} dz \ge 0 \tag{2.1}$$

for all $\boldsymbol{\phi} \in C_0^\infty(\mathbb{R}^n; \mathbb{R}^m)$.

(ii) The function $W(\boldsymbol{F})$ is called quasiconvex if it is quasiconvex at all points $\boldsymbol{F}$.

(iii) The quasiconvex envelope $QW(\boldsymbol{F})$ of $W(\boldsymbol{F})$ is the largest quasiconvex function satisfying $QW(\boldsymbol{F}) \le W(\boldsymbol{F})$ for all $\boldsymbol{F} \in \mathbb{M}$.

(iv) The set of points $\boldsymbol{F}$, such that $QW(\boldsymbol{F}) < W(\boldsymbol{F})$ is called the binodal region $\mathfrak{B}$. The boundary of the binodal region is called the binodal, $\mathfrak{B}\mathrm{in} = \partial\mathfrak{B}$.

There is also a formula for the quasiconvex envelope

$$QW(\boldsymbol{F}) = \inf_{\boldsymbol{\phi}\in C_0^\infty(\Omega;\mathbb{R}^m)} \fint_{\Omega} W(\boldsymbol{F} + \nabla\boldsymbol{\phi}) d\boldsymbol{x}, \tag{2.2}$$

due to Dacorogna [12].

We now describe our method from [20] for computing a subset of $\mathfrak{B}\mathrm{in}$, which in this paper will be shown to be the entire binodal. We start with the definition of the jump set, which lies in the closure of $\mathfrak{B}$.

**Definition 2.2** The jump set $\mathfrak{J}$ is the closure of the set of points $\boldsymbol{F} \in \mathbb{M}$ for which there exists $\boldsymbol{a} \in \mathbb{R}^m \setminus \{0\}$ and a unit vector $\boldsymbol{n} \in \mathbb{R}^n$, such that

$$\begin{cases} (W_{\boldsymbol{F}}(\boldsymbol{F}+\boldsymbol{a}\otimes\boldsymbol{n}) - W_{\boldsymbol{F}}(\boldsymbol{F}))\boldsymbol{n} = 0, \\ (W_{\boldsymbol{F}}(\boldsymbol{F}+\boldsymbol{a}\otimes\boldsymbol{n}) - W_{\boldsymbol{F}}(\boldsymbol{F}))^T\boldsymbol{a} = 0, \\ W(\boldsymbol{F}+\boldsymbol{a}\otimes\boldsymbol{n}) - W(\boldsymbol{F}) = W_{\boldsymbol{F}}(\boldsymbol{F})\boldsymbol{n}\cdot\boldsymbol{a}. \end{cases} \tag{2.3}$$

These equations have a dual nature. On the one hand they are necessary conditions on the surface of jump discontinuity of the gradient of any strong local minimizer of the energy $E[\boldsymbol{y}]$. On the other, these equations describe the set of homogeneous deformation gradients $\boldsymbol{F}$ that permit energy-neutral nucleation of layers of a new phase with a compatible deformation gradient $\boldsymbol{F}+\boldsymbol{a}\otimes\boldsymbol{n}$. If $\boldsymbol{F}$ is stable, i.e., if $QW(\boldsymbol{F}) = W(\boldsymbol{F})$, then such bifurcation property signals the boundary of stability, crossing which violates the quasiconvexity property. See [16] and [18] for the discussion.

We remark that if $\boldsymbol{F}$ solves (2.3), then $\boldsymbol{F}+\boldsymbol{a}\otimes\boldsymbol{n}$ also solves (2.3) with the same $\boldsymbol{n}$ and $\boldsymbol{a}$ replaced by $-\boldsymbol{a}$. This symmetry of the jump set makes it convenient to modify our notation and use $\boldsymbol{F}_-$ for $\boldsymbol{F}$ and $\boldsymbol{F}_+$ for $\boldsymbol{F}+\boldsymbol{a}\otimes\boldsymbol{n}$. We can also rewrite jump set equations in their canonical symmetric form

$$\begin{cases} \operatorname{rank}[\![\boldsymbol{F}]\!] = 1, \\ [\![W_{\boldsymbol{F}}]\!][\![\boldsymbol{F}]\!]^T = 0, \\ [\![W_{\boldsymbol{F}}]\!]^T[\![\boldsymbol{F}]\!] = 0, \\ [\![W]\!] - \langle W_{\boldsymbol{F}}^{\pm}, [\![\boldsymbol{F}]\!]\rangle = 0, \end{cases} \tag{2.4}$$

where the last equation is a shorthand for a pair of equations for each choice of the sign. It is easy to see that one of these equations is a consequence of the rest of the system. However, even if one discards it (breaking the symmetry of the system), the remaining equations are still not independent, since the inner product of the second with $[\![\boldsymbol{F}]\!]$ equals to the inner product of the third with $[\![\boldsymbol{F}]\!]^T$. For this reason, we prefer to keep the symmetry at the price of some unavoidable redundancy.

Vectors $\boldsymbol{a}$ and $\boldsymbol{n}$ from (2.3) are then defined up to the $\mathbb{Z}_2$ symmetry $(\boldsymbol{a},\boldsymbol{n}) \sim (-\boldsymbol{a},-\boldsymbol{n})$ via the first equation in (2.4):

$$[\![\boldsymbol{F}]\!] = \boldsymbol{a}\otimes\boldsymbol{n}. \tag{2.5}$$

We also remark that if $\boldsymbol{F}$ solves (2.3), it is possible that it does so with several, or even infinitely many choices of $\boldsymbol{a}\otimes\boldsymbol{n}$. It follows that in general $\mathfrak{J}$ is not a submanifold in $\mathbb{M}$, but rather a codimension one variety with self-intersections and singularities.

Since $\mathfrak{J} \subset \overline{\mathfrak{B}}$ (under very mild non-degeneracy assumptions), establishing quasiconvexity at a point of $\mathfrak{J}$ implies that this point belongs to the binodal. Moreover, this also gives a formula for the quasiconvex hull along the rank-one lines joining $\boldsymbol{F}_\pm$ [18].

**Theorem 2.3** *Suppose $\boldsymbol{F}_\pm$ is the corresponding pair of points on the jump set and*

$$QW(\boldsymbol{F}_\pm) = W(\boldsymbol{F}_\pm). \tag{2.6}$$

*Then for any $\lambda \in [0,1]$*

$$QW\big(\lambda\boldsymbol{F}_+ + (1-\lambda)\boldsymbol{F}_-\big) = \lambda W(\boldsymbol{F}_+) + (1-\lambda)W(\boldsymbol{F}_-). \tag{2.7}$$

If we prove that the entire jump set is quasiconvex, then formula (2.7) gives the explicit form of the quasiconvex envelope of $W(\boldsymbol{F})$. For the energy (1.1) this formula produces our main result: formula (5.2).

**Definition 2.4**

(i) The function is called rank-one convex if it is convex along all rank-one lines, i.e., when $\phi(t) = W(\boldsymbol{F} + t\boldsymbol{a} \otimes \boldsymbol{n})$ is convex for all $\boldsymbol{F}$, $\boldsymbol{a}$ and $\boldsymbol{n}$.
(ii) The rank-one convex envelope $RW$ of $W$ is the largest rank-one convex function, such that $RW(\boldsymbol{F}) \le W(\boldsymbol{F})$ for all $\boldsymbol{F} \in \mathbb{M}$.
(iii) We say that $W$ is rank-one convex at $\boldsymbol{F}_0$ if $W(\boldsymbol{F}_0) = RW(\boldsymbol{F}_0)$.
(iv) The set of points $\boldsymbol{F}$, such that $RW(\boldsymbol{F}) < W(\boldsymbol{F})$ is called the rank-one binodal region and its boundary is called the rank-one binodal.

It is well-known that every quasiconvex function is rank-one convex [23]. As a consequence we always have the implication

$$W(\boldsymbol{F}_0) = QW(\boldsymbol{F}_0) \quad \Rightarrow \quad W(\boldsymbol{F}_0) = RW(\boldsymbol{F}_0).$$

The converse is known to be false [25], when $n \ge 2$ and $m \ge 3$. However, it is not known if there is any difference between the two notions if $m = n = 2$. Since it is not known if $QW(\boldsymbol{F}) = RW(\boldsymbol{F})$, we cannot conclude a priori that the binodal agrees with the rank-one binodal everywhere.

In the two-dimensional example ($m = n = 2$) we consider in this paper we are going to show that the entire jump set is polyconvex by applying the method from [20]. The idea is for a given $\boldsymbol{F}_0$ to find a supporting null-Lagrangian $N(\boldsymbol{F})$, such that

(i) $N(\boldsymbol{F}) \le W(\boldsymbol{F})$ for all $\boldsymbol{F}$,
(ii) $N(\boldsymbol{F}_0) = W(\boldsymbol{F}_0)$.

This implies that $QW(\boldsymbol{F}_0) = W(\boldsymbol{F}_0)$. In general, finding the set of points $\boldsymbol{F}_0$, where such a null-Lagrangian exists is algebraically intractable for most nonlinear energies. However, and this is the main point of the method in [20], the search for such a null-Lagrangian simplifies to the point of feasibility in dimensions 2 and 3 when $\boldsymbol{F}_0 = \boldsymbol{F}_\pm \in \mathfrak{J}$. When $m = n = 2$ the desired null-Lagrangian has an explicit general formula, satisfying (ii). Stability of $\boldsymbol{F}_\pm$ is then a corollary of (i), which has the form

$$W(\boldsymbol{F}) \ge W(\boldsymbol{F}_\pm) + \big\langle W_{\boldsymbol{F}}(\boldsymbol{F}_\pm), \boldsymbol{F} - \boldsymbol{F}_\pm \big\rangle + \frac{\langle [\![W_{\boldsymbol{F}}]\!], \operatorname{cof}[\![\boldsymbol{F}]\!] \rangle}{|[\![\boldsymbol{F}]\!]|^2} \det(\boldsymbol{F} - \boldsymbol{F}_\pm), \tag{2.8}$$

for every $\boldsymbol{F} \in \mathbb{M}$. In other words, the energy is polyconvex at $\boldsymbol{F}_\pm \in \mathfrak{J}$ if and only if the minimum value of

$$\Psi(\boldsymbol{F}) = W(\boldsymbol{F}) - W(\boldsymbol{F}_\pm) - \big\langle W_{\boldsymbol{F}}(\boldsymbol{F}_\pm), \boldsymbol{F} - \boldsymbol{F}_\pm \big\rangle - \frac{\langle [\![W_{\boldsymbol{F}}]\!], \operatorname{cof}[\![\boldsymbol{F}]\!] \rangle}{|[\![\boldsymbol{F}]\!]|^2} \det(\boldsymbol{F} - \boldsymbol{F}_\pm) \tag{2.9}$$

is 0 (achieved at $\boldsymbol{F} = \boldsymbol{F}_\pm$). Our notation emphasizes the fact that either choice of sign in $\boldsymbol{F}_\pm$ results in one and the same function $\Psi(\boldsymbol{F})$.

## 3 The Jump Set

Failure of rank-one convexity implies that $W(\boldsymbol{F})$ has a non-trivial jump set. In [19] we have computed it and used our local bounds to identify an unstable portion of the jump set, when $\mu$ is not too large. Let us briefly recall its derivation. The simplest non-trivial case is when the non-convex function $h(d)$ is of "double-well" shape. The precise assumptions are not on the shape of the graph of $h(d)$ itself, but rather on the shape of the graph of $h'(d)$.

**Definition 3.1** We will say that the function $h(d)$ has a double-well shape if the following 3 properties are satisfied.

(i) There exists $d_0 > 0$, such that $h'(d)$ is concave on $(0, d_0)$ and has a strict local maximum there,
(ii) $h'(d)$ is convex on $(d_0, +\infty)$ and has a strict local minimum there
(iii) $\lim_{d\to 0^+} h'(d) = -\infty$.

Condition (iii) is consistent with the requirement that infinite compression costs infinite energy

$$\lim_{d\to 0^+} h(d) = +\infty. \tag{3.1}$$

We start with the kinematic compatibility equation (2.5) and take the determinant of $\boldsymbol{F}_+ = \boldsymbol{F}_- + \boldsymbol{a}\otimes\boldsymbol{n}$ to obtain

$$d_+ = d_- + \operatorname{cof}\boldsymbol{F}_-\boldsymbol{n}\cdot\boldsymbol{a}, \qquad d_\pm = \det\boldsymbol{F}_\pm. \tag{3.2}$$

Using the formula

$$\boldsymbol{P}(\boldsymbol{F}) = \mu\boldsymbol{F} + h'(\det\boldsymbol{F})\operatorname{cof}\boldsymbol{F}$$

for the Piola-Kirchhoff stress we compute

$$[\![\boldsymbol{P}]\!]\boldsymbol{n} = \mu\boldsymbol{a} + [\![h']\!]\operatorname{cof}\boldsymbol{F}_-\boldsymbol{n},$$

where we have used the well-known relation $\operatorname{cof}(\boldsymbol{F}_- + \boldsymbol{a}\otimes\boldsymbol{n})\boldsymbol{n} = (\operatorname{cof}\boldsymbol{F}_-)\boldsymbol{n}$. Similarly,

$$[\![\boldsymbol{P}]\!]^T\boldsymbol{a} = \mu|\boldsymbol{a}|^2\boldsymbol{n} + [\![h']\!]\operatorname{cof}\boldsymbol{F}_-^T\boldsymbol{a}.$$

Thus, the first two equations in (2.3) imply

$$\boldsymbol{a} = -\frac{[\![h']\!]}{\mu}\operatorname{cof}\boldsymbol{F}_-\boldsymbol{n}, \qquad [\![h']\!]^2\operatorname{cof}(\boldsymbol{C}_-)\boldsymbol{n} = \mu^2|\boldsymbol{a}|^2\boldsymbol{n}, \tag{3.3}$$

where $\boldsymbol{C}_\pm = \boldsymbol{F}_\pm^T\boldsymbol{F}_\pm$ is the Cauchy-Green strain tensor. We conclude that $\boldsymbol{n}$ must be an eigenvector of $\boldsymbol{C}_-$. Equations (3.3) permit us to find a relation between the two Cauchy-Green tensors $\boldsymbol{C}_\pm$. Using the kinematic compatibility equation we compute

$$\boldsymbol{C}_+ = \boldsymbol{C}_- + \boldsymbol{F}_-^T\boldsymbol{a}\otimes\boldsymbol{n} + \boldsymbol{n}\otimes\boldsymbol{F}_-^T\boldsymbol{a} + |\boldsymbol{a}|^2\boldsymbol{n}\otimes\boldsymbol{n}.$$

Applying $\boldsymbol{F}_-^T$ to the first equation in (3.3) we obtain $\boldsymbol{F}_-^T\boldsymbol{a} = -([\![h']\!]/\mu)d_-\boldsymbol{n}$, so that

$$[\![\boldsymbol{C}]\!] = \left(|\boldsymbol{a}|^2 - \frac{2[\![h']\!]d_-}{\mu}\right)\boldsymbol{n}\otimes\boldsymbol{n}. \tag{3.4}$$

It follows that the Cauchy-Green tensors $\boldsymbol{C}_+$ and $\boldsymbol{C}_-$ are simultaneously diagonalizable, since, by (3.3) $\boldsymbol{n}$ is an eigenvector of $\boldsymbol{C}_-$. According to Eq. (3.4) $\boldsymbol{C}_+$ and $\boldsymbol{C}_-$ have the same eigenvectors and the same eigenvalues corresponding to all eigenvectors orthogonal to $\boldsymbol{n}$. Hence $\boldsymbol{F}_\pm$ would have singular values $(\varepsilon_\pm, \varepsilon_2, \dots, \varepsilon_n)$, the first one corresponding to the eigenvector $\boldsymbol{n}$ of $\boldsymbol{C}_\pm$. Substituting the first equation in (3.3) into (3.2) we obtain

$$d_+ = d_- - \frac{[\![h']\!]}{\mu}\operatorname{cof}\boldsymbol{C}_-\boldsymbol{n}\cdot\boldsymbol{n} = d_- - \frac{[\![h']\!]d_-^2}{\mu\varepsilon_-^2},$$

which can be written in the more symmetric form as

$$\mu\frac{[\![d]\!]}{[\![h']\!]} = -\prod_{j=2}^{n}\varepsilon_j^2 = -\frac{d_\pm^2}{\varepsilon_\pm^2}. \tag{3.5}$$

This will be the equation for the jump set, when we determine $d_+$ as a function of $d_-$ from the Maxwell relation (the last equation in (2.4), which hasn't been used so far). It is well-known that the Maxwell relation does not change if we add any quadratic function in $\boldsymbol{F}$ to the energy. Thus, the term $\mu|\boldsymbol{F}|^2/2$ can be disregarded and the Maxwell relation becomes

$$[\![h]\!] = \{h'\operatorname{cof}\boldsymbol{F}\}\boldsymbol{n}\cdot\boldsymbol{a}.$$

Recalling that due to (3.2) $(\operatorname{cof}\boldsymbol{F}_+)\boldsymbol{n}\cdot\boldsymbol{a} = (\operatorname{cof}\boldsymbol{F}_-)\boldsymbol{n}\cdot\boldsymbol{a} = [\![d]\!]$ we obtain

$$[\![h]\!] = \{h'\}[\![d]\!]. \tag{3.6}$$

Equation (3.6) can be written in such a way as to highlight its geometric meaning. Observe that the straight line with equation

$$y = h'(d_-) + \frac{[\![h']\!]}{[\![d]\!]}(x - d_-)$$

is the secant line joining $(d_-, h'(d_-))$ and $(d_+, h'(d_+))$. Then the Maxwell relation (3.6) is equivalent to

$$\int_{d_-}^{d_+}\big\{h'(x) - y(x)\big\}dx = 0. \tag{3.7}$$

Now the relation (3.7) can be interpreted geometrically as equality of areas between the secant line and the graph of $h'(d)$. Hence, according to (3.5), the slope of the straight line orthogonal to the secant is the single common eigenvalue of $\operatorname{cof}\boldsymbol{C}_+$ and $\operatorname{cof}\boldsymbol{C}_-$, implying that the secant line has a negative slope.

It is now easy to show that under assumptions (i)–(iii) in Definition 3.1 there exists a single interval $(d_1, d_2)$ on which $h(d)$ differs from its convex hull, which on $(d_1, d_2)$ agrees with the common tangent line at $d_1$ and $d_2$ to the graph of $h(d)$. In terms of $h'(d)$ this double-tangency can also be interpreted geometrically as the horizontal "Maxwell line" with the equal area property. Then $d_- \in (d_1, d_0)$ and $d_+ \in (d_0, d_2)$. For every $d_- \in (d_1, d_0)$ there is a unique $d_+ = D(d_-)$ with equal area property (3.7) (see Fig. 2). (By continuity we can set $D(d_0) = d_0$.) For example if $h(d)$ is a quartic polynomial $(d - d_1)^2(d - d_2)^2$, then $D(d) = d_1 + d_2 - d$, and $d_0 = (d_1 + d_2)/2$. Regarding the function $D(d)$ as known, we obtain the explicit description of the jump set in terms of the singular values of $\boldsymbol{F}$. It can be viewed as

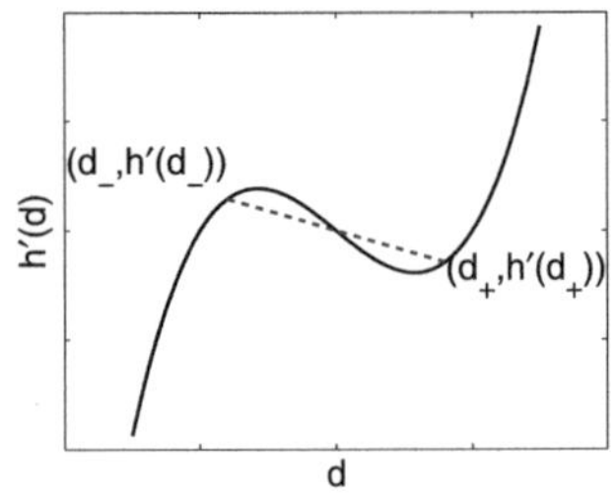


**Fig. 2** The map $d_\pm = D(d_\mp)$ is defined by the equal area condition (3.6)

a parametric hypersurface in $\mathbb{R}^n$ parametrized by $n-1$ parameters $d, \varepsilon_3, \ldots, \varepsilon_n$ via (3.5). Explicitly, choosing the ordering of eigenvalues where $\boldsymbol{n} = \boldsymbol{e}_1$ we obtain

$$\begin{cases} \varepsilon_1 = d\sqrt{-\dfrac{h'(D(d)) - h'(d)}{\mu(D(d)-d)}}, \\ \varepsilon_2 = \left(\displaystyle\prod_{j=3}^{n} \frac{1}{\varepsilon_j}\right)\sqrt{-\dfrac{\mu(D(d)-d)}{h'(D(d)) - h'(d)}}, \\ \varepsilon_3 = \varepsilon_3, \ldots, \varepsilon_n = \varepsilon_n. \end{cases} \tag{3.8}$$

The entire jump set is therefore a set of $\boldsymbol{F}$, whose singular values have the ordering, which satisfies (3.8). We note that the hypersurface (3.8) is invariant under all permutations of $(\varepsilon_2, \ldots, \varepsilon_n)$. Therefore, the entire jump set is the union of $n$ components: surface (3.8) and the ones obtained from it by swapping $\varepsilon_1$ and $\varepsilon_k$ in (3.8). When $n = 2$ we have

$$\begin{cases} \varepsilon_1 = d\sqrt{-\dfrac{h'(D(d)) - h'(d)}{\mu(D(d)-d)}}, \\ \varepsilon_2 = \sqrt{-\dfrac{\mu(D(d)-d)}{h'(D(d)) - h'(d)}}, \end{cases} \tag{3.9}$$

with the second component given by interchanging $\varepsilon_1$ and $\varepsilon_2$. Let us now restrict our attention to the 2D case. When $\mu$ is sufficiently large the jump set is a connected embedded submanifold of $\mathbb{R}^{2\times 2}$. For smaller values of $\mu$ the two symmetry-related components of the jump set in $(\varepsilon_1, \varepsilon_2)$ plane intersect and the jump set is no longer a manifold, as it has self-intersections. Since, the pair $(\varepsilon_1, \varepsilon_2)$ determines parameter $d = \varepsilon_1\varepsilon_2$ uniquely, the self-intersection happens only at the solutions of the equation $\varepsilon_1(d) = \varepsilon_2(d)$. Thus, if

$$\mu > \max_{d\in[d_1,d_2]} -d\frac{h'(D(d)) - h'(d)}{D(d)-d}, \tag{3.10}$$

then the jump set has no self-intersections (see Fig. 3).

If $\boldsymbol{F}_\pm$ is a corresponding pair on the jump set (i.e., the pair satisfying (2.4)), then there are frames in which both $\boldsymbol{F}_\pm$ are diagonal. The rank-1 lines connecting $\boldsymbol{F}_+$ and $\boldsymbol{F}_-$, i.e., the lines along the direction $\boldsymbol{n}$ are horizontal lines in representation (3.9).

*Remark 3.2* In [19] we have found necessary conditions for the points on the jump set to be stable. Using our formula from [19] we have the lower bound

$$\mu_*^2 = \max_{d\in[d_1,d_2]} R^4(d)\left(\frac{(D(d)-d)^2R^2(d)}{R^2(d)+h''(d)} - d^2 + 2dD(d)\right), \tag{3.11}$$

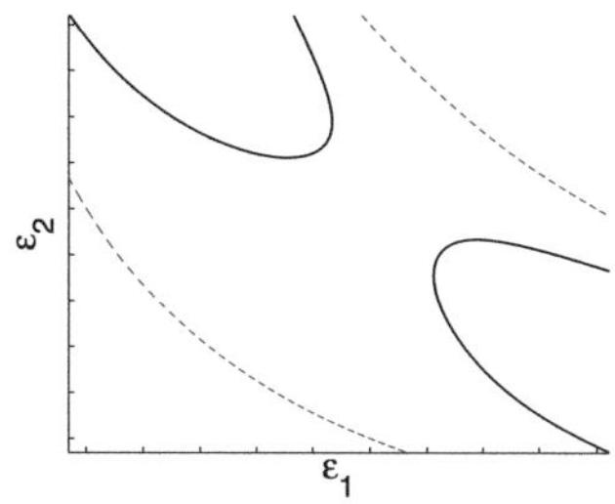


**Fig. 3** The jump set for the non-convex Hadamard material when $\mu$ satisfies (3.10)

where $R(d)^2 = -[\![h']\!]/[\![d]\!]$. If $\mu < \mu_*$ there will be a part of the jump set which fails rank-one convexity.

## 4 Polyconvexity

In this section we are going to prove that for sufficiently large $\mu > 0$ the function $\Psi(\boldsymbol{F})$, given by (2.9) is minimized only at $\boldsymbol{F} = \boldsymbol{F}_\pm$ for every corresponding pair $\boldsymbol{F}_\pm$ on the jump set.

The isotropy of the energy implies that it is sufficient to verify non-negativity of $\Psi(\boldsymbol{F})$ when $\boldsymbol{F}_\pm$ are both diagonal

$$\boldsymbol{F}_\pm = \begin{bmatrix} \varepsilon_\pm & 0 \\ 0 & \varepsilon_0 \end{bmatrix}, \tag{4.1}$$

where we have chosen the representative of the symmetry class on the jump set, given by (3.9), so that $\boldsymbol{n} = \boldsymbol{e}_1$ and therefore, $\boldsymbol{F}_\pm \boldsymbol{e}_2 = \varepsilon_0 \boldsymbol{e}_2$.

Let us now examine minima of the functions $\Psi(\boldsymbol{F})$, given by (2.9) when $\mu$ is sufficiently large. For the Hadamard material

$$\begin{aligned} \Psi(\boldsymbol{F}) &= \Psi(\boldsymbol{F};\boldsymbol{F}_\pm) \\ &= \frac{\mu}{2}|\boldsymbol{F}-\boldsymbol{F}_\pm|^2 + h(\det \boldsymbol{F}) - h_\pm - h'_\pm\langle \operatorname{cof}\boldsymbol{F}_\pm, \boldsymbol{F}-\boldsymbol{F}_\pm\rangle - m_0 \det(\boldsymbol{F}-\boldsymbol{F}_\pm), \end{aligned}$$

where

$$m_0 = \frac{\langle [\![W_F]\!], \operatorname{cof}[\![\boldsymbol{F}]\!]\rangle}{|[\![\boldsymbol{F}]\!]|^2} = \frac{[\![h'd]\!]}{[\![d]\!]}. \tag{4.2}$$

If $\boldsymbol{F}$ is a critical point of $\Psi(\boldsymbol{F};\boldsymbol{F}_\pm)$, then

$$\begin{gathered} \mu \boldsymbol{F} + \lambda \operatorname{cof}\boldsymbol{F} = \mu \begin{bmatrix} \varepsilon_\pm & 0 \\ 0 & \varepsilon_0 \end{bmatrix} + \lambda_\pm \begin{bmatrix} \varepsilon_0 & 0 \\ 0 & \varepsilon_\pm \end{bmatrix}, \\ \lambda = h'(\det \boldsymbol{F}) - m_0, \qquad \lambda_\pm = h'_\pm - m_0. \end{gathered} \tag{4.3}$$

In components we have

$$\begin{gathered} \mu F_{12} - \lambda F_{21} = 0, \qquad \mu F_{21} - \lambda F_{12} = 0, \qquad \mu F_{11} + \lambda F_{22} = \mu\varepsilon_\pm + \lambda_\pm \varepsilon_0, \\ \mu F_{22} + \lambda F_{11} = \mu \varepsilon_\pm + \lambda_\pm x_\pm. \end{gathered}$$

Using (4.2) we get

$$\lambda_\pm = h'_\pm - \frac{[\![h'd]\!]}{[\![d]\!]} = -\frac{[\![h']\!]}{[\![d]\!]} d_\mp = \mu \frac{d_\mp}{\varepsilon_0^2} = \mu \frac{\varepsilon_\mp}{\varepsilon_0}, \tag{4.4}$$

where the penultimate equality is due to (3.5).

If $\lambda \neq \pm\mu$, then $\boldsymbol{F}$ must also be diagonal. This is our main case. Before considering it, let us rule out the exceptional cases. If $\lambda = -\mu$ then $(\varepsilon_\pm + \varepsilon_0)(\mu + \lambda_\pm) = 0$, which is impossible, since $\varepsilon_\pm > 0$ and $\varepsilon_0 > 0$ and $\lambda_\pm > 0$, due to (4.4). If $\lambda = \mu$, then $(\varepsilon_\pm - \varepsilon_0)(\mu - \lambda_\pm) = 0$ and either $\varepsilon_\pm = \varepsilon_0$ or $\lambda_\pm = \mu$. However, according to (4.4), equation $\lambda_\pm = \mu$ implies $\varepsilon_\mp = \varepsilon_0$. Hence, if $\lambda = \mu$, then either $\boldsymbol{F}_+$ or $\boldsymbol{F}_-$ is a point of self-intersection of the jump set. If $\mu$ is sufficiently large, as to satisfy (3.10), then such points are not present, and in particular, combining inequality (3.10) with jump set equation (3.5) we obtain

$$\varepsilon_\pm < \varepsilon_0. \tag{4.5}$$

We now consider the only remaining possibility $\lambda \neq \pm\mu$. Then any critical point $\boldsymbol{F}$ of $\Psi$ must necessarily be diagonal in the frame in which $\boldsymbol{F}_\pm$ are diagonal. Denoting by $x$ and $y$ the two diagonal entries of $\boldsymbol{F}$ we obtain

$$\Psi(\boldsymbol{F}) = \Phi(x, y) + \text{const},$$

where

$$\Phi(x, y) = \frac{\mu}{2}(x^2 + y^2) + h(xy) - m_0 xy - \alpha x - \beta y, \quad x > 0,\ y > 0. \tag{4.6}$$

$$\alpha = \mu\varepsilon_\pm + \lambda_\pm\varepsilon_0 > 0, \qquad \beta = \mu\varepsilon_0 + \lambda_\pm\varepsilon_\pm > 0. \tag{4.7}$$

Emphasizing explicit dependence on $\mu$ the jump set equations (3.9) will be written as follows:

$$\begin{cases} \varepsilon_\pm = \dfrac{d_\pm R}{\sqrt{\mu}}, \\ \varepsilon_0 = \dfrac{\sqrt{\mu}}{R}, \end{cases} \qquad R = \sqrt{-\frac{[\![h']\!]}{[\![d]\!]}}. \tag{4.8}$$

In this notation parameters (4.7) are:

$$\alpha = 2\sqrt{\mu} R\{d\}, \qquad \beta = \frac{\mu^2 + R^4 d_+ d_-}{R\sqrt{\mu}}. \tag{4.9}$$

We compute

$$\beta - \alpha = \frac{(\mu - R^2 d_+)(\mu - R^2 d_-)}{R\sqrt{\mu}} > 0,$$

due to (3.10).

We are now ready to examine global minima of $\Phi(x, y)$. We begin by first minimizing $\Phi(x, y)$ over all pairs $(x, y)$, satisfying $xy = d$ for each fixed $d > 0$, and then minimizing

$$\phi(d) = h(d) - m_0 d + \phi_0(d), \qquad \phi_0(d) = \min_{xy=d} \Phi_0(x, y)$$

in $d>0$, where $m_0$ is given in (4.2) and

$$\Phi_0(x,y)=\frac{\mu}{2}\left(x^2+y^2\right)-\alpha x-\beta y.$$

As shown in Appendix B, the minimum of $\Phi_0(x,y)$ is achieved at $(d/y(d),y(d))$, where $y(d)$ is the largest root of

$$q(y,d)=\mu y^4-\beta y^3+d\alpha y-\mu d^2=0. \tag{4.10}$$

The function $y(d)$ is analytic in $d$, regarded as a complex variable. Singularities of such a function occur at values of $d$ for which $q_y(y(d),d)=0$. It is a routine algebraic calculation (solving the system $q=0$, $q_y=0$) to show than none of the singularities of $y(d)$ lie on the real $d$-axis.

Using formulas (4.9) we compute

$$y(d_\pm)=\frac{\sqrt{\mu}}{R}.$$

Hence,

$$z(d)=\frac{(y(d)-\sqrt{\mu}/R)\mu^{3/2}}{(d-d_-)(d-d_+)}, \tag{4.11}$$

is analytic in the same complex domain as $y(d)$, and in particular on the real $d$-axis. The factor $\mu^{3/2}$ in the numerator in (4.11) is for convenience, since as is easy to show

$$\lim_{\mu\to\infty}z(d)=R^3, \tag{4.12}$$

and convergence is uniform in $d$ on compact subsets of $\mathbb{R}$. When $d_\pm$ varies in $(d_1,d_2)$, the parameter $R=R(d_\pm)$ varies in $(0,\sqrt{-h''(d_0)}]$. Thus, $z(d)=z(d;\mu)$ is uniformly bounded, as $\mu\to+\infty$. Therefore, the asymptotics of $y(d)=y(d;\mu)$ as $\mu\to+\infty$ can be read off the representation

$$y(d;\mu)=\frac{\sqrt{\mu}}{R}+\mu^{-3/2}(d-d_-)(d-d_+)z(d;\mu). \tag{4.13}$$

It is a calculation best done using computer algebra system, such as Maple, to show that

$$\phi_0'(d)=\frac{\mu y(d)^2-\beta y(d)}{d}.$$

In a similar fashion one verifies that $\phi_0''(d)$ cannot vanish on $d>0$. Thus, $\phi_0'(d)$ is an increasing function of $d$ on $(0,+\infty)$, and the minima of $\phi(d)$ occur at critical points of $\phi(d)$, since $\phi(d)\to+\infty$, when $d\to+\infty$. Then, minimizers of $\phi(d)$ satisfy

$$\phi'(d)=\frac{\mu y(d)^2-\beta y(d)}{d}+\left(h'(d)-m_0\right)=0. \tag{4.14}$$

This implies that there is a compact interval $[d_{\min},d_{\max}]\subset(0,+\infty)$, where all solutions of (4.14) lie. Indeed, $d_{\min}$ (respectively $d_{\max}$), chosen to be independent of $d_\pm\in(d_1,d_2)$, is found from the condition that both terms in (4.14) are negative (respectively positive). The details are in Appendix B.

We note that $\phi(d_+) = \phi(d_-)$ due to (3.6). Our goal is to prove that

$$\psi(d) = \phi(d) - \phi(d_\pm) \geq 0, \quad \forall d > 0.$$

We compute (using Maple)

$$\lim_{\mu\to\infty} \psi(d) = \psi_\infty(d) = h(d) - \{h\} - \{h'\}(d - \{d\}) + \frac{R^2}{2}(d - d_+)(d - d_-).$$

The function $\psi_\infty(d)$ satisfies $\psi_\infty(d_\pm) = 0$ (due to (3.6)) and $\psi'_\infty(d_\pm) = 0$ (due to the value of $R$, given in (4.8)). We observe that if $h(d)$ is *any* quartic polynomial with leading coefficient $a$, then

$$\psi_\infty(d) = a(d - d_+)^2(d - d_-)^2. \tag{4.15}$$

The geometric meaning of $\psi'_\infty(d)$ is the difference between $h'(d)$ and the equal-area secant line passing through $(d_\pm, h'(d_\pm))$ (see Fig. 2). The assumption of the double-well shape of $h(d)$ implies that there are exactly 3 points of intersection of this secant line with the graph of $h'(d)$, unless $d_+ = d_- = d_0$. In every case (including the limiting one) the points $d_\pm$ correspond to local (and hence global) minima of $\psi_\infty(d)$, while the third point is a point of local maximum.

**Lemma 4.1** *There exists* $c_0 > 0$, *such that for all* $d_- \in (d_1, d_2)$ *(setting* $d_+ = D(d_-)$*) and for all* $d \in [d_{\min}, d_{\max}]$ *we have*

$$\frac{\psi_\infty(d; d_-)}{(d - d_+)^2(d - d_-)^2} \geq c_0. \tag{4.16}$$

*Proof* If inequality (4.16) fails, then there exist sequences $d_-^{(n)}$ and $d_n \in [d_{\min}, d_{\max}]$ such that

$$\lim_{n\to\infty} \frac{\psi_\infty(d_n; d_-^{(n)})}{(d_n - d_+^{(n)})^2(d_n - d_-^{(n)})^2} = 0. \tag{4.17}$$

Since both sequences are bounded we can extract convergent subsequences, not relabeled:

$$d_-^{(n)} \to d_-^\infty, \qquad d_+^{(n)} \to d_+^\infty = D(d_-^\infty), \qquad d_n \to d_\infty.$$

Since denominator in (4.17) is uniformly bounded we conclude that $\psi_\infty(d_\infty; d_-^\infty) = 0$. Thus, $d_\infty$ is either $d_-^\infty$ or $d_+^\infty$. To fix notation we assume that $d_\infty = d_-^\infty$.

If $d_-^\infty \neq d_0$ then $d_+^\infty \neq d_-^\infty$ and we conclude that

$$\lim_{n\to\infty} \frac{\psi_\infty(d_n; d_-^{(n)})}{(d_n - d_+^{(n)})^2(d_n - d_-^{(n)})^2} = \frac{\psi''_\infty(d_-^\infty; d_-^\infty)}{2[\![d^\infty]\!]^2} = \frac{h''(d_-^\infty) + R^2(d_-^\infty)}{2[\![d^\infty]\!]^2} > 0,$$

contradicting (4.17). The only possibility is therefore $d_\infty = d_-^\infty = d_+^\infty = d_0$. In this case

$$\lim_{n\to\infty} \frac{\psi_\infty(d_n; d_-^{(n)})}{(d_n - d_+^{(n)})^2(d_n - d_-^{(n)})^2} = \frac{h^{(4)}(d_0)}{24} > 0, \tag{4.18}$$

since for sufficiently large $n$ all sequences $d_n$ and $d_\pm^{(n)}$ lie in any arbitrarily small neighborhood of $d = d_0$. In that neighborhood the function $h(d)$ will be indistinguishable from its quartic Taylor polynomial with leading coefficient $h^{(4)}(d_0)/24 > 0$. The value of the limit in (4.18) then follows from (4.15). □

In order to show that $d_\pm$ are the only minimizers of $\psi(d;\mu)$ we use representation (4.13) for $y(d;\mu)$ and obtain (via a Maple calculation)

$$\lim_{\mu\to\infty} \mu^2 \frac{\psi(d;\mu) - \psi_\infty(d)}{(d-d_+)^2(d-d_-)^2} = -\frac{R^6}{2},$$

where the limit is uniform in $d \in [d_{\min}, d_{\max}]$ and in $d_- \in (d_1, d_2)$.

Now, recalling that $R^2 \le -h''(d_0)$ for all $d_- \in (d_1, d_2)$, we have for sufficiently large $\mu$

$$\mu^2 \frac{\psi(d;\mu) - \psi_\infty(d)}{(d-d_+)^2(d-d_-)^2} \ge h''(d_0)^3$$

for all $d \in [d_{\min}, d_{\max}]$ and all $d_- \in (d_1, d_2)$. Applying Lemma 4.1 we have

$$\frac{\psi(d;\mu)}{(d-d_+)^2(d-d_-)^2} \ge c_0 + \mu^{-2} h''(d_0)^3 > 0$$

when $\mu$ is sufficiently large, and where the lower bound on $\mu$ does not depend on neither $d$, nor $d_-$. We conclude that $\psi(d;\mu)$ is minimized only at $d_\pm$, since $\psi(d_\pm;\mu) = 0$. Therefore, the entire jump set satisfies the polyconvexity condition.

*Remark 4.2* The lowest value of $\mu$ for which the entire jump set becomes polyconvex is not given explicitly here. In our numerical calculations we observed that if the point on the jump set corresponding to $d_+ = d_- = d_0$ is polyconvex, then the entire jump set is polyconvex, at least when $h(d)$ is a quartic polynomial. We can then write equations corresponding to the appearance of $d_*$, such that $\phi(d_*) = \phi(d_0)$ and $\phi'(d_*) = \phi'(d_0) = 0$. This leads to explicit but very cumbersome equation in the single variable $d$, whose solution is $d = d_*$. To give a rough idea of the strength of our results we have evaluated this value of $\mu$ for $h(d) = (d-1)^2(d-3)^2$. In this case the entire jump set is polyconvex if and only if $\mu \ge 10.53$. Inequality (3.10) says that the jump set has no self-intersections, when $\mu > 8.45$ and inequality (3.11) implies that the jump set does not capture the entire binodal if $\mu < 9.98$. Thus, there is a small interval of values [9.98, 10.53] of the parameter $\mu$ for which the jump set contains a region whose stability status is unknown. This interval represents the gap between necessary and sufficient conditions.

## 5 The Quasiconvex Envelope

We have now proved that when $\mu$ is sufficiently large, the entire jump set satisfies the quasiconvexity condition. As a consequence, we obtain an explicit rule for finding $QW(\boldsymbol{F})$ for the two-well Hadamard energy (1.1).

We recall a theorem from [13] that the quasiconvex envelope of the objective isotropic energy is also objective and isotropic. Therefore, $QW(\boldsymbol{F})$ depends on $\boldsymbol{F}$ only through its singular values $0 < \varepsilon_1(\boldsymbol{F}) \le \varepsilon_2(\boldsymbol{F})$. In order to describe $QW(\varepsilon_1, \varepsilon_2)$, for $d \in (d_1, d_2)$, let

$$R(d) = \sqrt{-\frac{h'(D(d)) - h'(d)}{D(d) - d}}.$$

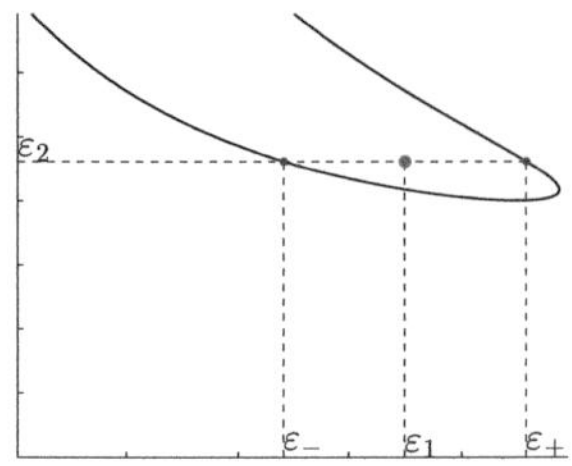


**Fig. 4** The convex combination rule (2.7) for computing the quasiconvex hull

Then,

$$\mathcal{B}_\mu = \left\{(\varepsilon_1, \varepsilon_2) : \varepsilon_1 > 0,\ d_1 < \varepsilon_1\varepsilon_2 < d_2,\ \varepsilon_2 R(\varepsilon_1\varepsilon_2) > \sqrt{\mu}\right\}$$

describes the region inside the curve in Fig. 4. If $(\varepsilon_1, \varepsilon_2) \in \mathcal{B}_\mu$, then there exists a uniquely determined pair $\varepsilon_-(\varepsilon_2) < \varepsilon_+(\varepsilon_2)$, such that $(\varepsilon_\pm, \varepsilon_2) \in \partial\mathcal{B}_\mu$ (see Fig. 4). Algebraically, $\varepsilon_\pm(\varepsilon_2)$ are found as the two solutions $\varepsilon = \varepsilon_\pm$ of

$$\varepsilon_2 R(\varepsilon\varepsilon_2) = \sqrt{\mu}. \tag{5.1}$$

We observe that for $(\varepsilon_1, \varepsilon_2) \in \mathcal{B}_\mu$ we always have

$$\varepsilon_-(\varepsilon_2) < \varepsilon_1 < \varepsilon_+(\varepsilon_2).$$

For any $\boldsymbol{F} \in \mathbb{R}_+^{2\times 2}$, let $0 < \varepsilon_1(\boldsymbol{F}) \le \varepsilon_2(\boldsymbol{F})$ be the ordered pair of its singular values. Then, according to (2.7)

$$QW(\varepsilon_1, \varepsilon_2) = \begin{cases} W(\varepsilon_1, \varepsilon_2), & (\varepsilon_1, \varepsilon_2) \notin \mathcal{B}_\mu, \\ \frac{\varepsilon_1-\varepsilon_-}{\varepsilon_+-\varepsilon_-} W(\varepsilon_+, \varepsilon_2) + \frac{\varepsilon_+-\varepsilon_1}{\varepsilon_+-\varepsilon_-} W(\varepsilon_-, \varepsilon_2), & (\varepsilon_1, \varepsilon_2) \in \mathcal{B}_\mu, \end{cases} \tag{5.2}$$

where $\varepsilon_\pm = \varepsilon_\pm(\varepsilon_2)$.

Let us examine the structure of $QW$ and how it is related to $W$. We first note that

$$\varepsilon_\pm(\varepsilon_2) = \frac{\varepsilon'_\pm(\frac{\varepsilon_2}{\sqrt{\mu}})}{\sqrt{\mu}},$$

where $\varepsilon'_\pm(\varepsilon'_2)$ are the two solutions of $\varepsilon'_2 R(\varepsilon'\varepsilon'_2) = 1$. Therefore,

$$\mathcal{B}_\mu = \left\{\left(\frac{\varepsilon'_1}{\sqrt{\mu}}, \varepsilon'_2\sqrt{\mu}\right) : (\varepsilon'_1, \varepsilon'_2) \in \mathcal{B}_1\right\}.$$

In other words $\mathcal{B}_\mu$ is a hyperbolic rotation of $\mathcal{B}_1$:

$$\mathcal{B}_\mu = \mathcal{R}_\mu \mathcal{B}_1, \qquad \begin{bmatrix} \varepsilon_1 \\ \varepsilon_2 \end{bmatrix} = \begin{bmatrix} \frac{1}{\sqrt{\mu}} & 0 \\ 0 & \sqrt{\mu} \end{bmatrix} \begin{bmatrix} \varepsilon'_1 \\ \varepsilon'_2 \end{bmatrix}.$$

We can describe $\mathcal{B}_1$ more explicitly in terms of functions $\varepsilon'_\pm(\varepsilon'_2)$:

$$\mathcal{B}_1 = \left\{(\varepsilon'_1, \varepsilon'_2) : \varepsilon'_2 > \frac{1}{\sqrt{-h''(d_0)}},\ \varepsilon'_-(\varepsilon'_2) < \varepsilon'_1 < \varepsilon'_+(\varepsilon'_2)\right\}, \tag{5.3}$$

where $d_0 \in (d_1, d_2)$ is the unique fixed point of the map $D(d)$.

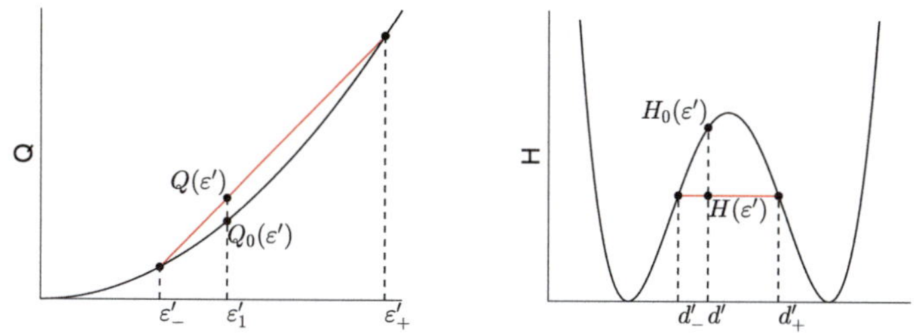


**Fig. 5** Geometric interpretation of functions $Q(\boldsymbol{\varepsilon}')$ and $H(\boldsymbol{\varepsilon}')$

Denoting $\boldsymbol{\varepsilon} = (\varepsilon_1, \varepsilon_2)$ and $\boldsymbol{\varepsilon}' = (\varepsilon'_1, \varepsilon'_2)$ we can also write

$$W(\boldsymbol{\varepsilon}) = \frac{\mu\varepsilon_2^2}{2} + \Psi_0(\mathcal{R}_\mu^{-1}\boldsymbol{\varepsilon}), \qquad QW(\boldsymbol{\varepsilon}) = \frac{\mu\varepsilon_2^2}{2} + \Psi(\mathcal{R}_\mu^{-1}\boldsymbol{\varepsilon}),$$

where

$$\Psi_0(\boldsymbol{\varepsilon}') = Q_0(\boldsymbol{\varepsilon}') + H_0(\boldsymbol{\varepsilon}'), \qquad \Psi(\boldsymbol{\varepsilon}') = Q(\boldsymbol{\varepsilon}') + H(\boldsymbol{\varepsilon}').$$

For $\boldsymbol{\varepsilon}' \notin \mathcal{B}_1$

$$Q(\boldsymbol{\varepsilon}') = Q_0(\boldsymbol{\varepsilon}') = \frac{(\varepsilon'_1)^2}{2}, \qquad H(\boldsymbol{\varepsilon}') = H_0(\boldsymbol{\varepsilon}') = h(\varepsilon'_1\varepsilon'_2).$$

For $\boldsymbol{\varepsilon}' \in \mathcal{B}_1$ $Q_0$ and $H_0$ retain their form above, while

$$Q(\boldsymbol{\varepsilon}') = \frac{\varepsilon'_1 - \varepsilon'_-}{\varepsilon'_+ - \varepsilon'_-}\frac{(\varepsilon'_+)^2}{2} + \frac{\varepsilon'_+ - \varepsilon'_1}{\varepsilon'_+ - \varepsilon'_-}\frac{(\varepsilon'_-)^2}{2},$$

$$H(\boldsymbol{\varepsilon}') = \frac{\varepsilon'_1 - \varepsilon'_-}{\varepsilon'_+ - \varepsilon'_-}h(\varepsilon'_+\varepsilon'_2) + \frac{\varepsilon'_+ - \varepsilon'_1}{\varepsilon'_+ - \varepsilon'_-}h(\varepsilon'_-\varepsilon'_2).$$

These formulas are deceptively messy. Functions $Q$ and $H$ have a simple geometric interpretation shown in Fig. 5. Denoting

$$d' = \varepsilon'_1\varepsilon'_2, \qquad d'_\pm = \varepsilon'_\pm\varepsilon'_2.$$

The point $(d', H(\boldsymbol{\varepsilon}'))$ lies on the chord connecting points $(d'_\pm, h(d'_\pm))$, while the point $(\varepsilon'_1, Q(\boldsymbol{\varepsilon}'))$ lies on the chord connecting points $(\varepsilon'_\pm, (\varepsilon'_\pm)^2/2)$. Comparing this formula to the naive first guess (1.2) we can write $U(\boldsymbol{\varepsilon})$ in the exact same form

$$U(\boldsymbol{\varepsilon}) = \frac{\mu\varepsilon_2^2}{2} + Q_0(\mathcal{R}_\mu^{-1}\boldsymbol{\varepsilon}) + \widetilde{H}(\mathcal{R}_\mu^{-1}\boldsymbol{\varepsilon}),$$

where the point $(d', \widetilde{H}(\boldsymbol{\varepsilon}'))$ lies on the Horizontal chord (common tangent) connecting points $(d_1, h(d_1))$ and $(d_2, h(d_2))$ in Fig. 5. We can have an idea of the relation between $W(\boldsymbol{\varepsilon})$, $QW(\boldsymbol{\varepsilon})$ and $U(\boldsymbol{\varepsilon})$ by observing that

$$U(\boldsymbol{\varepsilon}) \le QW(\boldsymbol{\varepsilon}) \le W(\boldsymbol{\varepsilon}).$$

Both inequalities above are strict when $\boldsymbol{\varepsilon} \in \mathcal{B}_\mu$. When $\varepsilon_1\varepsilon_2 \notin (d_1, d_2)$, then both inequalities become equalities, and when $\boldsymbol{\varepsilon} \notin \mathcal{B}_\mu$, but $\varepsilon_1\varepsilon_2 \in (d_1, d_2)$, then the left inequality remains strict, while the right inequality turns into equality. We note that the above analysis used simplified notation, where $\varepsilon_2$ denoted the larger of the two singular values of $\boldsymbol{F}$ and $\varepsilon_1$, the smaller.

## 6 Conclusions

The problem of computing quasi-convex envelopes for non-rank one convex energies is notoriously complex and only very few explicit constructions of this type are known in the literature [7, 24]. In this paper we presented a new explicit construction of this type for a subclass of double-well Hadamard energies which model materials undergoing isotropic-to-isotropic elastic phase transitions. The explicit calculations are performed in two space dimensions, but the technique works in 3D, as well, producing analogous results. Our example is non-trivial because the energy wells are not rank one connected and the construction does not reduce to convexification of any auxiliary energy. The possibility to obtain the explicit result is due to the fact that in the range of the dominance of the quadratic term in the Hadamard energy, the energy-minimizing microstructures are simple laminates which we prove by showing stability of the entire jump set representing deformation gradients that can coexist at jump discontinuities. When the relative size of the quadratic convex term decreases, the non-convex term in the Hadamard energy starts to dominate and the microstructures get more and more complex in some regions of the binodal. Yet, our method of identifying a laminate subset of the binodal remains effective, even though it no longer captures the entire binodal. When the rigidity modulus $\mu$ tends to zero and the material progressively fluidizes, the relaxation of Hadamard material should involve laminates of infinite rank, since the part of the binodal that is not captured by our method cannot be captured by laminates of any finite rank. We believe that such situations where either simple or infinite rank laminates are optimal must be generic, notwithstanding examples to the contrary for very special energies [3, 4, 9, 10].

**Acknowledgements** This work was started during the stay of YG at ESPCI, Paris supported by Chair Joliot. The work of both authors was supported by the National Science Foundation under Grant No. DMS-1714287. LT was supported additionally by the French Government under the Grant No. ANR-10-IDEX-0001-02 PSL.



## Appendix A

**Theorem A.1** *Suppose that $W(\boldsymbol{F})$ is given by* (1.1) *and that $h(d)$ is of class $C^1$, non-convex, has superlinear growth as $d \to +\infty$, and $h(d) = h^{**}(d)$ for all $d$ outside of a finite interval. Then $QW(\boldsymbol{F})$ cannot be written in the form $\mu'|\boldsymbol{F}|^2/2 + H(d)$ for any function $H$ and $\mu' > 0$.*

Let $d_0$ be such that $h^{**}(d_0) = h(d_0)$. Then for any $\boldsymbol{F}_0$, such that $\det \boldsymbol{F}_0 = d_0$ we have $QW(\boldsymbol{F}_0) = W(\boldsymbol{F}_0)$. Indeed, we have $W(\boldsymbol{F}_0) = U(\boldsymbol{F}_0)$, where $U(\boldsymbol{F})$ is defined in (1.2). But $U(\boldsymbol{F})$ is polyconvex and therefore quasiconvex, while satisfying $U(\boldsymbol{F}) \le W(\boldsymbol{F})$. It follows that

$$U(\boldsymbol{F}) \le QW(\boldsymbol{F}) \le W(\boldsymbol{F}). \tag{A.1}$$

Thus, $QW(\boldsymbol{F}_0) = W(\boldsymbol{F}_0)$ for any $\boldsymbol{F}_0$, such that $W(\boldsymbol{F}_0) = U(\boldsymbol{F}_0)$.

Theorem A.1 is proved by contradiction. Hence we assume that there exists a number $\mu'$ and a function $H(d)$, such that

$$QW(\boldsymbol{F}) = \frac{\mu'}{2}|\boldsymbol{F}|^2 + H(d). \tag{A.2}$$

We fix $d_0$, such that $h(d_0) = h^{**}(d_0)$. Then for diagonal $\boldsymbol{F}$ with components $f$ and $d_0/f$ on the diagonal we must have

$$\frac{\mu'}{2}\left(f^2 + \frac{d_0^2}{f^2}\right) + H(d_0) = \frac{\mu}{2}\left(f^2 + \frac{d_0^2}{f^2}\right) + h(d_0),$$

for every $f > 0$. It follows that $\mu = \mu'$ and $H(d) = h(d)$ for any $d$ for which $h(d) = h^{**}(d)$. But then inequality (A.1) implies $h^{**}(d) \le H(d) \le h(d)$ for all $d > 0$. However quasiconvexity of (A.2) implies that $H(d)$ has to be convex [6]. It follows that $H(d) = h^{**}(d)$ and we must conclude that if (A.2) is true then

$$QW(\boldsymbol{F}) = \frac{\mu}{2}|\boldsymbol{F}|^2 + h^{**}(d). \tag{A.3}$$

Showing that (A.3) is impossible is the main technical difficulty to be overcome. The idea is to prove that there exists $\boldsymbol{F}_0$, such that $h^{**}(\det \boldsymbol{F}_0) < h(\det \boldsymbol{F}_0)$, yet $W(\boldsymbol{F}_0) = QW(\boldsymbol{F}_0)$, which would contradict (A.3).

Let us consider the set

$$\mathcal{C} = \{d : h(d) > h^{**}(d)\},$$

which is open and bounded. In particular $\mathcal{C}$ is an at most countable union of finite intervals. Let $(d_*, d^*)$ be one such interval. Let us prove a geometrically "obvious" statement that the tangent line to the graph of $h(d)$ at $d_*$ must also be tangent to it at $d^*$.

**Lemma A.2** *Suppose that $h(d)$ is of class $C^1$ and has superlinear growth. Suppose that $h(d) > h^{**}(d)$ for all $d \in (d_*, d^*)$ and $h(d_*) = h^{**}(d_*)$ and $h(d^*) = h^{**}(d^*)$. Then $h'(d_*) = h'(d^*)$ and $h(d^*) - h(d_*) = h'(d_*)(d^* - d_*)$.*

*Proof* We assume, without loss of generality, that

$$h(d^*) = h'(d^*) = 0.$$

We will now show that $h(d_*) = h'(d_*) = 0$. It is enough to show that $h'(d_*) = 0$, since the graph of $h(d)$ can not be below the tangent line at either $d^*$ or $d_*$, we will then conclude (since both tangent lines are horizontal) that $h(d_*) = h(d^*) = 0$.

From the inequalities

$$0 = h(d^*) \ge h(d_*) + h'(d_*)(d^* - d_*) \ge h'(d_*)(d^* - d_*)$$

we conclude that $h'(d_*) < 0$, if it is not equal to 0. Now consider supporting lines to the graph of $h(d)$ with small and negative slope. In other words, for all $0 < \epsilon < -h'(d_*)$ we consider

$$\phi(\epsilon) = \min_d \big(h(d) + \epsilon d\big).$$

Let $d_\epsilon$ denote a minimizer, which must exist, since $h(d)$ has a superlinear growth. This minimizer must satisfy $h^{**}(d_\epsilon)=h(d_\epsilon)$, so that $d_\epsilon \notin (d_*, d^*)$. It is easy to see that $d_\epsilon$ cannot be larger than $d^*$. Indeed, if $d_\epsilon > d^*$

$$0=h\big(d^*\big)\ge h(d_\epsilon)+h'(d_\epsilon)\big(d^*-d_\epsilon\big)=h(d_\epsilon)+\epsilon\big(d_\epsilon-d^*\big)\ge\epsilon\big(d_\epsilon-d^*\big)>0,$$

which is a contradiction. Let us show that $d_\epsilon$ cannot be smaller that $d_*$ either. If $d_\epsilon < d_*$ then we have

$$h(d_\epsilon)\ge h(d_*)+h'(d_*)(d_\epsilon-d_*)$$

and

$$h(d_*)\ge h(d_\epsilon)-\epsilon(d_*-d_\epsilon)\ge h(d_*)-\big(h'(d_*)+\epsilon\big)(d_*-d_\epsilon)>h(d_*),$$

which is a contradiction. □

We will now show that for all sufficiently small $\epsilon>0$ we must have $QW(\boldsymbol{F})=W(\boldsymbol{F})$ for any $\boldsymbol{F}$ with $\det\boldsymbol{F}=d^*-\epsilon$. This would contradict formula (A.3).

**Lemma A.3** *For $\delta\in(d_*,d^*)$ we define*

$$\mathcal{D}(\delta)=\big\{d:h(d)<h(\delta)+h'(\delta)(d-\delta)\big\}\ne\emptyset,$$

*and $\theta(\delta)=\inf\mathcal{D}(\delta)$. Then there exists $\delta_n\to d_*$, such that $\theta(\delta_n)\to d^*$, as $n\to\infty$.*

*Proof* In our case $h(d)>0$ for all $d\in(d_*,d^*)$, while $h(d_*)=h(d^*)=0$. For any $\epsilon>0$ let $E_\epsilon=\{d\in(d_*,d^*):h(d)<\epsilon\}$. Let

$$d^\epsilon=\inf\left\{d\in E_\epsilon:d>\frac{d_*+d^*}{2}\right\},\qquad d_\epsilon=\sup\left\{d\in E_\epsilon:d<\frac{d_*+d^*}{2}\right\}$$

Then $h(d_\epsilon)=h(d^\epsilon)=\epsilon$, $h(d)>\epsilon$ for all $d\in(d_\epsilon,d^\epsilon)$ and $d_\epsilon\to d_*$ and $d^\epsilon\to d^*$, as $\epsilon\to 0$. Now consider the function

$$\phi_\epsilon(d)=h(d)-\frac{\epsilon(d-d_*)}{d^\epsilon-d_*}.$$

Let $d_\epsilon^*$ be a minimizer of $\phi_\epsilon$:

$$\phi_\epsilon\big(d_\epsilon^*\big)=\min_{d\in[d_*,d^\epsilon]}\phi_\epsilon(d)=m_\epsilon.$$

We observe that $m_\epsilon<0$. Indeed,

$$0=h'(d_*)=\lim_{d\to d_*}\frac{h(d)}{d-d_*}.$$

Thus, for all $d>d_*$ sufficiently close to $d_*$ we have

$$\frac{h(d)}{d-d_*}<\frac{\epsilon}{d^\epsilon-d_*}.$$

Therefore, $d_\epsilon^*\in(d_*,d_\epsilon)$. Indeed, for all $d\in(d_\epsilon,d^\epsilon)$ we have

$$\phi_\epsilon(d)\ge\epsilon-\frac{\epsilon(d-d_*)}{d^\epsilon-d_*}=\frac{\epsilon(d^\epsilon-d)}{d^\epsilon-d_*}>0.$$

We conclude that $d^*_\epsilon \to d_*$, as $\epsilon \to 0$ and $\phi'_\epsilon(d^*_\epsilon) = 0$, i.e.,

$$h'(d_\epsilon) = \frac{\epsilon}{d^\epsilon - d_*} > 0.$$

In addition we have $\phi_\epsilon(d) \geq \phi_\epsilon(d^*_\epsilon)$, i.e.,

$$h(d) \geq \frac{\epsilon(d - d^*_\epsilon)}{d^\epsilon - d_*} + h\big(d^*_\epsilon\big) = h\big(d^*_\epsilon\big) + h'\big(d^*_\epsilon\big)\big(d - d^*_\epsilon\big),$$

for all $d \in [d_*, d^\epsilon]$. This inequality will also hold for $d < d_*$. Indeed, since $\phi_\epsilon(d^*_\epsilon) < 0$ we have

$$h\big(d^*_\epsilon\big) < \frac{\epsilon(d^*_\epsilon - d_*)}{d^\epsilon - d_*}.$$

Therefore,

$$\frac{\epsilon(d - d^*_\epsilon)}{d^\epsilon - d_*} + h\big(d^*_\epsilon\big) < \frac{\epsilon(d - d_*)}{d^\epsilon - d_*} < 0 \leq h(d),$$

when $d < d_*$. We conclude that

$$d^* \in \mathcal{D}\big(d^*_\epsilon\big) \subset \big(d^\epsilon, +\infty\big). \qquad \square$$

Now, let $\delta_n$ be as in the lemma. The superlinear growth of $h(d)$ implies that there exists $\beta > 0$ so that $\mathcal{D}(\delta_n) \subset (\theta(\delta_n), \beta)$ for all sufficiently large $n$. Hence,

$$\lim_{n\to\infty} \sup_d \big\{h(\delta_n) + h'(\delta_n)(d - \delta_n) - h(d)\big\} = 0.$$

Let $\boldsymbol{F}_n = \sqrt{\delta_n}\boldsymbol{I}_2$. Then,

$$\lim_{n\to\infty} \inf_{\{\boldsymbol{F}:\det \boldsymbol{F}\in\mathcal{D}(\delta_n)\}} |\boldsymbol{F} - \boldsymbol{F}_n| \geq \operatorname{dist}\big(\sqrt{d_*}\boldsymbol{I}_2, \big\{\boldsymbol{F} : \det \boldsymbol{F} \geq d^*\big\}\big) > 0.$$

It follows that there exists $N > 0$, such that

$$\sup_d \big\{h(\delta_N) + h'(\delta_N)(d - \delta_N) - h(d)\big\} < \frac{\mu}{2}|\boldsymbol{F} - \boldsymbol{F}_N|^2$$

for every $\boldsymbol{F}$ with $\det \boldsymbol{F} \in \mathcal{D}(\delta_N)$. Then, for any $\boldsymbol{F}$, either $\det \boldsymbol{F} \notin \mathcal{D}(\delta_N)$ and hence

$$h(\det \boldsymbol{F}) - h(\delta_N) - h'(\delta_N)(\det \boldsymbol{F} - \delta_N) \geq 0,$$

or $\det \boldsymbol{F} \in \mathcal{D}(\delta_N)$ and thus

$$h(\delta_N) + h'(\delta_N)(\det \boldsymbol{F} - \delta_N) - h(\det \boldsymbol{F}) \leq \frac{\mu}{2}|\boldsymbol{F} - \boldsymbol{F}_N|^2.$$

In either case

$$\frac{\mu}{2}|\boldsymbol{F} - \boldsymbol{F}_N|^2 + h(\det \boldsymbol{F}) - h(\delta_N) - h'(\delta_N)(\det \boldsymbol{F} - \delta_N) \geq 0. \tag{A.4}$$

We observe that the function

$$\mathcal{N}(\boldsymbol{F}) = h(\delta_N) + h'(\delta_N)(\det \boldsymbol{F} - \delta_N)$$

is a null-Lagrangian and satisfies $W(\boldsymbol{F}) \geq \mathcal{N}(\boldsymbol{F})$ for all $\boldsymbol{F}$ and $W(\boldsymbol{F}_N) = \mathcal{N}(\boldsymbol{F}_N)$. Thus, since $\mathcal{N}(\boldsymbol{F})$ is quasiconvex, we have

$$W(\boldsymbol{F}) \leq QW(\boldsymbol{F}) \leq \mathcal{N}(\boldsymbol{F}),$$

which implies that $QW(\boldsymbol{F}_N) = W(\boldsymbol{F}_N)$, as desired, while $h^{**}(\delta_N) < h(\delta_N)$, since $\delta_N \in (d_*, d^*)$. Theorem A.1 is proved now.

## Appendix B: Minimization of $\Phi_0(x, y)$ over $xy = d$

We use the method of Lagrange multipliers to minimize

$$\Phi_0(x, y) = \frac{\mu}{2}\left(x^2 + y^2\right) - \alpha x - \beta y,$$

subject to the constraints $x > 0$, $y > 0$, $xy = d$. We note that $\Phi_0(x, y) \to +\infty$ when $x^2 + y^2 \to +\infty$, since $\mu > 0$. Therefore, the minimum of $\Phi_0(x, y)$ is achieved at a critical point on the branch of the hyperbola $xy = d$, $x > 0$. This critical point satisfies the following linear the system of equations:

$$\begin{cases} \mu x - \lambda y = \alpha, \\ \mu y - \lambda x = \beta. \end{cases}$$

Its solution is

$$x = x(\lambda) = \frac{\lambda\beta + \alpha\mu}{\mu^2 - \lambda^2}, \qquad y = y(\lambda) = \frac{\lambda\alpha + \beta\mu}{\mu^2 - \lambda^2}. \tag{B.1}$$

These equations describe both branches of the hyperbola

$$\mu\left(y^2 - x^2\right) + \alpha x - \beta y = 0. \tag{B.2}$$

The minimum value of $\Phi_0(x, y)$, subject to the constraints $x > 0$, $y > 0$, $xy = d$ is achieved at a point of intersection of the hyperbola (B.1), shown as a thick solid line on Fig. 6 with the hyperbola $xy = d$, shown as a dashed line. The $x$ and $y$ coordinates of these points of intersection solve the quartic equations

$$\mu x^4 - \alpha x^3 + d\beta x - \mu d^2 = 0, \qquad \mu y^4 - \beta y^3 + d\alpha y - \mu d^2 = 0, \tag{B.3}$$

respectively. When $d$ is very small there are 3 points of intersection of the two hyperbolas in the first quadrant and one in the third, as shown in Fig. 6. There is a critical value $d_* > 0$,

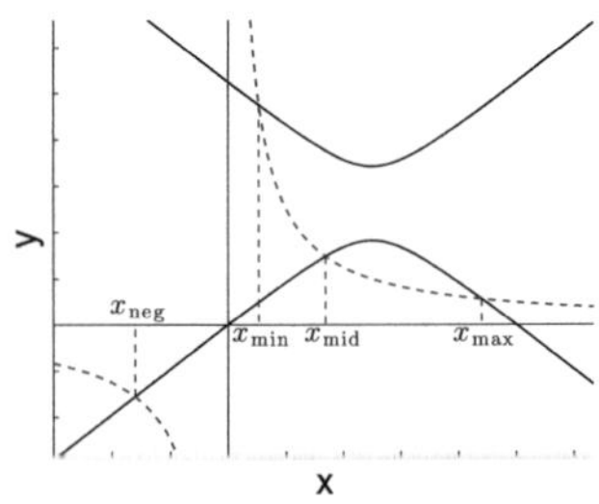


**Fig. 6** Critical points of $\Phi_0(x, y)$ subject to the constraint $xy = d$

such that for $d > d_*$ there will be only one point of intersection of the two hyperbolas in the first quadrant, located on the upper branch of the hyperbola (B.2). In this case this point must be the minimizer, since the minimum is achieved at a critical point on $xy = d$, $x > 0$. When $0 < d < d_*$, the other three points of intersection will lie on the lower branch of (B.2). Let us show that in this case the point of intersection with the upper branch corresponds to the minimizer. Let

$$x_{\text{neg}} < 0 < x_{\min}(d) < x_{\text{mid}}(d) < x_{\max}(d)$$

denote the 4 roots of (B.3). They are critical points of $f(x; d) = \Phi_0(x, d/x)$. Since $f(x, d) \to +\infty$ when $x \to 0^+$ and $x \to +\infty$, we conclude that $x_{\min}(d)$ and $x_{\max}(d)$ are points of local minimum, while $x_{\text{mid}}(d)$ is a point of local maximum of $f(x; d)$. It will also be convenient to denote $y_\gamma = d/x_\gamma$, where $\gamma$ is one of 'neg', 'min', 'mid' or 'max'. We note that

$$\lim_{d\to 0+} \big(x_{\min}(d), y_{\min}(d)\big) = \left(0, \frac{\beta}{\mu}\right), \qquad \lim_{d\to 0+} \big(x_{\max}(d), y_{\max}(d)\big) = \left(\frac{\alpha}{\mu}, 0\right).$$

Thus,

$$\lim_{d\to 0+} f\big(x_{\min}(d); d\big) = -\frac{\beta^2}{2\mu} < -\frac{\alpha^2}{2\mu} = \lim_{d\to 0+} f\big(x_{\max}(d); d\big).$$

This shows that

$$\Delta(d) = f\big(x_{\max}(d); d\big) - f\big(x_{\min}(d); d\big) > 0$$

when $d$ is sufficiently small. In order to prove that $\Delta(d) > 0$ for all $d \le d_*$ we note that

$$f\big(x_\gamma(d); d\big)' = \frac{\partial f}{\partial d}\big(x_\gamma(d); d\big), \quad \gamma \in \{\text{'neg'}, \text{'min'}, \text{'mid'}, \text{'max'}\}.$$

Then, after a straightforward calculation, we find that

$$\Delta'(d) = \big[f\big(x_{\max}(d); d\big) - f\big(x_{\min}(d); d\big)\big]' = \mu\frac{x_{\max} - x_{\min}}{x_{\max}x_{\min}}\left(\frac{\beta}{\mu} - (y_{\min} + y_{\max})\right).$$

Finding one more term of the asymptotic expansion of the roots of (B.3) in powers of $d$ we compute

$$\lim_{d\to 0+} \Delta'(d) = \frac{\mu(\alpha^2 - \beta^2)}{\alpha\beta} < 0.$$

Thus, $\Delta(d)$ is positive and decreasing when $d$ is sufficiently small. The decrease of $\Delta(d)$ will continue either until $d$ reaches $d_*$ or a value at which $\Delta'(d) = 0$, whichever occurs first. Let us show that $\Delta'(d) \neq 0$ for any $d \in [0, d_*]$. Indeed, if $\Delta'(d) = 0$ then we must have

$$y_{\min} + y_{\max} = \frac{\beta}{\mu}.$$

However, since the $y$'s are roots of the $y$-quartic in (B.3) we also have

$$y_{\text{neg}} + y_{\min} + y_{\text{mid}} + y_{\max} = \frac{\beta}{\mu}.$$

Hence, if $\Delta'(d) = 0$ we must have $y_{\mathrm{mid}} = -y_{\mathrm{neg}}$. However, if for some $y > 0$ both $y$ and $-y$ solve (B.3), then such $y$ must satisfy

$$\mu y^4 - \mu d^2 = 0, \qquad -\beta y^3 + d\alpha y = 0,$$

resulting in the consistency requirement $\alpha = \beta$, which is not true at all points satisfying (4.5). Thus, $\Delta(d)$ achieves its smallest value at $d = d_*$. However, at $d = d_*$ the function $f(x; d_*)$ has only one local minimum at $x_{\min}(d_*)$, while it is an increasing function in the neighborhood of $x_{\min}(d_*) = x_{\mathrm{mid}}(d_*)$. Thus, $\Delta(d_*) > 0$, and therefore, $\Delta(d) > 0$ for all $d \in [0, d_*]$. We conclude that for all $d > 0$

$$\phi(d) = \min_{xy=d} \Phi(x, y) = \Phi_0\big(x_{\min}(d), y_{\min}(d)\big) + h(d) - m_0 d, \tag{B.4}$$

where $x_{\min}(d)$ is the smallest positive root of the $x$-quartic in (B.3). ($y_{\min}(d) = d/x_{\min}(d)$ will be the largest root of the $y$-quartic in (B.3).) Moreover, Maple calculation shows

$$\phi'(d) = \frac{\mu y_{\min}(d)^2 - \beta y_{\min}(d)}{d} + h'(d) - m_0. \tag{B.5}$$

When $d$ is sufficiently large $\phi'(d) > 0$ and therefore the minimum of $\phi(d)$ is attained at a critical point of $\phi(d)$. We note that $y_{\min}(d)$ is an decreasing function of $d$ on $(0, d_{\min})$ and an increasing function on $(d_{\min}, +\infty)$, which is geometrically obvious, since hyperbola (B.2) is independent of $d$. The value of $d_{\min}$ is easily found from the condition that the hyperbola $xy = d$ passes through the upper vertex of hyperbola (B.2). Another observation is that

$$y_{\min}\left(\frac{\alpha\beta}{\mu^2}\right) = \frac{\beta}{\mu}.$$

The immediate consequence of this observation is that

$$\begin{cases} \phi'(d) > 0, & \text{if } h'(d) > m_0 \text{ and } d > \delta_{\max}(\mu) = \dfrac{\alpha\beta}{\mu^2}, \\[2ex] \phi'(d) < 0, & \text{if } h'(d) < m_0 \text{ and } d < \delta_{\min}(\mu) = \dfrac{\alpha(\beta + \sqrt{\beta^2 - \alpha^2})}{4\mu^2}. \end{cases}$$

We easily compute

$$\lim_{\mu\to\infty} \delta_{\max}(\mu) = 2\{d\}, \qquad \lim_{\mu\to\infty} \delta_{\min}(\mu) = \{d\}.$$

It follows that there exist $0 < d_{\min} < d_{\max}$ and $M_0 > 0$ that depend only on the choice of $h(d)$, so that for all $\mu > M_0$ all critical points of $\phi(d)$ are confined to the interval $(d_{\min}, d_{\max})$.

## References

1. Allaire, G., Kohn, R.V.: Explicit optimal bounds on the elastic energy of a two-phase composite in two space dimensions. Q. Appl. Math. **LI**(4), 675–699 (1993)
2. Allaire, G., Kohn, R.V.: Optimal bounds on the effective behavior of a mixture of two well-ordered elastic materials. Q. Appl. Math. **LI**(4), 643–674 (1993)
3. Antimonov, M.A., Cherkaev, A., Freidin, A.B.: Phase transformations surfaces and exact energy lower bounds. Int. J. Eng. Sci. **90**, 153–182 (2016)

4. Ball, J.M., James, R.: Proposed experimental tests of a theory of fine microstructure and two-well problem. Philos. Trans. R. Soc. Lond. **338A**, 389–450 (1992)
5. Ball, J.M., James, R.D.: Incompatible sets of gradients and metastability. Arch. Ration. Mech. Anal. **218**(3), 1363–1416 (2015)
6. Ball, J.M., Murat, F.: $W^{1,p}$-quasiconvexity and variational problems for multiple integrals. J. Funct. Anal. **58**(3), 225–253 (1984)
7. Benešová, B., Kružík, M.: Weak lower semicontinuity of integral functionals and applications. SIAM Rev. **59**(4), 703–766 (2017)
8. Cherepanov, G.P.: Inverse problems of the plane theory of elasticity. J. Appl. Math. Mech. **38**(6), 963–979 (1974)
9. Cherkaev, A., Kucuk, I.: Detecting stress fields in an optimal structure. I. Two-dimensional case and analyzer. Struct. Multidiscip. Optim. **26**(1–2), 1–15 (2004)
10. Cherkaev, A., Kucuk, I.: Detecting stress fields in an optimal structure. II. Three-dimensional case. Struct. Multidiscip. Optim. **26**(1–2), 16–27 (2004)
11. Dacorogna, B.: A relaxation theorem and its application to the equilibrium of gases. Arch. Ration. Mech. Anal. **77**(4), 359–386 (1981)
12. Dacorogna, B.: Quasiconvexity and relaxation of nonconvex problems in the calculus of variations. J. Funct. Anal. **46**(1), 102–118 (1982)
13. Dacorogna, B.: Direct Methods in the Calculus of Variations, 2nd edn. Springer, New York (2008)
14. Golubović, L., Lubensky, T.C.: Nonlinear elasticity of amorphous solids. Phys. Rev. Lett. **63**(10), 1082 (1989)
15. Grabovsky, Y., Kohn, R.V.: Microstructures minimizing the energy of a two phase elastic composite in two space dimensions. II: the Vigdergauz microstructure. J. Mech. Phys. Solids **43**(6), 949–972 (1995)
16. Grabovsky, Y., Truskinovsky, L.: Roughening instability of broken extremals. Arch. Ration. Mech. Anal. **200**(1), 183–202 (2011)
17. Grabovsky, Y., Truskinovsky, L.: Marginal material stability. J. Nonlinear Sci. **23**(5), 891–969 (2013)
18. Grabovsky, Y., Truskinovsky, L.: Normality condition in elasticity. J. Nonlinear Sci. **24**(6), 1125–1146 (2014)
19. Grabovsky, Y., Truskinovsky, L.: Legendre-Hadamard conditions for two-phase configurations. J. Elast. **123**(2), 225–243 (2016)
20. Grabovsky, Y., Truskinovsky, L.: When rank-one convexity meets polyconvexity: an algebraic approach to elastic binodal. J. Nonlinear Sci. (2018, in press). https://doi.org/10.1007/s00332-018-9485-7
21. Hadamard, J.: Leçons sur la propagation des ondes et les équations de l'hydrodynamique. Hermann, Paris (1903)
22. John, F.: Plane elastic waves of finite amplitude. Hadamard materials and harmonic materials. Commun. Pure Appl. Math. **19**(3), 309–341 (1966)
23. Morrey, C.B. Jr.: Quasi-convexity and the lower semicontinuity of multiple integrals. Pac. J. Math. **2**, 25–53 (1952)
24. Raoult, A.: Quasiconvex envelopes in nonlinear elasticity. In: Poly-, Quasi-and Rank-One Convexity in Applied Mechanics, pp. 17–51. Springer, Berlin (2010)
25. Šverák, V.: Rank-one convexity does not imply quasiconvexity. Proc. R. Soc. Edinb., Sect. A, Math. **120**(1–2), 185–189 (1992)
26. Tanaka, T.: Collapse of gels and the critical endpoint. Phys. Rev. Lett. **40**(12), 820 (1978)
27. Vigdergauz, S.B.: Two-dimensional grained composites of extreme rigidity. ASME J. Appl. Mech. **61**(2), 390–394 (1994)

J Elast (2019) 135:375–397
https://doi.org/10.1007/s10659-019-09724-0

# The State of Stress and Strain Adjacent to Notches in a New Class of Nonlinear Elastic Bodies

**Vojtěch Kulvait[1] · Josef Málek[1] · K.R. Rajagopal[2]**

*Dedicated to the memory of Professor Walter Noll*

Received: 9 July 2018 / Published online: 26 February 2019


**Abstract** In this paper we study the deformation of a body with a notch subject to an anti-plane state of stress within the context of a new class of elastic models. These models stem as approximations of constitutive response functions for an elastic body that is defined within the context of an implicit constitutive relation between the stress and the deformation gradient. Gum metal and many metallic alloys are described well by such constitutive relations. We consider the state of anti-plane stress of a body with a smoothened V-notch within the context of constitutive relations for the linearized strain in terms of a power-law for the stretch. The problem is solved numerically and the convergence and the stability of the solution is studied.

**Keywords** Implicit constitutive theory · Power law models · Small strain elasticity

**Mathematics Subject Classification** 74B20 · 74B15 · 35Q74

## 1 Introduction

Walter Noll [39–41] introduced the concept of a Simple Material[1] whose constitutive representation subsumes many classical constitutive expressions for elastic and viscoelastic

[1]Noll [42] later generalized the concept of his definition of a Simple Material but not to include implicit relationships. We shall not get into a discussion of the same here.

✉ K.R. Rajagopal
krajagopal@tamu.edu

V. Kulvait
vojtech.kulvait@ovgu.de

J. Málek
malek@karlin.mff.cuni.cz

1 Charles University, Faculty of Mathematics and Physics, Mathematical Institute, Sokolovská 83, 186 75 Prague, Czech Republic

2 Department of Mechanical Engineering, Texas A&M University, College Station, TX 77843, USA

solids and viscous and viscoelastic fluids, thus providing a framework within which one could study the response characteristics of a large class of materials. This notwithstanding, since the Cauchy stress in a Simple Material depends on the history of the density and the deformation gradient it precludes the possibility that the histories of the stress and the deformation gradient could be related by an implicit constitutive relation. Recently, Rajagopal [49, 50] has provided a compelling rationale for considering such implicit constitutive relations to describe the response of both fluids and solids. In general, one can have an implicit relationship between the history of the density, stress, deformation gradient, and possibly other relevant physical variables such as the temperature, electrical and magnetic fields, etc.[2]

As algebraic implicit constitutive relations are somewhat recent, we provide some related references to studies concerning the response of bodies described by implicit constitutive relations. Průša and Rajagopal [46] generalizing the approach of Coleman and Noll [19] were able to obtain constitutive response relations of the type due to Maxwell, Oldroyd, Rivlin and Ericksen, that are approximations that hold within the context of retarded motions. Perlácová and Průša [45], building on the earlier work of LeRoux and Rajagopal [31], generalizing it and using algebraic implicit constitutive relations (that is constitutive relations wherein only the stress and the symmetric part of the velocity gradient, and not any of their time derivatives, appear) used them describe the response of colloids and suspensions. Málek et al. [35] developed a generalization of the Navier-Stokes constitutive relation and introduced stress power-law fluids and Rajagopal [53] showed that had one started by expressing the velocity gradient as a function of the stress than vice-versa one would not arrive at erroneous assumptions such as the Stokes assumption in fluid mechanics.

Rigorous mathematical analyses for a subclass of fluids described by implicit constitutive relations (including activated fluids such as the Bingham fluid or activated Euler fluids) have been carried out for steady flows by Málek et al. [8, 33] and for unsteady flows by Bulíček et al. [5, 6]; see also a recent study by Blechta et al. [3] where the systematic classification of fluids described by implicit constitutive relation are carried out and the mathematical analysis of steady/unsteady flows of activated Euler fluids subject to various types of boundary conditions is presented. The analysis of unsteady flows of Bingham fluids subject to boundary conditions described by implicit relations is considered in [10] and flows of heat conducting fluids are analysed in [11, 37]. The convergence of numerical schemes proposed to approximate flows of incompressible fluids towards the (weak) solution of the original PDE problems was established in [22, 27, 62]. PDE analysis for a class of solids with bounded linearized strain is developed in [2, 13, 14], see also the survey paper [7], while a study regarding the properties of finite element approximations is presented in [4] (see also [28, 38] for relevant results of computer simulations). Large data and long time PDE analysis of unsteady motions of generalized Kelvin-Voigt solids is developed in [12]; see also the study [9]. Finally, the analysis of the equations governing the motion of a special class of compressible fluids with bounded divergence is developed in [24], see also [36] for the derivation of these models using a thermodynamical approach.

The response of solids described by implicit constitutive relations has been studied in the case of polymers by Rajagopal and Saccomandi [56], inelastic solids by Rajagopal and Srinivasa [59] and compressible fluids by Málek and Rajagopal [36]. The response of electroelastic bodies described by implicit constitutive relations have been studied in the articles by Bustamante and Rajagopal [15], [16], and the response of magnetoelastic bodies described by implicit constitutive relations has been analysed by Bustamante and Rajagopal

[2]Rajagopal [55] classified the material symmetry possessed by the sub-classes of bodies whose histories of the stress, deformation gradient, density, etc., are given in terms of an implicit constitutive relations.

[17]. A review of implicit constitutive theories can be found in the articles by Rajagopal and Saccomandi [57] and Málek and Průša [34].

When one confines attention to elastic solids, a special sub-class of the implicit constitutive relations that is relevant is the class of bodies whose response is described by an implicit relationship between the stress and the deformation gradient. Rajagopal and Srinivasa [58], [48] have provided a thermodynamic basis for implicit theories of elasticity. Cauchy elastic bodies wherein the stress is expressed as an explicit function of the deformation gradient are a special subclass of the above class of implicit relations. Another special subclass is defined by an explicit expression for the Cauchy-Green tensor $\boldsymbol{B}$ in terms of the Cauchy stress $\boldsymbol{T}$. Truesdell and Moon [63] studied conditions that guaranteed semi-invertibility of isotropic functions and thus is relevant to the conditions under which the relation between the Cauchy-Green tensor $\boldsymbol{B}$ and the Cauchy stress $\boldsymbol{T}$ is invertible. But not all functions of the Cauchy-Green tensor $\boldsymbol{B}$ as a function of the Cauchy stress $\boldsymbol{T}$ are invertible. The class of response functions that express the Cauchy-Green tensor $\boldsymbol{B}$ as a function of the Cauchy stress $\boldsymbol{T}$ includes response functions that are not Cauchy elastic (Cauchy [18]). Moreover, Truesdell and Moon [63] were not interested in issues of causality that would lead one to conclude that implicit models that relate the stress to kinematical quantities have a sounder philosophical underpinning. Moreover, since the study of Truesdell and Moon [63] was confined to isotropic response, it does not address any of the models that describe the relationship between the Cauchy-Green tensor $\boldsymbol{C}$ and the Cauchy stress in the case of the anisotropic response of elastic bodies.

There are several advantages to the use of implicit models in describing elastic response, one of them being the fact that the use of the linearization based on the displacement gradient being small leading to a nonlinear relationship between the linearized strain and the stress (see Rajagopal [51, 52]). Such responses are exhibited by a variety of metallic alloys as well as materials like concrete and Devindran et al. [21] used such implicit models to describe the response of several metallic alloys. Also, a sub-class of these models, those that have limited linearized strain[3] are particularly well suited for the study of problems such as fracture and the state of stresses and strains at notches within the context of brittle materials (see Rajagopal and Walton [60], Gou et al. [25], Kulvait et al. [28], Ortiz et al. [43, 44], Montero et al. [38], Itou et al. [26]) and unlike the classical linearized theory of elasticity the strains are not singular. Of course, this is not surprising as the constitutive relation has built into it the concept of bounded strains. The problem of anti-plane state of stress in a body with a V-notch that is pointed or sharply radiused has been studied by Zappolorto et al. [64] in which they obtain closed form solutions in the case of strain limiting models, and Rajagopal and Zappalorto [61] have studied the state of stress and strain adjacent to a crack tip in the case of a non-monotonic relationship between the stress and the strain.

Instead of strain limiting theories, one could have constitutive relations wherein the relationship between the linearized strain and the stress is given by a power-law. Such models have been found to be very useful in describing the response of Gum metal and several Titanium alloys (see Rajagopal [54], Kulvait et al. [30]). In such models, unlike strain limiting models, one finds that appropriate values of the power-law index, the strains grows as the stress grows, but much slower than that for a linearized elastic model. It would be interesting to examine the state of strain and stress at notches which have been smoothed (see Fig. 2) and it is towards the investigation of this question that this study is aimed. Two parameters, an angle $\alpha$ and a radius $r_c$ determine the nature of the smoothening that is carried out. We

[3]Bulíček et al. [7] study several mathematical aspects concerning strain limiting bodies.

study the anti-plane state of stress for a rectangle with a smoothened notch within the context of different materials. The material characterization for a variety of Titanium alloys has been carried out by Kulvait et al. [30] and we use these material parameters to characterize the different alloys that comprise the rectangular body with a notch that is subjected to anti-plane stress in the study.

In the present study we are interested in the class of constitutive relations, where the spherical (usually referred to as "hydrostatic part" or "isotropic part", but neither of these terminologies seem appropriate) and deviatoric parts of the response are modeled through

$$\begin{aligned} \operatorname{tr} \boldsymbol{\varepsilon} &= \sigma_1(\operatorname{tr} \boldsymbol{T}) \operatorname{tr} \boldsymbol{T}, \\ \boldsymbol{\varepsilon}^d &= \sigma_2(|\boldsymbol{T}^d|) \boldsymbol{T}^d, \end{aligned} \tag{1}$$

where $\sigma_1$ and $\sigma_2$ are scalar functions, with $\sigma_1(0) = 0$ and $\sigma_2(0) = 0$, $\boldsymbol{T}$ is a Cauchy stress, $\operatorname{tr} \boldsymbol{T}$ and $\boldsymbol{T}^d$ denote the trace and the deviatoric part of the stress $\boldsymbol{T}$, $\boldsymbol{\varepsilon}$ is the linearized strain tensor, $\operatorname{tr} \boldsymbol{\varepsilon}$ and $\boldsymbol{\varepsilon}^d$ denote the trace and deviatoric part of the tensor $\boldsymbol{\varepsilon}$. Of course, one could provide the form for the linearized strain in terms of the Cauchy stress that implies the assumed forms for spherical and deviatoric parts, but we have chosen to express it as above so that the structure of the two parts of the linearized strain are transparent. In particular we are interested in power-law models, wherein

$$\begin{aligned} \sigma_1(\operatorname{tr} \boldsymbol{T}) &= \frac{1}{3K} \left( \frac{\tau_K^2 + |\operatorname{tr} \boldsymbol{T}|^2}{\tau_K^2} \right)^{\frac{s'-2}{2}}, \\ \sigma_2(|\boldsymbol{T}^d|) &= \frac{1}{2\mu} \left( \frac{\tau_\mu^2 + |\boldsymbol{T}^d|^2}{\tau_\mu^2} \right)^{\frac{q'-2}{2}}, \end{aligned} \tag{2}$$

parameters $q' \in (1, \infty)$, $s' \in (1, \infty)$, $K > 0$, $\tau_K > 0$, $\mu > 0$, $\tau_\mu > 0$ are material moduli. One should not think of expressions obtained by inverting (1), with (2) inserted in the appropriate place, as constitutive models. This is because these models have the meaning of approximations only within the context of the expressions for the Cauchy-Green tensor $\boldsymbol{B}$ in terms of the stress $\boldsymbol{T}$. A detailed discussion concerning the relevant issues can be found in Rajagopal [47].

Recently, it has been shown that the model (2) is capable of capturing the nonlinear response characteristics of titanium alloys in the small strain range, see [30]. We have identified the parameters of the power-law model (2), for four different beta phase titanium alloys. Based on these identified parameters, we have studied the behavior of these alloys when subject to anti-plane shear stress.

This study is devoted to the anti-plane stress of a body containing a smoothed V-notch, which we refer to as the VC-notch. As it is an anti-plane stress problem, we shall assume that the stress and the displacement field depend only on plane coordinates ($x_1$ and $x_2$) and the only nonzero components of the stress tensor are $T_{13}$ and $T_{23}$. The boundary tractions on a body that is subject to the anti-plane stress setting is in a direction perpendicular to the $(x_1, x_2)$ plane, see Fig. 1. These assumptions allow us to rephrase the problem of finding the unknown stress and displacement of the body as a variational problem on a discrete space of finite elements to find an unknown Airy's stress function.

In [28] we have presented a numerical study of the behavior of the strain-limiting model in the anti-plane stress setting. Here, we are interested in the behavior of the class of materials defined through (1) and (2) in the geometry of a square plate with a smoothened V-shaped notch.

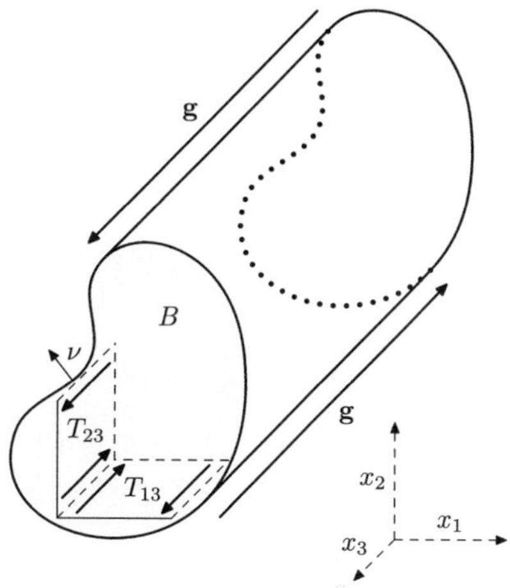


**Fig. 1** Anti-plane stress setting. The only nonzero components of the stress tensor are $T_{13}$ and $T_{23}$

The solver of the discretized problem on the finite element space uses the damped Newton method and utilizes FEniCS software library, see [32]. Further implementation details for solving such problems can be found in the doctoral thesis [29] and the Python scripts can be accessed from the public git repository https://bitbucket.org/kulvait/simulationsfenics.

## 2 Boundary Value Problem

The fundamental boundary value problem of elastostatics can be formulated as follows.

**Definition 1** (Problem (P)) Let $\Omega \subset \mathbb{R}^n$, $n = 2$ or $3$ be an open, bounded, connected set with the boundary $\partial\Omega$ consisting of two smooth disjoint parts $\Gamma_D$ and $\Gamma_N$ such that $\partial\Omega = \overline{\Gamma_D \cup \Gamma_N}$, where $\boldsymbol{\nu} = \boldsymbol{\nu}(\mathbf{x})$ denotes the unit outward normal vector at $\mathbf{x} \in \partial\Omega$. Let $\boldsymbol{G} : \mathbb{R}^{n\times n}_{\text{sym}} \to \mathbb{R}^{n\times n}_{\text{sym}}$, $\mathbf{u}_0 : \Omega \to \mathbb{R}^n$, $\mathbf{f} : \Omega \to \mathbb{R}^n$ and $\mathbf{g} : \Gamma_N \to \mathbb{R}^n$ be given. We say that the pair of functions $(\mathbf{u}, \boldsymbol{T}) : \Omega \to \mathbb{R}^n \times \mathbb{R}^{n\times n}_{\text{sym}}$ solves the Problem (P) if

$$-\operatorname{div} \boldsymbol{T} = \mathbf{f} \quad \text{in } \Omega, \tag{3a}$$

$$\boldsymbol{\varepsilon} = \boldsymbol{G}(\boldsymbol{T}) \quad \text{in } \Omega, \tag{3b}$$

$$\mathbf{u} = \mathbf{u}_0 \quad \text{on } \Gamma_D, \tag{3c}$$

$$\boldsymbol{T}\boldsymbol{\nu} = \mathbf{g} \quad \text{on } \Gamma_N, \tag{3d}$$

where

$$\boldsymbol{\varepsilon} = \frac{1}{2}\left(\nabla\mathbf{u} + (\nabla\mathbf{u})^T\right).$$

The above problem describes the state of the elastic body occupying the set $\Omega$ at equilibrium characterized by (3a), where $\boldsymbol{T}$ is the stress tensor and $\mathbf{f}$ stands for the external body forces. The body obeys the nonlinear constitutive equation (3b) that relates the linearized strain $\boldsymbol{\varepsilon}$, which is the symmetric part of the gradient of the displacement $\mathbf{u}$, and the stress tensor $\boldsymbol{T}$ by means of a nonlinear function $\boldsymbol{G}$. Equation (3c) prescribes the displacement $\mathbf{u}_0$ on the Dirichlet part of the boundary $\Gamma_D$, and (3d) prescribes the boundary traction $\mathbf{g}$ on the

Neumann part of the boundary $\Gamma_N$. The state of the body is subject to an additional assumption that the square of the norm of the displacement gradient can be neglected in comparison to the norm of the displacement gradient itself.

The geometry and the type of the deformation of anti-plane stress, see Fig. 1, allows us to assume that $\Omega = B \times \mathbb{R}$, $\mathbf{u} = (0, 0, u(x_1, x_2))$, $\partial B = \Gamma_N$, $\mathbf{g} = (0, 0, g)$ and $\mathbf{f} = \mathbf{0}$. The stress tensor $\boldsymbol{T}$ has two nontrivial components and can be represented by vector $\boldsymbol{T}_v = (T_{13}, T_{23})$. The trace and the norm of the deviatoric part of the stress tensor can be expressed as

$$\operatorname{tr} \boldsymbol{T} = 0, \quad \left|\boldsymbol{T}^d\right| = \sqrt{2}|\boldsymbol{T}_v|.$$

Also, the strain tensor $\boldsymbol{\varepsilon}$ has also only two nontrivial components $\varepsilon_{13}$ and $\varepsilon_{23}$ and can be represented by the vector $\boldsymbol{\varepsilon}_v = (\varepsilon_{13}, \varepsilon_{23})$. Constitutive relation (1) reduces to the form

$$\boldsymbol{\varepsilon}_v = \sigma_2\left(\sqrt{2}|\boldsymbol{T}_v|\right)\boldsymbol{T}_v \equiv \sigma\left(|\boldsymbol{T}_v|\right)\boldsymbol{T}_v,$$

and Problem (P) can be reformulated as:

**Definition 2** (Problem (P) in Anti-plane Stress Setting) Let $\Omega = B \times \mathbb{R}$, where $B \subset \mathbb{R}^2$ is an open, bounded, simply connected set. Let $\sigma : \mathbb{R} \to \mathbb{R}$ represent the constitutive response of the material and let $g : \partial B \to \mathbb{R}$ be a given function. We say that a pair of functions $(\boldsymbol{\varepsilon}_v, \boldsymbol{T}_v)$ is the solution of Problem (P) in the anti-plane stress setting [4] when the following is true

$$-\frac{\partial T_{13}}{\partial x_1} - \frac{\partial T_{23}}{\partial x_2} = 0 \quad \text{in } B, \tag{4a}$$

$$\varepsilon_{13} = \sigma\left(|\boldsymbol{T}_v|\right)T_{13}, \qquad \varepsilon_{23} = \sigma\left(|\boldsymbol{T}_v|\right)T_{23} \quad \text{in } B, \tag{4b}$$

$$T_{13}\nu_1 + T_{23}\nu_2 = g \quad \text{on } \partial B. \tag{4c}$$

In the remainder of this chapter, we use the following particular forms of the constitutive function $\sigma$. For the power-law model (2), we have the response

$$\sigma\left(|\boldsymbol{T}_v|\right) = \frac{1}{2\mu}\left(\frac{\tau_0^2 + 3|\boldsymbol{T}_v|^2}{\tau_0^2}\right)^{\frac{q'-2}{2}}. \tag{5}$$

The linear Hooke's law is characterized by the response (5), when $q' = 2$, that is

$$\sigma\left(|\boldsymbol{T}_v|\right) = \frac{1}{2\mu_L}. \tag{6}$$

In the case of strain-limiting model, see [28] we have

$$\sigma\left(|\boldsymbol{T}_v|\right) = \frac{\tau_\mu}{2\mu_l(\tau_\mu^a + (\sqrt{2}|\boldsymbol{T}_v|)^a)^{1/a}}. \tag{7}$$

[4] Since we consider constitutive relations of the type (4b), the anti-plane stress state is equivalent to the classical definition of anti-plane strain.

## 2.1 Compatibility Conditions

The Saint-Venant compatibility conditions for $\boldsymbol{\varepsilon}$, are reduced to the two nontrivial equations

$$\frac{\partial^2 \varepsilon_{13}}{\partial x_1 \partial x_2} - \frac{\partial^2 \varepsilon_{23}}{\partial x_1 \partial x_1} = 0, \qquad \frac{\partial^2 \varepsilon_{23}}{\partial x_1 \partial x_2} - \frac{\partial^2 \varepsilon_{13}}{\partial x_2 \partial x_2} = 0, \tag{8}$$

which implies the existence of a constant $C$ such that

$$\frac{\partial \varepsilon_{13}}{\partial x_2} - \frac{\partial \varepsilon_{23}}{\partial x_1} = C. \tag{9}$$

Expressing (9) in terms of $\mathbf{u}$, we obtain

$$\frac{\partial^2 u_1}{\partial x_3 \partial x_2} - \frac{\partial^2 u_3}{\partial x_1 \partial x_2} - \frac{\partial^2 u_2}{\partial x_3 \partial x_1} + \frac{\partial^2 u_3}{\partial x_2 \partial x_1} = C. \tag{10}$$

Since $\mathbf{u} = (0, 0, u(x_1, x_2))$, we conclude that $C = 0$ and thus (8) leads to

$$\frac{\partial \varepsilon_{13}}{\partial x_2} - \frac{\partial \varepsilon_{23}}{\partial x_1} = 0. \tag{11}$$

The condition (11) is the necessary and sufficient condition for the existence of the displacement $u(x_1, x_2)$ fulfilling

$$\varepsilon_{13} = \frac{1}{2}\frac{\partial u}{\partial x_1}, \qquad \varepsilon_{23} = \frac{1}{2}\frac{\partial u}{\partial x_2}.$$

## 2.2 Airy's Function

Airy's stress function $A$ is a scalar function, whose derivatives are components of $\boldsymbol{T}_\upsilon$ defined through

$$T_\upsilon^1 = T_{13} = \frac{\partial A}{\partial x_2}, \qquad T_\upsilon^2 = T_{23} = -\frac{\partial A}{\partial x_1}. \tag{12}$$

Every stress field of the type (12) fulfills the equilibrium equation (4a). From (12), we have that $|\boldsymbol{T}_\upsilon| = |\nabla A|$, and thus $\sigma(|\boldsymbol{T}_\upsilon|) = \sigma(|\nabla A|)$. Substituting the constitutive relation (4b) into the compatibility condition (11) leads to

$$-\frac{\partial}{\partial x_1}\left(\sigma\left(|\nabla A|\right)\frac{\partial A}{\partial x_1}\right) - \frac{\partial}{\partial x_2}\left(\sigma\left(|\nabla A|\right)\frac{\partial A}{\partial x_2}\right) = 0 \quad \text{in } B. \tag{13}$$

The boundary condition (4c) takes the form

$$\frac{\partial A}{\partial x_2}\nu_1 - \frac{\partial A}{\partial x_1}\nu_2 = \frac{\partial A}{\partial x_1}t_1 + \frac{\partial A}{\partial x_2}t_2 = g \quad \text{on } \partial B, \tag{14}$$

where $\mathbf{t} = (-\nu_2, \nu_1)$ represents the tangential vector to the boundary $\partial B$. We parametrize the boundary by the counterclockwise oriented closed curve $\boldsymbol{\xi} = (\xi_1, \xi_2)$, $\boldsymbol{\xi}(0) = \boldsymbol{\xi}(h)$, such that $\boldsymbol{\xi}([0, h]) = \partial B$ and

$$t_1\big(\boldsymbol{\xi}(b)\big) = \xi_1'(b), \qquad t_2\big(\boldsymbol{\xi}(b)\big) = \xi_2'(b). \tag{15}$$

Substituting (15) into (14) and using the chain rule, we conclude that $g$ is equal to the tangential derivative of $A$. We have that

$$g\big(\boldsymbol{\xi}(b)\big) = \nabla A \cdot \boldsymbol{\xi}'(b) = \frac{dA(\boldsymbol{\xi}(b))}{db}, \qquad b \in [0, h). \tag{16}$$

Integrating (16) along the boundary, we obtain the Dirichlet boundary condition for $A$

$$A(x) = A_0^g\big(\boldsymbol{\xi}(b)\big) = A\big(\boldsymbol{\xi}(0)\big) + \int_0^b g\big(\boldsymbol{\xi}(s)\big)\,\mathrm{d}s, \quad b \in [0, h),\ x = \boldsymbol{\xi}(b) \in \partial B. \tag{17}$$

By introducing Airy's stress function via (12), the boundary value problem (4a)–(4c) takes the form summarized in the following definition.

**Definition 3** (Airy's Function for Solving Problem (P) in the Anti-plane Stress Setting) Let $B \subset \mathbb{R}^2$ be an open, bounded, simply connected set and the functions $(g, \sigma)$ fulfill the assumptions of Definition 2. We say that the function $A$ is the Airy's function solving Problem (P) in the anti-plane stress setting if the following is true

$$\begin{aligned} -\operatorname{div}\big(\sigma\big(|\nabla A|\big)\nabla A\big) &= 0 \quad \text{in } B, \\ A(\mathbf{x}) &= A_0^g(\mathbf{x}) \quad \text{on } \partial B, \end{aligned} \tag{18}$$

where $A_0^g$ is the function introduced in (17).

## 3 Finite Element Simulations

In this section, we delineate the setting that we follow when performing computer simulations. First, we depict the geometries of the computational domains and describe the boundary conditions that we use. Then, we define the variational formulation of the problem on finite dimensional spaces. Finally, we tabulate the parameters of the models of the titanium alloys that we simulate.

### 3.1 Computational Domain

We use the geometry of the square with a smoothened V-notch, which we shall refer to as the VC-geometry, see Fig. 2. The tip of the V-notch is smoothened by the circle arc of radius $r_c$ that is tangent to the V-notch. The geometry is parametrized by angle $\alpha$ and the radius of arc $r_c$. When $\alpha < \pi/2$, then the left part of the boundary consists of five parts $\Gamma_4 = \Gamma_4^A \cup \Gamma_4^B \cup \Gamma_4^C \cup \Gamma_4^D \cup \Gamma_4^E$. There is a restriction that the arc fits inside the unit square, which imposes the condition

$$0 < r_c < \frac{0.5 \sin(\frac{\alpha}{2})}{\cos^2(\frac{\alpha}{2})}.$$

The center of the arc is at the point

$$C = \left(0.5 - \frac{r_c}{\sin(\frac{\alpha}{2})}, 0.5\right).$$

When $\pi/2 \le \alpha < \pi$, the left part of the boundary consists of three parts $\Gamma_4 = \Gamma_4^A \cup \Gamma_4^B \cup \Gamma_4^C$. The restriction that the arc fits into the unit square takes the form

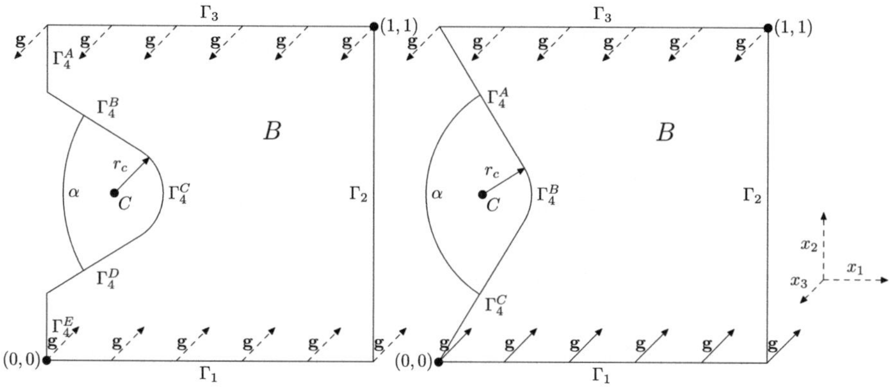


**Fig. 2** VC-geometry of a computational domain. On the left, the domain is depicted when $\alpha < \pi/2$. On the right, when $\pi/2 \le \alpha < \pi$

$$0 < r_c < \frac{0.5}{\cos\left(\frac{\alpha}{2}\right)}$$

and the center of the arc is at point

$$C = \left( \frac{0.5}{\tan(\frac{\alpha}{2})} - \frac{r_c}{\sin(\frac{\alpha}{2})}, 0.5 \right).$$

As we use SI units the edge length of the square is 1 m.

## 3.2 Boundary Conditions

We impose the same boundary conditions on all computational domains. The domain is loaded by the shearing force $F = 100$ MPa that acts on the top and on the bottom edge of the unit square. Therefore $g = F$ on $\Gamma_3$ (top), $g = -F$ on $\Gamma_1$ (down) and $g = 0$ on the remaining parts of the boundary. Boundary conditions (17) take the form

$$A_0^g(x_1, x_2) = \begin{cases} -Fx_1 & \text{on } \Gamma_1, \\ -F & \text{on } \Gamma_2, \\ -Fx_1 & \text{on } \Gamma_3, \\ 0 & \text{on } \Gamma_4, \end{cases}$$

where $F = 1 \times 10^8$ Pa.

## 3.3 Variational Problem on the Space of Finite Elements

Now we formulate the boundary value problem documented in Definition 3 as a variational problem on the space of finite elements. To derive the weak formulation, we multiply the first equation in (18) by a test function $\phi \in W_0^{1,q'}(B)$, integrate over $B$ and use integration by parts. For the power-law solid (5) where $\sigma$ has the form

$$\sigma(|\boldsymbol{T}_v|) \sim \left(1 + |\boldsymbol{T}_v|^2\right)^{\frac{q'-2}{2}},$$

we seek the function $A \in W^{1,q'}(B)$ that satisfies the formulation

$$\begin{aligned} &\int_B \sigma\left(|\nabla A|\right)\nabla A \cdot \nabla\phi \, \mathrm{d}x = 0 \quad \forall\phi \in W_0^{1,q'}(B), \\ &A = A_0^g \quad \text{on } \partial B, \end{aligned} \tag{19}$$

where $A_0^g$ is specified in (17). When the response is given by Hooke's law (6), where $\sigma$ is a constant, we set $q' = 2$ and seek Airy's function in the space $A \in W^{1,2}(B)$.

To find numerical solutions, we first construct a triangulation $\mathcal{T}_h$ of the computational domain $B$. We employ a finite dimensional space $V_h \subset W^{1,q'}(B)$ over $\mathcal{T}_h$ and its counterpart with zero trace

$$V_h^0 = \{u \in V_h, u|_{\partial B} = 0\}.$$

**Definition 4** (Airy's solution to the Discrete Problem (P) within the context of the Anti-plane Stress Setting) Let $B \subset \mathbb{R}^2$ be a computational domain described in Fig. 2. Let $\mathcal{T}_h$ be a triangulation over $\overline{B}$. Let the functions $(\sigma, g)$ represent the problem data. We say that the function $A_h \in V_h \subset W^{1,q'}(B)$ is an Airy's function that solves the discrete Problem (P) within the context of the anti-plane stress setting, if

$$\begin{aligned} &\sum_{T \in \mathcal{T}_h} \int_T \sigma\left(|\nabla A_h|\right)\nabla A_h \cdot \nabla\phi_h \, \mathrm{d}x = 0 \quad \forall\phi_h \in V_h^0, \\ &A_h = A_0^g \quad \text{on } \partial B, \end{aligned} \tag{20}$$

where $A_0^g$ is the function defined in (17).

As $V_h$ we use the space of piece-wise continuous second order polynomials $V_h = X_h^2$, where

$$X_h^2 = \left\{u_h \subset C^0(\overline{B}), u_h|_T \in P^2(T), \forall T \in \mathcal{T}_h\right\}.$$

We use Lagrange elements of second order so that the $X_h^2 \subseteq W^{1,q'}(B)$ and therefore we can omit the sum over elements in (20). For the problems of the type (20), the quasi-norm interpolation error estimates were established, see for example [23].

For solving the problem computationally, we have developed the software in Python that utilizes the FEniCS library, see [1]. To find a solution for nonlinear power-law problems, we use the damped Newton method, where the convergence criterion was based on the $L^2$ norm of the residuum. Computationally intensive tasks were performed using a cluster infrastructure supported by the Charles university, see http://cluster.karlin.mff.cuni.cz/.

### 3.4 Material Parameters Used for Simulations of Titanium Alloys

We study the response of three different titanium alloys, namely Gum Metal, Ti-30Nb-10Ta-5Zr alloy and Ti-24Nb-4Zr-7.9Sn alloy. For each alloy, we compare four different models

**Table 1** Parameters used in the computer simulations. Linear model LIN with the response (6) is parametrized by the shear modulus $\mu_L$. Power law models NLB and NLS with the response (5) are parametrized by $\tau_0$, $q'$ and $\mu$. Data for the models are based on previous experimental corroboration, see [30]

| Material | Model | $\mu_L$ [GPa] | $\tau_0$ [GPa] | $q'$ | $\mu$ [GPa] |
|---|---|---|---|---|---|
| Gum metal | LIN1 | 23.5 | – | – | – |
| Gum metal | NLB1 | – | 0.5 | 2.23 | 20.2 |
| Gum metal | NLS1 | – | 0.5 | 7.65 | 18668 |
| Ti-30Nb-10Ta-5Zr | LIN2 | 21.75 | – | – | – |
| Ti-30Nb-10Ta-5Zr | NLB2 | – | 0.5 | 2.49 | 22.3 |
| Ti-30Nb-10Ta-5Zr | NLS2 | – | 0.5 | 9.15 | 1001 |
| Ti-24Nb-4Zr-7.9Sn | LIN3 | 22.05 | – | – | – |
| Ti-24Nb-4Zr-7.9Sn | NLB3 | – | 0.5 | 2.99 | 16.5 |
| Ti-24Nb-4Zr-7.9Sn | NLS3 | – | 0.5 | 15.68 | 3378 |

derived in [30]. The parameters were estimated based on the general model (2) and subsequently they are applied to the model (5). The model labeled NLB is the power law model with parameters obtained as the best fit under the nonlinear bulk response condition and the model NLS is a model with parameters obtained under the nonlinear shear response condition, see [30].[5] The model LIN is a linear model (6), where the shear modulus is given by the Voigt-Reuss-Hill approximation. The original model (2) has the parameters $(\tau_0, s', K, q', \mu)$ that appear in its representation. In the anti-plane stress setting, the parameters $s'$ and $K$ do not enter the model (5), and therefore we work with a reduced set of parameters $(\tau_0, q', \mu)$. Parameters used in the computations are summarized in Table 1.

## 4 Results

We have performed the following set of simulations. We computed solutions for $\alpha \in \{1°, 2°, \dots 70°, 80° \dots 180°\}$. For each angle $\alpha$, we considered $r_c \in \{0.05, 0.01, 0.005, 0.001\}$. In total, we used 324 different geometries of the computational domain. For each geometry, we generate one basic mesh and used 5 adaptive refinements. In total, we use 1944 different meshes. Meshes were adaptively refined according to the local error estimator for the linear problem, this process is illustrated in Fig. 3. Mesh construction and refinement were performed using COMSOL Multiphysics software, version 3.5*a*, see [20]. On each mesh, we performed computer simulations for 9 different model settings described in Table 1, see Sect. 3.4.

### 4.1 Global Convergence of Solutions

The primary goal of computer simulations is to study distributions of strains and stresses for each model and to find differences between the linear and nonlinear solutions and between

[5]We are not using the same model and obtaining two different sets of values for the material parameters under two loading conditions. The NLB and NLS models are two different models that provide equally good fits. We used different experiments to obtain the material parameters for the two different models. These different sets of values fit the data equally well for the both the experiments. Which of these two models better explains the response of the body can only be determined by having other experiments against which these models can be corroborated. Such experiments are not available at the present time.

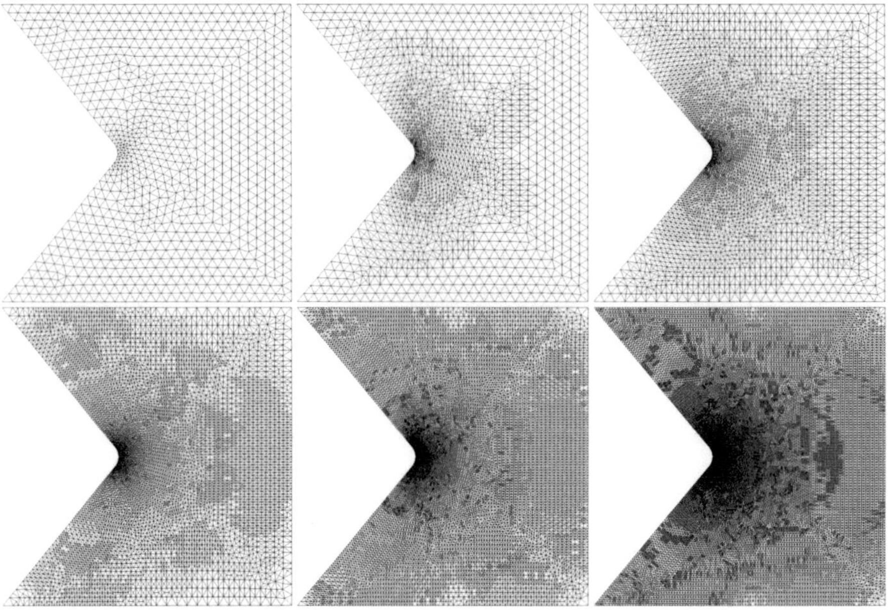

**Fig. 3** Visualization of the adaptive refinement of the mesh in 5 steps. In upper left, there is the unrefined mesh and in lower right, there is the mesh after 5 refinement steps. Visualization for the VC geometry, $\alpha = 100°$, $r_c = 0.05$

solutions of the problems for NLB and NLS that differ in the magnitude of the shear response exponent $q'$.

Prior to studying this, we investigate how accurate and reliable the solutions are. In this section, we study the global stability of the solutions with respect to a refinement level. To measure the relative error of the solution with respect to the reference solution, we define the relative error norm.

**Definition 5** (Relative Error Norm) Let $A_{\text{ref}}$ be a function (reference solution) on a finite dimensional space $V_h$. Let $A$ be a function (solution) on a finite dimensional space $Q_h \subseteq V_h$. Let $\|.\|$ denote a norm on $V_h$. Then the relative error norm of $A$ with respect to $A_{\text{ref}}$ is defined through

$$\|A\|^{\text{rel}} = \frac{\|A - A_{\text{ref}}\|}{\|A_{\text{ref}}\|}. \tag{21}$$

For a sequence of adaptively refined meshes, the triangulation on one level is a subset of triangulations on the levels above. Let $V_h^i$ be a finite dimensional space of solutions on the $i$th refinement level, then $V_h^0 \subseteq V_h^1 \cdots \subseteq V_h^5$. When analysing adaptively refined problems, we use a solution on the densest mesh, that is on the 5-th refinement level, as the reference solution. Then we compute the relative error norm for the solutions computed on croaser meshes. Our spaces $V_h^i$ consists of piece-wise continuous quadratic polynomials. Since $V_h^i \subseteq W^{1,2}(B)$ for $i \in \{0 \dots 5\}$, we use the norm $\|.\| = \|.\|_{1,2}$ in (21) for all the problems.

### 4.1.1 Tables of Mesh Properties and Error Norms

In Tables 2, 3 and 4 we list some important parameters with regard to the computations with respect to the mesh refinement. In particular, we list the number of elements, the number of degrees of freedom, the mesh size parameters $h_{\min}$ and $h_{\max}$ and the relative error norms for each problem listed in Table 1. For example, $\|A_{\mathrm{NLB2}}\|_{1,2}^{\mathrm{rel}}$ in a row in the table denotes the error norms with respect to the mesh refinement for the solution of the problem NLB2. In each table, the error norms for all the linear problems are compressed to the single row with

**Table 2** Mesh properties and error norms of the solutions with respect to the refinement level for $\alpha = 100°$ and $r_c = 0.001$. For a detailed description of row labels, see Sect. 4.1 and Table 1

| Refinement | 0 | 1 | 2 | 3 | 4 | 5 |
|---|---|---|---|---|---|---|
| Elements | 2504 | 6231 | 13 442 | 28 305 | 58 294 | 127 106 |
| DOFs | 5169 | 12 644 | 27 159 | 57 066 | 117 185 | 255 155 |
| $h_{\min}$ | 0.0002 | 0.0002 | 0.0002 | 0.0002 | $8 \times 10^{-5}$ | $6 \times 10^{-5}$ |
| $h_{\max}$ | 0.04 | 0.04 | 0.04 | 0.04 | 0.03 | 0.02 |
| $\|A_{\mathrm{LIN}}\|_{1,2}^{\mathrm{rel}}$ | 0.0006 | 0.0003 | 0.0002 | 0.0002 | 0.0002 | NA |
| $\|A_{\mathrm{NLB1}}\|_{1,2}^{\mathrm{rel}}$ | 0.0006 | 0.0003 | 0.0002 | 0.0002 | 0.0001 | NA |
| $\|A_{\mathrm{NLB2}}\|_{1,2}^{\mathrm{rel}}$ | 0.0005 | 0.0003 | 0.0002 | 0.0002 | 0.0001 | NA |
| $\|A_{\mathrm{NLB3}}\|_{1,2}^{\mathrm{rel}}$ | 0.0005 | 0.0002 | 0.0002 | 0.0001 | 0.0001 | NA |
| $\|A_{\mathrm{NLS1}}\|_{1,2}^{\mathrm{rel}}$ | 0.0004 | 0.0002 | 0.0001 | 0.000 08 | 0.000 06 | NA |
| $\|A_{\mathrm{NLS2}}\|_{1,2}^{\mathrm{rel}}$ | 0.0004 | 0.0002 | 0.0001 | 0.000 08 | 0.000 06 | NA |
| $\|A_{\mathrm{NLS3}}\|_{1,2}^{\mathrm{rel}}$ | 0.0004 | 0.0002 | 0.0001 | 0.000 08 | 0.000 05 | NA |

NA stands for Not Applicable

**Table 3** Mesh properties and error norms of the solutions with respect to the refinement level for $\alpha = 100°$ and $r_c = 0.01$. For a detailed description of row labels, see Sect. 4.1 and Table 1

| Refinement | 0 | 1 | 2 | 3 | 4 | 5 |
|---|---|---|---|---|---|---|
| Elements | 2028 | 4681 | 9667 | 20 025 | 40 747 | 87 058 |
| DOFs | 4197 | 9518 | 19 550 | 40 428 | 82 038 | 174 841 |
| $h_{\min}$ | 0.002 | 0.001 | 0.001 | 0.0006 | 0.0003 | 0.0002 |
| $h_{\max}$ | 0.04 | 0.04 | 0.04 | 0.04 | 0.04 | 0.02 |
| $\|A_{\mathrm{LIN}}\|_{1,2}^{\mathrm{rel}}$ | 0.001 | 0.001 | 0.0007 | 0.0003 | 0.0001 | NA |
| $\|A_{\mathrm{NLB1}}\|_{1,2}^{\mathrm{rel}}$ | 0.001 | 0.0009 | 0.0006 | 0.0002 | 0.000 09 | NA |
| $\|A_{\mathrm{NLB2}}\|_{1,2}^{\mathrm{rel}}$ | 0.001 | 0.0008 | 0.0006 | 0.0002 | 0.000 09 | NA |
| $\|A_{\mathrm{NLB3}}\|_{1,2}^{\mathrm{rel}}$ | 0.0009 | 0.0007 | 0.0005 | 0.0002 | 0.000 08 | NA |
| $\|A_{\mathrm{NLS1}}\|_{1,2}^{\mathrm{rel}}$ | 0.0007 | 0.0005 | 0.0004 | 0.0001 | 0.000 05 | NA |
| $\|A_{\mathrm{NLS2}}\|_{1,2}^{\mathrm{rel}}$ | 0.0006 | 0.0005 | 0.0003 | 0.0001 | 0.000 05 | NA |
| $\|A_{\mathrm{NLS3}}\|_{1,2}^{\mathrm{rel}}$ | 0.0006 | 0.0005 | 0.0003 | 0.0001 | 0.000 05 | NA |

NA stands for Not Applicable

**Table 4** Mesh properties and error norms of the solutions with respect to the refinement level for $\alpha = 100°$ and $r_c = 0.05$. For a detailed description of row labels, see Sect. 4.1 and Table 1

| Refinement | 0 | 1 | 2 | 3 | 4 | 5 |
|---|---|---|---|---|---|---|
| Elements | 1758 | 4153 | 8481 | 17 287 | 34 819 | 70 086 |
| DOFs | 3645 | 8460 | 17 172 | 34 920 | 70 138 | 140 827 |
| $h_{\min}$ | 0.01 | 0.002 | 0.002 | 0.001 | 0.0008 | 0.0006 |
| $h_{\max}$ | 0.04 | 0.04 | 0.04 | 0.04 | 0.04 | 0.03 |
| $\|A_{\mathrm{LIN}}\|_{1,2}^{\mathrm{rel}}$ | 0.003 | 0.001 | 0.0004 | 0.0003 | 0.0002 | NA |
| $\|A_{\mathrm{NLB1}}\|_{1,2}^{\mathrm{rel}}$ | 0.003 | 0.0009 | 0.0004 | 0.0003 | 0.0002 | NA |
| $\|A_{\mathrm{NLB2}}\|_{1,2}^{\mathrm{rel}}$ | 0.003 | 0.0009 | 0.0003 | 0.0003 | 0.0002 | NA |
| $\|A_{\mathrm{NLB3}}\|_{1,2}^{\mathrm{rel}}$ | 0.003 | 0.0008 | 0.0003 | 0.0003 | 0.0002 | NA |
| $\|A_{\mathrm{NLS1}}\|_{1,2}^{\mathrm{rel}}$ | 0.002 | 0.0006 | 0.0002 | 0.0002 | 0.0001 | NA |
| $\|A_{\mathrm{NLS2}}\|_{1,2}^{\mathrm{rel}}$ | 0.002 | 0.0006 | 0.0002 | 0.0002 | 0.0001 | NA |
| $\|A_{\mathrm{NLS3}}\|_{1,2}^{\mathrm{rel}}$ | 0.002 | 0.0006 | 0.0002 | 0.0002 | 0.0001 | NA |

NA stands for Not Applicable

a label $\|A_{\mathrm{LIN}}\|_{1,2}^{\mathrm{rel}}$, because for every mesh, the solution $A_{\mathrm{LIN}}$ to the linear problem does not depend on the shear modulus $\mu_L$. It means that all the linear problems have identical Airy's functions and consequently error norms are identical as well. We note that

$$\|A_{\mathrm{LIN}}\|_{1,2}^{\mathrm{rel}} = \|A_{\mathrm{LIN1}}\|_{1,2}^{\mathrm{rel}} = \|A_{\mathrm{LIN2}}\|_{1,2}^{\mathrm{rel}} = \|A_{\mathrm{LIN3}}\|_{1,2}^{\mathrm{rel}}. \tag{22}$$

We have chosen the following illustrative set of geometries to include in the tables. We consider data for $\alpha = 100°$ and $r_c \in \{0.001, 0.01, 0.05\}$, see Tables 2, 3 and 4.

## 4.2 Comparison of Solutions

Now, we focus on the geometry that has been smoothened by a circle or an arc of diameter $r_c$. We consider $\alpha = 90°$ and compare the behavior of the models LIN2, NLB2 and NLS2.

In Fig. 2, we consider $r_c = 0.01$. Visualizations of the stress and strain distributions are depicted in Fig. 4 and Fig. 5 respectively. We construct plots of stress and strain over the line from the smoothened tip, that is the point $C + (r_c, 0)$ to the right, which is denoted as $\varphi = 0°$ and to the top, which is denoted as $\varphi = 90°$, see Fig. 6.

We find that there is no singularity in the stresses or the strains for the various situations considered in the study.

## 4.3 Dependence on $\alpha$ and $r_c$

Thus far, we have been comparing the solutions when the geometrical setting is fixed. In this section, we study how the solutions depend on the parameters of the geometry. We consider the behavior of solutions when the diameter $r_c$ is decreasing and when the opening angle $\alpha$ is increasing.

In Figs. 7, 8 and 9, we plot the dependence of the highest value of $T_{23}$ and $\varepsilon_{23}$ in the whole computational domain with respect to $\alpha$ for all the models that were studied. Plots for LIN2 model in Fig. 7. Plots for NLB2 and NLS2 are provided, respectively in Figs. 8 and 9.

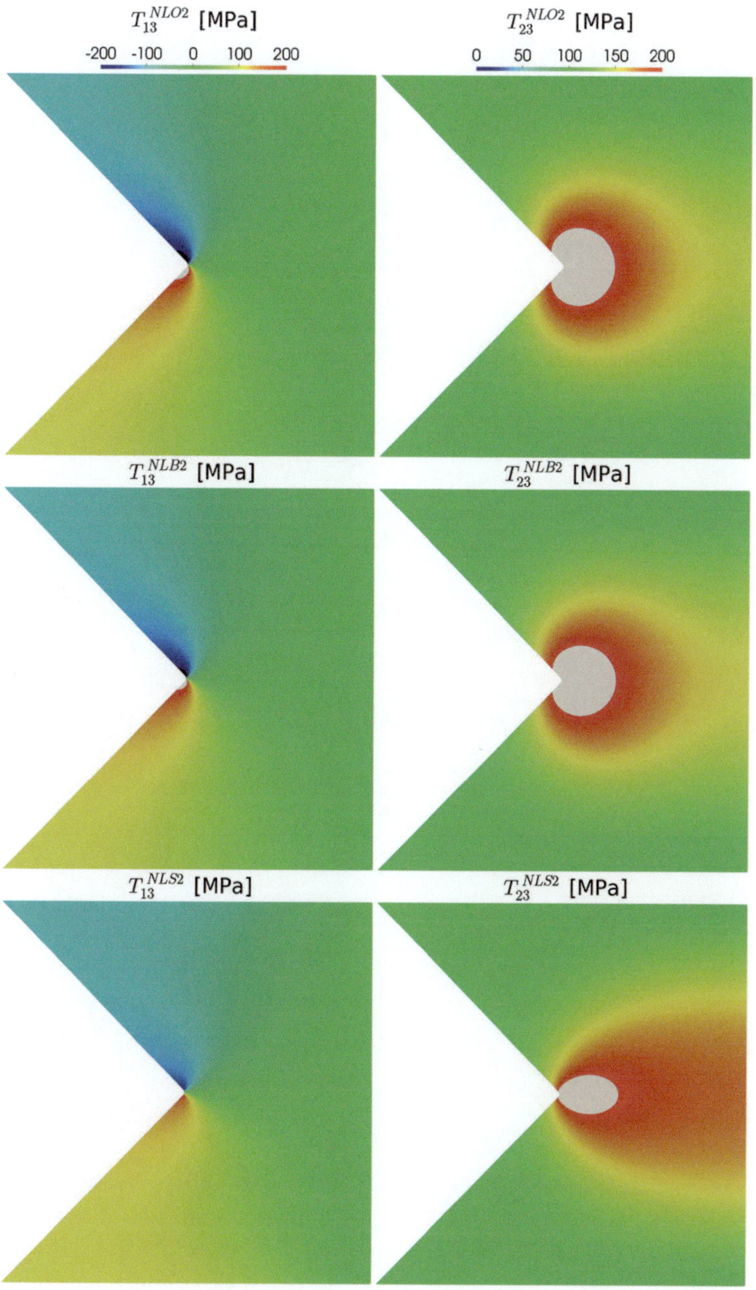


**Fig. 4** Comparison of stress distributions in VC geometry with $\alpha = 90^\circ$ and $r_c = 0.01$ for the models NLO2, NLB2, NLS2

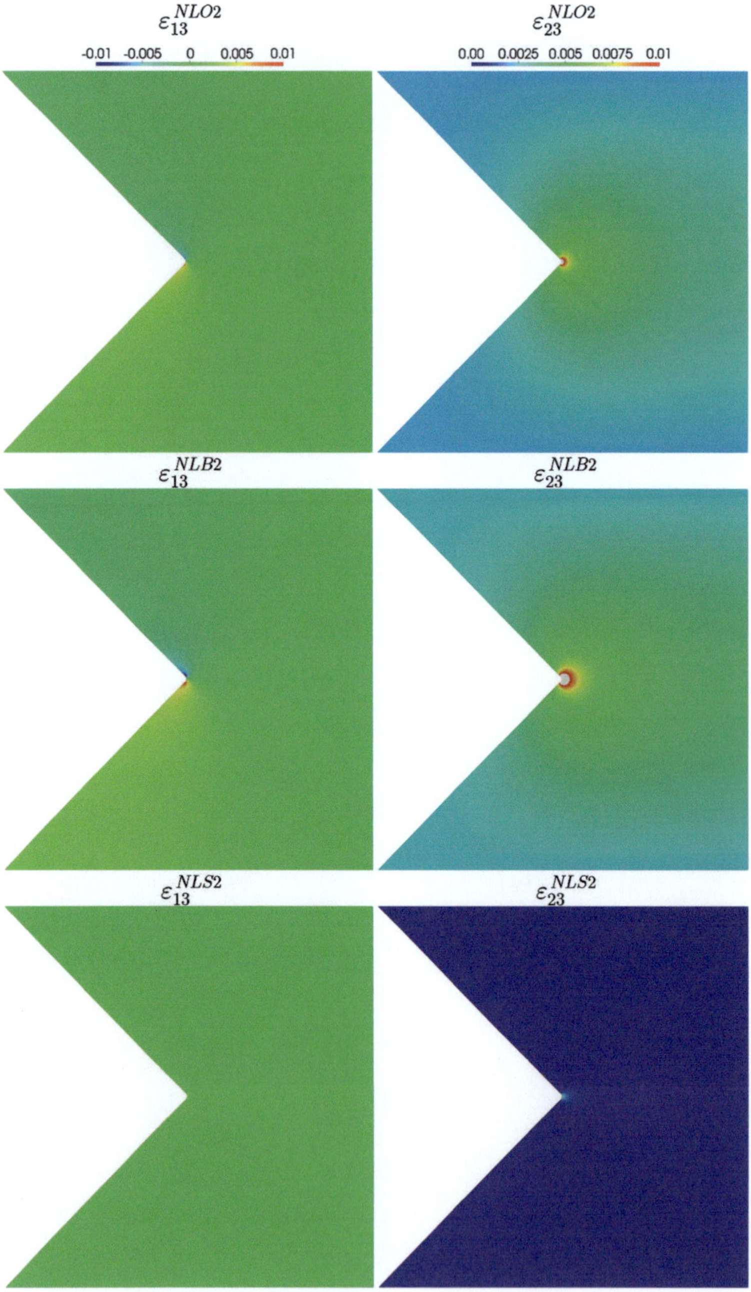


**Fig. 5** Comparison of strain distributions in VC geometry with $\alpha = 90°$ and $r_c = 0.01$ for the models NLO2, NLB2, NLS2

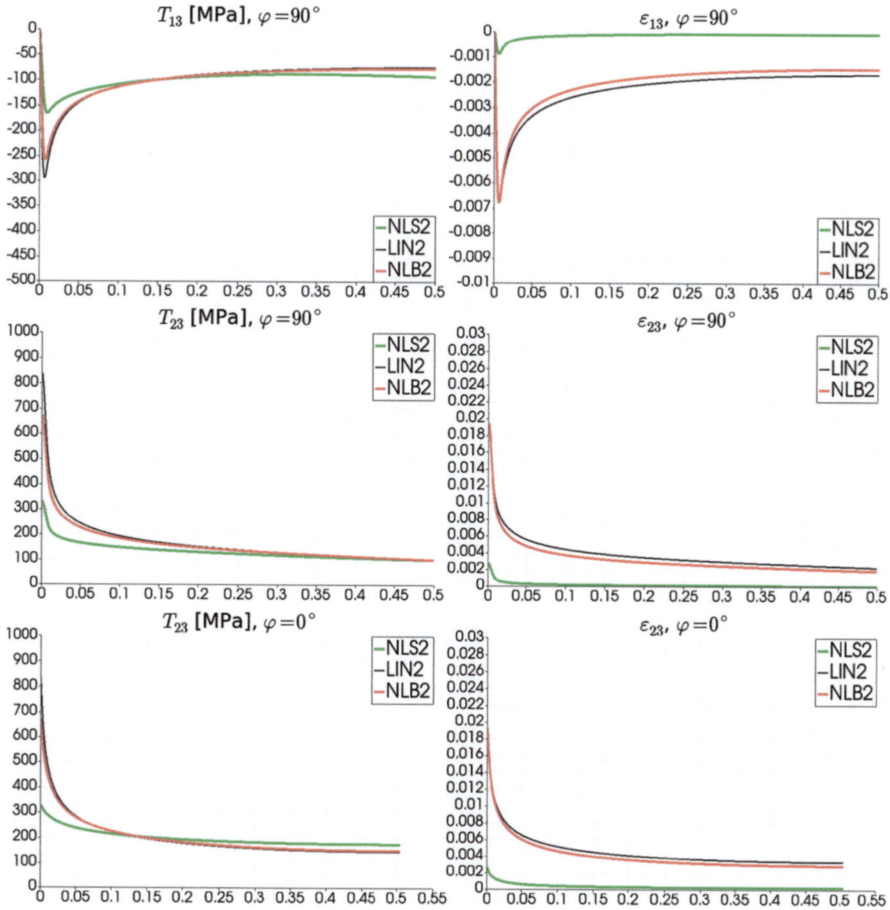


**Fig. 6** Plots of stress and strain for the models LIN2, NLO2, NLB2 and NLS2 in VC geometry with $\alpha = 90°$ and $r_c = 0.01$. The plots are created from the line starting at the point $C + (r_c, 0)$ of the geometry and ending either at point $(1, 0.5)$, which is denoted $\varphi = 0°$, or at point $C + (r_c, 0.5)$, which corresponds to $\varphi = 90°$

Next we define the distance from the notch tip to the point, where $T_{23}$ or $\varepsilon_{23}$ acquires a particular value.

**Definition 6** (Functions $\text{dst}^T$ and $\text{dst}^e$) Let $(r, \varphi)$ be the cylindrical coordinate system centered at point $C + (r_c, 0)$. We define the functions $\text{dst}^T : [0, 2\pi) \times \mathbb{R} \to \mathbb{R}_0^+$ and $\text{dst}^e : [0, 2\pi) \times \mathbb{R} \to \mathbb{R}_0^+$ as follows

$$\text{dst}^T(\varphi, T) = \begin{cases} \infty & \text{if } \forall r \geq 0 : T < T_{23}(r, \varphi), \\ -\infty & \text{if } \forall r \geq 0 : T > T_{23}(r, \varphi), \\ \min\{r \geq 0, T_{23}(r, \varphi) = T\} & \text{otherwise.} \end{cases}$$

$$\text{dst}^e(\varphi, e) = \begin{cases} \infty & \text{if } \forall r \geq 0 : e < \varepsilon_{23}(r, \varphi), \\ -\infty & \text{if } \forall r \geq 0 : e > \varepsilon_{23}(r, \varphi), \\ \min\{r \geq 0, \varepsilon_{23}(r, \varphi) = e\} & \text{otherwise.} \end{cases}$$

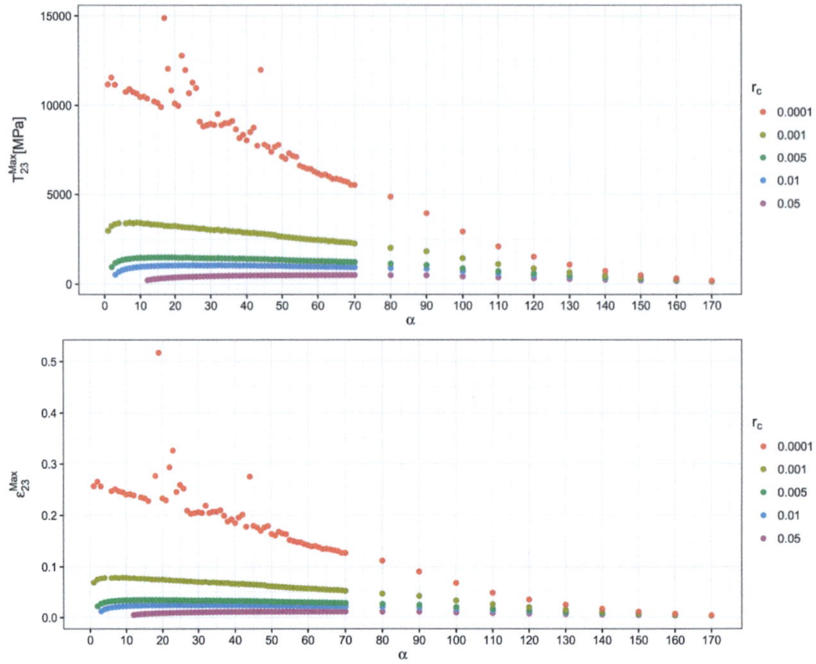


**Fig. 7** Maximal $T_{23}$ and $\varepsilon_{23}$ with respect to $\alpha$ for the model LIN2

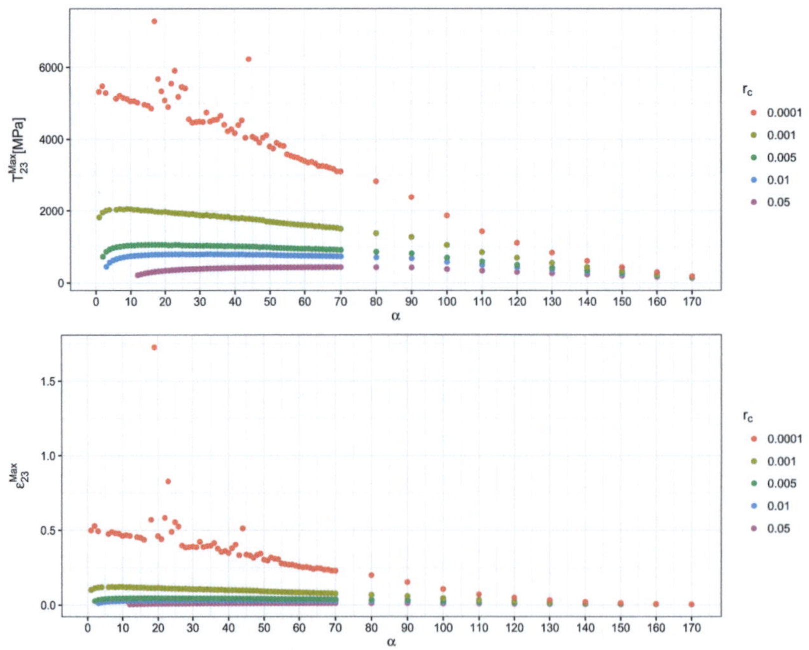


**Fig. 8** Maximal $T_{23}$ and $\varepsilon_{23}$ with respect to $\alpha$ for the model NLB2

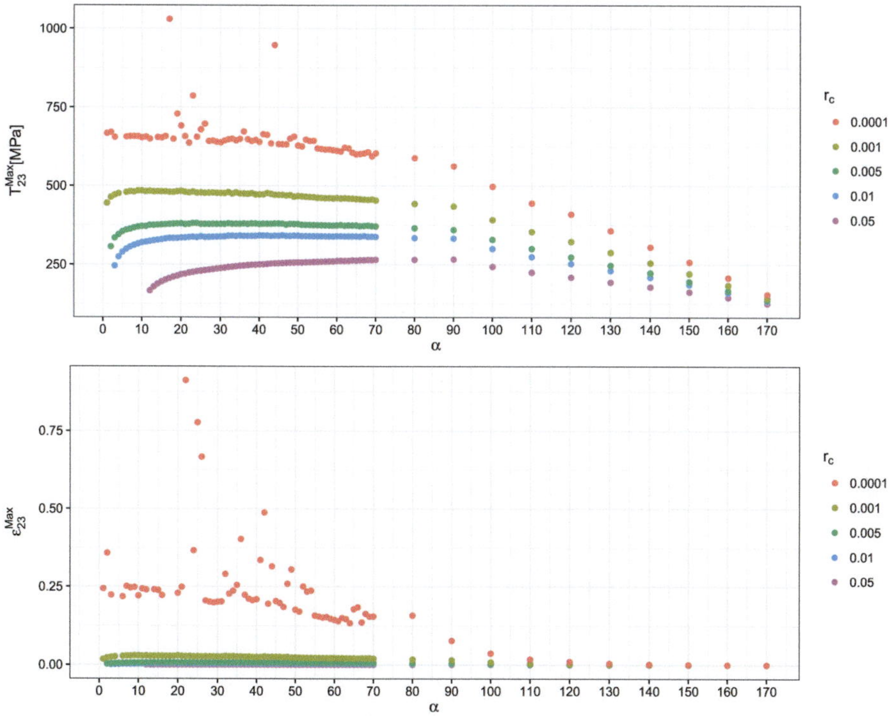


**Fig. 9** Maximal $T_{23}$ and $\varepsilon_{23}$ with respect to $\alpha$ for the model NLS2

For visualizations, we define the three following distances called DT2, DT3 and DE1 such that

$$\begin{aligned} \mathrm{DT2} &= \mathrm{dst}^{T}(0, 200\ \mathrm{MPa}), \\ \mathrm{DT3} &= \mathrm{dst}^{T}(0, 300\ \mathrm{MPa}), \\ \mathrm{DE1} &= \mathrm{dst}^{e}(0, 0.01). \end{aligned} \tag{23}$$

These distances are visualized for model LIN2 in Fig. 10, for model NLB2 in Fig. 11 and for model NLS2 in Fig. 12.

## 5 Discussion

We considered finite element approximations of the Problem (P) in the reduced geometrical setting of anti-plane stress. We simulated the response of three titanium alloys, described by the power-law models (5), with the material moduli previously identified in [30], see Table 1. Numerical simulations were performed on adaptively refined triangular meshes.

We have verified that the computed variational solutions are stable. In Tables 2, 3 and 4, we list relative error norms (21) with respect to the refinement level for all the problems studied for numerous different parameters of the geometry. The differences between the computed solutions on the two most refined meshes are typically of order lower than $10^{-4}$. The global stability measured by the error norms for the power-law models is comparable

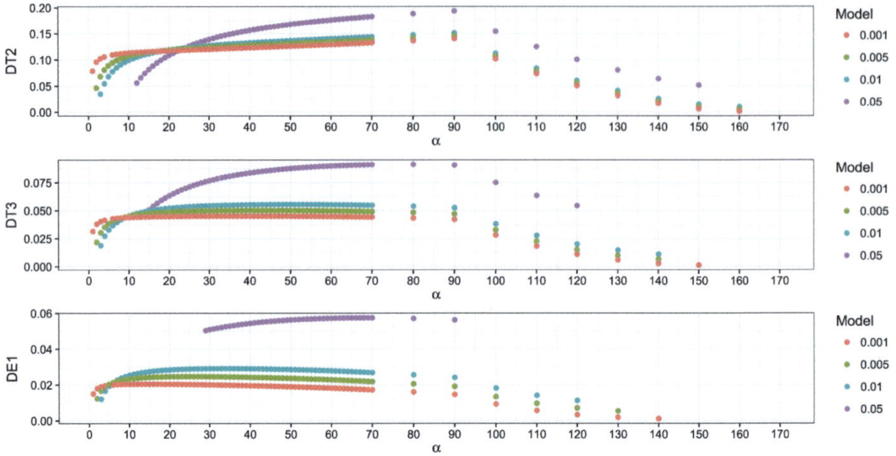


**Fig. 10** Distances DT2, DT3 and DE1, see (23), with respect to $\alpha$ for the model LIN2

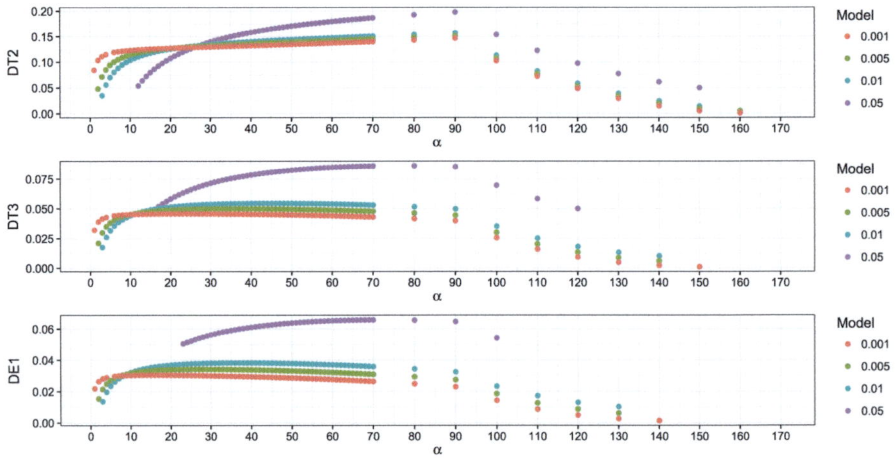


**Fig. 11** Distances DT2, DT3 and DE1, see (23), with respect to $\alpha$ for the model NLB2

to or even better than the global stability for the linear problem. There is no singularity of stress and strain observed, see Fig. 4 and Fig. 6.

It is evident that with increasing angle $\alpha$, the solution is becoming even more stable due to the decrease of the severity of the stress concentration. We observe that the solutions are more stable for higher values of the power law exponents $q'$. Namely the error norms for the model NLS2 with the highest $q'$ are consistently smaller than the error norms for the NLB2 model with the lower value of $q'$.

We measured the local error of the computed solutions with respect to the refinement by the norm of the difference in the stress.

When studying the behavior of the solutions with respect to parameters of the geometry, see Figs. 7–12, we found that there is substantial stability of the solution in comparison to the solutions across completely different meshes. We can clearly identify patterns in Figs. 7–12 that arise with respect to the notch opening angle $\alpha$.

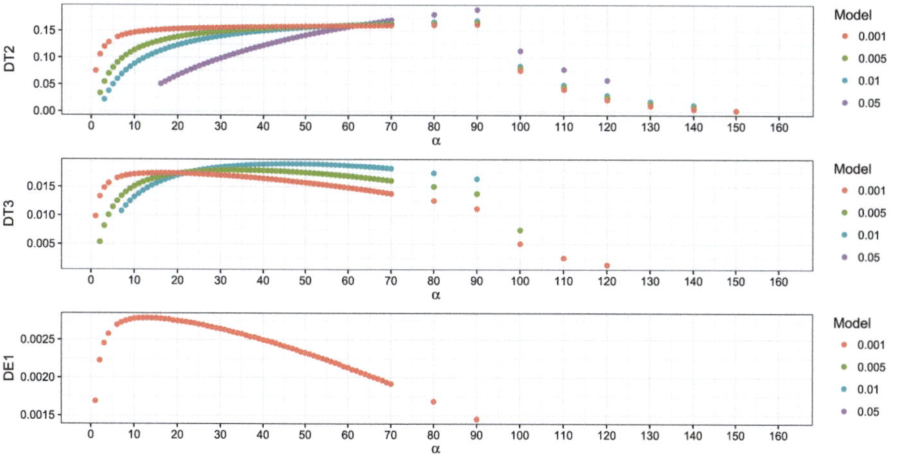


**Fig. 12** Distances DT2, DT3 and DE1, see (23), with respect to $\alpha$ for the model NLS2

**Acknowledgements** Josef Málek thanks the Czech Science Foundation for support through the project 18-12719S. K.R. Rajagopal thanks the Office of Naval Research for support of this work.



## References

1. Alnæs, M., Blechta, J., Hake, J., Johansson, A., Kehlet, B., Logg, A., Richardson, C., Ring, J., Rognes, M.E., Wells, G.N.: The FEniCS project version 1.5. Arch. Numer. Softw. **3**(100), 9–23 (2015). https://doi.org/10.11588/ans.2015.100.20553
2. Beck, L., Bulíček, M., Málek, J., Süli, E.: On the existence of integrable solutions to nonlinear elliptic systems and variational problems with linear growth. Arch. Ration. Mech. Anal. **225**(2), 717–769 (2017). https://doi.org/10.1007/s00205-017-1113-4
3. Blechta, J., Málek, J., Rajagopal, K.R.: On the classification of incompressible fluids (2019)
4. Bonito, A., Girault, V., Süli, E.: Finite element approximation of a strain-limiting elastic model. arXiv:1805.04006 [math.NA] (2018)
5. Bulíček, M., Gwiazda, P., Málek, J., Rajagopal, K.R., Świerczewska-Gwiazda, A.: On flows of fluids described by an implicit constitutive equation characterized by a maximal monotone graph. In: Robinson, J.C., Rodrigo, J.L., Sadowski, W. (eds.) Mathematical Aspects of Fluid Mechanics, pp. 23–51. Cambridge University Press, Cambridge (2012). https://doi.org/10.1017/cbo9781139235792.003
6. Bulíček, M., Gwiazda, P., Málek, J., Świerczewska-Gwiazda, A.: On unsteady flows of implicitly constituted incompressible fluids. SIAM J. Math. Anal. **44**(4), 2756–2801 (2012). https://doi.org/10.1137/110830289
7. Bulíček, M., Málek, J., Rajagopal, K.R., Süli, E.: On elastic solids with limiting small strain: modelling and analysis. EMS Surv. Math. Sci. **1**(2), 283–332 (2014). https://doi.org/10.4171/emss/7
8. Bulíček, M., Gwiazda, P., Málek, J., Świerczewska-Gwiazda, A.: On steady flows of incompressible fluids with implicit power-law-like rheology. Adv. Calc. Var. **2**(2), 109–136 (2009). https://doi.org/10.1515/ACV.2009.006
9. Bulíček, M., Gwiazda, P., Málek, J., Świerczewska-Gwiazda, A.: On scalar hyperbolic conservation laws with a discontinuous flux. Math. Models Methods Appl. Sci. **21**(1), 89–113 (2011). https://doi.org/10.1142/S021820251100499X
10. Bulíček, M., Málek, J.: On unsteady internal flows of Bingham fluids subject to threshold slip on the impermeable boundary. In: Advances in Mathematical Fluid Mechanics, pp. 135–156. Birkhäuser, Basel (2016). https://doi.org/10.1007/978-3-0348-0939-9_8

11. Bulíček, M., Málek, J.: Internal flows of incompressible fluids subject to stick-slip boundary conditions. Vietnam J. Math. **45**(1–2), 207–220 (2017). https://doi.org/10.1007/s10013-016-0221-z
12. Bulíček, M., Málek, J., Rajagopal, K.R.: On Kelvin-Voigt model and its generalizations. Evol. Equ. Control Theory **1**(1), 17–42 (2012). https://doi.org/10.3934/eect.2012.1.17
13. Bulíček, M., Málek, J., Rajagopal, K.R., Walton, J.: Existence of solutions for the anti-plane stress for a new class of "strain-limiting" elastic bodies. Calc. Var. Partial Differ. Equ. **54**(2), 2115–2147 (2015). https://doi.org/10.1007/s00526-015-0859-5
14. Bulíček, M., Málek, J., Süli, E.: Analysis and approximation of a strain-limiting nonlinear elastic model. Math. Mech. Solids **20**(1), 92–118 (2015). https://doi.org/10.1177/1081286514543601
15. Bustamante, R., Rajagopal, K.R.: On a new class of electroelastic bodies, I. Proc. R. Soc. A, Math. Phys. Eng. Sci. **469**(2149), 20120521 (2012). https://doi.org/10.1098/rspa.2012.0521
16. Bustamante, R., Rajagopal, K.R.: On a new class of electro-elastic bodies, II: boundary value problems. Proc. R. Soc. A, Math. Phys. Eng. Sci. **469**(2155), 20130106 (2013). https://doi.org/10.1098/rspa.2013.0106
17. Bustamante, R., Rajagopal, K.R.: Implicit constitutive relations for nonlinear magnetoelastic bodies. Proc. R. Soc. A, Math. Phys. Eng. Sci. **471**(2175), 20140959 (2015). https://doi.org/10.1098/rspa.2014.0959
18. Cauchy, A.L.B.: Recherches sur l'équilibre et le mouvement intérieur des corps solides ou fluides, élastiques ou non élastiques. In: Bulletin de la Société Philomatique, pp. 9–13 (1823)
19. Coleman, B.D., Noll, W.: An approximation theorem for functionals, with applications in continuum mechanics. Arch. Ration. Mech. Anal. **6**(1), 355–370 (1960)
20. COMSOL AB: Comsol multiphysics user's guide. http://www.comsol.com (2008)
21. Devendiran, V.K., Sandeep, R.K., Kannan, K., Rajagopal, K.R.: A thermodynamically consistent constitutive equation for describing the response exhibited by several alloys and the study of a meaningful physical problem. Int. J. Solids Struct. **108**, 1–10 (2017). https://doi.org/10.1016/j.ijsolstr.2016.07.036
22. Diening, L., Kreuzer, C., Süli, E.: Finite element approximation of steady flows of incompressible fluids with implicit power-law-like rheology. SIAM J. Numer. Anal. **51**(2), 984–1015 (2013)
23. Ebmeyer, C., Liu, S.W.: Quasi-norm interpolation error estimates for the piecewise linear finite element approximation of p-Laplacian problems. Numer. Math. **100**(2), 233–258 (2005). https://doi.org/10.1007/s00211-005-0594-5
24. Feireisl, E., Liao, X., Málek, J.: Global weak solutions to a class of non-Newtonian compressible fluids. Math. Methods Appl. Sci. **38**(16), 3482–3494 (2015). https://doi.org/10.1002/mma.3432
25. Gou, K., Mallikarjuna, M., Rajagopal, K.R., Walton, J.R.: Modeling fracture in the context of a strain-limiting theory of elasticity: a single plane-strain crack. Int. J. Eng. Sci. **88**, 73–82 (2015). https://doi.org/10.1016/j.ijengsci.2014.04.018
26. Itou, H., Kovtunenko, V.A., Rajagopal, K.R.: Nonlinear elasticity with limiting small strain for cracks subject to non-penetration. Math. Mech. Solids (2016). https://doi.org/10.1177/1081286516632380
27. Kreuzer, C., Süli, E.: Adaptive finite element approximation of steady flows of incompressible fluids with implicit power-law-like rheology. ESAIM: M2AN **50**(5), 1333–1369 (2016). https://doi.org/10.1051/m2an/2015085
28. Kulvait, V., Málek, J., Rajagopal, K.R.: Anti-plane stress state of a plate with a V-notch for a new class of elastic solids. Int. J. Fract. **179**(1–2), 59–73 (2013). https://doi.org/10.1007/s10704-012-9772-5
29. Kulvait, V.: Mathematical analysis and computer simulations of deformation of nonlinear elastic bodies in the small strain range. PhD thesis. Charles University, Faculty of Mathematics and Physics (2017)
30. Kulvait, V., Málek, J., Rajagopal, K.R.: Modeling gum metal and other newly developed titanium alloys within a new class of constitutive relations for elastic bodies. Arch. Mech. **69**(1), 223–241 (2017)
31. Le Roux, C., Rajagopal, K.R.: Shear flows of a new class of power-law fluids. Appl. Math. **58**(2), 153–177 (2013). https://doi.org/10.1007/s10492-013-0008-4
32. Logg, A., Mardal, K.A., Wells, G.: Automated Solution of Differential Equations by the Finite Element Method: The FEniCS Book. Springer, New York (2012)
33. Málek, J.: Mathematical properties of flows of incompressible power-law-like fluids that are described by implicit constitutive relations. Electron. Trans. Numer. Anal. **31**, 110–125 (2008)
34. Málek, J., Průša, V.: Derivation of equations for continuum mechanics and thermodynamics of fluids. In: Giga, Y., Novotny, A. (eds.) Handbook of Mathematical Analysis in Mechanics of Viscous Fluids, pp. 1–70. Springer, Cham (2016)
35. Málek, J., Průša, V., Rajagopal, K.R.: Generalizations of the Navier-Stokes fluid from a new perspective. Int. J. Eng. Sci. **48**(12), 1907–1924 (2010). https://doi.org/10.1016/j.ijengsci.2010.06.013
36. Málek, J., Rajagopal, K.R.: Compressible generalized Newtonian fluids. Z. Angew. Math. Phys. **61**(6), 1097–1110 (2010). https://doi.org/10.1007/s00033-010-0061-8
37. Maringová, E., Žabenský, J.: On a Navier-Stokes-Fourier-like system capturing transitions between viscous and inviscid fluid regimes and between no-slip and perfect-slip boundary conditions. Nonlinear Anal., Real World Appl. **41**, 152–178 (2018)

38. Montero, S., Bustamante, R., Ortiz-Bernardin, A.: A finite element analysis of some boundary value problems for a new type of constitutive relation for elastic bodies. Acta Mech. **227**(2), 601–615 (2016). https://doi.org/10.1007/s00707-015-1480-6
39. Noll, W.: On the continuity of the solid and fluid states. J. Ration. Mech. Anal. **4**, 3–81 (1955)
40. Noll, W.: On the Foundation of the Mechanics of Continuous Media. Technical Report Series. Books on Demand (1957)
41. Noll, W.: A mathematical theory of the mechanical behavior of continuous media. Arch. Ration. Mech. Anal. **2**(1), 197–226 (1958)
42. Noll, W.: A new mathematical theory of simple materials. Arch. Ration. Mech. Anal. **48**(1), 1–50 (1972). https://doi.org/10.1007/BF00253367
43. Ortiz, A., Bustamante, R., Rajagopal, K.R.: A numerical study of a plate with a hole for a new class of elastic bodies. Acta Mech. **223**(9), 1971–1981 (2012). https://doi.org/10.1007/s00707-012-0690-4
44. Ortiz, A., Bustamante, R., Rajagopal, K.R.: A numerical study of elastic bodies that are described by constitutive equations that exhibit limited strains. Int. J. Solids Struct. **51**(3), 875–885 (2014). https://doi.org/10.1016/j.ijsolstr.2013.11.014
45. Perlácová, T., Průša, V.: Tensorial implicit constitutive relations in mechanics of incompressible non-Newtonian fluids. J. Non-Newton. Fluid Mech. **216**, 13–21 (2015). https://doi.org/10.1016/j.jnnfm.2014.12.006
46. Průša, V., Rajagopal, K.R.: On implicit constitutive relations for materials with fading memory. J. Non-Newton. Fluid Mech. **181**, 22–29 (2012)
47. Rajagopal, K.: A note on the linearization of the constitutive relations of non-linear elastic bodies. Mech. Res. Commun. (2017)
48. Rajagopal, K., Srinivasa, A.: On a class of non-dissipative materials that are not hyperelastic. Proc. R. Soc. A, Math. Phys. Eng. Sci. **465**(2102), 493–500 (2009). https://doi.org/10.1098/rspa.2008.0319
49. Rajagopal, K.R.: On implicit constitutive theories. Appl. Math. **48**(4), 279–319 (2003)
50. Rajagopal, K.R.: On implicit constitutive theories for fluids. J. Fluid Mech. **550**, 243–249 (2006)
51. Rajagopal, K.R.: The elasticity of elasticity. Z. Angew. Math. Phys. **58**(2), 309–317 (2007)
52. Rajagopal, K.R.: Conspectus of concepts of elasticity. Math. Mech. Solids **16**(5), 536–562 (2011). https://doi.org/10.1177/1081286510387856
53. Rajagopal, K.R.: A new development and interpretation of the Navier–Stokes fluid which reveals why the "Stokes assumption" is inapt. Int. J. Non-Linear Mech. **50**, 141–151 (2013). https://doi.org/10.1016/j.ijnonlinmec.2012.10.007
54. Rajagopal, K.R.: On the nonlinear elastic response of bodies in the small strain range. Acta Mech. **225**(6), 1545–1553 (2014). https://doi.org/10.1007/s00707-013-1015-y
55. Rajagopal, K.R.: A note on the classification of anisotropy of bodies defined by implicit constitutive relations. Mech. Res. Commun. **64**, 38–41 (2015). https://doi.org/10.1016/j.mechrescom.2014.11.005
56. Rajagopal, K.R., Saccomandi, G.: The mechanics and mathematics of the effect of pressure on the shear modulus of elastomers. Proc. R. Soc. A, Math. Phys. Eng. Sci. **465**(2112), 3859–3874 (2009). https://doi.org/10.1098/rspa.2009.0416
57. Rajagopal, K.R., Saccomandi, G.: A novel approach to the description of constitutive relations. Frontiers in Materials **3** (2016). https://doi.org/10.3389/fmats.2016.00036
58. Rajagopal, K.R., Srinivasa, A.R.: On the response of non-dissipative solids. Proc. R. Soc. A, Math. Phys. Eng. Sci. **463**(2078), 357–367 (2006). https://doi.org/10.1098/rspa.2006.1760
59. Rajagopal, K.R., Srinivasa, A.R.: Inelastic response of solids described by implicit constitutive relations with nonlinear small strain elastic response. Int. J. Plast. **71**, 1–9 (2015). https://doi.org/10.1016/j.ijplas.2015.02.007
60. Rajagopal, K.R., Walton, J.R.: Modeling fracture in the context of a strain-limiting theory of elasticity: a single anti-plane shear crack. Int. J. Fract. **169**(1), 39–48 (2011). https://doi.org/10.1007/s10704-010-9581-7
61. Rajagopal, K.R., Zappalorto, M.: Bodies described by non-monotonic strain-stress constitutive equations containing a crack subject to anti-plane shear stress. Int. J. Mech. Sci. **149**, 494–499 (2018). https://doi.org/10.1016/j.ijmecsci.2017.07.060
62. Süli, E., Tscherpel, T.: Fully discrete finite element approximation of unsteady flows of implicitly constituted incompressible fluids. arXiv:1804.02264 (2018)
63. Truesdell, C., Moon, H.: Inequalities sufficient to ensure semi-invertibility of isotropic functions. J. Elas. **5**(34) (1975)
64. Zappalorto, M., Berto, F., Rajagopal, K.R.: On the anti-plane state of stress near pointed or sharply radiused notches in strain limiting elastic materials: closed form solution and implications for fracture assessments. Int. J. Fract. **199**(2), 169–184 (2016). https://doi.org/10.1007/s10704-016-0102-1

J Elast (2019) 135:399–408
https://doi.org/10.1007/s10659-018-9704-3

# A Personal Appreciation of Walter Noll

**A. Ian Murdoch**[1]

Received: 19 April 2018 / Published online: 30 October 2018


**Abstract** Walter Noll is personally acknowledged to have been an inspirational teacher and an exemplar in his analysis of fundamental concepts and relations in Continuum Mechanics. Attention is drawn to unfounded criticisms of two aspects of his work and how considerations of local spatial and temporal averaging of microscopic behaviour can complement and inform the continuum viewpoint. Remarks on such considerations briefly summarise the results of published works which both utilised Noll's direct notation and attempted to emulate his precision and clarity.

**Keywords** Reminiscences · Criticisms · Microscopic considerations

**Mathematics Subject Classification** 70 · 74A

## 1 Reminiscences

The inspiration and expertise of teachers and mentors form an invaluable basis for the development of any student. In this respect I was singularly fortunate to begin graduate studies in the Mathematics Department of Carnegie-Mellon University in the Autumn of 1970. Most significantly, the following Spring I took a first course in Continuum Mechanics from Mort Gurtin and was immediately attracted to the subject and stimulated by the enthusiasm and rigour with which he presented material. Prior to arriving in the Department I had been unaware of the subject, and that two of its leading exponents, Walter Noll and Bernard Coleman, were among its members. Walter was held in the highest regard by students and, *inter alia*, his development of direct notation[1] influenced a course I took from Bill Williams

[1] That is, at no stage are co-ordinates employed, but geometrical points (and displacements between pairs of such points) are treated *directly*.

✉ A.I. Murdoch
a.i.murdoch@maths.strath.ac.uk

1 Department of Mathematics and Statistics, University of Strathclyde, Livingstone Tower, 26 Richmond Street, Glasgow, G1 1XH, UK

on Multidimensional Algebra and Calculus in my first semester. However, it was only in Spring 1972 that I experienced Walter first-hand, sitting in on a course which formed the basis of his expository memoir [1]. It is not easy to convey the impact of his lectures to anyone who has only read his works and thus missed the careful motivation behind their content. In crystal clear fashion he discussed fundamental continuum concepts and relations in terms of axioms and a notation free of superfluous elements. His calm and measured delivery was precise, and gave a lasting impression of someone who had not only thought deeply about every item, but who also *instinctively* thought logically. Here was someone one might aspire to emulate, although his habit of smoking cigarillos is now no longer an option! Both in and out of class Walter was approachable and, when answering any of my queries, always displayed a level of understanding deeper than my own, no matter how long or hard I had considered the points in question. On a personal level Walter was always kind, calm, courteous, and in my experience never lost his temper nor disparaged the work of others.

Prior to undertaking doctoral research it was necessary to pass an oral comprehensive examination in two subjects. Mine took place in September 1972 with choices Continuum Mechanics and Complex Analysis, and examiners Walter, David Owen and Zeev Nehari. In preparation I had studied the derivation of the stress tensor beyond the level of coursework, and in some detail, but was not satisfied with my understanding. Passing Walter on a corridor some days before the exam, I confessed this apologetically. He replied that he did not understand either![2] In retrospect I guess it was only natural that, after being asked to present something of my choice for ten minutes (I spoke on the notions of observers and Euclidean point spaces), Walter asked me to talk about the stress tensor. Fortunately I managed to say something sensible. Having enjoyed (and passed) this exam I needed to sort out a doctoral supervisor. Somewhat intimidated by Walter's reputation, and guided by the grapevine of senior fellow graduate students, I was fortunate enough to be accepted as a student by Mort Gurtin. The following March, the 15th meeting of the Society for Natural Philosophy (addressing mathematical problems related to mixtures of interacting species) was held at Rochester, NY. As Chairman, Walter had expenses paid by the Society, but opted to stay with friends, and most kindly donated his hotel room etc to myself (otherwise beyond my means). This was both surprising and deeply appreciated, particularly since I knew nothing of the subject matter and was very much a rookie. It is still a pleasure to recall my first meeting with such luminaries as Truesdell and Müller.

By this time, after first having me look into time-reversal considerations for materials with internal state variables, Mort had suggested that my PhD topic should be 'surfaces', having noted that I had taken Walter's course in Differential Geometry in the Fall of 1972. Here I understood that Mort was thinking of surfaces of bodies, and my first reaction was to think about liquid boundaries and surface tension. The next day I queried whether he meant interfaces, since any material boundary such as that between two immiscible liquids may exhibit properties which depend upon both liquids. Receiving the answer 'I guess so' led to my consideration of bidimensional continua sandwiched between pairs of bulk phases. Here Walter's notes on surfaces proved invaluable by virtue of their clear geometric visualisations, which involve local bijections between surface points and corresponding tangent planes, and are expressed in direct notation. While it was necessary to develop a theory of bidimensional bodies ('material surfaces') via definitions of configurations, deformations, and delineation of appropriate divergence theorems for vector and tensor fields defined on surfaces, these were mainly direct analogues of standard three-dimensional concepts, and perceived as such via use of Walter's notation. The resulting thesis, owing significantly to Mort's guidance

[2] Stress remained one of his interests over the years: cf. [2–4].

and meticulous input, was examined by Professors Gurtin, Markovitz, Noll, and Williams, and published verbatim as [5]. Later work on symmetry considerations for elastic material surfaces [6, 7] (of both second-grade and Cosserat varieties) was likewise greatly indebted to Walter's approach.

## 2 Unfounded Criticisms

### 2.1 Preamble

Most researchers will never have been fortunate enough to have heard Walter either teach or talk, and be directly acquainted with his work and ideas only via his papers or articles. The concise nature of these cannot always convey viewpoint, motivation or reasoning behind his particular approach.[3] While the value and depth of his papers were widely advertised by his former adviser Clifford Truesdell, this sometimes proved to be less helpful than intended. Truesdell's writings were stimulating and greatly informative, but often their confrontational stance could alienate researchers with different scientific cultural backgrounds; here one only has to think of understandable responses to his designation of 'rational' mechanics in respect of his favoured approaches. Unfortunately, as a consequence of such alienation, and lack of acquaintance with contemporary mathematical concepts and notation, Noll's work has not been fully appreciated. In several cases criticism has come from within the Continuum Mechanics fraternity.

### 2.2 Simple Fluids

Rivlin took issue with Noll's definition of a simple fluid. Noll (cf. [9, 10]) had shown that the stress response functional for any simple fluid can be expressed via material frame-indifference[4] (mfi) in terms of a reduced functional whose domain consists only of positive-definite symmetric tensor histories which at current time take the value **1**, the identity tensor. The restriction imposed by mfi at this stage is that such reduced functional be invariant under the orthogonal group. Rivlin [11] argued that the definition of simple fluid for the functional in question mandated invariance under the unimodular group, a stronger restriction than that of Noll, given the larger nature of the unimodular group. However, such restriction for Rivlin involved histories which at current time take the form $\mathbf{H}^T\mathbf{H}$ with $\mathbf{H}$ unimodular. Since $\mathbf{H}^T\mathbf{H} = \mathbf{1}$ if and only if $\mathbf{H}$ is orthogonal, this criticism is unfounded: such histories do not lie in the domain of the reduced response functional unless $\mathbf{H}$ is orthogonal. This was pointed out by myself in [12] and the source of the error identified. This error stemmed from a failure to distinguish between the domain of a functional and that which corresponds to a reduced form of this functional. Rivlin did not accept this technicality and replied with a repetition of his mistake [13, 14]. This proved to be a salutary lesson in having to appreciate that, for whatever reason, one cannot always convince someone of their misconceptions or errors. Such a situation, of much greater significance, also arose concerning the status and interpretation of mfi.

[3] Perhaps with this in mind Walter drafted the collection of essays [8] and established the website www.math.cmu.edu/~wn0g/noll.

[4] Introduced by Truesdell and Noll in [10] as a principle of classical physics and variously phrased on pages 36, 44 and 296 therein.

## 2.3 Material Frame-Indifference

In a summer school held in 1982 at Noto, Sicily, I learned from Ingo Müller of his approximative constitutive relations for stress and heat flux fields in a rarefied gas derived on the basis of kinetic theory (cf. [15]). These he claimed to violate mfi because these fields were dependent upon spin, contradicting the impossibility of such dependence as argued in most texts on the basis of mfi (cf., e.g., [16], p. 156, [17], p. 251). However, what such texts accomplish is to rule out dependence upon spin as computed relative to the frame of a general observer. Crucially, however, Müller's spin is computed with respect to an (and hence any) inertial frame[5] and accordingly was a field upon which all observers could agree, since the notion of inertial frame is central to classical physics (in serving to establish the concept of force). However, from my understanding of mfi as expounded in Noll's lectures, such dependence upon inertial spin was indeed possible, and this I indicated to Ingo. Since he strongly disagreed, I thought long and hard on the matter and wrote [18] to explain my perspective. I also wrote to Walter to see if I had misunderstood the nature of mfi. In his reply of November, 1983, Walter wrote 'As to the matter concerning frame-indifference, I fully agree with your viewpoint, and I told Ingo Müller so when I recently met him at a meeting in Providence, RI.' Unfortunately Ingo never accepted this and the nature of mfi/objectivity continues to be controversial, particularly if it is identified with 'invariance under superposed rigid motions' (isrbm). However, the possibility that material response could be sensitive to spin relative to an inertial frame was accepted by both Noll and Müller and does not violate any fundamental tenet of Physics.

## 2.4 Comments on MFI, ISRBM and Spin-Dependence

The essence of the notions of mfi and isrbm stems from recognition that constitutive relations which purport to describe material behaviour as manifest to different observers must be subject to some restriction in order to make sense. Said differently, any two observers should be able to agree upon the nature and behaviour of any given material of interest, and to recognise how this imposes restrictions upon, and relations between, response functions. How to label, frame and phrase discussion of this fundamental aspect of modelling has long been problematic. As detailed on pp. 28–29 of [19], over the years Walter variously described the concept as 'isotropy of space' and 'objectivity' before finally settling on 'frame-indifference'.

It clarifies matters to discard once and for all the isrbm interpretation. While at first sight it is 'obvious' that rotation of a body without distortion will merely rotate (*inter alia*) the stress field therein in precisely the same way, a little thought suggests this might not be the case. If the body is initially at rest in an inertial frame, then subsequent rotation with respect to this frame will involve a motion which introduces non-uniform acceleration of points of the body. Accordingly, some form of external force system is required to preserve rigidity. Just how this might be achieved is problematic, not to say metaphysical. Of course, what isrbm immediately achieves by its very nature is to negate any constitutive dependence upon spin since no restriction is placed upon the rate at which rotation is performed. Since material behaviour rarely seems to exhibit spin dependence, its absence might be regarded as evidence of the veracity of isrbm. However, fundamental considerations indicate that spin dependence is to be expected quite generally: common experience merely indicates

[5] This was not entirely clear from the paper, but subsequently clarified via correspondence with, and correction by, the author.

that sensitivity thereto is usually negligible. To appreciate this somewhat subtle point, one must recognise that continuum fields describe physical behaviour which is monitored by measuring devices. In order to interpret recorded data it is necessary to relate localised measurement values to continuum field values. However, local measurements involve associated scales of both length and time, since no measuring device can register physical information that is either instantaneous or localised at a geometrical point.[6] Rather, measurement values represent local space-time averages of material behaviour at specific scales. Thus any measurement made on a body must be expected to be sensitive to rotation of the body (which occurs during the time lapse required to register any given measurement) relative to the measuring device employed.[7] Accordingly, fields such as stress and heat flux must *a priori* be expected to be sensitive[8] to rotation with respect to an (hence any) inertial frame. In [15] Müller had exhibited such sensitivity via kinetic theory. The foregoing considerations should suffice to put paid to isrbm, but it has proved a hard beast to slay and continues to attract believers.

It was helpful to talk privately with Rivlin, at Nottingham in 1985, on the matter of isrbm. In a few words he disposed of its standard interpretation by saying that of course one should not envisage rigidly rotating a body of interest relative to the observer, but rather rotating the observer relative to the body. Although such motions would appear identical to the observer (consider films taken of the body with a camera held by the observer in which any 'background' has been erased: both films reveal the same relative rotation) there is a physical difference between the two relative rotations. The state of the body has not changed as a consequence of rotation of the observer (for whom field values will merely be perceived to be rotated versions of those which correspond to absence of relative rotation). Rivlin [23] described this viewpoint as 'invariance under superposed rotation'. In practice this is almost identical with 'invariance under change of observer' (cf. [24], p. 206): here a change of observer, from $O$ to $O^*$ say, is envisaged in which the same response functions are employed by $O$ and $O^*$. This notion may be re-interpreted as a change in perspective of a single observer were he/she to have undergone a rotation. In so doing there can be no query (otherwise reasonable) as to why the same response functions should be involved: for example, in the case of elastic bodies, just why $O$ and $O^*$ should of necessity select the same reference configuration becomes immediately apparent.

Since isrbm continued to be regarded widely as an absolute statement, in 2002 I wrote [25] to point out that it was helpful to emphasise the nature of objectivity in its broad sense as stemming from observer consensus. Here I was influenced by comments made by Jerry Ericksen, in a letter of November, 1980, when discussing a model of heat conduction which I felt illuminated the isrbm/mfi controversy (cf. (2) in [26]): 'As I interpret the word, objectivity refers to what it is that different observers can agree upon. First, different observers can and do make their own choice of reference configuration'. However, I was ignoring the advice of Truesdell who, in October, 1980, wrote 'I hope you can avoid using "objectivity" because that term is often used (e.g. by Müller) to denote "invariance under superposed rigid motions".' In [25] attention was drawn to observer consensus implicit in usage of mathematical ideas and results, to physical concepts such as inertial frames, events, and, for classical

[6]For example, a 'snapshot' has an associated length scale of 'resolution' and a time scale of 'exposure'.

[7]Such device should be at rest in some inertial frame to obviate its own possible sensitivity to rotation.

[8]Of course, this sensitivity may be very slight, and detectable only at high spins. However, the point is that it is to be expected and is in the very nature of things. Indeed, such sensitivity has been suggested in studies of superfluid Helium [20, 21] and small-scale fluid turbulence [22]. Models which interpret large-scale measurements of atmospheric air movement or oceanographic currents would also seem to be candidates for inclusion of constitutive dependency upon inertial spin.

physics, to time lapses and distances between simultaneous events. The specific additional aspects of objectivity for continuum modelling of macroscopic behaviour were postulated to be: agreement upon the nature of any given ideal material, and upon all possible responses of an ideal material, no matter what the relative motion of the observers might be. Derivation of restrictions upon response functions involved precisely Rivlin's notion of behaviour conceived in terms of two separate motions of a single observer relative to the body. The argumentation employed a second observer in order to stress that different observers may choose different response functions, echoing Ericksen's above remark.

A more detailed discussion which includes microscopic considerations is to be found in Chap. 12 of [27].

## 3 Influence

Noll's legacy derives not only from his works but also from the example set by the precision and clarity he brought to the understanding, formulation and presentation of fundamental aspects of continuum modelling. In attempting to extend or complement his work, it is the latter that has been of most personal interest. Specifically, I have tried to examine the physical basis of primitive objects which appear in Noll's axiomatic approach and the general continuum viewpoint.

It is usual to regard a body to be a set of material points which at any instant occupy a definite region in space. While this may appear to be entirely natural, courtesy of our sensory perception of matter at the macroscopic level, Nature is more complex, and can require much finer investigation for its character to be revealed. Since macroscopic material properties ultimately derive from the behaviour of molecules, which in turn consist of assemblies of atomic nuclei and electrons, it is natural to explore how the macroscopic/continuum viewpoint can be related to discrete/microscopic considerations. If one starts from a model of a system of interacting point masses, it is necessary to implement an averaging procedure applied to discrete entities in order to establish fields which can be identified with those of continuum physics. This can be done via either adopting the ensemble averaging of Statistical Mechanics followed by further spatial (and possible temporal) averaging or direct application of spatial (with possible additional temporal) averaging.[9] The results of thirty years thinking about such issues form the content of [27]. The discussions therein are couched in direct notation and employ Noll's notion of each observer perceiving space in terms of an individual copy of Euclidean space: such considerations are addressed in detail in two appendices. Local spatial averaging of corpuscular behaviour is implemented via scale-dependent weighting functions, and results in fields that may be averaged further in time. Bodies (regarded as systems of point masses) whose corpuscular content change with time are treated via the notion of a membership function $e$: at any given time a point mass may belong to the system ($e = 1$) or not ($e = 0$). Such considerations lead to an appreciation of the physical interpretations of fields, and the relations they satisfy, which is deeper than those which appear intuitively clear from a macroscopic viewpoint (wherein the material of which a body consists at any instant is regarded to be smeared throughout a definite portion of space that it is considered to occupy). Some consequences of the foregoing microscopic viewpoint are now briefly outlined in a sequence of remarks. These remarks are intended to stimulate interest in study of the cited works.

[9]Cf., e.g., p. 821 of Irving and Kirkwood's classic contribution [28], [29] and Sect. 14.5 of [27].

*Remark 1* Spatial densities of mass $\rho$ and momentum $\mathbf{p}$ at any prescribed length scale $\varepsilon$ are defined in terms of a weighting function $w$ and satisfy (via a simple calculation: cf. [27], pp. 46–47) the continuity equation

$$\partial\rho/\partial t + \operatorname{div}\mathbf{p} = 0. \tag{1}$$

While the form (1) is independent of the specific form of $w$, the smoothness of fields defined thereby are inherited from that of $w$ and the physical interpretations thereof depend upon the choice of $w$. The simplest choice yields the values of $\rho$ and $\mathbf{p}$ at location $\mathbf{x}$ and time $t$ as the total mass and momentum of the point masses within a distance $\varepsilon$ of $\mathbf{x}$ at time $t$ divided by $4\pi\varepsilon^3/3$.

*Remark 2* The region occupied by the body at scale $\varepsilon$ is that in which $\rho > 0$ and is thus, for the simplest choice of $w$, the union of a set of spherical regions each of radius $\varepsilon$ and centred at a body point mass. The boundary of this region (to be regarded as the boundary of the body at this scale) is accordingly scale-dependent. This is helpful when modelling porous media: at sufficiently large scale pores are not evident, but at decreasing scales pore boundaries become increasingly well-resolved, and pore space increases. Porosity is then a spatial average, at large scale, of pore space defined in terms of the smaller scale, and is accordingly a concept which involves *two* length scales (cf. [27], Remark 4.3.5). It can be useful to consider two porosities, defined via selection of two different small scales, since the region available to any permeating fluid will depend on both its molecular size and phase (liquid or gas).

*Remark 3* Since $\rho$ and $\mathbf{p}$ are scale-dependent (see Remark 1) the velocity

$$\mathbf{v} := \mathbf{p}/\rho \tag{2}$$

is also scale-dependent (cf. [27], Sect. 3.6). From Remark 1 it follows that the simplest choice of $w$ yields $\mathbf{v}(\mathbf{x}, t)$ as the velocity of the mass centre of point masses within the radius $\varepsilon$ spherical averaging region centred at $\mathbf{x}$ at time $t$. The corresponding motion map, with respect to the situation at any selected reference time, is defined in terms of the solution to an initial-value problem. Roughly speaking, this corresponds to consideration of any point in the region occupied by the body at the reference time and subsequently moving from this point with a velocity which at any later time coincides with the instantaneous local value of $\mathbf{v}$. With each point of the reference-time region can be identified a *material point*. This notion is thus seen to be a mathematical artefact whose motion is governed by $\mathbf{v}$ (and hence is scale-dependent) and, for the simplest choice of $w$, represents a local mass centre velocity. Such visualisation renders a precise physical interpretation of a concept which is usually undefined, or loosely described in terms of 'marks on a body'. The scale-dependence of $\mathbf{v}$ upon context is exemplified by attempting to describe motion of air as it flows over an aerofoil, round a skyscraper, or in meteorological contexts.

*Remark 4* The motion of a point mass (which represents any fundamental discrete entity such as a molecule, atom or ion) in an inertial frame is governed by Newton's second law in which account is taken of interactions therewith from all other molecules of the body and also from external agencies. Spatial averaging of such equations of motion[10] yields directly

[10]Each equation of motion is multiplied by the weighting function evaluated at field point $\mathbf{x}$ and time $t$ for the point mass in question and the sum of such equations is taken over *all* point masses.

the local form (cf. [27], Sect. 5.5)

$$\mathbf{f}+\mathbf{b}=\partial/\partial t\{\mathbf{p}\}+\operatorname{div}\boldsymbol{\mathcal{D}} \tag{3}$$

of linear momentum balance. Here **f** denotes the interaction force density and **b** the resultant external force density ($\mathbf{b}=\rho\mathbf{g}$ if gravity is the only such agency). Tensor field $\boldsymbol{\mathcal{D}}$ takes symmetric values and involves thermal velocities.[11] Any tensor $\mathbf{T}^{-}$ (termed the interaction stress tensor) which satisfies

$$\operatorname{div}\mathbf{T}^{-}=\mathbf{f} \tag{4}$$

yields[12] the standard local form of balance

$$\operatorname{div}\mathbf{T}+\mathbf{b}=\rho\mathbf{a}\quad\text{with } \mathbf{T}:=\mathbf{T}^{-}-\boldsymbol{\mathcal{D}}. \tag{5}$$

Here **T** denotes the Cauchy stress tensor and is seen to have two distinct contributions. For ideal gases, in which interactions are absent (so $\mathbf{T}^{-}=\mathbf{0}$), $-\boldsymbol{\mathcal{D}}$ constitutes a pressure.

*Remark 5* In treating non-reacting mixtures the analogue of (5) for a species $\alpha$ is (cf. [27], Sects. 9.3-5)

$$\operatorname{div}\mathbf{T}_{\alpha}+\mathbf{f}_{\alpha}+\mathbf{b}_{\alpha}=\rho_{\alpha}\mathbf{a}_{\alpha}\quad\text{with } \mathbf{T}_{\alpha}:=\mathbf{T}_{\alpha}^{-}-\boldsymbol{\mathcal{D}}_{\alpha}. \tag{6}$$

Here partial stress $\mathbf{T}_{\alpha}^{-}$ involves only $\alpha-\alpha$ interactions, $\mathbf{f}_{\alpha}$ represents a body force density which derives from the effect of all *other* constituents upon species $\alpha$, and $\mathbf{b}_{\alpha}$ is the resultant external body force density for this species. These interpretations are immediate, and differ from those postulated in early works by Truesdell and others on mixtures (cf., e.g., [30, 31]) which were incorrect and led to a paradox, as pointed out by Gurtin, Oliver & Williams [32].

*Remark 6* Spatial averaging of the moments of individual equations of motion about an arbitrary point yields a local moment of momentum balance, and averaging applied to the scalar product of each such equation with the relevant corpuscular velocity leads to a local form of energy balance (cf. [27], Ch. 7 and 6, respectively). Here thermal quantities are distinguished by the presence of thermal velocities in their definitions. For interactions delivered in terms of separation-dependent pair potentials, the internal energy density is the sum of densities of energy of assembly and heat content.[13]

*Remark 7* Additional balances can be derived by averaging moments of mass, and tensor moments of individual equations of motion about an arbitrary point. These balances involve a local measure of fine-scale inhomogeneity[14] together with couple stress, and, if adopted, require appropriate book-keeping in derivation of the (modified) energy balance (cf. [27], Ch. 6, 7).

[11] That is, velocities of point masses relative to the local continuum velocity **v**.

[12] Of course, the physical/geometric interpretation of $\mathbf{T}^{-}$ depends upon which solution to (4) is chosen (cf. [27], Sect. 5.8).

[13] That is, kinetic energy associated with thermal motions.

[14] For the simplest form of weighting function, this measure at $(\mathbf{x},t)$ is the displacement from **x** of the mass centre associated with those point masses in the sphere centred at **x** with radius $\varepsilon$ at time $t$.

*Remark 8* Further averaging in time, at any temporal scale $\Delta$ (essentially a mean value over a time interval of duration $\Delta$ ending at current time $t$), is straightforward and balances remain formally unchanged (cf. [27], Ch. 8). Of course, field values are now identified as local averages jointly in space and time at prescribed scales $\varepsilon$ and $\Delta$, and as such have the same nature as measurement values. In computing any such value, at location $\mathbf{x}$ and time $t$, time averaging is effected keeping $\mathbf{x}$ fixed, and so, if this value is to be identified with a local measurement, then this measurement corresponds to a fixed point for the observer. This makes sense for measuring devices which deliver local information but remain stationary relative to the observer. Temporal averages taken over material point trajectories (and which thus yield information upon which all observers can agree) requires further study.

*Remark 9* Any material system $M$ whose microscopic content changes with time[15] (and accordingly exchanges matter with its environment) can be treated in terms of a membership function[16] $e$. When applied to reacting mixtures, the considerations of Remark 3 make it clear how material points which correspond to different constituents can 'occupy' the same locations, and how material points for a given constituent can suddenly vanish or appear (cf. [27], Ch. 8).

*Remark 10* Modelling electrons and atomic nuclei as point mass/charges, spatial averaging can be used to establish Maxwell's equations as they pertain to electrostatics and magnetostatics [33] and thereby elucidate the character of electromagnetic fields.

Over the years attempts to master concepts, arguments, and technicalities have incurred many instances of personal frustration and no little nervous tension. At such times it has always been helpful to pause, take stock, and ask myself: How would Walter have approached the problem? A profound legacy indeed.



## References

1. Noll, W.: Lectures on the foundations of continuum mechanics and thermodynamics. Arch. Ration. Mech. Anal. **52**, 62–92 (1973)
2. Noll, W.: Die Herleitung der Grundgleichungen der Thermomechanik der Kontinua aus der statistischen Mechanik. J. Ration. Mech. Anal. **4**, 627–646 (1955)
3. Noll, W.: On the theory of surface interactions (2005), 14 pp., published on the website www.math.cmu.edu/~wn0g/noll
4. Noll, W.: Thoughts on the concept of stress. J. Elast. **100**, 25–32 (2010)
5. Gurtin, M.E., Murdoch, A.I.: A continuum theory of elastic material surfaces. Arch. Ration. Mech. Anal. **57**, 291–323 (1975)

[15] E.g. A creature or vehicle which gains propulsion by ejection of fluid, constituent in a reacting mixture, or the set of molecules which define the exposed surface of a propagating crack.

[16] See text preceding Remark 1.

6. Murdoch, A.I., Cohen, H.: Symmetry considerations for material surfaces. Arch. Ration. Mech. Anal. **72**, 61–98 (1979)
7. Murdoch, A.I., Cohen, H.: Symmetry considerations for material surfaces: addendum. Arch. Ration. Mech. Anal. **76**, 393–400 (1981)
8. Noll, W.: Contributions to Natural Philosophy. Research Report 04-CAN-0187, Center for Nonlinear Analysis, Carnegie-Mellon University (2004)
9. Noll, W.: A mathematical theory of the mechanical behaviour of continuous media. Arch. Ration. Mech. Anal. **2**, 197–226 (1958)
10. Truesdell, C., Noll, W.: The non-linear field theories of mechanics. In: Flügge, S. (ed.) Handbuch der Physik III/3. Springer, Berlin (1965)
11. Rivlin, R.S.: A note on the simple fluid. J. Non-Newton. Fluid Mech. **11**, 209–213 (1982)
12. Murdoch, A.I.: On unfounded criticism of the simple fluid of Noll. J. Non-Newton. Fluid Mech. **12**, 387–392 (1983)
13. Rivlin, R.S., Smith, G.F.: On the fallacy in a paper by A.I. Murdoch. J. Non-Newton. Fluid Mech. **12**, 393–394 (1983)
14. Murdoch, A.I.: Reply to the note "On the fallacy in a paper by A.I. Murdoch" by R.S. Rivlin and G.F. Smith. J. Non-Newton. Fluid Mech. **15**, 247–248 (1984)
15. Müller, I.: On the frame dependence of stress and heat flux. Arch. Ration. Mech. Anal. **45**, 241–250 (1973)
16. Silhavy, M.: The Mechanics and Thermomechanics of Continuous Media. Springer, Berlin (1997)
17. Gurtin, M.E., Fried, E., Anand, L.: The Mechanics and Thermodynamics of Continua. Cambridge University Press, Cambridge (2010)
18. Murdoch, A.I.: On material frame-indifference, intrinsic spin, and certain constitutive relations motivated by the kinetic theory of gases. Arch. Ration. Mech. Anal. **83**, 185–194 (1983)
19. Noll, W.: The genesis of Truesdell's nonlinear field theories of mechanics. J. Elast. **70**, 23–30 (2003)
20. Roberts, P.H., Donnelly, R.J.: Superfluid mechanics. In: Van Dyke, M., Vincentini, W.G., Wehausen, J.V. (eds.) Annual Review of Fluid Mechanics, vol. 6. Annual Reviews, Palo Alto (1974)
21. Hills, R.N., Roberts, P.H.: Superfluid mechanics for a high density of vortex lines. Arch. Ration. Mech. Anal. **66**, 43–71 (1977)
22. Piquet, J.: Turbulent Flows, revised 2nd printing edn. Springer, Berlin (2001)
23. Rivlin, R.S.: Red herrings and sundry unidentified fish in nonlinear continuum mechanics. In: Kanninen, M.G., Adler, D., Rosenfield, A.R., Jaffee, R.I. (eds.) Inelastic Behaviour of Solids, pp. 117–134. McGraw Hill, New York (1970)
24. Jog, C.S.: Continuum Mechanics, 3rd edn. Cambridge University Press, Cambridge (2015)
25. Murdoch, A.I.: Objectivity in classical continuum physics: a rationale for discarding the 'principle of invariance under superposed rigid body motions' in favour of purely objective considerations. Contin. Mech. Thermodyn. **15**, 309–320 (2003)
26. Murdoch, A.I.: On material frame-indifference. Proc. R. Soc. Lond. A **380**, 417–426 (1983)
27. Murdoch, A.I.: Physical Foundations of Continuum Mechanics. Cambridge University Press, Cambridge (2012)
28. Irving, J.H., Kirkwood, J.: The statistical mechanical theory of transport processes. IV. The equations of hydrodynamics. J. Chem. Phys. **18**, 817–829 (1950)
29. Murdoch, A.I.: Some primitive concepts in continuum mechanics regarded in terms of objective space-time averaging: the key role played by inertial observers. J. Elast. **84**, 69–97 (2006)
30. Truesdell, C.: Rational Thermodynamics. McGraw Hill, New York (1969)
31. Bowen, R.M.: Theory of mixtures. In: Eringen, A.C. (ed.) Continuum Physics III. Academic Press, New York (1976)
32. Gurtin, M.E., Oliver, M.L., Williams, W.O.: On balance of forces for mixtures. Q. Appl. Math. **30**, 527–530 (1973)
33. Murdoch, A.I.: On spatially-averaged electrokinetics of point charges and Maxwell's equations. J. Elast. **131**, 75–109 (2018)

J Elast (2019) 135:409–433
https://doi.org/10.1007/s10659-018-09712-w

# Macroscopic and Microscopic Behavior of Narrow Elastic Ribbons

**Roberto Paroni**[1] · **Giuseppe Tomassetti**[2]

*Dedicated to the memory of Walter Noll*

Received: 29 June 2018 / Published online: 12 December 2018


**Abstract** A one-dimensional model for a narrow ribbon is derived from the plate theory of Kirchhoff by means of a power expansion in the width variable. The energy found coincides with the corrected Sadowsky's energy. Furthermore, we derive the Euler-Lagrange equations and use them to study an equilibrium configuration of a twisted ribbon. Within this example we also describe how to construct the fine scale oscillations that develop in the deformed configuration.

**Keywords** Ribbons · Elasticity · Relaxed energy

**Mathematics Subject Classification** 74K10 · 74K20 · 74B20 · 49S05

## 1 Introduction

Geometrically a ribbon is a body with a rectangular parallelepiped configuration having thickness $\varepsilon$ much smaller than the width $h$ which, in turn, is much smaller than the length $\ell$. Customarily, the theory of finite deformations of an extremely thin elastic ribbon is studied by associating to the two-dimensional rectangular region, of width $h$ and length $\ell$, the Kirchhoff bending energy, [3, 24].

No stretching energy is introduced by assuming the material to be inextensible. This is the starting point of the seminal work of Sadowsky [22, 32]. For $h$ extremely small Sadowsky, in [32], proposed to model an isotropic ribbon through a one-dimensional energy that depends on the curvature and the torsion of the deformed image of the interval $(0, \ell)$. Nowadays, this energy is known as Sadowsky's energy. A formal justification of the Sad-

✉ R. Paroni
roberto.paroni@unipi.it

G. Tomassetti
giuseppe.tomassetti@uniroma3.it

1 DICI, Università di Pisa, 56122 Pisa, Italy

2 Dipartimento di Ingegneria, Università Roma Tre, 00154 Roma, Italy

owsky's energy was given by Wunderlich [36, 40] also by means of a parametrization of the deformed configuration of the ribbon as a ruled surface. His ingenious argument is more geometrical than mechanical. Interesting generalizations and clarifications on the argument may be found in [1, 2, 4–8, 12, 13, 23, 33, 34], moreover a comprehensive collection of papers on ribbons is [15].

The deduction of a one-dimensional model for ribbons was recently obtained by means of $\Gamma$-convergence in [16] in the isotropic case and in [17] for a general material symmetry. The obtained asymptotic energy density turns out to display two regimes only one of which agrees with the functional proposed by Sadowsky. Another key difference to Sadowsky's theory is that the asymptotic functional does not depend only on the deformation of the midline, but rather on a curve endowed with a triad of directors satisfying several constraints.

The starting point in [16, 17] was the Kirchhoff bending energy as done by Sadowsky; in [18] analogous results were obtained starting from von Kàrmàn models. The same asymptotic models found in [16, 17] were also recently obtained from the nonlinear three-dimensional theory of elasticity in [19], i.e., by letting $\varepsilon$ and $h$ simultaneously to zero. Other models for ribbons deduced from the three-dimensional theory of nonlinear elasticity may be found in [20, 21].

The present paper is concerned with the derivation of a ribbon model starting from the Kirchhoff bending energy. We follow the typical mechanical approach used to derive plate and rod theories, namely, in our contest, on a power series expansion in the width variable. The power series of the energy is essentially truncated at order $h^2$. The one-dimensional energy obtained depends on the symmetric tensor $\mathbf{K}^\circ$ which represents the second fundamental form of the deformed surface restricted to the midline of the ribbon. This line is naturally endowed with three directors; three of the four components of the tensor $\mathbf{K}^\circ$ may be written in terms of these directors. Moreover the inextensibility of the ribbon leads to the constraint $\det \mathbf{K}^\circ = 0$, among others. Due to this constraint the energy is not convex and, as a consequence, the energy minimization problem generally has no solution. Indeed, what may happen is that a minimizing sequence of isometric deformation may have a limit that does not satisfy the constraint $\det \mathbf{K}^\circ = 0$; this, for instance, happens if the minimizing sequence develops fine scale oscillations. This difficulty is resolved by using a so-called relaxed energy in place of the original bending energy. The relaxed energy automatically accounts for the fine scale oscillations and, as a consequence, is defined also for symmetric tensors having determinant different from zero. Solutions for which the determinant is different from zero should then be interpreted as the limit of a sequence of finely oscillating deformations. Of course, microscopic oscillation effects are expected to be lost for moderately thin ribbons, where the membrane energy is important, and the ribbon cannot be considered inextensible. This situation is of no less importance since the competition and interplay between bending and stretching energy can produce exotic behaviors, such as for instance temperature-induced sharp shape transitions between helicoidal to twisted configurations [27, 35, 37–39].

The theory developed has some analogies with the theory of nets modeled as a continuum of fibers by Tchebychev and Rivlin [30, 31]. In Tchebychev nets two families of inextensible cords cross each other without sliding. Pipkin in [29], see also [28], observed that the minimization problem may not have a kinematically admissible minimizer. He therefore introduced the relaxed problem in which the inextensibility constraint is weakened: the cords can grow shorter but not longer. He named his model 'inextensible nets with slack'.

In the case of ribbons the $\det \mathbf{K}^\circ = 0$ constraint is completely absorbed in the relaxed bending energy density. The model that we obtain coincides with those found in [16, 17]. Our derivation, contrary to that of Wunderlich [40] makes no use of ruled surfaces.

The paper is organized as follows. In Sect. 2 we briefly review the Kirchhoff theory for plates and in the following section we derive the model for a narrow ribbon. In Sect. 4 we explicitly write the bending energy density of the ribbon. Several results used in this Section are proved in Sects. A.1 and A.2. The Euler-Lagrange equations are then derived in Sect. 5. In the last Section of the paper we study an equilibrium configuration of a twisted ribbon. This example is particularly interesting since the constraint $\det \mathbf{K}^\circ = 0$ does not hold. We therefore show how to construct a sequence of isometric deformations whose limit is the equilibrium configuration. In this construction we use, like Wunderlich, ruled surfaces.

Standard notation is used throughout. Vectors and tensors are denoted in boldface letters; Latin indices take values in $\{1, 2, 3\}$, while Greek in $\{1, 2\}$. We adopt Einstein's summation convention, and we denote with a dot, $\cdot$, the standard inner product and with the wedge, $\wedge$, the standard cross product. The transpose of the second order tensor $\mathbf{A}$ is denoted by $\mathbf{A}^\top$, while the cofactor by $\mathbf{A}^* = (\det \mathbf{A})\mathbf{A}^{-\top}$ for an invertible matrix $\mathbf{A}$. We recall that if $\mathbf{A}, \mathbf{B}$ are two tensors that can be represented by $2 \times 2$ matrices, then

$$\det(\mathbf{A} - \mathbf{B}) = \det \mathbf{A} - \mathbf{A}^* \cdot \mathbf{B} + \det \mathbf{B}. \tag{1}$$

Finally, the partial derivative with respect to the variable $x_i$ is simply denoted by $\partial_i$.

## 2 Plate Model

We consider deformations of an inextensible elastic plate that initially occupies a rectangular domain $\omega = (0, \ell) \times (-h/2, h/2)$, of length $\ell$ and width $h$, in the $x_1, x_2$ plane. Let $\mathbf{e}_\alpha$, $\alpha = 1, 2$, be an orthonormal system of base vectors oriented as the initial domain with $\mathbf{e}_1$ parallel to the $x_1$ axis. Deformations of the plate are specified by a mapping $\boldsymbol{\chi}$ that takes the point $x$ of $\omega$ to the point $\boldsymbol{\chi}(x)$ of the three-dimensional space. Let $\mathbf{e}_3$ be a unit vector orthogonal to $\mathbf{e}_1$ and $\mathbf{e}_2$ so that $\{\mathbf{e}_1, \mathbf{e}_2, \mathbf{e}_3\}$ is an orthonormal system of the three-dimensional space. The vectors $\mathbf{e}_\alpha$, $\alpha = 1, 2$, will be considered as two or three components vectors as convenient; the context should make it clear. The deformation gradient, $\nabla\boldsymbol{\chi}$, is denoted by $\mathsf{F}$ and is represented by a $3 \times 2$ matrix. Since the plate is inextensible we have that

$$\mathsf{F}\mathbf{e}_\alpha \cdot \mathsf{F}\mathbf{e}_\beta = \mathbf{e}_\alpha \cdot \mathbf{e}_\beta, \tag{2}$$

for $\alpha, \beta = 1, 2$. Equivalently, we may write

$$\partial_\alpha \boldsymbol{\chi} \cdot \partial_\beta \boldsymbol{\chi} = \mathbf{e}_\alpha \cdot \mathbf{e}_\beta. \tag{3}$$

Let $\boldsymbol{\nu}$ be the unit normal to the deformed plate defined by

$$\boldsymbol{\nu} = \mathsf{F}\mathbf{e}_1 \wedge \mathsf{F}\mathbf{e}_2, \tag{4}$$

and let

$$\mathbf{K} = -\mathsf{F}^\top \nabla \boldsymbol{\nu} \tag{5}$$

be the second fundamental form of $\boldsymbol{\chi}$. The symmetric tensor $\mathbf{K}$ is represented by a $2 \times 2$ matrix whose components, with respect to the $\mathbf{e}_\alpha$ basis, are $K_{\alpha\beta} = -\partial_\alpha \boldsymbol{\chi} \cdot \partial_\beta \boldsymbol{\nu} = \partial^2_{\alpha\beta} \boldsymbol{\chi} \cdot \boldsymbol{\nu}$. Since it is impossible to have deformations that simultaneously change both principal curvatures without stretching, by the inextensibility of the plate we must have that

$$\det \mathbf{K} = 0; \tag{6}$$

as follows from Gauss's Theorema Egregium.

As energy density of the plate we simply take a quadratic form in $\mathbf{K}$:

$$Q(x_1,\mathbf{K}) = \frac{1}{2}\mathbb{D}(x_1)\mathbf{K}\cdot\mathbf{K}$$

where the bending stiffness tensor $\mathbb{D}$ maps symmetric second order tensors to symmetric second order tensors. For simplicity, we assume that $\mathbb{D}$ depends only on the $x_1$ coordinate. We further assume that $\mathbb{D}$ is strictly positive definite and has the major symmetry, i.e., there exists a positive constant $d$ such that $\mathbb{D}(x_1)\mathbf{A}\cdot\mathbf{A} \geq d|\mathbf{A}|^2$ and $\mathbb{D}(x_1)\mathbf{A}\cdot\mathbf{B} = \mathbb{D}(x_1)\mathbf{B}\cdot\mathbf{A}$ for every symmetric tensors $\mathbf{A}$ and $\mathbf{B}$ and for every $x_1$. Hence the total bending energy in a given deformation $\boldsymbol{\chi}$ is

$$\mathcal{E}^h(\boldsymbol{\chi}) = \frac{1}{2}\int_\omega \mathbb{D}\mathbf{K}\cdot\mathbf{K}da = \frac{1}{2}\int_0^\ell \int_{-h/2}^{+h/2} \mathbb{D}\mathbf{K}\cdot\mathbf{K}dx_2dx_1, \tag{7}$$

where $\mathbf{K}$ is the curvature associated to $\boldsymbol{\chi}$.

## 3 A Model for a Narrow Ribbon

Hereafter, we assume that the width $h$ of the plate is much smaller than its length $\ell$. Under this assumption it appears natural to formally expand the deformation $\boldsymbol{\chi}$ in series with respect to $x_2$, by writing

$$\boldsymbol{\chi}(x_1,x_2) = \boldsymbol{\chi}^\circ(x_1) + \sum_{i=1}^\infty \boldsymbol{\chi}^{(i)}(x_1)x_2^i,$$

deduce the formal expansion

$$\mathbf{K}(x_1,x_2) = \mathbf{K}^\circ(x_1) + \sum_{i=1}^\infty \mathbf{K}^{(i)}(x_1)x_2^i,$$

and then substitute this last expansion into (7). If we content ourselves with $O(h^2)$ approximation of the energy, and if we reckon that $\mathbb{D}$ does not depend on $x_2$ (symmetry with respect to $x_2$ would indeed suffice), the resulting expression of the energy is

$$\mathcal{E}^h(\boldsymbol{\chi}) = \frac{h}{2}\int_0^\ell \mathbb{D}\mathbf{K}^\circ\cdot\mathbf{K}^\circ dx_1 + o(h^3). \tag{8}$$

We now render this expression more explicit. To do so we find it convenient to adopt the following notation: given any field $f(x_1,x_2)$, we set $f^\circ(x_1) := f(x_1,0)$ to denote the trace of $f$ on the midline of the ribbon. By making use of this notation, we introduce the *directors*:

$$\mathbf{d}_1 = (\mathsf{F}\mathbf{e}_1)^\circ, \qquad \mathbf{d}_2 = (\mathsf{F}\mathbf{e}_2)^\circ, \qquad \mathbf{d}_3 = \boldsymbol{\nu}^\circ, \tag{9}$$

and we observe that these vectors constitute an orthonormal basis:

$$\mathbf{d}_i\cdot\mathbf{d}_j = \mathbf{e}_i\cdot\mathbf{e}_j, \tag{10}$$

as can be checked by means of (2) and (4).

Keeping in mind that the operation of taking the trace on the midline commutes with that of performing differentiation with respect to $x_1$, we see that the first director is tangent to the midline of the ribbon

$$\mathbf{d}_1 = (\partial_1 \boldsymbol{\chi})^\circ = \partial_1 \boldsymbol{\chi}^\circ. \tag{11}$$

The second director

$$\mathbf{d}_2 = (\partial_2 \boldsymbol{\chi})^\circ$$

is tangent to the ribbon on the midline and "represents" the orientation of the cross section of the ribbon. The third director is perpendicular to the ribbon on the midline.

Taking $\alpha = \beta = 1$ in (3), differentiating with respect to $x_2$, and evaluating the result at $x_2 = 0$, we find

$$0 = (\partial_2 \partial_1 \boldsymbol{\chi})^\circ \cdot (\partial_1 \boldsymbol{\chi})^\circ = \partial_1 (\mathsf{F}\mathbf{e}_2)^\circ \cdot (\mathsf{F}\mathbf{e}_1)^\circ,$$

that is

$$\mathbf{d}_2' \cdot \mathbf{d}_1 = 0. \tag{12}$$

This is a constraint that must be satisfied by the directors. From the definition of the second fundamental form we find that

$$K^\circ_{1\beta} = \left(\partial^2_{1\beta} \boldsymbol{\chi}\right)^\circ \cdot \boldsymbol{\nu}^\circ = \partial_1 (\partial_\beta \boldsymbol{\chi})^\circ \cdot \boldsymbol{\nu}^\circ = \mathbf{d}_\beta' \cdot \mathbf{d}_3. \tag{13}$$

The components $K^\circ_{11} = \mathbf{d}_1' \cdot \mathbf{d}_3$ and $K^\circ_{12} = \mathbf{d}_2' \cdot \mathbf{d}_3$ are the curvatures of the ribbon on the midline. It is convenient to introduce the *curvature*

$$\mathbf{k} := \frac{1}{2} \sum_{i=1}^{3} \mathbf{d}_i \wedge \mathbf{d}_i',$$

which, as is easily seen, satisfies the identity

$$\mathbf{d}_i' = \mathbf{k} \wedge \mathbf{d}_i,$$

which in the components $k_i := \mathbf{k} \cdot \mathbf{d}_i$ rewrites as

$$k_1 = \mathbf{d}_2' \cdot \mathbf{d}_3 = K^\circ_{12}, \qquad k_2 = \mathbf{d}_3' \cdot \mathbf{d}_1 = -K^\circ_{11}, \qquad k_3 = \mathbf{d}_1' \cdot \mathbf{d}_2. \tag{14}$$

Then, $k_1$ measures torsion, $k_2$ measures flexure with respect to $\mathbf{d}_2$ and $k_3$ measure flexure with respect to $\mathbf{d}_3$. Equation (12) implies that the ribbon does not bend around $\mathbf{d}_3$:

$$k_3 = 0. \tag{15}$$

We shall simply refer to $k_2$ as the *curvature* of the midline curve $\boldsymbol{\chi}^\circ$ and to $k_1$ as the *twist* of the cross-section of the ribbon. The component $K^\circ_{22}$ can not be expressed in terms of only the directors. Hereafter, we set $\gamma := K^\circ_{22}$ so to write

$$K^\circ_{\alpha\beta} = \begin{pmatrix} \mathbf{d}_1' \cdot \mathbf{d}_3 & \mathbf{d}_2' \cdot \mathbf{d}_3 \\ \mathbf{d}_2' \cdot \mathbf{d}_3 & \gamma \end{pmatrix} = \begin{pmatrix} -k_2 & k_1 \\ k_1 & \gamma \end{pmatrix}. \tag{16}$$

By Gauss's Theorema Egregium, surfaces that satisfy (3) have Gaussian curvature everywhere equal to zero. Evaluating (6) at $x_2 = 0$ we deduce that

$$\det \mathbf{K}^\circ = 0. \tag{17}$$

This is a severe constraint on the deformations, indeed note that if $k_2 = 0$ then also $k_1 = 0$: in words, twist is not allowed if the midline is kept straight.

With (8) in mind, we set

$$\mathcal{E}^\circ(\mathbf{r}, \mathbf{d}_i, \gamma) := \frac{1}{2}\int_0^\ell \mathbb{D}\mathbf{K}^\circ \cdot \mathbf{K}^\circ dx_1 = \int_0^\ell Q\big(x_1, \mathbf{K}^\circ\big) dx_1, \tag{18}$$

where $\mathbf{K}^\circ$ is given in terms of $\mathbf{d}_i$ and $\gamma$ by (16), and

$$\mathbf{r} := \chi^\circ$$

is related to $\mathbf{d}_1$ by (11), that is

$$\mathbf{d}_1 = \mathbf{r}'. \tag{19}$$

Because of the constraint (17), minimizing the energy (18) is the same as minimizing the energy

$$\mathcal{E}_\circ(\mathbf{r}, \mathbf{d}_i, \gamma) := \int_0^\ell W\big(x_1, \mathbf{K}^\circ\big) dx_1, \tag{20}$$

where, the density $W$ is defined by

$$W(x_1, \mathbf{L}) = \begin{cases} Q(x_1, \mathbf{L}) & \text{if } \det \mathbf{L} = 0, \\ +\infty & \text{otherwise.} \end{cases} \tag{21}$$

We view (20) as the *microscopic energy*. The energy density $W$, when finite, is a strictly positive quadratic function over a non convex set. Indeed, as it is easy to check, the set of 2-by-2 matrices with zero determinant is not convex. Thus, $W$ is not convex and hence the microscopic energy (20) may not have a minimizer. What may go wrong is that a minimizing sequence $(\mathbf{r}_n, (\mathbf{d}_i)_n, \gamma_n)$, which determines (according to (16)) a $\mathbf{K}_n^\circ$ with null determinant, may approach a limit $\mathbf{K}^\circ$ with determinant not everywhere equal to zero. This can be achieved for instance when the $n$-th element of the sequence is obtained by alternating, on a scale of order $1/n$ or smaller, two deformations with finite energy, i.e., $\det \mathbf{K}_n^\circ = 0$ everywhere, but with an "average" (weak limit of the sequence) $\mathbf{K}^\circ$ with determinant different from zero. Hereafter, we shall say that $\mathbf{K}^\circ$ is *generated* by $\mathbf{K}_n^\circ$. Thus, we may extend the definition of energy so to make it finite also for deformations with $\det\mathbf{K}^\circ \neq 0$: the energy that we may associate to a $\mathbf{K}^\circ$ generated by $\mathbf{K}_n^\circ$ is the limit of the energies associated to $\mathbf{K}_n^\circ$. The *macroscopic energy* or *relaxed energy* is the least energy that is generated by a sequence of deformations with finite energy; this corresponds to, see for instance [11, 14],

$$\mathcal{E}_\circ^{**}(\mathbf{r}, \mathbf{d}_i, \gamma) := \int_0^\ell W^{**}\big(x_1, \mathbf{K}^\circ\big) dx_1,$$

where $W^{**}$ is the *convex envelope* of $W$.

This energy may be further reduced, since the field $\gamma$ is neither related to $\mathbf{r}$ nor to $\mathbf{d}_i$. We minimize once and for all over $\gamma$ to obtain

$$\min_{\gamma} \mathcal{E}_\circ^{**}(\mathbf{r}, \mathbf{d}_i, \gamma) = \int_0^\ell \overline{W}(x_1, k_1, k_2) dx_1,$$

where

$$\overline{W}(x_1, k_1, k_2) := \min_{\gamma} W^{**}\left(x_1, \begin{pmatrix} -k_2 & k_1 \\ k_1 & \gamma \end{pmatrix}\right). \tag{22}$$

We interpret

$$\overline{\mathcal{E}}(\mathbf{r}, \mathbf{d}_i) := \int_0^\ell \overline{W}(x_1, k_1, k_2) dx_1, \tag{23}$$

as the *macroscopic energy* of the ribbon. A key observation here is that the macroscopic energy is always less or equal to the microscopic energy; indeed,

$$\begin{aligned} \mathcal{E}_\circ(\mathbf{r}, \mathbf{d}_i, \gamma) &= \int_0^\ell W\big(x_1, \mathbf{K}^\circ\big) dx_1 \geq \int_0^\ell W^{**}\big(x_1, \mathbf{K}^\circ\big) dx_1 \\ &\geq \int_0^\ell \overline{W}(x_1, k_1, k_2) dx_1 = \overline{\mathcal{E}}(\mathbf{r}, \mathbf{d}_i). \end{aligned}$$

The energy of the ribbon (23) coincides with that obtained by means of $\Gamma$-convergence in [17].

## 4 Representation of the Energy Density of the Ribbon

In what follows, for brevity, we do not write the dependence of $W$ on the variable $x_1$. In order to have an explicit representation of the energy density $\overline{W}$ of the ribbon it is necessary to first evaluate the *convex envelope* $W^{**}$ of the energy $W$ defined in (21). It turns out that, given any symmetric 2-by-2 matrix $\mathbf{L}$,

$$W^{**}(\mathbf{L}) = Q(\mathbf{L}) + \alpha^+ (\det \mathbf{L})^+ + \alpha^- (\det \mathbf{L})^-, \tag{24}$$

where

$$\alpha^- = \min_{\det \mathbf{L}=1} Q(\mathbf{L}), \qquad \alpha^+ = \min_{\det \mathbf{L}=-1} Q(\mathbf{L}), \tag{25}$$

and where

$$(\det \mathbf{L})^+ = \begin{cases} \det \mathbf{L} & \text{if } \det \mathbf{L} > 0, \\ 0 & \text{else,} \end{cases} \qquad (\det \mathbf{L})^- = \begin{cases} -\det \mathbf{L} & \text{if } \det \mathbf{L} < 0, \\ 0 & \text{else.} \end{cases}$$

This representation has been obtained in [17] by means of convex duality. In Sect. A.1, besides giving a new proof, we prove that the convex envelope can be obtained by a single "lamination". That is: for every symmetric matrix $\mathbf{L}$ there exist two symmetric matrices $\mathbf{A}$ and $\mathbf{B}$ having null determinant and a $t \in [0, 1]$ such that

$$W^{**}(\mathbf{L}) = (1-t) Q(\mathbf{A}) + t Q(\mathbf{B}), \quad \text{with } \mathbf{L} = (1-t)\mathbf{A} + t\mathbf{B}. \tag{26}$$

The quantities $\mathbf{A}$, $\mathbf{B}$, and $t$ can be evaluated by using the symmetric matrices $\mathbf{D}^-$ and $\mathbf{D}^+$ that minimize problems (25). More precisely, $\mathbf{D}^-$ and $\mathbf{D}^+$ are two symmetric matrices with $\det\mathbf{D}^- = -1$ and $\det\mathbf{D}^+ = +1$ such that

$$\alpha^- = Q(\mathbf{D}^+), \qquad \alpha^+ = Q(\mathbf{D}^-). \tag{27}$$

If $\det\mathbf{L} > 0$ then the real number $t$ and the matrix $\mathbf{B}$ are given by

$$t = \frac{1}{2 - 2\lambda(\sqrt{\lambda^2+1} - \lambda)}, \qquad \mathbf{B} = \mathbf{L} - \eta\mathbf{D}^-$$

where $\lambda$ and $\eta$ are defined by

$$\lambda := \frac{\mathbf{L}^* \cdot \mathbf{D}^-}{2\sqrt{\det\mathbf{L}}}, \qquad \eta := \sqrt{\det\mathbf{L}}\big(\sqrt{\lambda^2+1} - \lambda\big).$$

While, if $\det\mathbf{L} < 0$ we have that

$$t = \frac{1}{2 + 2\lambda(\lambda - \sqrt{\lambda^2+1})}, \qquad \mathbf{B} = \mathbf{L} - \eta\mathbf{D}^+, \tag{28}$$

with

$$\lambda := \frac{\mathbf{L}^* \cdot \mathbf{D}^+}{2\sqrt{-\det\mathbf{L}}}, \qquad \eta := \sqrt{-\det\mathbf{L}}\big(\lambda - \sqrt{\lambda^2+1}\big). \tag{29}$$

In both cases, $\det\mathbf{L} > 0$ and $\det\mathbf{L} < 0$, the matrix $\mathbf{A}$ is given by

$$\mathbf{A} = \frac{\mathbf{L} - t\mathbf{B}}{1-t}. \tag{30}$$

The elastic constants $\alpha^\pm$ are computed in Sect. A.2, here we briefly outline the results. We denote by

$$\mathbb{D}_{\alpha\beta\gamma\delta} := \mathbf{e}_\alpha \otimes \mathbf{e}_\beta \cdot \mathbb{D}\mathbf{e}_\gamma \otimes \mathbf{e}_\delta.$$

For an orthotropic ribbon with respect to the basis $(\mathbf{e}_1, \mathbf{e}_2)$ it is found that

$$\alpha^- = \sqrt{\mathbb{D}_{1111}\mathbb{D}_{2222}} + \mathbb{D}_{1122}, \qquad \alpha^+ = \min\{\sqrt{\mathbb{D}_{1111}\mathbb{D}_{2222}} - \mathbb{D}_{1122}, 2\mathbb{D}_{1212}\}.$$

In particular, for the isotropic strain energy

$$Q(\mathbf{L}) = d_\mu|\mathbf{L}|^2 + \frac{d_\lambda}{2}(L_{11} + L_{22})^2,$$

where $d_\mu$ and $d_\lambda$ are two elastic constants, we find

$$\alpha^- = 2(d_\mu + d_\lambda), \qquad \alpha^+ = 2d_\mu.$$

In this case, the convex envelope of $W$, (24), turns out to be, see Sect. A.2 for details,

$$W^{**}(\mathbf{L}) = \left(d_\mu + \frac{d_\lambda}{2}\right)\big(|\mathbf{L}|^2 + 2|\det\mathbf{L}|\big). \tag{31}$$

One can also check, by a direct computation, that $\alpha^- = Q(\mathbf{D}^+)$, and $\alpha^+ = Q(\mathbf{D}^-)$ for

$$\mathbf{D}^+ = \begin{pmatrix} 1 & 0 \\ 0 & 1 \end{pmatrix}, \qquad \mathbf{D}^- = \begin{pmatrix} 1 & 0 \\ 0 & -1 \end{pmatrix}. \tag{32}$$

In the isotropic case, a simple computation leads to

$$\min_{L_{22}\in\mathbb{R}} W^{**}(\mathbf{L}) = \left(d_\mu + \frac{d_\lambda}{2}\right) \begin{cases} \frac{(L_{11}^2+L_{12}^2)^2}{L_{11}^2} & \text{if } |L_{11}| > |L_{12}|, \\ 4L_{12}^2 & \text{else if } |L_{11}| \le |L_{12}|, \end{cases}$$

and hence, the energy density of the ribbon (22) is

$$\overline{W}(k_1, k_2) = \left(d_\mu + \frac{d_\lambda}{2}\right) \begin{cases} \frac{(k_2^2+k_1^2)^2}{k_2^2} & \text{if } |k_2| > |k_1|, \\ 4k_1^2 & \text{else if } |k_2| \le |k_1|. \end{cases} \tag{33}$$

This is the corrected Sadowsky energy density found in [16]. Indeed, in the regime in which the absolute value of the curvature $k_2$ is larger than the absolute value of the twist, the energy density coincides with that proposed by Sadowsky [32].

## 5 Euler-Lagrange Equations

In order to derive the Euler-Lagrange equations we follow the approach of [26], based on incorporating the constraints into the energy functional through Lagrange multipliers. To this effect, we introduce the Lagrangian associated to the energy (23)

$$\mathcal{L}(\mathbf{r}, \mathbf{d}_1, \mathbf{d}_2, \mathbf{d}_3, \mathbf{n}, m_3) = \int_0^\ell \overline{W}(k_1, k_2) + m_3 k_3 dx_1 + \int_0^\ell \mathbf{n}\cdot(\mathbf{r}' - \mathbf{d}_1) dx_1.$$

For notational simplicity, we have dropped the dependence of $\overline{W}$ on the variable $x_1$. Here $m_3$ and $\mathbf{n}$ are the Lagrange multipliers associated to the constraints $k_3 = 0$ and $\mathbf{r}' = \mathbf{d}_1$, respectively, see (15) and (19). We consider an $\varepsilon$-parametrized family of admissible perturbations

$$\begin{aligned} \mathbf{r}_\varepsilon &= \mathbf{r} + \varepsilon\dot{\mathbf{r}}, \\ \mathbf{d}_{i\varepsilon} &= \exp(\varepsilon\dot{\boldsymbol{\Theta}})\mathbf{d}_i, \quad i = 1, 2, 3, \\ \mathbf{n}_\varepsilon &= \mathbf{n} + \varepsilon\dot{\mathbf{n}}, \\ m_{3,\varepsilon} &= \mathrm{m} + \varepsilon\dot{m}_3, \end{aligned} \tag{34}$$

of a putative stationary point $(\mathbf{r}, \mathbf{d}_1, \mathbf{d}_2, \mathbf{d}_3, \mathbf{n}, m_3)$. Here $\dot{\mathbf{n}}$, $\dot{m}_3$, $\dot{\mathbf{r}}$, and $\dot{\boldsymbol{\Theta}}$ are smooth functions taking values, respectively, in $\mathbb{R}^3$, $\mathbb{R}$, $\mathbb{R}^3$, and $\mathbb{R}^{3\times3}$, with $\dot{\boldsymbol{\Theta}}$ a skew-symmetric matrix. We leave it for later the determination of boundary conditions and, accordingly, we restrict attention to perturbations that vanish at the endpoints:

$$\dot{\mathbf{r}}(0) = \dot{\mathbf{r}}(L) = \mathbf{0}, \qquad \dot{\boldsymbol{\Theta}}(0) = \dot{\boldsymbol{\Theta}}(L) = \mathbf{0}.$$

We note on passing that, since

$$\mathbf{Q}_\varepsilon := \exp(\varepsilon\dot{\boldsymbol{\Theta}}) \tag{35}$$

is a proper rotation such that

$$\mathbf{Q}_0 = \mathbf{I}, \tag{36}$$

the ordered list of vectors $\{\mathbf{d}_{i\varepsilon}\}_{i=1,2,3}$ constitutes a positively-oriented orthonormal triad which depends smoothly on $\varepsilon$ and coincides with the triad $\{\mathbf{d}_i\}_{i=1,2,3}$ for $\varepsilon = 0$. We also recall, for later use, that the exponential matrix commutes with its derivative, namely, $\frac{d}{d\varepsilon}\mathbf{Q}_\varepsilon = \mathbf{Q}_\varepsilon\dot{\boldsymbol{\Theta}} = \dot{\boldsymbol{\Theta}}\mathbf{Q}_\varepsilon$. In particular, we have

$$\left.\frac{d}{d\varepsilon}\right|_{\varepsilon=0} \mathbf{Q}_\varepsilon = \dot{\boldsymbol{\Theta}}. \tag{37}$$

As a consequence, if $\dot{\boldsymbol{\theta}}$ is the axial vector of $\dot{\boldsymbol{\Theta}}$, then

$$\dot{\mathbf{d}}_i := \left.\frac{d}{d\varepsilon}\right|_{\varepsilon=0} \mathbf{d}_{i\varepsilon} = \dot{\boldsymbol{\Theta}}\mathbf{d}_i = \dot{\boldsymbol{\theta}} \wedge \mathbf{d}_i \tag{38}$$

for every $i = 1, 2, 3$.

The next step is to evaluate the functional $\mathcal{L}$ at the typical perturbed configuration. The result is a differentiable function of $\varepsilon$, whose derivative at $\varepsilon = 0$ we require to vanish. We set

$$m_\alpha := \partial_{k_\alpha} \overline{W}(k_1, k_2), \quad \alpha = 1, 2,$$

and we define the *referential moment vector* and the *referential curvature vector* as

$$\mathsf{m} := (m_1, m_2, m_3), \qquad \mathsf{k} := (k_1, k_2, k_3).$$

We denote by $k_{i\varepsilon}$ the quantities obtained by means of (14) but with $\mathbf{d}_i$ replaced by $\mathbf{d}_{i\varepsilon}$ and we set $\mathsf{k}_\varepsilon := (k_{1\varepsilon}, k_{2\varepsilon}, k_{3\varepsilon})$. On letting $\dot{\mathsf{k}} := \frac{d}{d\varepsilon}|_{\varepsilon=0}\mathsf{k}_\varepsilon$, we may write, on integrating by parts,

$$\begin{aligned}\dot{\mathcal{L}} &:= \left.\frac{d}{d\varepsilon}\right|_{\varepsilon=0} \mathcal{L}(\mathbf{r}_\varepsilon, \mathbf{d}_{1\varepsilon}, \mathbf{d}_{2\varepsilon}, \mathbf{d}_{3\varepsilon}, \mathbf{n}_\varepsilon, m_{3\varepsilon}) \\ &= \int_0^\ell \mathsf{m} \cdot \dot{\mathsf{k}} + \dot{m}_3 k_3 dx_1 + \int_0^\ell \dot{\mathbf{n}} \cdot \big(\mathbf{r}' - \mathbf{d}_1\big) - \mathbf{n}' \cdot \dot{\mathbf{r}} - \mathbf{n} \cdot \dot{\boldsymbol{\theta}} \wedge \mathbf{d}_1 dx_1.\end{aligned}$$

One can write the relation between the curvature vector and the directors in the perturbed configuration as:

$$k_{i\varepsilon} = \frac{1}{2}\epsilon_{ijk}\mathbf{d}'_{j\varepsilon} \cdot \mathbf{d}_{k\varepsilon},$$

an identity whose differentiation with respect to $\varepsilon$ yields

$$\begin{aligned}\dot{k}_i &= \frac{1}{2}\epsilon_{ijk}\dot{\mathbf{d}}'_j \cdot \mathbf{d}_k + \frac{1}{2}\epsilon_{ijk}\mathbf{d}'_j \cdot \dot{\mathbf{d}}_k \\ &= \frac{1}{2}\epsilon_{ijk}(\dot{\boldsymbol{\theta}} \wedge \mathbf{d}_j)' \cdot \mathbf{d}_k + \frac{1}{2}\epsilon_{ijk}\mathbf{d}'_j \cdot \dot{\boldsymbol{\theta}} \wedge \mathbf{d}_k \\ &= \frac{1}{2}\epsilon_{ijk}\dot{\boldsymbol{\theta}}' \wedge \mathbf{d}_j \cdot \mathbf{d}_k + \underbrace{\frac{1}{2}\epsilon_{ijk}\dot{\boldsymbol{\theta}} \wedge \mathbf{d}'_j \cdot \mathbf{d}_k + \frac{1}{2}\epsilon_{ijk}\mathbf{d}'_j \cdot \dot{\boldsymbol{\theta}} \wedge \mathbf{d}_k}_{=0}\end{aligned}$$

$$= \frac{1}{2}\epsilon_{ijk}\mathbf{d}_j \wedge \mathbf{d}_k \cdot \dot{\boldsymbol{\theta}}'$$

$$= \mathbf{d}_i \cdot \dot{\boldsymbol{\theta}}',$$

where we have used the identity $\mathbf{d}_i = \frac{1}{2}\epsilon_{ijk}\mathbf{d}_j \wedge \mathbf{d}_k$. Now, on letting

$$\mathbf{m} := m_i \mathbf{d}_i, \tag{39}$$

we can write

$$\mathbf{m} \cdot \dot{\mathbf{k}} = \sum_{i=1}^{3} m_i \dot{k}_i = \sum_{i=1}^{3} m_i \mathbf{d}_i \cdot \dot{\boldsymbol{\theta}}' = \mathbf{m} \cdot \dot{\boldsymbol{\theta}}',$$

and we arrive at the following expression:

$$\begin{aligned}\dot{\mathcal{L}} &= \int_0^\ell \mathbf{m} \cdot \dot{\boldsymbol{\theta}}' + \dot{m}_3 k_3 - \mathbf{n} \cdot \dot{\boldsymbol{\theta}} \wedge \mathbf{d}_1 dx_1 + \int_0^\ell \dot{\mathbf{n}} \cdot (\mathbf{r}' - \mathbf{d}_1) - \mathbf{n}' \cdot \dot{\mathbf{r}} dx_1 \\ &= -\int_0^\ell (\mathbf{m}' + \mathbf{d}_1 \wedge \mathbf{n}) \cdot \dot{\boldsymbol{\theta}} + \dot{m}_3 k_3 dx_1 + \int_0^\ell \dot{\mathbf{n}} \cdot (\mathbf{r}' - \mathbf{d}_1) - \mathbf{n}' \cdot \dot{\mathbf{r}} dx_1\end{aligned}$$

On requiring that $\dot{\mathcal{L}}$ vanish for every perturbation we obtain the *constraint equations*:

$$\mathbf{r}' = \mathbf{d}_1, \qquad k_3 = 0,$$

and the *equilibrium equations*:

$$\mathbf{n}' = \mathbf{0}, \qquad \mathbf{m}' + \mathbf{d}_1 \wedge \mathbf{n} = \mathbf{0}. \tag{40}$$

The equilibrium equations can be rendered in referential form by introducing the components

$$n_i := \mathbf{n} \cdot \mathbf{d}_i,$$

so that

$$\mathbf{n} = n_i \mathbf{d}_i. \tag{41}$$

On substituting (41) into the first equilibrium equation and (39) into the second equilibrium equation we obtain, in the order,

$$\begin{aligned} n_i' \mathbf{d}_i + n_i \mathbf{k} \wedge \mathbf{d}_i &= \mathbf{0}, \\ m_i' \mathbf{d}_i + m_i \mathbf{d}_i' + \mathbf{d}_1 \wedge \mathbf{n} &= \mathbf{0}. \end{aligned} \tag{42}$$

Now we introduce the rotation matrix

$$\mathbf{D} = \mathbf{d}_i \otimes \mathbf{e}_i, \tag{43}$$

and we observe that

$$\mathbf{D}^T \mathbf{d}_i = \mathbf{e}_i. \tag{44}$$

We also notice that, given two arbitrary vectors **a** and **b**, the following identity holds true:

$$\mathbf{D}^T (\mathbf{a} \wedge \mathbf{b}) = (\mathbf{D}^T \mathbf{a}) \wedge (\mathbf{D}^T \mathbf{b})$$

Thus, in particular, we have

$$\mathbf{D}^T(\mathbf{k}\wedge\mathbf{d}_i)=\mathbf{k}\wedge\mathbf{e}_i,$$

and

$$\mathbf{D}^T(\mathbf{d}_1\wedge\mathbf{n})=\mathbf{e}_1\wedge\mathbf{n}.$$

Thus on premultiplying both equations in (42) by $\mathbf{D}^T$, we arrive at

$$n_i'\mathbf{e}_i+n_i\mathbf{k}\wedge\mathbf{e}_i=\mathbf{0},$$
$$m_i'\mathbf{e}_i+m_i\mathbf{k}\wedge\mathbf{e}_i+\mathbf{e}_1\wedge\mathbf{n}=\mathbf{0},$$

which are equivalent to

$$\mathbf{n}'+\mathbf{k}\wedge\mathbf{n}=\mathbf{0},$$
$$\mathbf{m}'+\mathbf{k}\wedge\mathbf{m}+\mathbf{e}_1\wedge\mathbf{n}=\mathbf{0}.$$

## 6 Example: A Twisted Ribbon

A class of deformations that are of particular importance for the mechanics of a ribbon are the twisting deformations.

We consider an experiment where an isotropic ribbon, while being subject to a *tensile force* $N$, is twisted under controlled *terminal rotation* $\Theta$.

We model this experiment as follows. We identify the middle line of the ribbon with a straight line passing through the origin and parallel to the unit vector $\mathbf{e}_1$. We clamp the end $x_1=0$ to a fixed base support by asking

$$\mathbf{r}(0)=\mathbf{0},\quad\text{and}\quad\mathbf{d}_i(0)=\mathbf{e}_i,\quad\text{for } i=1,2,3.$$

The condition that the end $x_1=\ell$ be clamped and subject to a tensile force $N$ is rendered by the following set of equations:

$$\mathbf{n}(\ell)\cdot\mathbf{e}_1=N,\qquad\mathbf{r}(\ell)\cdot\mathbf{e}_2=0,\qquad\mathbf{r}(\ell)\cdot\mathbf{e}_3=0,$$
$$\mathbf{d}_1(\ell)=\mathbf{e}_1,$$
$$\mathbf{d}_2(\ell)=\cos\Theta\mathbf{e}_2+\sin\Theta\mathbf{e}_3,\qquad\mathbf{d}_3(\ell)=-\sin\Theta\mathbf{e}_2+\cos\Theta\mathbf{e}_3.$$

**The Boundary-Value Problem** For the reader's convenience we recapitulate the relevant field equations.

- The referential equilibrium equations:

$$\begin{aligned}&\mathbf{n}'+\mathbf{k}\wedge\mathbf{n}=\mathbf{0},\\&\mathbf{m}'+\mathbf{k}\wedge\mathbf{m}+\mathbf{e}_1\wedge\mathbf{n}=\mathbf{0}.\end{aligned}\tag{45a}$$

- The compatibility equation:

$$\mathbf{d}_i'=\mathbf{k}\wedge\mathbf{d}_i,\quad i=1,2,3.\tag{45b}$$

- The constitutive equations:

$$m_\alpha = \frac{\partial \overline{W}(k_1, k_2)}{\partial k_\alpha}, \quad \alpha = 1, 2. \tag{45c}$$

- The constraint equations:

$$\begin{aligned} &\mathbf{d}_i \cdot \mathbf{d}_j = \delta_{ij}, \\ &k_3 = 0, \\ &\mathbf{r}' = \mathbf{d}_1. \end{aligned} \tag{45d}$$

**The Reference State** A solution of the twist problem is the following

$$\begin{aligned} &\mathbf{r}(x_1) = x_1 \mathbf{e}_1, \\ &\mathbf{d}_1(x_1) = \mathbf{e}_1, \\ &\mathbf{d}_2(x_1) = +\cos(\Theta x_1/\ell)\mathbf{e}_2 + \sin(\Theta x_1/\ell)\mathbf{e}_3, \\ &\mathbf{d}_3(x_1) = -\sin(\Theta x_1/\ell)\mathbf{e}_2 + \cos(\Theta x_1/\ell)\mathbf{e}_3. \end{aligned} \tag{46}$$

The corresponding referential curvatures are

$$\begin{aligned} &k_1(x_1) = \Theta/\ell, \\ &k_2(x_1) = 0, \\ &k_3(x_1) = 0. \end{aligned} \tag{47}$$

The corresponding referential tension and moment are, respectively,

$$\begin{aligned} &\mathsf{n}(x_1) = (N, 0, 0), \\ &m_1(x_1) = \partial_{k_1} \overline{W}(k_1, k_2) =: GI_t \Theta, \\ &m_2(x_1) = \partial_{k_2} \overline{W}(k_1, k_2) = 0, \\ &m_3(x_1) = 0. \end{aligned} \tag{48}$$

The bending moments $m_1$ and $m_2$ are determined by the corresponding curvatures through the constitutive equations (33). Instead, the bending moment $m_3$ is computed by making use of the equilibrium equation $m_2' + k_3 m_1 - k_1 m_3 - n_3 = 0$. In the second equation of (48) we have set

$$GI_t := 4(2d_\mu + d_\lambda)\frac{1}{\ell}.$$

**Computation of $\overline{W}$ in Terms of $Q$** The solution found is the *macroscopic* solution, i.e., the solution of the macroscopic problem. We now reconstruct the *microscopic* solution. To do so we need to express the energy in terms of the quadratic form $Q$.

From (16) and (47) we find

$$\mathbf{K}^\circ = \begin{pmatrix} 0 & k_1 \\ k_1 & \gamma \end{pmatrix},$$

where $k_1 = \Theta/\ell$ is the amount of twist, see (47), and $\gamma$ has to be determined from (22). Since

$$W^{**}(\mathbf{L}) = \frac{GI_t\ell}{8}\big(|\mathbf{L}|^2 + 2|\det\mathbf{L}|\big),$$

see (31), we have that $W^{**}(\mathbf{K}^\circ) = GI_t\ell/8(4k_1^2+\gamma^2)$ and thence the minimum in (22) is achieved for

$$\gamma = 0.$$

We therefore have

$$\mathbf{K}^\circ = \begin{pmatrix} 0 & k_1 \\ k_1 & 0 \end{pmatrix}, \qquad \overline{W}(k_1,0) = W^{**}\big(\mathbf{K}^\circ\big). \tag{49}$$

Thus, we are in the case of negative Gaussian curvature ($\det\mathbf{K}^\circ < 0$). We look for $\mathbf{A}$, $\mathbf{B}$, and $t$ such that $\det\mathbf{A} = \det\mathbf{B} = 0$ and $\mathbf{K}^\circ = (1-t)\mathbf{A} + t\mathbf{B}$.

Since

$$\big(\mathbf{K}^\circ\big)^* = \begin{pmatrix} 0 & -k_1 \\ -k_1 & 0 \end{pmatrix},$$

on recalling (29) and (32) we have

$$\lambda = \frac{(\mathbf{K}^\circ)^*\cdot\mathbf{D}^+}{2\sqrt{-\det\mathbf{L}}} = 0. \tag{50}$$

By (28) and (29) we find

$$t = \frac{1}{2+2\lambda(\lambda-\sqrt{\lambda^2+1})} = \frac{1}{2}, \tag{51}$$

and

$$\eta = \sqrt{-\det\mathbf{K}^\circ}\big(\lambda - \sqrt{\lambda^2+1}\big) = -k_1, \tag{52}$$

so that by (28) and (30) we deduce that

$$\mathbf{B} = \mathbf{K}^\circ - \eta\mathbf{D}^+ = \begin{pmatrix} k_1 & k_1 \\ k_1 & k_1 \end{pmatrix}, \qquad \mathbf{A} = \frac{\mathbf{K}^\circ - t\mathbf{B}}{1-t} = \begin{pmatrix} -k_1 & k_1 \\ k_1 & -k_1 \end{pmatrix}.$$

According to (26) and (49), we have

$$\mathbf{K}^\circ = \frac{1}{2}\mathbf{A} + \frac{1}{2}\mathbf{B}, \quad \text{and} \quad \overline{W}(k_1,0) = W^{**}\big(\mathbf{K}^\circ\big) = \frac{1}{2}Q(\mathbf{A}) + \frac{1}{2}Q(\mathbf{B}). \tag{53}$$

**Microscopic Oscillations** The macroscopic curvature $\mathbf{K}^\circ$ has determinant different from zero and therefore it is microscopically inaccessible. Equation (53) shows that it can be generated by mixing, on a fine scale and in equal proportions, the "curvatures" $\mathbf{A}$ and $\mathbf{B}$, which have both null determinant. The oscillations occurring in a twisted ribbon are also studied in [9, 10, 25].

In a ribbon of finite width $h$, given a matrix with a null determinant, like $\mathbf{A}$ and $\mathbf{B}$, it is always possible to find a deformation whose second fundamental form is the given matrix, essentially by solving (5). Instead, given two matrices $\mathbf{A}$ and $\mathbf{B}$ with null determinant it is

**Fig. 1** Period of the function $f_n$

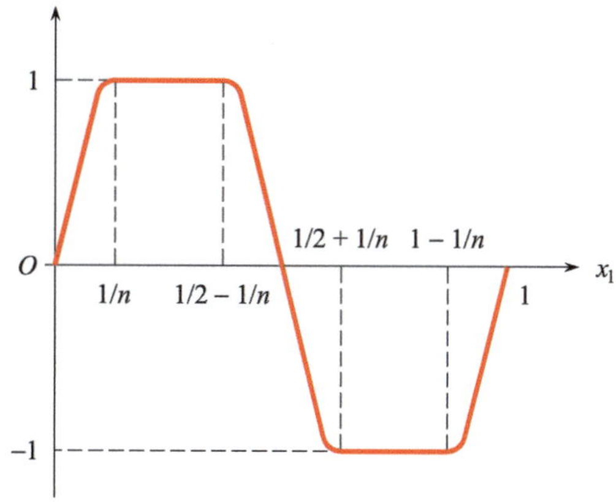


possible to find a (smooth) deformation whose fundamental form takes only the values **A** and **B** only if **A** and **B** are proportional, see Sect. A.3 for a proof of this statement. Since in the case under consideration the matrices **A** and **B** are not proportional, it is necessary to introduce a boundary layer between the regions where the curvature is equal to **A** and **B**. Since

$$\mathbf{A} = -k_1 \begin{pmatrix} -1 \\ 1 \end{pmatrix} \otimes \begin{pmatrix} -1 \\ 1 \end{pmatrix} \quad \text{and} \quad \mathbf{B} = k_1 \begin{pmatrix} 1 \\ 1 \end{pmatrix} \otimes \begin{pmatrix} 1 \\ 1 \end{pmatrix}$$

the boundary layer can be introduced as follows. Let $f_n : \mathbb{R} \to [-1, 1]$ be a periodic function whose period is similar to that represented in Fig. 1 and let

$$\mathbf{K}^{(n)}(x_1) = k_1 f_n(nx_1) \begin{pmatrix} f_n(nx_1) \\ 1 \end{pmatrix} \otimes \begin{pmatrix} f_n(nx_1) \\ 1 \end{pmatrix}.$$

Then, $\mathbf{K}^{(n)}$ is a periodic function, whose period size is $1/n$, having null determinant everywhere and with average (and weak limit) equal to $\mathbf{K}^\circ$. Moreover, the function $\mathbf{K}^{(n)}$ is equal to **A** and **B** except on $n$ boundary layers of size $2/n^2$. Since the energy on these layers is bounded they will contribute, to the total energy, by a constant times $n2/n^2$. Thus, the contribution of the layers will approach zero as the oscillations increase, i.e., as $n$ goes to $+\infty$. From the second equation of (53) we can therefore deduce that

$$\lim_{n\to\infty} \int_0^\ell Q(\mathbf{K}^{(n)})dx_1 = \int_0^\ell W^{**}(\mathbf{K}^\circ)dx_1 = \int_0^\ell \overline{W}(k_1, 0)dx_1.$$

**Reconstruction of the Microscopic Deformation from the Microscopic Curvature** With (14) in mind, we set

$$k_1^{(n)} := K_{12}^{(n)} = k_1 f_n^2(n\cdot), \qquad k_2^{(n)} := -K_{11}^{(n)} = -k_1 f_n^3(n\cdot), \tag{54}$$

and we construct a set of directors $\mathbf{d}_i^{(n)}$ such that

$$(\mathbf{d}_2^{(n)})' \cdot \mathbf{d}_3^{(n)} = k_1^{(n)}, \quad (\mathbf{d}_3^{(n)})' \cdot \mathbf{d}_1^{(n)} = k_2^{(n)}, \quad \text{and} \quad (\mathbf{d}_1^{(n)})' \cdot \mathbf{d}_2^{(n)} = 0. \tag{55}$$

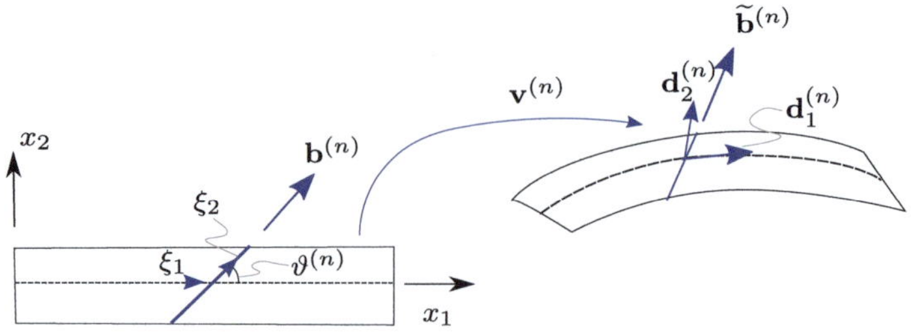


**Fig. 2** The parameterizations $(x_1, x_2)$ and $(\xi_1, \xi_2)$ of the reference configuration, and the deformation $\mathbf{v}^{(n)}$

This task is accomplished by introducing the skew-symmetric tensor

$$\widetilde{\mathbf{K}}^{(n)} = \begin{pmatrix} 0 & 0 & +k_2^{(n)} \\ 0 & 0 & -k_1^{(n)} \\ -k_2^{(n)} & +k_1^{(n)} & 0 \end{pmatrix}, \tag{56}$$

and by finding a proper rotation $\mathbf{D}^{(n)}(x_1)$ that solves the Cauchy problem:

$$\begin{cases} (\mathbf{D}^{(n)})' = \mathbf{D}^{(n)}\widetilde{\mathbf{K}}^{(n)} & \text{in } (0, \ell), \\ \mathbf{D}^{(n)}(0) = \mathbf{I}. \end{cases} \tag{57}$$

The set of directors is then found by setting

$$\mathbf{d}_i^{(n)} := \mathbf{D}^{(n)}\mathbf{e}_i,$$

compare with (43). We are now in a position, for small $h$, to construct a isometric deformation $\boldsymbol{\chi}^{(n)} : \omega \to \mathbb{R}^3$ by means of a ruled surface *à la* Wunderlich [40], see also [16]. We first determine the microscopic deformation of the midline

$$\mathbf{r}^{(n)}(x_1) := \int_0^{x_1} \mathbf{d}_1^{(n)}(s)ds, \tag{58}$$

compare the third equation of (45d), and then we set

$$\Phi^{(n)}(\xi_1, \xi_2) := \xi_1\mathbf{e}_1 + \xi_2\mathbf{b}^{(n)}(\xi_1), \qquad \mathbf{b}^{(n)} := \cos\vartheta^{(n)}\mathbf{e}_1 + \sin\vartheta^{(n)}\mathbf{e}_2,$$
$$\mathbf{v}^{(n)}(\xi_1, \xi_2) := \mathbf{r}^{(n)}(\xi_1) + \xi_2\tilde{\mathbf{b}}^{(n)}(\xi_1), \qquad \tilde{\mathbf{b}}^{(n)} := \cos\vartheta^{(n)}\mathbf{d}_1^{(n)} + \sin\vartheta^{(n)}\mathbf{d}_2^{(n)},$$

where $\mathbf{b}^{(n)}$ is the eigenvector associated to the null eigenvalue of $\mathbf{K}^{(n)}$. The ribbon therefore does not bend along $\mathbf{b}^{(n)}$. The map $(x_1, x_2) = \Phi^{(n)}(\xi_1, \xi_2)$ parametrizes the domain $\omega$, see Fig. 2, in the coordinates $\xi_1$ and $\xi_2$, where the latter coordinate parametrizes the lines that do not bend. The map $\mathbf{v}^{(n)}$ instead deforms the ribbon by imposing that:

(i) the deformation of the midline is $\mathbf{r}^{(n)}$;
(ii) the straight segments parallel to $\mathbf{b}^{(n)}$ are mapped to the straight segments parallel to $\tilde{\mathbf{b}}^{(n)}$.

Thus, the deformed surface is a ruled surface where the curve $\xi_1 \mapsto \mathbf{r}^{(n)}(\xi_1)$ is the *directrix* and the lines $\xi_2 \mapsto \mathbf{r}^{(n)}(\xi_1) + \xi_2\tilde{\mathbf{b}}^{(n)}(\xi_1)$ are the *rulings*.

Then, the microscopic deformation in the $x_1$ and $x_2$ variables is

$$\boldsymbol{\chi}^{(n)} := \mathbf{v}^{(n)} \circ \left(\boldsymbol{\Phi}^{(n)}\right)^{-1}.$$

Some calculations show that the second fundamental form of $\boldsymbol{\chi}^{(n)}$, hereafter denoted by $\mathbf{K}(\boldsymbol{\chi}^{(n)})$, is given by

$$\mathbf{K}\big(\boldsymbol{\chi}^{(n)}\big) = \big[\big(\nabla\boldsymbol{\Phi}^{(n)}\big)^{-\top}\nabla^2 v_i^{(n)}\big(\nabla\boldsymbol{\Phi}^{(n)}\big)^{-1} \nu_i^{(n)}\big] \circ \big(\boldsymbol{\Phi}^{(n)}\big)^{-1}, \tag{59}$$

where $\boldsymbol{\nu}^{(n)}$ is the normal to $\boldsymbol{\chi}^{(n)}$ that coincides with $\mathbf{d}_3^{(n)}$ on the midline of the surface. At $\xi_2 = 0$ we have that

$$\begin{aligned}(\nabla^2 v_i^{(n)})\nu_i^{(n)} &= \begin{pmatrix} (\mathbf{r}^{(n)})'' \cdot \mathbf{d}_3^{(n)} & \cos\vartheta^{(n)}(\mathbf{d}_1^{(n)})' \cdot \mathbf{d}_3^{(n)} + \sin\vartheta^{(n)}(\mathbf{d}_2^{(n)})' \cdot \mathbf{d}_3^{(n)} \\ \text{sym} & 0 \end{pmatrix} \\ &= \begin{pmatrix} K_{11}^{(n)} & \cos\vartheta^{(n)} K_{11}^{(n)} + \sin\vartheta^{(n)} K_{12}^{(n)} \\ \text{sym} & 0 \end{pmatrix},\end{aligned}$$

where to obtain the second equality we have used (54), (55), and (58). Also, at $\xi_2 = 0$,

$$\nabla\boldsymbol{\Phi}^{(n)} = \begin{pmatrix} 1 & \cos\vartheta^{(n)} \\ 0 & \sin\vartheta^{(n)} \end{pmatrix},$$

and hence from (59) we deduce that

$$\mathbf{K}\big(\boldsymbol{\chi}^{(n)}\big) \circ \boldsymbol{\Phi}^{(n)}(\xi_1, 0) = \begin{pmatrix} K_{11}^{(n)} & K_{12}^{(n)} \\ K_{12}^{(n)} & -K_{11}^{(n)} \frac{\cos^2\vartheta^{(n)}}{\sin^2\vartheta^{(n)}} - 2K_{12}^{(n)} \frac{\cos\vartheta^{(n)}}{\sin\vartheta^{(n)}} \end{pmatrix}.$$

By using that $(\cos\vartheta^{(n)}, \sin\vartheta^{(n)})$ is the eigenvector associated to the null eigenvalue of $\mathbf{K}^{(n)}$, we easily show that

$$K_{22}^{(n)} = -K_{11}^{(n)} \frac{\cos^2\vartheta^{(n)}}{\sin^2\vartheta^{(n)}} - 2K_{12}^{(n)},$$

and therefore deduce that on the midline the second fundamental form of $\boldsymbol{\chi}^{(n)}$ is exactly equal to $\mathbf{K}^{(n)}$.

**Acknowledgements** R. Paroni acknowledges support from the Università di Pisa through the project PRA_2018_61 "Modellazione multi-scala in ingegneria strutturale".

**Publisher's Note** Springer Nature remains neutral with regard to jurisdictional claims in published maps and institutional affiliations.

# Appendix

## A.1 Convexification

The aim of this appendix is to determine the convex envelope of

$$W(\mathbf{L}) = \begin{cases} Q(\mathbf{L}) & \text{if } \det\mathbf{L} = 0, \\ +\infty & \text{else,} \end{cases}$$

where $Q$ is a positive definite quadratic form defined on 2-by-2 symmetric matrices.

Consider the function

$$Q_{\det}(\mathbf{L};\alpha) = Q(\mathbf{L}) + \alpha \det \mathbf{L}$$

with $\alpha$ a real number. For fixed $\alpha$, $Q_{\det}(\cdot;\alpha)$ is a quadratic function—recall that $\mathbf{L}$ may be represented by a $2\times 2$ matrix—and hence it is convex if and only if $Q_{\det}(\mathbf{L};\alpha) \geq 0$ for every $\mathbf{L}$. This inequality is satisfied for all $\mathbf{L}$ with determinant equal to zero, since $Q$ is positive definite. If $\det \mathbf{L} > 0$, the condition $Q_{\det}(\mathbf{L};\alpha) \geq 0$ for every $\mathbf{L}$ holds if and only if

$$Q(\mathbf{L}/\sqrt{\det \mathbf{L}}) + \alpha \geq 0 \quad \text{for every } \mathbf{L},$$

that is equivalent to

$$\alpha \geq -Q(\mathbf{L}) \quad \text{for every } \mathbf{L} \text{ with } \det \mathbf{L} = 1,$$

or

$$\alpha \geq -\min_{\det \mathbf{L}=1} Q(\mathbf{L}) =: -\alpha^-. \tag{60}$$

Similarly, if $\det \mathbf{L} < 0$, the condition $Q_{\det}(\mathbf{L};\alpha) \geq 0$ for every $\mathbf{L}$ holds if and only if

$$\alpha \leq \min_{\det \mathbf{L}=-1} Q(\mathbf{L}) =: \alpha^+. \tag{61}$$

Note that $\alpha^-, \alpha^+ > 0$, since $Q$ is positive definite. Hence, $Q_{\det}(\cdot;\alpha)$ is convex if and only if $-\alpha^- \leq \alpha \leq \alpha^+$, and

$$W_c(\mathbf{L}) = \sup_{-\alpha^- \leq \alpha \leq \alpha^+} Q_{\det}(\mathbf{L};\alpha) \tag{62}$$

is also convex, for the point-wise supremum of convex functions is convex. It immediately follows that

$$W_c(\mathbf{L}) = Q(\mathbf{L}) + \alpha^+ (\det \mathbf{L})^+ + \alpha^- (\det \mathbf{L})^-, \tag{63}$$

where $(\det \mathbf{L})^+$ is the positive part of $\det \mathbf{L}$, and $(\det \mathbf{L})^-$ is the negative part, i.e., $(\det \mathbf{L})^- = -\det \mathbf{L}$ if $\det \mathbf{L} < 0$ and zero otherwise.

Clearly, $W(\mathbf{L}) \geq Q_{\det}(\mathbf{L};\alpha)$ for any $\mathbf{L}$ and for any $\alpha$, hence $W(\mathbf{L}) \geq W_c(\mathbf{L})$. Since $W_c$ is convex it also follows that

$$W^{**}(\mathbf{L}) \geq W_c(\mathbf{L}).$$

In the rest of the section we will show that the inequality above is indeed an equality, and that the lower bound can be achieved by a single lamination. More precisely, we show that given $\mathbf{L}$ we can find two symmetric tensors $\mathbf{A}$ and $\mathbf{B}$, and $0 < t < 1$ such that

$$(1-t)\mathbf{A} + t\mathbf{B} = \mathbf{L}, \quad \text{and} \quad (1-t)W(\mathbf{A}) + tW(\mathbf{B}) = W_c(\mathbf{L}).$$

By the definition of $W$, the statement above is equivalent to: given $\mathbf{L}$ there exist two symmetric tensors $\mathbf{A}$ and $\mathbf{B}$ with $\det \mathbf{A} = \det \mathbf{B} = 0$, and $0 < t < 1$ such that

$$(1-t)\mathbf{A} + t\mathbf{B} = \mathbf{L}, \quad \text{and} \quad (1-t)Q(\mathbf{A}) + tQ(\mathbf{B}) = W_c(\mathbf{L}).$$

In this statement we may eliminate the tensor $\mathbf{A}$ by noticing that: given $\mathbf{L}$, $\mathbf{B}$ and $t$, as above, there exists a symmetric tensor $\mathbf{A}$ with $\det \mathbf{A} = 0$ and satisfying the equation

$$(1-t)\mathbf{A} = \mathbf{L} - t\mathbf{B}$$

if and only if $\det(\mathbf{L} - t\mathbf{B}) = 0$. Since

$$\begin{aligned}(1-t)Q(\mathbf{A}) + tQ(\mathbf{B}) &= \frac{1}{1-t}Q\big((1-t)\mathbf{A}\big) + tQ(\mathbf{B}) \\ &= \frac{1}{1-t}Q(\mathbf{L} - t\mathbf{B}) + tQ(\mathbf{B}) \\ &= Q(\mathbf{L}) + \frac{t}{1-t}Q(\mathbf{L} - \mathbf{B}),\end{aligned}$$

the statement to be proven is: given $\mathbf{L}$ there exist a symmetric tensor $\mathbf{B}$ with $\det\mathbf{B} = 0$, and $0 < t < 1$ such that

$$\det(\mathbf{L} - t\mathbf{B}) = 0, \quad \text{and} \quad Q(\mathbf{L}) + \frac{t}{1-t}Q(\mathbf{L} - \mathbf{B}) = W_c(\mathbf{L}). \tag{64}$$

Recalling (60) and (61), let $\mathbf{D}^-, \mathbf{D}^+$ be the two symmetric matrices with $\det\mathbf{D}^- = -1$ and $\det\mathbf{D}^+ = 1$ such that

$$\alpha^- = Q\big(\mathbf{D}^+\big), \qquad \alpha^+ = Q\big(\mathbf{D}^-\big). \tag{65}$$

Note that the tensors $\mathbf{D}^-$ and $\mathbf{D}^+$ only depend on the bending stiffness tensor $\mathbb{D}$. We now show that if $\det\mathbf{L} > 0$ there exist a symmetric tensor $\mathbf{B}$ with null determinant and $0 < t < 1$ such that

$$\det(\mathbf{L} - t\mathbf{B}) = 0, \qquad \det(\mathbf{L} - \mathbf{B}) = -\frac{1-t}{t}\det\mathbf{L}, \tag{66}$$

and

$$\mathbf{D}^- = \frac{\mathbf{L} - \mathbf{B}}{\sqrt{-\det(\mathbf{L} - \mathbf{B})}}. \tag{67}$$

Assuming, for the moment, the validity of this statement we deduce (64) in the case $\det\mathbf{L} > 0$:

$$\begin{aligned}Q(\mathbf{L}) + \frac{t}{1-t}Q(\mathbf{L} - \mathbf{B}) &= Q(\mathbf{L}) + \frac{t}{1-t}\big(-\det(\mathbf{L} - \mathbf{B})\big)Q\big(\mathbf{D}^-\big) \\ &= Q(\mathbf{L}) + Q\big(\mathbf{D}^-\big)\det\mathbf{L} \\ &= Q(\mathbf{L}) + \alpha^+\det\mathbf{L} = W_c(\mathbf{L}),\end{aligned}$$

where we have used, in order, (67), (66), (65), and (63). Hence, assuming (66) and (67) we have shown that if $\det\mathbf{L} > 0$ then $W^{**}(\mathbf{L}) = W_c(\mathbf{L})$. A similar proof can be made for $\det\mathbf{L} < 0$, while the result is trivial for $\det\mathbf{L} = 0$.

We therefore only need to show the validity of (66) and (67). In doing it, we shall provide formulae that deliver $\mathbf{B}$ and $t$.

For $\det\mathbf{L} > 0$, set $\mathbf{B} = \mathbf{L} - \eta\mathbf{D}^-$ where the real number $\eta$ is chosen by imposing that $\det\mathbf{B} = 0$. With (1), this amount to solve the second order equation

$$\det\mathbf{L} - \eta\mathbf{L}^* \cdot \mathbf{D}^- - \eta^2 = 0, \tag{68}$$

where we used the fact that $\det\mathbf{D}^- = -1$. Of the two solutions we choose

$$\eta = \frac{-\mathbf{L}^* \cdot \mathbf{D}^- + \sqrt{(\mathbf{L}^* \cdot \mathbf{D}^-)^2 + 4\det\mathbf{L}}}{2}.$$

To keep the notation compact, we set

$$\lambda = \frac{\mathbf{L}^* \cdot \mathbf{D}^-}{2\sqrt{\det \mathbf{L}}}$$

so that

$$\eta = \sqrt{\det \mathbf{L}}\big(\sqrt{\lambda^2 + 1} - \lambda\big).$$

Now that $\mathbf{B}$ has been fixed, we find $t$ by solving $(66)_1$. We first write

$$\mathbf{L} - t\mathbf{B} = \mathbf{L} - t\big(\mathbf{L} - \eta\mathbf{D}^-\big) = (1-t)\mathbf{L} + t\eta\mathbf{D}^-,$$

and by using (1), we write $(66)_1$ as

$$(1-t)^2 \det \mathbf{L} + (1-t)t\eta\mathbf{L}^* \cdot \mathbf{D}^- - t^2\eta^2 = 0,$$

which simplifies, thanks to (68), to

$$\det \mathbf{L} + t\big(\eta\mathbf{L}^* \cdot \mathbf{D}^- - 2\det \mathbf{L}\big) = 0.$$

Thus

$$t = \frac{-\det \mathbf{L}}{\eta\mathbf{L}^* \cdot \mathbf{D}^- - 2\det \mathbf{L}} = \frac{1}{2 - 2\lambda(\sqrt{\lambda^2+1} - \lambda)}.$$

We may easily verify that $0 < t < 1$. Indeed, $t > 0$ if and only if $2\lambda(\sqrt{\lambda^2+1} - \lambda) < 2$, which clearly holds if $\lambda \le 0$, while for $\lambda > 0$ is equivalent to $\sqrt{\lambda^2+1} < \frac{1}{\lambda} + \lambda$ which is trivially verified. Similarly we may check that $t < 1$.

From the definition of $\mathbf{B}$, we have that $\eta\mathbf{D}^- = \mathbf{L} - \mathbf{B}$. Taking the determinant, we find $-\eta^2 = \det(\mathbf{L} - \mathbf{B})$ from which (67) follows. Finally, by applying (1) twice, we obtain that

$$\begin{aligned}\det(\mathbf{L} - \mathbf{B}) &= \det \mathbf{L} - \mathbf{L}^* \cdot \mathbf{B} = \det \mathbf{L} - \frac{\det \mathbf{L}}{t} + \frac{\det \mathbf{L} - \mathbf{L}^* \cdot (t\mathbf{B})}{t} \\ &= \frac{t-1}{t}\det \mathbf{L} + \frac{\det(\mathbf{L} - t\mathbf{B})}{t},\end{aligned}$$

and recalling $(66)_1$ we deduce $(66)_2$.

We close this appendix by summarizing the results found:

$$W^{**}(\mathbf{L}) = Q(\mathbf{L}) + \alpha^+(\det \mathbf{L})^+ + \alpha^-(\det \mathbf{L})^-,$$

where

$$\alpha^- = \min_{\det \mathbf{L}=1} Q(\mathbf{L}) = Q\big(\mathbf{D}^+\big), \qquad \alpha^+ = \min_{\det \mathbf{L}=-1} Q(\mathbf{L}) = Q\big(\mathbf{D}^-\big).$$

If $\det \mathbf{L} > 0$, set

$$\lambda = \frac{\mathbf{L}^* \cdot \mathbf{D}^-}{2\sqrt{\det \mathbf{L}}}, \qquad t = \frac{1}{2 - 2\lambda\big(\sqrt{\lambda^2+1} - \lambda\big)},$$

$$\eta = \sqrt{\det \mathbf{L}}\big(\sqrt{\lambda^2+1} - \lambda\big), \qquad \mathbf{B} = \mathbf{L} - \eta\mathbf{D}^-;$$

while, if $\det \mathbf{L} < 0$, set

$$\lambda = \frac{\mathbf{L}^* \cdot \mathbf{D}^+}{2\sqrt{-\det \mathbf{L}}}, \qquad t = \frac{1}{2 + 2\lambda\big(\lambda - \sqrt{\lambda^2+1}\big)},$$

$$\eta = \sqrt{-\det \mathbf{L}}\big(\lambda - \sqrt{\lambda^2+1}\big), \qquad \mathbf{B} = \mathbf{L} - \eta \mathbf{D}^+.$$

Then, with

$$\mathbf{A} = \frac{\mathbf{L} - t\mathbf{B}}{1-t},$$

we have that

$$(1-t)\mathbf{A} + t\mathbf{B} = \mathbf{L}, \quad \text{and} \quad (1-t)Q(\mathbf{A}) + tQ(\mathbf{B}) = W^{**}(\mathbf{L}).$$

## A.2 Evaluation of $\alpha^{\pm}$ for Various Symmetries

We first turn our attention to the computation of the constants $\alpha^+$ and $\alpha^-$ in the definition of the convexification $W^{**}$ in (24). According to (25) we must minimize the quadratic form $Q(\mathbf{L})$ over the manifolds $\{\mathbf{L} : \det \mathbf{L} = \pm 1\}$. We accomplish this task through the method of Lagrange multipliers by seeking stationary points of the *augmented functional* $(\mathbf{L}, \beta) \mapsto Q(\mathbf{L}) + \beta(\det \mathbf{L} \pm 1)$. We note that the determinant of a 2-by-2 matrix is a quadratic form. Thus, there exists a fourth-order tensor $\mathbb{E}$ such that

$$\det \mathbf{L} = L_{11}L_{22} - L_{12}^2 = \frac{1}{2}\mathbb{E}\mathbf{L} \cdot \mathbf{L}. \tag{69}$$

Thus, the augmented functional is

$$(\mathbf{L}, \beta) \mapsto \frac{1}{2}(\mathbb{D} + \beta\mathbb{E})\mathbf{L} \cdot \mathbf{L} \mp \beta. \tag{70}$$

We now argue that the minima $\alpha^{\pm}$ have the following property:

$$\mathbb{D} \mp \alpha^{\mp}\mathbb{E} \text{ is singular,} \tag{71}$$

that is $\alpha^{\mp}$ are *generalized eigenvalues of* $\mathbb{D}$ *with respect to* $\mathbb{E}$. To verify the last statement, let us recall the definition of $\mathbf{D}^{\pm}$, see (27), that is

$$\mathbf{D}^{\pm} = \operatorname{argmin}_{\det \mathbf{L} = \pm 1} Q(\mathbf{L}). \tag{72}$$

Then there exist $\beta^{\mp}$ such that the pair $(\mathbf{D}^{\pm}, \beta^{\mp})$ is a stationary point of the augmented functional defined in (70). The stationarity conditions for such pair are

$$\big(\mathbb{D} + \beta^{\mp}\mathbb{E}\big)\mathbf{D}^{\pm} = \mathbf{0}, \tag{73a}$$

$$\frac{1}{2}\mathbb{E}\mathbf{D}^{\pm} \cdot \mathbf{D}^{\pm} \mp 1 = 0. \tag{73b}$$

It follows from the second stationarity condition that $\mathbf{D}^{\pm} \neq \mathbf{0}$. This implies that the tensor $\mathbb{D} + \beta^{\mp}\mathbb{E}$ is singular. Moreover, the minimum $\alpha^{\mp}$ coincides with the Lagrange multiplier $\beta^{\mp}$, up to sign:

$$\alpha^{\mp} = Q\big(\mathbf{D}^{\pm}\big) = \frac{1}{2}\mathbb{D}\mathbf{D}^{\pm} \cdot \mathbf{D}^{\pm} \overset{(73a)}{=} -\frac{\beta^{\mp}}{2}\mathbb{E}\mathbf{D}^{\pm} \cdot \mathbf{D}^{\pm} \overset{(73b)}{=} \mp\beta^{\mp}. \tag{74}$$

To carry our calculation further on, we observe that

$$\mathbb{D}\mathbf{L}\cdot\mathbf{L}=\begin{pmatrix}\mathbb{D}_{1111} & \mathbb{D}_{1122} & \mathbb{D}_{1112}\\ \mathbb{D}_{1122} & \mathbb{D}_{2222} & \mathbb{D}_{1222}\\ \mathbb{D}_{1112} & \mathbb{D}_{1222} & \mathbb{D}_{1212}\end{pmatrix}\begin{pmatrix}L_{11}\\ L_{22}\\ 2L_{12}\end{pmatrix}\cdot\begin{pmatrix}L_{11}\\ L_{22}\\ 2L_{12}\end{pmatrix} \tag{75}$$

and that

$$\mathbb{E}\mathbf{L}\cdot\mathbf{L}=\begin{pmatrix}0 & 1 & 0\\ 1 & 0 & 0\\ 0 & 0 & -1/2\end{pmatrix}\begin{pmatrix}L_{11}\\ L_{22}\\ 2L_{12}\end{pmatrix}\cdot\begin{pmatrix}L_{11}\\ L_{22}\\ 2L_{12}\end{pmatrix}. \tag{76}$$

Thus, on denoting by $\mathbb{d}$ and $\mathbb{e}$ the matrices appearing in the definitions of the quadratic forms (75) and (76), we can write (71) as

$$\det\left(\mathbb{d}\mp\alpha^{\mp}\mathbb{e}\right)=0. \tag{77}$$

**Orthotropic Response** In the special case, when the material is orthotropic with respect to the basis $(\mathbf{e}_1,\mathbf{e}_2)$, we have

$$\mathbb{D}_{1112}=0, \quad \text{and} \quad \mathbb{D}_{1222}=0$$

(the other vanishing components of $\mathbb{D}$ are obtained by application of the standard symmetries: $\mathbb{D}_{\alpha\beta\gamma\delta}=\mathbb{D}_{\beta\alpha\gamma\delta}=\mathbb{D}_{\gamma\delta\alpha\beta}$). In this case, the characteristic polynomial

$$p(\beta):=\det(\mathbb{d}+\beta\mathbb{e})=\left(\mathbb{D}_{1212}-\frac{\beta}{2}\right)\left(\mathbb{D}_{1111}\mathbb{D}_{2222}-(\mathbb{D}_{1122}+\beta)^2\right)$$

has the following set of roots:

$$\{-\sqrt{\mathbb{D}_{1111}\mathbb{D}_{2222}}-\mathbb{D}_{1122},\sqrt{\mathbb{D}_{1111}\mathbb{D}_{2222}}-\mathbb{D}_{1122},2\mathbb{D}_{1212}\}.$$

The positive definiteness of $\mathbb{D}$ implies that the first root is negative, and that the remaining roots are positive. The negative root must coincide with $\beta^-=-\alpha^-$, that is

$$\alpha^-=\sqrt{\mathbb{D}_{1111}\mathbb{D}_{2222}}+\mathbb{D}_{1122}.$$

On the other hand, the minimality of $\alpha^+$ yields

$$\alpha^+=\min\{\sqrt{\mathbb{D}_{1111}\mathbb{D}_{2222}}-\mathbb{D}_{1122},2\mathbb{D}_{1212}\}.$$

We note that the foregoing calculation can be used also when the plate is orthotropic with respect to a basis $(\mathbf{e}'_1,\mathbf{e}'_2)$, not necessarily coincident with $(\mathbf{e}_1,\mathbf{e}_2)$. In this case, the relevant strain energy is $W'(\mathbf{L})=W(\mathbf{R}\mathbf{L}\mathbf{R}^T)$, where $\mathbf{R}=\mathbf{e}'_\alpha\otimes\mathbf{e}_\alpha$ is a rotation matrix. Then it can be easily shown that the convexification of $Q'$ is given by

$$\left(W'\right)^{**}(\mathbf{L})=W^{**}\left(\mathbf{R}\mathbf{L}\mathbf{R}^T\right).$$

**Isotropic Response** For the isotropic strain energy $Q(\mathbf{L}) = d_\mu|\mathbf{L}|^2 + \frac{d_\lambda}{2}(L_{11} + L_{22})^2$ the relevant components of $\mathbb{D}$ are $\mathbb{D}_{1111} = \mathbb{D}_{2222} = 2d_\mu + d_\lambda$, $\mathbb{D}_{1122} = d_\lambda$, $\mathbb{D}_{1212} = d_\mu$. Then

$$\alpha^- = 2(d_\mu + d_\lambda), \qquad \alpha^+ = 2d_\mu,$$

and so

$$\begin{aligned} W^{**}(\mathbf{L}) &= d_\mu|\mathbf{L}|^2 + \frac{d_\lambda}{2}(L_{11} + L_{22})^2 + 2d_\mu(\det\mathbf{L})^+ + 2(d_\mu + d_\lambda)(\det\mathbf{L})^- \\ &= d_\mu|\mathbf{L}|^2 + \frac{d_\lambda}{2}|\mathbf{L}|^2 + d_\lambda\det\mathbf{L} + 2d_\mu(\det\mathbf{L})^+ + 2(d_\mu + d_\lambda)(\det\mathbf{L})^- \\ &= \left(d_\mu + \frac{d_\lambda}{2}\right)|\mathbf{L}|^2 + 2d_\mu|\det\mathbf{L}| + d_\lambda\big(\det\mathbf{L} + 2(\det\mathbf{L})^-\big) \\ &= \left(d_\mu + \frac{d_\lambda}{2}\right)\big(|\mathbf{L}|^2 + 2|\det\mathbf{L}|\big). \end{aligned}$$

## A.3 Compatibility Among Curvatures

The aim of this Appendix is to prove the following statement.

Let $\omega$, $\omega_\mathbf{A}$ and $\omega_\mathbf{B}$ be three two-dimensional open sets, with $\omega = \omega_\mathbf{A} \cup \omega_\mathbf{B} \cup \Gamma$, where $\Gamma = \overline{\omega}_\mathbf{A} \cap \overline{\omega}_\mathbf{B}$ is a smooth curve and $\overline{\omega}_\mathbf{A}$ denotes the closure of $\omega_\mathbf{A}$. Let $\mathbf{A}$ and $\mathbf{B}$, with $\mathbf{A} \neq \mathbf{B}$, be two 2-by-2 symmetric matrices with null determinant. If there exists a deformation $\mathbf{y}$ continuously differentiable on $\omega$ and twice differentiable on $\overline{\omega}_\mathbf{A}$ and on $\overline{\omega}_\mathbf{B}$ such that

$$\mathbf{K}_\mathbf{y} = \mathbf{A} \text{ on } \overline{\omega}_\mathbf{A}, \quad \text{and} \quad \mathbf{K}_\mathbf{y} = \mathbf{B} \text{ on } \overline{\omega}_\mathbf{B},$$

where $\mathbf{K}_\mathbf{y}$ denotes the second fundamental form of $\mathbf{y}$, then the curve $\Gamma$ is straight and the matrices $\mathbf{A}$ and $\mathbf{B}$ are proportional, that is, there exists a constant $\sigma$ for which either $\mathbf{A} = \sigma\mathbf{B}$ or $\mathbf{B} = \sigma\mathbf{A}$.

We assume that a deformation $\mathbf{y}$ as described in the statement above exists and we study the consequences.

Since $\mathbf{A} \neq \mathbf{B}$, we may assume without loss of generality that $\mathbf{A} \neq \mathbf{0}$. Since $\mathbf{y}$ is continuously differentiable we have that $\nabla\mathbf{y}$ and the normal $\boldsymbol{\nu} = \partial_1\mathbf{y} \wedge \partial_2\mathbf{y}$ are continuous on $\omega$. Let $\mathbf{t}$ be the unit tangent to the curve $\Gamma$. Since $(\nabla\boldsymbol{\nu})\mathbf{t} = \partial_\mathbf{t}\boldsymbol{\nu}$ and since $\mathbf{K} = -(\nabla\mathbf{y})^\top\nabla\boldsymbol{\nu}$ we have that

$$\mathbf{At} = \mathbf{Bt}, \quad \text{on } \Gamma.$$

Thus $\det(\mathbf{A} - \mathbf{B}) = 0$, and the identity $\det(\mathbf{A} - \mathbf{B}) = \det\mathbf{A} - \mathbf{A}^* \cdot \mathbf{B} + \det\mathbf{B}$ implies that

$$\mathbf{A}^* \cdot \mathbf{B} = 0.$$

Let $(\lambda, \mathbf{a}_1)$ and $(0, \mathbf{a}_2)$ be the pairs of eigenvalues and eigenvectors of $\mathbf{A}$, so that

$$\mathbf{A} = \lambda\mathbf{a}_1 \otimes \mathbf{a}_1,$$

with $\lambda \neq 0$. Then $\mathbf{A}^* = \lambda\mathbf{a}_2 \otimes \mathbf{a}_2$ and $\mathbf{A}^* \cdot \mathbf{B} = 0$ implies that $\mathbf{Ba}_2 \cdot \mathbf{a}_2 = 0$. We may therefore write

$$\mathbf{B} = \beta\mathbf{a}_1 \otimes \mathbf{a}_1 + \tilde{\beta}(\mathbf{a}_1 \otimes \mathbf{a}_2 + \mathbf{a}_2 \otimes \mathbf{a}_1),$$

and, since $\det \mathbf{B} = 0$, conclude that

$$\mathbf{B} = \beta \mathbf{a}_1 \otimes \mathbf{a}_1.$$

Setting $\sigma = \beta/\lambda$ we obtain $\mathbf{B} = \sigma \mathbf{A}$. Since $\mathbf{A} \neq \mathbf{B}$, from the identity $\mathbf{At} = \mathbf{Bt}$ we deduce that $\mathbf{a}_1 \cdot \mathbf{t} = 0$, which implies that $\mathbf{t}$ is constant and, in turn, that the curve $\Gamma$ is a straight segment.

## References


1. Agostiniani, V., DeSimone, A., Koumatos, K.: Shape programming for narrow ribbons of nematic elastomers. J. Elast. **127**, 1–24 (2017)
2. Agostiniani, V., DeSimone, A.: Dimension reduction via $\Gamma$-convergence for soft active materials. Meccanica **52**, 3457–3470 (2017)
3. Antman, S.S.: Nonlinear Problems of Elasticity, 2nd edn. Applied Mathematical Sciences, vol. 107. Springer, New York (2005)
4. Audoly, B., Seffen, K.A.: Buckling of naturally curved elastic strips: the ribbon model makes a difference. J. Elast. **119**, 293–320 (2015)
5. Chen, Y.C., Fosdick, R., Fried, E.: Representation for a smooth isometric mapping from a connected planar domain to a surface. J. Elast. **119**, 335–350 (2015)
6. Chen, Y.C., Fosdick, R., Fried, E.: Representation of a smooth isometric deformation of a planar material region into a curved surface. J. Elast. **130**, 145–195 (2018)
7. Chen, Y.C., Fosdick, R., Fried, E.: Issues concerning isometric deformations of planar regions to curved surfaces. J. Elast. **132**, 1–42 (2018)
8. Chen, Y.C., Fried, E.: Möbius bands, unstretchable material sheets, and developable surfaces. Proc. R. Soc., Math. Phys. Eng. Sci. **472**, 20150760 (2016)
9. Chopin, J., Kudrolli, A.: Helicoids, wrinkles, and loops in twisted ribbons. Phys. Rev. Lett. **111**, 174302 (2013)
10. Chopin, J., Démery, V., Davidovitch, B.: Roadmap to the morphological instabilities of a stretched twisted ribbon. J. Elast. **119**, 137–189 (2015)
11. Dacorogna, B.: Direct Methods in the Calculus of Variations. Applied Math. Sciences, vol. 78. Springer, Berlin (1989)
12. Dias, M.A., Audoly, B.: "Wunderlich, Meet Kirchhoff": a general and unified description of elastic ribbons and thin rods. J. Elast. **119**, 49–66 (2015)
13. Efrati, E.: Non-Euclidean ribbons. J. Elast. **119**, 251–261 (2014)
14. Fonseca, I., Leoni, G.: Modern Methods in the Calculus of Variations: $L^p$ Spaces. Springer Monographs in Mathematics. Springer, New York (2007)
15. Fosdick, R., Fried, E.: The Mechanics of Ribbons and Möbius Bands. Springer, Dordrecht (2016)
16. Freddi, L., Hornung, P., Mora, M.G., Paroni, R.: A corrected Sadowsky functional for inextensible elastic ribbons. J. Elast. **123**, 125–136 (2016)
17. Freddi, L., Hornung, P., Mora, M.G., Paroni, R.: A variational model for anisotropic and naturally twisted ribbons. SIAM J. Math. Anal. **48**, 3883–3906 (2016)
18. Freddi, L., Hornung, P., Mora, M.G., Paroni, R.: One-dimensional von Kármán models for elastic ribbons. Meccanica **53**, 659–670 (2018)
19. Freddi, L., Hornung, P., Mora, M.G., Paroni, R.: Generalised Sadowsky theories for ribbons from three-dimensional nonlinear elasticity. Submitted
20. Freddi, L., Mora, M.G., Paroni, R.: Nonlinear thin-walled beams with a rectangular cross-section—Part I. Math. Models Methods Appl. Sci. **22**, 1150016 (2012)
21. Freddi, L., Mora, M.G., Paroni, R.: Nonlinear thin-walled beams with a rectangular cross-section—Part II. Math. Models Methods Appl. Sci. **23**, 743–775 (2013)
22. Hinz, D.F., Fried, E.: Translation of Michael Sadowsky's paper "An elementary proof for the existence of a developable Möbius band and the attribution of the geometric problem to a variational problem". J. Elast. **119**, 3–6 (2015)
23. Kirby, N.O., Fried, E.: Gamma-limit of a model for the elastic energy of an inextensible ribbon. J. Elast. **119**, 35–47 (2015)
24. Kirchhoff, G.: Über das Gleichgewicht und die Bewegung einer elastischen Scheibe. J. Reine Angew. Math. **40**, 51–88 (1850)
25. Kohn, R.V., O'Brien, E.: The wrinkling of a twisted ribbon. J. Nonlinear Sci. **28**, 1221–1249 (2018)

26. Moore, A., Healey, T.: Computation of elastic equilibria of complete Möbius bands and their stability. Math. Mech. Solids (2018). https://doi.org/10.1177/1081286518761789
27. Nardinocchi, P., Pezzulla, M., Teresi, L.: Anisotropic swelling of thin gel sheets. Soft Matter **11**, 1492–1499 (2015)
28. Paroni, R.: An existence theorem for inextensible nets with slack. Math. Mech. Solids **17**, 460–472 (2012)
29. Pipkin, A.C.: Inextensible networks with slack. Q. Appl. Math. **40**, 63–71 (1982/1983)
30. Rivlin, R.S.: Plane strain of a net formed by inextensible cords. J. Ration. Mech. Anal. **4**, 951–974 (1955)
31. Rivlin, R.S.: The deformation of a membrane formed by inextensible cords. Arch. Ration. Mech. Anal. **2**, 447–476 (1958/1959)
32. Sadowsky, M.: Ein elementarer Beweis für die Existenz eines abwickelbaren Möbiuschen Bandes und die Zurückführung des geometrischen Problems auf ein Variationsproblem. Sitz.ber. Preuss. Akad. Wiss. 412–415 (1930)
33. Starostin, E.L., van der Heijden, G.H.M.: The equilibrium shape of an elastic developable Möbius strip. Proc. Appl. Math. Mech. **7**, 2020115–2020116 (2007)
34. Starostin, E.L., van der Heijden, G.H.M.: Equilibrium shapes with stress localisation for inextensible elastic Möbius and other strips. J. Elast. **119**, 67–112 (2015)
35. Tomassetti, G., Varano, V.: Capturing the helical to spiral transitions in thin ribbons of nematic elastomers. Meccanica **52**, 3431–3441 (2017)
36. Todres, R.E.: Translation of W. Wunderlich's "On a developable Möbius band". J. Elast. **119**, 23–34 (2015)
37. Teresi, L., Varano, V.: Modeling helicoid to spiral-ribbon transitions of twist-nematic elastomers. Soft Matter **9**, 3081–3088 (2013)
38. Sawa, Y., Urayama, K., Takigawa, T., Gimenez-Pinto, V., Mbanga, B.L., Ye, F., Selinger, J.V., Selinger, R.L.: Shape and chirality transitions in off-axis twist nematic elastomer ribbons. Phys. Rev. E **88**, 022502 (2013)
39. Wan, G., Jin, C., Trase, I., Zhao, S., Chen, Z.: Helical structures mimicking chiral seedpod opening and tendril coiling. Sensors **18**, 2973 (2018)
40. Wunderlich, W.: Über ein abwickelbares Möbiusband. Monatshefte Math. **66**, 276–289 (1962)

J Elast (2019) 135:435–456
https://doi.org/10.1007/s10659-018-9709-y

# On the Mechanical Modeling of Matter, Molecular and Continuum

**Paolo Podio-Guidugli**[1,2]

*To Walter Noll, the alchemist who distilled the quintessence of Continuum Mechanics*

Received: 18 July 2018 / Published online: 12 December 2018


**Abstract** Any mechanical modeling of matter depends primarily on the chosen spatial and temporal observation scales. Roughly speaking, there are three such scales, microscopic for quantum, mesoscopic for statistical, and macroscopic for continuum mechanics. This by itself demands for adequate scale-bridging procedures. This paper focuses on the passage from molecular to continuum formulations of the basic balance laws of linear momentum and energy, kinetic, internal, and total. The procedure used is a modification of that introduced by Irving & Kirkwood (J. Chem. Phys. 18(6):817–829, 1950), as improved by Noll (Indiana Univ. Math. J. 4:627–646, 1955); alternative procedures are mentioned. The proposed modification consists primarily in equipping provisionally the target continuum balances with internal-source terms accounting at the macroscopic scale for microscopic motion randomness, when, e.g., a loosely aggregated material system evolves at relatively high temperature. Attention is also devoted to those continuum notions that either are hardly given a universally accepted molecular counterpart or even cannot and ultimately need not be given one, such as the notion of material point.

**Keywords** Molecular mechanics · Continuum mechanics · Scale bridging

**Mathematics Subject Classification** 70A05 · 74A05 · 74A99

✉ P. Podio-Guidugli
p.podioguidugli@gmail.com

1 Accademia Nazionale dei Lincei, Rome, Italy

2 Dipartimento di Matematica, Università di Roma TorVergata, Rome, Italy

## 1 Terminological Premiss

In this paper I call molecular mechanics (MM) the discipline opposed in nature to continuum mechanics (CM), regardless of how appealing the use of discrete mechanics might be.[1] Here are my two reasons why.

Matter has been modeled both as discrete and as continuous ever since, around 1820, Navier and Cauchy started the quest for a discrete counterpart of the continuous description they had just begun to create;[2] according to [3], the expression *mécanique moleculaire* was coined by Lamé in his *Leçons sur la Théorie Mathématique de l'Élasticité des Corps Solides*, appeared in 1852. This is my first reason to talk about molecular, rather than discrete, mechanics.

The great men just mentioned were convinced that matter the naked eye perceives as continuous is really made up of elementary particles. Needless to say, they all had not the faintest idea of the physical nature of those particles, just as Democritus had no idea about the atoms he talked about, apart for the indivisibility conveyed by the term's etymology. As for etymology, molecules are nothing but small masses (*moles*, in Latin), just as particles are nothing but small parts. In the years 1820s, postulating pairwise elastic interactions among molecules sufficed to attempt a mechanical modelization, although not to achieve it successfully some fifteen years before Hamilton and some fifty years before Gibbs and Boltzmann, when neither a Hamiltonian nor a statistical approach had been proposed. These days computer simulations are provided by molecular dynamics (MD), the *brachium saeculare* of all theories consisting of a clever sophistication of one or another Hamiltonian version of Newtonian equations in probabilistic sauce. This is my second reason to use MM as the antonym of CM.

## 2 Introduction

The purpose of this paper is to discuss whether and to what extent a CM consistent with MM can be constructed. It will be shown that such construction is indeed possible, albeit to some extent, in more than one way. The issue has been brought up just at the middle of the last century and pursued by many ever since. Its study appears to be inspired by the communal belief that, as expressed in the title of [5], the physical foundations of CM are to be found at scales much smaller than its own.

Not that, in my opinion, the standing of CM require such an implicit validation: as a mathematical theory, CM stands on its own feet. As is desirable, it allows both for controllable and useful predictions and for predictions that can be falsified; to me, this is enough to regard it consistent with physics. At first, CM was created to formalize a body of phenomenological evidences that can be captured by the naked eye and sequenced by the pulse; thanks to and alongside of the ever increasing availability of analytical tools it partly inspired, it has developed to the point that it can now describe efficiently a variety of microscopic and submicroscopic phenomena such as, say, displacive phase transitions, phase segregation and frustration.

Sure enough, CM cannot be pushed down to scales at which molecules have individual citizenship. Bottom-up procedures, none of them excluded, have their limitations as well, both conceptual and computational. Much of this paper is devoted to set the stage for and

[1] To my knowledge, nobody ever proposed to use indiscrete, instead of continuum, mechanics.

[2] See Love's "Historical Introduction" [1] and [2], Sect. 301.

present in detail—respectively, in Sects. 5.1 and 6—a version of my taste of a bottom-up procedure that has no computational value as is, the classic procedure introduced by Irving & Kirkwood in 1950 [6] and improved by Noll in 1955 [7] (see also [8]).

Computationally, the scale divide between MM and CM is passable upwards via MD. Ever since the papers by Andersen [9] and Parrinello and Rahman [10–12] appeared,[3] successful simulations of certain displacive phase transitions have been achieved, by the use of Lagrangians (or for that matter Hamiltonians) extended by inclusion of certain macroscopic degrees of freedom (DOFs) interpreted as either the stress or the strain inducing the transition in a minuscule speck of matter. Moreover, it seems possible to implement a conceptual use, at zero temperature, of an Andersen-Parrinello-Rahman (APR) approach to inspect the mutual coherency of macroscopic and microscopic constitutive assignments under form, respectively, of stored-energy mappings and intermolecular potentials [14]. It is also possible, as proposed in [15], to apply the Irving-Kirkwood procedure to an APR-extended Hamiltonian system, so as to arrive at CM balance equations featuring non-standard terms due the presence of the macroscopic DOFs, under form of additional fluxes and sources that influence the thermodynamic and transport properties. Being aware of such additional contributions may prove crucial when designing multi-scale computational schemes in which information is exchanged between the molecular and continuum levels, such as the scheme envisaged in [3], where an APR approach of both conceptual and computational value is sketched.

However, the current, perhaps intrinsic, limitations of MD are such that, although it can simulate well the bulk behavior of condensed matter, it cannot mimic convincingly the boundary of a body or the common boundary of two adjacent parts of it. In CM, the boundary between two adjacent subbodies is thought of as the result of an operation of mental surgery (a so-called *Euler's cut*) that assigns to it a purely geometrical nature. In fact, to my knowledge, so far MM offers no clue to simulate the physical contact of body parts, let alone offering hints to tackle a class of modeling problems which have not been solved satisfactorily even within CM, namely, the problems of modeling the physical contact between two bodies of different constitution (say, a Cosserat and a conventional body, or a liquid crystal and its container) or with different types of boundaries (say, smooth and fractal).

I think it appropriate to conclude this section by completing a summary of the paper's contents.

The passage from molecular to continuum mechanical models of matter is the subject of Sect. 3, where the two ways, statistical and deterministic, to arrive at CM balances via an indispensable space-time averaging are delineated and briefly put into reciprocal perspective. In Sect. 4, certain difficulties intrinsic to MM $\leftrightarrow$ CM bridging are listed, all of which emanate from two characteristic differences between the two disciplines, the one in the notion of material system, the other in the nature of material interactions, both differences impinging on how the constitution of matter chunks is typified in MM and in CM. Although these differences are ineliminable, it is suggested in Sect. 4.4 that they might be coped with better than it is done with the use of the available bottom-up procedures—the modified Irving-Kirkwood-Noll one presented in Sect. 6 included—by equating power expenditures; more about this issue is said in Sect. 5.2. My main conclusions are collected in Sect. 7. Some side material about APR MD is found in Appendix A, while Hellinger's 1914 notions of material body and material point are recapitulated in Appendix B.

[3] See [13] for a well documented account of developments up to the year 2010.

## 3 From Molecular to Continuum Balances

That mechanics had historically evolved into three disciplines, each of which standing on its own feet and having its own devotees, and that bridges, not walls, were to be built between any two of those disciplines, was clear to Walter Noll, as early as in 1957. We read in [17] (italics are mine): "...continuous matter is really made up of elementary particles. The basic laws governing the elementary particles are those of *quantum mechanics*. The science that provides the link between these basic laws and the laws describing the behavior of gross matter is *statistical mechanics*. At the present time this link is quite weak, partly because the mathematical difficulties are formidable, and partly because the basic laws themselves are not yet completely clear. A rigorous theory of *continuum mechanics* would give at least some precise information on what kind of gross behavior the basic laws ought to predict."

It appears that Noll here envisages three observation scales, the micro/meso-scopic scale of statistical mechanics (SM) been markedly different both from the truly microscopic scale of quantum mechanics and the macroscopic scale of continuum mechanics; that he assigns to SM the role of bridging between quantum and continuum mechanics; and that he suggests the adoption of a top-down view to direct and inspire the search for such quantum-mechanical predictions as to match the findings of a rigorously reformulated CM.

In fact, the years when Noll was making the above quoted statements were years when a new and authoritative narration of continuum mechanics was about to be made available through two masterfully written treatises, namely, "The Classical Field Theories" ([18], 1960), written by C. Truesdell and R.A. Toupin, with an appendix due to J.L. Ericksen, and the "The Nonlinear Field Theories of Mechanics" ([19], 1965), written by C. Truesdell and W. Noll himself.

Moreover, just a few years before, Noll had straightened and completed in [7] the path-breaking work of 1950 by J.H. Irving and J.G. Kirkwood [6], where local counterparts of three basic equations of CM—mass conservation, balance of linear momentum, and balance of energy—are constructed, under form of Liouvillean evolutions of the shrewdly defined SM observables that the authors associate with CM mass density, linear momentum, and energy (internal and kinetic); and where counterparts of the CM notions of stress and heat flux are identified.[4] The Liouvillean equation Irving & Kirkwood use is that induced by a standard Hamiltonian system of massy particles; they suggest that, beside the phase-space averaging per unit volume intrinsic to their notion of observables, a convenient space $\times$ time averaging of the counterpart equations they derive should be performed, so as to justify the conclusion that a standard Hamiltonian system of identical particles does serve as a discrete counterpart of a conventional continuous body.[5]

[4]From the abstract of [6]: "The equations of hydrodynamics–continuity equation, equation of motion, and equation of energy transport–are derived by means of the classical statistical mechanics. Thereby, expressions are obtained for the stress tensor and heat current density in terms of molecular variables. In addition to the familiar terms occurring in the kinetic theory of gases, there are terms depending upon intermolecular force...." It is kind of amusing for a continuum mechanist to see that to the Authors, in 1950, hydrodynamics is a synonym of all of CM; and that, on taking a reversed, indeed complementary perspective of CM's, terms involving characteristic and general CM notions such as stress and heat flux are regarded as additional to "the familiar terms occurring in the kinetic theory of gases". At variance with what is done in [6] and with common practice, kinetic terms under form of additional internal sources will be included in some of the CM balances derived *à la* Irving-Kirkwood-Noll in Sect. 6.

[5]The Irving & Kirkwood procedure is robust enough to be so adapted as to show that Hamiltonian systems of disparate and/or variously structured particles can serve as discrete counterparts of nonconventional continua such as Cosserat bodies and liquid crystals (see, e.g., [20–22]).

Counterparts of CM balances, as well as of CM notions such as stress and heat flux, can be arrived at directly by procedures of space × time averaging.[6] I shall term such procedures *deterministic*, Irving & Kirkwood's *statistical*.

It seems worth repeating that all averaging procedures, whatever their type, appear to be inspired by the belief that the physical foundations of CM are to be found at smaller scales. However, the choice of one such procedure is not simply a matter of taste or mental habit. Those who are in favour of a deterministic approach may prefer not to commit themselves to the implicit presumption that a macroscopic evolution consists in a sequence of states of statistical equilibrium, with the relevance to the outcome of the chosen statistical ensemble presumed to vanish only in an appropriate thermodynamic limit. And, those who consider the Irving-Kirkwood-Noll assessment of the matters basically definitive may question the relevance to the outcome of the chosen spatial and temporal weighting functions, when an averaging approach of the type in [5] is taken.

No weights are used in the deterministic approach of [3, 4], which is inspired by the APR method of MD. Within this framework, CM fields interpreted as velocity and velocity gradient at the macroscopic current place $x$ and time $t$ are found by minimisation of a quadratic functional under form of an integral over a mesoscopic space × time cell of dimensions $L_m \times T_m$ centred at $(x, t)$; such cell, just as the computational cell in MD, is presumed to encompass all particles of a given microscopic system, whose individual positions and velocities are supposed to be known; the integrand evaluates the difference in kinetic energy of the microscopic velocity distribution and an affine approximation of it; and it is proved in [4] that the sought-for minimisers yield a macroscopic affine representation of the velocity field which, at each instant of mesoscopic time $\tau \in (t - T_m/2, t + T_m/2)$, best restitutes an estimate of the particles' total kinetic energy.

## 4 Continuum Notions with Questionable Molecular Counterparts

In this section I discuss the two main reasons why, no matter which averaging procedure one adopts, statistical or deterministic, counterparts of some basic CM notions are hard to find on starting from MM: because MM and CM differ both *in the notion of material system* and *in the nature of material interactions*.

As to the first reason, in ordinary MM a material system is nothing but a generally finite collection of massy particles, perhaps of variable number. To decide what counterpart under form of a typical CM material system such a typical MM material system should have is a deceivingly simple task, which is taken up in Sect. 4.1 here below.[7]

As to the second reason, in a molecular picture of matter the only interactions are at a distance; contacts between molecules do not occur: when molecules happen to collide, their interaction is regarded as instantaneous and modeled as an exchange of momentum and kinetic energy, not force. *Distance forces*, inertial and not, are central to Newtonian mechanics, whence MD takes the system of ordinary differential equations governing the motion of a statistically significant number of particles. Since long, in an attempt to bridge

[6] See [5], where an account of [23–25] is given and, for two very recent contributions, [3, 4]; how estimates of mass density may depend on the chosen microscopic spatial scale is vividly illustrated in [26].

[7] CM notions of material body have been considered for one or another nonstandard applicative purpose, which defy individuation of adequate molecular counterparts: for one example, consider the construction of the current appearance of the spatial regions occupied by the CM material bodies considered in [27], in connection with the theory of structures of optimal shape. The study of such nonstandard material systems is beyond the scope of this paper.

between the scale of MD simulations and the macroscopic scale of CM, various sorts of conceptual post-processing procedures have been proposed to arrive at a molecular counterpart of the stress associated to those simulations. *Stress*, a notion central to CM, is germane to *contact forces* between a body and its environment and between adjacent body parts: in Noll's words, "The soul of continuum mechanics is the analysis of contact forces." [28]. It is via the representation that contact forces are given in terms of stress, reviewed in Sect. 4.3, that the CM balance equations acquire the divergence structure leading to their formulation under form of partial differential equations to hold at interior body points, consistent with a set of boundary conditions.

## 4.1 Material Bodies and Material Points

In standard CM, a body is made of *material points*, its indivisible elementary constituents, its 'atoms' (see Appendix B). It is customarily presumed that a body can be observed 'atom-wise' against the background of a chosen space-time structure; this presumption is customarily accompanied by the convenient limitation of restricting attention to globally invertible, smooth *deformations*, that is, changes in volume and shape; *motions* are families of deformations depending smoothly on time. Accordingly, once and for all a body is identified point-wise with a bounded regularly open space region $B$ with conveniently smooth measurable boundary $\partial B$, and a positive-valued mass-density field is defined over $B$. In so doing, each material point is given its own distinguishing label: whatever the body motion, neither material points can be created or annihilated nor the distinction between interior and boundary points is ever lost; moreover, each material point has specified velocity and acceleration at each instant of time. It follows that CM is a field theory in which, at any fixed time, the evaluation of a field of interest at a chosen space point is in fact an evaluation at the material point currently occupying that space point.

This conceptual setting implies an ineliminable difference between the interpretation of standard CM balances in local form and their MM-derived counterparts. In fact, there is not a univocal way to mimic a material point on starting from a collection of molecules, let alone a material body and its boundary. The molecules of a system in motion are presumed individually localizable in a space $\times$ time box; they have individual mass and velocity. The notion of material point one obtains, whatever the averaging procedure of choice, is somewhat fictitious, being essentially that of the current mass center of the molecule system.

## 4.2 Constitutive Notions

In CM, whichever notion is constitutive in nature is expressed as a property of a body's typical material part or rather, ultimately, of a body's material points. Therefore, in view of the considerations expressed in the preceding subsection, all counterparts of CM constitutive notions constructed on starting from MM are bound to be arguable.

One of such notions is that of aggregation state. In general, the molecules of a system in motion may both exchange mutual interactions and experience collisions (consideration of external actions is irrelevant to our present developments). The two most studied instances are when mutual interactions are negligible, as is the case for rarefied gases, and when collisions are negligible, as for crystalline systems. An assessment of the relative importance of these two mechanisms, interaction and collision, might perhaps serve as a MM gauge of aggregation state. But it would be unlikely to arrive at paralleling the mathematically precise

CM notions of, say, fluidity and solidity for the so-called 'simple' materials,[8] once again because those notions specify constitutive peculiarities of certain types of material points.

### 4.3 Contact Force and Stress

For $\mathcal{P}$ a regularly open bounded part with smooth boundary $\partial\mathcal{P}$ of a space region $\mathcal{R}$ and for $\boldsymbol{v}$ the restriction to the closure of $\mathcal{P}$ of a smooth vector field over $\mathcal{R}$, consider the identity

$$\begin{aligned}\int_{\mathcal{P}} \boldsymbol{A}\cdot \operatorname{grad}\boldsymbol{v} &= \int_{\mathcal{P}} (-\operatorname{div}\boldsymbol{A})\cdot\boldsymbol{v} + \int_{\partial\mathcal{P}} \boldsymbol{A}\boldsymbol{n}\cdot\boldsymbol{v} \\ &= \int_{\mathcal{P}} \boldsymbol{d}_A\cdot\boldsymbol{v} + \int_{\partial\mathcal{P}} \boldsymbol{c}_A\cdot\boldsymbol{v}, \quad \boldsymbol{d}_A := -\operatorname{div}\boldsymbol{A},\ \boldsymbol{c}_A := \boldsymbol{A}\boldsymbol{n}, \end{aligned} \tag{1}$$

where $\boldsymbol{n}$ denotes the outer unit normal to the boundary of $\mathcal{P}$; interpret $(\boldsymbol{d}_A, \boldsymbol{c}_A)$ as the (distance, contact) pair of vector fields associated with tensor $\boldsymbol{A}$; moreover, interpret (1) as the statement that, whatever $\boldsymbol{v}$, the formal power expenditure of $\boldsymbol{A}$ over $\mathcal{P}$ equals the sum of the formal power expenditures of $\boldsymbol{d}_A$ over $\mathcal{P}$ and of $\boldsymbol{c}_A$ over $\partial\mathcal{P}$; finally, note that, for $\boldsymbol{v}$ an arbitrary constant field, (1) implies the formal force balance

$$\boldsymbol{0} = \int_{\mathcal{P}} \boldsymbol{d}_A + \int_{\partial\mathcal{P}} \boldsymbol{c}_A. \tag{2}$$

Relationship (2) has the same form as the physical force balance

$$\boldsymbol{0} = \int_{\mathcal{P}} \boldsymbol{d} + \int_{\partial\mathcal{P}} \boldsymbol{c} \tag{3}$$

for a body part $P$ currently occupying region $\mathcal{P}$, where the total ($\equiv$ inertial + noninertial) *distance force*

$$\boldsymbol{d} = \boldsymbol{d}^{in} + \boldsymbol{d}^{ni} \tag{4}$$

and the *contact force* $\boldsymbol{c}$ enter. Recall a famous result due to Cauchy: for $\mathcal{P}$ under form of a right tetrahedron that is made omothetically smaller and smaller, whose orthonormal faces have outer normals $\boldsymbol{n}_i$; $(i = 1, 2, 3)$ and are acted upon by contact forces $\boldsymbol{c}_i$ and whose oblique face has outer normal $\boldsymbol{n}$, (3) implies that there is a second-order tensor $\boldsymbol{C}$, the *Cauchy stress*, such that

$$\boldsymbol{C} = \sum_{1}^{3} \boldsymbol{c}_i \otimes \boldsymbol{n}_i, \quad \boldsymbol{C}\boldsymbol{n} = \boldsymbol{c}, \tag{5}$$

where $\boldsymbol{c}$ is the *Cauchy contact force* acting upon the oblique face. The conceptual core of this doubly constructive result is that the stress and contact-force fields carry the same information content; moreover, it follows from (3) and (5) that

$$-\operatorname{div}\boldsymbol{C} = \boldsymbol{d}. \tag{6}$$

It is important to realize that *the CM notions of Cauchy contact force and Cauchy stress are independent of constitutive specifications*, except of course those implicit in the balance statements (3) and (4). Not so in MD, where particle interactions are constitutively prescribed *ab initio*, as inevitable within an approach of Hamiltonian type.

---

[8] See Sect. 28 of [19] for the standard algebraic notions in terms of material symmetries and [29] for a discussion of internal-power based notions.

### 4.4 Power Expenditures

On taking $\boldsymbol{A} = \boldsymbol{C}$ in (1), we see that the total power expenditure in terms of stress is expressible as the sum of a bulk power expenditure in terms of distance force and a boundary power expenditure in terms of contact force:

$$\int_{\mathcal{P}} \boldsymbol{C} \cdot \operatorname{grad} \boldsymbol{v} = \int_{\mathcal{P}} (-\operatorname{div} \boldsymbol{C}) \cdot \boldsymbol{v} + \int_{\partial\mathcal{P}} \boldsymbol{C}\boldsymbol{n} \cdot \boldsymbol{v} = \int_{\mathcal{P}} \boldsymbol{d} \cdot \boldsymbol{v} + \int_{\partial\mathcal{P}} \boldsymbol{c}. \tag{7}$$

Now, for a system of $N$ molecules, the instantaneous power expenditure is

$$\sum_{J=1}^{N} \boldsymbol{f}_J \cdot \boldsymbol{v}_J, \quad \boldsymbol{f}_J = \boldsymbol{f}_J^{\text{int}} + \boldsymbol{f}_J^{\text{ext}}, \tag{8}$$

where the total force at a distance $\boldsymbol{f}_J$ experienced by the $J$-th molecule is split into the *int*eraction force $\boldsymbol{f}_J^{\text{int}}$ and the *ext*ernal force $\boldsymbol{f}_J^{\text{ext}}$. I interpret $\boldsymbol{f}_J^{\text{int}}$ as short-range molecule interactions and ignore long-range ones; moreover, on indulging to a common practice with rare exceptions (for one, see [3, 4]), I omit range specifications. For simplicity, the only external forces here considered are inertial forces:

$$\boldsymbol{f}_J^{\text{ext}} = -m_J \dot{\boldsymbol{v}}_J. \tag{9}$$

Ideally, were $N$ an integer multiple of Avogadro's number and were the system regarded as a discrete mechanical analogue of a continuous body $P$, one should associate to the given molecule system such CM stress and velocity fields as to guarantee a perfect matching of power expenditures:

$$\int_{\mathcal{P}} \boldsymbol{C} \cdot \operatorname{grad} \boldsymbol{v} = \sum_{J=1}^{N} \boldsymbol{f}_J \cdot \boldsymbol{v}_J. \tag{10}$$

However, as we shall see when we review it in Sect. 6, the Irving-Kirkwood-Noll procedure is not based on equalising MM and CM power expenditures. Consequently, while it leads to an MM counterpart of $\boldsymbol{C}$ whose divergence does provide a fair evaluation of $\operatorname{div} \boldsymbol{C}$, it fails to account properly for the power expenditure at the boundary, appearing in the right side of equation (7). The procedure proposed by DiCarlo in [3, 4] and summarised in the closure of Sect. 3 appears preferable, although it too does not assign a pivotal role to power expenditures.

## 5 Statistical Molecular Balances. Power Expenditures

The general Liouvillean format of SM balance laws is the subject of Sect. 5.1, an indispensable preliminary to the outline, in Sect. 6, of the Irving-Kirkwood-Noll procedure. The remarks about molecular and continuum power expenditures in Sect. 5.2 are additional to those in Sect. 4.4.

### 5.1 The Format of SM Balance Laws

For $(q, p)$ a pair of $n$-strings of Hamiltonian coordinates, let $z = (q, p)$ denote a point in the state space $\mathcal{Z}$; moreover, let

$$\widetilde{\rho} = \widetilde{\rho}(z, t), \quad (z, t) \in \mathcal{Z} \times \mathbb{R},$$

be a chosen probability measure over $\mathcal{Z}$. A *micro-observable* is a function $z \mapsto o = \widehat{o}(z)$ defined over the whole state space; the corresponding *macro-observable* is defined to be the result of a phase-space average:

$$\langle o\rangle(t) := \int_{\mathcal{Z}} \widehat{o}(z)\widetilde{\rho}(z,t)dz, \tag{11}$$

and is referred to as the *expected value* of $\widehat{o}$ at time $t$ with respect to the chosen probability measure. Under reasonable assumptions, it can be proved that, for a system whose Hamiltonian is $H$,[9]

$$\partial_t\langle o\rangle = \big\langle \partial_q\widehat{o}[\partial_p H] - \partial_p\widehat{o}[\partial_q H]\big\rangle. \tag{12}$$

Whenever

$$H = \widehat{H}(q,p) = \widehat{K}(p) + \widehat{U}(q),$$

the general Liouvillean balance (12) becomes

$$\partial_t\langle o\rangle = \big\langle \partial_q\widehat{o}[\partial_p K] - \partial_p\widehat{o}[\partial_q U]\big\rangle. \tag{13}$$

Moreover, for the systems of molecules ($\equiv$ mass points) on which we focus,

$$q = (\boldsymbol{r}_1,\ldots,\boldsymbol{r}_N), \qquad p = (\boldsymbol{p}_1 = m_1\dot{\boldsymbol{r}}_1,\ldots,\boldsymbol{p}_N = m_N\dot{\boldsymbol{r}}_N), \tag{14}$$

where $\boldsymbol{r}_J = x_J - x_o$ is the current position vector of the $J$-th molecule with respect to a fixed origin $x_o$; the total kinetic energy reads

$$\widehat{K}(p) = \sum_J K_J, \quad K_J := \frac{|\boldsymbol{p}_J|^2}{2m_J}; \tag{15}$$

molecule interactions are ruled by a potential energy of the form

$$\widehat{U}(q) = \sum_J U_J, \qquad U_J := \frac{1}{2}\sum_K V(r_{JK}), \quad r_{JK} := |\boldsymbol{r}_J - \boldsymbol{r}_K|,\ V(0) = 0. \tag{16}$$

Consequently, the evolution of an SM macro-observable is ruled by the Liouvillean equation

$$\boxed{\partial_t\langle o\rangle = \sum_J \big\langle \partial_{\boldsymbol{r}_J}\widehat{o}[\boldsymbol{v}_j] + \partial_{\boldsymbol{p}_J}\widehat{o}\big[\boldsymbol{f}_J^{\text{int}}\big]\big\rangle}, \tag{17}$$

where we let

$$\boldsymbol{f}_J^{\text{int}} := \sum_K \boldsymbol{f}_{JK}, \quad \boldsymbol{f}_{JK} := -\partial_{\boldsymbol{r}_J}V(r_{JK}) = -V'(r_{JK})\frac{\boldsymbol{r}_J - \boldsymbol{r}_K}{|\boldsymbol{r}_J - \boldsymbol{r}_K|}, \tag{18}$$

and we identify $\boldsymbol{f}_{JK}$ as the force exerted by particle $K$ on particle $J$, $\boldsymbol{f}_J^{\text{int}}$ as the total force exerted on the $J$-th molecule by all other molecules, in accordance with the notation introduced in (8). Note that $\boldsymbol{f}_{JK} = -\boldsymbol{f}_{KJ}$ and that $\boldsymbol{f}_{JK}$ is attractive (repulsive) for $V'(r_{JK}) > 0$ (for $V'(r_{JK}) < 0$).

---

[9]Here $\partial_t$, $\partial_q$, and $\partial_p$, denote differentiation with respect to the suffix variables; moreover,

$$\partial_q\widehat{o}[\partial_p H] = \sum_J (\partial_{q_J}\widehat{o})(\partial_{p_J}H) \quad \text{et sim. for } \partial_p\widehat{o}[\partial_q H].$$

*Remark 1* For simplicity, in this paper I ignore both external contributions to the system's potential energy (consistent with what I did in Sect. 4.4) and internal contributions that are not reducible to pair-wise form (as to such matters, see [30], [31] and the comments in Sect. 2 of [3]). I also ignore certain pair-wise interactions considered instead in Dissipative Particle Dynamics, such as *dissipative interactions* of the form

$$\boldsymbol{f}^d_{JK} = -w^d(r_{JK})\theta^d_{JK}\frac{\boldsymbol{r}_J - \boldsymbol{r}_K}{|\boldsymbol{r}_J - \boldsymbol{r}_K|}, \qquad \theta^d_{JK} = \frac{\boldsymbol{r}_J - \boldsymbol{r}_K}{|\boldsymbol{r}_J - \boldsymbol{r}_K|}\cdot(\boldsymbol{v}_J - \boldsymbol{v}_K),$$

and *random interactions* of the form

$$\boldsymbol{f}^r_{JK} = -w^r(r_{JK})\theta^r_{JK}\frac{\boldsymbol{r}_J - \boldsymbol{r}_K}{|\boldsymbol{r}_J - \boldsymbol{r}_K|},$$

where $\theta^r_{JK}$ is a randomly fluctuating variable with Gaussian statistics [32], and where both weighting functions $w^d$ and $w^r$, not differently from the potential function $V$, may be taken to vanish beyond a specified cut-off distance.

### 5.2 More About Molecular and Continuum Power Expenditures

It follows from (14), (15), and (9), that

$$\sum_J \dot{K}_J = -\sum_J \boldsymbol{f}^{\text{ext}}_J\cdot\boldsymbol{v}_J; \tag{19}$$

from (16) and (18), that

$$\sum_J \dot{U}_j = -\sum_J \boldsymbol{f}^{\text{int}}_J\cdot\boldsymbol{v}_J. \tag{20}$$

Hence, for the total instantaneous power expenditure (8) we have that

$$\sum_J \boldsymbol{f}_J\cdot\boldsymbol{v}_J = -\Big(\sum_J \dot{K}_J + \sum_J \dot{U}_J\Big). \tag{21}$$

With a view toward satisfying the power-matching condition (10), an instantaneous, necessarily nondissipative CM counterpart of a given Hamiltonian molecule system would be an imagined continuous body $P$ currently occupying with mass density $\rho$ a region $\mathcal{P}$ encompassing all molecules of the system, constitutively described by a stored-energy mapping $w = \widehat{w}(\boldsymbol{F})$ such that, mass conservation being granted,

$$\int_{\mathcal{P}} \boldsymbol{C}\cdot\operatorname{grad}\boldsymbol{v} = \dot{K} + \dot{W},$$

with

$$K = \frac{1}{2}\int_{\mathcal{P}} \rho\boldsymbol{v}\cdot\boldsymbol{v} \tag{22}$$

and with

$$W = \int_{\mathcal{P}} (\det\boldsymbol{F})^{-1}\widehat{w}(\boldsymbol{F}), \qquad \boldsymbol{C} = (\det\boldsymbol{F})^{-1}\partial_{\boldsymbol{F}}\widehat{w}(\boldsymbol{F})\boldsymbol{F}^T. \tag{23}$$

In (23), $\boldsymbol{F}$ denotes the gradient of a deformation of $P$, which needs not be determined at this stage, whose time rate is the field $\boldsymbol{v}$ in (22); in fact, $\boldsymbol{F}$ is to be associated with the molecular motion via grad $\boldsymbol{v}$, by solving the evolution equation

$$\dot{\boldsymbol{F}} = (\operatorname{grad} \boldsymbol{v})\boldsymbol{F}. \tag{24}$$

We see that the crux of micro → macro bridging is to derive from the molecular motion appropriate evaluations of the CM velocity field $\boldsymbol{v}$ and its space and time derivatives grad $\boldsymbol{v}$ and $\dot{\boldsymbol{v}}$: while grad $\boldsymbol{v}$ decides the internal power expenditure once a choice of $\widehat{w}$ is made, the CM inertial force is proportional to $\dot{\boldsymbol{v}}$. A satisfactory power-expenditure based bridging would be achieved in case the result of this process of choice and evaluation would be such that

$$\dot{W} \simeq -\sum_J \dot{U}_J \quad \text{and} \quad \dot{K} \simeq -\sum_J \dot{K}_J.$$

## 6 A Reasoned Résumé of the Irving-Kirkwood-Noll Procedure

The material in this section updates the account given in [33] of Irving-Kirkwood paper [6] and of Noll papers [7] and [8]; the results presented in Sects. 6.3 and 6.4 are new.

### 6.1 IK Micro- and Macro-observables

The key idea of the Irving-Kirkwood paper is to choose the following common form for all micro-observables:

$$\widehat{o}(z, x) = \sum_J \widehat{o}_J(z)\, \delta(\boldsymbol{r}_J - \boldsymbol{r}), \qquad \boldsymbol{r} = x - x_o, \tag{25}$$

(here $\delta(\cdot)$ denotes the Dirac measure). According to (11), all associated macro-observables have the common form

$$\langle o\rangle(x, t) = \left\langle \sum_J \widehat{o}_J(z)\delta(\boldsymbol{r}_J - \boldsymbol{r}) \right\rangle, \tag{26}$$

or rather, equivalently,

$$\langle o\rangle(x, t) = \sum_J \langle o_J\rangle(x, t), \qquad \langle o_J\rangle(x, t) = \int_{\mathcal{Z}} \widehat{o}_J(z)\, \delta(\boldsymbol{r}_J - \boldsymbol{r})\, \widetilde{\rho}(z, t)\, dz \tag{27}$$

(as noted in [6],

$$\big\langle \delta(\boldsymbol{r}_J - \boldsymbol{r})\big\rangle(x, t) = \int_{\mathcal{Z}} \delta(\boldsymbol{r}_J - \boldsymbol{r})\, \widetilde{\rho}(z, t)\, dz$$

is the expectation of finding the $J$-th molecule at $x$ at time $t$).

### 6.2 CM Counterparts of IK Macro-observables

Irving & Kirkwood introduce the CM counterparts of four SM macroscopic fields of type (26), namely,

– for $\widehat{o}_J(z) = m_J$, the *mass density* field

$$\rho(x,t) \equiv \Big\langle \sum_J m_J \delta(\boldsymbol{r}_J - \boldsymbol{r}) \Big\rangle; \tag{28}$$

– for $\widehat{o}_J(z) = \boldsymbol{p}_J$, the *momentum* field

$$\boldsymbol{p}(x,t) \equiv \sum_J \langle \boldsymbol{p}_J \delta(\boldsymbol{r}_J - \boldsymbol{r}) \rangle; \tag{29}$$

– for $\widehat{o}_J(z) = K_J$, the *kinetic-energy* field

$$(\rho\kappa)(x,t) \equiv \sum_J \langle K_J \delta(\boldsymbol{r}_J - \boldsymbol{r}) \rangle. \tag{30}$$

– for $\widehat{o}_J(z) = U_J$, the *internal-energy* field

$$(\rho\epsilon)(x,t) \equiv \sum_J \langle U_J \delta(\boldsymbol{r}_J - \boldsymbol{r}) \rangle. \tag{31}$$

The CM *motion velocity* field

$$\boldsymbol{v}(x,t)\big(= (\rho^{-1}\boldsymbol{p})(x,t)\big) \equiv \Big(\Big\langle \sum_J m_J \delta(\boldsymbol{r}_J - \boldsymbol{r}) \Big\rangle\Big)^{-1} \sum_J \langle \boldsymbol{p}_J \delta(\boldsymbol{r}_J - \boldsymbol{r}) \rangle \tag{32}$$

is a derived notion, obtained by combining (28) and (29).

*Remark 2* Recall that in standard CM the mass densities of kinetic energy and inertia force read, respectively,

$$\kappa = \frac{1}{2}\boldsymbol{v}\cdot\boldsymbol{v} \quad \text{and} \quad \boldsymbol{d}^{in} = -\dot{\boldsymbol{v}}, \tag{33}$$

and satisfy the consistency relation[10]

$$\dot{\kappa} + \boldsymbol{d}^{in}\cdot\boldsymbol{v} = 0. \tag{34}$$

A use of (28), (30), and (32) yields the following molecular version of $(33)_1$:

$$\Big\langle \sum_J m_J \delta(\boldsymbol{r}_J - \boldsymbol{r}) \Big\rangle \sum_J \langle K_J \delta(\boldsymbol{r}_K - \boldsymbol{r}) \rangle = \frac{1}{2} \sum_J \langle \boldsymbol{p}_J \delta(\boldsymbol{r}_J - \boldsymbol{r}) \rangle \cdot \sum_K \langle \boldsymbol{p}_K \delta(\boldsymbol{r}_K - \boldsymbol{r}) \rangle. \tag{35}$$

*Remark 3* Consider $\widehat{o}(z,x) = \sum_J \widehat{m}_J \boldsymbol{r}_J \delta(\boldsymbol{r}_j - \boldsymbol{r})$, the collective IK micro-observable corresponding to the individual micro-observables $\widehat{o}_J(z) = m_J \boldsymbol{r}_J$ $(J = 1, \ldots, N)$; for the asso-

[10]That for whatever chosen class of massy continuous bodies the constitutive choices for the two manifestations of inertial interactions, kinetic energy and inertia force, should be consistent in the following sense:

$$\frac{d}{dt}\,(\text{kinetic energy}) + (\text{inertial power}) = 0$$

was pointed out in [34]; the discovery power intrinsic to this requirement has been exemplified in [35–38].

ciated macro-observable of type (26), introduce both the CM *first-order moment field*

$$m(x,t) \equiv \left\langle \sum_J m_J r_J \delta(r_J - r) \right\rangle;$$

and the CM *mass-center field*

$$c_M(x,t) \equiv \left\langle \sum_J m_J \delta(r_J - r) \right\rangle^{-1} \left\langle \sum_J m_J r_J \delta(r_J - r) \right\rangle;$$

the latter, if one so wishes, can be thought of as delivering the position vector of the virtual *IK material point* associable with the molecule system at hand.

### 6.3 IK-SM Balances of Mass and Momentum

All CM balances I hereafter consider do not include non-inertial external sources, because they are irrelevant to my developments to come. However, at variance with [6], I include specific additional sources in the momentum balance as well as in the balances of kinetic, internal, and total, energy, with a view to facilitate the identification of the microscopic counterparts of the macroscopic fields occurring in those balances.

#### *6.3.1 Mass*

The target CM conservation equation is

$$\partial_t \rho + \operatorname{div}(\rho v) = 0. \tag{36}$$

For $\widehat{o}(z,x) = \sum_K m_K\, \delta(r_K - r)$ (cf. (28)),

$$\partial_{p_J} \widehat{o} = \mathbf{0} \quad \text{and} \quad \partial_{r_J} \widehat{o} = -m_J \partial_r \delta(r_J - r),$$

since $\partial_x f(x-y) = -\partial_y f(x-y)$. Consequently, the general evolution equation (17) takes the IK form boxed here below:

$$\partial_t \rho \equiv \boxed{\partial_t \left\langle \sum_J m_J\, \delta(r_J - r) \right\rangle = -\partial_r \left\langle \sum_J p_J \delta(r_J - r) \right\rangle} \equiv -\operatorname{div}(\rho v) \tag{37}$$

(cf. [6], p. 821, first equation in the right column). In (37), the CM counterpart $\partial_t \rho$ of the left SM term is identified with the use of (28), that of the right term, which accounts for the convective transport mechanism responsible for mass balance, with the use of (29) and (32). Had we included an internal source in (36), we would have found no molecular justification of IK type for such an inclusion.

#### *6.3.2 Momentum*

I write the target CM balance equation as follows:

$$\partial_t(\rho v) + \operatorname{div}(\rho v \otimes v) = \operatorname{div} C + s_p^{\text{add}}, \tag{38}$$

where, as it is customary, the convective and diffusive currents $\rho\boldsymbol{v}\otimes\boldsymbol{v}$ and $-\boldsymbol{C}$ appear, which determine the respective momentum inflows at the boundary of a body part; and where, as it is not customary, I have provisionally included an additional internal momentum source denoted by $\boldsymbol{s}_p^{\mathrm{add}}$.

For $\widehat{o}(z,x)=\sum_K \boldsymbol{p}_K\,\delta(\boldsymbol{r}_K-\boldsymbol{r})$ (cf. (29)),

$$\partial_{\boldsymbol{r}_J}\widehat{o}[\boldsymbol{v}_J]=-\partial_{\boldsymbol{r}}\big(\boldsymbol{p}_J\delta(\boldsymbol{r}_J-\boldsymbol{r})\big)[\boldsymbol{v}_J]\quad\text{and}\quad\partial_{\boldsymbol{p}_J}\widehat{o}[\boldsymbol{f}_J]=\boldsymbol{f}_J\delta(\boldsymbol{r}_J-\boldsymbol{r}).$$

Accordingly, the Liouvillean evolution equation (17) takes the following IK form:

$$\boxed{\partial_t\Big\langle\sum_J \boldsymbol{p}_J\,\delta(\boldsymbol{r}_J-\boldsymbol{r})\Big\rangle=-\sum_J\big\langle\partial_{\boldsymbol{r}}\big(\boldsymbol{p}_J\delta(\boldsymbol{r}_J-\boldsymbol{r})\big)[\boldsymbol{v}_J]+\boldsymbol{f}_J\delta(\boldsymbol{r}_J-\boldsymbol{r})\big\rangle};\tag{39}$$

it remains for us to show, after Irving & Kirkwood [6] and Noll [7, 8], what are the CM counterparts of its terms.

For the left side, this is an easy job: in view of definition (29),

$$\partial_t(\rho\boldsymbol{v})\equiv\partial_t\Big\langle\sum_J \boldsymbol{p}_J\,\delta(\boldsymbol{r}_J-\boldsymbol{r})\Big\rangle.$$

The first term on the right side requires more work. We observe that

$$\partial_{\boldsymbol{r}}\big(\boldsymbol{p}_J\delta(\boldsymbol{r}_J-\boldsymbol{r})\big)[\boldsymbol{v}_J]=\partial_{\boldsymbol{r}}\big(m_J\delta(\boldsymbol{r}_J-\boldsymbol{r})\boldsymbol{v}_J\otimes\boldsymbol{v}_J\big)$$

and that

$$(\boldsymbol{v}_J-\boldsymbol{v})\otimes(\boldsymbol{v}_J-\boldsymbol{v})=\boldsymbol{v}_J\otimes\boldsymbol{v}_J+\boldsymbol{v}\otimes\boldsymbol{v}-\boldsymbol{v}\otimes\boldsymbol{v}_J-\boldsymbol{v}_J\otimes\boldsymbol{v},$$

where $(\boldsymbol{v}_J-\boldsymbol{v})$ is the *velocity fluctuation* of the $J$-th molecule; hence,

$$\Big\langle\sum_J m_J\delta(\boldsymbol{r}_J-\boldsymbol{r})\boldsymbol{v}_J\otimes\boldsymbol{v}_J\Big\rangle=\Big\langle\sum_J m_J\delta(\boldsymbol{r}_J-\boldsymbol{r})(\boldsymbol{v}_J-\boldsymbol{v})\otimes(\boldsymbol{v}_J-\boldsymbol{v})\Big\rangle+\rho\,\boldsymbol{v}\otimes\boldsymbol{v},$$

where the last term is the momentum's convective current. We are left with the task of showing in what sense

$$\operatorname{div}\boldsymbol{C}+\boldsymbol{s}_p^{\mathrm{add}}\equiv-\partial_{\boldsymbol{r}}\Big\langle\sum_J m_J\delta(\boldsymbol{r}_J-\boldsymbol{r})(\boldsymbol{v}_J-\boldsymbol{v})\otimes(\boldsymbol{v}_J-\boldsymbol{v})\Big\rangle+\Big\langle\sum_J \boldsymbol{f}_J\delta(\boldsymbol{r}_J-\boldsymbol{r})\Big\rangle.\tag{40}$$

The identification of continuum counterparts for the terms on the right side of (40) is not univocal. My present proposal differs from the one given by Irving & Kirkwood in [6] and accepted in [7, 8] and [33].

Firstly, I set

$$\boldsymbol{K}=-\Big\langle\sum_J m_J\,\delta(\boldsymbol{r}_J-\boldsymbol{r})(\boldsymbol{v}_J-\boldsymbol{v})\otimes(\boldsymbol{v}_J-\boldsymbol{v})\Big\rangle.\tag{41}$$

Note that $\boldsymbol{K}$ is a Galilean-invariant symmetric tensor field. In [6], this tensor is regarded as a kinetic part of the stress. At variance with this, I propose to regard $\boldsymbol{K}$ as responsible for an internal momentum source depending solely on velocity fluctuations:

$$-\partial_{\boldsymbol{r}}\Big\langle\sum_J m_J\delta(\boldsymbol{r}_J-\boldsymbol{r})(\boldsymbol{v}_J-\boldsymbol{v})\otimes(\boldsymbol{v}_J-\boldsymbol{v})\Big\rangle\equiv\boldsymbol{s}_p^{\mathrm{add}}.\tag{42}$$

Such an internal source of kinetic origin would be all-important in the case of an ideal gas (see the comments in the introductory section of [6]). I expect it to be at times negligible (see the discussion in Sect. 8.1.3 of [31]) in the case of liquid-like or solid-like matter, especially at low temperatures.

Secondly, on relying on Noll's construction in [7, 8] of a so-called reaching stress $\boldsymbol{R}$ in terms of given pair-wise molecular interactions of type (18):

$$\left.\begin{aligned} \boldsymbol{R} &= \int_{\mathcal{V}} \underbrace{-\Big(\frac{1}{2}\int_0^1 \gamma\big(x-(1-\alpha)\boldsymbol{v}, x+\alpha\boldsymbol{v}\big)d\alpha\Big)}_{\cong \boldsymbol{c}(\boldsymbol{n}_i)} \boldsymbol{v}\otimes\boldsymbol{v}\, d\boldsymbol{v}, \\ \gamma(x,y) &= -|x-y|^{-1}\sum_J\sum_K \big\langle V'(r_{JK})\,\delta(x_J-x)\delta(x_K-y)\big\rangle \end{aligned}\right\}, \tag{43}$$

I set

$$(\boldsymbol{f} \equiv)\Big\langle \sum_J \boldsymbol{f}_J \delta(\boldsymbol{r}_J - \boldsymbol{r})\Big\rangle \equiv \operatorname{div} \boldsymbol{R}, \tag{44}$$

I stipulate that

$$\boldsymbol{R} \equiv \boldsymbol{C}, \tag{45}$$

and, finally, I interpret $-\boldsymbol{R}$ as the molecular counterpart of the diffusive current, constitutively determined as specified by (16) and (18).

My present interpretation is strongly suggested by the—to me—striking similarity in structure between (43) and the standard representation $(5)_1$ of the Cauchy stress. A side merit of this interpretation is that it identifies the stress *directly* and not, as is often pointed out after [6], through its divergence, the derived construct appearing in (38). The molecular interpretation for the Cauchy stress given by Irving & Kirkwood, who did not include any source term in the CM balance (38), is $\boldsymbol{C} \equiv (\boldsymbol{R} + \boldsymbol{K})$. Needless to say, whichever interpretation is chosen for stress it affects the determination of the SM balances of the IKN macro-observables regarded as molecular counterparts of the CM densities of energy, total, internal, and kinetic. Therefore, my interpretations of energy terms to come are bound to be different from those in [6].

## 6.4 IK-SM Balances of Energy

### *6.4.1 Prelims*

A customary point-wise expression for the CM balance of total energy is

$$\partial_t(\rho\tau) + \operatorname{div}(\rho\tau\boldsymbol{v} + \boldsymbol{q}) = \operatorname{div}\big(\boldsymbol{C}^T\boldsymbol{v}\big), \quad \tau = \epsilon + \kappa, \tag{46}$$

where $\tau$ and $\epsilon$ denote total and internal energy per unit mass (the kinetic-energy density $\kappa$ has been introduced in (33)). Likewise, customary point-wise balances of internal energy and of kinetic energy read

$$\partial_t(\rho\epsilon) + \operatorname{div}(\rho\epsilon\boldsymbol{v} + \boldsymbol{q}) = \boldsymbol{C}\cdot \operatorname{grad}\boldsymbol{v}, \tag{47}$$

and

$$\partial_t(\rho\kappa) + \operatorname{div}(\rho\kappa\boldsymbol{v}) = \boldsymbol{v}\cdot\operatorname{div}\boldsymbol{C}; \tag{48}$$

in view of the identity

$$\operatorname{div}\big(\boldsymbol{C}^T\boldsymbol{v}\big)=\boldsymbol{C}\cdot\operatorname{grad}\boldsymbol{v}+\boldsymbol{v}\cdot\operatorname{div}\boldsymbol{C},$$

member-wise summation of (47) and (48) restitutes (46).

The convective contributions to the boundary inflow of total, internal, and kinetic, energy are, respectively, $\rho\tau\boldsymbol{v}$, $\rho\epsilon\boldsymbol{v}$, and $\rho\kappa\boldsymbol{v}$; we see from (47) that the thermal flux $\boldsymbol{q}$ is indeed a diffusive current of internal energy, and from (48) that there is no diffusive contribution to the balance of kinetic energy. Interestingly, a glance at (46) reveals that $\operatorname{div}(\boldsymbol{C}^T\boldsymbol{v})$, an internal source term of mixed nature, constitutive and kinetic, is needed to achieve balance of total energy. This source splits into a *source of internal energy*, $\boldsymbol{C}\cdot\operatorname{grad}\boldsymbol{v}$, equal to the total power expenditure per unit current volume in terms of Cauchy stress, plus a *source of kinetic energy*, $\boldsymbol{v}\cdot\operatorname{div}\boldsymbol{C}$, which equals the negative of the power expended by the inertial distance force:[11]

$$\boldsymbol{v}\cdot\operatorname{div}\boldsymbol{C}=-\boldsymbol{d}^{in}\cdot\boldsymbol{v}. \tag{49}$$

As anticipated, in preparation for the identifications the terms occurring in their SM counterparts the target energy balances are hereafter restated in forms which provisionally include additional internal source terms; two out of three balances suffice, and I choose

– (*kinetic-energy balance*)

$$\partial_t(\rho\kappa)+\operatorname{div}(\rho\kappa\boldsymbol{v})=\boldsymbol{v}\cdot\operatorname{div}\boldsymbol{C}+s_\kappa^{\rm add}, \tag{50}$$

– (*internal-energy balance*)

$$\partial_t(\rho\epsilon)+\operatorname{div}(\rho\epsilon\boldsymbol{v}+\boldsymbol{q})=\boldsymbol{C}\cdot\operatorname{grad}\boldsymbol{v}+s_\epsilon^{\rm add}. \tag{51}$$

Under the condition that

$$s_\tau^{\rm add}=s_\epsilon^{\rm add}+s_\kappa^{\rm add}, \tag{52}$$

the identifications pertaining to the total-energy balance

$$\partial_t(\rho\tau)+\operatorname{div}(\rho\tau\boldsymbol{v}+\boldsymbol{q})=\operatorname{div}\big(\boldsymbol{C}^T\boldsymbol{v}\big)+s_\tau^{\rm add} \tag{53}$$

will be dictated by those made apropos of kinetic and internal energy balances.

### 6.4.2 Kinetic-Energy Balance

For $\widehat{o}(z,x)=\sum_K K_K\,\delta(\boldsymbol{r}_K-\boldsymbol{r})$ (cf. (30) and $(15)_2$),

$$\partial_{\boldsymbol{r}_J}\widehat{o}=-\partial_{\boldsymbol{r}}\big(K_J\delta(\boldsymbol{r}_j-\boldsymbol{r})\big)\quad\text{and}\quad\partial_{\boldsymbol{p}_J}\widehat{o}=\delta(\boldsymbol{r}_j-\boldsymbol{r})\boldsymbol{v}_J.$$

[11] In fact, provided that no source term is included in the momentum balance, as is customarily done in CM, and provided that mass is conserved, (38) can be written as

$$\big(\partial_t(\rho\boldsymbol{v})+\operatorname{div}(\rho\boldsymbol{v}\otimes\boldsymbol{v})\big)=\rho\dot{\boldsymbol{v}}=\operatorname{div}\boldsymbol{C}.$$

It is easy to see that, in the present circumstances, the balance (48) is equivalent to condition (34) of inertial consistency.

The related Liouvillean evolution equation (17) is

$$\boxed{\partial_t\Big\langle\sum_J K_J\delta(\boldsymbol{r}_J-\boldsymbol{r})\Big\rangle+\partial_r\Big\langle\sum_J K_J\delta(\boldsymbol{r}_j-\boldsymbol{r})\boldsymbol{v}_J\Big\rangle=\Big\langle\sum_J\delta(\boldsymbol{r}_j-\boldsymbol{r})\boldsymbol{f}_J\cdot\boldsymbol{v}_J\Big\rangle} \tag{54}$$

(cf. equation (6.4) in [6]). With a view to show under what assumptions equation (54) qualifies as a counterpart of the CM balance (50), we begin by introducing velocity fluctuations, so that

$$\Big\langle\sum_J K_J\delta(\boldsymbol{r}_j-\boldsymbol{r})\boldsymbol{v}_J\Big\rangle=\Big\langle\Big(\sum_J K_J\delta(\boldsymbol{r}_j-\boldsymbol{r})\Big)\Big\rangle\boldsymbol{v}+\Big\langle\sum_J K_J\delta(\boldsymbol{r}_j-\boldsymbol{r})(\boldsymbol{v}_J-\boldsymbol{v})\Big\rangle$$

and

$$\Big\langle\sum_J\delta(\boldsymbol{r}_j-\boldsymbol{r})\boldsymbol{f}_J\cdot\boldsymbol{v}_J\Big\rangle=\Big\langle\sum_J\delta(\boldsymbol{r}_j-\boldsymbol{r})\boldsymbol{f}_J\cdot(\boldsymbol{v}_J-\boldsymbol{v})\Big\rangle+\Big\langle\sum_J\delta(\boldsymbol{r}_j-\boldsymbol{r})\boldsymbol{f}_J\Big\rangle\cdot\boldsymbol{v}.$$

Next, we juxtapose (50) and (54) piece by piece on taking into account the last two relations here above:

$$\begin{aligned}
&\partial_t(\rho\kappa)+\mathrm{div}(\rho\kappa\boldsymbol{v})\\
&\quad\equiv\partial_t\Big\langle\sum_J K_J\delta(\boldsymbol{r}_J-\boldsymbol{r})\Big\rangle+\partial_r\Big(\Big\langle\sum_J K_J\delta(\boldsymbol{r}_j-\boldsymbol{r})\Big\rangle\boldsymbol{v}\Big)=-\partial_r\Big\langle\sum_J K_J\delta(\boldsymbol{r}_j-\boldsymbol{r})(\boldsymbol{v}_j-\boldsymbol{v})\Big\rangle\\
&\qquad+\Big\langle\sum_J\delta(\boldsymbol{r}_j-\boldsymbol{r})\boldsymbol{f}_J\cdot(\boldsymbol{v}_j-\boldsymbol{v})\Big\rangle+\Big\langle\sum_J\delta(\boldsymbol{r}_j-\boldsymbol{r})\boldsymbol{f}_J\Big\rangle\cdot\boldsymbol{v}\\
&\quad\equiv\boldsymbol{v}\cdot\mathrm{div}\,\boldsymbol{C}+s_\kappa^{\mathrm{add}}.
\end{aligned}$$

On recalling that, in view of (44) and (45),

$$\Big\langle\sum_J\boldsymbol{f}_J\delta(\boldsymbol{r}_J-\boldsymbol{r})\Big\rangle\cdot\boldsymbol{v}\equiv\boldsymbol{v}\cdot\mathrm{div}\,\boldsymbol{C}, \tag{55}$$

we are now in a position to conclude that

$$-\partial_r\Big\langle\sum_J K_J\delta(\boldsymbol{r}_j-\boldsymbol{r})(\boldsymbol{v}_j-\boldsymbol{v})\Big\rangle+\Big\langle\sum_J\delta(\boldsymbol{r}_j-\boldsymbol{r})\boldsymbol{f}_J\cdot(\boldsymbol{v}_j-\boldsymbol{v})\Big\rangle\equiv s_\kappa^{\mathrm{add}}. \tag{56}$$

Interestingly, under the reasonable assumption that the force field $\boldsymbol{f}$ is equipowerful to the force system $\{\boldsymbol{f}_J\,|\,J=1,2,\ldots,N\}$:

$$\Big\langle\sum_J\delta(\boldsymbol{r}_j-\boldsymbol{r})\boldsymbol{f}_J\cdot\boldsymbol{v}_j\Big\rangle=\boldsymbol{f}\cdot\boldsymbol{v}, \tag{57}$$

identification (56) reduces to

$$-\partial_r\Big\langle\sum_J K_J\delta(\boldsymbol{r}_j-\boldsymbol{r})(\boldsymbol{v}_j-\boldsymbol{v})\Big\rangle\equiv s_\kappa^{\mathrm{add}}. \tag{58}$$

### 6.4.3 Internal-Energy Balance

For $\widehat{o}(z,x)=\sum_J\langle U_J\delta(\boldsymbol{r}_J-\boldsymbol{r})\rangle$ (cf. (31) and $(16)_2$), $\partial_{\boldsymbol{p}_J}\widehat{o}=\boldsymbol{0}$; the relative specialization of the general Liouvillean evolution equation (17) is

$$\partial_t\langle o\rangle=\Big\langle\sum_J \boldsymbol{v}_J\cdot\partial_{\boldsymbol{r}_J}\widehat{o}\Big\rangle,\quad \langle o\rangle=\Big\langle\sum_J U_J\delta(\boldsymbol{r}_J-\boldsymbol{r})\Big\rangle. \tag{59}$$

After a few manipulations (see Sect. 6.4 of [33]), a convenient expression for $\partial_{\boldsymbol{r}_J}\widehat{o}$ is found:

$$\partial_{\boldsymbol{r}_J}\widehat{o}=-U_J\,\partial_{\boldsymbol{r}}\delta(\boldsymbol{r}_J-\boldsymbol{r})-\frac{1}{2}\bigg(\Big(\sum_K \boldsymbol{f}_{JK}\Big)\delta(\boldsymbol{r}_J-\boldsymbol{r})+\sum_K \boldsymbol{f}_{JK}\delta(\boldsymbol{r}_K-\boldsymbol{r})\bigg).$$

Furthermore,

(i) $\langle\sum_J \boldsymbol{v}_J\cdot U_J\partial_{\boldsymbol{r}}\delta(\boldsymbol{r}_J-\boldsymbol{r})\rangle=$

$$\partial_{\boldsymbol{r}}\Big\langle\sum_J U_J\delta(\boldsymbol{r}_J-\boldsymbol{r})(\boldsymbol{v}_J-\boldsymbol{v})\Big\rangle+\partial_{\boldsymbol{r}}\Big\langle\sum_J U_J\delta(\boldsymbol{r}_J-\boldsymbol{r})\boldsymbol{v}\Big\rangle;$$

(ii) $\langle\sum_J \boldsymbol{v}_J\cdot((\sum_K \boldsymbol{f}_{JK})\delta(\boldsymbol{r}_J-\boldsymbol{r})+\sum_K \boldsymbol{f}_{JK}\delta(\boldsymbol{r}_K-\boldsymbol{r}))\rangle=$

$$\Big\langle\sum_J\sum_K \boldsymbol{v}_J\cdot\boldsymbol{f}_{JK}\big(\delta(\boldsymbol{r}_J-\boldsymbol{r})+\delta(\boldsymbol{r}_K-\boldsymbol{r})\big)\Big\rangle;$$

(iii) $\langle\sum_J\sum_K \boldsymbol{f}_{JK}(\delta(\boldsymbol{r}_J-\boldsymbol{r})+\delta(\boldsymbol{r}_K-\boldsymbol{r}))\rangle=\boldsymbol{0}$.

On taking all this into account, equation (59) becomes:

$$\boxed{\begin{aligned}&\partial_t\Big\langle\sum_J U_J\delta(\boldsymbol{r}_J-\boldsymbol{r})\Big\rangle+\partial_{\boldsymbol{r}}\Big\langle\sum_J U_J\delta(\boldsymbol{r}_J-\boldsymbol{r})\boldsymbol{v}\Big\rangle+\partial_{\boldsymbol{r}}\Big\langle\sum_J U_J\delta(\boldsymbol{r}_J-\boldsymbol{r})(\boldsymbol{v}_J-\boldsymbol{v})\Big\rangle\\&\quad+\frac{1}{2}\Big\langle\sum_J\sum_K(\boldsymbol{v}_J-\boldsymbol{v})\cdot\boldsymbol{f}_{JK}\big(\delta(\boldsymbol{r}_J-\boldsymbol{r})+\delta(\boldsymbol{r}_K-\boldsymbol{r})\big)\Big\rangle=0.\end{aligned}} \tag{60}$$

With a view toward paralleling equation (51), we begin by setting

$$\partial_t(\rho\epsilon)+\operatorname{div}(\rho\epsilon\boldsymbol{v})\equiv\partial_t\Big\langle\sum_J U_J\delta(\boldsymbol{r}_J-\boldsymbol{r})\Big\rangle+\partial_{\boldsymbol{r}}\Big\langle\sum_J U_J\delta(\boldsymbol{r}_J-\boldsymbol{r})\boldsymbol{v}\Big\rangle. \tag{61}$$

We remain with the task of choosing for the CM thermal flux $\boldsymbol{q}$ one of the identifications compatible both with the following relation:

$$\begin{aligned}\operatorname{div}\boldsymbol{q}-\boldsymbol{C}\cdot\operatorname{grad}\boldsymbol{v}-s_\epsilon^{\rm add}\overset{?}{\equiv}\;&\partial_{\boldsymbol{r}}\Big\langle\sum_J U_J\delta(\boldsymbol{r}_J-\boldsymbol{r})(\boldsymbol{v}_J-\boldsymbol{v})\Big\rangle\\&+\frac{1}{2}\Big\langle\sum_J\sum_K(\boldsymbol{v}_J-\boldsymbol{v})\cdot\boldsymbol{f}_{JK}\big(\delta(\boldsymbol{r}_J-\boldsymbol{r})+\delta(\boldsymbol{r}_K-\boldsymbol{r})\big)\Big\rangle\end{aligned} \tag{62}$$

and with the identification made for $\boldsymbol{C}$ by means of (43) and (45). I envisage various ways to proceed, of which I select the one that seems to me the least cumbersome of all.[12]

I stipulate that $s_\tau^{\text{add}} = 0$, or rather as per (52) that $s_\epsilon^{\text{add}} = -s_\kappa^{\text{add}}$, and I choose $s_\kappa^{\text{add}}$ as in (56), so that (62) becomes

$$\operatorname{div}\boldsymbol{q} - \boldsymbol{C}\cdot\operatorname{grad}\boldsymbol{v} \overset{?}{\equiv} \partial_r\left(\left\langle\sum_J (K_J + U_J)\delta(\boldsymbol{r}_J - \boldsymbol{r})(\boldsymbol{v}_J - \boldsymbol{v})\right\rangle\right)$$
$$+\frac{1}{2}\left\langle\sum_J\sum_K (\boldsymbol{v}_J - \boldsymbol{v})\cdot\boldsymbol{f}_{JK}\big(\delta(\boldsymbol{r}_J - \boldsymbol{r}) + \delta(\boldsymbol{r}_K - \boldsymbol{r})\big)\right\rangle$$

I move further by setting:

$$\begin{aligned}
\boldsymbol{q} - \boldsymbol{q}^{\text{add}} &\equiv \left\langle\sum_J (K_J + U_J)\delta(\boldsymbol{r}_J - \boldsymbol{r})(\boldsymbol{v}_J - \boldsymbol{v})\right\rangle,\\
\operatorname{div}\boldsymbol{q}^{\text{add}} - \boldsymbol{R}\cdot\operatorname{grad}\boldsymbol{v} &\equiv \frac{1}{2}\left\langle\sum_J\sum_K (\boldsymbol{v}_J - \boldsymbol{v})\cdot\boldsymbol{f}_{JK}\big(\delta(\boldsymbol{r}_J - \boldsymbol{r}) + \delta(\boldsymbol{r}_K - \boldsymbol{r})\big)\right\rangle;
\end{aligned} \tag{63}$$

these juxtapositions make patent the twofold, constitutive and kinetic, nature of both $\boldsymbol{q}$ and $\boldsymbol{q}^{\text{add}}$.

## 7 Conclusions

A version of the IKN construction of molecular counterparts for the CM balances of mass, momentum, and energy, has been given. The key differences from the original construction are encountered when, in Sect. 6.3.2, momentum balance is dealt with: (i) the Cauchy stress is given the purely constitutive molecular interpretation furnished by relations (43)–(45); (ii) a momentum source is introduced, depending solely on molecules' velocity fluctuations as specified by equation (42). As recalled in Sect. 6.4.1, the CM balances of kinetic, internal and total energies all feature diffusive sources depending on Cauchy stress. Therefore, when MM counterparts of energy balances are sought for *à la* Irving-Kirkwood-Noll, they cannot be brought into term-wise correspondence with the corresponding CM balances, unless the latter are provisionally equipped with additional source terms. The molecular identification of these additional sources, as well as of another basic CM notion, that of heat flux, must be made in a mutually consistent manner. This can be done in more than one way. My preference is for letting the additional CM sources of kinetic and internal energies to balance each other, so as to let the total energy balance keep the standard form; accordingly, as argued in Sect. 6.4.3, the heat flux receives the twofold molecular interpretation, both constitutive and kinematical, specified by (63).

In my view, there can be circumstances—say, when a loosely aggregated material system evolves at relatively high temperatures—when a purely kinematical contribution to momentum balance may turn out to be a nonnegligible *macroscopic consequence of microscopic motion randomness*. Such a contribution can be interpreted as an *internal source of momentum* that CM could not predict were it not IKN-hybridized with SM, in the manner I propose in this paper or otherwise. Quite similarly, kinematical contributions to any of CM energy balances should be expected whenever velocity fluctuations count.

[12] My procedure differs from Irving & Kirkwood's (in [6], see the text from the second column of p. 825 to the first column of p. 827) and has a different outcome.

**Acknowledgements** Over the years, many of the matters I here dealt with have been the subject of a number of useful discussions with Antonio DiCarlo, which I gratefully acknowledge.

**Publisher's Note** Springer Nature remains neutral with regard to jurisdictional claims in published maps and institutional affiliations.

## Appendix A: Andersen-Parrinello-Rahman's Molecular Dynamics

In Andersen-Parrinello-Rahman MD, the additional degrees of freedom enter the expression of the transition-inducing macroscopic power expenditure on the computational box. The modifier 'macroscopic' is here used in the sense of SM: the system under study is small at the CM scale and yet microscopically large. In fact, at the CM scale, an MD space-time region is to be regarded as a (point, instant) pair. If, at that point and instant, $\boldsymbol{P}$ and $\dot{\boldsymbol{F}}$ denote, respectively, the Piola stress and the time rate of the deformation gradient $\boldsymbol{F}$, so that the macroscopic *internal* power expenditure per unit volume is $\boldsymbol{P} \cdot \dot{\boldsymbol{F}}$, then, in an MD simulation of APR type, that power expenditure is regarded as an *external* power expenditure over the computational box.[13] When the macroscopic stress plays the role of an external control—the case of stress-induced phase transitions—then $\dot{\boldsymbol{F}}$ accounts for the fluctuations of the computational box during an MD run; the additional DOFs are just the components of $\boldsymbol{F}$, the tensor accounting for changes in volume and shape of the computational box.

An extended Lagrangian of APR type implies a non-Hamiltonian formulation of the evolution problem of a given molecule system. Moreover, the associated ensemble averages are not microcanonical, as is customary when energy, volume and molecule number are conserved, but rather *isenthalpic* and *isobaric* when the transitions under study are stress-induced, because energy is replaced by enthalpy, and volume by scalar pressure or, more generally, applied stress.

## Appendix B: Hellinger's Notions of Material Body and Material Point

Here is, reproduced from [16], the CM notion of a material body as stated by E. Hellinger in 1914:

> "[t]he general *three-dimensionally extended continuous medium* ... – by abstraction from all individual properties of matter[14] – consists in a collection of material points that ... must be *distinguishable* the one for the other...".[15]

As was noted in [39], while a Hellinger-type presumption appears acceptable in the case of solid bodies, whose referential shape typically evolves in space × time in a way that we

---

[13]That $\boldsymbol{P} \cdot \dot{\boldsymbol{F}} \propto \boldsymbol{C} \cdot \operatorname{grad} \boldsymbol{v}$ is a well-known direct consequence of the relationship between Piola's and Cauchy's stress measures ($\boldsymbol{P} = (\det \boldsymbol{F})\boldsymbol{C}\boldsymbol{F}^{-T}$).

[14]That is to say, as I reckon it, whatever the constitutive response.

[15]**2. Der Begriff des Kontinuums**. (The Notion of Continuum.) **2.a. Das Kontinuum and seine Deformation**. (The Continuum and its Deformation.) Das allgemeine *dreidimensionalen ausgedehnte kontinuirliche Medium*, auf das sich die folgenden Betrachtungen beziehen (which is the object of the following considerations), bedeutet - unter Abstraktion von allen individuellen Eigenschaften der Materie - eine Gesamtheit von materiellen Teilchbar, die erstens voneinander *unterscheidbar* sein und zweitens den Raum bzw. einem stetig begrenzten Raumteil *stetig ausfüllen* sollen (and, secondly, must continuously fill the whole space or, respectively, a continuous and bounded portion of space).

can track point-wise at each instant of the observation interval, this is not the case for liquids and, worse, gases.[16]

## References

1. Love, A.E.H.: A Treatise on the Mathematical Theory of Elasticity. Dover, New York (1927)
2. Truesdell, C., Toupin, R.A.: The classical field theories. In: Handbuch der Physik, III/1. Springer, Berlin (1960)
3. DiCarlo, A.: A major serendipitous contribution to continuum mechanics. Mech. Res. Commun. **93**, 41–46 (2018). https://doi.org/10.1016/j.mechrescom.2017.10.002
4. DiCarlo, A.: Continuum mechanics as a computable coarse-grained picture of molecular dynamics (2018, this volume)
5. Murdoch, A.I.: Physical Foundations of Continuum Mechanics. Cambridge University Press, Cambridge (2012)
6. Irving, J.H., Kirkwood, J.G.: The statistical mechanical theory of transport processes. IV. The equations of hydrodynamics. J. Chem. Phys. **18**(6), 817–829 (1950). https://doi.org/10.1063/1.1747782
7. Noll, W.: Die Herleitung der Grundgleichungen der Thermomechanik der Kontinua aus der Statistischen Mechanik. Indiana Univ. Math. J. **4**, 627–646 (1955). https://doi.org/10.1512/iumj.1955.4.54022 and, J. Ration. Mech. Anal. **4**, 627–646 (1955). English translation: Derivation of the fundamental equations of continuum thermodynamics from statistical mechanics. J. Elast. **100**, 5–24 (2010)
8. Noll, W.: Thoughts on the concept of stress. J. Elast. **100**, 25–32 (2010)
9. Andersen, H.C.: Molecular dynamics simulations at constant pressure and/or temperature. J. Chem. Phys. **72**, 2384–2393 (1980)
10. Parrinello, M., Rahman, A.: Crystal structure and pair potentials: a molecular-dynamics study. Phys. Rev. Lett. **45**, 1196–1199 (1980)
11. Parrinello, M., Rahman, A.: Polymorphic transitions in single crystals: a new molecular dynamics method. J. Appl. Phys. **52**, 7182–7190 (1981)
12. Parrinello, M., Rahman, A.: Strain fluctuations and elastic constants. J. Chem. Phys. **76**, 2662–2666 (1982)
13. Tuckerman, M.: Statistical Mechanics: Theory and Molecular Simulation. Oxford University Press, London (2010)
14. Podio-Guidugli, P.: On microscopic and macroscopic notions of stress. In: Elasticity and Inelasticity. Proceedings of "Problems of Mechanics of Deformable Bodies", pp. 279–285, Moscow, Jan 20–21, 2011. Moscow University Press, Moscow (2011)
15. Giusteri, G.G., Podio-Guidugli, P., Fried, E.: Continuum balances from extended Hamiltonian dynamics. J. Chem. Phys. **146**, 224101 (2017)
16. Hellinger, E.: Die Allgemeinen Ansatze der Mechanik der Kontinua. Enzyclopaedie der Matematischen Wissenschaften, Art. IV. 30 (1914)
17. Noll, W.: The foundations of classical mechanics in the light of recent advances in continuum mechanics. In: Noll, W. (ed.) Proceedings of an International Symposium on the Axiomatic Method, University of California, Berkeley, Dec. 27, 1957–Jan. 4, 1958, pp. 265–281. Springer, Berlin (1959). Reprinted in: The Foundations of Mechanics and Thermodynamics: Selected Papers by W. Noll, pp. 32–47. Springer, Berlin (1974). https://doi.org/10.1007/978-3-642-65817-4_2.
18. Truesdell, C., Toupin, R.A.: The classical field theories. In: Flügge, S. (ed.) Handbuch der Physik III/1. Springer, Berlin (1960)
19. Truesdell, C., Noll, W.: The nonlinear field theories of mechanics. In: Flügge, S. (ed.) Handbuch der Physik III/3. Springer, Berlin (1965)
20. Pitteri, M.: Continuum equations of balance in classical statistical mechanics. Arch. Ration. Mech. Anal. **94**, 291–305 (1986). Erratum: **100**, 315–316 (1988)
21. Seguin, B., Fried, E.: Statistical foundations of liquid-crystal theory. I: Discrete systems of rod-like molecules. Arch. Ration. Mech. Anal. **206**, 1039–1072 (2012)
22. Seguin, B., Fried, E.: Statistical foundations of liquid-crystal theory II: Macroscopic balance laws. Arch. Ration. Mech. Anal. **207**, 1–37 (2013)

[16]"The qualifier 'ephemeral' was proposed for continuous models of bodies, such as gases, for which the generally tacit axiom of permanence of material elements fails to apply. Consequently, to their scrutiny, a Eulerian (local) approach is mandatory, such as one adopted, e.g., in molecular dynamics." (from the abstract of [40], where references to other relevant papers by G. Capriz and coworkers can be found).

23. Murdoch, A.I.: A corpuscular approach to continuum mechanics: basic considerations. Arch. Ration. Mech. Anal. **88**(4), 291–321 (1985)
24. Murdoch, A.I.: Some primitive concepts in continuum mechanics regarded in terms of objective space-time molecular averaging: the key rôle played by inertial observers. J. Elast. **84**, 69–97 (2006)
25. Murdoch, A.I.: On molecular modelling and continuum concepts. J. Elast. **100**(1), 33–61 (2010)
26. Roberts, A.J.: A One-dimensional Introduction to Continuum Mechanics. World Scientific, Singapore (1994)
27. Capriz, G., Mazzini, G.: A $\sigma$-algebra and a concept of limit for bodies. Math. Models Methods Appl. Sci. **10**(6), 801–813 (2000)
28. Noll, W.: On the concept of force. https://www.math.cmu.edu/~wn0g/Force.pdf (2007)
29. Podio-Guidugli, P.: On the aggregation state of simple materials. In: Šilhavý, M. (ed.) Mathematical Modeling of Bodies with Complicated Bulk and Boundary Behavior. Quaderni di Matematica, vol. 20, pp. 159–168 (2008)
30. Admal, N.C., Tadmor, E.B.: A unified interpretation of stress in molecule systems. J. Elast. **100**(1), 63–143 (2010)
31. Tadmor, E.B., Miller, R.E.: Modeling Materials: Continuum, Atomistic and Multiscale Techniques. Cambridge University Press, Cambridge (2011)
32. Groot, R.D., Warren, P.B.: Dissipative particle dynamics: bridging the gap between atomistic and mesoscopic simulation. J. Chem. Phys. **107**(11), 4423 (1997)
33. Podio-Guidugli, P.: On the modeling of transport phenomena in continuum and statistical mechanics. Discrete Contin. Dyn. Syst. **10**(6), 1393–1411 (2017)
34. Podio-Guidugli, P.: Inertia and invariance. Ann. Mat. Pura Appl. (4) **CLXXII**, 103–124 (1997)
35. Gurtin, M.E., Podio-Guidugli, P.: Configurational forces and the basic laws for crack propagation. J. Mech. Phys. Solids **44**(6), 905–927 (1996)
36. Gurtin, M.E., Podio-Guidugli, P.: On configurational inertial forces at a phase interface. J. Elast. **44**, 255–269 (1996)
37. Gurtin, M.E., Podio-Guidugli, P.: Configurational forces and a constitutive theory for crack propagation that allows for kinking and curving. J. Mech. Phys. Solids **46**(2), 1–36 (1998)
38. Podio-Guidugli, P.: La scelta dei termini inerziali per i continui con microstruttura. Rend. Lincei, Mat. Appl. Ser. IX **XIV**(4), 319–326 (2003)
39. Capriz, G., Podio-Guidugli, P.: Whence the boundary conditions in modern continuum physics? In: Atti dei Convegni Lincei, vol. 210, pp. 19–42. Accademia dei Lincei, Rome (2004)
40. Capriz, G., Giovine, P.: Classes of ephemeral continua. Math. Methods Appl. Sci. **41**(3), 1175–1196 (2018)

J Elast (2019) 135:457–483
https://doi.org/10.1007/s10659-018-9691-4

# Hyper-stresses in $k$-Jet Field Theories

**Reuven Segev**[1] · **Jędrzej Śniatycki**[2,3]

Received: 4 March 2018 / Published online: 18 September 2018


**Abstract** For high-order continuum mechanics and classical field theories configurations are modeled as sections of general fiber bundles and generalized velocities are modeled as variations thereof. Smooth stress fields are considered and it is shown that three distinct mathematical stress objects play the roles of the traditional stress tensor of continuum mechanics in Euclidean spaces. These objects are referred to as the variational hyper-stress, the traction hyper-stress and the non-holonomic stress. The properties of these three stress objects and the relations between them are studied.

**Keywords** Continuum mechanics · Classical field theories · Fiber bundle · Hyper-stress · High order continuum mechanics

**Mathematics Subject Classification** 74A10 · 70S10 · 53Z05 · 58A32

## 1 Introduction

This manuscript is concerned with hyper-stresses for theories formulated on general fiber bundles over differentiable manifolds. Although the study of higher order continuum mechanics entered the focus of attention early in the second half of the 20th century (*e.g.*, [16, 17, 30, 31] it is still the subject of current research from both theoretical (*e.g.*, [6–8, 19, 21]) and practical (*e.g.*, [1–3, 22]) interests.

In parallel, analogous questions engaged the mathematical physics literature considering classical field theories (*e.g.*, [4, 5, 9, 10, 12]), although the variational approach has been

✉ R. Segev
rsegev@bgu.ac.il

J. Śniatycki
sniatycki@gmail.com

1 Department of Mechanical Engineering, Ben-Gurion University of the Negev, Beer-Sheva, Israel

2 Departments of Mathematics and Statistics, University of Calgary, Calgary, Alberta, Canada

3 University of Victoria, Victoria, B.C., Canada

much more pronounced. While in continuum mechanics fields are usually defined on a three-dimensional body manifold or the three-dimensional physical space, classical field theories are mostly formulated as sections of fiber bundles over a space-time manifold.

Following Walter Noll's [18], it is by now accepted that the body object of continuum mechanics should be modeled as a manifold devoid of a metric or a connection. Furthermore, fields defined on the body, such as various order parameters or internal degrees of freedom, cannot always be expected to be valued in Euclidean spaces. Thus, formulations in the setting of general manifolds are relevant in continuum mechanics as much as they are for physical field theories.

While traditionally, $k$-th order hyper-stresses emerge in continuum mechanics as derivatives of a strain energy density with respect to derivatives of order $k$ of the configuration, in [24, 27] their existence follows from a representation theorem for forces which are conjugate to the space of $C^k$-velocity fields. This setting has the advantage that it applies to the geometry of general manifolds and to hyper-stresses that may be as irregular as Borel measures. It is noted that for a theory of order $k$ on a differentiable manifold, the derivatives of a particular order $0 < r \leq k$, do not give rise to an invariant geometric object. Rather, one has to use the $k$-jet of the generalized velocity field combining all derivatives of order less or equal to $k$ into a single invariant object. (See [23] as a general reference on jet bundles.)

Among the peculiarities of continuum mechanics on manifolds, even in the smooth case considered in this paper and $k = 1$, we emphasize the distinction between variational stresses and traction stresses [25]. For classical continuum mechanics in a Euclidean space, the same mathematical object determines the traction on subbodies—the traction stress—and acts on velocity gradients to produce power—the variational stress, in our terminology. Yet, for the general geometric setting, two distinct mathematical objects play these roles. While the variational stress determines the traction stress, there is a class of traction stresses for each variational stress field. A unique variational stress density is determined when a traction stress and a body force density are given.

As we show below, it turns out that for high order theories, a variational hyper-stress does not determine a unique traction hyper-stress. In fact, a traction hyper-stress field together with the body hyper-force encode more information than a variational hyper-stress field. Now, a third object is needed, which we refer to as the non-holonomic stress field. While the non-holonomic stress has been introduced in [26] for the case $k = 2$ as a generalization of the variational stress enabling one to apply integral transformations, here we show that non-holonomic stresses are the objects that determine traction hyper-stresses as defined in [29].

Specifically, let $W$ denote the vector bundle over the body $\mathscr{X}$ where generalized velocity field assume their values and let $J^k W$ denote it $k$-the jet bundle. Then, a variational hyper-stress density is a section $S$ of the vector bundle

$$L\Big(J^k W, \bigwedge^n T^* \mathscr{X}\Big), \tag{1.1}$$

so that for a vector field $w$, $S \cdot j^k w$ is the power density that the hyper-stress expends for the $k$-th jet of the generalized velocity. The traction hyper-stress, $\sigma$, determines the hyper-traction, $\boldsymbol{t}_{\mathscr{R}}$—a section of $L(J^{k-1} W, \bigwedge^{n-1} T^* \partial \mathscr{R})$—on the boundary of an arbitrary subbody $\mathscr{R}$. Thus, $\sigma$ is a section of

$$L\Big(J^{k-1} W, \bigwedge^{n-1} T^* \mathscr{X}\Big) \tag{1.2}$$

and the corresponding hyper-traction is determined by the condition that $\boldsymbol{t}_{\mathscr{R}} \cdot j^{k-1} w$ is the $(n-1)$-form obtained by restricting $\sigma \cdot j^{k-1} w$ to $T \partial \mathscr{R}$. The non-holonomic stress density,

$P$, is a section of

$$L\Big(J^1(J^{k-1}W), \textstyle\bigwedge^n T^*\mathscr{X}\Big). \tag{1.3}$$

The term "non-holonomic" stress has been chosen in order to reflect the fact that it acts on the first jet of sections $A$ of $J^{k-1}W$. A section of $J^{k-1}W$ need not be holonomic in the sense that it need not be given as $j^{k-1}w$, for some vector field $w$. Hence the distinction between the non-holonomic stresses and the variational stresses.

In order to introduce the terminology and exhibit the various roles played by these three distinct mathematical objects in the simplest terms, consider the case where the body $\mathscr{R}$ is identified with its current configuration in $\mathbb{R}^3$. Let $w^j$ be the components of a velocity field and let $w^j_{,i_1\cdots i_k}$ be its $k$-th partial derivatives. The components of the variational stress of highest order $S_j^{i_1\cdots i_k}$ act on the derivatives of the velocity in the form

$$I = \int_{\mathscr{R}} S_j^{i_1\cdots i_k} w^j_{,i_1\cdots i_k}\, \mathrm{d}V. \tag{1.4}$$

Since the partial derivatives are symmetric relative to permutations of the indices $i_1 \cdots i_k$, the components of the variational stress satisfy the same symmetry conditions. We remark that this symmetry condition for the components of the variational stress emerges naturally in the hyper-elastic case where the components of the variational stress are given in terms of a potential function $\varphi$ and the symmetric components of the $k$-th partial derivatives of the configuration $\kappa^j_{,i_1\cdots i_k}$ by

$$S_j^{i_1\cdots i_k} = \frac{\partial \varphi}{\partial \kappa^j_{,i_1\cdots i_k}}. \tag{1.5}$$

Returning to the expression for the action of the variational stress, one has

$$\begin{aligned} I &= \int_{\mathscr{R}} \big[\big(S_j^{i_1\cdots i_k} w^j_{,i_1\cdots i_{k-1}}\big)_{,i_k} - S_{j,i_k}^{i_1\cdots i_k} w^j_{,i_1\cdots i_{k-1}}\big]\mathrm{d}V, \\ &= \int_{\partial\mathscr{R}} S_j^{i_1\cdots i_k} n_{i_k} w^j_{,i_1\cdots i_{k-1}}\, \mathrm{d}A - \int_{\mathscr{R}} S_{j,i_k}^{i_1\cdots i_k} w^j_{,i_1\cdots i_{k-1}}\, \mathrm{d}V, \end{aligned} \tag{1.6}$$

where $n$ is the unit normal to the boundary. We will refer to a tensor such as $\boldsymbol{t}_j^{i_1\cdots i_{k-1}} = S_j^{i_1\cdots i_k} n_{i_k}$ as a hyper-traction and to $\boldsymbol{b}_j^{i_1\cdots i_{k-1}} = S_{j,i_k}^{i_1\cdots i_k}$ as a body hyper-force. Thus, in the expression

$$\boldsymbol{t}_j^{i_1\cdots i_{k-1}} = S_j^{i_1\cdots i_k} n_{i_k}, \tag{1.7}$$

the variational hyper-stress $S$ acts as a traction hyper-stress by determining the hyper-traction on the boundary. That is, a traction hyper-stress $\sigma_j^{i_1\cdots i_k}$ determines hyper-tractions by

$$\boldsymbol{t}_j^{i_1\cdots i_{k-1}} = \sigma_j^{i_1\cdots i_k} n_{i_k}. \tag{1.8}$$

Nevertheless, if the role of the traction hyper-stress is exhibited in the last equation, there is no a-priori reason to assume that $\sigma_j^{i_1\cdots i_k}$ is symmetric relative to permutations of all indices $i_1 \cdots i_k$ and not merely $i_1 \cdots i_{k-1}$. In other words a traction hyper-stress need not be determined by a variational hyper-stress but rather, by a more general object to which we refer as the non-holonomic stress.

One could postulate that traction hyper-stresses are indeed symmetric relative to all the indices $i_1 \cdots i_k$. However, it turns out that this condition cannot be formulated invariantly in the general geometric case. Indeed, on general fiber bundles, a traction hyper-stress field together with a body hyper-force determine a unique non-holonomic hyper-stress field. A non-holonomic hyper-stress field does not act on $k$-jets of velocity fields but rather on the first jets of sections of the $(k-1)$-jet bundle. Such sections of the $(k-1)$-jet bundle need not be holonomic—compatible—hence the terminology. It follows that non-holonomic hyper-stress fields obey less restrictive symmetry conditions in comparison with variational stresses. Conversely, a non-holonomic hyper-stress field determines a unique traction hyper-stress field. Furthermore, a non-holonomic hyper-stress determines a unique variational stress that may be used to compute the force on each subbody. However, this variational hyper-stress does not encode enough information as to determine the traction hyper-stress on each subbody. We find it somewhat intriguing that for higher-order continuum mechanics—originally based on the variational approach—it turns out that the generalization of the Cauchy construction plays such a crucial role.

Section 2 introduces the notation and some of the terminology used in the sequel, mainly, that corresponding to symmetric tensors. Section 3 reviews the fundamentals of continuum mechanics and field theories on $k$-jet bundles of fiber bundles. In particular, variational hyper-stress densities are introduced. Section 4 outlines higher-order theories and introduces traction hyper-stresses using a framework which generalizes the classical Cauchy approach. Section 5 considers non-holonomic stresses and their relations with traction stresses. Next, the variational stress induced by a non-holonomic stress is presented in Sect. 6. Finally, in Sect. 7, we describe the geometric setting in which elastic constitutive relations are formulated.

## 2 Notation and Preliminaries

Multi-index notation, as used in this manuscript, will help make the algebraic expressions below more compact. In this section we introduce the notation adopted here, with particular application to symmetric tensors. (For additional details, see [29].)

### 2.1 Multi-indices

A collection of indices $i_1 \cdots i_k$, $i_r = 1, \dots, n$ will be represented as a multi-index $I$ and we will write $|I| = k$, the length of the multi-index. Multi-indices will be denoted by upper-case roman letters and the associated indices will be denoted by the corresponding lower-case letters. In what follows, we will use the summation convention for repeated indices as well as repeated multi-indices. Whenever the syntax is violated, *e.g.*, when a multi-index appears more than twice in a term, it is understood that summation does not apply.

A multi-index $I$ induces a sequence $(I_1, \dots, I_n)$ in which $I_r$ is the number of times the index $r$ appears in the sequence $i_1 \cdots i_k$. Thus, $|I| = \sum_r I_r$. Multi-indices may be concatenated naturally such that $|IJ| = |I| + |J|$. Additional convenient notation is introduced by

$$I! := I_1! \cdots I_n!, \qquad \delta^I_J := \delta^{i_1}_{j_1} \cdots \delta^{i_k}_{j_k}. \tag{2.1}$$

Greek letters will be used for strictly increasing multi-indices used in the representation of alternating tensors and forms, *e.g.*,

$$\omega = \omega_\lambda \mathrm{d}x^\lambda := \omega_{\lambda_1 \cdots \lambda_{|\lambda|}} \mathrm{d}x^{\lambda_1} \wedge \cdots \wedge \mathrm{d}x^{\lambda_{|\lambda|}}. \tag{2.2}$$

To simplify the notation, the local volume element induced by a coordinate system will be denoted by $\mathrm{d}x$, *i.e.*,

$$\mathrm{d}x = \mathrm{d}x^1 \wedge \cdots \wedge \mathrm{d}x^n. \tag{2.3}$$

## 2.2 Permutations

We denote by $\mathscr{P}_l$ the group of permutations of $(1, \ldots, l)$. For a multi-index $I = i_1 \cdots i_l$ and a permutation $p \in \mathscr{P}_l$,

$$p(I) = I \circ p = i_{p(1)} \cdots i_{p(l)}. \tag{2.4}$$

An $l$-permutation $p$ acts on an $l$-dimensional array by $p(T)_I = T_{p(I)}$.

We observe that for a multi-index $I$, of all $|I|!$ permutations $p(I)$, there are $I!$ permutations that leave $I$ invariant. Thus, there are $|I|!/I!$ elements in the collection, $\mathscr{P}_I$, containing permutations that give distinct multi-indices $J = p(I)$.

We will also use the notation

$$|\varepsilon|^I_J = \begin{cases} 1, & \text{if } I \text{ may be obtained by a permutation of } J, \\ 0, & \text{otherwise.} \end{cases} \tag{2.5}$$

## 2.3 Symmetric Tensors

We will view an $l$-tensor as an $l$-multilinear mapping, an element of $L^l(\boldsymbol{V}, \boldsymbol{W}) \simeq (\bigotimes^l \boldsymbol{V}^*) \otimes \boldsymbol{W}$ for two vector spaces $\boldsymbol{V}$ and $\boldsymbol{W}$. In order to simplify the notation, for the rest of this section, we will use $\boldsymbol{W} = \mathbb{R}$. For other finite dimensional vector spaces, the extension is straightforward. A permutation $p$ acts on an $l$-tensor $T$ by

$$p(T)(v_1, \ldots, v_l) = T(v_{p(1)}, \ldots, v_{p(l)}) \tag{2.6}$$

which implies that the array of $p(T)$ is indeed $p(T)_I = T_{p(I)}$. A symmetric tensor $T$ satisfies the condition

$$p(T) = T, \quad \text{or equivalently,} \quad T_{p(I)} = T_I, \tag{2.7}$$

for all permutations $p \in \mathscr{P}_l$. A symmetric array is uniquely determined by the elements of the form $T_{\boldsymbol{I}}$ where we use the convention that multi-indices denoted by bold faced characters are non-decreasing, that is $i_1 \le \cdots \le i_l$, and alternatively, when a multi-index appears inside angle brackets, it is implied that it is permuted so that it is non-decreasing. When the summation convention should be applied to such multi-indices, it is understood that the summation is carried out only for non-decreasing occurrences.

For an $n$-dimensional vector space $\boldsymbol{V}$, the subspace of symmetric tensors, $L^l_S(\boldsymbol{V}, \mathbb{R})$, has the dimension (see [14]),

$$\dim L^l_S(\boldsymbol{V}, \mathbb{R}) = \frac{(n + l - 1)!}{(n - 1)! l!}. \tag{2.8}$$

## 2.4 Symmetrization

We will use the symmetrization operator

$$\mathscr{S} : L^l(\boldsymbol{V}, \boldsymbol{W}) \longrightarrow L^l_S(\boldsymbol{V}, \mathbb{R}), \quad \text{whereby,} \quad \mathscr{S}(T) = \frac{1}{l!} \sum_{p \in \mathscr{P}_l} p(T), \tag{2.9}$$

which implies that the array of the symmetrized tensor is the symmetrized array, *i.e.*,

$$T_{(I)} := \mathscr{S}(T)_I = \frac{1}{l!}\sum_{p\in\mathscr{P}_l} T_{p(I)}. \tag{2.10}$$

The symmetrized tensor product is $T \odot R = \mathscr{S}(T \otimes R)$. Using the notation

$$\boldsymbol{e}^J := \bigotimes^J \boldsymbol{e}^J = \boldsymbol{e}^{j_1} \otimes \cdots \otimes \boldsymbol{e}^{j_l}, \qquad \boldsymbol{e}^{(J)} := \bigodot^J \boldsymbol{e}^J = \boldsymbol{e}^{j_1} \odot \cdots \odot \boldsymbol{e}^{j_l}, \tag{2.11}$$

a basis for the space of symmetric tensors may be formed by the elements $\{\boldsymbol{e}^{(\boldsymbol{I})}\}$ (where we observe the $\boldsymbol{I}$ multi-indices are non-decreasing). Thus, one has the identification

$$L^l_S(\boldsymbol{V}, \mathbb{R}) \simeq \bigodot^l \boldsymbol{V}^*. \tag{2.12}$$

Consider the inclusion

$$\iota_S : \bigodot^l \boldsymbol{V}^* \longrightarrow \bigotimes^l \boldsymbol{V}^*. \tag{2.13}$$

For a symmetric tensor $T$, one has

$$\begin{aligned} \iota_S(T) = T &= T_J \boldsymbol{e}^J = T_J \boldsymbol{e}^{(J)}, \\ &= \sum_{\boldsymbol{I}} \sum_{p\in\mathscr{P}_{\boldsymbol{I}}} T_{p(\boldsymbol{I})} \boldsymbol{e}^{(p(\boldsymbol{I}))}, \\ &= \sum_{\boldsymbol{I}} \frac{|\boldsymbol{I}|!}{\boldsymbol{I}!} T_{\boldsymbol{I}} \boldsymbol{e}^{(\boldsymbol{I})} \quad \text{(no sum)}. \end{aligned} \tag{2.14}$$

If we want the components of $T$ relative to the basis in $\odot^l \boldsymbol{V}^*$ to be equal to the corresponding components of the array of $\iota_S(T) = T$, we should use the modified basis

$$\underset{\leftarrow}{\overleftarrow{\boldsymbol{e}}}^{(\boldsymbol{I})} := \frac{|\boldsymbol{I}|!}{\boldsymbol{I}!} \boldsymbol{e}^{(\boldsymbol{I})}, \qquad \text{so that } T = T_{\boldsymbol{I}} \underset{\leftarrow}{\overleftarrow{\boldsymbol{e}}}^{(\boldsymbol{I})}. \tag{2.15}$$

Similarly, one can easily show that for the primal basis $\{\boldsymbol{e}_i\}$,

$$\underset{\leftarrow}{\overleftarrow{\boldsymbol{e}}}^{(\boldsymbol{I})}(\boldsymbol{e}_{(\boldsymbol{J})}) = \delta^{\boldsymbol{I}}_{\boldsymbol{J}}, \qquad \boldsymbol{e}^{(\boldsymbol{I})}(\underset{\leftarrow}{\overleftarrow{\boldsymbol{e}}}_{(\boldsymbol{J})}) = \delta^{\boldsymbol{I}}_{\boldsymbol{J}}, \tag{2.16}$$

and so $\{\underset{\leftarrow}{\overleftarrow{\boldsymbol{e}}}^{(\boldsymbol{I})}\}$ may serve as the dual basis of $\{\boldsymbol{e}_{(\boldsymbol{J})}\}$ and vice versa.

For a basis of the tangent space to a manifold induced by the coordinates $(x^i)$, we will use the basis consisting of the elements

$$\partial_{\boldsymbol{I}} := \partial_{i_1} \odot \cdots \odot \partial_{i_l}, \quad \partial_i := \frac{\partial}{\partial x^i}, \tag{2.17}$$

for the space of symmetric tensors, while the dual basis consists of the elements

$$\underset{\leftarrow}{\overleftarrow{\mathrm{d}x}}^{\boldsymbol{I}} := \frac{|\boldsymbol{I}|!}{\boldsymbol{I}!} \mathrm{d}x^{i_1} \odot \cdots \odot \mathrm{d}x^{i_l}. \tag{2.18}$$

It is noted that in both cases, only non-decreasing indices are used.

It follows that $\bigodot^l V^*$ may be identified with $(\bigodot^l V)^*$. In analogy with the above, for $R = R^I e_{(I)} \in \bigodot^l V$, $T = T_J \underleftarrow{e}^J$, $T(R) = T_I R^I$, and

$$T_J R^J = \sum_I \frac{|I|!}{I!} T_I R^I \quad \text{(no sum on } I\text{)}. \tag{2.19}$$

So setting,

$$\underleftarrow{R}^I = \frac{|I|!}{I!} R^I, \qquad \underrightarrow{R}^J = \frac{J!}{|J|!} R^J, \tag{2.20}$$

one has

$$T_J R^J = T_I \underleftarrow{R}^I, \qquad T_J \underrightarrow{R}^J = T_I R^I = T(R). \tag{2.21}$$

The last expression may be interpreted as a statement that the components $\underrightarrow{R}^J$ represent the inclusion of the symmetric tensor $R$ in the space of all tensors.

We will refer to elements of $(\bigodot^{l-1} V^*) \otimes V^*$ as *almost symmetric* tensors. It can be shown [14] that for an almost symmetric tensor $T$,

$$\mathscr{S}(T) = \frac{1}{l} \sum_{p \in \tilde{\mathscr{P}}_l} p(T), \tag{2.22}$$

where $\tilde{\mathscr{P}}_l$ contains permutations $p$ such that $p(l) = 1, \dots, l$, and $p(1) < \dots < p(l-1)$, that is, $p$ switches $l$ with another number and then orders the first $l-1$ numbers.

For a symmetric array $w_I$ and an array $T^J$, $|I| = |J|$, we will encounter below expressions such as

$$T^I w_I = T^{(I)} w_I, \quad \text{and} \quad T^{Jj} w_{Jj} = T^{(Jj)} w_{Jj}. \tag{2.23}$$

We wish to find the relation between these two expressions. Note that each pair $Jj$, determines by ordering a unique $I$ represented by $(J_1, \dots, J_j + 1, \dots, J_n)$.

Consider a particular $I$ represented by $(I_1, \dots, I_n)$. Let

$$\{I\} = \{r \mid I_r > 0\}, \quad c(I) = \text{cardinality}\{I\}, \tag{2.24}$$

*i.e.*, $c(I)$ is the number of indices that appear in $I$ one time or more. Since a pair $Jj$ determines a unique $I$, we may also write $c(Jj)$. Let $j \in \{I\}$, then,

$$Jj, \quad \text{with } J = (I_1, \dots, I_j - 1, \dots, I_n) \tag{2.25}$$

will give $I$ upon rearranging. Thus, for each $I$ there are exactly $c(I)$ pairs $Jj$ that may be rearranged to give $I$. It follows that

$$T^{Jj} w_{Jj} = \sum_{j \in \{I\}} T^{Jj} w_{Jj} = c(I) T^I w_I \quad \text{(no sum on } I\text{)}. \tag{2.26}$$

Define the arrays $\overrightarrow{T}^{Jj}$ and $\overleftarrow{T}^I$ by

$$\overrightarrow{T}^{Jj} = \frac{1}{c(Jj)} T^{Jj}, \quad \text{and} \quad \overleftarrow{T}^I = c(I) T^I. \tag{2.27}$$

It is concluded that

$$\overrightarrow{T}^{Jj} w_{Jj} = T^I w_I, \quad \text{and} \quad T^{Jj} w_{Jj} = \overleftarrow{T}^I w_I. \tag{2.28}$$

## 2.5 Jets

For a fiber bundle $\xi : \mathscr{Y} \to \mathscr{X}$, we will denote by $\xi^k : J^k(\mathscr{X}, \mathscr{Y}) \to \mathscr{X}$ the corresponding $k$-jet bundle of sections of $\xi$. When no ambiguity may occur, we will often use the simpler notation $\xi^k : J^k\mathscr{Y} \to \mathscr{X}$. One has the additional natural projections $\xi^k_l : J^k(\mathscr{X}, \mathscr{Y}) \to J^l(\mathscr{X}, \mathscr{Y})$, $l < k$, and in particular $\xi^k_0 : J^k(\mathscr{X}, \mathscr{Y}) \to \mathscr{Y}$, [23].

Let $\eta : W \to \mathscr{X}$ be a vector bundle and let $\eta^k : J^k W \to \mathscr{X}$ be the corresponding $k$-jet bundle. For $l < k$, the $l$-vertical subbundle of the jet bundle is defined by

$$V^l J^k W = \text{Kernel}\, \eta^k_l = \big\{ \dot{A} \in J^k W \mid \eta^k_l(\dot{A}) = 0 \big\}. \tag{2.29}$$

Evidently, $V^l J^k W$ is a vector subbundle of the jet bundle and we denote the inclusion by

$$\iota_{V^l_k} : V^l J^k W \longrightarrow J^k W. \tag{2.30}$$

## 2.6 Pullback of Forms

Here we use "$\sharp$" to indicate the pullback of a form by duality, rather than using "$*$". The latter is used to indicate the pullback of a form viewed as a section of the pullback bundle.

# 3 Continuum Mechanics and Field Theories on $k$-Jet Bundles

The fundamental object we consider is a fiber bundle

$$\xi : \mathscr{Y} \longrightarrow \mathscr{X} \tag{3.1}$$

where $\mathscr{X}$ is an $n$-dimensional orientable manifold and the typical fiber is an $m$-dimensional manifold. For the sake of simplicity, it is assumed that $\mathscr{X}$ is compact. This fiber bundle may have the following interpretations.

In Lagrangian continuum mechanics, it is usually assumed that the total space $\mathscr{Y}$ is trivial, *i.e.*,

$$\mathscr{Y} = \mathscr{X} \times \mathscr{S} \tag{3.2}$$

where the manifold $\mathscr{X}$ is interpreted as the body manifold and the manifold $\mathscr{S}$ is interpreted as the space manifold. In this case, a section $\kappa : \mathscr{X} \to \mathscr{Y}$ may be identified with a mapping $\mathscr{X} \to \mathscr{S}$. Such a mapping is interpreted as a configuration of the body in the physical space.

In Eulerian continuum mechanics, $\mathscr{X}$ is interpreted as the physical space and the fibers of $\mathscr{Y}$ are interpreted as the possible values that fields over the space may have. For classical field theories, $\mathscr{X}$ is interpreted as space-time. (See [13] for further motivation, references and examples.)

### 3.1 The Configuration Space

A configuration of the system, or a field, is a section $\kappa : \mathscr{X} \to \mathscr{Y}$. In accordance with the paradigm of analytical mechanics, a central role is played by the configuration space of the system, an infinite dimensional object in the case of a field theory. We are motivated by the case of continuum mechanics where one requires traditionally that configurations of a body in space be embeddings, and the fact that the subset of embeddings is open in the manifold of all $C^k$-mappings between two manifolds, for $k \geq 1$. (See [11, 15].) Thus, for $k \geq 1$, we consider the collection of $C^k$-sections of $\xi$, equipped with the $C^k$-topology, and define the configuration space $\mathscr{Q}$ to be an open subset of the manifold of sections, *i.e.*,

$$\mathscr{Q} \subset C^k(\xi) := C^k(\mathscr{X}, \mathscr{Y}), \quad \text{is open.} \tag{3.3}$$

Note that by the notation $C^k(\mathscr{X}, \mathscr{Y})$ we refer only to sections of the fiber bundle rather than all mappings $\mathscr{X} \to \mathscr{Y}$.

### 3.2 Generalized Velocities

Generalized velocities (virtual velocities, virtual displacements) at the configuration $\kappa$, or variations of the configuration $\kappa$, are elements of the tangent space $T_\kappa \mathscr{Q}$. Since $\mathscr{Q}$ is assumed to be open,

$$T_\kappa \mathscr{Q} = T_\kappa C^k(\mathscr{X}, \mathscr{Y}). \tag{3.4}$$

Furthermore, let

$$V\xi = V\mathscr{Y} := \text{Kernel}\, T\xi \tag{3.5}$$

be the vertical subbundle of $T\mathscr{Y}$. Then (see [20, p. 51]),

$$T_\kappa \mathscr{Q} = T_\kappa C^k(\mathscr{X}, \mathscr{Y}) \simeq C^k\big(\mathscr{X}, \kappa^* V\mathscr{Y}\big), \tag{3.6}$$

where $\kappa^* V\mathscr{Y}$ denotes the pullback of the vertical bundle onto $\mathscr{X}$. Thus, every generalized velocity may be identified uniquely with a $C^k$-section

$$w : \mathscr{X} \longrightarrow \kappa^* V\mathscr{Y}, \tag{3.7}$$

in accordance with the traditional interpretation in continuum mechanics. (It is noted that in [20], $V\mathscr{Y}$ is denoted by $TF(\mathscr{Y})$ and $\kappa^* V\mathscr{Y}$ is denoted by $T_\kappa \mathscr{Y}$.)

Thus,

$$TC^k(\mathscr{X}, \mathscr{Y}) \simeq C^k(\mathscr{X}, V\mathscr{Y}) \tag{3.8}$$

and a section $w$ represents an element of $T_\kappa \mathscr{Q}$ if

$$\tau_\mathscr{Y}|_{V\mathscr{Y}} \circ w = \kappa, \tag{3.9}$$

where $\tau_\mathscr{Y} : T\mathscr{Y} \longrightarrow \mathscr{Y}$ is the natural tangent bundle projection. In the sequel, in order to simplify the notation, we will denote the vector bundle

$$\xi \circ \tau_\mathscr{Y}|_{V\mathscr{Y}} : \kappa^* V\mathscr{Y} \longrightarrow \mathscr{X} \quad \text{by } \pi : W \longrightarrow \mathscr{X}. \tag{3.10}$$

Using local coordinates $(x^i, y^\alpha)$ in $\mathscr{Y}$, an element of $T\mathscr{Y}$ is represented locally in the form $\dot{x}^i \partial_i + \dot{y}^\alpha \partial_\alpha$. For an element of $V\mathscr{Y}$, $\dot{x}^i = 0$, and so, an element of $\kappa^* V\mathscr{Y} = W$ is represented locally in the form $\dot{y}^\alpha \kappa^* \partial_\alpha$, which we may write also simply as $\dot{y}^\alpha \boldsymbol{e}_\alpha$, $\boldsymbol{e}_\alpha := \kappa^* \partial_\alpha$. The dual bases will be denoted accordingly. However, in what follows we often identify $\kappa^* \partial_\alpha$ with $\partial_\alpha$ in the notation.

### 3.3 Generalized Forces and Their Representations by Variational Hyper-stresses

A generalized force at a configuration $\kappa$ is an element of $T^*_\kappa \mathscr{Q}$. As such, a force $F$ at $\kappa$ is a continuous linear functional in $C^k(\mathscr{X}, \kappa^* V\mathscr{Y})^*$. Henceforth, we will use the concise notation and will write

$$T_\kappa \mathscr{Q} = C^k\big(\kappa^* V\mathscr{Y}\big) = C^k(W), \quad F \in C^k\big(\kappa^* V\mathscr{Y}\big)^* = C^k(W)^*. \tag{3.11}$$

We observe that the jet extension mapping

$$j^k : C^k(W) \longrightarrow C^0\big(J^k W\big), \quad w \longmapsto j^k w, \tag{3.12}$$

is an embedding. (Local representations of the various variables and mappings are provided below.) That is, if an atlas is used in order to define compatible Banach space structures on $C^k(W)$ and $C^0(J^k W)$, then $j^k$ is an isometry (see [20]). Note that $j^k$ is not surjective as non-holonomic sections are not included in the image. It follows from the Hahn-Banach theorem that its dual mapping,

$$j^{k*} : C^0\big(J^k W\big)^* \longrightarrow C^k(W)^*, \tag{3.13}$$

is surjective. One concludes that for a given force $F$, there is some element $\varsigma \in C^0(\mathscr{X}, J^k W)^*$ such that

$$F = j^{k*}\varsigma \quad \text{or} \quad F(w) = \varsigma\big(j^k w\big), \tag{3.14}$$

for every $w \in C^k(W)$. In their form, (3.14) generalize the equation of equilibrium and the expression for the principle of virtual work of continuum mechanics. Specifically, it follows that the power expended by the stress on the jet of a velocity field is equal to the power expended by the force for that velocity field. We obtained equations that are usually considered as representing laws of mechanics from a mathematical representation theorem. It is noted, however, that the stress object contains terms that are not included in the standard treatment. For example, we did not require (nor can we require without an additional geometric structure) that the power vanishes for some particular classes of vector fields (*e.g.*, translations and rotations). In addition, we have no notions such as "internal force" and "external forces".

The fact that $j^k$ is not surjective, implies that the representation is not unique. This non-uniqueness of the stress representation of a given force is a generalization of the well-known static indeterminacy of continuum mechanics.

Since $\varsigma$ is a linear functional on the space of continuous sections of the vector bundle $\pi^k : J^k W \to \mathscr{X}$, it may be represented by a measure $\mu$ valued in $(J^k W)^*$ in the form

$$\varsigma(\dot{A}) = \int_{\mathscr{X}} \dot{A} \cdot \mathrm{d}\mu \tag{3.15}$$

for any section $\dot{A}$ of $\pi^k$.

For a chart in $\mathscr{X}$ with coordinates $(x^i)$ and a local basis $\{\boldsymbol{e}_\alpha\}$ for $W$, a section of $W$ is represented locally in the form

$$w = w^\alpha \boldsymbol{e}_\alpha \tag{3.16}$$

and its jet is represented in the form

$$j^k w = w^\alpha_{,\boldsymbol{I}} \boldsymbol{e}^{\boldsymbol{I}}_\alpha, \quad 0 \le |\boldsymbol{I}| \le k, \tag{3.17}$$

where

$$\boldsymbol{e}^{\boldsymbol{I}}_\alpha := \overleftarrow{\underleftarrow{\mathrm{d}x}}^{\boldsymbol{I}} \otimes \boldsymbol{e}_\alpha. \tag{3.18}$$

Thus, a section $\dot{A}$ of the jet bundle is represented in the form

$$\dot{A} = \dot{A}^\alpha_{\boldsymbol{I}} \boldsymbol{e}^{\boldsymbol{I}}_\alpha, \quad 0 \le |\boldsymbol{I}| \le k. \tag{3.19}$$

Locally, $\mu$ is represented by a collection of real valued measures $\mu^{\boldsymbol{I}}_\alpha$, $|\boldsymbol{I}| \le k$ so that

$$\dot{A} \cdot \mathrm{d}\mu = \dot{A}^\alpha_{\boldsymbol{I}} \cdot \mathrm{d}\mu^{\boldsymbol{I}}_\alpha. \tag{3.20}$$

The representing elements $\varsigma \in C^0(J^k W)^*$ are the *hyper-stresses* as implied by the principle of virtual work above. Furthermore, their role as a means for restriction of forces results from their representation by measures. Thus, a hyper-stress $\varsigma$ represented by the measure $\mu$, induces for any $n$-dimensional submanifold $\mathscr{R}$, a force $F_{\mathscr{R}}$ by

$$F_{\mathscr{R}}(w) = \int_{\mathscr{R}} j^k w \cdot \mathrm{d}\mu. \tag{3.21}$$

Hence, while forces cannot be restricted to submanifolds, the importance of the representation by hyper-stresses follows from the fact that measures may be restricted to Borel sets.

### 3.4 Smooth Variational Hyper-stresses

In what follows, we consider a smooth configuration $\kappa$ and set $W = \kappa^* V\mathscr{Y}$. Furthermore, the focus of this work is placed on smooth force distributions and hyper-stresses, in other words, forces and hyperstresses represented by smooth densities. Consider the vector bundle of linear mappings

$$L\left(J^k W, \textstyle\bigwedge^n T^*\mathscr{X}\right) \simeq \left(J^k W\right)^* \otimes_{\mathscr{X}} \textstyle\bigwedge^n T^*\mathscr{X}. \tag{3.22}$$

A *smooth variational hyper-stress field* is a smooth section $S$ of $L(J^k W, \bigwedge^n T^*\mathscr{X})$. Alternatively, a variational hyper-stress field may be viewed as a smooth vector bundle morphism

$$S : J^k W \longrightarrow \textstyle\bigwedge^n T^*\mathscr{X}, \tag{3.23}$$

where both bundles have $\mathscr{X}$ as the base manifold. A variational hyper-stress field $S$ induces a stress $\varsigma$ by

$$\varsigma(\dot{A}) = \int_{\mathscr{X}} S \cdot \dot{A}. \tag{3.24}$$

The integral on the right hand side makes sense as $S \cdot \dot{A}$ is an $n$-form on $\mathscr{X}$. In particular, the power that a force $F$ expends for the generalized velocity $w$ is given by

$$F(w) = \varsigma\left(j^k w\right) = \int_{\mathscr{X}} S \cdot j^k w, \tag{3.25}$$

and the restriction of the force, induced by the stress, to the subbody $\mathscr{R}$ is given by

$$F_{\mathscr{R}}(w) = \int_{\mathscr{R}} S \cdot j^k w. \tag{3.26}$$

To consider the local representation of the variational hyper-stress densities, we first note that as the dual basis to $\{\overleftarrow{\underleftarrow{\mathrm{d}x}}^{\boldsymbol{I}}\}$ is $\{\partial_{\boldsymbol{I}}\}$, the dual basis corresponding to $\{\boldsymbol{e}^{\boldsymbol{I}}_{\alpha}\}$ is

$$\left\{\boldsymbol{e}^{\alpha}_{\boldsymbol{I}} := \partial_{\boldsymbol{I}} \otimes \boldsymbol{e}^{\alpha}\right\}. \tag{3.27}$$

It follows that a stress density is represented locally in the form

$$S^{\boldsymbol{I}}_{\alpha} \boldsymbol{e}^{\alpha}_{\boldsymbol{I}} \otimes \mathrm{d}x, \quad 0 \leq |\boldsymbol{I}| \leq k, \tag{3.28}$$

and the action of the variational stress density on a section $\dot{A}$ of the jet bundle is given by

$$S \cdot \dot{A} = S^{\boldsymbol{I}}_{\alpha} \dot{A}^{\alpha}_{\boldsymbol{I}} \mathrm{d}x, \quad \text{in particular,} \quad S \cdot j^k w = S^{\boldsymbol{I}}_{\alpha} w^{\alpha}_{,\boldsymbol{I}} \mathrm{d}x. \tag{3.29}$$

Consider next the transformation rule for the components $S^{\boldsymbol{I}}_{\alpha}$. Let $x^{i'}(x^i)$ be a transformation of coordinates in a subset of $\mathscr{X}$ and $w^{\alpha'} = \dot{A}^{\alpha'}_{\alpha}(x) w^{\alpha}$ be the coordinate transformation in $W$. It is noted that by definition, the transformation rule for the components of elements of $J^k W$, may be written in the form

$$w^{\alpha'}_{,\boldsymbol{I}'} = G^{\alpha' \boldsymbol{I}}_{\boldsymbol{I}' \alpha} w^{\alpha}_{,\boldsymbol{I}}, \quad |\boldsymbol{I}| \leq \left|\boldsymbol{I}'\right|, \tag{3.30}$$

where $G^{\alpha' \boldsymbol{I}}_{\boldsymbol{I}' \alpha}$ contains derivatives of the aforementioned transformations.

Using the transformation rules (3.30) and denoting the Jacobian determinant of the transformation by $\mathscr{J}$, we may write

$$\begin{aligned} S^{\boldsymbol{I}}_{\alpha} w^{\alpha}_{,\boldsymbol{I}} \mathrm{d}x &= S^{\boldsymbol{I}'}_{\alpha'} w^{\alpha'}_{,\boldsymbol{I}'} \mathrm{d}x', \quad |\boldsymbol{I}| \leq \left|\boldsymbol{I}'\right|, \\ &= S^{\boldsymbol{I}'}_{\alpha'} G^{\alpha' \boldsymbol{I}}_{\boldsymbol{I}' \alpha} w^{\alpha}_{,\boldsymbol{I}} \mathrm{d}x\, \mathscr{J}. \end{aligned} \tag{3.31}$$

The independence of the point values of the derivatives $w^{\alpha}_{,\boldsymbol{I}}$ imply that the transformation rule for the components $S^{\boldsymbol{I}}_{\alpha}$ is

$$S^{\boldsymbol{I}}_{\alpha} = \mathscr{J} S^{\boldsymbol{I}'}_{\alpha'} G^{\alpha' \boldsymbol{I}}_{\boldsymbol{I}' \alpha}, \quad |\boldsymbol{I}| \leq \left|\boldsymbol{I}'\right| \leq k. \tag{3.32}$$

We conclude that the transformation of the components of the variational stress densities of order $l$ involves all components of equal or higher order $l \leq r \leq k$. In particular, the statement that all the components of order $r > l$ vanish, is invariant. This, of course, gives meaning to the statement that a material is of order $k$ and not any higher order.

# 4 Hyper-stresses, the Cauchy Approach

## 4.1 Body Hyper-forces

A *body hyper-force density* is an element of $L(J^{k-1}W, \bigwedge^n T^*\mathscr{X})$. A *body hyper-force field* is a section of $L(J^{k-1}W, \bigwedge^n T^*\mathscr{X})$. Given a body force field $\boldsymbol{b}$ and a subbody $\mathscr{R} \subset \mathscr{X}$, the total power of the body hyper-force is

$$F_b(\dot{A}) = \int_{\mathscr{R}} \boldsymbol{b} \cdot \dot{A}, \tag{4.1}$$

for any section $\dot{A}$ of $J^{k-1}W$. In particular, for a section $w$ of $W$,

$$F_b(j^{k-1}w) = \int_{\mathscr{R}} \boldsymbol{b} \cdot j^{k-1}w. \tag{4.2}$$

Using the notation introduced above, a body hyper-force field is represented locally in the form

$$\boldsymbol{b} = \boldsymbol{b}^{\boldsymbol{J}}_\alpha \boldsymbol{e}^\alpha_{\boldsymbol{J}} \otimes \mathrm{d}x, \quad 0 \le |\boldsymbol{J}| \le k-1. \tag{4.3}$$

The action of a body force is

$$\boldsymbol{b} \cdot \dot{A} = \boldsymbol{b}^{\boldsymbol{J}}_\alpha \dot{A}^\alpha_{\boldsymbol{J}} \mathrm{d}x, \quad \text{and in particular,} \quad \boldsymbol{b} \cdot j^k w = \boldsymbol{b}^{\boldsymbol{J}}_\alpha w^\alpha_{,\boldsymbol{J}} \mathrm{d}x, \quad 0 \le |\boldsymbol{J}| \le k-1. \tag{4.4}$$

Consider next the transformation rule for the components $\boldsymbol{b}^{\boldsymbol{J}}_\alpha$. Using (3.30), one has

$$\begin{aligned} \boldsymbol{b}^{\boldsymbol{J}}_\alpha w^\alpha_{,\boldsymbol{J}} \mathrm{d}x &= \boldsymbol{b}^{\boldsymbol{J}'}_{\alpha'} w^{\alpha'}_{,\boldsymbol{J}'} \mathrm{d}x', \quad |\boldsymbol{J}| \le |\boldsymbol{J}'| \le k-1, \\ &= \boldsymbol{b}^{\boldsymbol{J}'}_{\alpha'} G^{\alpha' \boldsymbol{J}}_{\boldsymbol{J}'\alpha} w^\alpha_{,\boldsymbol{J}} \mathrm{d}x \, \mathscr{J}. \end{aligned} \tag{4.5}$$

The independence of the point values of the derivatives $w^\alpha_{,\boldsymbol{I}}$ imply that the transformation rule for the components $S^{\boldsymbol{I}}_\alpha$ is

$$\boldsymbol{b}^{\boldsymbol{J}}_\alpha = \mathscr{J} \boldsymbol{b}^{\boldsymbol{J}'}_{\alpha'} G^{\alpha' \boldsymbol{J}}_{\boldsymbol{J}'\alpha}, \quad |\boldsymbol{J}| \le |\boldsymbol{J}'| \le k-1. \tag{4.6}$$

Similarly to the variational stress, we conclude that the transformation of the components of the body hyper-force densities of order $l$ involves all components of equal or higher order $l \le r \le k$. In particular, the statement that all the components of order $r > l$ vanish, is invariant.

Thus, for $k$-th order continuum mechanics, body hyper-forces are like variational hyper-stresses for continuum mechanics of order $k-1$. All the analysis concerning variational hyper-stresses applies to body hyper-forces. It observed that in case a metric structure is provided, body hyper-forces may include densities of couples.

## 4.2 Hyper-traction Fields

Let $\mathscr{R} \subset \mathscr{X}$ be a subbody. A *surface hyper-force density* or a *hyper-traction* on $\partial\mathscr{R}$ is an element of $L(J^{k-1}W, \bigwedge^{n-1} T^*\partial\mathscr{R})$. Here, and in what follows, we view $L(J^{k-1}W, \bigwedge^{n-1} T^*\partial\mathscr{R})$ as a vector bundle over $\partial\mathscr{R}$ and in our notation we omit the indication that $J^{k-1}W$ is restricted to $\partial\mathscr{R}$. A *hyper-traction field* on $\partial\mathscr{R}$ is a section of

$L(J^{k-1}W, \bigwedge^{n-1}T^*\partial\mathscr{R})$. Given a subbody $\mathscr{R} \subset \mathscr{X}$ and a hyper-traction field $\boldsymbol{t}_{\mathscr{R}}$ and, the total power of the hyper-traction is the action of the functional

$$F_t(\dot{A}) = \int_{\partial\mathscr{R}} \boldsymbol{t}_{\mathscr{R}} \cdot \dot{A}, \tag{4.7}$$

for any section $\dot{A}$ of $(J^{k-1}W)|_{\partial\mathscr{R}}$. In particular, for a section $w$ of $W$,

$$F_t\left(j^{k-1}w\right) = \int_{\partial\mathscr{R}} \boldsymbol{t}_{\mathscr{R}} \cdot j^{k-1}w. \tag{4.8}$$

It is noted that the jet of $w$ is taken relative to $\mathscr{X}$ and not $\partial\mathscr{R}$.

Let $(z^p)$, $p = 1, \ldots, n-1$ be local coordinates in $\partial\mathscr{R}$, so that

$$\mathrm{d}z = \mathrm{d}z^1 \wedge \cdots \wedge \mathrm{d}z^{n-1} \tag{4.9}$$

is a local volume form. Then, in an adapted coordinate system, a hyper-traction field is represented locally in the form

$$\boldsymbol{t}_{\mathscr{R}} = \boldsymbol{t}^{\boldsymbol{J}}_{\mathscr{R}\alpha} \boldsymbol{e}^{\alpha}_{\boldsymbol{J}} \otimes \mathrm{d}z, \quad 0 \le |\boldsymbol{J}| \le k-1. \tag{4.10}$$

The action of a hyper-traction is

$$\boldsymbol{t}_{\mathscr{R}} \cdot \dot{A} = \boldsymbol{t}^{\boldsymbol{J}}_{\mathscr{R}\alpha} \dot{A}^{\alpha}_{\boldsymbol{J}} \mathrm{d}z, \quad \text{and in particular,} \quad \boldsymbol{t}_{\mathscr{R}} \cdot j^k w = \boldsymbol{t}^{\boldsymbol{J}}_{\mathscr{R}\alpha} w^{\alpha}_{,\boldsymbol{J}} \mathrm{d}z, \quad 0 \le |\boldsymbol{J}| \le k-1. \tag{4.11}$$

## 4.3 Smooth Force Functionals and Hyper-force Systems

We will say that a force functional on a subbody $\mathscr{R}$ is smooth if it is induced by a body hyper-force field $\boldsymbol{b}$ and a hyper-traction field $\boldsymbol{t}_{\mathscr{R}}$ in the form

$$F_{\mathscr{R}}(w) = \int_{\mathscr{R}} \boldsymbol{b} \cdot j^{k-1}w + \int_{\partial\mathscr{R}} \boldsymbol{t}_{\mathscr{R}} \cdot j^{k-1}w \tag{4.12}$$

for every virtual displacement $w : \mathscr{X} \to W$. Henceforth, we will be interested only in smooth force functionals.

Let $\{\boldsymbol{t}_{\mathscr{R}}\}$ be a collection of sections $\boldsymbol{t}_{\mathscr{R}}$ of $L(U|_{\partial\mathscr{R}}, \bigwedge^{n-1}T^*\partial\mathscr{R})$ for all subbodies $\mathscr{R} \subset \mathscr{X}$. We will refer to $\{\boldsymbol{t}_{\mathscr{R}}\}$ as a *system of hyper-tractions*. The collection $(\boldsymbol{b}, \{\boldsymbol{t}_R\})$ will be referred to as a *smooth system of hyper-forces*, where each force functional in $\{F_{\mathscr{R}}\}$ is represented as in the equation above.

## 4.4 Traction Hyper-stresses

A *traction hyper-stress* is an element

$$\sigma_0 \in L\left(J^{k-1}W, \bigwedge^{n-1}T^*\mathscr{X}\right) \tag{4.13}$$

and a *traction hyper-stress field* is a smooth section $\sigma$ of $L(J^{k-1}W, \bigwedge^{n-1}T^*\mathscr{X})$.

Let $\mathscr{R} \subset \mathscr{X}$ be an $n$-dimensional submanifold with boundary $\partial\mathscr{R}$. The inclusion

$$\iota_{\partial\mathscr{R}} : \partial\mathscr{R} \longrightarrow \mathscr{X}, \tag{4.14}$$

induces the following diagram where $(\iota_{\partial\mathscr{R}})^*T\mathscr{X} \simeq T\mathscr{X}|_{\partial\mathscr{R}}$ is the pullback of the tangent bundle and $\delta\iota_{\partial\mathscr{R}}$ is the induced vector bundle morphism over the boundary that makes the diagram commutative—an inclusion of tangent vectors.

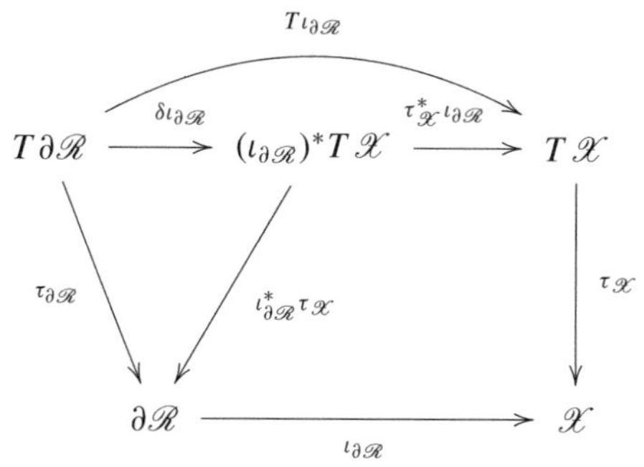


The dual vector bundle morphism,

$$\rho_{\partial\mathscr{R}} = (\delta\iota_{\partial\mathscr{R}})^* : \left(\bigwedge^{n-1}T^*\mathscr{X}\right)\Big|_{\partial\mathscr{R}} \simeq \bigwedge^{n-1}\left(\iota^*_{\partial\mathscr{R}}T\mathscr{X}\right)^* \longrightarrow \bigwedge^{n-1}T^*\partial\mathscr{R}, \tag{4.15}$$

restricts alternating tensors to vectors tangent to the boundary and it induces the pullback of forms.

Let $z \in \partial\mathscr{R}$, and $\sigma_0 \in L(J^{k-1}W, \bigwedge^{n-1}T^*\mathscr{X})_z$. Then,

$$\boldsymbol{t}_0 = \rho_{\partial\mathscr{R}} \circ \sigma_0 \in L\left(J^{k-1}W, \bigwedge^{n-1}T^*\partial\mathscr{R}\right)_z \tag{4.16}$$

is a hyper-traction induced at $z$ by $\sigma_0$. Similarly, a traction hyper-stress field $\sigma$ induces a hyper-traction field

$$\boldsymbol{t} = \rho_{\partial\mathscr{R}} \circ \sigma = \iota^{\sharp}_{\partial\mathscr{R}}\sigma. \tag{4.17}$$

Thus, (4.17) is a generalization of the traditional Cauchy formula.

Let $v \lrcorner \omega$ denote the contraction (inner product) of the form $\omega$ with the tangent vector $v$. As $\{\partial_i \lrcorner \mathrm{d}x\}$ may serve as a basis for $\bigwedge^{n-1}T^*\mathscr{X}$, in analogy with (3.28), a traction hyper-stress field may be represented in the form

$$\sigma = \sigma_\alpha^{\boldsymbol{J}i}\boldsymbol{e}^\alpha_{\boldsymbol{J}} \otimes (\partial_i \lrcorner \mathrm{d}x), \quad |\boldsymbol{J}| \leq k-1. \tag{4.18}$$

For a section $\dot{B}$ of $J^{k-1}W$, one has therefore,

$$\sigma \cdot \dot{B} = \sigma_\alpha^{\boldsymbol{J}i}\dot{B}^\alpha_{\boldsymbol{J}}(\partial_i \lrcorner \mathrm{d}x), \tag{4.19}$$

and in particular,

$$\sigma \cdot j^{k-1}w = \sigma_\alpha^{\boldsymbol{J}i} w^\alpha_{,\boldsymbol{J}}(\partial_i \lrcorner \mathrm{d}x), \quad |\boldsymbol{J}| \leq k-1. \tag{4.20}$$

Consider next the transformation rule for the components $\sigma_\alpha^{\boldsymbol{J}i}$. Using the transformations (3.30) and the invariance of the action, we have

$$\begin{aligned} \sigma_\alpha^{\boldsymbol{I}i} w^\alpha_{,\boldsymbol{I}}\partial_i \lrcorner \mathrm{d}x &= \sigma_{\alpha'}^{\boldsymbol{I}'i'} w^{\alpha'}_{,\boldsymbol{I}'}\partial_{i'} \lrcorner \mathrm{d}x', \quad |\boldsymbol{I}| \leq |\boldsymbol{I}'|, \\ &= \sigma_{\alpha'}^{\boldsymbol{I}'i'} G^{\alpha'\boldsymbol{I}}_{\boldsymbol{I}'\alpha} w^\alpha_{,\boldsymbol{I}} x^i_{,i'}\partial_i \lrcorner \mathrm{d}x\,\mathscr{J}. \end{aligned} \tag{4.21}$$

Thus,

$$\sigma_{\alpha}^{Ii} = \mathscr{J} \sigma_{\alpha'}^{I'i'} G_{I'\alpha}^{\alpha' I} x_{,i'}^{i}, \quad |I| \leq |I'|. \tag{4.22}$$

The last equation is a generalization of the classical relation between the Cauchy stress and the first Piola-Kirchhoff stress, which as presented here, is just a transformation of coordinates of the right geometric object. Once again, the transformation of the components of the traction hyper-stress densities of order $l$ involves all components of equal or higher order $l \leq r \leq k-1$.

Finally, given a traction hyper-stress, $\sigma$, and a body hyper-force, $\boldsymbol{b}$, one may write the force functional on each subbody as

$$\begin{aligned} F_{\mathscr{R}}(w) &= \int_{\mathscr{R}} \boldsymbol{b} \cdot j^{k-1} w + \int_{\partial \mathscr{R}} \sigma \cdot j^{k-1} w, \\ &= \int_{\mathscr{R}} \boldsymbol{b} \cdot j^{k-1} w + \int_{\mathscr{R}} \mathrm{d}(\sigma \cdot j^{k-1} w). \end{aligned} \tag{4.23}$$

## 4.5 Cauchy's Theorem and Traction Hyper-stresses

We review here Cauchy's theorem on manifolds as in [25, 28]. The Cauchy theorem asserts that under certain boundedness and locality conditions, a system of forces on all subbodies induce a unique traction stress field. The theorem is formulated for a general vector bundle $U$ and the terminology used is for the case of standard continuum mechanics, $k = 1$. We will use it later for the case $U = J^{k-1} W$.

It is assumed in the sequel that the collection of subbodies of $\mathscr{X}$ includes $n$-dimensional chains and in particular simplices. Furthermore, it is assumed that a particular orientation has been chosen for the manifold $\mathscr{X}$.

**Definition 1** We will say that a system $\{\boldsymbol{t}_{\mathscr{R}}\}$ of tractions is *consistent* if the following conditions are satisfied.

(1) **Boundedness.** There is a section $\xi$ of $L(J^1 U, \bigwedge^n T^* \mathscr{X})$ such that for each $\mathscr{R}$,

$$\left| \int_{\partial \mathscr{R}} \boldsymbol{t}_{\mathscr{R}}(u|_{\partial \mathscr{R}}) \right| \leqslant \int_{\mathscr{R}} \left| \xi(j^1 u) \right|, \tag{4.24}$$

for every smooth section $u$ of $U$.

(2) **Cauchy's postulate of locality.** Let $v_1, \ldots, v_{n-1} \in T_x \mathscr{X}$ be a collection of tangent vector at $x \in \mathscr{X}$. Let $\mathscr{R}$ be any subbody such that $v_1, \ldots, v_{n-1} \in T_x \partial \mathscr{R}$, and the collection of vectors is positively oriented relative to the orientation of $\partial \mathscr{R}$. Then, $\boldsymbol{t}_{\mathscr{R}}(u)(v_1, \ldots, v_{n-1})$ is independent of $\mathscr{R}$. Thus, there is an alternating multilinear mapping

$$\Sigma_x : (T_x \mathscr{X})^{n-1} \longrightarrow U_x^* \tag{4.25}$$

such that

$$\boldsymbol{t}_{\mathscr{R}}(u)(v_1, \ldots, v_{n-1}) = \Sigma_x(v_1, \ldots, v_{n-1})(u), \quad \text{for all } u \in U_x. \tag{4.26}$$

(3) **Regularity.** The mapping $x \mapsto \Sigma_x$ is smooth.

A smooth system of forces $(\boldsymbol{b}, \{\boldsymbol{t}_{\mathscr{R}}\})$ is said to be consistent if $\{\boldsymbol{t}_{\mathscr{R}}\}$ is a consistent system of tractions.

**Theorem 2** *Let $\{t_{\mathscr{R}}\}$ be a consistent force system, then, there is a unique traction stress field $\sigma$ such that for each subbody $\mathscr{R}$, the surface force is given by the Cauchy formula*

$$t_{\mathscr{R}} = \rho_{\partial\mathscr{R}} \circ \sigma. \tag{4.27}$$

Cauchy's theorem, in the generalized setting outlined above, applies readily to traction hyper-stresses by setting $U = J^{k-1}W$. Assuming that the system of hyper-tractions satisfies the conditions above, it follows from Cauchy's theorem that there is a unique traction hyper-stress that induces it using (4.27).

## 5 Non-holonomic Variational Stresses

Unlike the simple case, $k = 1$, variational hyper-stresses do not determine uniquely traction hyper-stresses for higher order theories. Another mathematical object is needed—the non-holonomic stress.

### 5.1 The Exterior Jet and Non-holonomic Variational Stresses

**Proposition 3** *There is a natural linear, first order-differential operator, the exterior jet $\mathfrak{d}$, taking sections of $L(J^{k-1}W, \bigwedge^{n-1}T^*\mathscr{X})$ into sections of $L(J^1(J^{k-1}W), \bigwedge^n T^*\mathscr{X})$, defined by the condition*

$$\mathfrak{d}\sigma \cdot j^1\dot{A} = d(\sigma \cdot \dot{A}) \tag{5.1}$$

*for every section $\dot{A}$ of $J^{k-1}W$.*

*Proof* Let $\sigma$ be a traction hyper-stress field represented as in (4.18) and let a generic section $\dot{A}$ of $J^{k-1}W$ be given locally by $\dot{A} = \dot{A}^\alpha_J e^J_\alpha$. One has

$$\mathrm{d}(\sigma \cdot \dot{A}) = \left(\sigma^{Ji}_{\alpha,i}\dot{A}^\alpha_J + \sigma^{Ji}_\alpha \dot{A}^\alpha_{J,i}\right)\mathrm{d}x, \quad |J| \leq k-1. \tag{5.2}$$

It follows that $\mathrm{d}(\sigma \cdot \dot{A})$ is an $n$-form that depends pointwise on the values of the representatives of $\dot{A}$ and their first derivatives, *i.e.*, it depends on $j^1\dot{A}$. Hence, there is a unique section $\mathfrak{d}\sigma$ of $L(J^1(J^{k-1}W), \bigwedge^n T^*\mathscr{X})$, for which condition (5.1) holds. The value of the section $\mathfrak{d}\sigma$ at a point $x \in \mathscr{X}$ depends linearly on the first jet at $x$ of the section $\sigma : \mathscr{X} \to L(J^{k-1}W, \bigwedge^{n-1}T^*\mathscr{X})$. Hence, $\mathfrak{d}$ is a first order linear differential operator. □

**Definition 4** An element of $L(J^1(J^{k-1}W), \bigwedge^n T^*\mathscr{X})$ will be referred to as a *non-holonomic (variational hyper-) stress.* A section of $L(J^1(J^{k-1}W), \bigwedge^n T^*\mathscr{X})$ is a *non-holonomic stress field.*

Thus, $\mathfrak{d}\sigma$ is a non-holonomic stress field. As noted, the terminology follows from the fact that an element of $J^1(J^{k-1}W)$ need not be associated with the value of a $k$-th jet of a section of $W$.

## 5.2 Local Representation of Non-holonomic Stresses and Exterior Jets

An element $\dot{B} \in J^1(J^{k-1}W)$ is represented locally in the form

$$\dot{B} = \dot{B}^\alpha_{\boldsymbol{J}} \boldsymbol{e}^{\boldsymbol{J}}_\alpha + \dot{B}^\alpha_{\boldsymbol{J}j} \boldsymbol{e}^{\boldsymbol{J}}_\alpha \otimes \mathrm{d}x^j, \tag{5.3}$$

where it is noted that while $\dot{B}^\alpha_{\boldsymbol{J}}$ and $\dot{B}^\alpha_{\boldsymbol{J}j}$ are symmetric under permutations of $\boldsymbol{J}$, $\dot{B}^\alpha_{\boldsymbol{J}j}$ need not be symmetric under permutations of $\boldsymbol{J}j$. For a section $\dot{A}$ of $J^{k-1}W$, represented locally in the form $\dot{A} = \dot{A}^\alpha_{\boldsymbol{J}} \boldsymbol{e}^{\boldsymbol{J}}_\alpha$, one has locally

$$j^1 \dot{A} = \dot{A}^\alpha_{\boldsymbol{J}} \boldsymbol{e}^{\boldsymbol{J}}_\alpha + \dot{A}^\alpha_{\boldsymbol{J},j} \boldsymbol{e}^{\boldsymbol{J}}_\alpha \otimes \mathrm{d}x^j, \tag{5.4}$$

and for a section $w$ of $W$,

$$j^1(j^{k-1}w) = w^\alpha_{,\boldsymbol{J}} \boldsymbol{e}^{\boldsymbol{J}}_\alpha + w^\alpha_{,\boldsymbol{J}j} \boldsymbol{e}^{\boldsymbol{J}}_\alpha \otimes \mathrm{d}x^j. \tag{5.5}$$

Evidently, $j^1(j^{k-1}w)$ represents a holonomic element of $J^1(J^{k-1}W)$ for which complete symmetry holds.

It follows that a non-holonomic stress should be locally of the form

$$P = \left(P^{\boldsymbol{J}}_\alpha \boldsymbol{e}^\alpha_{\boldsymbol{J}} + \underline{P}^{\boldsymbol{J}j}_\alpha \boldsymbol{e}^\alpha_{\boldsymbol{J}} \otimes \partial_j\right) \otimes \mathrm{d}x. \tag{5.6}$$

Its action is given by

$$P(\dot{B}) = \left(P^{\boldsymbol{J}}_\alpha \dot{B}^\alpha_{\boldsymbol{J}} + \underline{P}^{\boldsymbol{J}j}_\alpha \dot{B}^\alpha_{\boldsymbol{J}j}\right)\mathrm{d}x, \qquad P(j^1\dot{A}) = \left(P^{\boldsymbol{J}}_\alpha \dot{A}^\alpha_{\boldsymbol{J}} + \underline{P}^{\boldsymbol{J}j}_\alpha \dot{A}^\alpha_{\boldsymbol{J},j}\right)\mathrm{d}x, \tag{5.7}$$

and

$$P\left(j^1(j^{k-1}w)\right) = \left(P^{\boldsymbol{J}}_\alpha w^\alpha_{,\boldsymbol{J}} + \underline{P}^{\boldsymbol{J}j}_\alpha w^\alpha_{,\boldsymbol{J}j}\right)\mathrm{d}x. \tag{5.8}$$

Given a traction hyper-stress field $\sigma$, the exterior jet field should be of the form

$$\mathfrak{d}\sigma = \left((\mathfrak{d}\sigma)^{\boldsymbol{J}}_\alpha \boldsymbol{e}^\alpha_{\boldsymbol{J}} + \underline{(\mathfrak{d}\sigma)}^{\boldsymbol{J}j}_\alpha \boldsymbol{e}^\alpha_{\boldsymbol{J}} \otimes \partial_j\right) \otimes \mathrm{d}x. \tag{5.9}$$

The definition of the exterior jet implies that for any section $\dot{A}$ of $J^{k-1}W$, $\mathfrak{d}\sigma \cdot j^1\dot{A} = \mathrm{d}(\sigma \cdot \dot{A})$, and so

$$(\mathfrak{d}\sigma)^{\boldsymbol{J}}_\alpha \dot{A}^\alpha_{\boldsymbol{J}} + \underline{(\mathfrak{d}\sigma)}^{\boldsymbol{J}j}_\alpha \dot{A}^\alpha_{\boldsymbol{J},j} = \sigma^{\boldsymbol{J}j}_{\alpha,j} \dot{A}^\alpha_{\boldsymbol{J}} + \sigma^{\boldsymbol{J}j}_\alpha \dot{A}^\alpha_{\boldsymbol{J},j}. \tag{5.10}$$

At any point $x$, the values of the components $\dot{A}^\alpha_{\boldsymbol{J}}$ and $\dot{A}^\alpha_{\boldsymbol{J},j}$ are independent. Thus,

$$(\mathfrak{d}\sigma)^{\boldsymbol{J}}_\alpha = \sigma^{\boldsymbol{J}j}_{\alpha,j}, \qquad \underline{(\mathfrak{d}\sigma)}^{\boldsymbol{J}j}_\alpha = \sigma^{\boldsymbol{J}j}_\alpha, \tag{5.11}$$

$$\mathfrak{d}\sigma = \left(\sigma^{\boldsymbol{J}j}_{\alpha,j} \boldsymbol{e}^\alpha_{\boldsymbol{J}} + \sigma^{\boldsymbol{J}j}_\alpha \boldsymbol{e}^\alpha_{\boldsymbol{J}} \otimes \partial_j\right) \otimes \mathrm{d}x. \tag{5.12}$$

## 5.3 The Non-holonomic Stress Induced by a Consistent Hyper-force System

We now consider systems consisting of body hyper-forces and hyper-tractions as in (4.12). We will say that the hyper-force system is consistent if it satisfies the conditions of consistency of the generalized Cauchy theorem for the case $U = J^{k-1}W$.

**Proposition 5** *A smooth consistent hyper-force system* $(\boldsymbol{b}, \{\boldsymbol{t}_{\mathscr{R}}\})$ *induces a unique non-holonomic stress field* $P$ *which represents it by*

$$F_{\mathscr{R}}(w) = \int_{\mathscr{R}} P \cdot j^1\big(j^{k-1}w\big). \tag{5.13}$$

*Proof* A consistent hyper-force system induces, by the generalized Cauchy theorem for the case $U = J^{k-1}W$, a unique traction hyper-stress $\sigma$. Thus, (4.23) applies, and using the definition of the exterior jet, we may rewrite it as

$$F_{\mathscr{R}}(w) = \int_{\mathscr{R}} \boldsymbol{b} \cdot j^{k-1}w + \eth\sigma \cdot j^1\big(j^{k-1}w\big). \tag{5.14}$$

Observing (5.14), the value of the integrand at any $x \in \mathscr{X}$ is an $n$-alternating tensor that is evidently a linear function of $j^1(j^{k-1}w)(x)$. For a section $\dot{A}$ of $J^{k-1}W$, set

$$P \cdot j^1\dot{A} = \boldsymbol{b} \cdot \dot{A} + \eth\sigma \cdot j^1\dot{A}. \tag{5.15}$$

Thus, $P$ is a section of $L(J^1(J^{k-1}W), \bigwedge^n T^*\mathscr{X})$ such that

$$F_{\mathscr{R}}(w) = \int_{\mathscr{R}} \big(\boldsymbol{b} \cdot j^{k-1}w + \eth\sigma \cdot j^1\big(j^{k-1}w\big)\big) = \int_{\mathscr{R}} P \cdot j^1\big(j^{k-1}w\big). \tag{5.16}$$

In fact, $P$ induces a linear functional $\widehat{P}_{\mathscr{R}}$, which extends $F_{\mathscr{R}}$ to act on all sections of $J^{k-1}W$ by

$$\widehat{P}_{\mathscr{R}}(\dot{A}) = \int_{\mathscr{R}} \big(\boldsymbol{b} \cdot \dot{A} + \eth\sigma \cdot j^1\dot{A}\big) = \int_{\mathscr{R}} P \cdot j^1\dot{A}. \tag{5.17}$$

The local representation of $P$ may be readily obtained from (5.12) as

$$P = \big(\big(\sigma^{\boldsymbol{J}j}_{\alpha,j} + \boldsymbol{b}^{\boldsymbol{J}}_{\alpha}\big)\boldsymbol{e}^{\alpha}_{\boldsymbol{J}} + \sigma^{\boldsymbol{J}j}_{\alpha}\boldsymbol{e}^{\alpha}_{\boldsymbol{J}} \otimes \partial_j\big) \otimes \mathrm{d}x, \quad 0 \le |\boldsymbol{J}| \le k-1, \tag{5.18}$$

$$P^{\boldsymbol{J}}_{\alpha} = \sigma^{\boldsymbol{J}j}_{\alpha,j} + \boldsymbol{b}^{\boldsymbol{J}}_{\alpha}, \qquad \underline{P}^{\boldsymbol{J}j}_{\alpha} = \sigma^{\boldsymbol{J}j}_{\alpha}. \tag{5.19}$$

□

**Proposition 6** *There is a natural mapping*

$$p_\sigma : L\Big(J^1\big(J^{k-1}W\big), \textstyle\bigwedge^n T^*\mathscr{X}\Big) \longrightarrow L\Big(J^{k-1}W, \textstyle\bigwedge^{n-1} T^*\mathscr{X}\Big), \tag{5.20}$$

*whereby a non-holonomic stress field* $P$ *induces a unique traction hyper-stress* $\sigma$.

*Proof* The local expression for the action of $P$ is

$$P \cdot j^1\dot{A} = \big(P^{\boldsymbol{J}}_{\alpha}\dot{A}^{\alpha}_{\boldsymbol{J}} + \underline{P}^{\boldsymbol{J}j}_{\alpha}\dot{A}^{\alpha}_{\boldsymbol{J},j}\big) \otimes \mathrm{d}x, \quad 0 \le |\boldsymbol{J}| \le k-1. \tag{5.21}$$

Restricting $P(x)$ to sections for which $\dot{A}^{\alpha}_{\boldsymbol{J}}(x) = 0$ for all $\boldsymbol{J}$, $|\boldsymbol{J}| \le k-1$, *i.e.*, to elements of

$$V^0 J^1\big(J^{k-1}W\big) = \mathrm{Kernel}\,\pi^1_0, \tag{5.22}$$

the action of $P$ determines $\sigma$ uniquely. In other words, the inclusion $\iota_1^0$ of the vertical bundle induces

$$\begin{aligned}\iota_1^{0*} : L\Big(J^1(J^{k-1}W), \textstyle\bigwedge^n T^*\mathscr{X}\Big) &\longrightarrow L\Big(VJ^1(J^{k-1}W), \textstyle\bigwedge^n T^*\mathscr{X}\Big),\\ &\simeq L\Big(L(T\mathscr{X}, J^{k-1}W), \textstyle\bigwedge^n T^*\mathscr{X}\Big),\\ &\simeq T\mathscr{X} \otimes (J^{k-1}W)^* \otimes \textstyle\bigwedge^n T^*\mathscr{X},\\ &\simeq (J^{k-1}W)^* \otimes \textstyle\bigwedge^{n-1} T^*\mathscr{X},\\ &\simeq L\Big(J^{k-1}W, \textstyle\bigwedge^{n-1}\mathscr{X}\Big),\end{aligned} \tag{5.23}$$

where we used the natural isomorphism $VJ^1U \simeq L(T\mathscr{X}, U)$ for any vector bundle $U$ over $\mathscr{X}$, and the isomorphism

$$T\mathscr{X} \otimes \textstyle\bigwedge^n T^*\mathscr{X} \longrightarrow \textstyle\bigwedge^{n-1} T^*\mathscr{X}, \quad (v, \theta) \longmapsto v \lrcorner \theta. \tag{5.24}$$

Thus, locally,

$$\sigma_\alpha^{\boldsymbol{J}j} = \underline{P}_\alpha^{\boldsymbol{J}j}. \tag{5.25}$$

□

**Proposition 7** *There is a natural linear differential operator* div *mapping jets of non-holonomic stress fields into body hyper-force fields. The differential operator* div *is defined by*

$$\operatorname{div} P = \partial \circ p_\sigma(P) - P. \tag{5.26}$$

*The non-holonomic stress $P$ represents the force system $(\boldsymbol{b}, \{\boldsymbol{t}_{\mathscr{R}}\})$ if an only if*

$$p_\sigma \circ P = \sigma, \quad \textit{and} \quad \operatorname{div} P + \boldsymbol{b} = 0. \tag{5.27}$$

*Proof* To prove that $\operatorname{div} P$ is a section of $L(J^{k-1}W, \bigwedge^n T^*\mathscr{X})$, we use (5.25) and (5.12). Thus, locally, for $0 \le |\boldsymbol{J}| \le k-1$,

$$\begin{aligned}\operatorname{div} P &= \big(\underline{P}_{\alpha,j}^{\boldsymbol{J}j} \boldsymbol{e}_{\boldsymbol{J}}^\alpha + \underline{P}_\alpha^{\boldsymbol{J}j} \boldsymbol{e}_{\boldsymbol{J}}^\alpha \otimes \partial_j - P_\alpha^{\boldsymbol{J}} \boldsymbol{e}_{\boldsymbol{J}}^\alpha - \underline{P}_\alpha^{\boldsymbol{J}j} \boldsymbol{e}_{\boldsymbol{J}}^\alpha \otimes \partial_j\big) \otimes \mathrm{d}x,\\ &= \big(\underline{P}_{\alpha,j}^{\boldsymbol{J}j} - P_\alpha^{\boldsymbol{J}}\big) \boldsymbol{e}_{\boldsymbol{J}}^\alpha \otimes \mathrm{d}x.\end{aligned} \tag{5.28}$$

In view of Propositions 5 and 6, it remains to show that $\boldsymbol{b} = -\operatorname{div} P$. This follows, however, from (5.15) and (5.26). □

## 6 Induced Variational Stresses

One could naively assume that just as in the case $k = 1$, a variational hyper-stress, a section of $L(J^k W, \bigwedge^n T^*\mathscr{X})$ will induce a unique traction hyper-stress. This, however, is not the case. It is sufficient to observe that in the local expression for the variational hyper-stress $S_\alpha^{\boldsymbol{I}} \boldsymbol{e}_{\boldsymbol{I}}^\alpha \otimes \mathrm{d}x$, $|\boldsymbol{I}| \le k$, the components $S_\alpha^{\boldsymbol{I}}$ are symmetric under permutations of $\boldsymbol{I}$. On the other hand, in the local expression for the traction hyper-stress $\sigma_\alpha^{\boldsymbol{J}i} \boldsymbol{e}_{\boldsymbol{J}}^\alpha \otimes (\partial_i \lrcorner \mathrm{d}x)$, $|\boldsymbol{J}| \le k-1$, the arrays $\sigma^{Ji}$, which are of the same order as $S_\alpha^{\boldsymbol{I}}$, need not be symmetric under all permutations of $\boldsymbol{J}j$. In other words, unlike the non-holonomic stress that is in one-to-one correspondence with a force system, the variational hyper-stress encodes less information and cannot be used to determine the hyper-traction on the boundaries of subbodies. The present section studies these issues with some detail.

## 6.1 The Variational Hyper-stress Induced by a Non-holonomic Stress

Let

$$\iota_{1,k-1}^{k} : J^k W \longrightarrow J^1\big(J^{k-1}W\big), \qquad j^k w(x) \longmapsto j^1\big(j^{k-1}w\big)(x), \tag{6.1}$$

be the natural inclusion of holonomic elements. Then, the dual mapping

$$\iota_{1,k-1}^{k*} : L\Big(J^1\big(J^{k-1}W\big), \textstyle\bigwedge^n T^*\mathscr{X}\Big) \longrightarrow L\Big(J^k W, \textstyle\bigwedge^n T^*\mathscr{X}\Big), \quad P \longmapsto P \circ \iota_{1,k-1}^{k}, \tag{6.2}$$

assigns variational hyper-stresses to non-holonomic stresses. Let $S = \iota_{1,k-1}^{k*}(P)$. The local representations of the actions of $S$ and $P$ imply that for any section $w$ of $W$,

$$\begin{aligned}
S_\alpha^{\boldsymbol{I}} w^\alpha_{,\boldsymbol{I}} &= P_\alpha^{\boldsymbol{J}} w^\alpha_{,\boldsymbol{J}} + \underline{P}_\alpha^{\boldsymbol{J}j} w_{,\boldsymbol{J}j}, && 0 \le |\boldsymbol{I}| \le k, \quad 0 \le |\boldsymbol{J}| \le k-1,\\
&= P_\alpha w^\alpha + \overrightarrow{P}_\alpha^{\boldsymbol{L}j} w^\alpha_{,\boldsymbol{L}j} + \underline{P}_\alpha^{\boldsymbol{L}j} w^\alpha_{,\boldsymbol{L}j} + \overleftarrow{\underline{P}}_\alpha^{\boldsymbol{K}} w^\alpha_{,\boldsymbol{K}}, && 0 \le |\boldsymbol{L}| \le k-2, \quad |\boldsymbol{K}| = k,\\
&= P_\alpha w^\alpha + \big(\overrightarrow{P}_\alpha^{\boldsymbol{L}j} + \underline{P}_\alpha^{\boldsymbol{L}j}\big) w^\alpha_{,\boldsymbol{L}j} + \overleftarrow{\underline{P}}_\alpha^{\boldsymbol{K}} w^\alpha_{,\boldsymbol{K}},\\
&= P_\alpha w^\alpha + \big(P_\alpha^{\boldsymbol{J}} + \overleftarrow{\underline{P}}_\alpha^{\boldsymbol{J}}\big) w^\alpha_{,\boldsymbol{J}} + \overleftarrow{\underline{P}}_\alpha^{\boldsymbol{K}} w^\alpha_{,\boldsymbol{K}}, && 1 \le |\boldsymbol{J}| \le k-1, \quad |\boldsymbol{K}| = k.
\end{aligned}$$

As the relations above should hold for any point $x \in \mathscr{X}$ and since compatibility imposes no restrictions on the point values of the various derivatives, we conclude that $S = \iota_{1,k-1}^{k*}(P)$ is represented by

$$\begin{aligned}
S_\alpha^{\boldsymbol{K}} &= \overleftarrow{\underline{P}}_\alpha^{(\boldsymbol{K})}, && |\boldsymbol{K}| = k,\\
S_\alpha^{\boldsymbol{J}} &= P_\alpha^{\boldsymbol{J}} + \overleftarrow{\underline{P}}_\alpha^{(\boldsymbol{J})}, && 1 \le |\boldsymbol{J}| < k,\\
S_\alpha &= P_\alpha.
\end{aligned} \tag{6.3}$$

*Remark 8* It is emphasized that without additional geometric structure there is no projection $J^1(J^{k-1}W) \to J^k W$, an inverse to the inclusion above. (See, for example [23, p. 169].) That is, a non-holonomic section does not determine a holonomic section. As a consequence, a variational hyper-stress cannot determine by duality a unique non-holonomic stress. Accepting this somewhat disappointing obstacle, one might consider the following alternative to the procedure of representation of forces by variational stresses in Sect. 3.3. Let the completely non-holonomic, iterated jet bundle

$$\hat{\pi}^k : \hat{J}^k W \longrightarrow \mathscr{X} \tag{6.4}$$

be defined by

$$\hat{J}^1 W = J^1 W, \qquad \hat{J}^r W = J^1\big(\hat{J}^{r-1} W\big), \tag{6.5}$$

and let

$$\hat{j}^k : C^k(W) \longrightarrow C^0\big(\hat{J}^k W\big) \tag{6.6}$$

be the natural repeated jet extension. Then, the analog of the construction in Sect. 3.3 may be carried out replacing $j^k$ by $\hat{j}^k$. This will lead to representation of forces by completely non-holonomic stresses, elements $\Pi$ of $C^0(\hat{J}^k W)^*$, in the form

$$F = \hat{j}^{k*}(\Pi). \tag{6.7}$$

## 6.2 The Reduced Exterior Jet

The *reduced exterior jet* of a traction hyper-stress $\sigma$ is defined as

$$\eth_k \sigma := \iota^{k*}_{1,k-1}(\eth\sigma), \tag{6.8}$$

that is,

$$\eth_k \sigma \cdot j^k w = \eth\sigma \cdot j^1\big(j^{k-1} w\big). \tag{6.9}$$

Using (6.3) and (5.12) one obtains immediately,

$$\begin{aligned} (\eth_k\sigma)^{\boldsymbol{K}}_\alpha &= \overleftarrow{\sigma}^{(\boldsymbol{K})}_\alpha, & |\boldsymbol{K}| = k,\\ (\eth_k\sigma)^{\boldsymbol{J}}_\alpha &= \sigma^{\boldsymbol{J}j}_{\alpha,j} + \overleftarrow{\sigma}^{(\boldsymbol{J})}_\alpha, & 1 \le |\boldsymbol{J}| < k,\\ (\eth_k\sigma)_\alpha &= \sigma^{j}_{\alpha,j}. \end{aligned} \tag{6.10}$$

Evidently, one can obtain these relations directly from (6.9).

## 6.3 The Variational Hyper-stress Induced by a Consistent Hyper-force System

Let $(\boldsymbol{b}, \{\boldsymbol{t}_{\mathscr{R}}\})$ be a consistent hyper-force system corresponding to the traction hyper-stress $\sigma$. Then, the corresponding unique non-holonomic stress, as in (5.18), determines through (6.3) a unique variational hyper-stress $S$ given locally by

$$\begin{aligned} S^{\boldsymbol{K}}_\alpha &= \overleftarrow{\sigma}^{(\boldsymbol{K})}_\alpha, & |\boldsymbol{K}| = k,\\ S^{\boldsymbol{J}}_\alpha &= \sigma^{\boldsymbol{J}j}_{\alpha,j} + \boldsymbol{b}^{\boldsymbol{J}}_\alpha + \overleftarrow{\sigma}^{(\boldsymbol{J})}_\alpha, & 1 \le |\boldsymbol{J}| < k,\\ S_\alpha &= \sigma^{j}_{\alpha,j} + \boldsymbol{b}_\alpha. \end{aligned} \tag{6.11}$$

As mentioned above, only the symmetrized components of $\sigma$ determine $S$. Conversely, specifying $S$ and $\boldsymbol{b}$ is not sufficient in order to determine a unique traction hyper-stress.

One may be tempted to postulate a-priori that the components of $\sigma$ be symmetric. However, as shown below, symmetry of the components of elements of $L(J^{k-1}W, \bigwedge^{n-1} T^*\mathscr{X})$ is not an invariant property.

*Example 9* Consider the case where $W = \mathscr{X} \times \mathbb{R}$ and $k - 1 = 3$. Let $x^{i'} = x^{i'}(x^i)$ be a local coordinate transformation. Then,

$$\begin{aligned} w_{,i'} &= w_{,i} x^i_{,i'},\\ w_{,i'j'} &= w_{,ij} x^i_{,i'} x^j_{,j'} + w_{,i} x^i_{,i'j'},\\ w_{,i'j'k'} &= w_{,ijk} x^i_{,i'} x^j_{,j'} x^k_{,k'} + w_{,ij} x^i_{,i'k'} x^j_{,j'} + w_{,ij} x^i_{,i'} x^j_{,j'k'} + w_{,ik} x^i_{,i'j'} x^k_{,k'} + w_{,i} x^i_{,i'j'k'}. \end{aligned} \tag{6.12}$$

Invariance of the action imposes the condition that

$$
\begin{aligned}
\sigma^{Il} w_{,I} \partial_l \lrcorner \mathrm{d}x &= \sigma^{I'l'} w_{,I'} \partial_{l'} \lrcorner \mathrm{d}x', \quad |I| \leq 3, \\
&= \sigma^{I'l'} w_{,I'} x^l_{,l'} \partial_l \lrcorner \mathrm{d}x \mathcal{J},
\end{aligned}
\tag{6.13}
$$

where $\mathcal{J}$ is the Jacobian determinant. After substitution of the transformed derivatives, and in view of the arbitrariness of the values of $w_{,I}$, we obtain

$$
\begin{aligned}
\sigma^{ijkl}/\mathcal{J} &= \sigma^{i'j'k'l'} x^i_{,i'} x^j_{,j'} x^k_{,k'} x^l_{,l'}, \\
\sigma^{ijl}/\mathcal{J} &= \sigma^{i'j'l'} x^i_{,i'} x^j_{,j'} x^l_{,l'} + \sigma^{i'j'k'l'} \left( x^i_{,i'k'} x^j_{,j'} x^l_{,l'} + x^i_{,i'} x^j_{,j'k'} x^l_{,l'} + x^i_{,i'j'} x^k_{,k'} x^l_{,l'} \right), \\
\sigma^{il}/\mathcal{J} &= \sigma^{i'l'} x^i_{,i'} x^l_{,l'} + \sigma^{i'j'l'} x^i_{,i'j'} x^l_{,l'} + \sigma^{i'j'k'l'} x^i_{,i'j'k'} x^l_{,l'}, \\
\sigma^l/\mathcal{J} &= \sigma^{l'} x^l_{,l'}.
\end{aligned}
\tag{6.14}
$$

We note that in case $\sigma^{i'j'k'l'}$ is symmetric, so is $\sigma^{ijkl}$. However, even if all the components $\sigma^{I'}$ are symmetric, this need not hold for $\sigma^{ijl}$ and $\sigma^{il}$.

We conclude that symmetry of the components of a traction hyper-stress is not an invariant property. One does not encounter this issue in the formulations of continuum mechanics in a Euclidean space (or whenever a connection is given) because one assumes, invariantly in this case, that the hyper-stress contains only the components of the tensor of highest order. Then, as noted above, these components preserve their symmetry under a transformation of coordinates.

*Example 10* It is observed that for the case $k = 2$, so that $\sigma^{ijl} = \sigma^{ijkl} = 0$ in the last equation, $\sigma^{il}$ is symmetric if and only if $\sigma^{i'l'}$ is symmetric. Hence, for $k = 2$, one may impose the condition that $\sigma^{ij}$ is symmetric. This, together with the condition $\boldsymbol{b}^i_\alpha = 0$, imply that the variational stress determines a unique traction hyper-stress and a unique body hyper-force.

In light of the foregoing discussion we make the following

**Definition 11** A variational hyper-stress $S$ is said to be *consistent* with a non-holonomic stress $P$ if $S = \iota^{k*}_{1,k-1}(P)$.

**Proposition 12** *If a variational hyper-stress $S$ is consistent with a non-holonomic stress $P$ representing a hyper-force system $\{F_{\mathcal{R}}\} = (\boldsymbol{b}, \{\boldsymbol{t}_{\mathcal{R}}\})$, then, although $\boldsymbol{b}$ and $\{\boldsymbol{t}_{\mathcal{R}}\}$ cannot be determined by $S$, the force functionals $\{F_{\mathcal{R}}\}$ can be determined by $S$ in the form*

$$
F_{\mathcal{R}}(w) = \int_{\mathcal{R}} S \cdot j^k w. \tag{6.15}
$$

*In other words, while $S$ represents the force system $\{F_{\mathcal{R}}\}$ as in Sects.* 3.3, 3.4, *it does not determine the corresponding body hyper-force and hyper-tractions.*

*Proof* The assertion follows immediately from the fact that consistency implies that

$$
F_{\mathcal{R}}(w) = \int_{\mathcal{R}} P \cdot j^1\left(j^{k-1} w\right) = \int_{\mathcal{R}} S \cdot j^k w. \tag{6.16}
$$

□

# 7 Spaces of Hyper-stresses and Constitutive Relations

Hyper-stresses have been defined above in the context of some particular configuration $\kappa$. To formulate constitutive relations, one has to identify the appropriate spaces to which hyper-stresses belong for all possible configurations. The present section is concerned with these issues.

## 7.1 The Space of Variational Hyper-stress Density Values

We first recall the following two natural isomorphisms. As fiber bundles over $\mathscr{X}$,

$$V J^k \mathscr{Y} \simeq J^k V \mathscr{Y}, \tag{7.1}$$

and for any section $\kappa : \mathscr{X} \to \mathscr{Y}$,

$$\big(j^k\kappa\big)^* V J^k \mathscr{Y} \simeq J^k\big(\kappa^* V \mathscr{Y}\big) \tag{7.2}$$

(see [23, Theorem 4.4.1] and [20, Theorem 17.1, p. 82]).

For a configuration $\kappa$ and a point $x \in \mathscr{X}$, consider the value $S(x)$ of some variational hyper-stress field at the configuration $\kappa$. Using the isomorphism above and recalling (3.10), one has,

$$\begin{aligned}
S(x) &\in L\Big(\big(J^k W\big)_x, \Big(\textstyle\bigwedge^n T^* \mathscr{X}\Big)_x\Big), \\
&= L\Big(\Big(J^k\big(\kappa^* V \mathscr{Y}\big)\Big)_x, \Big(\textstyle\bigwedge^n T^* \mathscr{X}\Big)_x\Big), \\
&= L\Big(\Big(\big(j^k\kappa\big)^* V J^k \mathscr{Y}\Big)_x, \Big(\textstyle\bigwedge^n T^* \mathscr{X}\Big)_x\Big), \\
&= L\Big(\big(V J^k \mathscr{Y}\big)_{j^k\kappa(x)}, \pi^{k*}\Big(\textstyle\bigwedge^n T^* \mathscr{X}\Big)_{j^k\kappa(x)}\Big), \\
&= L\Big(V J^k \mathscr{Y}, \pi^{k*} \textstyle\bigwedge^n T^* \mathscr{X}\Big)_{j^k\kappa(x)}, \\
&= \Big[\big(j^k\kappa\big)^* L\Big(V J^k \mathscr{Y}, \pi^{k*} \textstyle\bigwedge^n T^* \mathscr{X}\Big)\Big]_x.
\end{aligned} \tag{7.3}$$

One concludes that the values of variational hyper-stress fields may always be identified with elements of the vector bundle

$$\pi_{\mathfrak{S}} : \mathfrak{S} := L\Big(V J^k \mathscr{Y}, \pi^{k*} \textstyle\bigwedge^n T^* \mathscr{X}\Big) \longrightarrow J^k \mathscr{Y}, \tag{7.4}$$

independently of the configuration under consideration. For any configuration $\kappa$, variational hyper-stress densities are valued in

$$\big(j^k\kappa\big)^* \mathfrak{S} \simeq L\Big(J^k\big(\kappa^* V \mathscr{Y}\big), \textstyle\bigwedge^n T^* \mathscr{X}\Big). \tag{7.5}$$

Intuitively, sections of $J^k \mathscr{Y}$ may be thought of as "local configurations" and sections of $V J^k \mathscr{Y}$ may be conceived as variations thereof or as "local velocity fields". Thus, following this line of thought, variational hyper-stresses may be thought of as "local forces".

## 7.2 The Space of Traction Hyper-stress Values

In analogy with Sect. 7.1, we identify here the space where traction hyper-stress fields assume their values, independently of a particular configuration. Thus, for a configuration $\kappa$ and a point $x \in \mathscr{X}$, we consider the value $\sigma(x)$ of some traction hyper-stress field at the configuration $\kappa$. One has,

$$
\begin{aligned}
\sigma(x) &\in L\Big(\big(J^{k-1}W\big)_x, \Big(\bigwedge\nolimits^{n-1}T^*\mathscr{X}\Big)_x\Big),\\
&= L\Big(\big(J^{k-1}(\kappa^* V\mathscr{Y})\big)_x, \Big(\bigwedge\nolimits^{n-1}T^*\mathscr{X}\Big)_x\Big),\\
&= L\Big(\big((j^{k-1}\kappa)^* V J^{k-1}\mathscr{Y}\big)_x, \Big(\bigwedge\nolimits^{n-1}T^*\mathscr{X}\Big)_x\Big),\\
&= L\Big(\big(V J^{k-1}\mathscr{Y}\big)_{j^{k-1}\kappa(x)}, \pi^{k-1*}\Big(\bigwedge\nolimits^{n-1}T^*\mathscr{X}\Big)_{j^{k-1}\kappa(x)}\Big),\\
&= L\Big(V J^{k-1}\mathscr{Y}, \pi^{k-1*}\bigwedge\nolimits^{n-1}T^*\mathscr{X}\Big)_{j^{k-1}\kappa(x)},\\
&= \Big[(j^{k-1}\kappa)^* L\Big(V J^{k-1}\mathscr{Y}, \pi^{k-1*}\bigwedge\nolimits^{n-1}T^*\mathscr{X}\Big)\Big]_x.
\end{aligned}
\tag{7.6}
$$

It is concluded that the values of traction hyper-stress fields may be identified with elements of the vector bundle

$$
\pi_{\mathfrak{T}} : \mathfrak{T} := L\Big(V J^{k-1}\mathscr{Y}, \pi^{k-1*}\bigwedge\nolimits^{n-1}T^*\mathscr{X}\Big) \longrightarrow J^{k-1}\mathscr{Y}. \tag{7.7}
$$

For a given configuration $\kappa$, traction hyper-stresses are valued in

$$
(j^{k-1}\kappa)^*\mathfrak{T} \simeq L\Big((j^{k-1}\kappa)^* V J^{k-1}\mathscr{Y}, \bigwedge\nolimits^{n-1}T^*\mathscr{X}\Big). \tag{7.8}
$$

## 7.3 The Space of Values of Non-holonomic Stresses

Let

$$
A : \mathscr{X} \longrightarrow J^{k-1}\mathscr{Y} \tag{7.9}
$$

be a smooth section and let $x \in \mathscr{X}$. Than, the value of a non-holonomic stress field $P$ at $x$ satisfies

$$
\begin{aligned}
P(x) &\in L\Big(\big(J^1(J^{k-1}W)\big)_x, \Big(\bigwedge\nolimits^{n}T^*\mathscr{X}\Big)_x\Big),\\
&= L\Big(\big(J^1(A^* V J^{k-1}\mathscr{Y})\big)_x, \Big(\bigwedge\nolimits^{n}T^*\mathscr{X}\Big)_x\Big),\\
&= L\Big(\big((j^1A)^*(V J^1(J^{k-1}\mathscr{Y}))\big)_x, \Big(\bigwedge\nolimits^{n}T^*\mathscr{X}\Big)_x\Big),\\
&= L\Big(\big(V J^1(J^{k-1}\mathscr{Y})\big)_{j^1A(x)}, \pi^{1*}\Big(\bigwedge\nolimits^{n}T^*\mathscr{X}\Big)_{j^1A(x)}\Big),\\
&= L\Big(\big(V J^1(J^{k-1}\mathscr{Y})\big), \pi^{1*}\Big(\bigwedge\nolimits^{n}T^*\mathscr{X}\Big)\Big)_{j^1A(x)},\\
&= \Big[(j^1A)^* L\Big(\big(V J^1(J^{k-1}\mathscr{Y})\big), \pi^{1*}\Big(\bigwedge\nolimits^{n}T^*\mathscr{X}\Big)\Big)\Big]_x.
\end{aligned}
\tag{7.10}
$$

Here, we view $\pi^{k-1}: J^{k-1}\mathscr{Y} \to \mathscr{X}$ as a fiber bundle over $\mathscr{X}$, and so $\pi^1 : J^1(J^{k-1}\mathscr{Y}) \to \mathscr{X}$ is the natural projection. Hence, values of non-holonomic hyper-stress fields may be viewed as elements of the vector bundle

$$\pi_{\mathscr{P}} : \mathscr{P} := L\Big(\big(VJ^1(J^{k-1}\mathscr{Y})\big), \pi^{1*}\Big(\bigwedge^n T^*\mathscr{X}\Big)\Big) \longrightarrow J^1\big(J^{k-1}\mathscr{Y}\big). \tag{7.11}$$

For a given section $A : \mathscr{X} \to J^{k-1}\mathscr{Y}$, non-holonomic stresses at $A$ are valued in

$$\big(j^1 A\big)^* \mathscr{P} = L\Big(J^1\big(A^* V J^{k-1}\mathscr{Y}\big), \bigwedge^n T^*\mathscr{X}\Big). \tag{7.12}$$

## 7.4 Elastic Constitutive Relations

Once the spaces of hyper-stresses have been identified, one may introduce constitutive relations for the variational hyper-stresses and for the non-holonomic stresses.

An elastic constitutive relation for the variational hyper-stress is a section

$$\Psi : J^k\mathscr{Y} \longrightarrow \mathfrak{S} = L\Big(VJ^k\mathscr{Y}, \pi^{k*}\bigwedge^n T^*\mathscr{X}\Big) \tag{7.13}$$

of $\pi_{\mathfrak{S}}$. Thus, a constitutive relation assigns a value of a variational hyper-stress at a point $x$ to a $k$-jet of a section at $x$. In particular, for a section $\kappa$,

$$\Psi \circ j^k\kappa : \mathscr{X} \longrightarrow \mathfrak{S} \tag{7.14}$$

is identified with a variational hyper-stress field according to (7.3).

Similarly, an elastic constitutive relation for the non-holonomic stress is a section

$$\Phi : J^1\big(J^{k-1}\mathscr{Y}\big) \longrightarrow \mathscr{P} = L\Big(\big(VJ^1(J^{k-1}\mathscr{Y})\big), \pi^{1*}\Big(\bigwedge^n T^*\mathscr{X}\Big)\Big) \tag{7.15}$$

of $\pi_{\mathscr{P}}$. For a section $A : \mathscr{X} \to J^{k-1}\mathscr{Y}$,

$$\Phi \circ j^1 A : \mathscr{X} \longrightarrow \mathscr{P} \tag{7.16}$$

is identified with a non-holonomic stress field according to (7.10). In particular, for a section $\kappa : \mathscr{X} \to \mathscr{Y}$, one has the induced non-holonomic stress field $\Phi \circ j^1(j^{k-1}\kappa)$.

**Acknowledgements** Both authors are grateful to BIRS for sponsoring the Banff Workshop on Material Evolution, June 11–18, 2017, which led to this collaboration. R.S.'s work has been partially supported by H. Greenhill Chair for Theoretical and Applied Mechanics and the Pearlstone Center for Aeronautical Engineering Studies at Ben-Gurion University.



## References

1. Aifantis, E.C.: On the role of gradients in the localization of deformation and fracture. Int. J. Eng. Sci. **30**, 1279–1299 (1992)
2. Askes, H., Aifantis, E.C.: Gradient elasticity in statics and dynamics: an overview of formulations, length scale identification procedures, finite element implementations and new results. Int. J. Solids Struct. **48**, 1962–1990 (2011)

3. Bertram, A.: Compendium on Gradient Materials (2017). http://www.ifme.ovgu.de/ifme_media/FL/Publikationen/Compendium+on+Gradient+Materials_May+2017.pdf
4. Binz, E., Śniatycki, J., Fischer, H.: Geometry of Classical Fields. North-Holland, Amsterdam (1988)
5. de León, M., Rodrigues, P.R.: Generalized Classical Mechanics and Field Theory: A Geometrical Approach of Lagrangian and Hamiltonian Formalisms Involving Higher Order Derivatives. North-Holland, Amsterdam (1985)
6. dell'Isola, F., Seppecher, P., Madeo, A.: How contact interactions may depend on the shape of Cauchy cuts in $N$th gradient continua: approach "á la d'Alembert". Z. Angew. Math. Phys. **63**, 1119–1141 (2012)
7. dell'Isola, F., Seppecher, P., della Corte, A.: The postulations á la d'Alembert and á la Cauchy for higher gradient continuum theories are equivalent: a review of existing results. Proc. R. Soc. A **471**, 20150415 (2015)
8. Fosdick, R.: A generalized continuum theory with internal corner and surface contact interactions. Contin. Mech. Thermodyn. **28**, 275–292 (2016)
9. Giachetta, G., Mangiarotti, L., Sardanashvily, G.: Advanced Classical Field Theory. World Scientific, Singapore (2009)
10. Gotay, M.J., Isenberg, J., Marsden, J.E., Montgomery, R.: Momentum maps and classical fields. Part I: Covariant field theory (2003). arXiv:physics/9801019 [math-ph]
11. Hirsch, M.: Differential Topology. Springer, Berlin (1976)
12. Kijowski, J., Tuczyjew, W.M.: A Symplectic Framework for Field Theories. Lecture Notes in Physics, vol. 107. Springer, Berlin (1979)
13. Kupferman, R., Olami, E., Segev, R.: Stress theory for classical fields. Math. Mech. Solids (2017). https://doi.org/10.1177/1081286517723697
14. Lewis, A.: Notes on Global Analysis. Appendix F: Multilinear Algebra (2014). http://www.mast.queensu.ca/~andrew/teaching/math942/pdf/1appendixf.pdf
15. Michor, P.W.: Manifolds of Differentiable Mappings. Shiva, Chandigarh (1980)
16. Mindlin, R.D.: Micro-structure in linear elasticity. Arch. Ration. Mech. Anal. **16**, 51–78 (1964)
17. Mindlin, R.D.: Second gradient of strain and surface tension in linear elasticity. Int. J. Solids Struct. **1**, 417–438 (1965)
18. Noll, W.: The foundations of classical mechanics in the light of recent advances in continuum mechanics. In: Henkin, L., Suppes, P., Tarski, A. (eds.) The Axiomatic Method, with Special Reference to Geometry and Physics, pp. 266–281. North-Holland, Amsterdam (1959)
19. Noll, W., Virga, E.G.: On edge interactions and surface tension. Arch. Ration. Mech. Anal. **111**, 1–31 (1990)
20. Palais, R.S.: Foundations of Global Non-Linear Analysis. Benjamin, Elmsford (1968)
21. Podio-Guidugli, P.: Cauchy's construction for flat complex bodies. J. Elast. **118**, 101–107 (2015)
22. Ru, C.Q., Aifantis, E.C.: A simple approach to solve boundary-value problems in gradient elasticity. Acta Mech. **101**, 59–68 (1993)
23. Saunders, D.J.: The Geometry of Jet Bundles. London Mathematical Society: Lecture Notes Series, vol. 142. Cambridge University Press, Cambridge (1989)
24. Segev, R.: Forces and the existence of stresses in invariant continuum mechanics. J. Math. Phys. **27**, 163–170 (1986)
25. Segev, R.: Notes on metric independent analysis of classical fields. Math. Methods Appl. Sci. **36**, 497–566 (2013)
26. Segev, R.: Geometric analysis of hyper-stresses. Int. J. Eng. Sci. **120**, 100–118 (2017)
27. Segev, R., DeBotton, G.: On the consistency conditions for force systems. Int. J. Non-Linear Mech. **26**, 47–59 (1991)
28. Segev, R., Rodnay, G.: Cauchy's theorem on manifolds. J. Elast. **56**(2), 129–144 (1999)
29. Segev, R., Śniatycki, J.: On jets, almost symmetric tensors, and traction hyper-stresses. Math. Mech. Complex Syst. **6**, 101–124 (2018)
30. Toupin, R.A.: Elastic materials with couple-stresses. Arch. Ration. Mech. Anal. **11**, 385–414 (1962)
31. Toupin, R.A.: Theories of elasticity with couple-stress. Arch. Ration. Mech. Anal. **17**, 85–112 (1964)

J Elast (2019) 135:485–507
https://doi.org/10.1007/s10659-018-09721-9

# Multi-Component Multiphase Flow Through a Poroelastic Medium

**Brian Seguin**[1] · **Noel J. Walkington**[2]

*In memory of Walter Noll*

Received: 20 June 2018 / Published online: 6 February 2019


**Abstract** An axiomatic development for a continuum description of a multi-component multiphase porous flow in an elastic medium is developed. The Coleman–Noll procedure is used to derive constitutive restrictions which guarantee that the resulting model satisfies an appropriate statement of the second law of thermodynamics and a corresponding dissipation inequality. Many of the models and formulations appearing in the engineering literature are shown to be special cases of the model developed here.

**Keywords** Poroelasticity · Multiphase flow · Biot theory · Thermodynamics

**Mathematics Subject Classification** 76S05 · 74F10 · 80A17

## 1 Introduction

Coleman–Noll theory is utilized to develop a continuum description of flow through an elastic porous medium which satisfies the statement of the second law of thermodynamics developed in their pioneering paper [12]. Geological problems involve fluids and media for which a detailed description of their properties is not available and are too complex to simulate precisely. Models of the gross properties of these systems utilize mixture theory with substantial constitutive postulates to model the dissipation, phase formation, and interaction between the phases and medium. In another paper the authors developed an axiomatic

N.J. Walkington was supported in part by National Science Foundation Grants DMS-1418991 and DMREF-1729478. This work was also supported by the NSF through the Center for Nonlinear Analysis.

✉ B. Seguin
bseguin@luc.edu

N.J. Walkington
noelw@andrew.cmu.edu

1 Department of Mathematics and Statistics, Loyola University Chicago, Chicago, IL 60660, USA

2 Department of Mathematical Sciences, Carnegie Mellon University, Pittsburgh, PA 15213, USA

formulation for flows in a rigid media [26]; extension to include elastic media requires additional constitutive postulates to model how the fluid pressure(s) enter into the macroscopic stress–strain relation of the combination of medium and fluids. Classical poroelasticity models utilize Biot theory to develop the macroscopic stress-strain relation. Biot theory is typically developed in the context of linear elasticity with a single fluid pressure [10]; in particular, the reference and current configurations of the medium and fluid(s) are not distinct in this theory. The poroelastic model below is developed in the current configuration; while this is less convenient for linearization of the elastic response of the medium, it significantly simplifies the constitutive theory for the fluid(s) in the pores.

## 1.1 Porous Flow Models

Porous flows of interest typically involve multiple fluids in a tortuous medium with interactions occurring over multiple spatial and temporal scales. Moreover, only gross properties of the spatial and compositional makeup of the medium and fluids are available, so models based on first principles can only realize macroscopic balance laws for mass, force, and energy, along with an entropy inequality. Modeling the kinematics, dynamics, and thermodynamics, required to close the systems of conservation laws has a long history. Darcy's experimental results of 1857 resulted in his eponymous law relating the mass flow and pressure gradient [13], and the series of papers by Biot and Willis [3–6] dating from the 1940's integrated elasticity of the medium into the modeling. The monograph by Bear [2] presents a comprehensive review of the classical theory, and the more recent monograph by Chen, Huan, and Ma [9] includes a detailed discussion of numerical methods to simulate solutions. The monograph by Cheng [10] provides an extensive overview of poroelastic models.

Due to their importance in the oil industry, a majority of the work on porous flow focuses on incompressible one or two phase problems [1, 7, 8, 14, 16, 17, 19, 22, 27, 31]. When multiple components are present a separation of time scales is assumed so that classical thermodynamics can be used to determine the formation and evolution of the various phases [11, 15, 18, 20, 23]. Integrating classical statements of thermodynamics into a continuum description of flow in a rigid medium consistent with the second law formulated by Coleman and Noll [12] was undertaken by the authors in [26]. Classical elasticity theory assumes that stresses can be determined from the deformation, but this is not so for a poroelastic medium since the stress in the medium also depends upon the pressure in the pores. In addition, deformation of the medium results in changes of pore volume and hence the fluid pressure. Homogenization theory for a periodic linear elastic medium with pore pressure $p$ [25] shows that there exists a homogenized elasticity tensor $\mathbb{C}$ and two Biot parameters, a tensor $\mathbf{B}$ and a scalar $\beta$, for which the homogenized stress $\mathbf{T}_0$ in the solid and a volumetric strain $\xi$ take the form

$$\mathbf{T}_0 = \mathbb{C}(\nabla \mathbf{u}) + p(\mathbf{B} - s_0\mathbf{1}) \quad \text{and} \quad \xi = \mathbf{B} \cdot \nabla \mathbf{u} + \beta p, \tag{1}$$

where $\mathbf{u}$ is the macroscopic displacement. If $V_s \subset V$ is the solid portion of a representative volume of the porous medium and $|\tilde{V}_s|$ is the volume of the deformed solid, then setting $s_0 = |V_s|/|V|$ and $s = |\tilde{V}_s|/|V|$ the volumetric strain is characterized by

$$s = s_0 + \xi + O(\epsilon),$$

where $\epsilon$ is the ratio of the pore size (period) of the medium to a characteristic length scale of the domain. These constitutive laws are comparable to those appearing in [10, equations (5.5) and (5.6)], except there the total stress is used, while in (1) only the stress in the solid phase is present. In the absence of detailed geometric information on the pores required to determine the homogenized coefficients $(\mathbb{C}, \mathbf{B}, \beta)$, the Biot theory appearing in the engineering literature typically assumes isotropic response so that $\mathbb{C}(\mathbf{E}) = 2\mu\mathbf{E} + \lambda\,\mathrm{tr}(\mathbf{E})\mathbf{1}$ and $\mathbf{B}$ is a multiple of the identity [24]. In Sect. 4.1 we show that formulae for the Biot parameters are developed from a linearization of the formulation presented here.

## 1.2 Overview

Currently it is accepted that computationally tractable simulations of porous flows through an elastic medium with acceptable predictive capability can be realized with a system of partial differential equations consisting of (i) a scalar equation representing the balance of mass for each conserved component (chemical species), (ii) a vector valued equation representing a macroscopic momentum balance, and (iii) a scalar equation representing the balance of energy. The model presented below fits into this paradigm, which results in a system of coupled partial differential equations for the mass densities $\{m_c\}_{c=1}^{N_c}$ of the conserved components,[1] the displacement of the solid $\mathbf{u}$, and the internal energy $e$. All other physical processes are modeled with algebraic equations. These include Darcy laws which enter as force balances for the fluid phases, and constitutive equations describing the formation of the phases.

The balances of mass take the form,

$$\frac{\partial}{\partial t}m_c + \mathrm{div}\left(m_c\mathbf{v}_0 - \mathbf{K}_c\nabla\mu_c + \sum_{\pi=1}^{N_p} M_{c\pi}\mathbf{v}_\pi^r\right) = h_c,$$

where the quantities inside the divergence represent the mass flux of the component arising from the motion of the medium and each phase, and the right-hand side the mass sinks and sources. Denoting the deformation gradient of the solid by $\mathbf{F} := \mathbf{1} + \nabla\mathbf{u}$, the balance of forces for the solid/fluid system takes the form

$$\sum_{\pi=0}^{N_p} s_\pi\mathbf{b}_\pi + \mathrm{div}\left[\left(\frac{\partial e_I}{\partial\mathbf{F}} - \theta\frac{\partial\eta_0}{\partial\mathbf{F}}\right)\mathbf{F}^\top + \left(e_0 - \theta\eta_0 - \sum_{\pi=1}^{N_p} s_\pi p_\pi\right)\mathbf{1}\right] = \mathbf{0}.$$

This equation neglects the inertial term since the motion is slow in relation to the time scales of interest. The term inside the divergence is the (total) stress, and the remaining term is the (total) body force. The balance of energy becomes

$$\begin{aligned}\dot{e} = r + \dot{\mathbf{m}}\cdot\boldsymbol{\mu} + \mathrm{div}(\mathbf{K}_\theta\nabla\theta) + \sum_{c=1}^{N_c}\nabla\mu_c\cdot\mathbf{K}_c\nabla\mu_c + \sum_{\pi=1}^{N_p}\left(\mathbf{v}_\pi^r\cdot\sum_{\substack{\pi'=0\\ \pi'\neq\pi}}^{N_p}\mathbf{A}_{\pi\pi'}(\mathbf{v}_\pi^r - \mathbf{v}_{\pi'}^r)\right)\\ + \nabla\mathbf{v}_0\cdot\left[\left(\frac{\partial e_I}{\partial\mathbf{F}} - \theta\frac{\partial\eta_0}{\partial\mathbf{F}}\right)\mathbf{F}^\top - \theta\eta\mathbf{1}\right] - \sum_{\pi=1}^{N_p}\theta\,\mathrm{div}(\eta_\pi\mathbf{v}_\pi^r).\end{aligned}$$

The left-hand side of this equation is the rate of change of the internal energy and the right-hand side represents the change due to external supply, chemical energy carried by the mass, diffusion of heat and mass, viscous dissipation due to the relative motion of the fluids, elastic energy, and transport due to the motion of the fluids relative to the solid.

[1] In porous flow contexts $\{m_c\}_{c=1}^{N_c}$ denote mass densities of components and $\{\rho_\pi\}_{\pi=1}^{N_p}$ are the mass densities of the phases.

Constitutive relations are required to close this system of partial differential equations. The development of constitutive relations to model phase formation and corresponding energy balance is non-trivial and is one of the main novelties of this work. Below we utilize the methodology introduced by Coleman and Noll [12], whereby the second law of thermodynamics is exploited to develop constitutive relations which guarantee a dissipation inequality for the resulting model. These relations naturally divide into two distinct types. The first type characterize the state of the fluids and phases at the pore scale and are introduced in Sect. 3.1. It is assumed that these processes evolve to an equilibrium at much shorter time scales than the macroscopic flow processes so that classical thermodynamics can be used to characterize the state of the fluids at the pore scale. This principle is referred to as local thermodynamic equilibrium, and while equilibrium has been assumed in poroelastic theories before [10], the novel formulation of this assumption we present provides a seamless integration of classical and continuum statements of the second law for multi-phase multicomponent fluids. In Sect. 3.2 these constitutive assumptions are introduced into the balance laws and the Coleman–Noll procedure is utilized to identify constitutive restrictions on the remaining variables which guarantee satisfaction of the second law of thermodynamics. The short Sect. 3.3 discusses the consequences of these restrictions. The second set of constitutive assumptions determine the dissipative processes, such as the Darcy law(s), and are discussed in Sect. 3.4. Section 3.5 discusses the simplifications that result when the temperature is assumed to be constant or is determined by that of the solid medium.

The next section introduces the notation required to describe systems with multiple components and phases, and develops the statements of the physical laws these quantities obey. The continuum model developed in Sect. 3 models a broad class of flows in poroelastic media. In Sect. 4 connections with other theories and the present theory are made. Finally, Sect. 5 summaries the proposed model by recapping the constitutive assumptions and balance laws.

## 2 Balance Laws

Balance laws will be formulated for a porous medium in which multiple phases of fluids flow through a porous solid phase. The fluid phases are formed by different components; that is, species which are conserved. There will be a total of $N_c$ components that make up $N_p$ phases. The formulation of the balance laws will be motivated by mixture theory and results from homogenization involving poroelastic materials.

$\mathcal{B}_R$ will denote a reference configuration of the body (medium and fluids) in a three-dimensional Euclidean space $\mathcal{E}$ with associated vector space $\mathcal{V}$, and $\mathcal{I}$ an interval of time. Unless stated otherwise, *all densities are taken per unit volume of space in the present configuration.*

### 2.1 Introduction of Relevant Quantities

A large number of quantities are required to characterize the evolution of the porous medium under consideration. The variables used below are gathered here for easy reference.

The motion and volume fraction of each phase are described by the following.

- $\chi : \mathcal{B}_R \times \mathcal{I} \to \mathcal{E}$ is the motion of the medium, and is identified as the motion of the solid phase. Let

$$\mathcal{T} := \{(\chi(X,t),t) \in \mathcal{E} \times \mathcal{I} \mid (X,t) \in \mathcal{B}_R \times \mathcal{I}\}$$

be the trajectory of the motion.

- The deformation gradient of this motion is denoted by $\mathbf{F} = \nabla \chi$.
- A superscript dot will denote the material time-derivative relative to the motion of the solid; if $\phi$ is a spatial field, then

$$\dot{\phi} := \frac{\partial \phi}{\partial t} + \mathbf{v}_0 \cdot \nabla \phi. \tag{2}$$

- $\mathbf{v}_\pi : \mathcal{T} \to \mathcal{V}$ is the velocity of phase $\pi$, for $0 \le \pi \le N_p$, where phase $\pi = 0$ corresponds to the solid phase. We write $\mathbf{v}_\pi^r := \mathbf{v}_\pi - \mathbf{v}_0$ for the velocity of phase $\pi$ relative to the solid phase.
- $s_\pi : \mathcal{T} \to [0, 1]$ represents the volume fraction of phase $0 \le \pi \le N_p$. We write $\mathbf{s} := (s_0, s_1, \ldots, s_{N_p})$ and assume that $\sum_{\pi=0}^{N_p} s_\pi = 1$.

Laws will be postulated for the balance of the mass of each component, force balance for each phase, a single energy balance, and a single entropy imbalance. The mass balances involve the following quantities.

- $m_0 : \mathcal{T} \to [0, \infty)$ is the mass density of the solid phase.
- $m_c : \mathcal{T} \to [0, \infty)$ is the mass density of component $c$, for $1 \le c \le N_c$. The vector of fluid mass densities is denoted by $\mathbf{m} := (m_1, \ldots, m_{N_c})$.
- $M_{c\pi} : \mathcal{T} \to [0, \infty)$ represents the mass density of component $c$ in phase $\pi$ for $1 \le c \le N_c$ and $1 \le \pi \le N_p$. The matrix of mass densities is denoted by $M = [M_{c\pi}]$. Clearly $m_c = \sum_{\pi=1}^{N_p} M_{c\pi}$.
- $\mathbf{h}_c : \mathcal{T} \to \mathcal{V}$ represents the mass flux of component $c$ due to diffusion.
- $h_c : \mathcal{T} \to \mathbb{R}$ is the supply (sinks and sources) of component $c$.

Next we list the quantities appearing in the balances of forces. It is assumed that the solid phase is elastic, while the fluid phases are viscous. In the fluid phases, the viscous forces will be modeled as body forces, dependent on the velocity of the fluid relative to the other phases, rather than surface forces as is customary. In this way, the force balance for a fluid phase will take the form of a Darcy Law. This formulation of viscous forces in porous flow is motivated by results in homogenization theory [25]. Also, the phases will interact with each other through body forces, as is common in mixture theory.

- $\mathbf{T}_\pi : \mathcal{T} \to \mathrm{Sym}$[2] is the elastic stress for phase $0 \le \pi \le N_p$. For the fluid phases, $\pi \ne 0$, we assume $\mathbf{T}_\pi = -s_\pi p_\pi \mathbf{1}$, where $p_\pi$ is the pressure in the fluid phase. The term $s_\pi p_\pi$ is sometimes referred to as the partial pressure in phase $\pi$.
- $\mathbf{f}_{\pi\pi'} : \mathcal{T} \to \mathcal{V}$ with $\pi \ne \pi'$ is the force density that phase $\pi'$ exerts on phase $\pi$. We assume that this can be additively decomposed into the sum of an elastic force $\mathbf{f}_{E\pi\pi'}$ and a viscous force $\mathbf{f}_{V\pi\pi'}$ so that $\mathbf{f}_{\pi\pi'} = \mathbf{f}_{E\pi\pi'} + \mathbf{f}_{V\pi\pi'}$.
- $\mathbf{b}_\pi : \mathcal{T} \to \mathcal{V}$ is the body force density acting on phase $\pi$ per unit volume of phase $\pi$.

The quantities appearing in the balance of energy are the following.

- $e : \mathcal{T} \to \mathbb{R}$ is the total internal energy density of the entire continuum.
- $e_\pi : \mathcal{T} \to \mathbb{R}$ is the internal energy density of phase $0 \le \pi \le N_p$. The vector of internal energy densities is denoted by $\mathbf{e} = (e_0, \ldots, e_{N_p})$.
- $\mathbf{q} : \mathcal{T} \to \mathcal{V}$ represents the heat flux.

[2]Here we assume that the stresses are symmetric. In more general mixture theory, it is possible for stresses to be nonsymmetric. In this case one must postulate a torque balance for each phase. However, if for each phase there is no external supply of torques, then torque balance implies that the stresses are symmetric. See Truesdell [29] for details.

- $\mu_c : \mathcal{T} \to \mathbb{R}$ is the chemical potential of component $1 \le c \le N_c$. The vector of chemical potentials is denoted by $\boldsymbol{\mu} := (\mu_1, \ldots, \mu_{N_c})$.
- $r : \mathcal{T} \to \mathbb{R}$ is the supply of energy.

Finally, we list the quantities appearing in the entropy imbalance.

- $\eta : \mathcal{T} \to \mathbb{R}$ is the total entropy density of the entire continuum.
- $\eta_\pi : \mathcal{T} \to \mathbb{R}$ is the entropy density of phase $0 \le \pi \le N_p$.
- $\theta : \mathcal{T} \to (0, \infty)$ is the (absolute) temperature.
- $\psi_\pi = e_\pi - \theta\eta_\pi$ is the (Helmholtz) free energy of phase $0 \le \pi \le N_p$. The free energy is used in place of the entropy in the isothermal setting.

### 2.2 Balances for Inertialess Continua

All of the balances will first be formulated globally using a spatial region $V(t)$ that is convecting with the motion of the solid phase. We will assume that the fluid phases are transported by the porous solid so that their velocities take the form $\mathbf{v}_\pi = \mathbf{v}_0 + \mathbf{v}_\pi^r$ $(1 \le \pi \le N_p)$, and that inertial effects are negligible.

1. *Mass balance*: The global form of mass balance for the solid states that

$$\frac{d}{dt}\int_{V(t)} m_0\, dv = 0. \tag{3}$$

The local form of this balance is

$$\frac{\partial}{\partial t} m_0 + \operatorname{div}(m_0 \mathbf{v}_0) = 0. \tag{4}$$

For fluid component $c$, global mass balance reads

$$\frac{d}{dt}\int_{V(t)} m_c\, dv = \int_{V(t)} h_c\, dv - \int_{\partial V(t)} \left(\mathbf{h}_c + \sum_{\pi=1}^{N_p} M_{c\pi}\mathbf{v}_\pi^r\right)\cdot \mathbf{n}\, da, \tag{5}$$

where $\mathbf{n}$ is the exterior unit normal. Localizing this balance yields

$$\frac{\partial}{\partial t} m_c + \operatorname{div}\left(m_c\mathbf{v}_0 + \mathbf{h}_c + \sum_{\pi=1}^{N_p} M_{c\pi}\mathbf{v}_\pi^r\right) = h_c. \tag{6}$$

2. *Force balance*: Assume for each subcollection $P \subset \{0, \ldots, N_p\}$ of phases that

$$\sum_{\pi\in P}\left[\int_{V(t)} s_\pi \mathbf{b}_\pi\, dv + \int_{\partial V(t)} \mathbf{T}_\pi \mathbf{n}\, da + \sum_{\pi'\in P^c}\int_{V(t)} \mathbf{f}_{\pi\pi'}\, dv\right] = \mathbf{0}, \tag{7}$$

where $P^c$ is the complement of $P$ in $\{0, \ldots, N_p\}$. The local form is

$$\sum_{\pi\in P}\left[s_\pi \mathbf{b}_\pi + \operatorname{div}(\mathbf{T}_\pi) + \sum_{\pi'\in P^c} \mathbf{f}_{\pi\pi'}\right] = \mathbf{0}. \tag{8}$$

3. *Energy balance*: The global form is

$$\frac{d}{dt}\int_{V(t)} e\,dv = \int_{V(t)} r\,dv - \int_{\partial V(t)} \mathbf{q}\cdot\mathbf{n}\,da + \sum_{c=1}^{N_c}\left[\int_{V(t)} \mu_c h_c\,dv - \int_{\partial V(t)} \mu_c \mathbf{h}_c\cdot\mathbf{n}\,da\right] + \sum_{\pi=0}^{N_p}\left[\int_{V(t)} s_\pi \mathbf{b}_\pi\cdot\mathbf{v}_\pi\,dv + \int_{\partial V(t)} \left(\mathbf{T}_\pi\mathbf{v}_\pi - e_\pi\mathbf{v}_\pi^r\right)\cdot\mathbf{n}\,da\right]. \tag{9}$$

The local form reads

$$\frac{\partial}{\partial t}e + \operatorname{div}\left[e\mathbf{v}_0 + \mathbf{q} + \sum_{c=1}^{N_c}\mu_c\mathbf{h}_c + \sum_{\pi=0}^{N_p}\left(e_\pi\mathbf{v}_\pi^r - \mathbf{T}_\pi\mathbf{v}_\pi\right)\right] = r + \sum_{c=1}^{N_c}\mu_c h_c + \sum_{\pi=0}^{N_p} s_\pi\mathbf{b}_\pi\cdot\mathbf{v}_\pi. \tag{10}$$

4. *Entropy imbalance*: The global form of entropy imbalance is

$$\frac{d}{dt}\int_{V(t)}\eta\,dv \geq \int_{V(t)}\frac{r}{\theta}\,dv - \int_{\partial V(t)}\frac{1}{\theta}\mathbf{q}\cdot\mathbf{n}\,da - \sum_{\pi=1}^{N_p}\int_{\partial V(t)}\eta_\pi\mathbf{v}_\pi^r\cdot\mathbf{n}\,da. \tag{11}$$

The local form is

$$\frac{\partial}{\partial t}\eta + \operatorname{div}\left(\eta\mathbf{v}_0 + \frac{\mathbf{q}}{\theta} + \sum_{\pi=1}^{N_p}\eta_\pi\mathbf{v}_\pi^r\right) \geq \frac{r}{\theta}. \tag{12}$$

## 2.3 Basic Consequences of the Balances

The aforementioned balances have a few consequences that will be useful in our analysis. First of all, using results in Noll's theory of interactions [21], force balance yields the law of mutual action:

$$\mathbf{f}_{\pi\pi'} = -\mathbf{f}_{\pi'\pi} \quad \text{for all } \pi,\pi' \in \{0,\ldots,N_p\}. \tag{13}$$

The energy balance can be rewritten in a different form which will be useful below. Eliminating the mass supplies $h_c$ and the external body forces $s_\pi\mathbf{b}_\pi$ from the energy balance using (6) and (8), and using (13), the energy balance becomes (recall $\mathbf{T}_\pi = -s_\pi p_\pi\mathbf{1}$ for $\pi \neq 0$ and the notation (2))

$$\dot{e} - \dot{\mathbf{m}}\cdot\boldsymbol{\mu} = r - \operatorname{div}\mathbf{q} - \sum_{c=1}^{N_c}\nabla\mu_c\cdot\mathbf{h}_c - \sum_{\pi=1}^{N_p}\left(\mathbf{v}_\pi^r\cdot\sum_{\substack{\pi'=0\\ \pi'\neq\pi}}^{N_p}\mathbf{f}_{V\pi\pi'}\right) + \nabla\mathbf{v}_0\cdot\left[\mathbf{T}_0 + \left(\boldsymbol{\mu}\cdot\mathbf{m} - e - \sum_{\pi=1}^{N_p}s_\pi p_\pi\right)\mathbf{1}\right]$$

$$+\sum_{\pi=1}^{N_p}\mathbf{v}_\pi^r\cdot\left[\sum_{c=1}^{N_c}\mu_c\nabla M_{c\pi}-\nabla e_\pi-\sum_{\substack{\pi'=0\\ \pi'\neq\pi}}^{N_p}\mathbf{f}_{E\pi\pi'}\right]$$

$$+\sum_{\pi=1}^{N_p}\operatorname{div}(\mathbf{v}_\pi^r)\left[\sum_{c=1}^{N_c}\mu_c M_{c\pi}-e_\pi-s_\pi p_\pi\right]. \tag{14}$$

This equation can be used to eliminate the radiation from the entropy imbalance to obtain

$$\theta\dot{\eta}-\dot{e}+\dot{\mathbf{m}}\cdot\boldsymbol{\mu}\geq\frac{1}{\theta}\nabla\theta\cdot\mathbf{q}+\sum_{c=1}^{N_c}\nabla\mu_c\cdot\mathbf{h}_c+\sum_{\pi=1}^{N_p}\left(\mathbf{v}_\pi^r\cdot\sum_{\substack{\pi'=0\\ \pi'\neq\pi}}^{N_p}\mathbf{f}_{V\pi\pi'}\right)$$

$$-\nabla\mathbf{v}_0\cdot\left[\mathbf{T}_0+\left(\boldsymbol{\mu}\cdot\mathbf{m}-e-\sum_{\pi=1}^{N_p}s_\pi p_\pi+\theta\eta\right)\mathbf{1}\right]$$

$$-\sum_{\pi=1}^{N_p}\mathbf{v}_\pi^r\cdot\left[\sum_{c=1}^{N_c}\mu_c\nabla M_{c\pi}-\nabla e_\pi-\sum_{\substack{\pi'=0\\ \pi'\neq\pi}}^{N_p}\mathbf{f}_{E\pi\pi'}+\theta\nabla\eta_\pi\right]$$

$$-\sum_{\pi=1}^{N_p}\operatorname{div}(\mathbf{v}_\pi^r)\left[\sum_{c=1}^{N_c}\mu_c M_{c\pi}-e_\pi-s_\pi p_\pi+\theta\eta_\pi\right]. \tag{15}$$

This last relation is sometimes referred to as the reduced entropy inequality and does not involve any external influences. It is the form of the entropy imbalance that is most useful when applying the Coleman–Noll procedure.

## 3 Constitutive Relations

### 3.1 Local Thermodynamic Equilibrium

It is necessary to specify how the mass of each component $m_c$, total internal energy $e$, and entropy $\eta$, are partitioned among the various phases, and to determine how much of each phase is present at any particular point. Classical thermodynamics was developed to determine macroscopic properties of equilibrated systems from their constituents (energy, volume, mass). Porous flow models postulate a separation of time scales so that this theory can be used to model the state of the fluids at the pore scale. This is realized by assuming that at each point of the continuum the phases are formed by maximizing the total entropy subject to constraints on the total amount of internal energy and mass available, and the deformation $\mathbf{F}$. This characterizes equilibria of classical thermodynamic ensembles [28]. This assumption has been made in poroelasticity before [10], but a clear statement of the second law has not been formulated precisely in the multi-component multiphase fluid setting.

We assume that the entropy of a fluid phase, $\pi\neq 0$, is specified by the constitutive law

$$\eta_\pi=\hat{\eta}_\pi\big(e_\pi,s_\pi,\{M_{c\pi}\}_{c=1}^{N_c}\big), \tag{16}$$

and the entropy for the solid phase is determined by

$$\eta_0=\hat{\eta}_0(e_0,s_0,\mathbf{F}). \tag{17}$$

The entropy for the entire continuum is determined by the local equilibrium assumption introduced below. The total internal energy at the pore scale is assumed to consist of the internal energy of the phases and the energy concentrated at the interfaces between the phases associated with surface tension and wetting. This additional contribution to the internal energy depends upon the phases present, and in the porous flow model is constitutively specified by a function of the form

$$e_I = \hat{e}_I(\mathbf{e}, \mathbf{s}, M, \mathbf{F}), \tag{18}$$

where, recall, $\mathbf{e} = (e_0, \ldots, e_{N_p})$.

**Assumption 1** (Local Equilibrium) The entropy $\eta$ of the continuum is specified by

$$\begin{aligned}\eta &= \hat{\eta}(e, \mathbf{m}, \mathbf{F})\\ &:= \sup_{(\mathbf{e},\mathbf{s},M)} \left\{ \hat{\eta}_0(e_0, s_0, \mathbf{F}) + \sum_{\pi=1}^{N_p} \hat{\eta}_\pi\left(e_\pi, s_\pi, \{M_{c\pi}\}_{c=1}^{N_c}\right) \mid \hat{e}_I(\mathbf{e}, \mathbf{s}, M, \mathbf{F}) \right.\\ &\qquad \left. + \sum_{\pi=0}^{N_p} e_\pi = e,\ \sum_{\pi=0}^{N_p} s_\pi = 1,\ \sum_{\pi=1}^{N_p} M_{c\pi} = m_c \right\}.\end{aligned} \tag{19}$$

Moreover, for each $(e, \mathbf{m}, \mathbf{F})$ the maximum is attained at a unique point

$$\begin{aligned}(\mathbf{e}, \mathbf{s}, M) = \arg\max_{(\mathbf{e},\mathbf{s},M)} &\left\{ \hat{\eta}_0(e_0, s_0, \mathbf{F}) + \sum_{\pi=1}^{N_p} \hat{\eta}_\pi\left(e_\pi, s_\pi, \{M_{c\pi}\}_{c=1}^{N_c}\right) \mid \hat{e}_I(\mathbf{e}, \mathbf{s}, M, \mathbf{F}) \right.\\ &\qquad \left. + \sum_{\pi=0}^{N_p} e_\pi = e,\ \sum_{\pi=0}^{N_p} s_\pi = 1,\ \sum_{\pi=1}^{N_p} M_{c\pi} = m_c \right\}.\end{aligned} \tag{20}$$

The supremum in these expressions is taken over $\mathbf{e} \in \mathbb{R}^{N_p+1}$, $\mathbf{s} \in [0, 1]^{N_p+1}$, and $M \in [0, \infty)^{N_c \times N_p}$.

By the uniqueness part of this assumption, we obtain constitutive laws for the energies, volume fractions, and mass densities:

$$(\mathbf{e}, \mathbf{s}, M) = \big(\hat{\mathbf{e}}(e, \mathbf{m}, \mathbf{F}), \hat{\mathbf{s}}(e, \mathbf{m}, \mathbf{F}), \hat{M}(e, \mathbf{m}, \mathbf{F})\big).$$

Under the assumption that the entropy and interfacial energy functions are smooth, the local thermodynamic equilibrium assumption implies the existence of Lagrange multipliers $(\lambda^e, \lambda^s, \boldsymbol{\lambda}^m) \in \mathbb{R} \times \mathbb{R} \times \mathbb{R}^{N_c}$ and KKT (Karush–Kuhn–Tucker) multipliers

$$\left(\boldsymbol{\lambda}^0, \boldsymbol{\lambda}^1, \Lambda\right) \in [0, \infty)^{1+N_\pi} \times [0, \infty)^{1+N_\pi} \times [0, \infty)^{N_\pi N_c}$$

such that

$$\frac{\partial \eta_\pi}{\partial e_\pi} = \lambda^e\left(1 + \frac{\partial e_I}{\partial e_\pi}\right), \qquad \pi \in \{0, \ldots, N_p\}, \tag{21}$$

$$\frac{\partial \eta_\pi}{\partial s_\pi} + \lambda^0_\pi - \lambda^1_\pi = \lambda^s + \lambda^e \frac{\partial e_I}{\partial s_\pi}, \qquad \pi \in \{0, \ldots, N_p\}, \tag{22}$$

$$\frac{\partial \eta_\pi}{\partial M_{c\pi}} + \Lambda_{c\pi} = \lambda_c^m + \lambda^e \frac{\partial e_I}{\partial M_{c\pi}}, \qquad \pi \in \{1, \ldots, N_p\}, \tag{23}$$

$$s_\pi \lambda_\pi^0 = 0, \quad (1 - s_\pi)\lambda_\pi^1 = 0, \qquad \pi \in \{0, \ldots, N_p\}, \tag{24}$$

$$M_{c\pi} \Lambda_{c\pi} = 0, \qquad \pi \in \{1, \ldots, N_p\}, \tag{25}$$

with $c \in \{1, \ldots, N_c\}$. We shall see that the Lagrange multipliers $\lambda^e$, $\lambda^s$, and $\lambda_c^m$ are related to the temperature $\theta$, a pressure $p$, and the chemical potentials $\mu_c$, respectively.

Let $(\bar{e}, \bar{\mathbf{m}}, \bar{\mathbf{F}})$ be a parameter dependent family of states at which the entropy $\eta$ can be considered. By the assumption of local equilibrium, there corresponds a parameter dependent family $(\bar{\mathbf{e}}, \bar{\mathbf{s}}, \bar{M})$ such that

$$\hat{\eta}(\bar{e}, \bar{\mathbf{m}}, \bar{\mathbf{F}}) = \hat{\eta}_0(\bar{e}_0, \bar{s}_0, \bar{\mathbf{F}}) + \sum_{\pi=1}^{N_p} \hat{\eta}_\pi \left(\bar{e}_\pi, \bar{s}_\pi, \{\bar{M}_{c\pi}\}_{c=1}^{N_c}\right).$$

Differentiating this equation with respect to the parameter, using the chain rule, and (21)–(25) yields

$$\eta' = \frac{\partial \eta}{\partial e}\bar{e}' + \frac{\partial \eta}{\partial \mathbf{m}} \cdot \bar{\mathbf{m}}' + \frac{\partial \eta}{\partial \mathbf{F}} \cdot \bar{\mathbf{F}}' = \lambda^e \bar{e}' + \boldsymbol{\lambda}^m \cdot \bar{\mathbf{m}}' + \left(\frac{\partial \eta_0}{\partial \mathbf{F}} - \lambda^e \frac{\partial e_I}{\partial \mathbf{F}}\right) \cdot \bar{\mathbf{F}}', \tag{26}$$

where prime denotes the derivative with respect to the parameter. Since the process $(\bar{e}, \bar{\mathbf{m}}, \bar{\mathbf{F}})$ was arbitrary, it follows that

$$\frac{\partial \eta}{\partial e} = \lambda^e, \quad \frac{\partial \eta}{\partial \mathbf{m}} = \boldsymbol{\lambda}^m, \quad \text{and} \quad \frac{\partial \eta}{\partial \mathbf{F}} = \frac{\partial \eta_0}{\partial \mathbf{F}} - \lambda^e \frac{\partial e_I}{\partial \mathbf{F}}. \tag{27}$$

### 3.2 Restrictions Due to the Second Law

Coleman and Noll's interpretation [12] of the second law of thermodynamics is that the entropy imbalance (12) must hold for all thermodynamic processes. Moreover, they stipulated that external influences can be specified arbitrarily, so for any thermodynamic process one can choose $h_c$, $\mathbf{b}_\pi$, and $r$ such that the balances of mass, force, and energy hold. Since the second law does not restrict the class of processes, it should be interpreted as a restriction on constitutive laws. Here, rather than specifying all of the constitutive laws and use the second law to find if and only if restrictions on them, we only find sufficient conditions for the second law to hold and use these to motivate additional constitutive laws that are consistent with the second law of thermodynamics.

**Proposition 1** *Assume that local equilibrium holds so that the entropy $\eta$ is determined by* (19) *and that the energies, volume fractions, and densities of the different phases are determined by* (20). *Under these constitutive assumptions, the entropy imbalance holds for all thermodynamics processes if*

1. *The entropy of fluid phase $\pi \neq 0$ is given by*

$$\eta_\pi = \frac{1}{\theta}\left(e_\pi + s_\pi p_\pi - \sum_{c=1}^{N_c} M_{c\pi}\mu_c\right). \tag{28}$$

2. *The stress in the solid phase is given by*

$$\mathbf{T}_0 = (e_0 - \theta\eta_0 + e_I)\mathbf{1} - \theta\frac{\partial \eta_0}{\partial \mathbf{F}}\mathbf{F}^\top + \frac{\partial e_I}{\partial \mathbf{F}}\mathbf{F}^\top. \tag{29}$$

3. *The interfacial energy only depends upon the volume fractions and the deformation gradient; that is,*

$$e_I = \hat{e}_I(\mathbf{s}, \mathbf{F}). \tag{30}$$

4. *The temperature, phase pressures, and the chemical potential of the components are related to the Lagrange multipliers through*

$$\theta = 1/\lambda^e, \qquad p_\pi = p + \frac{\partial e_I}{\partial s_\pi}, \qquad \mu_c = -\theta\lambda_c^m, \tag{31}$$

*where* $p := \lambda^s\theta$.

5. *The diffusive mass fluxes, heat flux, and viscous forces satisfy*

$$\sum_{c=1}^{N_c} \mathbf{h}_c \cdot \nabla\mu_c \le 0, \qquad \mathbf{q}\cdot\nabla\theta \le 0, \qquad \sum_{\pi=1}^{N_p}\left(\mathbf{v}_\pi^r \cdot \sum_{\substack{\pi'=0\\ \pi'\neq\pi}}^{N_p} \mathbf{f}_{V\pi\pi'}\right) \le 0. \tag{32}$$

6. *The elastic forces between the phases satisfy*

$$p_\pi \nabla s_\pi = \sum_{\substack{\pi'=0\\ \pi'\neq\pi}}^{N_p} \mathbf{f}_{E\pi\pi'}, \quad \pi \neq 0. \tag{33}$$

*Proof* A chain rule calculation similar to (26) in the reduced entropy inequality (15) shows

$$\begin{aligned}
0 \le\ & (\lambda^e\theta - 1)\dot{e} + (\boldsymbol{\mu} + \theta\boldsymbol{\lambda}^m)\cdot\dot{\mathbf{m}} \\
& - \frac{1}{\theta}\nabla\theta\cdot\mathbf{q} - \sum_{c=1}^{N_c}\nabla\mu_c\cdot\mathbf{h}_c - \sum_{\pi=1}^{N_p}\left(\mathbf{v}_\pi^r\cdot\sum_{\substack{\pi'=0\\ \pi'\neq\pi}}^{N_p}\mathbf{f}_{V\pi\pi'}\right) \\
& + \nabla\mathbf{v}_0\cdot\left[\mathbf{T}_0 - \left(\frac{\partial e_I}{\partial\mathbf{F}} - \theta\frac{\partial\eta_0}{\partial\mathbf{F}}\right)\mathbf{F}^\top + \left(\boldsymbol{\mu}\cdot\mathbf{m} - e - \sum_{\pi=1}^{N_p} s_\pi p_\pi + \theta\eta\right)\mathbf{1}\right] \\
& + \sum_{\pi=1}^{N_p}\mathbf{v}_\pi^r\cdot\left[\sum_{c=1}^{N_c}\mu_c\nabla M_{c\pi} - \nabla e_\pi - \sum_{\substack{\pi'=0\\ \pi'\neq\pi}}^{N_p}\mathbf{f}_{E\pi\pi'} + \theta\nabla\eta_\pi\right] \\
& + \sum_{\pi=1}^{N_p}\operatorname{div}(\mathbf{v}_\pi^r)\left[\sum_{c=1}^{N_c}\mu_c M_{c\pi} - e_\pi - s_\pi p_\pi + \theta\eta_\pi\right].
\end{aligned}$$

We must use Items 1–6 to show that this inequality holds regardless of the state of the continuum. It follows from Items 1, 4, 5, and 6 that it suffices to show that

$$\mathbf{T}_0 = \left(\frac{\partial e_I}{\partial \mathbf{F}} - \theta\frac{\partial \eta_0}{\partial \mathbf{F}}\right)\mathbf{F}^\top - \left(\boldsymbol{\mu}\cdot\mathbf{m} - e - \sum_{\pi=1}^{N_p} s_\pi p_\pi + \theta\eta\right)\mathbf{1} \tag{34}$$

and

$$\sum_{c=1}^{N_c} \mu_c \nabla M_{c\pi} - \nabla e_\pi - p_\pi \nabla s_\pi + \theta\nabla\eta_\pi = \mathbf{0}. \tag{35}$$

To obtain the first of these, sum (28) from $\pi = 1$ to $N_p$ and use the facts

$$\eta = \sum_{\pi=0}^{N_p} \eta_\pi \quad \text{and} \quad e = e_I + \sum_{\pi=0}^{N_p} e_\pi$$

to find that

$$e_0 - \theta\eta_0 + e_I = e - \theta\eta - \mathbf{m}\cdot\boldsymbol{\mu} + \sum_{\pi=1}^{N_p} s_\pi p_\pi.$$

Using this equation and Item 2 yields (34). To obtain (35), begin by taking the gradient of (16) and use (21)–(25) to find that

$$\theta\nabla\eta_\pi = \theta\lambda^e\left(1 + \frac{\partial e_I}{\partial e_\pi}\right)\nabla e_\pi + \left(\theta\lambda^s + \theta\lambda^e\frac{\partial e_I}{\partial s_\pi}\right)\nabla s_\pi + \sum_{c=1}^{N_c}\left(\theta\lambda_c^m + \theta\lambda^e\frac{\partial e_I}{\partial M_{c\pi}}\right)\nabla M_{c\pi}.$$

Combining this with Items 3 and 4 results in (35), which completes the proof. □

## 3.3 Comments and Consequences of the Restrictions

Item 1 of the Proposition 1 is consistent with the common assumption in classical thermodynamics that $\eta_\pi$ is a homogeneous function of $e_\pi$, $s_\pi$, and $M_{c\pi}$; see, for example, [28].

Item 6 specifies the net elastic forces on a fluid phase $\pi$ by the other phases. The net elastic force on the solid phase coming from the fluid phases can be computed from the fact that (13) implies

$$\mathbf{0} = \sum_{\pi=0}^{N_p}\sum_{\substack{\pi'=0\\ \pi'\neq\pi}}^{N_p} \mathbf{f}_{E\pi\pi'} = \sum_{\pi'=1}^{N_p} \mathbf{f}_{E0\pi'} + \sum_{\pi=1}^{N_p}\sum_{\substack{\pi'=0\\ \pi'\neq\pi}}^{N_p} \mathbf{f}_{E\pi\pi'}$$

and hence by $(31)_2$

$$\begin{aligned}\sum_{\pi'=1}^{N_p} \mathbf{f}_{E0\pi'} &= -\sum_{\pi=1}^{N_p} p_\pi\nabla s_\pi = -\sum_{\pi=1}^{N_p}\left(p + \frac{\partial e_I}{\partial s_\pi}\right)\nabla s_\pi \\ &= p\nabla s_0 - \nabla e_I + \frac{\partial e_I}{\partial s_0}\nabla s_0 + \frac{\partial e_I}{\partial \mathbf{F}}\nabla\mathbf{F}.\end{aligned} \tag{36}$$

This formula together (33) specifies net elastic force $\sum_{\substack{\pi'=0\\ \pi'\neq\pi}}^{N_p} \mathbf{f}_{E\pi\pi'}$ on any phase $\pi$ in terms of the given constitutive laws.

If (33) is used in the force balance for fluid phase $\pi$ and the divergence of the stress is expanded, then one obtains

$$s_\pi \mathbf{b}_\pi - s_\pi \nabla p_\pi + \sum_{\substack{\pi'=0\\ \pi'\neq\pi}}^{N_p} \mathbf{f}_{V\pi\pi'} = \mathbf{0}. \tag{37}$$

Thus, the net elastic force acting on the fluid phase from other phases cancels with part of the term associated with the divergence of the stress.

The results of Proposition 3.2 can also be used to rewrite the energy balance (14) in the form

$$\begin{aligned} \dot{e} - \boldsymbol{\mu}\cdot\dot{\mathbf{m}} = r - \operatorname{div}\mathbf{q} - \sum_{c=1}^{N_c} \nabla\mu_c \cdot \mathbf{h}_c - \sum_{\pi=1}^{N_p}\Bigg(\mathbf{v}_\pi^r \cdot \sum_{\substack{\pi'=0\\ \pi'\neq\pi}}^{N_p} \mathbf{f}_{V\pi\pi'}\Bigg) \\ + \nabla\mathbf{v}_0 \cdot \left(\frac{\partial e_I}{\partial \mathbf{F}}\mathbf{F}^\top - \theta\frac{\partial \eta_0}{\partial \mathbf{F}}\mathbf{F}^\top - \theta\eta\mathbf{1}\right) - \sum_{\pi=1}^{N_p} \theta \operatorname{div}\big(\eta_\pi \mathbf{v}_\pi^r\big). \end{aligned} \tag{38}$$

This form of the energy balance can be used to obtain a dissipation relation. Namely, using (27) and (31) we have

$$\theta\dot{\eta} = \dot{e} - \boldsymbol{\mu}\cdot\dot{\mathbf{m}} + \left(\frac{\partial e_I}{\partial \mathbf{F}}\mathbf{F}^\top - \theta\frac{\partial \eta_0}{\partial \mathbf{F}}\mathbf{F}^\top\right)\cdot\nabla\mathbf{v}_0.$$

Using this in (38) and then integrating the term involving the heat flux by parts, one obtains for any region $V(t)$ convecting with the solid phase that

$$\begin{aligned} -\frac{d}{dt}\int_{V(t)} \eta\, dv - \int_{V(t)} \frac{1}{\theta}\Bigg\{\frac{1}{\theta}\mathbf{q}\cdot\nabla\theta + \sum_{c=1}^{N_c}\nabla\mu_c\cdot\mathbf{h}_c + \sum_{\pi=1}^{N_p}\Bigg(\mathbf{v}_\pi^r \cdot \sum_{\substack{\pi'=0\\ \pi'\neq\pi}}^{N_p} \mathbf{f}_{V\pi\pi'}\Bigg)\Bigg\}\, dv \\ = -\int_{V(t)} \frac{r}{\theta}\, dv + \int_{\partial V(t)} \left(\frac{\mathbf{q}}{\theta} + \sum_{\pi=1}^{N_p}\eta_\pi \mathbf{v}_\pi\right)\cdot\mathbf{n}\, da. \end{aligned}$$

This equation can be used to establish bounds upon the gradients of the temperature, chemical potentials, and pressure. Such estimates are important for the development of stable numerical schemes.

### 3.4 Fluxes and Darcy's Law

Prototypical constitutive laws for the mass and heat flux are provided by Fick and Fourier laws. Here, these take the form

$$\mathbf{h}_c = -\hat{\mathbf{K}}_c(\mathbf{e}, \mathbf{s}, M)\nabla\mu_c \quad \text{and} \quad \mathbf{q} = -\hat{\mathbf{K}}_\theta(\mathbf{e}, \mathbf{s}, M)\nabla\theta, \tag{39}$$

where it is assumed that $\hat{\mathbf{K}}_c(\mathbf{e}, \mathbf{s}, M)$ and $\hat{\mathbf{K}}_\theta(\mathbf{e}, \mathbf{s}, M)$ are positive semidefinite symmetric tensors, which ensures that $(32)_{1,2}$ hold.

The specification of the viscous forces $\mathbf{f}_{V\pi\pi'}$ is related to Darcy's law, which is fundamental to the theory of porous flow and is discussed in detail in all of the monographs on the subject [2, 9, 10, 25]. These forces should depend on the velocities of the phases $\pi$ and $\pi'$ and, possibly, the volume fractions. For this reason, we assume that

$$\mathbf{f}_{V\pi\pi'} = -\hat{\mathbf{A}}_{\pi\pi'}(\mathbf{s})(\mathbf{v}_\pi - \mathbf{v}_{\pi'}), \tag{40}$$

where the $\hat{\mathbf{A}}_{\pi\pi'}(\mathbf{s})$ are assumed to be positive semidefinite symmetric tensors and satisfy $\hat{\mathbf{A}}_{\pi\pi'}(\mathbf{s}) = \hat{\mathbf{A}}_{\pi'\pi}(\mathbf{s})$. These properties ensure that the inequality $(32)_3$ holds. Substituting (40) into the force balance for phase $\pi$ (37) results in

$$s_\pi \mathbf{b}_\pi - s_\pi \nabla p_\pi - \sum_{\substack{\pi'=0 \\ \pi'\neq\pi}}^{N_p} \hat{\mathbf{A}}_{\pi\pi'}(\mathbf{s})(\mathbf{v}_\pi - \mathbf{v}_{\pi'}) = \mathbf{0}. \tag{41}$$

In the case of a single fluid phase, if $\hat{\mathbf{A}}_{10}(s_0, s_1)$ is invertible, the above equation can be written in the form

$$\mathbf{v}_1^r = s_1 \hat{\mathbf{A}}_{10}(s_0, s_1)^{-1}(\mathbf{b}_1 - \nabla p_1),$$

which is the traditional form of Darcy's law. In this way one can see that Darcy's law can be viewed as a consequence of force balance, in the absence of inertial forces, together with an appropriate constitutive law for the viscous forces. Moreover, the tensors $\mathbf{A}_{\pi\pi'}$ correspond to the (pseudo) inverses of the permeability tensors that appear in the engineering and experimental literature.

It is possible to formulate Darcy's law as the Euler–Lagrange equation of a (convex) dissipation function. Set

$$I_{\mathbf{s}}\left(\{\mathbf{v}_\pi^r\}_{\pi=1}^{N_p}\right) := \frac{1}{2}\left\|\{\mathbf{v}_\pi^r\}_{\pi=1}^{N_p}\right\|_{\mathbf{A}}^2 - \sum_{\pi=1}^{N_p} s_\pi(\mathbf{b}_\pi - \nabla p_\pi)\cdot \mathbf{v}_\pi^r,$$

where $\|.\|_{\mathbf{A}}$ is the norm associated with the inner-product

$$\left(\{\mathbf{v}_\pi^r\}_{\pi=1}^{N_p}, \{\mathbf{w}_\pi^r\}_{\pi=1}^{N_p}\right)_{\mathbf{A}} := \sum_{\pi=1}^{N_p}\left(\sum_{\substack{\pi'=0 \\ \pi'<\pi}}^{N_p} \mathbf{A}_{\pi\pi'}\left(\mathbf{v}_\pi^r - \mathbf{v}_{\pi'}^r\right)\cdot\left(\mathbf{w}_\pi^r - \mathbf{w}_{\pi'}^r\right)\right). \tag{42}$$

The associated Euler–Lagrange equation is

$$\left(\{\mathbf{v}_\pi^r\}_{\pi=1}^{N_p}, \{\mathbf{w}_\pi^r\}_{\pi=1}^{N_p}\right)_{\mathbf{A}} = \sum_{\pi=1}^{N_p} s_\pi(\mathbf{b}_\pi - \nabla p_\pi)\cdot\mathbf{w}_\pi^r \quad \text{for all } \{\mathbf{w}_\pi^r\}_{\pi=1}^{N_p},$$

which is the variational formulation of Darcy's law (41).

### 3.5 Mechanical and Isothermal Case

Often it is assumed that the temperature is constant or that the thermal energy in the solid phase dominates that of the fluid phases so that the temperature of the fluids become that of the solid much faster than the other processes taking place. In this case the energy balance reduces to an equation for the temperature of the solid that is decoupled from the other balances. For example, the temperature could satisfy a classic heat equation

$$\frac{\partial}{\partial t}(c\theta) - \operatorname{div}(\mathbf{K}_\theta \nabla\theta) = r,$$

where $c$ and $\mathbf{K}_\theta$ are the specific heat and conductivity of the solid phase, respectively.

To specify a constant temperature it is preferable to use the Helmholtz free-energy rather than the entropy in the local thermodynamic equilibrium assumption. Letting $\psi_\pi = e_\pi - \theta\eta_\pi$ denote the free-energy density of phase $\pi$, the local equilibrium assumption requires that

$$\psi = \hat{\psi}(\theta, \mathbf{m}, \mathbf{F}) := \inf_{(\mathbf{s}, M)} \left\{ \hat{\psi}_0(\theta, s_0, \mathbf{F}) + \sum_{\pi=1}^{N_p} \hat{\psi}_\pi\left(\theta, s_\pi, \{M_{c\pi}\}_{c=1}^{N_c}\right) + \hat{e}_I(\theta, \mathbf{s}, M, \mathbf{F}) \;\Big|\; \sum_{\pi=0}^{N_p} s_\pi = 1,\ \sum_{\pi=1}^{N_p} M_{c\pi} = m_c \right\}. \tag{43}$$

Moreover, for each $(\theta, \mathbf{m}, \mathbf{F})$ the minimum is attained at a unique point

$$(\mathbf{s}, M) = \operatorname*{arg\,min}_{(\mathbf{s}, M)} \left\{ \hat{\psi}_0(\theta, s_0, \mathbf{F}) + \sum_{\pi=1}^{N_p} \hat{\psi}_\pi\left(\theta, s_\pi, \{M_{c\pi}\}_{c=1}^{N_c}\right) + \hat{e}_I(\theta, \mathbf{s}, M, \mathbf{F}) \;\Big|\; \sum_{\pi=0}^{N_p} s_\pi = 1,\ \sum_{\pi=1}^{N_p} M_{c\pi} = m_c \right\}. \tag{44}$$

The infimum in these expressions is taken over $\mathbf{s} \in [0, 1]^{N_p+1}$ and $M \in [0, \infty)^{N_c \times N_p}$.

As in Sect. 3.1, there are Lagrange multipliers $(\lambda^s, \boldsymbol{\lambda}^m)$ and KKT multipliers $(\boldsymbol{\lambda}^0, \boldsymbol{\lambda}^1, \Lambda)$ such that

$$\frac{\partial \psi_\pi}{\partial s_\pi} + \lambda_\pi^0 - \lambda_\pi^1 = \lambda^s - \frac{\partial e_I}{\partial s_\pi}, \qquad \pi \in \{0, \ldots, N_p\}, \tag{45}$$

$$\frac{\partial \psi_\pi}{\partial M_{c\pi}} + \Lambda_{c\pi} = \lambda_c^m - \frac{\partial e_I}{\partial M_{c\pi}}, \qquad \pi \in \{1, \ldots, N_p\}, \tag{46}$$

as well as (24) and (25) with $c \in \{1, \ldots, N_c\}$.

Replacing the entropy with the free-energy in the reduced entropy inequality results in

$$\begin{aligned}
-\dot{\psi} - \eta\dot{\theta} + \dot{\mathbf{m}} \cdot \boldsymbol{\mu} \geq{}& \frac{1}{\theta}\nabla\theta \cdot \mathbf{q} + \sum_{c=1}^{N_c} \nabla\mu_c \cdot \mathbf{h}_c + \sum_{\pi=1}^{N_p} \left( \mathbf{v}_\pi^r \cdot \sum_{\substack{\pi'=0 \\ \pi'\neq\pi}}^{N_p} \mathbf{f}_{V\pi\pi'} \right) \\
&- \nabla\mathbf{v}_0 \cdot \left[ \mathbf{T}_0 + \left( \boldsymbol{\mu} \cdot \mathbf{m} - \psi - \sum_{\pi=1}^{N_p} s_\pi p_\pi \right) \mathbf{1} \right] \\
&- \sum_{\pi=1}^{N_p} \mathbf{v}_\pi^r \cdot \left[ \sum_{c=1}^{N_c} \mu_c \nabla M_{c\pi} - \nabla\psi_\pi - \sum_{\substack{\pi'=0 \\ \pi'\neq\pi}}^{N_p} \mathbf{f}_{E\pi\pi'} - \eta_\pi \nabla\theta \right] \\
&- \sum_{\pi=1}^{N_p} \operatorname{div}\left(\mathbf{v}_\pi^r\right) \left[ \sum_{c=1}^{N_c} \mu_c M_{c\pi} - \psi_\pi - s_\pi p_\pi \right],
\end{aligned}$$

which is sometimes referred to as the free-energy imbalance. The following result is analogous to Proposition 1, and is presented without proof.

**Proposition 2** *Assume that local equilibrium holds so that the Helmholtz free-energy* $\psi$ *is determined by* (43) *and that the volume fractions and densities of the different phases are determined by* (44). *Under these constitutive assumptions, the free-energy imbalance holds for all thermodynamic processes if*

1. *The entropy of each phase is given by* $\eta_\pi = -\frac{\partial \psi_\pi}{\partial \theta}$ *for* $0 \le \pi \le N_p$.
2. *The free-energy of fluid phase* $\pi \neq 0$ *is given by*

$$\psi_\pi = -s_\pi p_\pi + \sum_{c=1}^{N_c} M_{c\pi} \mu_c. \tag{47}$$

3. *The stress in the solid phase is given by*

$$\mathbf{T}_0 = (\psi_0 + e_I)\mathbf{1} + \frac{\partial \psi_0}{\partial \mathbf{F}} \mathbf{F}^\top + \frac{\partial e_I}{\partial \mathbf{F}} \mathbf{F}^\top. \tag{48}$$

4. *The interfacial energy only depends upon the volume fractions and the deformation gradient; that is,* $e_I = \hat{e}_I(\mathbf{s}, \mathbf{F})$.
5. *The phase pressures and the chemical potential of the components are related to the Lagrange multipliers through*

$$p_\pi = p + \frac{\partial e_I}{\partial s_\pi} \quad \textit{and} \quad \mu_c = \lambda_c^m, \tag{49}$$

   *where* $p := -\lambda^s$.
6. *The diffusive mass fluxes, heat flux, and viscous forces satisfy*

$$\sum_{c=1}^{N_c} \mathbf{h}_c \cdot \nabla \mu_c \le 0, \qquad \mathbf{q} \cdot \nabla \theta \le 0, \qquad \sum_{\pi=1}^{N_p} \left( \mathbf{v}_\pi^r \cdot \sum_{\substack{\pi'=0 \\ \pi' \neq \pi}}^{N_p} \mathbf{f}_{V\pi\pi'} \right) \le 0. \tag{50}$$

7. *The elastic forces between the phases satisfy*

$$p_\pi \nabla s_\pi = \sum_{\substack{\pi'=0 \\ \pi' \neq \pi}}^{N_p} \mathbf{f}_{E\pi\pi'} \qquad \pi \neq 0. \tag{51}$$

This result can be used to obtain an equation that describes the rate of change of the total Helmholtz free-energy. To see this, take the gradient of $\psi_\pi = \hat{\psi}_\pi(\theta, s_\pi, \{M_{c\pi}\}_{c=1}^{N_c})$ and use the chain rule to find

$$\nabla \psi_\pi = -\eta_\pi \nabla \theta - p_\pi \nabla s_\pi + \sum_{c=1}^{N_c} \mu_c \nabla M_{c\pi}.$$

Combining this with the gradient of (47) results in

$$\mathbf{0} = \eta_\pi \nabla \theta - s_\pi \nabla p_\pi + \sum_{c=1}^{N_c} M_{c\pi} \nabla \mu_c.$$

Now, using the chain rule to compute the time derivative of $\psi$, mass balance, and the previous equation, one can obtain

$$\begin{aligned}\frac{d}{dt}\int_{V(t)}\psi\,dv &= \int_{V(t)}\left(\mathbf{T}_0-\sum_{\pi=1}^{N_p}s_\pi p_\pi\mathbf{1}\right)\cdot\nabla\mathbf{v}_0+\sum_{\pi=1}^{N_p}s_\pi\nabla p_\pi\cdot\mathbf{v}_\pi\,dv\\ &\quad-\int_{V(t)}\left(\eta\dot{\theta}+\sum_{\pi=1}^{N_p}\eta_\pi\mathbf{v}_\pi^r\cdot\nabla\theta\right)dv+\sum_{c=1}^{N_c}\int_{V(t)}(\mu_c h_c+\nabla\mu_c\cdot\mathbf{h}_c)\,dv\\ &\quad-\int_{\partial V(t)}\mu_c\left(\mathbf{h}_c+\sum_{\pi=1}^{N_p}M_{c\pi}\mathbf{v}_\pi^r\right)\cdot\mathbf{n}\,da,\end{aligned}$$

where $V(t)$ is a region convecting with the motion of the solid. Integrating the term involving $\nabla\mathbf{v}_0$ by parts and then using the force balance (8), under isothermal conditions ($\dot{\theta}=0$ and $\nabla\theta=\mathbf{0}$) the previous equation becomes

$$\begin{aligned}&\frac{d}{dt}\int_{V(t)}\psi\,dv-\int_{V(t)}\left\{\sum_{\pi=1}^{N_p}\left(\sum_{\substack{\pi'=0\\ \pi'\neq\pi}}^{N_p}\mathbf{f}_{V\pi\pi'}\cdot\mathbf{v}_\pi^r\right)+\sum_{c=1}^{N_c}\nabla\mu_c\cdot\mathbf{h}_c\right\}dv\\ &=\int_{V(t)}\left\{\sum_{\pi=0}^{N_p}s_\pi\mathbf{b}_\pi\cdot\mathbf{v}_\pi+\sum_{c=1}^{N_c}\mu_c h_c\right\}dv\\ &\quad+\int_{\partial V(t)}\left\{\left(\mathbf{T}_0-\sum_{\pi=1}^{N_p}s_\pi p_\pi\mathbf{1}\right)\mathbf{v}_0-\mu_c\left(\mathbf{h}_c+\sum_{\pi=1}^{N_p}M_{c\pi}\mathbf{v}_\pi^r\right)\right\}\cdot\mathbf{n}\,da.\end{aligned}$$

## 4 Connection with Other Theories

### 4.1 Biot Constants

One of the most dominant theories for poroelastic materials in the case of linear elastic response for the solid is Biot's theory [3–6]. See the book by Cheng [10] for a recent overview of the subject or Sanchez-Palencia [25] for how the theory can be obtained via homogenization. Within this theory, only one fluid phase is considered and a global force balance is used. Thus, the stress in this theory is the total stress $\mathbf{T}$ given as the sum the stresses in the solid and fluid phases:

$$\mathbf{T}=\mathbf{T}_0+\mathbf{T}_1=\mathbf{T}_0-s_1p_1\mathbf{1}.$$

Constitutive laws for the total stress and the global pressure $p$ are determined by Biot constants. Perhaps surprisingly, one of these constants appears in the constitutive laws for both the stress and the pressure. This fact has been reproduced using homogenization theory. Here we show that under certain simplifying assumptions, our theory reproduces these constitutive laws and the Biot constants. In the following calculations, assume that $0<s_0<1$ so that we are not in a degenerate case. This implies that the KKT multipliers $\lambda_0^1$ and $\lambda_0^1$ are both zero by (24).

To obtain the constitutive laws appearing in Biot's theory we make two simplifying assumptions:

– there is no interfacial energy; that is, $e_I = 0$,
– only small displacements and small changes in volume fraction occur.

In Biot's theory of poroelasticity, there is no mention of any interfacial energy, so the first of these assumptions is reasonable. Moreover, Biot's theory is linear, so the second assumption is also justified.

Notice that by $(31)_2$ the first assumption implies that $p_\pi = p$ for all $\pi$, so there is only one pressure. Since we will be working with small displacements, it makes sense to work with a referential description. For $J := \det \mathbf{F}$, set

$$e_{0R} := Je_0, \qquad s_{0R} := Js_0, \qquad p_R := Jp, \qquad \eta_{0R} := J\eta_0.$$

The constitutive law $\hat{\eta}_{0R}$ for $\eta_{0R}$ is related to $\hat{\eta}_0$ by

$$\hat{\eta}_{0R}(Je_0, Js_0, \mathbf{F}) = J\hat{\eta}_0(e_0, s_0, \mathbf{F}).$$

Differentiating this equation with respect to $\mathbf{F}$ and using the fact that $\frac{\partial J}{\partial \mathbf{F}} = J\mathbf{F}^{-\top}$ results in

$$\theta \frac{\partial \eta_{0R}}{\partial \mathbf{F}} + s_0 p_R \mathbf{F}^{-\top} = \left( \theta \frac{\partial \eta_0}{\partial \mathbf{F}} \mathbf{F}^\top - e_0 \mathbf{1} + \theta \eta_0 \mathbf{1} \right) J\mathbf{F}^\top,$$

and hence, from (29), the Piola stress $\mathbf{T}_{0R} = J\mathbf{T}_0\mathbf{F}^{-\top}$ for the solid is given by

$$\mathbf{T}_{0R} = -\theta \frac{\partial \eta_{0R}}{\partial \mathbf{F}} - s_0 p_R \mathbf{F}^{-\top}.$$

Since the Piola stress for the fluid phase is given by $\mathbf{T}_{1R} = -s_1 p_R \mathbf{F}^{-\top}$, it follows that the total Piola stress is

$$\mathbf{T}_R = \theta \frac{\partial \eta_{0R}}{\partial \mathbf{F}} - p_R \mathbf{F}^{-\top}. \tag{52}$$

Based on the small displacements and small changes in volume fraction assumption, the parameter

$$\epsilon := \sqrt{|\nabla \mathbf{u}_0|^2 + |s_{0R} - s_{0R\circ}|^2}$$

is small, where $\mathbf{u}_0$ is the displacement of the solid and $s_{0R\circ}$ is a given reference volume fraction for the solid from which we are considering small perturbations. Assuming that $\eta_{0R}$ has a local maximum at $(e_{0R}, s_{0R}, \mathbf{F}) = (e_{0R\circ}, s_{0R\circ}, \mathbf{1})$, an asymptotic expansion in $\epsilon$ of the constitutive law for $\eta_{0R}$ reads[3]

$$\eta_{0R} = \nabla \mathbf{u}_0 \cdot \mathbb{A} \nabla \mathbf{u}_0 + 2(s_{0R} - s_{0R\circ}) \mathbf{A} \cdot \nabla \mathbf{u}_0 - a^2 (s_{0R} - s_{0R\circ})^2 + o(\epsilon^2),$$

where

$$\mathbb{A} := \frac{1}{2} \frac{\partial^2 \hat{\eta}_{0R}}{\partial \mathbf{F}^2}(e_{0R\circ}, s_{0R\circ}, \mathbf{1}),$$

$$\mathbf{A} := \frac{1}{2} \frac{\partial^2 \hat{\eta}_{0R}}{\partial s_{0R} \partial \mathbf{F}}(e_{0R\circ}, s_{0R\circ}, \mathbf{1}),$$

[3] One could also consider small changes in $e_R$, but here this is not done so we can compare the result with the traditional Biot theory, which is isothermal.

$$-a^2 := \frac{1}{2}\frac{\partial^2 \hat{\eta}_{0R}}{\partial s_{0R}^2}(e_{0R\circ}, s_{0R\circ}, \mathbf{1}).$$

Taking this expansion as a constitutive law for $\eta_{0R}$, from (22) and (31), we obtain

$$\frac{p_R}{\theta} = J\big(2\mathbf{A}\cdot\nabla\mathbf{u}_0 - 2a^2(s_{0R} - s_{0R\circ})\big) = 2\mathbf{A}\cdot\nabla\mathbf{u}_0 - 2a^2(s_{0R} - s_{0R\circ}) + o(\epsilon).$$

This motivates completing the square in the quadratic asymptotic expansion for $\eta_{0R}$ in a particular way so as to obtain a term like the right-hand side of the previous equation. Namely, we can write

$$\eta_{0R} = \nabla\mathbf{u}_0\cdot\tilde{\mathbb{A}}\nabla\mathbf{u}_0 - \frac{1}{a^2}\big(\mathbf{A}\cdot\nabla\mathbf{u}_0 - a^2(s_{0R} - s_{0R\circ})\big)^2,$$

where $\tilde{\mathbb{A}}\nabla\mathbf{u}_0 := \mathbb{A}\nabla\mathbf{u}_0 + \frac{1}{a^2}(\mathbf{A}\cdot\nabla\mathbf{u}_0)\mathbf{A}$. Now, if we set

$$\mathbb{C} := -2\theta\tilde{\mathbb{A}}, \qquad \mathbf{H} := \frac{1}{a^2}\mathbf{A}, \qquad h := -\frac{1}{2a^2\theta}$$

we can write

$$-\theta\eta_{0R} = \frac{1}{2}\nabla\mathbf{u}_0\cdot\mathbb{C}\nabla\mathbf{u}_0 - \frac{1}{2h}\big(\mathbf{H}\cdot\nabla\mathbf{u}_0 - (s_{0R} - s_{0R\circ})\big)^2.$$

Using this in (52) and linearizing the term $p_R\mathbf{F}^{-\top}$ on the right-hand side yields

$$\mathbf{T}_R = \mathbb{C}\nabla\mathbf{u}_0 + p_R(\mathbf{H} - \mathbf{1}) \quad \text{and} \quad s_{0R} - s_{0R\circ} = \mathbf{H}\cdot\nabla\mathbf{u}_0 + hp_R. \tag{53}$$

The material constant $\mathbf{H}$ appears in both the formula for the stress and for the pressure. This is consistent with what appears in [10] and [25]. These formulas differ slightly from those found in the literature since the measure of strain $s_{0R} - s_{0R\circ}$ used here differs from those used elsewhere. The fourth-order tensor $\mathbb{C}$ has the usual major symmetry since it came from the tensor $\mathbb{A}$, which is a second derivative of the entropy with respect to $\mathbf{F}$. The minor symmetry of $\mathbb{C}$ will follow from the principle of material frame-indifference.

### 4.2 Alternate Form of Mass Balance

In some formulations of poroelasticity involving only one fluid phase made up of one component, the mass balance for this component is replaced by an equation involving the pressure and the volume fraction of the phase. Here this equation is derived under certain assumptions and its relation with mass balance is examined.

The assumptions we will make in the context of the single fluid phase made up of a single component are the following:

- the fluid phase is incompressible,
- there is no interfacial energy,
- only small displacements and small changes in volume fraction can occur.

By the definition of incompressibility, the fluid phase can only undergo isochoric motions and thus

$$\frac{d}{dt}\int_{V_f(t)} s_1\, dv = 0,$$

where $V_f(t)$ is any volume *convecting with the fluid phase*. Localizing this balance yields

$$\frac{\partial}{\partial t}s_1 + \operatorname{div}(s_1\mathbf{v}_1) = 0. \tag{54}$$

Writing this in a reference configuration and using the assumption of small displacements results in

$$\frac{\partial}{\partial t}s_{1R} + \operatorname{div}(s_{1R}\mathbf{v}_1) = 0, \tag{55}$$

where $s_{1R} := Js_1$ is the volume fraction of the fluid per unit volume in the reference configuration. Using the fact that the volume fraction of the fluid and solid must add to one and that the reference volume fraction is constant, with the help of $(53)_2$ the previous equation can be written as

$$-\frac{\partial}{\partial t}(\mathbf{H}\cdot\nabla\mathbf{u}_0 + hp_R) + \operatorname{div}(s_{1R}\mathbf{v}_1) = 0, \tag{56}$$

which is sometimes referred to as the storage equation [30].

A relationship between (54), and hence (56), and mass balance becomes clear in the case where the fluid phase has no mass diffusion and no sources or sinks of mass, in which case mass balance (6) reads

$$\frac{\partial}{\partial t}m_1 + \operatorname{div}(m_1\mathbf{v}_1) = 0.$$

If we let $\bar{m}_1$ denote the mass of the fluid per unit volume of the fluid so that $m_1 = s_1\bar{m}_1$, then (54) and the previous equation imply that

$$\frac{\partial}{\partial t}\bar{m}_1 + \nabla\bar{m}_1\cdot\mathbf{v}_1 = 0, \tag{57}$$

which says that $\bar{m}_1$ is constant on particle paths. This means that once the fluid velocity is known, $\bar{m}_1$ can be found by integrating along such a path. Some authors take the previous equation as the definition of incompressibility. When this is done, (57) together with mass balance yields (54). For this reason, (54), or (56) if there are small displacements and small changes in volume fraction, are sometimes referred to as an alternate form of mass balance. See, for example, the book by Bear [2].

### 4.3 Alternative Statements of the Force Balances

The force balance for the fluids used by the authors in [26] to develop a continuum description of porous flow in a rigid medium took the form

$$\mathbf{0} = \int_{V(t)} \sum_{\substack{\pi'=0\\ \pi'\neq\pi}}^{N_p} \mathbf{f}_{V\pi\pi'}\,dv + \int_{V_\pi(t)} \mathbf{b}_\pi\,dv_\pi - \int_{\partial V_\pi(t)} p_\pi\mathbf{n}\,da_\pi,$$

where $dv_\pi$ and $da_\pi$ are the volume and area elements associated with the volume occupied by phase $\pi$. With the identification $dv_\pi = s_\pi\,dv$ the local form of this law becomes

$$s_\pi\mathbf{b}_\pi - s_\pi\nabla p_\pi + \sum_{\substack{\pi'=0\\ \pi'\neq\pi}}^{N_p} \mathbf{f}_{V\pi\pi'} = \mathbf{0},$$

which is identical to Eq. (37) which resulted from the more general force balance postulated in Sect. 2 and subsequent restrictions due to the second law. A slightly simpler derivation of the model proposed here is obtained if this force balance is assumed a priori for the fluids, and the force balance for the solid takes the form

$$\int_{V(t)} (s_0 \mathbf{b}_0 + \mathbf{f}_E)\, dv + \int_{\partial V(t)} \mathbf{T}_0 \mathbf{n}\, da + \sum_{\pi=1}^{N_p} \int_{V(t)} \mathbf{f}_{V0\pi}\, dv = \mathbf{0},$$

where $\mathbf{f}_E$ is the force on the solid due to elastic stresses (pressures) in the fluid. The local form is

$$s_0 \mathbf{b}_0 + \operatorname{div}(\mathbf{T}_0) + \mathbf{f}_E + \sum_{\pi=1}^{N_p} \mathbf{f}_{V0\pi} = \mathbf{0},$$

and the formula for $\mathbf{f}_E$ appearing on the right hand side of Eq. (36) follows directly from Coleman–Noll procedure.

## 5 Summary

In summary, the constitutive assumptions for the model are determined by

- the entropy $\eta_\pi$ for the fluid phases (16) and the solid phase (17),
- the interfacial energy $e_I$ specified by (30),
- local thermodynamics equilibrium, Assumption 1,
- the stress relation (29) for the solid phase,
- the Fick and Fourier laws (39),
- (33) and (36), which along with (13), determine $\sum_{\substack{\pi'=0 \\ \pi' \neq \pi}}^{N_p} \mathbf{f}_{E\pi\pi'}$,
- the formula (40) for $\mathbf{f}_{V\pi\pi'}$.

Substituting these constitutive laws into the balances yields a system of partial differential equations that describe the poroelastic continuum. In this list of balances we use the total force balance for the system in place of the force balance for the solid phase. The reasoning for this will be explained at the end of this section.

The mass balance for component $c$ is

$$\frac{\partial}{\partial t} m_c + \operatorname{div}\left( m_c \mathbf{v}_0 - \mathbf{K}_c \nabla \mu_c + \sum_{\pi=1}^{N_p} M_{c\pi} \mathbf{v}_\pi^r \right) = h_c.$$

Force balance for fluid phase $\pi$ is

$$s_\pi \mathbf{b}_\pi - s_\pi \nabla p_\pi - \sum_{\substack{\pi'=0 \\ \pi' \neq \pi}}^{N_p} \mathbf{A}_{\pi\pi'} \mathbf{v}_{\pi\pi'}^r = \mathbf{0},$$

and for the entire continuum is

$$\sum_{\pi=0}^{N_p} s_\pi \mathbf{b}_\pi + \operatorname{div}\left[ \left( \frac{\partial e_I}{\partial \mathbf{F}} - \theta \frac{\partial \eta_0}{\partial \mathbf{F}} \right) \mathbf{F}^\top + \left( e_0 - \theta \eta_0 - \sum_{\pi=1}^{N_p} s_\pi p_\pi \right) \mathbf{1} \right] = \mathbf{0}. \tag{58}$$

The balance of energy becomes

$$\dot{e} - \dot{\mathbf{m}} \cdot \boldsymbol{\mu} = r + \operatorname{div}(\mathbf{K}_\theta \nabla\theta) + \sum_{c=1}^{N_c} \nabla\mu_c \cdot \mathbf{K}_c \nabla\mu_c + \sum_{\pi=1}^{N_p} \left( \mathbf{v}_\pi^r \cdot \sum_{\substack{\pi'=0 \\ \pi'\neq\pi}}^{N_p} \mathbf{A}_{\pi\pi'} (\mathbf{v}_\pi^r - \mathbf{v}_{\pi'}^r) \right)$$

$$+ \nabla\mathbf{v}_0 \cdot \left[ \left( \frac{\partial e_I}{\partial \mathbf{F}} - \theta \frac{\partial \eta_0}{\partial \mathbf{F}} \right) \mathbf{F}^\top - \theta\eta\mathbf{1} \right] - \sum_{\pi=1}^{N_p} \theta \operatorname{div}(\eta_\pi \mathbf{v}_\pi^r).$$

The mass balance for the solid phase is not listed here since the solid mass does not appear in any of the other balances and is essentially decoupled from them. Also, the deformation gradient is determined from the kinematic equation $\dot{\mathbf{F}} = (\nabla\mathbf{v}_0)\mathbf{F}$.

The quantities to be solved for are the velocity $\mathbf{v}_0$ and deformation gradient $\mathbf{F}$ of the solid, the relative velocities $\mathbf{v}_\pi^r$ of the fluid phases $1 \leq \pi \leq N_p$, the masses of the components $m_c$, and the energy $e$. The other quantities are either determined by constitutive laws or are given a priori, such as the external influences $h_c$, $\mathbf{b}_\pi$, and $r$. Notice that the force balances for the fluid phases are algebraic in the velocities.

The force balance for just the solid phase reads

$$s_0 \mathbf{b}_0 + \operatorname{div}\left[ \left( \frac{\partial e_I}{\partial \mathbf{F}} - \theta \frac{\partial \eta_0}{\partial \mathbf{F}} \right) \mathbf{F}^\top + (e_0 - \theta\eta_0)\mathbf{1} \right] + p\nabla s_0$$

$$+ \frac{\partial e_I}{\partial s_0} \nabla s_0 + \frac{\partial e_I}{\partial \mathbf{F}} \nabla\mathbf{F} + \sum_{\pi=1}^{N_p} \mathbf{A}_{0\pi} \mathbf{v}_\pi^r = \mathbf{0}.$$

Since we are assuming force balance for all of the fluid phases, using a total force balance for the continuum or just using the force balance for the solid phase are equivalent. In the literature on poroelastic continuum, the total force balance (58) is used rather than the solid force balance. Comparing the two balances it is not hard to see why. The total force balance has a clear divergence structure and does not involve the relative velocities.



## References

1. Alt, H.W., DiBenedetto, E.: Nonsteady flow of water and oil through inhomogeneous porous media. Ann. Sc. Norm. Super. Pisa, Cl. Sci. **12**(3), 335–392 (1985). http://www.numdam.org/item?id=ASNSP_1985_4_12_3_335_0
2. Bear, J.: Dynamics of Fluids in Porous Media. Dover, New York (1972)
3. Biot, M.: General theory of three-dimensional consolidation. J. Appl. Phys. **12**, 155–164 (1941)
4. Biot, M.: Theory of elasticity and consolidation for a porous anisotropic solid. J. Appl. Phys. **26**, 182–185 (1955)
5. Biot, M.: General solutions of the equations of elasticity and consolidation for a porous material. J. Appl. Mech. **78**, 91–96 (1956)
6. Biot, M., Willis, D.: The elastic coefficients of the theory of consolidation. J. Appl. Mech. **24**, 594–601 (1957)
7. Chen, Z.: Degenerate two-phase incompressible flow. I. Existence, uniqueness and regularity of a weak solution. J. Differ. Equ. **171**(2), 203–232 (2001). https://doi.org/10.1006/jdeq.2000.3848
8. Chen, Z., Ewing, R.E.: Degenerate two-phase incompressible flow. III. Sharp error estimates. Numer. Math. **90**(2), 215–240 (2001). https://doi.org/10.1007/s002110100291

9. Chen, Z., Huan, G., Ma, Y.: Computational Methods for Multiphase Flows in Porous Media. Computational Science & Engineering SIAM, Philadelphia (2006). https://doi.org/10.1137/1.9780898718942
10. Cheng, A.H.: Poroelasticity. Theory and Applications of Transport in Porous Media, vol. 27. Springer, Berlin (2016)
11. Cimmelli, V.A., Jou, D., Ruggeri, T., Ván, P.: Entropy principle and recent results in non-equilibrium theories. Entropy **16**(3), 1756–1807 (2014). https://doi.org/10.3390/e16031756
12. Coleman, B.D., Noll, W.: The thermodynamics of elastic materials with heat conduction and viscosity. Arch. Ration. Mech. Anal. **13**, 167–178 (1963)
13. Darcy, H.: Recherches expérimentales relatives au mouvement de l'eau dans les tuyaux. Mallet-Bachelier, Paris (1857)
14. DiBenedetto, E., Showalter, R.E.: Implicit degenerate evolution equations and applications. SIAM J. Math. Anal. **12**(5), 731–751 (1981). https://doi.org/10.1137/0512062
15. Dinariev, O., Evseev, N.: Multiphase flow modeling with density functional method. Comput. Geosci. **20**(4), 835–856 (2016). https://doi.org/10.1007/s10596-015-9527-2
16. Ewing, R.: The Mathematics of Reservoir Simulation. Frontiers in Applied Mathematics. SIAM, Philadelphia (1983). https://books.google.com/books?id=k_4Y-RBOK9oC
17. Ganis, B., Kumar, K., Pencheva, G., Wheeler, M.F., Yotov, I.: A global Jacobian method for mortar discretizations of a fully implicit two-phase flow model. Multiscale Model. Simul. **12**(4), 1401–1423 (2014). https://doi.org/10.1137/140952922
18. Gibson, N.L., Medina, F.P., Peszynska, M., Showalter, R.E.: Evolution of phase transitions in methane hydrate. J. Math. Anal. Appl. **409**(2), 816–833 (2014). https://doi.org/10.1016/j.jmaa.2013.07.023
19. Kroener, D., Luckhaus, S.: Flow of oil and water in a porous medium. J. Differ. Equ. **55**(2), 276–288 (1984). https://doi.org/10.1016/0022-0396(84)90084-6
20. Lauser, A., Hager, C., Helmig, R., Wohlmuth, B.: A new approach for phase transitions in miscible multi-phase flow in porous media. Adv. Water Resour. **34**, 957–966 (2011). https://doi.org/10.1016/j.advwatres.2011.04.021
21. Noll, W.: Lectures on the foundations of continuum mechanics and thermodynamics. Arch. Ration. Mech. Anal. **52**, 62–92 (1973)
22. Peszyńska, M., Lu, Q., Wheeler, M.F.: Coupling different numerical algorithms for two phase fluid flow. In: The Mathematics of Finite Elements and Applications, X, MAFELAP 1999, Uxbridge, pp. 205–214. Elsevier, Oxford (2000). https://doi.org/10.1016/B978-008043568-8/50012-2
23. Peszynska, M., Showalter, R.E., Webster, J.T.: Advection of methane in the hydrate zone: model, analysis and examples. Math. Methods Appl. Sci. **38**(18), 4613–4629 (2015). https://doi.org/10.1002/mma.3401
24. Rice, J.R., Cleary Michael, P.: Some basic stress diffusion solutions for fluid saturated elastic porous media with compressible constituents. Rev. Geophys. **14**(2), 227–241 (1976). https://doi.org/10.1029/RG014i002p00227
25. Sanchez-Palencia, E.: Nonhomogeneous Media and Vibration Theory. Lecture Notes in Physics, vol. 127. Springer, Berlin (1980)
26. Seguin, B., Walkington, N.J.: Multi-component multiphase flow. Arch. Ration. Mech. Anal. (2017, submitted)
27. Showalter, R.E.: Diffusion in poro-elastic media. J. Math. Anal. Appl. **251**(1), 310–340 (2000). https://doi.org/10.1006/jmaa.2000.7048
28. Swendsen, R.: An Introduction to Statistical Mechanics and Thermodynamics. Oxford Graduate Texts. Oxford University Press, Oxford (2012). https://books.google.com/books?id=hbW609uEobsC
29. Truesdell, C.: Rational Thermodynamics. McGraw-Hill, New York (1969). A course of lectures on selected topics, with an appendix on the symmetry of the heat-conduction tensor by C.C. Wang
30. Verruijt, A.: An Introduction to Soil Mechanics. Springer, Berlin (2010)
31. Visintin, A.: Models of Phase Transitions. Progress in Nonlinear Differential Equations and Their Applications, vol. 28. Birkhäuser, Boston (1996). https://doi.org/10.1007/978-1-4612-4078-5

J Elast (2019) 135:509–542
https://doi.org/10.1007/s10659-018-09711-x

# Solvent-Vapor Induced Spherulitic Growth in Multicomponent Elastic Films

**Ding Zhao**[1] · **Chi-Sing Man**[1]

*Dedicated to Walter Noll, Thinker, Natural Philosopher, Gentleman, in Memoriam*

Received: 14 June 2018 / Published online: 18 December 2018


**Abstract** Herein we study isothermal spherulitic crystallization in multicomponent thin films under solvent-vapor annealing. Spherulitic crystallization is taken as a coherent phase transition between an amorphous phase and a crystalline phase. Both of the two phases of the film are modeled as (possibly prestressed) elastic solids, and no *a priori* assumption is made on the stresses in the two phases. The interface between the two bulk phases is treated as an evolving one-dimensional open thermodynamic system defined by various thermodynamic variables. In particular, each chemical component of the interface is generally endowed with a mass density and a chemical potential. The interface, as a continuum system, is composed of particles. We assign each particle of an evolving interface a well-defined interface velocity so that a moving interface with mass density carries linear momentum and kinetic energy. By appealing to free energy imbalance for isothermal processes and balance of mass and momentum, we derive dissipation inequalities for the bulk phases and for the interface, with which we obtain constitutive restrictions through the Coleman–Noll procedure. The driving force behind spherulitic crystallization is identified as the quantity conjugate to the normal velocity in the interfacial dissipation inequality. Finally we investigate the effect of prestress on the directional dependence of spherulitic growth.

**Keywords** Spherulitic crystallization in elastic films · Interface velocity · Prestress effect on spherulitic growth · Phase transition · Solvent-vapor annealing

**Mathematics Subject Classification (2010)** 74A20 · 74A50 · 74N20 · 80A17 · 82B26 · 82C26

✉ C.-S. Man
cman@uky.edu

D. Zhao
ding.zhao@uky.edu

[1] Department of Mathematics, University of Kentucky, Lexington, KY 40506-0027, USA

## 1 Introduction

Organic small-molecule semiconductors have attractive features for applications in organic photovoltaics (OPVs), organic thin-film transistors (OTFTs), and organic light-emitting diodes (OLEDs) (see [19] and the references therein). In particular, binary blends of organic small-molecule semiconductors have been actively studied as the host–guest pair in OTFTs and as the donor–acceptor pair in bulk heterojunction OPVs [20].

Solvent-vapor annealing is known to improve device performance as measured by the power conversion efficiency of OPVs and field-effect mobility of OTFTs. For thin-film blends of small-molecule semiconductors which are formed by spin coating and are largely amorphous in the as-spun state, restructuring in film morphology upon exposure to solvent vapor often proceeds by crystallization through heterogeneous nucleation and spherulitic growth (see [17] and the references therein). However, as pointed out by Hegde et al. [15] in a 2012 paper, the thermodynamic driving force behind the solvent-vapor induced crystallization was by then not explicitly addressed in the literature. Clarifying the physics of solvent-vapor induced spherulitic growth in organic semiconductor thin films is therefore of crucial importance for further development of the subject.

Continuum sharp-interface theories of phase transitions in solids have been studied extensively in the past few decades. Abeyaratne and Knowles [1] investigate the surface of strain discontinuity in a continuum medium but no interfacial structure is considered. What they call the "driving traction" is identified to be the quantity conjugate to the normal velocity and the product of the two gives the entropy production at the interface. With configurational force balance as a main tool, Gurtin and Struthers [13], Gurtin [11, 12], and Fried and Gurtin [8, 9] study phase interfaces of various kinds, which include massless solid–solid interface and solid–vapor interface with adatom density. Jabbour, Man and Paroni [17] consider isothermal solvent-vapor induced spherulitic crystallization in binary thin films, which is characterized by a constant growth rate under certain experimental conditions (see Remark 4.1 for more details). Their basic assumption is that diffusion plays no role in spherulitic growth, which is in direct contrast to the theory of Lifschitz and Slyozov [23] (see also [22, Sect. 100]) on the diffusion-driven growth of a spherical grain in supersaturated solid solution. Jabbour et al. make further assumptions that include: (i) the interface be massless; (ii) the motion pertaining to spherulitic growth be smooth; (iii) the stresses in both the amorphous and crystalline phases be spherical. Under their assumptions and without invoking configurational forces, they derive an expression for the thermodynamic driving force that depends explicitly on the chemical potential of the vapor and delivers a constant growth rate of the crystalline phase.

In this paper we will revisit the same problem of isothermal spherulite crystallization studied by Jabbour et al. [17]. Specifically, we consider an elastic thin-film, modeled as a planar object, which at all times consists of $N$ components and of two phases, namely a crystalline phase and an amorphous phase. The region occupied by the crystalline phase is assumed to be bounded and simply connected while that of the amorphous phase can be either bounded or unbounded. The film is placed in a chamber and is exposed to a gas stream that carries vapor of the same components of the film (except the involatile ones), and each of those components can undergo evaporation and condensation. The vapor/gas stream is modeled both in [17] and here as that of a mixture of ideal gases. We will keep the basic assumption of Jabbour et al. that there is no diffusion of any species during spherulitic crystallization. Moreover, in our discussion we will not appeal to the theory of configurational forces and configuration force balance in particular. On the other hand, the three aforementioned further assumptions of Jabbour et al. will be relaxed as follows:

(i) We model the interface $\Sigma$ between the two phases as a one-dimensional open thermodynamic system. The $i$-th chemical component of the interface is generally endowed with a mass density $\Gamma_i$ and a chemical potential $\mu_i^{\Sigma}$. As the interface $\Sigma$ evolves as a continuum system, each of its particles has a well-defined interface velocity; see Sect. 3.1. A moving interface with mass density carries linear momentum and kinetic energy.

(ii) The interface $\Sigma$ is the seat of jump discontinuities of various thermomechanical variables of the thin film. In particular, spherulitic crystallization is a coherent phase transition; the velocity field and the deformation gradient in the thin film suffer jump discontinuities across the interface that satisfy Hadamard's compatibility conditions (6).

(iii) No *a priori* restriction is imposed on stresses in the amorphous and crystalline phases as elastic solids.

In Sect. 4, after writing down the equations for balance of mass and momentum, we will combine the energy balance and entropy imbalance to obtain for isothermal processes the free energy imbalance, which delivers the dissipation inequalities for both bulk phases and the interface upon localization. Then we will follow the Coleman–Noll procedure [3] to derive constitutive restrictions from the dissipation inequalities. The driving force behind the crystallization will be identified as the quantity conjugate to the normal velocity in the interfacial dissipation inequality, a practice that can be traced back to classical irreversible thermodynamics [4, 7]. The main difference between the present treatment and the existing ones on similar topics, such as [1, 8, 9, 11–13] among others, is that we treat the interface as an open thermodynamic system endowed with mass densities and chemical potentials of its chemical components as well as a well-defined interface velocity. The product of the interface velocity and the interfacial stress will give the power expended by the interface. We will end Sect. 4 by a comparison between the driving force we get and its counterpart in [17], where an example will illustrate their difference.

In Sect. 5, we will specialize our theory to investigate the effect of prestress on the shape of the spherulite in an elastic film. We derive a formula for the normal velocity as a function of the direction of the outward normal. We show that if the initial interface $\Sigma(0)$ is the boundary of a convex region, then the crystalline region that $\Sigma(t)$ encloses will be convex for all time $t$. Using the aforementioned formula and fact, we estimate the shape of the spherulite as it evolves from the initial shape of a disc.

In this paper we adopt the summation convention for repeated indices. The range of a repeated index depends on the context. For a property pertaining to a mixture of $N$ components, a repeated index runs from 1 to $N$. For components of two-dimensional vectors or tensors, a repeated index runs from 1 to 2.

## 2 Preliminaries on Kinematics

In this paper we consider the isothermal crystallization of an isolated spherulite in a multicomponent elastic thin film under solvent-vapor annealing. We model the thin film as a two-dimensional object, which lies at all times in some fixed Euclidean plane $\mathbb{E}$. We choose a reference configuration $\mathcal{B}$ for the film and consider its motion and phase transition in a time interval $[0, d]$ when the film is exposed to a vapor of its constituents and a carrier gas, during which each component (except the involatile ones) undergoes evaporation or condensation. We assume that at all times $t$ in $[0, d]$ the film $\mathcal{B}$ consists of an isolated spherulite in crystalline phase plus a surrounding amorphous phase, and we model the phase interface

$\Sigma(t)$ as a smooth simple closed curve (see Sect. 3 for further discussions on the modeling of the interface). Since every smooth simple closed curve is diffeomorphic to the unit circle (see [31], p. 55), $\Sigma(t)$ separates a bounded and simply-connected region $\mathcal{B}_t^s$ in crystalline phase from a bounded or unbounded region $\mathcal{B}_t^a$ in amorphous phase.

## 2.1 Evolving Interface, Normal Velocity, and Normal Time-Derivative

Let the phase interface $\Sigma(t)$ be parametrized by arc length $s$. The parametrization is chosen so that $\mathbf{t}(s,t)$, the unit tangent vector of $\Sigma(t)$, falls to the right of $\mathbf{m}(s,t)$, the unit normal that points away from the enclosed crystalline phase. The (signed) curvature $\kappa$ of $\Sigma(t)$ is given by the Frenet relation

$$\frac{\partial \mathbf{t}}{\partial s} = \kappa \mathbf{m}. \tag{1}$$

Under this choice of $\mathbf{t}$ and $\mathbf{m}$, $\kappa \le 0$ on $\Sigma(t)$ whenever the enclosed region is convex. We assume that the unit tangent $\mathbf{t}(\mathbf{X},t)$, outward unit normal $\mathbf{m}(\mathbf{X},t)$, and curvature $\kappa(\mathbf{X},t)$ are smooth functions on $\{(\mathbf{X},t) \in \Sigma(t) : t \in [0,d]\}$.

Let $t_o \in [0,d]$. Since $\Sigma(t_o)$ is compact, there exists (see [6], p. 113) a tubular neighborhood $W(t_o)$ of $\Sigma(t_o)$ in $\mathcal{B}$ such that all normal lines of $\Sigma(t_o)$ are disjoint in $W(t_o)$. We further assume that the evolving interface $\Sigma(t)$, $t \in [0,d]$, satisfies what follows:

> For each $t_o \in [0,d]$, there exists a neighborhood $U(t_o;\epsilon) \subset [0,d]$ of $t_o$ such that for each $t \in U(t_o;\epsilon)$ the interface $\Sigma(t)$ lies in $W(t_o)$ and meets each normal line of $\Sigma(t_o)$ at one and only one point. Let $(\mathbf{X}_o,t_o) \in \Sigma(t_o)$, and let $\mathbf{X}_o + \alpha(t)\mathbf{m}(\mathbf{X}_o,t_o)$ be the point at which $\Sigma(t)$ meets the normal line of $\Sigma(t_o)$ at $\mathbf{X}_o$. The function $\alpha(t)$ is differentiable at $t = t_o$, and
>
> $$V(\mathbf{X}_o,t_o) = \left.\frac{d\alpha}{dt}\right|_{t=t_o} \tag{2}$$
>
> is called the *normal velocity* of the interface at $(\mathbf{X}_o,t_o)$. The field of normal velocities $V(\mathbf{X},t)$ is smooth on $\{(\mathbf{X},t) \in \Sigma(t) : t \in [0,d]\}$.

Let $\mathbf{X}_o$ be a point on $\Sigma(t_o)$, the interface at time $t_o$. The *normal trajectory* through $\mathbf{X}_o$ at time $t_o$ is a curve $\mathbb{X}(t)$ which is the solution to the initial value problem

$$\frac{d\mathbb{X}(t)}{dt} = V\big(\mathbb{X}(t),t\big)\mathbf{m}\big(\mathbb{X}(t),t\big), \qquad \mathbb{X}(t_o) = \mathbf{X}_o. \tag{3}$$

Given a scalar, vector, or tensor field $\varphi(\mathbf{X},t)$ for $\mathbf{X} \in \Sigma(t)$, the *normal time-derivative* of $\varphi$ following $\Sigma(t)$ is defined by

$$\mathring{\varphi}(\mathbf{X},t) = \frac{d\varphi(\mathbb{X}(t),t)}{dt}, \tag{4}$$

which represents the rate of change of $\varphi$ following the normal trajectory $\mathbb{X}(t)$.

## 2.2 Hadamard's Compatibility Conditions

Let $\varphi$ be a scalar, vector, or tensor field that is smooth up to the phase interface $\Sigma(t)$. For each $\mathbf{X} \in \Sigma(t)$, let $\varphi^-(\mathbf{X},t)$ and $\varphi^+(\mathbf{X},t)$ denote the values of $\varphi$ when $\mathbf{X}$ is a limit point from the region enclosed by $\Sigma(t)$ and from outside of $\Sigma(t)$, respectively. We use

$$[\![\varphi]\!](\mathbf{X},t) = \varphi^+(\mathbf{X},t) - \varphi^-(\mathbf{X},t) \tag{5}$$

to denote the jump of the field $\varphi$ across $\Sigma(t)$.

A coherent phase transition is manifested by the propagation of the phase interface $\Sigma(t)$ in the reference configuration $\mathcal{B}$ during which the motion $\chi(\mathbf{X},t)$ is continuous while the velocity $\dot{\chi} = \partial\chi/\partial t$ and deformation gradient $\mathbf{F} = \text{Grad}\,\chi$ suffer a jump discontinuity across $\Sigma(t)$, respectively. For such a motion there holds Hadamard's compatibility conditions [14]:

$$[\![\dot{\chi}]\!] = -[\![\mathbf{F}]\!]\mathbf{m}V \quad \text{and} \quad [\![\mathbf{F}]\!]\mathbf{t} = \mathbf{0}, \tag{6}$$

where the unit tangent $\mathbf{t}$, outward unit normal $\mathbf{m}$, and normal velocity $V$ of the interface $\Sigma(t)$ are defined in Sect. 2.1.

Note that the normal time derivative of the motion is, by the chain rule,

$$\mathring{\chi} = \frac{d}{dt}\chi(\mathbb{X}(t),t) = \dot{\chi}^{\pm} + \mathbf{F}^{\pm}\mathbf{m}V = \langle\!\langle\dot{\chi}\rangle\!\rangle + \langle\!\langle\mathbf{F}\rangle\!\rangle\mathbf{m}V, \tag{7}$$

where $\langle\!\langle f\rangle\!\rangle = \frac{1}{2}(f^+ + f^-)$ for any continuous scalar, vector, or tensor field that is smooth up to the interface.

For later use, we mention two algebraic identities, both of which can be verified by direct computation:

$$[\![f_1 f_2]\!] = [\![f_1]\!]\langle\!\langle f_2\rangle\!\rangle + \langle\!\langle f_1\rangle\!\rangle[\![f_2]\!], \tag{8}$$

$$\langle\!\langle f_1 f_2\rangle\!\rangle = \langle\!\langle f_1\rangle\!\rangle\langle\!\langle f_2\rangle\!\rangle + \frac{1}{4}[\![f_1]\!][\![f_2]\!]; \tag{9}$$

here $f_1$ is a scalar field and $f_2$ a scalar, vector, or tensor field. For two vector fields $\mathbf{f}_1$ and $\mathbf{f}_2$, we easily obtain from (8) and (9) the formulas

$$[\![\mathbf{f}_1\cdot\mathbf{f}_2]\!] = [\![\mathbf{f}_1]\!]\cdot\langle\!\langle\mathbf{f}_2\rangle\!\rangle + \langle\!\langle\mathbf{f}_1\rangle\!\rangle\cdot[\![\mathbf{f}_2]\!], \tag{10}$$

$$\langle\!\langle\mathbf{f}_1\cdot\mathbf{f}_2\rangle\!\rangle = \langle\!\langle\mathbf{f}_1\rangle\!\rangle\cdot\langle\!\langle\mathbf{f}_2\rangle\!\rangle + \frac{1}{4}[\![\mathbf{f}_1]\!]\cdot[\![\mathbf{f}_2]\!], \tag{11}$$

respectively.

### 2.3 Transport Identities

Let $\mathcal{S}(t)$ be an evolving subcurve of $\Sigma(t)$ and $\mathbb{X}_a(t)$, $\mathbb{X}_b(t)$ be the endpoints of $\mathcal{S}(t)$. For any scalar or vector interfacial field $\varphi$, we write $\varphi_a(t) = \varphi(\mathbb{X}_a(t),t)$ and $\varphi_b(t) = \varphi(\mathbb{X}_b(t),t)$. We have the following transport theorem for line integrals.

**Theorem 2.1** (Transport theorem for line integrals [11]) *For $\mathcal{S}(t)$ a smoothly evolving subcurve of $\Sigma(t)$ and $\varphi$ a smooth interfacial field,*

$$\frac{d}{dt}\int_{\mathcal{S}(t)}\varphi\,ds = \int_{\mathcal{S}(t)}(\mathring{\varphi} - \varphi\kappa V)\,ds + \int_{\partial\mathcal{S}(t)}\varphi W, \tag{12}$$

*where*

$$\int_{\partial\mathcal{S}(t)}\varphi W = \varphi_b W_b - \varphi_a W_a, \tag{13}$$

*and* $W = \frac{d\mathbb{X}}{dt}\cdot\mathbf{t}$ *denotes the tangential component of the velocity of the endpoints of* $\mathcal{S}(t)$.

We also introduce the transport relation in the bulk with singular curve:

**Theorem 2.2** (Transport theorem with singular curve [14]) *Let $\mathcal{R}$ be any fixed region in $\mathcal{B}$ and $\Sigma(t) \subset \mathcal{B}$ be a curve such that a continuous scalar, vector or tensor field $\varphi$ is smooth up to $\Sigma(t)$, then we have*

$$\frac{d}{dt}\int_{\mathcal{R}} \varphi\, da = \int_{\mathcal{R}} \frac{\partial \varphi}{\partial t}\, da - \int_{\mathcal{R}\cap\Sigma(t)} [\![\varphi]\!] V\, ds, \tag{14}$$

*where $V$ is the normal velocity of $\Sigma(t)$.*

## 3 The Sharp Interface

In Gibbsian thermostatics the thin interfacial region, where densities of extensive thermodynamic variables change their values rapidly from those of one phase to their counterparts of another phase, is replaced by a sharp dividing surface endowed with corresponding "surface excesses" of the extensive variables ([10, pp. 219–331]; [5]). The Gibbs dividing surface is, in Gibbs' description, an "imaginary" construct, and a slight change in its rather arbitrary location may lead to huge changes in the values of the surfaces excesses (see, e.g., [5], p. 25). Nevertheless useful formulas that relate measurable quantities and are independent of the location of the dividing surface can be derived by the method of Gibbs. Exploring whether and how this Gibbsian approach can be extended to cover our present dynamical context, however, is out of the scope of the present paper.[1]

Just as we represent the thin film itself as a two-dimensional object, we model the interface between two phases in the film as a one-dimensional thermodynamic system $\Sigma$. Physically the density of each extensive thermodynamic variable $\Phi_i$ (e.g., density $\rho_i$ of the mass $M_i$ of the $i$-th component) typically varies rapidly through the thickness of a thin two-dimensional interfacial region $\mathcal{R}_i(t)$ at time $t$. A segment $\mathcal{S}$ of $\Sigma(t)$, our one-dimensional model of the interface, will carry the total amount of $\Phi_i$ in the part $\mathcal{P}_i \subset \mathcal{R}_i(t)$ it represents. Thus the extensive thermodynamic variables of the one-dimensional interface $\Sigma(t)$ are not excesses as in the Gibbsian approach. For example, in our model the interface usually carries mass, and mass densities are always non-negative. Note that the interfacial regions $\mathcal{R}_i(t)$ and $\mathcal{R}_j(t)$ that pertain to two different extensive variables $\Phi_i$ and $\Phi_j$, respectively, may be different.[2] For intensive thermodynamic variables, we assume that we can assign to each point in $\Sigma(t)$ a temperature, a chemical potential for each component, etc. (To make this assumption more palatable, some readers may wish to add the further assumption that each intensive variable does not change its value in the thickness direction of the thin interfacial regions of the extensive variables.) A legitimate objection can be raised that the locations of the two boundaries of each interfacial region, which according to the aforementioned scheme will determine the value of the extensive thermodynamic variable in question to be assigned to $\Sigma(t)$, are themselves ill-defined. This difficulty is similar to that concerning the location of

[1] One possibility is that all the formulas and equations in Sects. 3 and 4 would remain valid if we interpret the densities of extensive thermodynamic variables we assign to the interface as those of excesses. But we shall not pursue this possibility in this article.

[2] Take, for instance, the case of a mixture of two chemical species $A$ and $B$. Let $\rho_A$ and $\rho_B$ be the mass density of $A$ and $B$, respectively. The interfacial region pertaining to $\rho_A$ may be much thinner than that pertaining to $\rho_B$. Under special circumstances (see Remark 4.3 for an example), $\mathcal{R}_A$ may degenerate to one with zero thickness and $\mathcal{S}$ does not carry any mass of species $A$.

the dividing surface in the Gibbsian approach. But, likewise, we can see if we could circumvent this difficulty by deriving useful relations that no extensive thermodynamic variable of the interface appears.

Even after all these, a question remains. As the interface $\Sigma$ evolves, it sweeps past different material points in the thin film $\mathcal{B}$. At each instant $t$, $\Sigma(t)$ is a thermodynamic system. How do we link these systems together so that they constitute the transitory configurations of one evolving thermodynamic system? In particular, for a continuum theory each thermodynamic system is a collection of particles. How do we relate the particles of $\Sigma(t_1)$ to those of $\Sigma(t_2)$ for $t_1 \neq t_2$? We will present our answer to this question in the next subsection.

### 3.1 Interface as an Evolving Open Thermodynamic System, Interface Velocity

We consider the growth of an isolated spherulite in a thin film in the time interval $[0, d]$. Let $\mathcal{B}$ be the reference configuration of the film. We assume that at each instant $t$ in $[0, d]$ the film $\mathcal{B}$ consists of a bounded region $\mathcal{B}_t^s$ in crystalline phase surrounded by another (bounded or unbounded) region $\mathcal{B}_t^a$ in amorphous phase, and that the boundary $\Sigma(t)$ between the two regions is a smooth simple closed curve that represents the phase interface.

We assume that spherulitic crystallization is a coherent phase transition. Let $\mathbb{E}$ be a fixed two-dimensional Euclidean space, and let $\chi : \mathcal{B} \times [0, d] \to \mathbb{E}$ be the motion of the film that accompanies the phase transition in question. We assume that the motion $\chi$ satisfies the following three conditions:

(i) $\chi$ is continuous on $\mathcal{B} \times [0, d]$;
(ii) $\chi$ is smooth in $\overline{\mathcal{B}_t^s}$, (i.e. the closure of $\mathcal{B}_t^s$), and $\overline{\mathcal{B}_t^a}$;
(iii) $\chi$ satisfies the compatibility conditions (6).

We treat the interface $\Sigma(t)$, for each $t$ in $[0, d]$ as a continuum thermodynamic system in itself. As a continuum system, each $\Sigma(t)$ is composed of particles. Before we can apply the principles and laws of continuum mechanics and thermodynamics to the evolving interface, we must relate, for two different instants $t_1$ and $t_2$, the particles that constitute $\Sigma(t_1)$ and $\Sigma(t_2)$, respectively.[3] In the kinematics of evolving curves, only the normal velocity is intrinsic. Hence we relate the particles of $\Sigma(t)$ to those of $\Sigma(0)$ as follows.

Fix a point $P_0$ on $\Sigma(0)$ and take it as the starting point for measuring arc length. If $\hat{\mathbf{X}}(s)$ is the arc-length parametrization of the initial curve $\Sigma(0)$, then a generic point $\hat{\mathbf{X}}(s_0)$ in $\Sigma(0)$ is specified by a number $s_0$, which is its distance from $P_0$ as measured along $\Sigma(0)$. The parametrization (by time $t$) of the normal trajectory $\mathcal{N}_{s_0}$ through $\hat{\mathbf{X}}(s_0)$ is the solution of the initial value problem:

$$\frac{d\mathbb{X}(t)}{dt} = V\big(\mathbb{X}(t), t\big)\mathbf{m}\big(\mathbb{X}(t), t\big), \qquad \mathbb{X}(0) = \hat{\mathbf{X}}(s_0); \tag{15}$$

see (3). We can use the same name $s_0$ to specify the points in $\Sigma(t)$. Indeed we can plot the normal trajectory $\mathcal{N}_{s_0}$ that starts from a given point $s_0$ in $\Sigma(0)$ and use the same name "$s_0$" for the point in $\Sigma(t)$ where $\mathcal{N}_{s_0}$ meets $\Sigma(t)$; see Fig. 1.

We will use the name "$s_0$" to label the particles of $\Sigma(t)$. The location of the particle labeled by $s_0$ at some time $t$ is none other than the point in $\Sigma(t)$ with the same label.

[3] In continuum mechanics the same particle of a body, at the atomic scale, need not consist of the same molecules at different times; cf. gas dynamics for example.

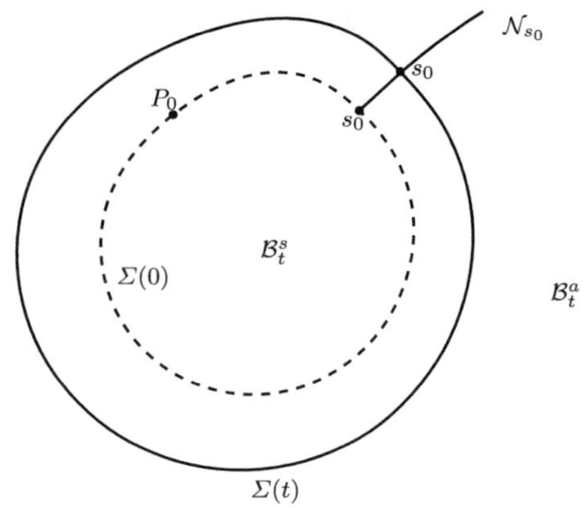


**Fig. 1** The initial curve $\Sigma(0)$ serving as a "reference"

Recall the normal time derivative of the motion in Sect. 2.1. If $\mathbb{X}(t)$ in (15) is a parametrization of the normal trajectory $\mathcal{N}_{s_0}$, then $\mathring{\chi}$ in (7) gives the velocity in physical space of the interfacial particle labeled by $s_0$. Hence we will call the quantity $\mathring{\chi}$ the interface velocity. As noted earlier, the interface is generally endowed with mass. When an interface $\Sigma$ with mass moves in physical space with velocity $\mathring{\chi}$, it carries linear momentum and kinetic energy. In particular, by (7) we have

$$\mathring{\chi} = \dot{\chi}^+ + V\mathbf{F}^+\mathbf{m} = \dot{\chi}^- + V\mathbf{F}^-\mathbf{m}. \tag{16}$$

Thus even in the special case where the bulk phases are always at rest (i.e., $\dot{\chi}^+ = \dot{\chi}^- = \mathbf{0}$, $\mathbf{F}^+ = \mathbf{F}^- = \mathbf{I}$, where $\mathbf{I}$ is the second-order identity tensor) during the phase transition, the propagating phase boundary $\Sigma$ still carries linear momentum and kinetic energy.

What we describe above can be taken as a Lagrangian description of the evolution of the interface: $\Sigma(0)$ is the "reference configuration" of the interface in the reference configuration of the film and $\mathring{\chi}$ is the "material velocity" of the particle named $s_0$.

*Remark 3.1* Note that it is not our intention to use $\Sigma(0)$ as the reference configuration for the constitutive relation of the interface. The role of $\Sigma(0)$ here is to provide names for a labeling of the normal trajectories that link the particles of $\Sigma(t_1)$ to those of $\Sigma(t_2)$ for $t_1 \neq t_2$. Once the particles of the interface $\Sigma$ are specified, the interface velocity $\mathring{\chi}$ is well defined.

*Remark 3.2* In our model, the interface $\Sigma$ is mathematically a curve of singularities across which the velocity $\dot{\chi}$, the deformation gradient $\mathbf{F}$, and other thermodynamic variables of the film suffer jump discontinuity. A particle of $\Sigma$ is a point of singularity.

*Remark 3.3* In this paper we consider films $\mathcal{B}$ exposed to a mixture of its components in vapor phase (plus a carrier gas), and exchange of matter between the vapor mixture and $\mathcal{B}$ (including the interface $\Sigma$) is allowed. Hence the bulk phases in $\mathcal{B}_t^a$, $\mathcal{B}_t^s$ and the interface $\Sigma$ are all open systems.

*Remark 3.4* As a kinematical quantity what we call "interface velocity" here is not new. For example, it appears as "motion velocity following [the interface] $\mathcal{S}$" in Gurtin's book [12, p. 64] on configurational forces. One noted difference between our treatment and the works of Gurtin and coworkers is that we take the interface as an evolving open thermodynamic system, which has mass and chemical potential. Thus an interface with a non-zero interface velocity will carry linear momentum and kinetic energy.

### 3.2 Thermodynamic Relations

We assume that the sharp interface $\Sigma$ behaves as if it is a one-dimensional elastic solid. For the present purpose of discussing thermodynamic relations, consider a sample of such a solid, which we still call $\Sigma$, and it has $N$ components and a homogeneous reference configuration that occupies the interval $[0, L]$. We parametrize the reference configuration by arc length $s$ so that the points in $[0, L]$ are labeled $\mathbf{X}(s)$. Let $\Sigma$ undergo a deformation $\chi : [0, L] \to \mathbb{E}$. We call

$$\mathbf{f} = \frac{\partial \chi}{\partial s} \tag{17}$$

the deformation gradient of the one-dimensional elastic solid $\Sigma$. The magnitude of $\mathbf{f}$, $j := \|\mathbf{f}\|$, is the stretch ratio (i.e., increase in length per unit length) of the deformation $\chi$. Let $\boldsymbol{\sigma}$ be the stress vector in $\Sigma$. Its value at $\mathbf{X}$ gives the force exerted across $\mathbf{X}$ by the material that the unit tangent $\mathbf{t}$ in the reference configuration points to on the material that $\mathbf{t}$ points from.

We now consider [27] the Gibbs relations of the reversible system $\Sigma$ in classical thermodynamics. Let $\Sigma$ undergo a homogeneous deformation $\chi$ and homogeneous changes in densities of the components. The work form of $\Sigma$ is:

$$w = -L\boldsymbol{\sigma} \cdot d\mathbf{f} - \mu_i^{\Sigma} dM_i^{\Sigma}, \tag{18}$$

where $M_i^{\Sigma}$ and $\mu_i^{\Sigma}$ ($i = 1, \ldots, N$) are the total mass and chemical potential of the $i$-th component in $\Sigma$, and the summation convention for repeated indices is in force.

Let $q$ be the heat form of the system $\Sigma$. By the First and Second Law of thermodynamics, for a reversible system there exist two functions of state called the internal energy $U$ and the entropy $S$ such that

$$dU = q - w, \qquad TdS = q, \tag{19}$$

where $T$ is the absolute temperature. Let $F := U - TS$ be the Helmholtz free energy. From (18), (19) and the definition of $F$, we obtain

$$\begin{aligned} dF &= d(U - TS) \\ &= dU - TdS - SdT \\ &= -SdT + L\boldsymbol{\sigma} \cdot d\mathbf{f} + \mu_i^{\Sigma}\, dM_i^{\Sigma}. \end{aligned} \tag{20}$$

In this paper we restrict attention to isothermal systems and processes. Henceforth $T$ is always a constant in all our discussions, and it will be suppressed in all functions for which it should be an independent variable. It follows from $dT = 0$ that we have

$$dF = L\boldsymbol{\sigma} \cdot d\mathbf{f} + \mu_i^{\Sigma}\, dM_i^{\Sigma}. \tag{21}$$

In terms of the Helmholtz free energy per unit length $\psi = F/L$, (21) can be recast as

$$d\psi = \boldsymbol{\sigma} \cdot d\mathbf{f} + \mu_i^{\Sigma}\, d\Gamma_i, \tag{22}$$

where $\Gamma_i := M_i^{\Sigma}/L$ is the mass density of the $i$-th component in $\Sigma$.

Note that by (21), we have

$$F = F\big(L\mathbf{f}, \mathbf{M}^{\Sigma}\big), \tag{23}$$

where $\mathbf{M}^{\Sigma} = \langle M_1^{\Sigma}, \ldots, M_N^{\Sigma}\rangle$. Assuming that $F$ is a homogeneous function of degree 1, we have, with the help of (21),

$$\begin{aligned} F &= \frac{\partial F}{\partial (L\mathbf{f})} \cdot (L\mathbf{f}) + \frac{\partial F}{\partial M_i^{\Sigma}} M_i^{\Sigma} \\ &= \boldsymbol{\sigma} \cdot (L\mathbf{f}) + \mu_i^{\Sigma} M_i^{\Sigma}, \end{aligned} \tag{24}$$

and thence

$$\psi = \boldsymbol{\sigma} \cdot \mathbf{f} + \mu_i^{\Sigma} \Gamma_i. \tag{25}$$

Dividing both sides of (23) by $L$, we obtain

$$\psi = \psi(\mathbf{f}, \boldsymbol{\Gamma}), \tag{26}$$

where $\boldsymbol{\Gamma} = \langle \Gamma_1, \ldots, \Gamma_N\rangle$. It follows from (23) and (25) that

$$\frac{\partial \psi}{\partial \mathbf{f}} = \boldsymbol{\sigma}, \qquad \frac{\partial \psi}{\partial \Gamma_i} = \mu_i^{\Sigma}. \tag{27}$$

### 3.3 Some Useful Formulas

We end our discussion on the interface with several useful formulas, which we present in a list of remarks. In what follows we return to the setting that $\Sigma(t)$, $t \in [0, d]$, denotes the evolving interface that separates the crystalline phase in $\mathcal{B}_t^s$ from the amorphous phase in $\mathcal{B}_t^a$, and $\boldsymbol{\chi} : \mathcal{B} \times [0, d] \to \mathbb{E}$ is the motion pertaining to the spherulitic crystallization in question, which is assumed to be a coherent phase transition (see Sect. 3.1).

*Remark 3.5* By the compatibility relation $(6)_2$,

$$\mathbf{f} = \frac{\partial \boldsymbol{\chi}}{\partial s} = \mathbf{F}\mathbf{t}. \tag{28}$$

Therefore it is parallel to the tangent vector of the interface in the current configuration.

*Remark 3.6* By the commutator relation [9, equation (14.15)], for any smooth scalar, vector or tensor field $\varphi$ on $\Sigma(t)$, we have

$$\frac{\partial \mathring{\varphi}}{\partial s} = \overset{\circ}{\overline{\left(\frac{\partial \varphi}{\partial s}\right)}} - \kappa V \frac{\partial \varphi}{\partial s}. \tag{29}$$

Replacing $\varphi$ by $\boldsymbol{\chi}$, we obtain the formula

$$\mathring{\mathbf{f}} = \frac{\partial \mathring{\boldsymbol{\chi}}}{\partial s} + \kappa V \mathbf{f}. \tag{30}$$

*Remark 3.7* We shall show in Sect. 4.3 that the stress vector $\boldsymbol{\sigma}$ is tangential to the interface in the current configuration, i.e.

$$\boldsymbol{\sigma} = \sigma \hat{\mathbf{t}}, \tag{31}$$

where $\hat{\mathbf{t}}$ denotes the unit tangent in the current configuration and $\sigma$ is, depending on its sign, the line tension or compression. Hence equation (25) can be recast as

$$\psi = \sigma j + \mu_i^{\Sigma} \Gamma_i. \tag{32}$$

Therefore the free energy per unit length in the current configuration is

$$\hat{\psi} = \sigma + \mu_i^{\Sigma} \hat{\Gamma}_i \tag{33}$$

where $\hat{\Gamma}_i = \Gamma_i / j$; see [18].

## 4 Driving Force for Spherulitic Growth

This section will be devoted to the derivation of the thermodynamic driving force for spherulitic crystallization.

We will proceed by writing down the equations of balance of mass, balance of linear and angular momentum, the First Law and Second Law of thermodynamics, and eventually the imbalance of free energy for isothermal systems and processes, which will deliver the bulk and the interfacial dissipation inequalities upon localization. The driving force will be identified as the term that is conjugate to the normal velocity in the interfacial dissipation inequality.

*Remark 4.1* For later reference, we summarize here some basic facts and findings of the experiments reported in [30]. These experiments followed a protocol similar to those reported in [20, 21]. All the experiments were performed at room temperature on thin films of the same thickness deposited on the same substrate. All samples consisted of a main component $A$ blended with some additive $B$ in various concentrations from zero onwards. The samples were divided into different groups $G_k$ $(k = 1, \ldots, K)$. All samples in the same group $G_k$ had the same concentration of $B$. Both components $A$ and $B$ in the samples were taken as involatile. In each experiment a sample which consisted of only components $A$ and $B$ was put in a chamber and was exposed to a stream of solvent vapor mixed with nitrogen as the filler gas, and the concentration (or equivalently the partial pressure $p_S$) of the solvent vapor $S$ in the gas/vapor mixture was kept constant. Time-lapse optical micrographs of the film were taken during solvent-vapor annealing, and spherulite radius $R$ was tracked as a function of time. Measurements were made for several values of $p_S$, but the flow rate of the gas/vapor stream and the total pressure of the gas/vapor mixture were the same for all experiments. The rate of spherulitic growth $V := dR/dt$ was found to be constant in each experiment but vary with $p_S$ and the group $G_k$ to which the sample in question belonged. When the measured growth rate $V$ is plotted against $p_S$ for samples in the same group $G_k$, it is found that $V$ decreases linearly with $p_S$ and is extrapolated to zero as $p_S$ approaches zero. Also, for a fixed $p_S$, $V$ decreases linearly as the concentration of the additive $B$ (in mole-percent) increases.

**Fig. 2** A sketch of the region $\mathcal{R}$

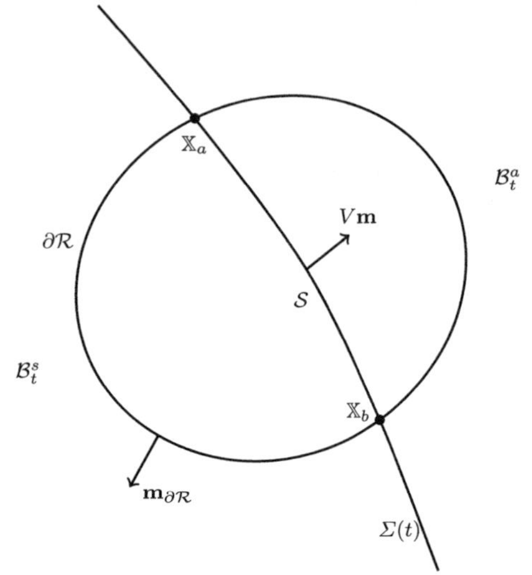


*Remark 4.2* Each thin film that we consider has an interface with the substrate and an interface with the gas/vapor mixture. As discussed in [30] these two interfaces and the thin film itself can be taken together as a composite thermodynamic system if the temperature and the chemical potentials of each chemical component of these three are always described by the same functions of $\mathbf{X}$ and $t$. On the assumption that the aforementioned conditions are valid, we can work exclusively with the thermodynamic functions of the composite thermodynamic system and simply call them the corresponding thermodynamic functions of the thin film, without even mentioning the two interfaces with the substrate and the gas/vapor mixture. This is what we shall henceforth do. Thus if a thin film $\mathcal{B}$ carries prestress that results from mismatch strains with the substrate, we shall just say that $\mathcal{B}$ is a film with prestress, without mentioning its cause.

## 4.1 Balance of Mass

For the $i$-th component, let $\rho_i$, $\Gamma_i$, $\mathcal{F}_i$ and $\mathfrak{f}_i$ be the bulk mass density, the mass density on the interface, the mass flux from the vapor into the bulk and into the interface, respectively, measured per unit reference area or length. Thus $\mathcal{F}_i > 0$, $\mathfrak{f}_i > 0$ (resp. $\mathcal{F}_i < 0$, $\mathfrak{f}_i < 0$) during condensation (resp. evaporation). Let $\mathcal{R}$ be a fixed region in $\mathcal{B}$, and $\mathcal{S}(t) = \mathcal{R} \cap \Sigma(t)$. We let the region $\mathcal{R}$ be such that the tangential component of the velocity of the endpoints of $\mathcal{S}(t)$ is 0, i.e. $W = 0$ in (12), but otherwise arbitrary. We make this choice because we find the tangential component $W$ irrelevant to our discussion when configurational forces are not considered. For simplicity, we will suppress the arguments of all integrands and write $\mathcal{S}$ for $\mathcal{S}(t)$, the part of the interface $\Sigma(t)$ that lies in $\mathcal{R}$; see Fig. 2.

The balance of mass requires that, for each component,

$$\frac{d}{dt}\left(\int_{\mathcal{R}} \rho_i \, da + \int_{\mathcal{S}} \Gamma_i \, ds\right) = \int_{\mathcal{R}} \mathcal{F}_i \, da + \int_{\mathcal{S}} \mathfrak{f}_i \, ds. \tag{34}$$

By applying (12) with $W = 0$ and (14), (34) becomes

$$\int_{\mathcal{R}} \dot{\rho}_i \, da - \int_{\mathcal{S}} [\![\rho_i]\!] V \, ds + \int_{\mathcal{S}} (\mathring{\Gamma}_i - \Gamma_i \kappa V) \, ds = \int_{\mathcal{R}} \mathcal{F}_i \, da + \int_{\mathcal{S}} \mathfrak{f}_i \, ds, \tag{35}$$

where the overhead dot represents the partial derivative with respect to time $t$.

We first consider the case where $\mathcal{R}$ lies entirely in $\mathcal{B}_t^s$ or $\mathcal{B}_t^a$, i.e. $\mathcal{R} \cap \Sigma = \emptyset$. In this case, (35) reduces to

$$\int_{\mathcal{R}} (\dot{\rho}_i - \mathcal{F}_i) \, da = 0. \tag{36}$$

By the arbitrariness of $\mathcal{R}$, we conclude that

$$\dot{\rho}_i = \mathcal{F}_i \tag{37}$$

in both of the crystalline phase $\mathcal{B}_t^s$ and the amorphous phase $\mathcal{B}_t^a$.

Next we take the limit of (35) by shrinking $\mathcal{R}$ to a chosen point in $\Sigma(t)$ in two steps [24]. First, $\partial\mathcal{R}$ is shrinked down to $\mathcal{S}$ in such a way that the area of $\mathcal{R}$ tends to 0 while the length of $\mathcal{S}$ remains unchanged. The area integrals will vanish in such a process, which leads to

$$-\int_{\mathcal{S}} [\![\rho_i]\!] V \, ds + \int_{\mathcal{S}} (\mathring{\Gamma}_i - \Gamma_i \kappa V) \, ds = \int_{\mathcal{S}} \mathfrak{f}_i \, ds. \tag{38}$$

Then $\mathcal{S}$ is shrinked to a point in $\Sigma(t)$ to yield the equation

$$\mathring{\Gamma}_i - \Gamma_i \kappa V - [\![\rho_i]\!] V = \mathfrak{f}_i \quad (\Gamma_i \geq 0) \tag{39}$$

at that point for each component.

Let

$$\rho = \sum_{i=1}^{N} \rho_i, \qquad \Gamma = \sum_{i=1}^{N} \Gamma_i, \qquad \mathcal{F} = \sum_{i=1}^{N} \mathcal{F}_i, \qquad \mathfrak{f} = \sum_{i=1}^{N} \mathfrak{f}_i \tag{40}$$

denote the total mass density in the bulk, the total mass density on the interface, the net mass flux into the bulk, the net mass flux into the interface, respectively. Summing over all components in (37) as well as in (39), we then have the mass balance relations

$$\dot{\rho} = \mathcal{F} \tag{41}$$

in both $\mathcal{B}_t^s$ and $\mathcal{B}_t^a$ and

$$\mathring{\Gamma} - \Gamma \kappa V - [\![\rho]\!] V = \mathfrak{f} \quad (\Gamma \geq 0) \tag{42}$$

on the interface $\Sigma(t)$.

*Remark 4.3* For an involatile $i$-th component and a reference configuration $\mathcal{B}$ with uniform $\rho_i$, we have $\mathfrak{f}_i = 0$ and $[\![\rho_i]\!] = 0$. By (39), an initial condition of $\Gamma_i = 0$ will lead to the solution of $\Gamma_i = 0$ for all $t$. In fact, under the present assumptions there is no interfacial region for the chemical component in question. Hence we will take $\Gamma_i = 0$ for an involatile component if we choose a reference configuration with uniform $\rho_i$.

### 4.2 Balance of Linear Momentum

Endowed with mass density, the interface $\Sigma$ carries momentum while it propagates. By the notion of interface velocity introduced in Sect. 3.1, the linear momentum associated with the interface is then

$$\int_{\Sigma} \Gamma \mathring{\chi}\, ds.$$

In discussing momentum balance and the First Law of thermodynamics, we shall consider the following force system:

$$\begin{aligned} &\mathbf{S} \ \text{first Piola stress}\\ &\boldsymbol{\sigma} \ \text{interfacial stress}\\ &\mathbf{B} \ \text{body force in the bulk phases}\\ &\mathbf{b} \ \text{body force on the interface.}\end{aligned} \tag{43}$$

The first Piola stress [14]

$$\mathbf{S} = (\det \mathbf{F})\mathbf{T}\mathbf{F}^{-\top}, \tag{44}$$

where $\mathbf{T}$ is the Cauchy stress and $\mathbf{F}$ the deformation gradient, is well known. The interfacial stress vector $\boldsymbol{\sigma}$ is introduced in Sect. 3.2. We will regard the body forces $\mathbf{B}$ in the bulk phases and $\mathbf{b}$ on the interface, which are measured per unit area and per unit length, respectively, as assignable.[4]

Let $\mathbf{u}_i$ denote the velocity of the $i$-th component that is condensing or evaporating. Then

$$\mathbf{u}_i = \begin{cases} \mathbf{v}^v & \text{during condensation} \\ \mathring{\chi} & \text{during evaporation from interface} \\ \dot{\chi} & \text{during evaporation from bulk phases,} \end{cases} \tag{45}$$

where $\mathbf{v}^v$ is the velocity of the gas/vapor flow. The balance of linear momentum requires that

$$\begin{aligned} &\frac{d}{dt}\left(\int_{\mathcal{R}} \rho \dot{\chi}\, da + \int_{\mathcal{S}} \Gamma \mathring{\chi}\, ds\right) \\ &\quad = \int_{\partial\mathcal{R}} \mathbf{S}\mathbf{m}_{\partial\mathcal{R}}\, ds + \int_{\partial\mathcal{S}} \boldsymbol{\sigma} + \int_{\mathcal{R}} \mathcal{F}_i \mathbf{u}_i\, da + \int_{\mathcal{S}} \mathfrak{f}_i \mathbf{u}_i\, ds + \int_{\mathcal{R}} \mathbf{B}\, da + \int_{\mathcal{S}} \mathbf{b}\, ds, \end{aligned} \tag{46}$$

where $\mathbf{m}_{\partial\mathcal{R}}$ is the outward unit normal of $\partial\mathcal{R}$. In the equation above

$$\int_{\partial\mathcal{S}} \boldsymbol{\sigma} = \boldsymbol{\sigma}(\mathbb{X}_b, t) - \boldsymbol{\sigma}(\mathbb{X}_a, t),$$

where $\mathbb{X}_a$ and $\mathbb{X}_b$ are the endpoints of $\mathcal{S}$; see Fig. 2.

By applying the transport identities (12), (14), and the divergence theorem with the presence of a discontinuity as well as invoking the mass balance relation (41), we obtain

[4] More discussion on this view of body forces can be found in [3], [34, p. 198], and [12, p. 53].

from (46)

$$\int_{\mathcal{R}} (\mathcal{F}\dot{\chi} + \rho\ddot{\chi})\, da - \int_{\mathcal{S}} [\![\rho\dot{\chi}]\!] V\, ds + \int_{\mathcal{S}} (\overline{\Gamma\mathring{\chi}}^{\circ} - \Gamma\mathring{\chi}\kappa V)\, ds$$
$$= \int_{\mathcal{R}} \operatorname{div}\mathbf{S}\, da + \int_{\mathcal{S}} [\![\mathbf{S}]\!]\mathbf{m}\, ds + \int_{\mathcal{S}} \frac{\partial \sigma}{\partial s}\, ds$$
$$+ \int_{\mathcal{R}} \mathcal{F}_i \mathbf{u}_i\, da + \int_{\mathcal{S}} \mathfrak{f}_i \mathbf{u}_i\, ds + \int_{\mathcal{R}} \mathbf{B}\, da + \int_{\mathcal{S}} \mathbf{b}\, ds. \tag{47}$$

In the case where $\mathcal{R}$ lies entirely in the bulk phases, i.e. $\mathcal{R} \cap \Sigma = \emptyset$, (47) reduces to

$$\int_{\mathcal{R}} (\mathcal{F}\dot{\chi} + \rho\ddot{\chi} - \operatorname{div}\mathbf{S} - \mathcal{F}_i \mathbf{u}_i - \mathbf{B})\, da = \mathbf{0}, \tag{48}$$

which leads to

$$\rho\ddot{\chi} = \operatorname{div}\mathbf{S} + \sum_{i=1}^{N} \mathcal{F}_i(\mathbf{u}_i - \dot{\chi}) + \mathbf{B} \tag{49}$$

upon localization.

By shrinking $\mathcal{R}$ to a point in $\Sigma(t)$ through the two-step process described in Sect. 4.1, we obtain the equation of balance of linear momentum on the interface, i.e.,

$$\overline{\Gamma\mathring{\chi}}^{\circ} - \Gamma\mathring{\chi}\kappa V = [\![\rho\dot{\chi}]\!] V + [\![\mathbf{S}]\!]\mathbf{m} + \frac{\partial \sigma}{\partial s} + \mathbf{u}_i \mathfrak{f}_i + \mathbf{b}. \tag{50}$$

### 4.3 Balance of Angular Momentum

Let $\{\mathbf{e}_1, \mathbf{e}_2\}$ be an orthonormal basis in the plane of the film, and let $\mathbf{e}_3$ be the unit vector normal to the plane of the film defined by $\mathbf{e}_3 = \mathbf{e}_1 \times \mathbf{e}_2$, where "$\times$" denotes the usual cross product of two vectors in space. For two vectors $\mathbf{a}$ and $\mathbf{b}$ in the plane of the film, we have $\mathbf{a} = a_j \mathbf{e}_j$, $\mathbf{b} = b_j \mathbf{e}_j$, where the repeated suffix $j$ means summation of terms with $j$ running from 1 to 2, and

$$\mathbf{a} \times \mathbf{b} = \epsilon_{3jk} a_j b_k \mathbf{e}_3 = (a_1 b_2 - a_2 b_1)\mathbf{e}_3,$$

where $\epsilon_{ijk}$ denotes the alternating symbol defined by

$$\epsilon_{ijk} = \begin{cases} 1, & \text{if } \{i, j, k\} \text{ is an even permutation of } \{1, 2, 3\} \\ -1, & \text{if } \{i, j, k\} \text{ is an odd permutation of } \{1, 2, 3\} \\ 0, & \text{if an index is repeated.} \end{cases} \tag{51}$$

Balance of angular momentum requires that

$$\frac{d}{dt}\left(\int_{\mathcal{R}} \chi \times \rho\dot{\chi}\, da + \int_{\mathcal{S}} \chi \times \Gamma\mathring{\chi}\, ds\right)$$
$$= \int_{\partial\mathcal{R}} \chi \times \mathbf{S}\mathbf{m}_{\partial\mathcal{R}}\, ds + \int_{\partial\mathcal{S}} \chi \times \sigma + \int_{\mathcal{R}} \chi \times \mathcal{F}_i \mathbf{u}_i\, da + \int_{\mathcal{S}} \chi \times \mathfrak{f}_i \mathbf{u}_i\, ds$$
$$+ \int_{\mathcal{R}} \chi \times \mathbf{B}\, da + \int_{\mathcal{S}} \chi \times \mathbf{b}\, ds; \tag{52}$$

here the vectors all lie in the plane of the film and $\mathbf{S} = S_{jk}\mathbf{e}_j \otimes \mathbf{e}_k$, where the summation convention for two-dimensional vectors and tensors is in force and each repeated index runs from 1 to 2. Using the divergence theorem, we have

$$\begin{aligned}
\int_{\partial\mathcal{R}} \chi \times \mathbf{S}\mathbf{m}\,ds &= \int_{\partial\mathcal{R}} \chi_j S_{kl} m_l \epsilon_{3jk}\mathbf{e}_3\,ds \\
&= \int_{\mathcal{R}} \frac{\partial(S_{kl}\chi_j)}{\partial X_l}\epsilon_{3jk}\mathbf{e}_3\,da + \int_{\mathcal{S}} \chi \times [\![\mathbf{S}]\!]\mathbf{m}\,ds \\
&= \int_{\mathcal{R}} \epsilon_{3jk}\frac{\partial S_{kl}}{\partial X_l}\chi_j\mathbf{e}_3\,da + \int_{R} \epsilon_{3jk}S_{kl}\frac{\partial \chi_j}{\partial X_l}\mathbf{e}_3\,da + \int_{\mathcal{S}} \chi \times [\![\mathbf{S}]\!]\mathbf{m}\,ds \\
&= \int_{\mathcal{R}} \chi \times \operatorname{div}\mathbf{S}\,da + \int_{\mathcal{R}} \epsilon_{3jk}S_{kl}\frac{\partial \chi_j}{\partial X_l}\mathbf{e}_3\,da + \int_{\mathcal{S}} \chi \times [\![\mathbf{S}]\!]\mathbf{m}\,ds. \qquad (53)
\end{aligned}$$

By invoking the transport identities (12), (14), and the balance of mass (41), we can rewrite (52) as

$$\begin{aligned}
&\int_{\mathcal{R}} (\chi \times \mathcal{F}\dot{\chi} + \chi \times \rho\ddot{\chi})\,da - \int_{\mathcal{S}} \chi \times [\![\rho\dot{\chi}]\!]V\,ds + \int_{\mathcal{S}} (\chi \times \overset{\circ}{\overline{\Gamma\mathring{\chi}}} - \chi \times \Gamma\mathring{\chi}\kappa V)\,ds \\
&= \int_{\mathcal{R}} \left(\chi \times \operatorname{div}\mathbf{S} + \epsilon_{3jk}S_{kl}\frac{\partial \chi_j}{\partial X_l}\mathbf{e}_3\right)da + \int_{\mathcal{S}} \left(\frac{\partial \chi}{\partial s} \times \sigma + \chi \times [\![\mathbf{S}]\!]\mathbf{m} + \chi \times \frac{\partial \sigma}{\partial s}\right)ds \\
&\quad + \int_{\mathcal{R}} \chi \times \mathcal{F}_i\mathbf{u}_i\,da + \int_{\mathcal{S}} \chi \times \mathfrak{f}_i\mathbf{u}_i\,ds + \int_{\mathcal{R}} \chi \times \mathbf{B}\,da + \int_{\mathcal{S}} \chi \times \mathbf{b}\,ds. \qquad (54)
\end{aligned}$$

With the help of balance of linear momentum (49), we conclude that away from the interface,

$$\epsilon_{3jk}S_{kl}\frac{\partial \chi_j}{\partial X_l}\mathbf{e}_3 = \mathbf{0}, \qquad (55)$$

which implies

$$\mathbf{S}\mathbf{F}^{\top} = \mathbf{F}\mathbf{S}^{\top} \qquad (56)$$

in the bulk phases where $\mathbf{F}^{\top}$ denotes the transpose of $\mathbf{F}$. By shrinking $\partial\mathcal{R}$ to $\mathcal{S}$ and invoking the balance of linear momentum on the interface (50), we obtain

$$\frac{\partial \chi}{\partial s} \times \sigma = \mathbf{0} \qquad (57)$$

on the phase interface. Note that by (28), the interfacial stress vector is parallel to the phase interface in the current configuration.

## 4.4 Free Energy Imbalance

Let $\mathcal{E}$ and $\varepsilon$ be the internal energy density in the bulk and on the interface, respectively. By the thermodynamical analysis in [17] and the First Law of thermodynamics, we have

$$\frac{d}{dt}\left(\int_{\mathcal{R}}\left(\mathcal{E} + \frac{1}{2}\rho|\dot{\chi}|^2\right)da + \int_{\mathcal{S}}\left(\varepsilon + \frac{1}{2}\Gamma|\mathring{\chi}|^2\right)ds\right)$$

$$= \underbrace{\int_{\partial\mathcal{R}} \mathbf{Sm}_{\partial\mathcal{R}} \cdot \dot{\chi}\, ds + \int_{\partial\mathcal{S}} \boldsymbol{\sigma} \cdot \mathring{\chi} + \sum_{i=1}^{N} \Big( \int_{\mathcal{R}} \frac{p_i^v}{\rho_i^v} \mathcal{F}_i\, da + \int_{\mathcal{S}} \frac{p_i^v}{\rho_i^v} \mathfrak{f}_i\, ds \Big)}_{I_1}$$
$$+ \underbrace{\int_{\mathcal{R}} \mathbf{B} \cdot \dot{\chi}\, da + \int_{\mathcal{S}} \mathbf{b} \cdot \mathring{\chi}\, ds}_{I_2}$$
$$+ \underbrace{\int_{\mathcal{R}} \Big( \mathcal{E}_i^v \mathcal{F}_i + \frac{1}{2} \mathcal{F}_i |\mathbf{u}_i|^2 \Big) da + \int_{\mathcal{S}} \Big( \mathcal{E}_i^v \mathfrak{f}_i + \frac{1}{2} \mathfrak{f}_i |\mathbf{u}_i|^2 \Big) ds}_{II}$$
$$+ \underbrace{\int_{\mathcal{R}} \mathfrak{R}\, da + \int_{\mathcal{S}} \mathfrak{r}\, ds}_{III}, \tag{58}$$

where $p_i^v$, $\rho_i^v$ and $\mathcal{E}_i^v$ denote the partial pressure, mass density, and specific internal energy of the $i$-th component in the vapor, respectively, and

$$\int_{\partial\mathcal{S}} \boldsymbol{\sigma} \cdot \mathring{\chi} = (\boldsymbol{\sigma} \cdot \mathring{\chi})(\mathbb{X}_b, t) - (\boldsymbol{\sigma} \cdot \mathring{\chi})(\mathbb{X}_a, t) \tag{59}$$

is the rate that the interfacial stress $\boldsymbol{\sigma}$ is working on $\mathcal{S}(t)$. The sum $I_1 + I_2$ represents the power expended on the thin film. The energy flow from the vapor to the film is expressed by the term $II$. The term $III$ describes the heat supply to the film, where $\mathfrak{R}$ and $\mathfrak{r}$ denote the heat supply density of the bulk phases and of the interface, respectively.

The Second Law asserts that the rate at which the entropy of a system is increasing is not smaller than the entropy inflow. Let $\eta$ and $\varsigma$ denote the entropy density of the bulk and of the interface, respectively. As we restrict our discussion to isothermal systems and processes at absolute temperature $T$, we have

$$\frac{d}{dt}\Big( \int_{\mathcal{R}} \eta\, da + \int_{\mathcal{S}} \varsigma\, ds \Big) \ge \frac{1}{T} \int_{\mathcal{R}} \mathfrak{R}\, da + \frac{1}{T} \int_{\mathcal{S}} \mathfrak{r}\, ds + \int_{\mathcal{R}} \eta_i^v \mathcal{F}_i\, da + \int_{\mathcal{S}} \eta_i^v \mathfrak{f}_i\, ds, \tag{60}$$

where $\eta_i^v$ is the specific entropy of the vapor of the $i$-th component.

By combining (58) and (60), we have the free energy inequality

$$\frac{d}{dt}\Big( \int_{\mathcal{R}} \Big( \Psi + \frac{1}{2}\rho|\dot{\chi}|^2 \Big) da + \int_{\mathcal{S}} \Big( \psi + \frac{1}{2}\Gamma|\mathring{\chi}|^2 \Big) ds \Big)$$
$$\le \int_{\partial\mathcal{R}} \mathbf{Sm}_{\partial\mathcal{R}} \cdot \dot{\chi}\, ds + \int_{\partial\mathcal{S}} \boldsymbol{\sigma} \cdot \mathring{\chi} + \int_{\mathcal{R}} \mathbf{B} \cdot \dot{\chi}\, da + \int_{\mathcal{S}} \mathbf{b} \cdot \mathring{\chi}\, ds$$
$$+ \int_{\mathcal{R}} \frac{1}{2} |\mathbf{u}_i|^2 \mathcal{F}_i\, da + \int_{\mathcal{S}} \frac{1}{2} |\mathbf{u}_i|^2 \mathfrak{f}_i\, ds + \int_{\mathcal{R}} \mu_i^v \mathcal{F}_i\, da + \int_{\mathcal{S}} \mu_i^v \mathfrak{f}_i\, ds, \tag{61}$$

where $\Psi^\alpha = \mathcal{E}^\alpha - T\eta^\alpha$ $(\alpha = a, s)$ and $\psi = \varepsilon - T\varsigma$ are the Helmholtz free energy density of the bulk phases and of the interface, respectively, and

$$\mu_i^v = \mathcal{E}_i^v - \eta_i^v T + \frac{p_i^v}{\rho_i^v} \tag{62}$$

is the chemical potential of the $i$-th component in the vapor.

### 4.5 Dissipation Inequalities

If the $j$-th component of the film is involatile, we have $\mathcal{F}_j \equiv 0$, $\mathfrak{f}_j \equiv 0$, and $\mu_j^v$, $\mathbf{u}_j$ are undefined. In what follows we shall continue to take advantage of the summation convention and let repeated indices indicate summation over all $N$ components but with the understanding that any term which contains $\mathcal{F}_j$ or $\mathfrak{f}_j$ of an involatile component should be dropped.

Upon localization of the free energy imbalance (61), we obtain by appealing to (37), (49), and (50), which pertain to balance of mass and of linear momentum, the dissipation inequality

$$\dot{\Psi}^\alpha - \mathbf{S} : \dot{\mathbf{F}} - \mu_i^v \mathcal{F}_i - \sum_{i=1}^{N} \frac{1}{2} |\dot{\chi} - \mathbf{u}_i|^2 \mathcal{F}_i \leq 0 \tag{63}$$

in both the crystalline phase and the amorphous phase and its counterpart

$$\begin{aligned} &-[\![\Psi]\!]V - \frac{1}{2}[\![\rho|\dot{\chi}|^2]\!]V + \mathring{\psi} - \psi\kappa V + \overline{\frac{1}{2}\Gamma|\mathring{\chi}|^2}^{\circ} - \frac{1}{2}\Gamma|\mathring{\chi}|^2\kappa V \\ &- [\![\mathbf{Sm}\cdot\dot{\chi}]\!] - \frac{\partial\boldsymbol{\sigma}}{\partial s}\cdot\mathring{\chi} - \boldsymbol{\sigma}\cdot\frac{\partial\mathring{\chi}}{\partial s} - \frac{1}{2}|\mathbf{u}_i|^2\mathfrak{f}_i - \mu_i^v\mathfrak{f}_i - \mathbf{b}\cdot\mathring{\chi} \leq 0 \end{aligned} \tag{64}$$

at the interface.

In this paper we restrict attention to isothermal spherulitic growth at some constant absolute temperature $T$. As mentioned before, we will suppress the explicit dependence on temperature in all constitutive functions. We assume that both the amorphous and the crystalline phase are elastic with constitutive relations

$$\Psi^\alpha = \Psi^\alpha(\mathbf{F}, \boldsymbol{\rho}), \qquad \frac{\partial \Psi^\alpha}{\partial \mathbf{F}} = \mathbf{S}, \qquad \frac{\partial \Psi^\alpha}{\partial \rho_i} = \mu_i \quad (\alpha = a, s) \tag{65}$$

where $\mathbf{F} = \operatorname{Grad}\chi$ and $\boldsymbol{\rho} = \langle \rho_1, \ldots, \rho_N \rangle$ denotes the list of mass densities.

By invoking the balance of mass (37) and the constitutive relations (65), dissipation inequality (63) reduces to

$$(\mu_i - \mu_i^v)\mathcal{F}_i - \sum_{i=1}^{N} \frac{1}{2} |\dot{\chi} - \mathbf{u}_i|^2 \mathcal{F}_i \leq 0. \tag{66}$$

The reduced dissipation inequality (66) will put restrictions on the constitutive relation that we will prescribe for $\mathcal{F}_i$.

Next, we recast the interfacial dissipation inequality (64) to a more transparent form. We model the phase interface $\Sigma$ as a one-dimensional elastic solid. Its thermodynamic relations are presented in Sect. 3.2. In what follows we will use a time-dependent reference configuration for $\Sigma$. At each instant $t$ we take $\Sigma(t) \subset \mathcal{B}$, the reference configuration of the film, as the reference configuration for $\Sigma$. To emphasize this choice, we write the free energy density of the interface as

$$\psi = \psi_{\Sigma(t)}(\mathbf{f}, \boldsymbol{\Gamma}). \tag{67}$$

Nevertheless thermodynamic relations that involve $\psi$, i.e., (22), (25), and (27), remain valid irrespective of the reference configuration chosen.

By (7), (10), (42), and (50) we have

$$
\begin{aligned}
&-[\![\mathbf{Sm}\cdot\dot{\chi}]\!] - \frac{\partial\boldsymbol{\sigma}}{\partial s}\cdot\mathring{\chi} \\
&= -[\![\mathbf{Sm}]\!]\cdot(\mathring{\chi} - \langle\!\langle\mathbf{Fm}\rangle\!\rangle V) - \langle\!\langle\mathbf{Sm}\rangle\!\rangle\cdot[\![\dot{\chi}]\!] - \frac{\partial\boldsymbol{\sigma}}{\partial s}\cdot\mathring{\chi} \\
&= [\![\mathbf{Sm}\cdot\mathbf{Fm}]\!]V - \left(\Gamma\mathring{\mathring{\chi}} - \sum_{i=1}^{N}\mathfrak{f}_i(\mathbf{u}_i - \mathring{\chi}) - \langle\!\langle\rho\rangle\!\rangle[\![\dot{\chi}]\!]V + [\![\rho]\!]\langle\!\langle\mathbf{Fm}\rangle\!\rangle V^2 - \mathbf{b}\right)\cdot\mathring{\chi}.
\end{aligned}
\tag{68}
$$

On the other hand, by (6)–(8) and the vector identities (10), (11), we observe that

$$
\begin{aligned}
&-\frac{1}{2}[\![\rho|\dot{\chi}|^2]\!]V + \langle\!\langle\rho\rangle\!\rangle[\![\dot{\chi}]\!]\cdot\mathring{\chi}V - [\![\rho]\!]\langle\!\langle\mathbf{Fm}\rangle\!\rangle\cdot\mathring{\chi}V^2 \\
&= -\frac{1}{2}\langle\!\langle\rho\rangle\!\rangle[\![\mathbf{Fm}\cdot\mathbf{Fm}]\!]V^3 - \frac{1}{2}[\![\rho]\!]\langle\!\langle\dot{\chi}\cdot\dot{\chi}\rangle\!\rangle V - [\![\rho]\!]\langle\!\langle\mathbf{Fm}\rangle\!\rangle\cdot\langle\!\langle\dot{\chi}\rangle\!\rangle V^2 \\
&\quad - [\![\rho]\!]\langle\!\langle\mathbf{Fm}\rangle\!\rangle\cdot\langle\!\langle\mathbf{Fm}\rangle\!\rangle V^3 \\
&= -\frac{1}{2}[\![\rho]\!]\mathring{\chi}\cdot\mathring{\chi}V - \frac{1}{2}[\![\rho\mathbf{Fm}\cdot\mathbf{Fm}]\!]V^3.
\end{aligned}
\tag{69}
$$

By invoking balance of mass at the interface (39), we obtain

$$
-\Gamma\mathring{\mathring{\chi}}\cdot\mathring{\chi} - \frac{1}{2}[\![\rho]\!]\mathring{\chi}\cdot\mathring{\chi}V + \overset{\circ}{\overline{\frac{1}{2}\Gamma|\mathring{\chi}|^2}} - \frac{1}{2}\Gamma|\mathring{\chi}|^2\kappa V = \frac{1}{2}\mathfrak{f}\mathring{\chi}\cdot\mathring{\chi}
\tag{70}
$$

and

$$
\frac{1}{2}\mathfrak{f}\mathring{\chi}\cdot\mathring{\chi} + \sum_{i=1}^{N}\mathfrak{f}_i(\mathbf{u}_i - \mathring{\chi})\cdot\mathring{\chi} - \frac{1}{2}|\mathbf{u}_i|^2\mathfrak{f}_i = -\frac{1}{2}\sum_{i=1}^{N}|\mathbf{u}_i - \mathring{\chi}|^2\mathfrak{f}_i.
\tag{71}
$$

By (7) we recast the last term on the right hand side of (69) as

$$
\frac{1}{2}[\![\rho\mathbf{Fm}\cdot\mathbf{Fm}]\!]V^3 = \frac{1}{2}[\![\rho|\mathring{\chi} - \dot{\chi}|^2]\!]V,
\tag{72}
$$

which represents the jump in the relative kinetic energy [12, p. 75] times $V$.

With (68)–(72), the interfacial dissipation inequality (64) can be rewritten as

$$
\begin{aligned}
&-\left(\left[\!\!\left[\Psi + \frac{1}{2}\rho|\mathring{\chi} - \dot{\chi}|^2\right]\!\!\right] - [\![\mathbf{Sm}\cdot\mathbf{Fm}]\!]\right)V - \frac{1}{2}\sum_{i=1}^{N}|\mathbf{u}_i - \mathring{\chi}|^2\mathfrak{f}_i \\
&+ \mathring{\psi} - \psi\kappa V - \boldsymbol{\sigma}\cdot\frac{\partial\mathring{\chi}}{\partial s} - \mu_i^v\mathfrak{f}_i \le 0.
\end{aligned}
\tag{73}
$$

With (26) replaced by (67), we gather (25)–(27) and the interfacial mass balance (39) to obtain

$$\begin{aligned}\mathring{\psi} - \psi\kappa V - \boldsymbol{\sigma}\cdot\frac{\partial\mathring{\boldsymbol{\chi}}}{\partial s} - \mu_i^v\mathfrak{f}_i &= \boldsymbol{\sigma}\cdot\mathring{\mathbf{f}} + \mu_i^\Sigma\mathring{\Gamma}_i - \psi\kappa V - \boldsymbol{\sigma}\cdot(\mathring{\mathbf{f}} - \kappa V\mathbf{f}) - \mu_i^v\mathfrak{f}_i \\ &= \mu_i^\Sigma\big(\mathfrak{f}_i + \kappa V\Gamma_i + [\![\rho_i]\!]V\big) - (\psi - \boldsymbol{\sigma}\cdot\mathbf{f})\kappa V - \mu_i^v\mathfrak{f}_i \\ &= \big(\mu_i^\Sigma - \mu_i^v\big)\mathfrak{f}_i - \big(\psi - \boldsymbol{\sigma}\cdot\mathbf{f} - \mu_i^\Sigma\Gamma_i\big)\kappa V + \mu_i^\Sigma[\![\rho_i]\!]V \\ &= \big(\mu_i^\Sigma - \mu_i^v\big)\mathfrak{f}_i + \mu_i^\Sigma[\![\rho_i]\!]V. \end{aligned} \tag{74}$$

Hence (73) can be rewritten as

$$\begin{aligned} &-\left([\![\Psi + \frac{1}{2}\rho|\mathring{\boldsymbol{\chi}} - \dot{\boldsymbol{\chi}}|^2]\!] - [\![\mathbf{Sm}\cdot\mathbf{Fm}]\!] - \mu_i^\Sigma[\![\rho_i]\!]\right)V + \big(\mu_i^\Sigma - \mu_i^v\big)\mathfrak{f}_i \\ &-\frac{1}{2}\sum_{i=1}^{N}|\mathbf{u}_i - \mathring{\boldsymbol{\chi}}|^2\mathfrak{f}_i \le 0. \end{aligned} \tag{75}$$

### 4.6 Constitutive Restrictions Imposed by Bulk Dissipation Inequality

For simplicity we will focus our discussion in this subsection and the next on the special case where all components in the film, except one, i.e., the solvent, are taken as involatile. We will index the involatile components from 1 to $N-1$ and specify the solvent by the subscript "$S$".

We will follow Coleman and Noll [3] and adopt as basic hypothesis the requirement that all admissible processes (or, in the terminology of [14], constitutive processes) be consistent with the dissipation inequalities. Specifically, all admissible processes should be consistent with inequalities (66) and (75). To start with, we should define what we mean by an admissible process in the present context. Since we assume that there is no diffusion of any species, the mass densities $\rho_i$ ($i = 1, \ldots, N-1$) of the involatile components remain unchanged in the reference configuration. Moreover, we choose a reference configuration $\mathcal{B}$ where the densities $\rho_i$ are uniform, and we take $\Gamma_i = 0$ for all the involatile components (cf. Remark 4.3).

In what follows an (isothermal) admissible process $\mathbb{P}$ of the isothermal thermodynamic system in question is defined by

- $\boldsymbol{\chi} : \mathcal{B}\times[0,d] \to \mathbb{E}$, which satisfies the three conditions laid out at the beginning of Sect. 3.1 and specifies also the evolution of the interface $\Sigma(t)$, $t\in[0,d]$,
- $\rho_S(\mathbf{X},t)$, where $\mathbf{X}\in\mathcal{B}\setminus\Sigma(t)$ for each $t$,
- $\Gamma_S(\mathbf{X},t)\ge 0$, where $\mathbf{X}\in\Sigma(t)$ for each $t$,
- the fields $\mathbf{B}$, $\mathbf{b}$, $\mathcal{F}_S$, $\mathfrak{f}_S$, and $\mathbf{v}^v$,

which for each $t\in[0,d]$ satisfy the balance equations

$$\dot{\rho} = \mathcal{F}_S \quad \text{in } \mathcal{B}\setminus\Sigma(t), \tag{76}$$

$$\mathring{\Gamma}_S - \Gamma_S\kappa V - [\![\rho_S]\!]V = \mathfrak{f}_S \quad \text{on } \Sigma(t), \tag{77}$$

$$\rho\ddot{\boldsymbol{\chi}} = \operatorname{div}\mathbf{S} + \mathcal{F}_S(\mathbf{u}_S - \dot{\boldsymbol{\chi}}) + \mathbf{B} \quad \text{in } \mathcal{B}\setminus\Sigma(t), \tag{78}$$

$$\overline{\Gamma_S\mathring{\boldsymbol{\chi}}}^{\circ} - \Gamma_S\mathring{\boldsymbol{\chi}}\kappa V = [\![\rho\dot{\boldsymbol{\chi}}]\!]V + [\![\mathbf{S}]\!]\mathbf{m} + \frac{\partial\boldsymbol{\sigma}}{\partial s} + \mathbf{u}_S\mathfrak{f}_S + \mathbf{b} \quad \text{on } \Sigma(t). \tag{79}$$

We assume that the fields $\mathbf{B}$, $\mathbf{b}$, and $\mathbf{v}^v$ are arbitrarily assignable.

In our present context the reduced bulk dissipation inequality asserts that

$$(\mu_S - \mu_S^v)\mathcal{F}_S - \frac{1}{2}|\dot{\chi} - \mathbf{u}_S|^2\mathcal{F}_S \le 0 \tag{80}$$

for every admissible process $\mathbb{P}$.

Note that when the solvent is evaporating, i.e. $\mathcal{F}_S < 0$ and $\mathbf{u}_S = \dot{\chi}$ by (45), we have

$$-\frac{1}{2}|\dot{\chi} - \mathbf{u}_S|^2\mathcal{F}_S = 0 \tag{81}$$

for each admissible process $\mathbb{P}$. When the solvent is condensing, i.e., $\mathcal{F}_S > 0$ and $\mathbf{u}_S = \mathbf{v}^v$, the same term will be non-positive, i.e.,

$$-\frac{1}{2}|\dot{\chi} - \mathbf{u}_S|^2\mathcal{F}_S \le 0. \tag{82}$$

Hence a sufficient condition for the bulk dissipation inequality (66) to be observed is that the inequality

$$(\mu_S - \mu_S^v)\mathcal{F}_S \le 0 \tag{83}$$

holds for all admissible processes.

We proceed to prove that condition (83) is also necessary. Indeed, suppose there is an admissible process $\mathbb{P}$ for which

$$(\mu_S(\mathbf{X}_o, t_o) - \mu_S^v(\mathbf{X}_o, t_o))\mathcal{F}_S(\mathbf{X}_o, t_o) > 0 \tag{84}$$

for some $\mathbf{X}_o \in \mathcal{B}\setminus\Sigma(t)$ and $t_o \in [0, d]$. Note that $\mathcal{F}_S(\mathbf{X}_o, t_o) > 0$, because $\mathcal{F}_S(\mathbf{X}_o, t_o) \le 0$ will lead to (81). Inequality (80) will then imply (83) so that (84) cannot hold. Let us construct a new admissible process $\tilde{\mathbb{P}}$ as follows: Replace the field $\mathbf{v}^v$ (that pertains to the process $\mathbb{P}$) by the constant velocity field

$$\tilde{\mathbf{v}}^v = \dot{\chi}(\mathbf{X}_o, t_o). \tag{85}$$

Let $\tilde{\mathbf{u}}_S$ be the modified $\mathbf{u}_S$ with $\mathbf{v}^v$ replaced by $\tilde{\mathbf{v}}^v$; cf. (45). Let

$$\tilde{\mathbf{B}} = \mathbf{B} + \mathcal{F}_S\mathbf{u}_S - \mathcal{F}_S\tilde{\mathbf{u}}_S, \tag{86}$$

$$\tilde{\mathbf{b}} = \mathbf{b} + \mathfrak{f}_S\mathbf{u}_S - \mathfrak{f}_S\tilde{\mathbf{u}}_S. \tag{87}$$

We make no change on $\mathcal{F}_S$ and $\mathfrak{f}_S$. Then the motion $\chi$ and the fields $\rho_S$, $\Gamma_S$ that pertain to process $\mathbb{P}$ will satisfy the balance equations for the process $\tilde{\mathbb{P}}$, i.e., (76), (77), and

$$\rho\ddot{\chi} = \operatorname{div}\mathbf{S} + \mathcal{F}_S(\tilde{\mathbf{u}}_S - \dot{\chi}) + \tilde{\mathbf{B}}, \tag{88}$$

$$\overline{\Gamma_S\mathring{\chi}} - \Gamma_S\mathring{\chi}\kappa V = [\![\rho\dot{\chi}]\!]V + [\![\mathbf{S}]\!]\mathbf{m} + \frac{\partial\sigma}{\partial s} + \tilde{\mathbf{u}}_S\mathfrak{f}_S + \tilde{\mathbf{b}}. \tag{89}$$

Note that except for the assignable $\tilde{\mathbf{B}}$, $\tilde{\mathbf{b}}$, $\tilde{\mathbf{v}}^v$, and $\tilde{\mathbf{u}}_S$, which differs from $\mathbf{u}_S$ in having $\mathbf{v}^v$ replaced by $\tilde{\mathbf{v}}^v$ for condensing solvent molecules, all fields of the process $\tilde{\mathbb{P}}$ remain the same as their respective counterpart that pertains to $\mathbb{P}$. For the process $\tilde{\mathbb{P}}$, however, we have

$$(\mu_S(\mathbf{X}_o, t_o) - \mu_S^v(\mathbf{X}_o, t_o))\mathcal{F}_S(\mathbf{X}_o, t_o) > 0, \tag{90}$$

$$\frac{1}{2}\left|\dot{\chi}(\mathbf{X}_o,t_o)-\tilde{\mathbf{u}}_S(\mathbf{X}_o,t_o)\right|^2\mathcal{F}_S(\mathbf{X}_o,t_o)=0. \tag{91}$$

Thus the admissible process $\tilde{\mathbb{P}}$ violates the bulk dissipation inequality. Hence a necessary condition for the bulk dissipation inequality to hold is:

$$\left(\mu_S-\mu_S^v\right)\mathcal{F}_S\leq 0 \tag{92}$$

for all admissible processes $\mathbb{P}$.

In summary, a necessary and sufficient condition for (80) to hold for all admissible process is:

$$\left(\mu_S^v-\mu_i\right)\mathcal{F}_S>0\quad \text{if } \mathcal{F}_S\neq 0 \text{ and } \mu_S^v-\mu_i\neq 0. \tag{93}$$

In the product $(\mu_S-\mu_S^v)\mathcal{F}_S$, $(\mu_S-\mu_S^v)$ is the thermodynamic driving force for the evaporation or condensation conjugate to the mass flux of the solvent $\mathcal{F}_S$. If we further assume that the flux $\mathcal{F}_S$ is a smooth function of $\mu_S-\mu_S^v$ and other parameters which are suppressed for simplicity, i.e.,

$$\mathcal{F}_S=\hat{\mathcal{F}}_S\left(\mu_S-\mu_S^v\right), \tag{94}$$

then condition (93) and continuity dictates that $\mathcal{F}_S(0)=0$. Moreover, linearization of (94) will lead to

$$\mathcal{F}_S=-K_S\left(\mu_S-\mu_S^v\right)\quad (K_S\geq 0) \tag{95}$$

when the driving force is small. The preceding constitutive relation has the same theoretical status as Fourier's law of heat conduction and Fick's law of diffusion. It should be emphasized that (95) is a special case of the general necessary and sufficient condition (93). One has to appeal to experimental approaches to single out the constitutive relation for $\mathcal{F}_S$ for a specific material.

## 4.7 Constitutive Restrictions Imposed by Interfacial Dissipation Inequality

In the context discussed at the beginning of Sect. 4.6 the interfacial dissipation inequality assumes the form

$$\begin{aligned}&-\left(\left[\!\left[\Psi+\frac{1}{2}\rho|\mathring{\chi}-\dot{\chi}|^2\right]\!\right]-[\![\mathbf{Sm}\cdot\mathbf{Fm}]\!]-\mu_S^\Sigma[\![\rho_S]\!]\right)V\\&+\left(\mu_S^\Sigma-\mu_S^v\right)\mathfrak{f}_S-\frac{1}{2}|\mathbf{u}_S-\mathring{\chi}|^2\mathfrak{f}_S\leq 0.\end{aligned} \tag{96}$$

By an argument similar to that which we use to prove (92) as a necessary and sufficient condition for the reduced bulk dissipation inequality (80) to hold for all admissible processes, it is easy to show that a necessary and sufficient condition for the reduced interfacial dissipation inequality (96) to be observed in all admissible processes is that the inequality

$$-\left(\left[\!\left[\Psi+\frac{1}{2}\rho|\mathring{\chi}-\dot{\chi}|^2\right]\!\right]-[\![\mathbf{Sm}\cdot\mathbf{Fm}]\!]-\mu_S^\Sigma[\![\rho_S]\!]\right)V+\left(\mu_S^\Sigma-\mu_S^v\right)\mathfrak{f}_S\leq 0 \tag{97}$$

holds for all admissible processes. However, it remains a question whether the two inequalities

$$-\left(\left[\!\left[\Psi+\frac{1}{2}\rho|\mathring{\chi}-\dot{\chi}|^2\right]\!\right]-[\![\mathbf{Sm}\cdot\mathbf{Fm}]\!]-\mu_S^\Sigma[\![\rho_S]\!]\right)V\leq 0 \tag{98}$$

and

$$\left(\mu_S^{\Sigma} - \mu_S^{v}\right)\mathfrak{f}_S \leq 0 \tag{99}$$

would each hold for all admissible processes.

A comparison of (99) with (92) and the fact that $\mu_S^{\Sigma} = \mu_S^{v}$ is a necessary condition for thermodynamic equilibrium [35, Chap. 3] suggest that inequality (99) would be observed in all admissible processes.

For admissible processes where $\mu_S^{\Sigma} = \mu_S^{v}$, inequality (98) reduces to

$$-\left( [\![\Psi + \frac{1}{2}\rho|\mathring{\chi} - \dot{\chi}|^2]\!] - [\![\mathbf{Sm}\cdot\mathbf{Fm}]\!] - \mu_S^{v}[\![\rho_S]\!] \right) V \leq 0, \tag{100}$$

which by condition (97) is observed in all such processes. In fact, if we follow Jabbour et al. [17] and impose $\Gamma_S = 0$ as *a priori* constraint, calculations similar to what we did above lead to the conclusion that (100) is a necessary and sufficient condition for the interfacial dissipation inequality to hold for all admissible processes.

*Remark 4.4* In the experiments reported in [20], [21], and [30], the measured values of $V$ for the thin films studied were less than $3 \times 10^{-5}$ m/s. By (72) the jump in relative kinetic energy is given by

$$\frac{1}{2}[\![\rho|\mathring{\chi} - \dot{\chi}|^2]\!] = \frac{1}{2}[\![\rho\mathbf{Fm}\cdot\mathbf{Fm}]\!]V^2 = \frac{1}{2}[\![\rho]\!]V^2 + [\![\rho\mathbf{m}\cdot\mathbf{Em}]\!]V^2, \tag{101}$$

where $\mathbf{E}$ is the Lagrangian strain. Hence this term would be very small with the measured values of $V$, which is consistent with the fact that motion in both bulk phases was unnoticed in those experiments. If motion in both bulk phases is negligible, then so is the jump in relative kinetic energy, which we shall neglect in our further discussions.

Let us assume the validity of inequalities (98) and (99), and let

$$f = [\![\Psi]\!] - [\![\mathbf{Sm}\cdot\mathbf{Fm}]\!] - \mu_S^{\Sigma}[\![\rho_S]\!] \tag{102}$$

be the driving force for spherulitic growth. Then we have

$$\begin{aligned} fV > 0 &\quad \text{if } f \neq 0 \text{ and } V \neq 0, \\ \mathfrak{f}_S(\mu_S^{\Sigma} - \mu_S^{v}) < 0 &\quad \text{if } \mathfrak{f}_S \neq 0 \text{ and } \mu_S^{\Sigma} - \mu_S^{v} \neq 0. \end{aligned} \tag{103}$$

If we further assume that $V$ and $\mathfrak{f}_S$ be smooth functions of $f$ and of $\mu_i^{\Sigma} - \mu_i^{v}$, respectively, i.e.,

$$V = \hat{\mathcal{V}}(f), \qquad \mathfrak{f}_S = \hat{\mathfrak{f}}_S\left(\mu_S^{\Sigma} - \mu_S^{v}\right), \tag{104}$$

where other arguments of the two functions are suppressed, continuity dictates that $\mathcal{V}(0) = 0$ and $\mathfrak{f}_S(0) = 0$. Then with small driving forces $f$ and small $\mu_S^{\Sigma} - \mu_S^{v}$, we have

$$V = \frac{f}{\beta}, \tag{105}$$

and by linearization

$$\mathfrak{f}_S = -k_S\left(\mu_S^{\Sigma} - \mu_S^{v}\right) \quad (k_S \geq 0) \tag{106}$$

where $\beta = 1/\mathcal{V}'(0)$ is referred to as the kinetic coefficient.

*Remark 4.5* For admissible processes where $\mu_S^\Sigma = \mu_S^v$, the driving force for spherulitic growth assumes the form

$$\tilde{f} = [\![\Psi]\!] - [\![\mathbf{Sm} \cdot \mathbf{Fm}]\!] - \mu_S^v [\![\rho_S]\!]. \tag{107}$$

In the experiments reported in [30], the kinetic coefficient was found to be a function of $p_S$, the partial pressure of solvent vapor in the gas/vapor stream, and other variables (e.g., the concentration of the additive $B$ in the film). As the films were thin (100 nm in thickness), while they were initially devoid of any solvent, it is plausible that once they were exposed to the gas/vapor stream, solvent got into the films so quickly that the system reached chemical equilibrium with $\mu_S = \mu_S^\Sigma = \mu_S^v$ even before nucleation began. Under this scenario, the driving force would be given by (107).

If we consider the special case where the Cauchy stress is spherical, i.e., $\mathbf{T} = -p\mathbf{I}$, then $\mathbf{Sm} \cdot \mathbf{Fm} = -Jp$, where $J = \det \mathbf{F}$. Thus

$$\begin{aligned} f &= [\![\Psi + Jp - \mu_i^\Sigma \rho_i]\!] \\ &= [\![(\mu_i - \mu_i^\Sigma)\rho_i]\!]. \end{aligned} \tag{108}$$

One can compare (108) with the driving force

$$\tilde{f} = [\![(\mu_i - \mu_i^v)\rho_i]\!] \tag{109}$$

derived in [17], where the interface is assumed to be massless.

*Remark 4.6* Let us consider a simple example to illustrate the difference between $f$ and $\tilde{f}$. Assume that the amorphous phase of a certain solid film with a single involatile component $A$ can form a solution with molecules of solvent $S$, while the crystallized phase rejects all molecules of $S$. Consider the growth of a spherulite in a film where the amorphous phase initially carries "dissolved" $S$ at concentration $\rho_0$. Since there is no diffusion, during spherulitic crystallization the molecules of $S$ are rejected into the interface and then evaporate. Indeed, after an initial buildup, the density of solvent $S$ in the interface will come to a steady state, i.e. $\mathring{\Gamma}_S$ is practically 0. If the curvature $\kappa$ is small enough that the term $\Gamma_S \kappa V$ can be neglected, then by (39), we have $0 > -\rho_0 V \approx \mathfrak{f}_S$ which implies $\mu_S^\Sigma > \mu_S^v$. Therefore

$$-\mu_S^\Sigma \rho_0 < -\mu_S^v \rho_0. \tag{110}$$

If we assume all other quantities be the same, then $f$ will deliver a smaller driving force than $\tilde{f}$.

## 5 Effect of Prestress on Directional Dependence of Growth Rate

In this section we study how the presence of a prestress affects the growth of an isolated spherulite in a thin film.

Consider a homogeneous amorphous thin film modeled as a two dimensional planar object. It is kept at a fixed temperature $T$ and carries an initial homogeneous Cauchy stress $\mathbf{T}^a$. We will assume that the amorphous film is isotropic[5] and linearly elastic.

The following assumptions are crucial to our discussion:

[5] We adopt the definition given in [34], i.e. a material is *isotropic* if there is a reference placement $\boldsymbol{\kappa}$ such that the peer group $\mathcal{G}_{\boldsymbol{\kappa}} \supset O(2)$.

- the amorphous film possesses a homogeneous stress-free natural configuration;
- the natural configuration, denoted by $\mathcal{B}$, will be taken as reference at all time (before and after crystallization);
- after spherulitic crystallization of a material point, the chosen reference configuration for that material point carries an initial stress;
- the amorphous phase can be taken at all time as in a state of a fixed homogeneous Cauchy stress $\mathbf{T}^a$;
- the crystalline phase is also isotropic.

In the experiments reported in [20, 21, 30], the thin film when under solvent-vapor annealing is composed of one major chemical component $A$, one minor component $B$ of less than 10 mole-percent in concentration, and a solvent $S$ of minuscule amount, where components $A$ and $B$ can be taken as involatile. Since components $A$ and $B$ are involatile (which implies $\mathcal{F}_i = 0$ for $i = A, B$) and non-diffusive, there is no mechanism in our model that may cause the density to change in a homogeneous reference configuration before and after the phase transition, which implies $[\![\rho_i]\!] = 0$ for $i = A, B$. Furthermore, since the amount of component $S$ is very small, for the sake of simplicity, we will put $[\![\rho_S]\!] = 0$ in formula (102) for the driving force $f$. Hence formula (105) assumes the form[6]

$$\beta V = [\![\Psi - \mathbf{Sm} \cdot \mathbf{Fm}]\!], \tag{111}$$

where $\beta > 0$ is the kinetic coefficient. In the discussion that follows, we will neglect inertia, "as it is generally unimportant in solid-state problems involving the interaction of composition and stress." [8, p. 1367].

### 5.1 Preliminaries on Helmholtz Free Energy

Consider a material point with three components $A$, $B$ and $S$ at the reference configuration. The discussion that follows can be extended to any number of components easily. Henceforth we shall use a superposed $\star$ to specify the quantities pertaining to the material point at the reference configuration. Suppose that the constitutive equations of a material point depend also on a list of microstructural variables $\boldsymbol{\xi} = (\xi_1^\star, \ldots, \xi_k^\star)$.

With the assumption that the component $S$ be of very small amount, the effect of the change in density of $S$ can be ignored; thus, the film can be treated as a closed system, i.e. $\rho_i = \rho_i^\star$ for $i = A, B, S$.

Let $\mathsf{E} = \frac{1}{2}(\mathbf{H} + \mathbf{H}^\top)$ be the infinitesimal strain tensor, where $\mathbf{H} = \mathbf{F} - \mathbf{I}$ is the displacement gradient. The relationship between the infinitesimal strain tensor and the (finite) Lagrange strain tensor is

$$\mathbf{E} = \frac{1}{2}(\mathbf{F}^\top\mathbf{F} - \mathbf{I}) = \frac{1}{2}(\mathbf{H} + \mathbf{H}^\top + \mathbf{H}^\top\mathbf{H}) = \mathsf{E} + o(\|\mathbf{H}\|). \tag{112}$$

For a closed system with fixed temperature $T$, if the second Piola stress $\mathbf{P}(\,\cdot\,; \boldsymbol{\xi}) = \mathbf{F}^{-1}\mathbf{S}$ is a smooth function of $(T, \mathsf{E}, \rho_A, \rho_B, \rho_S)$, we have [28, 33]

$$\mathbf{P}(T, \mathsf{E}, \rho_A, \rho_B, \rho_S; \boldsymbol{\xi}) = \overset{\star}{\mathbf{T}} + \mathbb{L}(T, \mathbf{0}, \rho_A^\star, \rho_B^\star, \rho_S^\star;\ \boldsymbol{\xi})[\mathsf{E}], \tag{113}$$

[6] At this point the reader may query the role of the solvent. As briefly summarized in Remark 4.1, the experiments reported in [30] suggest that the reciprocal of the kinetic coefficient, i.e., $1/\beta$, goes linearly to zero as the concentration of solvent vapor in the gas/vapor stream approaches zero. In other words, spherulitic crystallization of the amorphous phase is triggered by the presence of the solvent, without which the film will remain in its amorphous phase in a state of unstable equilibrium.

where

$$\overset{\star}{\mathbf{T}} = \mathbf{P}(T, \mathbf{0}, \rho_A^\star, \rho_B^\star, \rho_S^\star;\ \boldsymbol{\xi}) \tag{114}$$

denotes the initial stress and

$$\mathbb{L}(T, \mathbf{0}, \rho_A^\star, \rho_B^\star, \rho_S^\star;\ \boldsymbol{\xi}) = D_2\mathbf{P}(T, \mathbf{0}, \rho_A^\star, \rho_B^\star, \rho_S^\star;\ \boldsymbol{\xi}). \tag{115}$$

In (113), the $o(\|\mathsf{E}\|)$ terms are dropped. For notational simplicity, we will suppress the dependence of $\mathbb{L}$ on $T$, $\rho_i^\star$ (for $i = A, B, S$) and $\boldsymbol{\xi}$. Note that with fixed $T$, $\rho_i^\star$ and $\boldsymbol{\xi}$, $\mathbb{L}$ is a constant fourth-order tensor with major and minor symmetries.

In what follows we derive, for the amorphous and for the crystalline phase, respectively, an expression that delineates the dependence of the free energy density on $\mathsf{E}$ when all the other independent variables remain fixed. The empirical Hartig's Law [2, 25, 26] suggests that for many materials the initial stress $\overset{\star}{\mathbf{T}}$ should be treated as a microstructural variable, and that $\mathbb{L}$ can be written as [25, 26, 33]

$$\mathbb{L}[\mathsf{E}] = \mathbb{C}[\mathsf{E}] + \mathbb{D}[\overset{\star}{\mathbf{T}}, \mathsf{E}], \tag{116}$$

where $\mathbb{C}$ is a fourth-order tensor and $\mathbb{D}$ is a sixth-order tensor which will vanish when $\overset{\star}{\mathbf{T}} = \mathbf{0}$.

By integrating the thermodynamic relation [27]

$$d\Psi = \mathbf{P} : d\mathsf{E} + \mu_i d\rho_i \tag{117}$$

with constant $T$, $\rho_i$ and $\boldsymbol{\xi}$, we obtain

$$\Psi = \overset{\star}{\Psi} + \overset{\star}{\mathbf{T}} : \mathsf{E} + \frac{1}{2}\mathbb{L}[\mathsf{E}] : \mathsf{E}, \tag{118}$$

where $\Psi^\star$ is the free energy density pertaining to the reference configuration.

For our application, prior to crystallization the reference configuration $\mathcal{B}$ is natural, i.e. $\overset{\star}{\mathbf{T}}{}^a = \mathbf{0}$. The free energy in the amorphous phase is then, by letting $\overset{\star}{\mathbf{T}} = \mathbf{0}$ in (118),

$$\Psi^a = \overset{\star}{\Psi}{}^a + \frac{1}{2}\mathbb{C}^a[\mathsf{E}^a] : \mathsf{E}^a. \tag{119}$$

After spherulitic crystallization, the material points in the crystalline phase will carry an initial stress $\overset{\star}{\mathbf{T}}{}^s$. Then it follows from (113) and (116) that

$$\mathbf{P}^s(\mathsf{E}^s) = \overset{\star}{\mathbf{T}}{}^s + \mathbb{C}^s[\mathsf{E}^s] + \mathbb{D}[\overset{\star}{\mathbf{T}}{}^s, \mathsf{E}^s]. \tag{120}$$

The free energy density in the crystalline phase is then

$$\Psi^s = \overset{\star}{\Psi}{}^s + \overset{\star}{\mathbf{T}}{}^s : \mathsf{E}^s + \frac{1}{2}\mathbb{L}^s[\mathsf{E}^s] : \mathsf{E}^s. \tag{121}$$

A comparison of (118) with the integrated fundamental equation ([27, equation (21)]; see also [32, equation (54.15), where the case of a natural reference configuration is considered])

$$\Psi = \mathbf{P} : \mathsf{E} + \rho_i\mu_i \tag{122}$$

may raise the question whether the two formulas are compatible. The answer is affirmative. In fact by (117) and (122), we can get, for constant temperature $T$, the Gibbs–Duhem equation

$$\mathsf{E} : d\mathbf{P} + \rho_i d\mu_i = 0. \tag{123}$$

Since the incremental tensor $\mathbb{L}$ is a constant fourth-order tensor with major symmetry,

$$\mathsf{E} : d\mathbf{P} = \mathsf{E} : d\big(\overset{\star}{\mathbf{T}} + \mathbb{L}[\mathsf{E}]\big) = \mathbb{L}[\mathsf{E}] : d\mathsf{E}. \tag{124}$$

By integrating (123) from state $(T, \mathbf{0}, \rho_A^\star, \rho_B^\star, \rho_S^\star; \boldsymbol{\xi})$ to state $(T, \mathsf{E}, \rho_A^\star, \rho_B^\star, \rho_S^\star; \boldsymbol{\xi})$, we get

$$\rho_i^\star \mu_i = \rho_i^\star \mu_i^\star - \frac{1}{2}\mathbb{L}[\mathsf{E}] : \mathsf{E}. \tag{125}$$

Substituting (125) into (122), we obtain

$$\Psi = \rho_i^\star \mu_i^\star + \overset{\star}{\mathbf{T}} : \mathsf{E} + \frac{1}{2}\mathbb{L}[\mathsf{E}] : \mathsf{E}, \tag{126}$$

which agrees with (118) with $\overset{\star}{\Psi} = \rho_i^\star \mu_i^\star$.

## 5.2 Derivation of Growth Rate

Let $\sigma_1$ and $\sigma_2$ be the principal stresses of the prestress tensor $\mathbf{S}^a = \mathbf{T}^a$, where we have used the fact [14, p. 301] that with reference to a natural configuration the Cauchy and Piola stresses coincide to within an error of $o(\|\mathbf{H}\|)$. Let

$$\bar{\sigma} = \frac{1}{2}(\sigma_1 + \sigma_2), \qquad \tau = \frac{1}{2}(\sigma_1 - \sigma_2). \tag{127}$$

We assume that $\tau \neq 0$. Let $\mathbf{e}_1$ and $\mathbf{e}_2$ be orthonormal vectors that agree with the principal stress directions. Under the Cartesian coordinate system defined by the basis $\mathbf{e}_i$ $(i = 1, 2)$, the stress $\mathbf{S}^a$ is given by

$$\mathbf{S}^a = \bar{\sigma}\mathbf{I} + \begin{pmatrix} \tau & 0 \\ 0 & -\tau \end{pmatrix}. \tag{128}$$

By a judicious choice of $\mathbf{e}_1$ and $\mathbf{e}_2$, we can make $\tau$ positive or negative as we please.

By a representation theorem for isotropic linear tensor functions [29], we have

$$\mathbf{S}^a = \lambda^a \big(\operatorname{tr} \mathsf{E}^a\big)\mathbf{I} + 2\mu^a \mathsf{E}^a, \tag{129}$$

where $\lambda^a$ and $\mu^a$ are elastic constants. Under the assumption that the elasticity tensor $\mathbb{C}^a$ be positive definite, we have

$$\mu^a > 0, \quad \lambda^a + \mu^a > 0. \tag{130}$$

Moreover, we assume that the initial infinitesimal rotation

$$\mathbf{W}^a = \frac{1}{2}\big(\mathbf{H}^a - \mathbf{H}^{a\top}\big) = \mathbf{0}. \tag{131}$$

It follows from (128) and (129) that under the basis $\mathbf{e}_i$ $(i = 1, 2)$ the strain $\mathsf{E}^a$ is given by

$$\mathsf{E}^a = \frac{\bar{\sigma}}{2(\lambda^a + \mu^a)}\mathbf{I} + \frac{1}{2\mu^a}\begin{pmatrix} \tau & 0 \\ 0 & -\tau \end{pmatrix}. \tag{132}$$

Suppose at some instant a nucleus of spherulitic crystallization is formed and at time $t$ the spherulite occupies a region $\mathcal{B}_t^s$ in the film. In what follows we consider the growth of the spherulitic region and adopt the physical assumption that outside $\mathcal{B}_t^s$, the amorphous phase can be taken at all time as in a state of fixed homogeneous stress $\mathbf{S}^a$.

Let $\overset{\star}{\mathbf{T}}{}^s$ be the initial stress carried by the part of the reference configuration where spherulitic crystallization has occurred. As shown in [25, 29], with the presence of the initial stress and in two-dimensional case,

$$\mathbb{D}\big[\overset{\star}{\mathbf{T}}{}^s, \mathsf{E}^s\big] = \beta_1\big(\operatorname{tr}\mathsf{E}^s\big)(\operatorname{tr}\overset{\star}{\mathbf{T}}{}^s)\mathbf{I} + \beta_2(\operatorname{tr}\overset{\star}{\mathbf{T}}{}^s)\mathsf{E}^s + \beta_3\big(\big(\operatorname{tr}\mathsf{E}^s\big)\overset{\star}{\mathbf{T}}{}^s + \big(\operatorname{tr}\mathsf{E}^s\overset{\star}{\mathbf{T}}{}^s\big)\mathbf{I}\big), \tag{133}$$

where $\beta_1$, $\beta_2$, $\beta_3$ are material constants. Since we assume that the material point in the crystallized phase remains isotropic, we have $\overset{\star}{\mathbf{T}}{}^s = -p_0\mathbf{I}$ [16]. By (116), (133), and the formula [28]

$$\mathbf{S} = \overset{\star}{\mathbf{T}} + \mathbf{H}\overset{\star}{\mathbf{T}} + \mathbb{L}[\mathsf{E}] \tag{134}$$

in linear elasticity with initial stress, the stress tensor in the crystalline phase can be written as

$$\mathbf{S}^s = -p_0\mathbf{I} - p_0\mathbf{H}^s + \tilde{\lambda}^s\big(\operatorname{tr}\mathsf{E}^s\big)\mathbf{I} + 2\tilde{\mu}^s\mathsf{E}^s, \tag{135}$$

where

$$\tilde{\lambda}^s = \lambda^s - 2p_0\beta_1 - 2p_0\beta_3, \tag{136}$$

$$\tilde{\mu}^s = \mu^s - p_0\beta_2, \tag{137}$$

and $\lambda^s$ and $\mu^s$ are the elastic constants of the prestressed crystalline phase.

Since $\mathbf{F} = \mathbf{I} + \mathbf{H}$, we have

$$[\![\mathbf{F}]\!] = [\![\mathbf{H}]\!]. \tag{138}$$

Then by (131), $\mathbf{H}^{a^+} = \mathsf{E}^a$. The kinematic compatibility condition (6) requires that

$$\mathbf{H}^{a^+} - \mathbf{H}^{s^-} = \mathsf{E}^a - \mathbf{H}^{s^-} = \mathbf{a} \otimes \mathbf{m} \tag{139}$$

for some vector $\mathbf{a}$. Let

$$\mathbf{m} = (\cos\theta, \sin\theta) \tag{140}$$

under the basis $\mathbf{e}_i$ $(i = 1, 2)$. We choose a new orthonormal basis $\mathbf{f}_1$, $\mathbf{f}_2$ such that $\mathbf{f}_1 = \mathbf{m}$. Under this basis we have

$$\mathbf{a} \otimes \mathbf{m} = \begin{pmatrix} a_1 & 0 \\ a_2 & 0 \end{pmatrix}, \tag{141}$$

and the stress and the strain tensor in the amorphous phase under the new basis become

$$\mathbf{S}^a = \bar{\sigma}\mathbf{I} + \tau\begin{pmatrix} \cos 2\theta & -\sin 2\theta \\ -\sin 2\theta & -\cos 2\theta \end{pmatrix}, \tag{142}$$

$$\mathsf{E}^a = \frac{\bar{\sigma}}{2(\lambda^a + \mu^a)}\mathbf{I} + \frac{\tau}{2\mu^a}\begin{pmatrix} \cos 2\theta & -\sin 2\theta \\ -\sin 2\theta & -\cos 2\theta \end{pmatrix}, \tag{143}$$

respectively. It follows that

$$\begin{aligned} \mathbf{H}^{s^-} &= \mathsf{E}^a - \mathbf{a} \otimes \mathbf{m} \\ &= \frac{\bar{\sigma}}{2(\lambda^a + \mu^a)}\mathbf{I} + \frac{\tau}{2\mu^a}\begin{pmatrix} \cos 2\theta & -\sin 2\theta \\ -\sin 2\theta & -\cos 2\theta \end{pmatrix} - \begin{pmatrix} a_1 & 0 \\ a_2 & 0 \end{pmatrix} \end{aligned} \tag{144}$$

and

$$\begin{aligned} \mathsf{E}^{s^-} &= \frac{1}{2}(\mathbf{H}^{s^-} + (\mathbf{H}^{s^-})^\top) \\ &= \frac{\bar{\sigma}}{2(\lambda^a + \mu^a)}\mathbf{I} + \frac{\tau}{2\mu^a}\begin{pmatrix} \cos 2\theta & -\sin 2\theta \\ -\sin 2\theta & -\cos 2\theta \end{pmatrix} - \begin{pmatrix} a_1 & a_2/2 \\ a_2/2 & 0 \end{pmatrix}. \end{aligned} \tag{145}$$

Then by (135), we have

$$\begin{aligned} \mathbf{S}^{s^-} &= \left(\frac{\tilde{\lambda}^s + \tilde{\mu}^s - p_0/2}{\lambda^a + \mu^a}\bar{\sigma} - \tilde{\lambda}^s a_1 - p_0\right)\mathbf{I} \\ &\quad + \frac{\tilde{\mu}^s - p_0/2}{\mu^a}\tau\begin{pmatrix} \cos 2\theta & -\sin 2\theta \\ -\sin 2\theta & -\cos 2\theta \end{pmatrix} + \begin{pmatrix} (p_0 - 2\tilde{\mu}^s)a_1 & -\tilde{\mu}^s a_2 \\ (p_0 - \tilde{\mu}^s)a_2 & 0 \end{pmatrix}. \end{aligned} \tag{146}$$

By neglecting inertia effects, (50) reduces to

$$[\![\mathbf{Sm}]\!] + \frac{\partial \boldsymbol{\sigma}}{\partial s} = \mathbf{0}. \tag{147}$$

When the radius of the spherulite is sufficiently large, the effect of the interfacial stress $\boldsymbol{\sigma}$ can be neglected. Thus we have $[\![\mathbf{Sm}]\!] = \mathbf{0}$, i.e., $\mathbf{S}^{a^+}\mathbf{m} = \mathbf{S}^{s^-}\mathbf{m}$ on the interface. With (142) and (146), we obtain the equations

$$\begin{aligned} &\frac{\tilde{\lambda}^s + \tilde{\mu}^s - p_0/2}{\lambda^a + \mu^a}\bar{\sigma} - p_0 - \tilde{\lambda}^s a_1 + \frac{\tilde{\mu}^s - p_0/2}{\mu^a}\tau \cos 2\theta + (p_0 - 2\tilde{\mu}^s)a_1 \\ &= \bar{\sigma} + \tau \cos 2\theta, \end{aligned} \tag{148}$$

$$-\frac{\tilde{\mu}^s - p_0/2}{\mu^a}\tau \sin 2\theta + (p_0 - \tilde{\mu}^s)a_2 = -\tau \sin 2\theta, \tag{149}$$

from which we get

$$a_1 = \frac{1}{p_0 - \tilde{\lambda}^s - 2\tilde{\mu}^s}\left(p_0 + \left(1 - \frac{\tilde{\lambda}^s + \tilde{\mu}^s - p_0/2}{\lambda^a + \mu^a}\right)\bar{\sigma} + \tau\left(1 - \frac{\tilde{\mu}^s - p_0/2}{\mu^a}\right)\cos 2\theta\right), \tag{150}$$

$$a_2 = \frac{(\frac{\tilde{\mu}^s - p_0/2}{\mu^a} - 1)\tau \sin 2\theta}{p_0 - \tilde{\mu}^s}. \tag{151}$$

Note that

$$[\![\mathbf{Sm}\cdot\mathbf{Fm}]\!] = a_1(\bar{\sigma}+\tau\cos 2\theta) - a_2\tau\sin 2\theta. \tag{152}$$

The growth rate $V$ is given by (111) and the Helmholtz free energy densities for the amorphous phase and the crystalline phase are give by (119) and (121), respectively. We calculate

$$\mathbf{S}^a : \mathsf{E}^a = \frac{\bar{\sigma}^2}{\lambda^a+\mu^a} + \frac{\tau^2}{\mu^a}, \tag{153}$$

and

$$\begin{aligned}\mathbb{L}^s[\mathsf{E}^{s^-}] : \mathsf{E}^{s^-} = {} & \frac{\tilde{\lambda}^s+\tilde{\mu}^s}{(\lambda^a+\mu^a)^2}\bar{\sigma}^2 - \frac{2(\tilde{\lambda}^s+\tilde{\mu}^s)}{\lambda^a+\mu^a}a_1\bar{\sigma} \\ & + \frac{\tau^2\tilde{\mu}^s}{(\mu^a)^2} - \frac{2\tilde{\mu}^s\tau}{\mu^a}(a_1\cos 2\theta - a_2\sin 2\theta) \\ & + \tilde{\lambda}^s a_1^2 + 2\tilde{\mu}^s a_1^2 + \tilde{\mu}^s a_2^2,\end{aligned} \tag{154}$$

and therefore with (111) and (152)–(154) we obtain

$$\begin{aligned}\beta V = {} & \overset{\star}{\Psi}{}^a - \overset{\star}{\Psi}{}^s \\ & + \frac{1}{2}\Bigg(\frac{\bar{\sigma}^2}{\lambda^a+\mu^a}\left(1-\frac{\tilde{\lambda}_s+\tilde{\mu}_s}{\lambda^a+\mu^a}\right) + \frac{\tau^2}{\mu^a}\left(1-\frac{\tilde{\mu}_s}{\mu^a}\right) \\ & + 2\frac{\tilde{\lambda}_s+\tilde{\mu}^s}{\lambda^a+\mu^a}\bar{\sigma}a_1 + \frac{2\tilde{\mu}^s\tau}{\mu^a}(a_1\cos 2\theta - a_2\sin 2\theta) \\ & - \tilde{\lambda}^s a_1^2 - 2\tilde{\mu}^s a_1^2 - \tilde{\mu}^s a_2^2\Bigg) + \frac{p_0\bar{\sigma}}{\lambda^a+\mu^a} - p_0 a_1 \\ & - \big(a_1(\bar{\sigma}+\tau\cos 2\theta) - a_2\tau\sin 2\theta\big).\end{aligned} \tag{155}$$

From the expressions of $a_1$ and $a_2$ in (150), we observe that the growth-rate formula is in the form

$$\beta V = A + B\cos 2\theta + C\cos^2 2\theta, \tag{156}$$

where $A$, $B$, $C$ are all constants. A straightforward calculation shows that

$$\begin{aligned}B = {} & \frac{\tau}{p_0-\tilde{\lambda}^s-2\tilde{\mu}^s}\left(1-\frac{\tilde{\mu}^s-p_0/2}{\mu^a}\right)\Bigg(\left(\frac{\tilde{\mu}^s+\tilde{\lambda}^s}{\lambda^a+\mu^a}-1\right)\bar{\sigma} - p_0 \\ & + \left(p_0+\bar{\sigma}\left(1-\frac{\tilde{\lambda}^s+\tilde{\mu}^s-p_0/2}{\lambda^a+\mu^a}\right)\right)\left(\frac{\tilde{\mu}^s-\mu^a}{\mu^a-\tilde{\mu}^s-p_0/2} - \frac{p_0}{p_0-\tilde{\lambda}^s-2\tilde{\mu}^s}\right)\Bigg),\end{aligned} \tag{157}$$

and

$$\begin{aligned}C = {} & \tau^2\left(1-\frac{\tilde{\mu}^s-p_0/2}{\mu^a}\right)^2\Bigg(\frac{\tilde{\mu}^s-\mu^a}{\mu^a-\tilde{\mu}^s-p_0/2}\left(\frac{1}{p_0-\tilde{\lambda}^s-2\tilde{\mu}^s} - \frac{1}{p_0-\tilde{\mu}^s}\right) \\ & \frac{\tilde{\lambda}^s/2+\tilde{\mu}^s}{(p_0-\tilde{\lambda}^s-2\tilde{\mu}^s)^2} + \frac{\tilde{\mu}^s}{2(p_0-\tilde{\mu}^s)^2}\Bigg).\end{aligned} \tag{158}$$

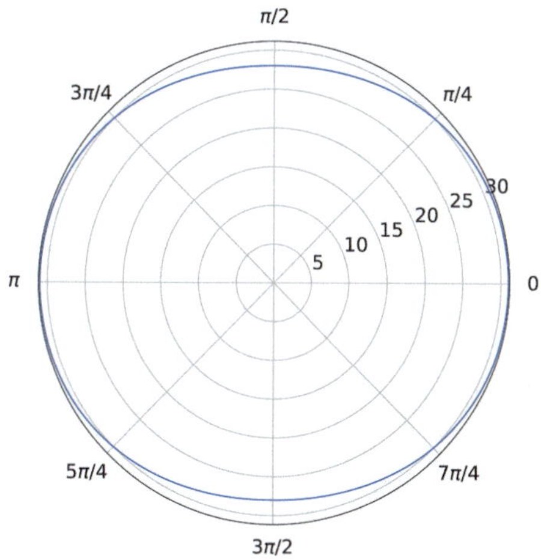


**Fig. 3** $A = 30$, $B = 1.5$, $C = -0.5$

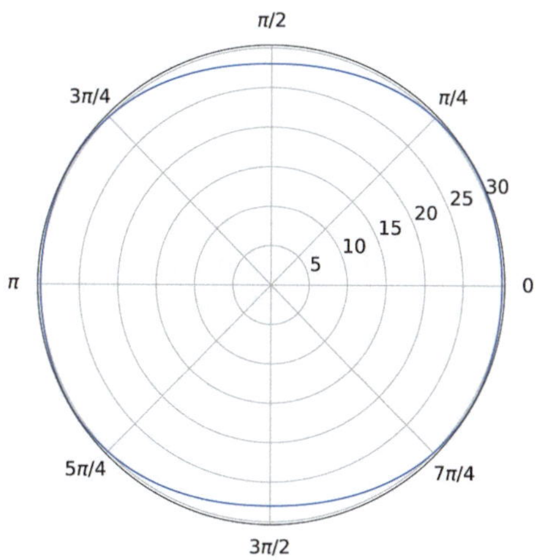


**Fig. 4** $A = 30$, $B = 1.0$, $C = -1.0$

It is obvious from the explicit expression of $B$ and $C$ that when $\tau = 0$, i.e. the prestress is spherical, we will recover the feature of constant growth rate obtained in [17].

A rotation of the Cartesian coordinate system by $\pi/2$ about the normal direction of the planar film changes $\theta$ to $\theta + \pi/2$ and the growth-rate formula (156) is recast as

$$\beta V = A - B\cos 2\theta + C\cos^2 2\theta. \tag{159}$$

Hence, without loss of generality, we may restrict our attention to (156) with $B \geq 0$. We are primarily interested in situations defined by $A > 0$, $A \gg B$ and $A \gg |C|$.

Examples of polar plots of $\beta V$ versus $\theta$ for growth-rate formula (156) are given in Fig. 3 and Fig. 4. It should be emphasized that these polar plots show the dependence of $\beta V$ on the direction of the unit outward normal **m** of the interface.

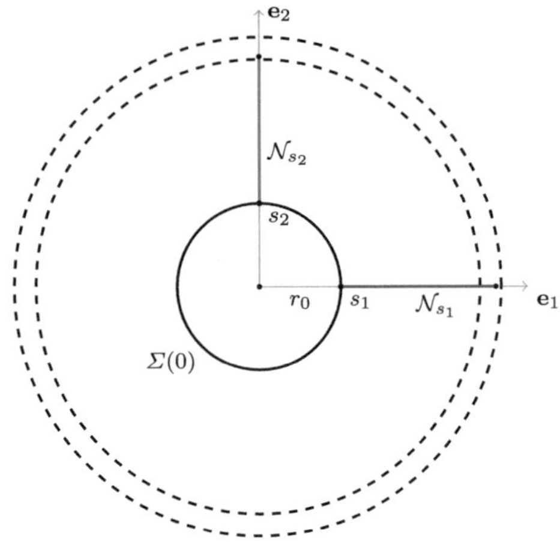


**Fig. 5** A sketch of shape estimation (not to scale)

### 5.3 Implication on the Shape of the Phase Boundary

Let an arc-length map $s = s(p,t)$ be prescribed [11] for the evolving interface $\Sigma(t)$, where $s(\cdot,t)$ is the arc-length for the curve $\Sigma(t)$ at time $t$ and $p$ is the parameter. Then

$$\frac{d\theta}{ds} = \kappa, \tag{160}$$

where $\theta$ is defined in (140). As noted in the paragraph containing equation (1), under our choice of **t** and **m**, $\kappa < 0$ for the curve that is the boundary of a strictly convex region.

The rate of change of curvature following a normal trajectory of the evolving curve is given by [9]

$$\mathring{\kappa} = V_{ss} + \kappa^2 V, \tag{161}$$

where

$$V_{ss} = V_{\theta\theta}\kappa^2 + V_\theta \kappa_\theta \kappa. \tag{162}$$

It is clear from (161) and (162) that $\mathring{\kappa} = 0$ when $\kappa = 0$. Thus if the initial curve $\Sigma(0)$ is the boundary of a convex region, then the region that $\Sigma(t)$ encloses will be convex for all $t$.

Consider the special case where the initial curve $\Sigma(0)$ is a circle of radius $r_0$; see Fig. 5. As noted in the beginning of Sect. 4.2, we choose a fixed point on the initial curve and measure from it the arc-length to the intersection of $\Sigma(0)$ with the axes $\mathbf{e}_1$ and $\mathbf{e}_2$, which we call $s_1$ and $s_2$, respectively. By symmetry, the normal trajectories starting from $s_1$ and $s_2$ will always stay on the axes; see $\mathcal{N}_{s_1}$ and $\mathcal{N}_{s_2}$ in Fig. 5. Therefore, at time $t$, the evolving curve $\Sigma(t)$ will intersect the axis $\mathbf{e}_1$ at the point $(r_0 + t(A+B+C)/\beta, 0)$ and the axis $\mathbf{e}_2$ at the point $(0, r_0 + t(A-B+C)/\beta)$. Note that $(A - B - |C|)/\beta \le V \le (A + B + |C|)/\beta$. The curve will always enclose a convex region and stay between two dashed circles of radius $r_0 + t(A+B+|C|)/\beta$ and $r_0 + t(A-B-|C|)/\beta$.

## 6 Closing Remarks

In deriving an expression for the driving force of spherulitic crystallization, we have followed a classical scheme. Indeed, writing the relevant terms of the dissipation inequalities as products of conjugate pairs of thermodynamic fluxes and forces can be traced back to classical irreversible thermodynamics [4, 7]. And our derivation is entirely analogous to that of Abeyaratne and Knowles [1], albeit in different contexts. What distinguishes our derivation from theirs and from other works in continuum sharp-interface theories of phase transition that we are aware of lies in our treatment of the interface. We take the sharp interface as a full-fledged open continuum thermodynamic system, whose particles have a well-defined interface velocity and whose components generally carry mass and chemical potential. Unlike the Gibbsian dividing interface in thermostatics, the thermodynamic variables in our treatment of the interface are not excesses. In particular, mass densities here are always non-negative. The foundations of this approach, its relationship with the Gibbsian dividing interface, and how it would fare in other more complicated problems are all topics that remain to be explored.

**Acknowledgements** D. Zhao would like to thank Prof. Michel Jabbour for helpful discussions and for the mechanics courses he offered at University of Kentucky. The research reported here was supported in part by the SOLAR Initiative of the National Science Foundation, USA, through Grant No. DMR-1035257.



## References

1. Abeyaratne, R., Knowles, J.K.: On the driving traction acting on a surface of strain discontinuity in a continuum. J. Mech. Phys. Solids **38**, 345–360 (1990)
2. Bell, J.F.: The experimental foundations of solid mechanics. In: Truesdell, C. (ed.) Mechanics of Solids, Vol. 1. Encyclopedia of Physics, vol. VIa/1, pp. 1–813. Springer, Berlin (1973)
3. Coleman, B.D., Noll, W.: The thermodynamics of elastic materials with heat conduction and viscosity. Arch. Ration. Mech. Anal. **13**, 167–178 (1963)
4. de Groot, S.R., Mazur, P.: Non-equilibrium Thermodynamics. North-Holland, Amsterdam (1962)
5. Defay, R., Prigogine, I., Bellemans, A.: Surface Tension and Adsorption. Wiley, New York (1966)
6. do Carmo, M.: Differential Geometry of Curves and Surfaces. Prentice-Hall, Englewood Cliffs (1976)
7. Fitts, D.D.: Nonequilibrium Thermodynamics: A Phenomenological Theory of Irreversible Processes in Fluid Systems. McGraw-Hill, New York (1962)
8. Fried, E., Gurtin, M.E.: Coherent solid-state phase transitions with atomic diffusion: a thermomechanical treatment. J. Stat. Phys. **95**, 1361–1427 (1999)
9. Fried, E., Gurtin, M.E.: A unified treatment of evolving interfaces accounting for small deformations and atomic transport with emphasis on grain-boundaries and epitaxy. Adv. Appl. Mech. **40**, 1–177 (2004)
10. Gibbs, J.W.: The Scientific Papers of J. Willard Gibbs, Volume 1, Thermodynamics. Dover, New York (1961)
11. Gurtin, M.E.: Thermomechanics of Evolving Phase Boundaries in the Plane. Clarendon Press, Oxford (1993)
12. Gurtin, M.E.: Configurational Forces as Basic Concepts of Continuum Physics. Springer, New York (2000)
13. Gurtin, M.E., Struthers, A.: Multiphase thermomechanics with interfacial structure. Arch. Ration. Mech. Anal. **112**, 97–160 (1990)
14. Gurtin, M.E., Fried, E., Anand, L.: The Mechanics and Thermodynamics of Continua. Cambridge University Press, Cambridge (2010)
15. Hegde, R., Henry, N., Whittle, B., Zang, H., Hu, B., Chen, J., Xiao, K., Dadmun, M.: The impact of controlled solvent exposure on the morphology, structure and function of bulk heterojunction solar cells. Sol. Energy Mater. Sol. Cells **107**, 112–124 (2012)
16. Hoger, A.: On the determination of residual stress in an elastic body. J. Elast. **16**, 303–324 (1986)
17. Jabbour, M., Man, C.-S., Paroni, R.: Spherulitic crystallization in binary thin films under solvent-vapor annealing. I. A sharp-interface theory. J. Chem. Phys. **139**, 144704 (2013). https://doi.org/10.1063/1.4823739
18. Johnson, R.E.: Conflicts between Gibbsian thermodynamics and recent treatments of interfacial energies in solid-liquid-vapor. J. Phys. Chem. **63**, 1655–1658 (1959)

19. Lee, S.S., Loo, Y.-L.: Structural complexities in the active layers of organic electronics. Annu. Rev. Chem. Biomol. Eng. **1**, 59–78 (2010)
20. Lee, S.S., Muralidharan, S., Woll, A.R., Loth, M.A., Li, Z., Anthony, J.E., Haataja, M., Loo, Y.-L.: Understanding heterogeneous nucleation in binary, solution-processed, organic semiconductor thin films. Chem. Mater. **24**, 2920–2928 (2012)
21. Lee, S.S., Tang, S.B., Smilgies, D.-M., Woll, A.R., Loth, M.A., Mativetsky, J.M., Anthony, J.E., Loo, Y.-L.: Guiding crystallization around bends and sharp corners. Adv. Mater. **24**, 2692–2698 (2012)
22. Lifshitz, E.M., Pitaevskii, L.P.: Physical Kinetics. Elsevier, Amsterdam (1981)
23. Lifshitz, I.M., Slyozov, V.V.: The kinetics of precipitation from supersaturated solid solutions. J. Phys. Chem. Solids **19**, 35–50 (1961)
24. Liu, I-S.: Continuum Mechanics. Springer, Berlin (2002)
25. Man, C.-S.: Hartig's law and linear elasticity with initial stress. Inverse Probl. **14**, 313–319 (1998)
26. Man, C.-S.: Effects of crystallographic texture on the acoustoelastic coefficients of polycrystals. Nondestruct. Test. Eval. **15**, 191–214 (1999)
27. Man, C.-S.: Classical thermodynamics of elastic solids as open systems. J. Elast. **126**, 272–280 (2017)
28. Man, C.-S., Carlson, D.E.: On the traction problem of dead loading in linear elasticity with initial stress. Arch. Ration. Mech. Anal. **128**, 223–247 (1994)
29. Man, C.-S., Schanding, J.B.: Differentiability of the scalar coefficients in two representation formulae for isotropic tensor functions in two dimensions. Q. Appl. Math. **54**, 121–132 (1996)
30. Man, C.-S., Jabbour, M.E., Paroni, R., Loo, Y.-L., Hailey, A.K., Lee, S.S., Anthony, J.E., Grimminger, M.A.: Spherulitic growth under solvent-vapor annealing: theory and experiments on thin-film blends of organic small-molecule semiconductors. Preprint
31. Milnor, J.W.: Topology from the Differentiable Viewpoint. University Press of Virginia, Charlottesville (1965)
32. Münster, A.: Classical Thermodynamics. Wiley, Chichester, New York (1970)
33. Paroni, R., Man, C.-S.: Constitutive equations of elastic polycrystalline materials. Arch. Ration. Mech. Anal. **150**, 153–177 (1999)
34. Truesdell, C.A.: A First Course in Rational Continuum Mechanics, 2nd edn. Academic Press, Boston (1991)
35. Zhao, D.: Spherulitic growth and thermodynamic equilibrium in multicomponent elastic films under solvent-vapor annealing. Doctoral dissertation, University of Kentucky, Lexington (2018)

MIX
Papier aus verantwortungsvollen Quellen
Paper from responsible sources
FSC® C105338

If you have any concerns about our products,
you can contact us on
**ProductSafety@springernature.com**

In case Publisher is established outside the EU,
the EU authorized representative is:
**Springer Nature Customer Service Center GmbH**
**Europaplatz 3, 69115 Heidelberg, Germany**

Printed by Libri Plureos GmbH
in Hamburg, Germany